STUDENT SOLUTIONS MANUAL FOR QUANTA, MATTER, AND CHANGE:

A MOLECULAR APPROACH TO PHYSICAL CHEMISTRY

STUDENT SOLUTIONS MANUAL FOR QUANTA, MATTER, AND CHANGE:

A MOLECULAR APPROACH TO PHYSICAL CHEMISTRY

C. A. Trapp

Professor of Chemistry, University of Louisville,
Louisville, Kentucky, USA

M. P. Cady

Professor of Chemistry,
Indiana University Southeast, New Albany Indiana, USA

C. Giunta

Professor of Chemistry,
Le Moyne College, Syracuse, NY, USA

OXFORD
UNIVERSITY PRESS

OXFORD
UNIVERSITY PRESS

Oxford University Press, Great Clarendon Street, Oxford,
OX2 6DP, United Kingdom

First printing

ISBN: 978-0-1995-5907-7

Published in the United States and Canada by
W. H. Freeman and Company
41 Madison Avenue
New York, NY 10010
www.whfreeman.com

ISBN-10: 1-4292-2375-8 ISBN-13: 978-1-4292-2375-1

British Library Cataloguing in Publication Data
Data available

Preface

This manual provides detailed solutions to all of the end-of-chapter (a) Exercises, and to the odd-numbered Discussion Questions and Problems in *Quanta, Matter, and Change*.

The solutions to some of the Exercises and many of the Problems in this manual relied heavily on the mathematical, graphical, and molecular modeling software that is now generally accessible to physical chemistry students. The availability of the software makes it possible to create and solve problems that can realistically mimic scientific research. Many of the problems specifically requested the use of such software, and, indeed, would have been almost unsolvable otherwise. We used the following software for many of the solutions in this manual: Excel™ for spreadsheet calculations and graphing, and Mathcad™ for mathematical calculations and the plotting of the results. When a quantum chemical calculation or molecular modeling process was called for, we usually provided the solution with PC Spartan™ because of its common availability. However, the majority of the Exercises and many of the Problems can still be solved with a modern hand-held scientific calculator.

In general we adhered rigorously to the rules for significant figures in displaying the final answers. However, when intermediate answers are shown, they are often given with one more figure than would be justified by the data. These excess figures are indicated with an overline.

The solutions were carefully cross-checked for errors not only by us, but very thoroughly by Valerie Walters, who also made many helpful suggestions for improving the solutions. We would be grateful to any readers who bring any remaining errors to our attention.

We warmly thank our publishers, especially Jonathan Crowe and Jessica Fiorillo, and also Samantha Calamari, for their patience in guiding this complex, detailed project to completion.

C. T.
M. C.
C. G.

Contents

STUDENT SOLUTIONS MANUAL FOR QUANTA, MATTER, AND CHANGE:

A MOLECULAR APPROACH TO PHYSICAL CHEMISTRY

Fundamentals

Exercises

F.1 Atoms

F1.1(a) The **nuclear atomic model** consists of atomic number *Z* protons concentrated along with all atomic neutrons within the nucleus, an extremely small central region of the atom. *Z* electrons occupy **atomic orbitals**, which are voluminous regions of the atom that describe where electrons are likely to be found with no more than two electrons in any orbital. The electrostatic attraction binds the negatively charged electrons to the positively charged nucleus, and the so-called strong interaction binds the protons and neutrons within the nucleus.

The atomic orbitals are arranged in shells around the nucleus, each shell being characterized by a **principal quantum number,** n = 1, 2, 3, 4.... A shell consists of n^2 individual orbitals, which are grouped together into n subshells. The **subshells**, and the orbitals they contain, are denoted s, p, d, and f. For all neutral atoms other than hydrogen, the subshells of a given shell have slightly different energies.

F1.2(a)

	Example	Element	Ground-state Electronic Configuration
(a)	Group 2	Ca, calcium	$[Ar]4s^2$
(b)	Group 7	Mn, manganese	$[Ar]3d^54s^2$
(c)	Group 15	As, arsenic	$[Ar]3d^{10}4s^24p^3$

F1.3(a) (a) chemical formula and name: $MgCl_2$, magnesium chloride

ions: Mg^{2+} and Cl^-

oxidation numbers of the elements: magnesium, +2; chlorine, –1

(b) chemical formula and name: FeO, iron(II) oxide

ions: Fe^{2+} and O^{2-}

oxidation numbers of the elements: iron, +2; oxygen, –2

(c) chemical formula and name: Hg_2Cl_2, mercury(I) chloride

ions: Cl^- and Hg_2^{2+} (a polyatomic ion)

oxidation numbers of the elements: mercury, +1; chlorine, –1

F1.4(a)

(light grey)	Metals conduct electricity, have luster, and are malleable and ductile.
(dark grey)	Nonmetals do not conduct electricity and are neither malleable nor ductile.
(white)	Metalloids typically have the appearance of metals but behave chemically like nonmetals.

1 IA	2 IIA	3 IIIB	4 IVB	5 VB	6 VIB	7 VIIB	8 VIIIB	9 VIIIB	10 VIIIB	11 IB	12 IIB	13 IIIA	14 IVA	15 VA	16 VIA	17 VIIA	18 VIIIA
1 H 1.008	Periodic Table of the Elements																2 He 4.003
3 Li 6.941	4 Be 9.012											5 B 10.81	6 C 12.01	7 N 14.01	8 O 16.00	9 F 19.00	10 Ne 20.18
11 Na 22.99	12 Mg 24.31											13 Al 26.98	14 Si 28.09	15 P 30.97	16 S 32.07	17 Cl 35.45	18 Ar 39.95
19 K 39.10	20 Ca 40.08	21 Sc 44.96	22 Ti 47.88	23 V 50.94	24 Cr 52.00	25 Mn 54.94	26 Fe 55.85	27 Co 58.93	28 Ni 58.69	29 Cu 63.55	30 Zn 65.38	31 Ga 69.72	32 Ge 72.59	33 As 74.92	34 Se 78.96	35 Br 79.90	36 Kr 83.80
37 Rb 85.47	38 Sr 87.62	39 Y 88.91	40 Zr 91.22	41 Nb 92.91	42 Mo 95.94	43 Tc (98)	44 Ru 101.1	45 Rh 102.9	46 Pd 106.4	47 Ag 107.9	48 Cd 112.4	49 In 114.8	50 Sn 118.7	51 Sb 121.8	52 Te 127.6	53 I 126.9	54 Xe 131.3
55 Cs 132.9	56 Ba 137.3	57 La 138.9	72 Hf 178.5	73 Ta 180.9	74 W 183.9	75 Re 186.2	76 Os 190.2	77 Ir 192.2	78 Pt 195.1	79 Au 197.0	80 Hg 200.6	81 Tl 204.4	82 Pb 207.2	83 Bi 209.0	84 Po (209)	85 At (210)	86 Rn (222)
87 Fr (223)	88 Ra 226	89 Ac (227)															
				58 Ce 140.1	59 Pr 140.9	60 Nd 144.2	61 Pm 145	62 Sm 150.4	63 Eu 152.0	64 Gd 157.3	65 Tb 158.9	66 Dy 162.5					
				90 Th 232.0	91 Pa (231)	92 U 238.0	93 Np 237	94 Pu (244)	95 Am (243)	96 Cm (247)	97 Bk (247)	98 Cf (251)					

F.2 Molecules

F2.1(a) A **single bond** is a shared pair of electrons between adjacent atoms within a molecule while a **multiple bond** involves the sharing of either two pairs of electrons (a double bond) or three pairs of electrons (a triple bond).

F2.2(a) (a) Sulfite anion, SO_3^{2-}

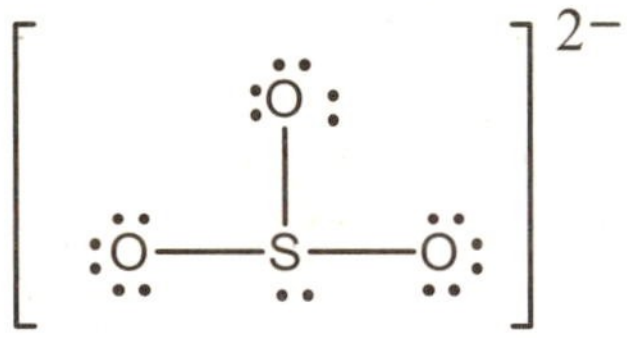

Alternatively, resonance structures may be drawn and, if desired, formal charges (shown in circles below) may be indicated.

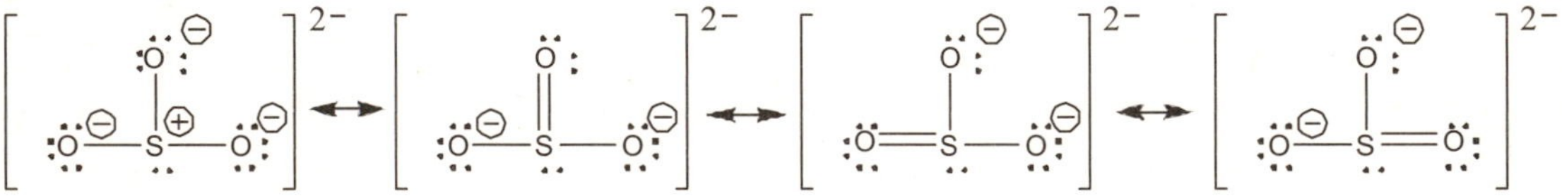

(b) Xenon tetrafluoride, XeF_4

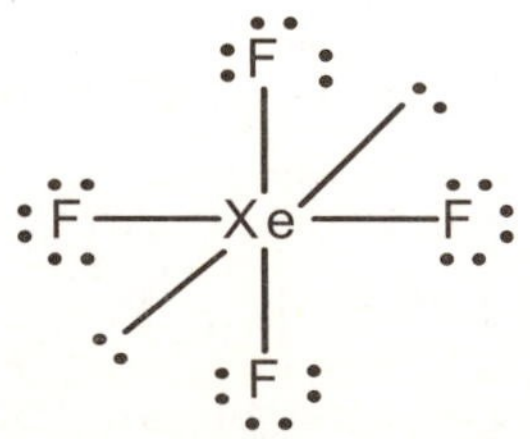

(c) White phosphorus, P_4

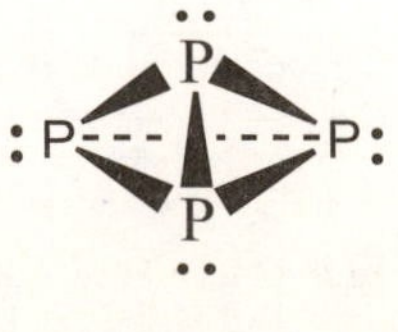

F2.3(a) **Valence-shell electron pair repulsion theory (VSEPR theory)** predicts molecular shape with the concept that regions of high electron density (as represented by single bonds, multiple bonds, and lone pairs) take up orientations around the central atom that maximize their separation. The resulting positions of attached atoms (not lone pairs) are used to classify the shape of the molecule. When the central atom has two or more lone pairs, the molecular geometry must minimize repulsion between the relatively diffuse orbitals of each lone pair. Furthermore, it is assumed that the repulsion between a lone pair and a bonding pair is stronger than the repulsion between two bonding pairs, thereby making bond angles smaller than the idealized bond angles that appear in the absence of a lone pair.

F2.4(a) Molecular shape and polyatomic ion shape are predicted by drawing the Lewis structure and applying the concepts of VSEPR theory.

(a) PCl_3

Lewis structure:

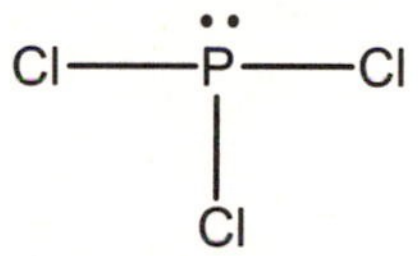

Orientations caused by repulsions between one lone pair and three bonding pairs:

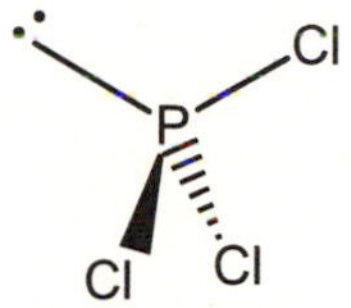

Molecular shape: trigonal pyramidal and bond angles somewhat smaller than 109.5°

(b) PCl_5

Lewis structure:

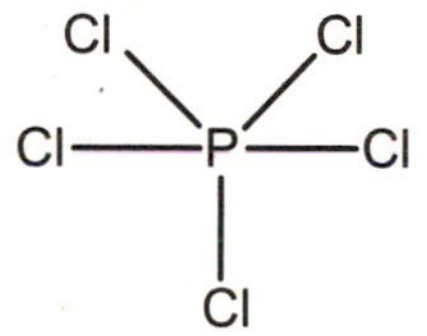

Orientations caused by repulsions between five bonding pairs (no lone pair):

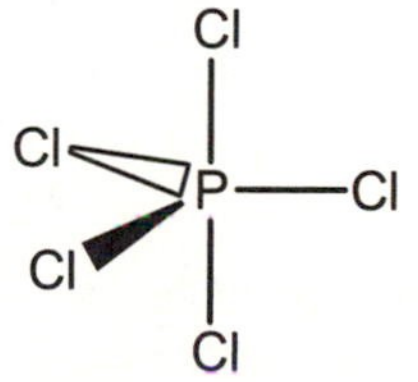

Molecular shape: trigonal bipyramidal with equatorial bond angles of 120° and axial bond angles of 90°

(c) XeF_2

Lewis structure:

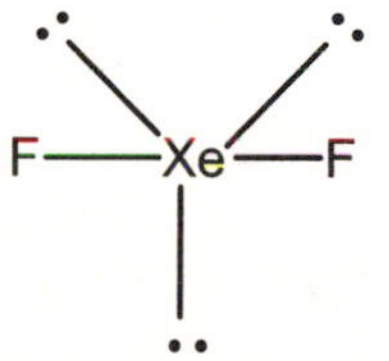

Orientations caused by repulsions between three lone pairs and two bonding pairs:

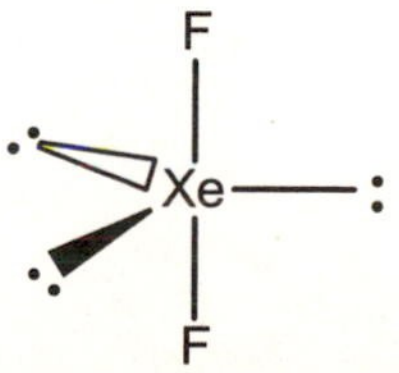

Molecular shape: linear with a 180° bond angle

(d) XeF4

Lewis structure: 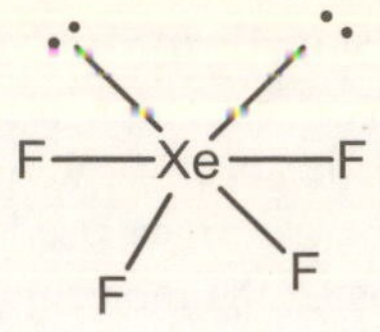

Orientations caused by repulsions between two lone pairs and four bonding pairs:

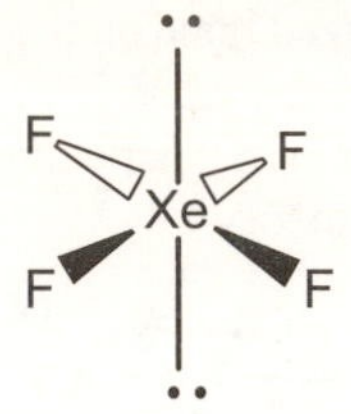

Molecular shape: square planar with 90° bond angles

F2.5(a) (a) δ^+ C—Cl δ^- (b) P—H Nonpolar or weakly polar. (c) δ^+ N—O δ^-

F2.6(a)
(a) CO_2 is a linear, nonpolar molecule.
(b) SO_2 is a bent, polar molecule.
(c) N_2O is a linear, polar molecule.
(d) SF_4 is a seesaw molecule, and it is a polar molecule.

F2.7(a) In the order of increasing dipole moment: CO_2, N_2O, SF_4, SO_2

F.3 Bulk matter

F3.1(a) The solid phase of matter has a shape that is independent of the container it occupies. It has a density compatible with the close proximity of either its constituent elemental atoms or its constituent molecules, and consequently it has low compressibility. Constituent atoms, or molecules, are held firmly at specific lattice sites by relatively strong net forces of attraction between neighboring constituents. Solids may be characterized by terms such as *brittle, ductile, tensile strength, toughness,* and *hardness.*

A liquid adopts the shape of the part of the container that it occupies; it can flow under the influence of gravitational attraction to occupy any shape. Like a solid, it has a density caused by the close proximity of either its constituent elemental atoms or its constituent molecules, and it has low compressibility. Liquids can flow because the constituent atoms or molecules have enough average kinetic energy to overcome the attractive forces between neighboring constituents, thereby making it possible for constituents to slip past immediate neighbors. This causes constituents to be placed randomly in contrast to the orderly array in crystals. Liquids are characterized by terms such as *surface tension, viscosity,* and *capillary action.* Liquids within a vertical, narrow tube exhibit a meniscus that is either concave up or concave down depending on the nature of the attractive or repulsive forces between the liquid and the material of the tube.

Gases have no fixed shape or volume. They expand to fill the container. The constituent molecules move freely and randomly. A perfect gas has a total molecular volume that is negligibly small compared to the container volume, and because of the relatively large average distance between molecules, intermolecular attractive forces are negligibly small. Gases are compressible.

F3.2(a)
(a) Mass is an extensive property.
(b) Mass density is an intensive property.
(c) Temperature is an intensive property.
(d) Number density is an intensive property.

F3.3(a) (a) $n = \dfrac{m}{M} = 25.0\ \text{g}\left(\dfrac{1\ \text{mol}}{46.069\ \text{g}}\right) = \boxed{0.543\ \text{mol}}$ [F.1]

(b) $N = nN_A = 0.543\ \text{mol}\left(\dfrac{6.0221\times10^{23}\ \text{molecules}}{\text{mol}}\right) = \boxed{3.27\times10^{23}\ \text{molecules}}$

F3.4(a) (a) $m = n\,M = 10.0\ \text{mol}\left(\dfrac{18.015\ \text{g}}{\text{mol}}\right) = \boxed{180.\ \text{g}}$ [F.1]

(b) weight $= F_{\text{gravity on Earth}} = m\,g_{\text{Earth}}$

$$= (180.\ \text{g})\times(9.81\ \text{m s}^{-2})\times\left(\frac{1\ \text{kg}}{1000\ \text{g}}\right) = 1.77\ \text{kg m s}^{-2} = \boxed{1.77\ \text{N}}$$

F3.5(a) $p = \dfrac{F}{A} = \dfrac{mg}{A}$

$$= \frac{(65\ \text{kg})\times(9.81\ \text{m s}^{-2})}{150\ \text{cm}^2}\left(\frac{1\ \text{cm}^2}{10^{-4}\ \text{m}^2}\right) = 4.3\times10^4\ \text{N m}^{-2} = 4.3\times10^4\ \text{Pa}\left(\frac{1\ \text{bar}}{10^5\ \text{Pa}}\right)$$

$$= \boxed{0.43\ \text{bar}}$$

F3.6(a) $0.43\ \text{bar}\left(\dfrac{1\ \text{atm}}{1.01325\ \text{bar}}\right) = \boxed{0.42\ \text{atm}}$

F3.7(a) (a) $1.45\ \text{atm}\left(\dfrac{1.01325\times10^5\ \text{Pa}}{1\ \text{atm}}\right) = \boxed{1.47\times10^5\ \text{Pa}}$

(b) $1.45\ \text{atm}\left(\dfrac{1.01325\ \text{bar}}{1\ \text{atm}}\right) = \boxed{1.47\ \text{bar}}$

F3.8(a) $T/\text{K} = \theta/°\text{C} + 273.15 = 37.0 + 273.15 = 310.2$ [F.2]

$\boxed{T = 310.2\ \text{K}}$

F3.9(a) To devise an equation relating the Fahrenheit and Celsius scales requires that consideration be given to both the degree size and a common reference point in each scale. Between the normal freezing point of water and its normal boiling point, there is a degree scaling of 100°C and 180°F. Thus, the scaling ratio is 5°C per 9°F. A convenient reference point is provided by the normal freezing point of water, which is 0°C and 32°F. So to calculate the Celsius temperature from a given Fahrenheit temperature, 32 must be subtracted from the Fahrenheit temperature (θ_F) followed by scaling to the Celsius temperature (θ) with the ratio 5°C / 9°F.

$$\boxed{\theta/°\text{C} = \tfrac{5}{9}\times(\theta_F/°\text{F} - 32) \quad \text{or} \quad \theta_F/°\text{F} = \tfrac{9}{5}\times\theta/°\text{C} + 32}$$

$$\theta_F/°\text{F} = \tfrac{9}{5}\times\theta/°\text{C} + 32 = \tfrac{9}{5}\times78.5 + 32 = 173$$

$\boxed{\theta_F = 173\ °\text{F}}$

F3.10(a)

$$110\text{ kPa}\times\left(\frac{(7.0+273.15)\text{ K}}{(20.0+273.15)\text{ K}}\right)=\boxed{105\text{ kPa}}$$

F3.11(a)

$$pV=nRT\text{ [F.3]}=\frac{mRT}{M}$$

$$M=\frac{mRT}{pV}=\frac{\rho RT}{p}\quad\text{where }\rho\text{ is the mass density}$$

$$=\frac{(3.710\text{ kg m}^{-3})(8.314\text{ J K}^{-1}\text{ mol}^{-1})(773.15\text{ K})}{93.2\times10^{3}\text{ Pa}}=0.256\text{ kg mol}^{-1}=256\text{ g mol}^{-1}$$

The molecular mass is eight times as large as the atomic mass of sulfur (32.06 g mol^{-1}), so the molecular formula is $\boxed{S_8}$.

F3.12(a)

$$n=22\text{ g}\times\left(\frac{1\text{ mol}}{30.07\text{ g}}\right)=0.73\text{ mol}$$

$$p=\frac{nRT}{V}\text{ [F.3]}=\frac{(0.73\text{ mol})(8.314\text{ J K}^{-1}\text{ mol}^{-1})(298.15\text{ K})}{1000.\text{ cm}^{3}}\left(\frac{\text{cm}^{3}}{10^{-6}\text{ m}^{3}}\right)$$

$$=1.8\times10^{6}\text{ Pa}=\boxed{1.8\text{ MPa}}$$

F3.13(a)

$$n_{N_2}=1.0\text{ mole}\quad\text{and}\quad n_{H_2}=2.0\text{ mole}$$

$$p_{N_2}=\frac{n_{N_2}RT}{V}\text{ [F.3]}=\frac{(1.0\text{ mol})(8.314\text{ J K}^{-1}\text{ mol}^{-1})(278.15\text{ K})}{10.0\text{ dm}^{3}}\left(\frac{\text{dm}^{3}}{10^{-3}\text{ m}^{3}}\right)=\boxed{2.3\times10^{5}\text{ Pa}}$$

Since there are twice as many moles of hydrogen as nitrogen, the hydrogen partial pressure must be twice as large.

$$p_{H_2}=\boxed{4.6\times10^{5}\text{ Pa}}\qquad p=p_{N_2}+p_{H_2}\text{ [F.4]}=\boxed{6.9\times10^{5}\text{ Pa}}$$

F.4 Thermodynamic properties

F4.1(a)

$$\Delta T=\frac{\Delta U}{C}\text{ [F.5]}=\frac{100.\text{ J}}{3.67\text{ J K}^{-1}}=\boxed{27.2\text{ K or }27.2\text{ °C}}$$

F4.2(a)

$$n=100.\text{ g}\times\left(\frac{1\text{ mol}}{207.2\text{ g}}\right)=0.483\text{ mol}$$

$$\Delta U=C\Delta T\text{ [F.5]}=nC_{m}\Delta T$$

$$=(0.483\text{ mol})(26.44\text{ J K}^{-1}\text{ mol}^{-1})(10.0\text{ K})=\boxed{128\text{ J}}$$

F4.3(a)

$$C_{s}=C_{m}/M=(111.46\text{ J K}^{-1}\text{ mol}^{-1})\times\left(\frac{1\text{ mol}}{46.069\text{ g}}\right)=\boxed{2.4194\text{ J K}^{-1}\text{ g}^{-1}}$$

F4.4(a)

$$C_{m}=C_{s}M=(4.18\text{ J K}^{-1}\text{ g}^{-1})\times\left(\frac{18.02\text{ g}}{\text{mol}}\right)=\boxed{75.3\text{ J K}^{-1}\text{ mol}^{-1}}$$

F4.5(a) Dividing eqn. F.7 by the number of moles n and using the molar properties (intensive) H_m, U_m, and V_m yields the equation

$$H_m - U_m = pV_m$$

Substitution of the perfect gas eos, $pV_m = RT$, yields

$$H_m - U_m = RT = (8.3145 \text{ J mol}^{-1} \text{ K}^{-1}) \times (298 \text{ K}) = \boxed{2.48 \text{ kJ mol}^{-1}}$$

F4.6(a)
$$H_m - U_m = pV_m \text{ [F.7]} = \frac{pM}{\rho} = \frac{(1.00\times10^5 \text{ Pa})\times(207.2 \text{ g mol}^{-1})}{11.350 \text{ g cm}^{-3}}\left(\frac{10^{-6} \text{ m}^3}{\text{cm}^3}\right) = \boxed{1.826 \text{ J mol}^{-1}}$$

F4.7(a)
$$S_{H_2O(g)} > S_{H_2O(l)}$$

F4.8(a)
$$S_{Fe(3000 \text{ K})} > S_{Fe(300 \text{ K})}$$

F4.9(a) The second law of thermodynamics states that any spontaneous (that is, natural) change in an isolated system is accompanied by an increase in the entropy of the system. This tendency is expressed commonly by saying that the natural direction of change is accompanied by dispersal of energy from a localized region to a less organized form.

F4.10(a) In a state of dynamic equilibrium, which is the character of all chemical equilibria, the forward and reverse reactions are occurring at the same rate and there is no net tendency to change in either direction. Examples:

$2\ SO_2(g) + O_2(g) \rightleftharpoons 2\ SO_3(g)$ Addition of oxygen shifts the equilibrium to the right. An increase in pressure also shifts it to the right so as to reduce the number of moles of gas.

$CaCO_3(s) \rightleftharpoons CaO(s) + CO_2(g)$ Addition of carbon dioxide shifts the equilibrium to the left. An increase in pressure also shifts it to the left so as to reduce the number of moles of gas.

$CO(g) + H_2O(g) \rightleftharpoons CO_2(g) + H_2(g)$ Addition of carbon monoxide shifts the equilibrium to the right. An increase in pressure has no effect on the equilibrium because there are equal numbers of moles of gas on left and right.

F.5 The relation between molecular and bulk properties

F5.1(a) Quantized energies are certain discrete values that are permitted for particles confined to a region of space.

F5.2(a)
$$\Delta E = E_{\text{upper}} - E_{\text{lower}} = 1 \text{ eV} = 1.6022\times10^{-19} \text{ J}$$

(a)
$$\frac{N_{\text{upper}}}{N_{\text{lower}}} = e^{-\Delta E/kT} \text{ [F.9]} = e^{-(1.6022\times10^{-19} \text{ J})/\{(1.381\times10^{-23} \text{ J K}^{-1})\times(300 \text{ K})\}} = \boxed{1.602\times10^{-17}}$$

(b)
$$\frac{N_{\text{upper}}}{N_{\text{lower}}} = e^{-\Delta E/kT} = e^{-(1.6022\times10^{-19} \text{ J})/\{(1.381\times10^{-23} \text{ J K}^{-1})\times(3000 \text{ K})\}} = \boxed{2.092\times10^{-2}}$$

F5.3(a)
$$\lim_{T\to0}\left(\frac{N_{\text{upper}}}{N_{\text{lower}}}\right) = \lim_{T\to0}\left(e^{-\Delta E/kT}\right) \text{ [F.9]} = e^{-\infty} = 0$$

In the limit of the absolute zero of temperature all particles occupy the lower state. The upper state is empty.

F5.4(a)
$$\Delta E = E_{\text{upper}} - E_{\text{lower}} = \tilde{\nu}hc = (2500 \text{ cm}^{-1})(6.626\times10^{-34} \text{ J s})(3.000\times10^{10} \text{ cm s}^{-1}) = 4.970\times10^{-20} \text{ J}$$

$$\frac{N_{\text{upper}}}{N_{\text{lower}}} = e^{-\Delta E/kT} \text{ [F.9]} = e^{-(4.970\times10^{-20} \text{ J})/\{(1.381\times10^{-23} \text{ J K}^{-1})\times(293 \text{ K})\}} = \boxed{4.631\times10^{-6}}$$

The ratio $N_{\text{upper}}/N_{\text{lower}}$ is so small that the population of the upper level is approximately zero.

F5.5(a) Kinetic molecular theory, a model for a perfect gas, assumes that the molecules, imagined as particles of negligible size, are in ceaseless, random motion and do not interact except during their brief collisions.

F5.6(a) Molecules can survive for long periods without undergoing chemical reaction at low temperatures when few molecules have the requisite speed and corresponding kinetic energy to promote excitation and bond breakage during collisions.

F5.7(a)

$$v_{\text{mean}} \propto (T/M)^{1/2} \quad \text{[F.11]}$$

$$\frac{v_{\text{mean}}(T_2)}{v_{\text{mean}}(T_1)} = \frac{(T_2/M)^{1/2}}{(T_1/M)^{1/2}} = \left(\frac{T_2}{T_1}\right)^{1/2}$$

$$\frac{v_{\text{mean}}(313\text{ K})}{v_{\text{mean}}(273\text{ K})} = \left(\frac{313\text{ K}}{273\text{ K}}\right)^{1/2} = \boxed{1.07}$$

F5.8(a)

$$v_{\text{mean}} \propto (T/M)^{1/2} \quad \text{[F.11]}$$

$$\frac{v_{\text{mean}}(M_2)}{v_{\text{mean}}(M_1)} = \frac{(T/M_2)^{1/2}}{(T/M_1)^{1/2}} = \left(\frac{M_1}{M_2}\right)^{1/2}$$

$$\frac{v_{\text{mean}}(\text{N}_2)}{v_{\text{mean}}(\text{CO}_2)} = \left(\frac{44.0\text{ g mol}^{-1}}{28.0\text{ g mol}^{-1}}\right)^{1/2} = \boxed{1.25}$$

F5.9(a) A gaseous argon atom has three translational degrees of freedom (the components of motion in the x, y, and z directions). Consequently, the equipartition theorem assigns a mean energy of $^3/_2kT$ to each atom. The molar internal energy is

$$U_{\text{m}} = \tfrac{3}{2}N_{\text{A}}kT = \tfrac{3}{2}RT \text{ [F.10]} = \tfrac{3}{2}(8.3145\text{ J mol}^{-1}\text{ K}^{-1})(298\text{ K}) = 3.72\text{ kJ mol}^{-1}$$

$$U = nU_{\text{m}} = mM^{-1}U_{\text{m}} = (5.0\text{ g})\left(\frac{1\text{ mol}}{39.95\text{ g}}\right)\left(\frac{3.72\text{ kJ}}{\text{mol}}\right) = \boxed{0.47\text{ kJ}}$$

F5.10(a) (a) A gaseous linear carbon dioxide molecule has three quadratic translational degrees of freedom (the components of motion in the x, y, and z directions) but it has only two rotational quadratic degrees of freedom because there is no rotation along the internuclear line. There is a total of five quadratic degrees of freedom for the molecule. Consequently, the equipartition theorem assigns a mean energy of $^5/_2kT$ to each molecule. The molar internal energy is

$$U_{\text{m}} = \tfrac{5}{2}N_{\text{A}}kT = \tfrac{5}{2}RT \text{ [F.10]} = \tfrac{5}{2}(8.3145\text{ J mol}^{-1}\text{ K}^{-1})(293\text{ K}) = 6.09\text{ kJ mol}^{-1}$$

$$U = nU_{\text{m}} = mM^{-1}U_{\text{m}} = (10.0\text{ g})\left(\frac{1\text{ mol}}{44.0\text{ g}}\right)\left(\frac{6.09\text{ kJ}}{\text{mol}}\right) = \boxed{1.38\text{ kJ}}$$

(b) A gaseous, nonlinear methane molecule has three quadratic translational degrees of freedom (the components of motion in the x, y, and z directions) and three quadratic rotational degrees of freedom. Consequently, the equipartition theorem assigns a mean energy of $^6/_2kT$ to each molecule. The molar internal energy is

$$U_{\text{m}} = \tfrac{6}{2}N_{\text{A}}kT = 3RT \text{ [F.10]} = 3(8.3145\text{ J mol}^{-1}\text{ K}^{-1})(293\text{ K}) = 7.31\text{ kJ mol}^{-1}$$

$$U = nU_{\text{m}} = mM^{-1}U_{\text{m}} = (10.0\text{ g})\left(\frac{1\text{ mol}}{16.04\text{ g}}\right)\left(\frac{7.31\text{ kJ}}{\text{mol}}\right) = \boxed{4.56\text{ kJ}}$$

F5.11(a) See Exercise F5.9(a) for the description of the molar internal energy of argon.

$$C_{V,\mathrm{m}} = \frac{\partial U_\mathrm{m}}{\partial T} = \frac{\partial(\frac{3}{2}RT)}{\partial T} = \tfrac{3}{2}R = \tfrac{3}{2}(8.3145\ \mathrm{J\ mol^{-1}\ K^{-1}}) = \boxed{12.47\ \mathrm{J\ mol^{-1}\ K^{-1}}}$$

F5.12(a) See Exercise F5.10(a) for the description of the molar internal energies of carbon dioxide and methane.

(a) $U_\mathrm{m} = \frac{5}{2}RT$ for carbon dioxide

$$C_{V,\mathrm{m}} = \frac{\partial U_\mathrm{m}}{\partial T} = \frac{\partial(\frac{5}{2}RT)}{\partial T} = \tfrac{5}{2}R = \tfrac{5}{2}(8.3145\ \mathrm{J\ mol^{-1}\ K^{-1}}) = \boxed{20.79\ \mathrm{J\ mol^{-1}\ K^{-1}}}$$

(b) $U_\mathrm{m} = 3RT$ for methane

$$C_{V,\mathrm{m}} = \frac{\partial U_\mathrm{m}}{\partial T} = \frac{\partial(3RT)}{\partial T} = 3R = 3(8.3145\ \mathrm{J\ mol^{-1}\ K^{-1}}) = \boxed{24.94\ \mathrm{J\ mol^{-1}\ K^{-1}}}$$

F.6 Particles

F6.1(a) $a = \mathrm{d}v/\mathrm{d}t = g$ so $\mathrm{d}v = g\,\mathrm{d}t$. The acceleration of free fall is constant near the surface of the Earth.

$$\int_{v=0}^{v(t)} \mathrm{d}v = \int_{t=0}^{t=t} g\,\mathrm{d}t$$

$$v(t) = gt$$

(a) $v(1.0\ \mathrm{s}) = (9.81\ \mathrm{m\ s^{-2}})\times(1.0\ \mathrm{s}) = \boxed{9.81\ \mathrm{m\ s^{-1}}}$

$$E_\mathrm{k} = \tfrac{1}{2}mv^2 = \tfrac{1}{2}(0.0010\ \mathrm{kg})\times(9.81\ \mathrm{m\ s^{-1}})^2 = \boxed{48\ \mathrm{mJ}}$$

(b) $v(3.0\ \mathrm{s}) = (9.81\ \mathrm{m\ s^{-2}})\times(3.0\ \mathrm{s}) = \boxed{29.4\ \mathrm{m\ s^{-1}}}$

$$E_\mathrm{k} = \tfrac{1}{2}mv^2 = \tfrac{1}{2}(0.0010\ \mathrm{kg})\times(29.4\ \mathrm{m\ s^{-1}})^2 = \boxed{0.43\ \mathrm{J}}$$

F6.2(a) The terminal velocity occurs when there is a balance between the force exerted by the electric field and the force of frictional drag. It will be in the direction of the field and have the magnitude s_terminal.

$$ze\mathcal{E} = 6\pi\eta R s_\mathrm{terminal}$$

$$\boxed{s_\mathrm{terminal} = \frac{ze\mathcal{E}}{6\pi\eta R}}$$

F6.3(a)

$$x(t) = A\sin(\omega t) + B\cos(\omega t)$$

$$\frac{\mathrm{d}x}{\mathrm{d}t} = A\frac{\mathrm{d}\sin(\omega t)}{\mathrm{d}t} + B\frac{\mathrm{d}\cos(\omega t)}{\mathrm{d}t} = A\times(\omega\cos(\omega t)) + B\times(-\omega\sin(\omega t))$$

$$= A\omega\cos(\omega t) - B\omega\sin(\omega t)$$

$$\frac{d^2x}{dt^2} = A\omega\frac{d\cos(\omega t)}{dt} - B\omega\frac{d\sin(\omega t)}{dt}$$
$$= A\omega\times(-\omega\sin(\omega t)) - B\omega\times(\omega\cos(\omega t))$$
$$= -\omega^2\left(A\sin(\omega t) + B\cos(\omega t)\right) = -\omega^2 x \quad \text{where } \omega = (k/m)^{1/2}$$
$$m\frac{d^2x}{dt^2} = -m\omega^2 x = -m\times\left(\frac{k}{m}\right)x = -kx$$

This confirms that $x(t)$ satisfies the harmonic oscillator equation of motion.

F6.4(a) The harmonic oscillator solution $x(t) = A\sin(\omega t)$ has the characteristics

$$x(t) = A\sin(\omega t) + B\cos(\omega t)$$

$$v(t) = \frac{dx}{dt} = A\omega\cos(\omega t)$$

$$x_{min} = x(t = n\pi/\omega,\ n = 0,1,2...) = 0 \quad \text{and} \quad x_{max} = x(t = (n+\tfrac{1}{2})\pi/\omega,\ n = 0,1,2...) = A$$

At x_{min} the harmonic oscillator restoration force (Hooke's law, $-kx$) is zero, and consequently, the harmonic potential energy, V, is a minimum that is taken to equal zero while kinetic energy, E_k, is a maximum. As kinetic energy causes movement away from x_{min}, kinetic energy continually converts to potential energy until no kinetic energy remains at x_{max} where the restoration force changes the direction of motion and the conversion process reverses. We may easily find an expression for the total energy $E(A)$ by examination of either x_{min} or x_{max}.

Analysis using x_{min}:

$$E = E_{k,max} = \tfrac{1}{2}m v_{max}{}^2 = \tfrac{1}{2}mA^2\omega^2 \quad \text{where } \omega = (k/m)^{1/2} \text{ or } m\omega^2 = k$$

$$\boxed{E = \tfrac{1}{2}kA^2}$$

We begin the analysis that uses x_{max} by deriving the expression for the harmonic potential energy.

$$dV = -F_x dx\,[\text{F.20}] = kx\,dx \quad (\text{i.e., } F_x = -kx;\ \text{Exercise F6.3a})$$

$$\int_0^{V(x)} dV = \int_0^x kx\,dx$$

$$V(x) = \tfrac{1}{2}kx^2 \quad \text{Thus, } V_{max} = V(x_{max}) = \tfrac{1}{2}kA^2 \text{ and } \boxed{E = V_{max} = \tfrac{1}{2}kA^2}$$

F6.5(a) The electron acceleration is caused by the centripetal electrostatic force of attraction of the positively charged nucleus for the negatively charged electron. The force and its resultant acceleration always point from the electron to the nucleus. A path of constant radius r is possible only when the electron speed v creates a centrifugal force, mv^2/r, that exactly balances the centripetal force. The magnitude of the centripetal acceleration is found by equating the force of Newton's second law of motion and the centrifugal force.

$$F = ma = \frac{mv^2}{r}$$

$$a = \frac{v^2}{r} = \frac{(2188\times10^3\ \text{m s}^{-1})^2}{53\times10^{-12}\ \text{m}} = 9.0\times10^{22}\ \text{m s}^{-2}$$

F6.6(a) The relationship between angular velocity ω and the speed of an electron in a stable circular path (i.e., an "orbit" of radius r) is found by recognizing that, when the electron travels through one orbit, it traverses 2π radians while traveling the distance $2\pi r$. Thus,

$$\omega = v\times(2\pi/2\pi r) = v/r$$

$$J = I\omega = (mr^2)\times\left(\frac{v}{r}\right) = mrv \quad \text{[F.15]}$$

$$= (9.10938\times10^{-31}\ \text{kg})\times(53\times10^{-12}\ \text{m})\times(2188\times10^{3}\ \text{m s}^{-1})$$

$$= 1.1\times10^{-34}\ \text{kg m}^2\ \text{s}^{-1} = 1.1\times10^{-34}\ \text{J s}$$

$$= (1.1\times10^{-34}\ \text{J s})\times\left(\frac{2\pi}{h}\right)\hbar \quad \text{where } \hbar = h/2\pi$$

$$= (1.1\times10^{-34}\ \text{J s})\times\left(\frac{2\pi}{6.63\times10^{-34}\ \text{J s}}\right)\hbar = \boxed{1.0\ \hbar}$$

F6.7(a) $w = \frac{1}{2}kx^2$ where $x = R - R_e$ is the displacement from equilibrium

(a) $$w = \tfrac{1}{2}(450\ \text{N m}^{-1})\times(10\times10^{-12}\ \text{m})^2 = 2.25\times10^{-20}\ \text{N m} = \boxed{2.25\times10^{-20}\ \text{J}}$$

(b) $$w = \tfrac{1}{2}(450\ \text{N m}^{-1})\times(20\times10^{-12}\ \text{m})^2 = 9.00\times10^{-20}\ \text{N m} = \boxed{9.00\times10^{-20}\ \text{J}}$$

F6.8(a) $$E_k = e\Delta\varphi$$

$$\tfrac{1}{2}mv^2 = e\Delta\varphi \quad \text{or} \quad v = \left(\frac{2e\Delta\varphi}{m}\right)^{1/2}$$

$$v = \left(\frac{2(1.6022\times10^{-19}\ \text{C})\times(100\times10^{3}\ \text{V})}{9.10938\times10^{-31}\ \text{kg}}\right)^{1/2} = 1.88\times10^{8}\left(\frac{\text{C V}}{\text{kg}}\right)^{1/2} = 1.88\times10^{8}\left(\frac{\text{J}}{\text{kg}}\right)^{1/2}$$

$$= 1.88\times10^{8}\left(\frac{\text{kg m}^2\ \text{s}^{-2}}{\text{kg}}\right)^{1/2} = \boxed{1.88\times10^{8}\ \text{m s}^{-1}}$$

$$E = E_k = e\Delta\varphi = e\times(100\ \text{kV}) = \boxed{100\ \text{keV}}$$

F6.9(a) The work needed to separate two ions to infinity is identical to the Coulomb potential drop that occurs when the two ions are brought from an infinite separation, where the interaction potential equals zero, to a separation of r.

In a vacuum:

$$w = -V = -\left(\frac{q_1q_2}{4\pi\varepsilon_0 r}\right)\ \text{[F.24]} = -\left(\frac{(e)\times(-e)}{4\pi\varepsilon_0 r}\right) = \frac{e^2}{4\pi\varepsilon_0 r}$$

$$= \frac{(1.6022\times10^{-19}\ \text{C})^2}{4\pi(8.85419\times10^{-12}\ \text{J}^{-1}\ \text{C}^2\ \text{m}^{-1})\times(200\times10^{-12}\ \text{m})} = \boxed{1.15\times10^{-18}\ \text{J}}$$

In water:

$$w = -V = -\left(\frac{q_1q_2}{4\pi\varepsilon r}\right)\ \text{[F.24]} = -\left(\frac{(e)\times(-e)}{4\pi\varepsilon r}\right) = \frac{e^2}{4\pi\varepsilon r} = \frac{e^2}{4\pi\varepsilon_r\varepsilon_0 r}\ \text{[F.25] where } \varepsilon_r = 78 \text{ for water at 25°C}$$

$$= \frac{(1.6022\times10^{-19}\ \text{C})^2}{4\pi(78)\times(8.85419\times10^{-12}\ \text{J}^{-1}\ \text{C}^2\ \text{m}^{-1})\times(200\times10^{-12}\ \text{m})} = \boxed{1.48\times10^{-20}\ \text{J}}$$

F6.10(a) We will model a solution by assuming that the LiH pair consists of the two point charge ions Li^+ and H^-. The electric potential will be calculated along the line of the ions.

$$\varphi = \varphi_{Li^+} + \varphi_{H^-} \text{ [F.27]} = \frac{e}{4\pi\varepsilon_0 r_{Li^+}} + \frac{(-e)}{4\pi\varepsilon_0 r_{Cl^-}} \text{ [F.26]} = \frac{e}{4\pi\varepsilon_0}\left(\frac{1}{r_{Li^+}} - \frac{1}{r_{Cl^-}}\right)$$

$$\varphi = \frac{1.6022\times10^{-19}\ \text{C}}{4\pi\left(8.85419\times10^{-12}\ \text{J}^{-1}\ \text{C}^2\ \text{m}^{-1}\right)}\left(\frac{1}{200\times10^{-12}\ \text{m}} - \frac{1}{150\times10^{-12}\ \text{m}}\right)$$

$$= -2.40\ \text{J C}^{-1} = -2.40\ \text{C V C}^{-1} = \boxed{-2.40\ \text{V}}$$

F6.11(a) We will assume that the electric circuit has a negligibly small heat capacity so that the water receives all of the electrically generated heat energy.

$$\Delta U_{H_2O} = \text{energy dissipated by the electric circuit}$$

$$= I\Delta\varphi\,\Delta t \text{ [F.30]}$$

$$= (2.23\ \text{A})\times(15.0\ \text{V})\times(720\ \text{s}) = 24.1\times10^3\ \text{C s}^{-1}\ \text{V s} = \boxed{24.1\ \text{kJ}}$$

$$\Delta U_{H_2O} = (nC\Delta T)_{H_2O} \text{ [F.5]}$$

$$\Delta T = \frac{\Delta U_{H_2O}}{(nC)_{H_2O}} = \frac{\Delta U_{H_2O}}{(mC/M)_{H_2O}} = \frac{24.1\times10^3\ \text{J}}{(200\ \text{g})\times\left(75.3\ \text{J K}^{-1}\ \text{mol}^{-1}\right)/\left(18.02\ \text{g mol}^{-1}\right)}$$

$$= 28.8\ \text{K} = \boxed{28.8\ °\text{C}}$$

F.7 Waves

F7.1(a)

$$\tilde{\nu} = \frac{1}{\lambda} \text{ [F.33]} = \frac{1}{590\times10^{-9}\ \text{m}}\left(\frac{10^{-2}\ \text{m}}{\text{cm}}\right) = \boxed{1.69\times10^4\ \text{cm}^{-1}}$$

$$\nu = \frac{c}{\lambda} \text{ [F.31]} = \frac{3.00\times10^8\ \text{m s}^{-1}}{590\times10^{-9}\ \text{m}} = 5.08\times10^{14}\ \text{s}^{-1} = \boxed{5.08\times10^{14}\ \text{Hz}}$$

F7.2(a)

$$c_{H_2O} = \frac{c}{n_r} \text{ [F.32]} = \frac{3.00\times10^8\ \text{m s}^{-1}}{1.33} = \boxed{2.26\times10^8\ \text{m s}^{-1}}$$

F7.3(a)

$$\lambda = \frac{1}{\tilde{\nu}} \text{ [F.33]} = \frac{1}{2500\ \text{cm}^{-1}}\left(\frac{10^6\ \mu\text{m}}{10^2\ \text{cm}}\right) = \boxed{4.00\ \mu\text{m}}$$

$$\nu = \frac{c}{\lambda} \text{ [F.31]} = \frac{3.00\times10^8\ \text{m s}^{-1}}{4.00\times10^{-6}\ \text{m}} = 7.50\times10^{13}\ \text{s}^{-1} = \boxed{7.50\times10^{13}\ \text{Hz}}$$

F7.4(a) (a) $f(x) = \cos(2\pi x/\lambda)$ where $\lambda = 1$ cm (See Figure F.1: ______________.)

(b) $f(x) = \cos(2\pi x/\lambda + \varphi)$ where $\varphi = \pi/3$ (See Figure F.1: - - - - - - - - - - - - - - - .)

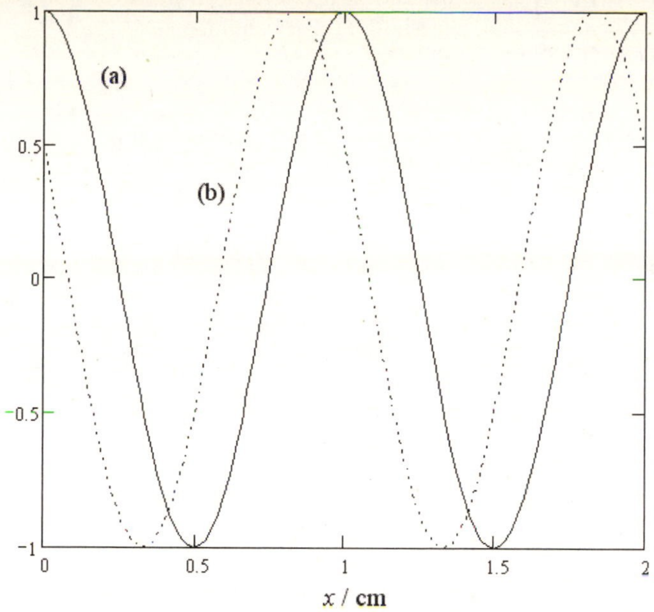

Figure F.1

F7.5(a)

$$f(x,t) = \cos\{2\pi\nu t - (2\pi/\lambda)x\}$$

$$\frac{\partial f}{\partial t} = -2\pi\nu \sin\{2\pi\nu t - (2\pi/\lambda)x\}$$

$$\frac{\partial^2 f}{\partial t^2} = -(2\pi\nu)^2 \cos\{2\pi\nu t - (2\pi/\lambda)x\}$$

$$= -(2\pi\nu)^2 f$$

$$\frac{\partial f}{\partial x} = -(-2\pi/\lambda)\sin\{2\pi\nu t - (2\pi/\lambda)x\} = (2\pi/\lambda)\sin\{2\pi\nu t - (2\pi/\lambda)x\}$$

$$\frac{\partial^2 f}{\partial x^2} = -(2\pi/\lambda)^2 \cos\{2\pi\nu t - (2\pi/\lambda)x\}$$

$$= -(2\pi/\lambda)^2 f$$

$$c^2\frac{\partial^2 f}{\partial x^2} = -(2\pi c/\lambda)^2 f$$

$$= -(2\pi\nu)^2 f$$

Inspection of the simplified equations for $\partial^2 f/\partial t^2$ and $c^2(\partial^2 f/\partial x^2)$ reveals that they are equal. Consequently, we conclude that $f(x,t)$ satisfies the wave equation.

F.8 Units

F8.1(a)

$$1.45\ \text{cm}^3 = 1.45\ (10^{-2}\ \text{m})^3 = \boxed{1.45\times10^{-6}\ \text{m}^3}$$

F8.2(a)

$$\left(11.2\ \frac{\text{g}}{\text{cm}^3}\right)\times\left(\frac{1\ \text{kg}}{10^3\ \text{g}}\right)\times\left(\frac{1\ \text{cm}}{10^{-2}\ \text{m}}\right)^3 = \boxed{11.2\times10^3\ \text{kg m}^{-3}}$$

F8.3(a)

$$\frac{\text{Pa}}{\text{J}} = \frac{\text{N m}^{-2}}{\text{J}} = \frac{(\text{N m})\ \text{m}^{-3}}{\text{J}} = \frac{\text{J m}^{-3}}{\text{J}} = \boxed{\text{m}^{-3}}$$

F8.4(a)

$$\frac{kT}{hc} = \frac{(1.381\times10^{-23}\ \mathrm{J\,K^{-1}})\times(298\ \mathrm{K})}{(6.626\times10^{-34}\ \mathrm{J\,s})\times(2.998\times10^{10}\ \mathrm{cm\,s^{-1}})} = \boxed{207.2\ \mathrm{cm^{-1}}}$$

F8.5(a)

$$\left(\frac{8.3144\ \mathrm{J}}{\mathrm{K\,mol}}\right)\times\left(\frac{\mathrm{Pa\,m^3}}{\mathrm{J}}\right)\times\left(\frac{1\ \mathrm{atm}}{101325\ \mathrm{Pa}}\right)\times\left(\frac{\mathrm{dm}}{10^{-1}\ \mathrm{m}}\right)^3 = \boxed{0.08206\ \mathrm{atm\,dm^3\,K^{-1}\,mol^{-1}}}$$

F8.6(a)

$$(1\ \mathrm{dm^3\ atm})\times\left(\frac{10^{-1}\ \mathrm{m}}{\mathrm{dm}}\right)^3\times\left(\frac{101325\ \mathrm{Pa}}{1\ \mathrm{atm}}\right)\times\left(\frac{\mathrm{J}}{\mathrm{Pa\,m^3}}\right) = \boxed{101.325\ \mathrm{J}}$$

F8.7(a) (a) Base unit of $\dfrac{e^2}{\varepsilon_0 r^2} = \dfrac{\mathrm{C^2}}{(\mathrm{C^2\,J^{-1}\,m^{-1}})\times(\mathrm{m^2})} = (\mathrm{J\,m^{-1}})\times\left(\dfrac{\mathrm{kg\,m^2\,s^{-2}}}{\mathrm{J}}\right) = \boxed{\mathrm{kg\,m\,s^{-2}}}$

(b) Unit of $\dfrac{e^2}{\varepsilon_0 r^2} = \dfrac{\mathrm{C^2}}{(\mathrm{C^2\,J^{-1}\,m^{-1}})\times(\mathrm{m^2})} = (\mathrm{J\,m^{-1}})\times\left(\dfrac{\mathrm{kg\,m^2\,s^{-2}}}{\mathrm{J}}\right) = \mathrm{kg\,m\,s^{-2}} = \boxed{\mathrm{N}}$

PART 1 Quantum theory

1 The principles of quantum theory

Answers to discussion questions

D1.1 At the end of the nineteenth century and the beginning of the twentieth, there were many experimental results on the properties of matter and radiation that could not be explained on the basis of established physical principles and theories. Here we list only some of the most significant.

(1) The photoelectric effect revealed that electromagnetic radiation, classically considered to be a wave, also exhibits the particle-like behavior of photons. Each photon is a discrete unit, or quantum, of energy that is absorbed during collisions with electrons. Photons are never partially absorbed. They either completely give up their energy or they are not absorbed. The energy of a photon can be calculated if either the radiation frequency or wavelength is known: $\boldsymbol{E_{\text{photon}} = h\nu = hc/\lambda}$.
(2) Absorption and emission spectra indicated that atoms and molecules can only absorb or emit discrete packets of energy (i.e., photons). This means that an atom or molecule has specific, allowed energy levels, and we say that their energies are quantized. During a spectroscopic transition the atom or molecule gains or loses the energy ΔE by either absorption of a photon or emission of a photon, respectively. Thus, spectral lines must satisfy the **Bohr frequency condition:** $\boldsymbol{\Delta E = h\nu}$.
(3) Neutron and electron diffraction studies indicated that these particles also possess wave-like properties of constructive and destructive interference. The joint particle and wave character of matter and radiation is called **wave-particle duality**. The **de Broglie relation**, $\lambda_{\text{de Broglie}} = h/p$, connects the wave character of a particle ($\lambda_{\text{de Broglie}}$) with its particulate momentum (p).

Evidence that resulted in the development of quantum theory also included:

(4) The energy density distribution of blackbody radiation as a function of wavelength.
(5) The heat capacities of monatomic solids such as copper metal.

D1.3 If the wavefunction describing the linear momentum of a particle is precisely known, the particle has a definite state of linear momentum; but then according to the uncertainty principle (eqn. 1.19a), the position of the particle is completely unknown. Conversely, if the position of a particle is precisely known, its linear momentum cannot be described by a single wavefunction. Rather, the wavefunction is a superposition of many wavefunctions, each corresponding to a different value for the linear momentum. All knowledge of the linear momentum of the particle is lost when its position is specified exactly. In the limit of an infinite number of superposed wavefunctions, the wavepacket illustrated in text Fig. 1.17 turns into the sharply spiked packet shown in Fig. 1.16. But the requirement of the superposition of an infinite number of momentum wavefunctions in order to locate the particle means a complete lack of knowledge of the momentum.

D1.5 By wave-particle duality we mean that in some experiments an entity behaves as a wave while in other experiments the same entity behaves as a particle. Electromagnetic radiation behaves as a wave in reflection and refraction experiments but it behaves as particulate photons in absorption and emission spectroscopy. Electrons behave as waves in diffraction experiments but as particles in the photoelectric effect. Consequences are especially important for small fundamental particles like electrons, atoms, and molecules. It is not possible to precisely specify complementary observables like position and momentum for fundamental particles. It is also impossible to specify the simultaneous energy and timing of an event. Rather, the multiplied uncertainties of complementary observables must always be greater than, or equal to, $\hbar/2$ (i.e., the Heisenberg uncertainty principle of eqns. 1.19a and 1.23). Quantum theory shows that, because of wave-particle duality, it is necessary to specify the wavefunction of fundamental particles and to use Postulates I–V of the text to interpret their behavior and observable properties.

D1.7 (a) The eigenvalue equation $\hat{\Omega}\psi = \omega\psi$ provides the relation between the operator for an observable ($\hat{\Omega}$) and the value of the observable (ω). Whenever the system is described by a wavefunction ψ, which is an eigenfunction of $\hat{\Omega}$, the outcome of a measurement of the observable Ω will be the eigenvalue ω. This is Postulate IV of the text.

(b) Should the system wavefunction be a superposition of eigenfunctions (eqn. 1.14, $\psi = \sum_k c_k \psi_k$) the probability of observing the eigenvalue ω_k is proportional to $|c_k|^2$. This is Postulate V of the text.

Solutions to exercises

E1.1(a)

$$\Delta E = h\nu\ [1.1] = \frac{hc}{\lambda}$$

$$= \frac{(6.626\times10^{-34}\ \text{J s})\times(2.998\times10^{8}\ \text{m s}^{-1})}{590\times10^{-9}\ \text{m}} = \boxed{3.37\times10^{-19}\ \text{J}}$$

E1.2(a) $\Delta E = h\nu = h/T$ where the period T equals $1/\nu$ $(T = 1/\nu)$

(a) $$\Delta E = \frac{6.626\times10^{-34}\ \text{J s}}{20\times10^{-15}\ \text{s}} = 3.3\times10^{-20}\ \text{J} = \boxed{33\ \text{zJ}}$$

This corresponds to $N_A \times (3.3\times10^{-20}\ \text{J}) = \boxed{20.\ \text{kJ mol}^{-1}}$

(b) $$\Delta E = \frac{6.626\times10^{-34}\ \text{J s}}{2.0\ \text{s}} = \boxed{3.3\times10^{-34}\ \text{J}},\ \boxed{0.20\ \text{nJ mol}^{-1}}$$

This is much too small to be measurable, thereby, demonstrating that for practical purposes the energy of a macroscopic object is a non-quantized, continuous variable.

E1.3(a)

$$E_k = \frac{1}{2}m_e v^2 = h\nu - \Phi = \frac{hc}{\lambda} - \Phi\ [1.2] \quad \text{and} \quad v = \left\{\frac{2}{m_e}\times E_k\right\}^{1/2}$$

$$\Phi = 2.14\,\text{eV} = (2.14)\times(1.602\times10^{-19}\ \text{J}) = 3.43\times10^{-19}\ \text{J}$$

(a) $$\frac{hc}{\lambda} = \frac{(6.626\times10^{-34}\ \text{J s})\times(2.998\times10^{8}\ \text{m s}^{-1})}{580\times10^{-9}\ \text{m}} = 3.42\times10^{-19}\ \text{J}$$

The photon energy is very nearly equal to the value of the work function.
Consequently, if photoejection occurs, the electrons will have $\boxed{\text{no kinetic energy and zero speed}}$.

(b) $$\frac{hc}{\lambda} = \frac{(6.626\times10^{-34}\ \text{J s})\times(2.998\times10^{8}\ \text{m s}^{-1})}{250\times10^{-9}\ \text{m}} = 7.95\times10^{-19}\ \text{J}$$

$$E_k = \tfrac{1}{2}mv^2 = (7.95-3.43)\times10^{-19}\ \text{J} = 4.52\times10^{-19}\ \text{J} = \boxed{0.452\ \text{aJ}}$$

$$v = \left(\frac{2E_k}{m}\right)^{1/2} = \left(\frac{(2)\times(4.52\times10^{-19}\ \text{J})}{9.109\times10^{-31}\,\text{kg}}\right)^{1/2} = \boxed{996\ \text{km s}^{-1}}$$

E1.4(a) $$E_k = \frac{1}{2}m_e v^2 = h\nu - \Phi = \frac{hc}{\lambda} - \Phi \;[1.2] \quad \text{or} \quad \Phi = \frac{hc}{\lambda} - E_k$$

$$\Phi = \frac{hc}{\lambda} - E_k = \frac{(6.626\times10^{-34}\ \text{J s})\times(2.998\times10^{8}\ \text{m s}^{-1})}{465\times10^{-9}\ \text{m}} - (2.11\ \text{eV})\times\left(\frac{1.602\times10^{-19}\ \text{J}}{\text{eV}}\right)$$
$$= 8.92\times10^{-20}\ \text{J}$$

The maximum wavelength needed for photoejection leaves the electron with zero kinetic energy. If the wavelength is longer, the radiation has insufficient energy for photoejection. If the absorbed wavelength is shorter, the electron will have a non-zero kinetic energy after photoejection. In the former case,

$$\lambda_{\text{max}} = \frac{hc}{\Phi} = \frac{(6.626\times10^{-34}\ \text{J s})\times(2.998\times10^{8}\ \text{m s}^{-1})}{8.92\times10^{-20}\ \text{J}} = 2.23\times10^{-6}\ \text{m} = \boxed{2.23\ \mu\text{m}}$$

E1.5(a) $$E_k = \frac{1}{2}m_e v^2 = h\nu - \Phi = \frac{hc}{\lambda} - \Phi \;[1.2] \quad \text{or} \quad \Phi = \frac{hc}{\lambda} - \frac{1}{2}m_e v^2 \quad \text{or} \quad v = \left\{\left(\frac{2}{m_e}\right)\times\left(\frac{hc}{\lambda} - \Phi\right)\right\}^{1/2}$$

$$\Phi = \frac{hc}{\lambda} - \frac{1}{2}m_e v^2 = \frac{(6.626\times10^{-34}\ \text{J s})\times(2.998\times10^{8}\ \text{m s}^{-1})}{165\times10^{-9}\ \text{m}} - \frac{1}{2}(9.109\times10^{-31}\ \text{kg})\times(1.24\times10^{6}\ \text{m s}^{-1})^2$$
$$= 5.04\times10^{-19}\ \text{J}$$

$$v = \left\{\left(\frac{2}{m_e}\right)\times\left(\frac{hc}{\lambda} - \Phi\right)\right\}^{1/2} = \left\{\left(\frac{2}{9.109\times10^{-31}\,\text{kg}}\right)\times\left(\frac{(6.626\times10^{-34}\,\text{J s})\times(2.998\times10^{8}\,\text{m s}^{-1})}{265\times10^{-9}\,\text{m}} - 5.04\times10^{-19}\,\text{J}\right)\right\}^{1}$$
$$= \boxed{7.35\times10^{5}\,\text{m s}^{-1}}$$

E1.6(a) $$E_{\text{binding}} = E_{\text{photon}} - E_k = h\nu - \frac{1}{2}m_e v^2 = \frac{hc}{\lambda} - \frac{1}{2}m_e v^2$$

$$E_{\text{binding}} = \frac{hc}{\lambda} - \frac{1}{2}m_e v^2 = \frac{(6.626\times10^{-34}\ \text{J s})\times(2.998\times10^{8}\ \text{m s}^{-1})}{150\times10^{-12}\ \text{m}} - \frac{1}{2}(9.109\times10^{-31}\ \text{kg})\times(2.14\times10^{7}\ \text{m s}^{-1})^2$$

$$= (1.12\times10^{-15}\ \text{J})\times\left(\frac{1\,\text{eV}}{1.602\times10^{-19}\ \text{J}}\right) = \boxed{6.96\ \text{keV}}$$ without a relativist mass correction.

Note: The photoelectron is moving at 7.1% of the speed of light. So, in order to calculate a more accurate value of the binding energy, it would be necessary to use the relativistic mass in place of the rest mass.

$$m = \frac{m_e}{\left(1-(v/c)^2\right)^{1/2}} = \frac{9.109\times10^{-31}\ \text{kg}}{\left(1-(2.14\times10^{7}\ \text{m s}^{-1}/2.998\times10^{8}\ \text{m s}^{-1})^2\right)^{1/2}} = 9.13\times10^{-31}\ \text{kg}$$

$$E_{\text{binding}} = \frac{hc}{\lambda} - \frac{1}{2}m_e v^2 = \frac{(6.626\times10^{-34}\ \text{J s})\times(2.998\times10^{8}\ \text{m s}^{-1})}{150\times10^{-12}\ \text{m}} - \frac{1}{2}(9.13\times10^{-31}\ \text{kg})\times(2.14\times10^{7}\ \text{m s}^{-1})^2$$

$$= (1.12\times10^{-15}\ \text{J})\times\left(\frac{1\ \text{eV}}{1.602\times10^{-19}\ \text{J}}\right) = \boxed{6.96\ \text{keV}}$$ with the relativistic mass correction.

The relativistic mass correction did not make a difference in this exercise.

E1.7(a) $$E = h\nu = \frac{hc}{\lambda}, \qquad E(\text{per mole}) = N_A E = \frac{N_A hc}{\lambda}$$

$$hc = (6.62608\times10^{-34}\ \text{J s})\times(2.99792\times10^{8}\ \text{m s}^{-1}) = 1.986\times10^{-25}\ \text{J m}$$

$$N_A hc = (6.02214\times10^{23}\ \text{mol}^{-1})\times(1.986\times10^{-25}\ \text{J m}) = 0.1196\ \text{J m mol}^{-1}$$

Thus, $E = \dfrac{1.986\times10^{-25}\ \text{J m}}{\lambda}$; $\quad E(\text{per mole}) = \dfrac{0.1196\ \text{J m mol}^{-1}}{\lambda}$

We can therefore draw up the following table.

λ/nm	E/J	E/(kJ mol^{-1})
(a) 620	3.20×10^{-19}	193
(b) 570	3.49×10^{-19}	210
(c) 380	5.23×10^{-19}	315

E1.8(a) Power is energy per unit time; hence

$$\frac{N}{\Delta t} = \frac{P}{h\nu}[P = \text{power in watts, } 1\text{ W} = 1\text{ J s}^{-1}] = \frac{P\lambda}{hc}$$

$$= \frac{P\lambda}{(6.626\times10^{-34}\text{ J s})\times(2.998\times10^{8}\text{ ms}^{-1})} = \frac{(P/\text{W})\times(\lambda/\text{nm})\text{s}^{-1}}{1.99\times10^{-16}}$$

$$= 5.03\times10^{15}(P/\text{W})\times(\lambda/\text{nm})\text{s}^{-1}$$

(a) $N/\Delta t = (5.03\times10^{15})\times(10)\times(590\text{s}^{-1}) = \boxed{3.0\times10^{19}\text{ s}^{-1}}$

(b) $N/\Delta t = (5.03\times10^{15})\times(250)\times(590\text{s}^{-1}) = \boxed{7.4\times10^{20}\text{ s}^{-1}}$

E1.9(a) $$\lambda = \frac{h}{p} = \frac{h}{mv} \text{ [1.3]}$$

(a) $$\lambda = \frac{6.626\times10^{-34}\text{ J s}}{(2.0\times10^{-3}\text{ kg})\times(1.0\times10^{-2}\text{ ms}^{-1})} = \boxed{3.3\times10^{-29}\text{ m}}$$

For a macroscopic object (e.g., 2 g) the de Broglie wavelength is much too small to be observed. Its particulate character predominates and energy levels are extraordinarily close, thereby producing an apparent continuum of possible energies.

(b) $$\lambda = \frac{6.626\times10^{-34}\text{ J s}}{(2.0\times10^{-3}\text{ kg})\times(2.5\times10^{5}\text{ m s}^{-1})} = \boxed{1.3\times10^{-36}\text{ m}}$$

(c) $$\lambda = \frac{6.626\times10^{-34}\text{ J s}}{(4.003)\times(1.6605\times10^{-27}\text{ kg})\times(1000\text{ m s}^{-1})} = \boxed{99.7\text{ pm}}$$

COMMENT. The wavelengths in (a) and (b) are smaller than the dimensions of any known particle, whereas the wavelength in (c) is comparable to atomic dimensions.

Question. For stationary particles, $v = 0$, corresponding to an infinite wavelength. What meaning can be ascribed to this result?

E1.10(a) $$\lambda = \frac{h}{p} = \frac{h}{mv} \text{ [1.3]}$$

$$v = \frac{h}{m_e\lambda} = \frac{6.626\times10^{-34}\text{ J s}}{(9.109\times10^{-31}\text{ kg})\times(100\times10^{-12}\text{ m})} = \boxed{7.27\times10^{6}\text{ m s}^{-1}}$$

The accelerating potential difference needed to provide this speed is

$$V=\frac{E_{\mathrm{k}}}{e}=\frac{\frac{1}{2}m_{\mathrm{e}}v^2}{e}$$

$$=\frac{\frac{1}{2}(9.109\times10^{-31}\ \mathrm{kg})\times(7.27\times10^{6}\ \mathrm{m\ s^{-1}})^2}{1.602\times10^{-19}\ \mathrm{C}}\left(\frac{1\ \mathrm{C\ V}}{\mathrm{J}}\right)=\boxed{150\ \mathrm{V}}$$

Low-energy electron diffraction (LEED) uses a beam of electrons of a well-defined low energy (typically in the range 20–200 eV) incident to a single crystal with a well-ordered surface structure in order to generate a back-scattered electron diffraction pattern.

E1.11(a)

$$E_{\mathrm{k}}=\frac{1}{2}m_{\mathrm{e}}v^2=h\nu=\frac{hc}{\lambda}\quad\text{or}\quad v=\left(\frac{2hc}{m_{\mathrm{e}}\lambda}\right)^{1/2}$$

$$v=\left(\frac{2(6.626\times10^{-34}\ \mathrm{J\ s})\times(2.998\times10^{8}\ \mathrm{m\ s^{-1}})}{(9.109\times10^{-31}\ \mathrm{kg})\times(150\times10^{-9}\ \mathrm{m})}\right)^{1/2}=\boxed{1.71\times10^{6}\ \mathrm{m\ s^{-1}}}$$

E1.12(a) The time-independent wavefunction in three-dimensional space is a function of x, y, and z so we write $\psi(x, y, z)$ or $\psi(\boldsymbol{r})$. The infinitesimal space element is $d\tau = dxdydz$ with each variable ranging from $-\infty$ to $+\infty$. The time-independent wavefunction is said to be a **stationary state**.

It is reasonable to expect that in some special cases the probability densities in each of the three independent directions should be mutually independent. This implies that the probability density for the time-independent wavefunction $\psi(\boldsymbol{r})$ should be the product of three probability densities, one for each coordinate: $|\psi(\boldsymbol{r})|^2 \propto |X(x)|^2 \times |Y(y)|^2 \times |Z(z)|^2$. Subsequently, we see that the wavefunction is the product of three independent wavefunctions in such a special case and we write $\psi(\boldsymbol{r}) \propto X(x) \times Y(y) \times Z(z)$. Such a wavefunction is said to exhibit the **separation of variables**.

For the special case of a particle free to move in a cube of volume L^3, we may generalize the wavefunction provided in Example 1.4 of the text, $X(x)\ \mu\ \sin(\pi x/L)$, to $Y(y) \propto \sin(\pi y/L)$ and $Z(z) \propto \sin(\pi z/L)$ and conclude that $\psi(\boldsymbol{r}) \propto \sin(\pi x/L) \times \sin(\pi y/L) \times \sin(\pi z/L)$. Alternatively, for a particle free to move in a rectangular parallelepiped of sides L_x, L_y and L_z, $\psi(\boldsymbol{r}) \propto \sin(\pi x/L_x) \times \sin(\pi y/L_y) \times \sin(\pi z/L_z)$.

Remarkably, when the potential energy term of the hamiltonian is either zero or a constant value throughout space, the time-independent wavefunction does not depend upon the particle mass! Mass does appear in both the kinetic energy operator and eigenvalues of operators that contain the kinetic energy operator. Similarly, electrical charge does not appear in the time-independent wavefunction in this particular case.

E1.13(a) An isolated, freely moving hydrogen atom is expected to have a translational, time-independent wavefunction that is a function of the center-of-mass coordinates x_{cm}, y_{cm}, and z_{cm}, so we write $\psi_{\mathrm{cm}}(x_{\mathrm{cm}}, y_{\mathrm{cm}}, z_{\mathrm{cm}})$ or $\psi_{\mathrm{cm}}(\boldsymbol{r}_{\mathrm{cm}})$ with each variable ranging from $-\infty$ to $+\infty$. The infinitesimal space element for the center of mass variables is $d\tau_{\mathrm{cm}} = dx_{\mathrm{cm}}dy_{\mathrm{cm}}dz_{\mathrm{cm}}$.

The hydrogen atom also has variables x, y, and z, which are the Cartesian coordinates of the electron with respect to the center-of-mass of the hydrogen atom. The electronic wavefunction can be written as $\psi_{\mathrm{el}}(x, y, z)$ or $\psi_{\mathrm{el}}(\boldsymbol{r})$ with each variable ranging from $-\infty$ to $+\infty$.

In general we expect that the total wavefunction is the product $\psi_{\mathrm{cm}}(\boldsymbol{r}_{\mathrm{cm}}) \times \psi_{\mathrm{el}}(\boldsymbol{r})$. (See E1.12a.) Furthermore, we expect that there are special cases for which the translational wavefunction exhibits the separation of variables: $\psi_{\mathrm{cm}}(\boldsymbol{r}_{\mathrm{cm}}) \propto X_{\mathrm{cm}}(x_{\mathrm{cm}}) \times Y(y_{\mathrm{cm}}) \times Z_{\mathrm{cm}}(z_{\mathrm{cm}})$. The electronic wavefunction does not exhibit the separation of Cartesian variables because the electrostatic potential between the electron and nucleus is proportional to $1/r$, which cannot be written as a sum of separate terms in the variables x, y, and z.

E1.14(a) An isolated, freely moving hydrogen atom is expected to have a translational, time-independent wavefunction that is a function of the center-of-mass coordinates x_{cm}, y_{cm}, and z_{cm}, so we write $\psi_{\mathrm{cm}}(x_{\mathrm{cm}}, y_{\mathrm{cm}}, z_{\mathrm{cm}})$ or $\psi_{\mathrm{cm}}(\boldsymbol{r}_{\mathrm{cm}})$ with each variable ranging from $-\infty$ to $+\infty$. The infinitesimal space element for the center of mass variables is $d\tau_{\mathrm{cm}} = dx_{\mathrm{cm}}dy_{\mathrm{cm}}dz_{\mathrm{cm}}$.

The hydrogen atom also has variables r, θ, and φ. These are the spherical polar coordinates of the electron with respect to the center of mass of the hydrogen atom. The electronic wavefunction can be written as $\psi_{el}(r, \theta, \varphi)$ with the variables lying in the ranges $0 \le \varphi \le 2\pi$, $0 \le \theta \le \pi$, and $0 \le r \le \infty$ (see text Figs. 3.12 and 3.13). Furthermore, the infinitesimal space element is $d\tau = dxdydz = r^2 \sin\theta\, drd\theta d\varphi$.

In general we expect that the total wavefunction is the product $\psi_{cm}(\boldsymbol{r}_{cm}) \times \psi_{el}(r, \theta, \varphi)$. (See E1.12a.) Furthermore, we expect that there are special cases for which the translational wavefunction exhibits the separation of variables: $\psi_{cm}(\boldsymbol{r}_{cm}) \propto X_{cm}(x_{cm}) \times Y(y_{cm}) \times Z_{cm}(z_{cm})$. In Chapter 4 you will discover that the electronic wavefunction also exhibits the separation of spherical polar coordinates because the electrostatic potential between the electron and nucleus is proportional to $1/r$.

E1.15(a) The normalized wavefunction has the form $\psi(\varphi) = N\,e^{i\varphi}$ where N is the normalized constant.

$$\int_0^{2\pi} \psi^* \psi \, d\varphi = 1 \; [1.6]$$

$$N^2 \int_0^{2\pi} e^{-i\varphi} e^{i\varphi} \, d\varphi = N^2 \int_0^{2\pi} d\varphi = 2\pi N^2 = 1$$

$$\boxed{N = \left(\frac{1}{2\pi}\right)^{1/2}}$$

E1.16(a) $\psi(\varphi) = (1/2\pi)^{1/2}\, e^{i\varphi}$ so $|\psi(\varphi)|^2 = (1/2\pi)\, e^{-i\varphi}\, e^{i\varphi} = 1/2\pi$. Thus, the probability of finding the atom in an infinitesimal volume element at any angle is $\boxed{(1/2\pi)\, d\varphi}$.

E1.17(a) The normalized wavefunction is $\psi = \left(\frac{1}{2\pi}\right)^{1/2} e^{i\varphi}$

Probability that $\pi/2 \le \varphi \le 3\pi/2 = \int_{\pi/2}^{3\pi/2} \psi^* \psi \, d\varphi$ [Postulate II]

$$= \left(\frac{1}{2\pi}\right) \int_{\pi/2}^{3\pi/2} e^{-i\varphi} e^{i\varphi} \, d\varphi = \left(\frac{1}{2\pi}\right) \int_{\pi/2}^{3\pi/2} d\varphi$$

$$= \left(\frac{1}{2\pi}\right) \varphi \Big|_{\varphi=\pi/2}^{\varphi=3\pi/2} = \left(\frac{1}{2\pi}\right) \times \left(\frac{3\pi}{2} - \frac{\pi}{2}\right) = \boxed{\frac{1}{2}}$$

E1.18(a)

$$\hat{E}_k = \frac{\hat{p}^2}{2m} = \frac{\hat{p}_x^2 + \hat{p}_y^2}{2m} = -\frac{\hbar^2}{2m}\left(\frac{\partial^2}{\partial x^2}\right)_y - \frac{\hbar^2}{2m}\left(\frac{\partial^2}{\partial y^2}\right)_x \quad [1.11]$$

E1.19(a) Let a and b be any real or complex number and let $f(x)$ be defined as $f(x) = ae^{2ix} + be^{-2ix}$. Then

$$\frac{df}{dx} = 2iae^{2ix} - 2ibe^{-2ix}$$

$$\frac{d^2 f}{dx^2} = (2i)^2 ae^{2ix} + (2i)^2 be^{-2ix} = -4\left(ae^{2ix} + be^{-2ix}\right)$$

$$= -4f(x)$$

$f(x)$ is an eigenfunction of the operator d^2/dx^2. The eigenvalue is $\boxed{-4}$.

E1.20(a) (a) $\frac{de^{ikx}}{dx} = (ik)e^{ikx}$ Thus, e^{ikx} is an eigenfunction of d/dx. The eigenvalue is $\boxed{ik}$.

(b) $\frac{de^{ax^2}}{dx} = (2ax)e^{ax^2}$ Thus, e^{ax^2} is not an eigenfunction of d/dx.

(c) $\frac{dx}{dx} = 1$ Thus, x is not an eigenfunction of d/dx.

(d) $\frac{dx^2}{dx} = 2x$ Thus, x^2 is not an eigenfunction of d/dx.

(e) $\frac{d(ax+b)}{dx} = a$ Thus, $ax + b$ is not an eigenfunction of d/dx.

(f) $\frac{d\sin(x+3a)}{dx} = \cos(x+3a)$ Thus, $\sin(x+3a)$ is not an eigenfunction of d/dx.

E1.21(a) ψ_i and ψ_j are orthogonal if $\int \psi_i^* \psi_j \, d\tau = 0$ [1.16].

Where $n \neq m$ and both n and m are integers,

$$\int_0^L \sin(n\pi x/L)\times\sin(m\pi x/L)\,dx = \left[\frac{\sin(\pi(n-m)x/L)}{2\pi(n-m)/L} - \frac{\sin((n+m)x/L)}{2\pi(n+m)/L}\right]_{x=0}^{x=L}$$

$$= \frac{\sin(\pi(n-m))}{2\pi(n-m)/L} - \frac{\sin(\pi(n+m))}{2\pi(n+m)/L} - \left\{\frac{\sin(0)}{2\pi(n-m)/L} - \frac{\sin(0)}{2\pi(n+m)/L}\right\}$$

$= 0$ because the sine of an integer multiple of π equals zero.

Thus, the functions $\sin(n\pi x/L)$ and $\sin(m\pi x/L)$ are orthogonal in the region $0 \le x \le L$.

E1.22(a) ψ_i and ψ_j are orthogonal if $\int \psi_i^* \psi_j \, d\tau = 0$ [1.16].

$$\int_0^{2\pi} (e^{i\varphi})^* \times e^{2i\varphi}\,d\varphi = \int_0^{2\pi} e^{-i\varphi}\times e^{2i\varphi}\,d\varphi = \int_0^{2\pi} e^{i\varphi}\,d\varphi = \frac{1}{i} e^{i\varphi}\Big|_{\varphi=0}^{\varphi=2\pi}$$

$$= \frac{1}{i}(e^{2\pi i} - e^0) = \frac{1}{i}(e^{2\pi i} - 1) = \frac{1}{i}(\cos(2\pi) - i\sin(2\pi) - 1) = \frac{1}{i}(1-0-1) = 0$$

(The Euler identity $e^{ai} = \cos(a) - i\sin(a)$ has been used in the math manipulations.)

Thus, the functions $e^{i\varphi}$ and $e^{2i\varphi}$ are orthogonal in the region $0 \le \varphi \le 2\pi$.

E1.23(a) The normalized form of this wavefunction is

$$\psi(x) = \left(\frac{2}{L}\right)^{1/2} \sin(2\pi x/L) \qquad \text{(See Self-test 1.4)}$$

The expectation value of the electron position is

$$\langle x\rangle = \int_0^L \psi^* x\,\psi\,dx\ [1.15b] = \left(\frac{2}{L}\right)\int_0^L x\sin^2\left(\frac{2\pi x}{L}\right)dx$$

$$= \left(\frac{2}{L}\right)\times\left[\frac{x^2}{4} - \frac{x\sin\left(\frac{4\pi x}{L}\right)}{8\pi/L} - \frac{\cos\left(\frac{4\pi x}{L}\right)}{8(2\pi/L)^2}\right]_{x=0}^{x=L}$$

$$= \left(\frac{2}{L}\right)\times\left[\frac{L^2}{4} - \frac{L\sin(4\pi)}{8\pi/L} - \frac{\cos(4\pi)}{8(2\pi/L)^2} - \left\{\frac{0^2}{4} - \frac{0\times\sin(0)}{8\pi/L} - \frac{\cos(0)}{8(2\pi/L)^2}\right\}\right] = \left(\frac{2}{L}\right)\times\left(\frac{L^2}{4}\right)$$

$$= \boxed{\frac{L}{2}}$$

E1.24(a) The normalized form of this wavefunction is

$$\psi(x) = \left(\frac{2}{L}\right)^{1/2} \sin(\pi x / L) \qquad \text{(See Example 1.4)}$$

$$\frac{d\psi}{dx} = \left(\frac{2}{L}\right)^{1/2} \left(\frac{\pi}{L}\right) \cos(\pi x / L)$$

The expectation value of the electron momentum is

$$\langle p_x \rangle = \int_0^L \psi^* \hat{p}_x \psi \, dx \text{ [1.15b]} = \int_0^L \psi^* \left(\frac{\hbar}{i}\frac{d}{dx}\right) \psi \, dx = \left(\frac{\hbar}{i}\right) \int_0^L \psi^* \left(\frac{d\psi}{dx}\right) dx$$

$$= \left(\frac{\hbar}{i}\right)\left(\frac{2}{L}\right)^{1/2}\left(\frac{\pi}{L}\right)\int_0^L \psi^* \cos\left(\frac{\pi x}{L}\right) dx = \left(\frac{\hbar}{i}\right)\left(\frac{2}{L}\right)\left(\frac{\pi}{L}\right)\int_0^L \sin\left(\frac{\pi x}{L}\right)\cos\left(\frac{\pi x}{L}\right) dx$$

$$= \left(\frac{h}{iL^2}\right)\int_0^L \sin\left(\frac{\pi x}{L}\right)\cos\left(\frac{\pi x}{L}\right) dx$$

$$= \left(\frac{h}{iL^2}\right) \times \left[\frac{\sin^2\left(\frac{\pi x}{L}\right)}{2\pi / L}\right]_{x=0}^{x=L} = \left(\frac{h}{iL^2}\right) \times \left(\frac{\sin^2(\pi)}{2\pi/L} - \frac{\sin^2(0)}{2\pi/L}\right) = \left(\frac{h}{iL^2}\right) \times (0+0) = \boxed{0}$$

E1.25(a) Let f and g be functions of x and examine the integral $\int_{-\infty}^{\infty} f^* \left(-\frac{\hbar^2}{2m}\frac{d^2}{dx^2}\right) g \, dx$. Integrate successively by parts (see Justification 1.1) and use the fact that these functions must be well behaved at the boundaries (i.e., they equal zero at infinity in either direction).

$$\int_{-\infty}^{\infty} f^* \left(-\frac{\hbar^2}{2m}\frac{d^2}{dx^2}\right) g \, dx = \left(-\frac{\hbar^2}{2m}\right)\int_{-\infty}^{\infty} f^* \left(\frac{d^2}{dx^2}\right) g \, dx = \left(-\frac{\hbar^2}{2m}\right)\int_{-\infty}^{\infty} f^* \left(\frac{d}{dx}\right)\frac{dg}{dx} dx$$

$$= \left(-\frac{\hbar^2}{2m}\right) \times \left\{\left[f^* \frac{dg}{dx}\right]_{-\infty}^{\infty} - \int_{-\infty}^{\infty} \frac{df^*}{dx} \times \frac{dg}{dx} dx\right\} = -\left(-\frac{\hbar^2}{2m}\right) \times \left\{\int_{-\infty}^{\infty} \frac{df^*}{dx} \times \frac{dg}{dx} dx\right\}$$

$$= -\left(-\frac{\hbar^2}{2m}\right) \times \left\{\left[g \frac{df^*}{dx}\right]_{-\infty}^{\infty} - \int_{-\infty}^{\infty} g \times \frac{d^2 f^*}{dx^2} dx\right\} = \left(-\frac{\hbar^2}{2m}\right) \times \left\{\int_{-\infty}^{\infty} g \times \frac{d^2 f^*}{dx^2} dx\right\}$$

$$= \int_{-\infty}^{\infty} g \left(-\frac{\hbar^2}{2m}\frac{d^2}{dx^2}\right) f^* \, dx = \left\{\int_{-\infty}^{\infty} g^* \left(-\frac{\hbar^2}{2m}\frac{d^2}{dx^2}\right) f \, dx\right\}^*$$

Thus, $\int_{-\infty}^{\infty} f^* \left(-\frac{\hbar^2}{2m}\frac{d^2}{dx^2}\right) g \, dx = \left\{\int_{-\infty}^{\infty} g^* \left(-\frac{\hbar^2}{2m}\frac{d^2}{dx^2}\right) f \, dx\right\}^*$

This is exactly the criterion that a hermitian operator must satisfy [1.13], so we conclude that the kinetic energy operator is hermitian.

E1.26(a) A quick examination of its expectation value reveals that it is complex because the expectation values of both position and momentum are real. All observables must be real.

$$\int \psi^* (\hat{x} + ia\hat{p}_x) \psi \, d\tau = \int \psi^* (\hat{x}) \psi \, d\tau + ia \int \psi^* (\hat{p}_x) \psi \, d\tau = \langle x \rangle + ia \langle p_x \rangle$$

Also, x and p_x are complementary observables so they cannot be simultaneously known.

The hermiticity of the operator may also be checked using the fact the both the position and linear momentum operators are hermitian.

$$\int g(\hat{x}+\mathrm{i}\hat{p}_x)\, f\,\mathrm{d}\tau = \int g(\hat{x})\, f\,\mathrm{d}\tau + \mathrm{i}\int g(\hat{p}_x)\, f\,\mathrm{d}\tau$$
$$= \left\{\int f(\hat{x})\, g\,\mathrm{d}\tau\right\}^* + \mathrm{i}\left\{\int g(\hat{p}_x)\, f\,\mathrm{d}\tau\right\}^*$$
$$= \left\{\int f(\hat{x})\, g\,\mathrm{d}\tau\right\}^* - \mathrm{i}^*\left\{\int g(\hat{p}_x)\, f\,\mathrm{d}\tau\right\}^*$$
$$= \left\{\int f(\hat{x}-\mathrm{i}\hat{p}_x)\, g\,\mathrm{d}\tau\right\}^*$$

This is not the relationship required by eqn. 1.13 so we conclude that the operator is not hermitian and cannot correspond to an observable.

E1.27(a)

$$\Delta p \approx 0.0100\,\text{percent of}\, p_0 = p_0\times(1.00\times10^{-4}) = m_p v\times(1.00\times10^{-4}) \quad (p_0 = m_p v)$$

$$\Delta q \approx \frac{\mathrm{h}}{2\Delta p}[1.19\mathrm{a}] \approx \frac{1.055\times10^{-34}\,\mathrm{J\,s}}{(2)\times(1.673\times10^{-27}\,\mathrm{kg})\times(6.1\times10^{6}\,\mathrm{m\,s^{-1}})\times(1.00\times10^{-4})}$$
$$\approx 5.2\times10^{-11}\,\mathrm{m} \approx \boxed{52\,\mathrm{pm}}$$

E1.28(a) The minimum uncertainty in position and momentum is given by the uncertainty principle in the form $\Delta p\Delta q \geq \frac{1}{2}\hbar$ [1.19a] with the choice of the equality. The uncertainty in momentum is $\Delta p = m\Delta v$ so

(a) $$\Delta v_{\min} = \frac{\hbar}{2m\Delta q} = \frac{1.055\times10^{-34}\,\mathrm{J\,s}}{(2)\times(0.500\,\mathrm{kg})\times(1.0\times10^{-6}\,\mathrm{m})} = \boxed{1.1\times10^{-28}\,\mathrm{m\,s^{-1}}}$$

(b) $$\Delta q_{\min} = \frac{\hbar}{2m\Delta v} = \frac{1.055\times10^{-34}\,\mathrm{J\,s}}{(2)\times(5.0\times10^{-3}\,\mathrm{kg})\times(1\times10^{-5}\,\mathrm{m\,s^{-1}})} = \boxed{1\times10^{-27}\,\mathrm{m}}$$

COMMENT. These uncertainties are extremely small; thus, the ball and bullet are effectively classical particles.

Question. If the ball were stationary (no uncertainty in position) the uncertainty in speed would be infinite. Thus, the ball could have a very high speed, contradicting the fact that it is stationary. What is the resolution of this apparent paradox?

E1.29(a) (a) $$[\hat{x},\hat{y}] = xy - yx = xy - xy = \boxed{0}$$

(b) $$[\hat{p}_x,\hat{p}_y] = \left(\frac{\hbar}{\mathrm{i}}\frac{\partial}{\partial x}\right)\left(\frac{\hbar}{\mathrm{i}}\frac{\partial}{\partial y}\right) - \left(\frac{\hbar}{\mathrm{i}}\frac{\partial}{\partial y}\right)\left(\frac{\hbar}{\mathrm{i}}\frac{\partial}{\partial x}\right) = \left(\frac{\hbar}{\mathrm{i}}\right)^2\left(\frac{\partial^2}{\partial x\partial y} - \frac{\partial^2}{\partial y\partial x}\right)$$

$$= \boxed{0} \quad \text{because} \quad \frac{\partial^2}{\partial x\partial y} = \frac{\partial^2}{\partial y\partial x} \quad \text{for all well-behaved functions.}$$

(c) $$[\hat{x},\hat{p}_x] = x\left(\frac{\hbar}{\mathrm{i}}\frac{\partial}{\partial x}\right) - \left(\frac{\hbar}{\mathrm{i}}\frac{\partial}{\partial x}\right)x = \cancel{x\left(\frac{\hbar}{\mathrm{i}}\frac{\partial}{\partial x}\right)} - \left(\frac{\hbar}{\mathrm{i}}\right) - \cancel{\left(\frac{\hbar}{\mathrm{i}}x\frac{\partial}{\partial x}\right)} = \boxed{\mathrm{i}\hbar}$$

(d) $$[\hat{x}^2,\hat{p}_x] = x^2\left(\frac{\hbar}{\mathrm{i}}\frac{\partial}{\partial x}\right) - \left(\frac{\hbar}{\mathrm{i}}\frac{\partial}{\partial x}\right)x^2 = \cancel{x^2\left(\frac{\hbar}{\mathrm{i}}\frac{\partial}{\partial x}\right)} - \frac{2x\hbar}{\mathrm{i}} - \cancel{x^2\left(\frac{\hbar}{\mathrm{i}}\frac{\partial}{\partial x}\right)} = \boxed{2\mathrm{i}x\hbar}$$

(e) $$[\hat{x}^n,\hat{p}_x] = x^n\left(\frac{\hbar}{\mathrm{i}}\frac{\partial}{\partial x}\right) - \left(\frac{\hbar}{\mathrm{i}}\frac{\partial}{\partial x}\right)x^n = \cancel{x^n\left(\frac{\hbar}{\mathrm{i}}\frac{\partial}{\partial x}\right)} - \frac{nx^{n-1}\hbar}{\mathrm{i}} - \cancel{x^n\left(\frac{\hbar}{\mathrm{i}}\frac{\partial}{\partial x}\right)} = \boxed{n\mathrm{i}x^{n-1}\hbar}$$

Solutions to problems

Solutions to numerical problems

P1.1

$$\psi=\left(\frac{2}{L}\right)^{1/2}\sin\frac{\pi x}{L} \quad \text{and} \quad \psi^2=\frac{2}{L}\sin^2\frac{\pi x}{L}$$

The probability P that the particle will be found in the region between a and b is the integral summation of all the probabilities of finding the particle within infinitesimally small volume elements within the region ($\psi^2\mathrm{d}x$ according to Postulate II).

$$P(a,b)=\int_a^b \psi^2\mathrm{d}x \quad \text{[Section 1.5, Postulate II]}$$

$$=\frac{2}{L}\int_a^b \sin^2\frac{\pi x}{L}\mathrm{d}x=\frac{2}{L}\times\left(\frac{x}{2}-\frac{L}{4\pi}\sin\frac{2\pi x}{L}\right)\Bigg|_{x=a}^{x=b}=\left(\frac{x}{L}-\frac{1}{2\pi}\sin\frac{2\pi x}{L}\right)\Bigg|_{x=a}^{x=b}$$

$$=\frac{b-a}{L}-\frac{1}{2\pi}\left(\sin\frac{2\pi b}{L}-\sin\frac{2\pi a}{L}\right)$$

$L=10.0\text{ nm}$

(a) $$P(4.95\text{ nm},5.05\text{ nm})=\frac{0.10}{10.0}-\frac{1}{2\pi}\left(\sin\frac{(2\pi)\times(5.05)}{10.0}-\sin\frac{(2\pi)\times(4.95)}{10.0}\right)$$

$$=0.010+0.010=\boxed{0.020}$$

(b) $$P(7.95\text{ nm},9.05\text{ nm})=\frac{1.10}{10.0}-\frac{1}{2\pi}\left(\sin\frac{(2\pi)\times(9.05)}{10.0}-\sin\frac{(2\pi)\times(7.95)}{10.0}\right)$$

$$=0.110-0.063=\boxed{0.047}$$

(c) $$P(9.90\text{ nm},10.0\text{ nm})=\frac{0.10}{10.0}-\frac{1}{2\pi}\left(\sin\frac{(2\pi)\times(10.0)}{10.0}-\sin\frac{(2\pi)\times(9.90)}{10.0}\right)$$

$$=0.010-0.009993=\boxed{7\times10^{-6}}$$

(d) $$P(0\text{ nm},5.0\text{ nm})=P(5.0\text{ nm},10.0\text{ nm})=\boxed{0.5} \quad \text{[by symmetry]}$$

(e) $$P\left(\frac{1}{3}L,\frac{2}{3}L\right)=\frac{1}{3}-\frac{1}{2\pi}\left(\sin\frac{4\pi}{3}-\sin\frac{2\pi}{3}\right)=\boxed{0.61}$$

P1.3 The normalization constant for this wavefunction is

$$N=\left(\frac{1}{\int_0^\pi x^2\times x^2\,\mathrm{d}x}\right)^{1/2}\text{[1.4]}=\left(\frac{5}{\pi^5}\right)^{1/2}$$

so the normalized wavefunction is

$$\psi=\left(\frac{5}{\pi^5}\right)^{1/2}x^2$$

$$P=\int_0^a\psi^2\,\mathrm{d}x=\left(\frac{5}{\pi^5}\right)\int_0^a x^4\,\mathrm{d}x=\left(\frac{a}{\pi}\right)^5$$

Consequently, $P=½$ when $\boxed{a=\pi/2^{1/5}}$

P1.5 The normalization constant for this wavefunction is

$$N=\left(\frac{1}{\int_0^{2\pi} e^{im\varphi}\times e^{-im\varphi}\,d\varphi}\right)^{1/2}[1.4]=\left(\frac{1}{\int_0^{2\pi} d\varphi}\right)^{1/2}=\left(\frac{1}{2\pi}\right)^{1/2}$$

so the normalized wavefunction is

$$\psi=(2\pi)^{-1/2}e^{-im\varphi}$$

$$\langle\varphi\rangle=\int_0^{2\pi}\psi^*\varphi\psi\,d\varphi=(2\pi)^{-1}\int_0^{2\pi}\varphi e^{im\varphi}e^{-im\varphi}\,d\varphi=(2\pi)^{-1}\int_0^{2\pi}\varphi\,d\varphi=(2\pi)^{-1}\left[\frac{\varphi^2}{2}\right]_{\varphi=0}^{\varphi=2\pi}$$

$$=\left(\frac{1}{2\pi}\right)\times\left(\frac{4\pi^2}{2}\right)=\boxed{\pi}$$

P1.7 The most probable location occurs when the probability density, $|\psi|^2$, is a maximum. Thus, we wish to find the value $x=x_{\max}$ such that $d|\psi|^2/dx=0$.

$$\psi(x)=Nxe^{-x^2/2a^2}$$

$$|\psi^2|=N^2x^2e^{-x^2/a^2}$$

$$d|\psi^2|/dx=N^2\left\{2xe^{-x^2/a^2}-(2x^3/a^2)e^{-x^2/a^2}\right\}=2N^2x\left\{1-x^2/a^2\right\}e^{-x^2/a^2}$$

The above derivative equals zero when the factor $1-x^2/a^2$ equals zero so we conclude that $\boxed{x_{\max}=a}$.

P1.9 The expectation value of the commutator is

$$\langle[\hat{x},\hat{p}]\rangle=\int\psi^*[\hat{x},\hat{p}]\psi\,d\tau$$

First evaluate the commutator acting on the wavefunction. The commutator of the position and momentum operators is defined as

$$[\hat{x},\hat{p}]=\hat{x}\hat{p}-\hat{p}\hat{x}=x\times\frac{\hbar}{i}\frac{d}{dx}-\frac{\hbar}{i}\frac{d}{dx}x,$$

so the commutator acting on the wavefunction is

$$[\hat{x},\hat{p}]\psi=x\times\frac{\hbar}{i}\frac{d\psi}{dx}-\frac{\hbar}{i}\frac{d}{dx}(x\psi),$$

where $\psi(x)=a^{1/2}e^{-ax/2}$

Evaluating this expression yields

$$[\hat{x},\hat{p}]\psi=\frac{x\hbar}{i}(a)^{1/2}(-a/2)e^{-ax/2}-\frac{\hbar}{i}[(a)^{1/2}e^{-ax/2}+x(-a/2)(a)^{1/2}e^{-ax/2}],$$

$$[\hat{x},\hat{p}]\psi=i\hbar(a)^{1/2}e^{-ax/2}$$

which is just $i\hbar$ times the original wavefunction. Putting this result into the expectation value yields

$$\langle[\hat{x},\hat{p}]\rangle=\int_0^{\infty}(a)^{1/2}e^{-ax/2}(i\hbar)(a)^{1/2}e^{-ax/2}dx=ia\hbar\int_0^{\infty}e^{-ax}dx$$

$$\langle[\hat{x},\hat{p}]\rangle=ia\hbar\times\left.\frac{e^{-ax}}{-a}\right|_0^{\infty}=\boxed{i\hbar}.$$

Note: Although the commutator is a well-defined and useful operator in quantum mechanics, it does not correspond to an observable quantity. Thus one need not be concerned about obtaining an imaginary expectation value.

Solutions to theoretical problems

P1.11 Time-independent Schrödinger equation of a single particle: $\left(-\frac{\hbar^2}{2m}\frac{d^2}{dx^2}+V\right)\psi = E\psi$ [1.7]

(a) $$\left(-\frac{\hbar^2}{2m_e}\frac{d^2}{dx^2}-\frac{e^2}{4\pi\varepsilon_0 x}\right)\psi = E\psi$$

(b) $$\left(-\frac{\hbar^2}{2m}\frac{d^2}{dx^2}\right)\psi = E\psi$$

(c) $F = c$ (a constant) implies that $V = -cx$ because $F = -dV/dx$.

$$\left(-\frac{\hbar^2}{2m}\frac{d^2}{dx^2}-cx\right)\psi = E\psi$$

P1.13 (i) $$\left[\hat{A},\hat{B}\right]=\hat{A}\hat{B}-\hat{B}\hat{A}=-\left(\hat{B}\hat{A}-\hat{A}\hat{B}\right)=-\left[\hat{B},\hat{A}\right]$$

(ii) $$\left[\hat{A},\hat{B}+\hat{C}\right]=\hat{A}\hat{B}+\hat{A}\hat{C}-\hat{B}\hat{A}-\hat{C}\hat{A}=\hat{A}\hat{B}-\hat{B}\hat{A}+\hat{A}\hat{C}-\hat{C}\hat{A}=\left[\hat{A},\hat{B}\right]+\left[\hat{A},\hat{C}\right]$$

(iii) $$\left[\hat{A}^2,\hat{B}\right]=\hat{A}^2\hat{B}-\hat{B}\hat{A}^2=\left(\hat{A}^2\hat{B}-\hat{A}\hat{B}\hat{A}\right)+\left(\hat{A}\hat{B}\hat{A}-\hat{B}\hat{A}^2\right)=\hat{A}\left(\hat{A}\hat{B}-\hat{B}\hat{A}\right)+\left(\hat{A}\hat{B}-\hat{B}\hat{A}\right)\hat{A}$$
$$=\hat{A}\left[\hat{A},\hat{B}\right]+\left[\hat{A},\hat{B}\right]\hat{A}$$

P1.15 (a) Since the relevant Heisenberg uncertainty principle is $\Delta p_x\Delta x \geq \tfrac{1}{2}\hbar$ [1.19a], knowledge of the values of both positions x and p_x simultaneously is restricted to non-zero uncertainties. However, if the uncertainty in one of these observables is very large, the uncertainty in the other can be small. The value of an observable can be known exactly in the case for which its complementary observable is completely uncertain. Non-complementary observables, like p_x and y, can be simultaneously measured without the restrictions that the uncertainty principle implies.

(b) The three components of momentum (p_x, p_y, p_z) commute with each other, so they are not complementary and they may be simultaneously measured without the restrictions that the uncertainty principle implies.

(c) The operators for kinetic energy and potential energy do not commute (see Self-test 1.13), so in general they cannot be measured without the restrictions that the uncertainty principle implies.

$$\left[\hat{V},\hat{E}_k\right]=\frac{\hbar^2}{2m}\left(\frac{d^2V}{dx^2}+2\frac{dV}{dx}\frac{d}{dx}\right)$$

However, the general form of the uncertainty principle, $\Delta\Omega_1\Delta\Omega_2 \geq \tfrac{1}{2}\left|\left\langle\left[\hat{\Omega}_1,\hat{\Omega}_2\right]\right\rangle\right|$ [1.23], indicates special cases for which uncertainty restrictions will not occur. These include the case for which

(i) $dV/dx = 0$ and (ii) $\left\langle\frac{d^2V}{dx^2}+2\frac{dV}{dx}\frac{d}{dx}\right\rangle = 0$

(d) The general form of the uncertainty principle indicates that $\Delta\mu\Delta E = -\frac{\hbar e}{2m_e}\langle\hat{p}_x\rangle$ (see Example 1.10) so in general, electric dipole moment and energy cannot be measured without uncertainty restrictions. However, in the special case for which $\langle\hat{p}_x\rangle = 0$ there will be no restriction on the simultaneous determination of them even though the corresponding operators are complementary.

(e) Since $\left[\hat{x},\hat{E}_k\right]=\frac{i\hbar}{m}\hat{p}_x$ (see Example 1.10), the general uncertainty principle indicates that

$$\Delta x\Delta E \geq \tfrac{1}{2}\left|\left\langle\frac{i\hbar}{m}\hat{p}_x\right\rangle\right| = \tfrac{1}{2}\left|\frac{i\hbar}{m}\langle\hat{p}_x\rangle\right| = \tfrac{1}{2}\left|\frac{i\hbar}{m}\right|\langle\hat{p}_x\rangle = \tfrac{1}{2}\left\{\left(\frac{-i\hbar}{m}\right)\left(\frac{i\hbar}{m}\right)\right\}^{1/2}\langle\hat{p}_x\rangle$$
$$\geq \frac{\hbar}{2m}\langle\hat{p}_x\rangle$$

Position and kinetic energy are complementary observables. However, in the special case for which $\langle \hat{p}_x \rangle = 0$ there will be no restriction on the simultaneous determination of them.

P1.17 (a) $\hat{i}\left(x^3 - kx\right) = (-x)^3 - k(-x) = -x^3 + kx = -1\times\left(x^3 - kx\right)$

$x^3 - kx$ is an eigenfunction of $\hat{i}$. Its eigenvalue is $\boxed{-1}$

(b) $\hat{i}\left(\cos(kx)\right) = \cos(-kx) = \cos(kx)$

$\cos(kx)$ is an eigenfunction of $\hat{i}$. Its eigenvalue is $\boxed{+1}$

(c) $\hat{i}\left(x^2 + 3x - 1\right) = (-x)^2 + 3(-x) - 1 = x^2 - 3x - 1$

$x^2 + 3x - 1$ is not an eigenfunction of $\hat{i}$

P1.19

$$\hat{E}_{\mathrm{k}} = \frac{\hat{p}_x^2}{2m_{\mathrm{e}}} = \frac{1}{2m_{\mathrm{e}}}\left(\frac{\hbar}{\mathrm{i}}\frac{\mathrm{d}}{\mathrm{d}x}\right)\left(\frac{\hbar}{\mathrm{i}}\frac{\mathrm{d}}{\mathrm{d}x}\right) = -\frac{\hbar^2}{2m_{\mathrm{e}}}\frac{\mathrm{d}^2}{\mathrm{d}x^2}$$

and the unnormalized wavefuction is $\psi = (\cos\chi)\mathrm{e}^{ikx} + (\sin\chi)\mathrm{e}^{-ikx} = c_1\mathrm{e}^{ikx} + c_2\,\mathrm{e}^{-ikx}$

$$\langle \hat{E}_{\mathrm{k}} \rangle = N^2\int\psi^*\left(\frac{\hat{p}_x^2}{2m_{\mathrm{e}}}\right)\psi\,\mathrm{d}\tau = \frac{\int\psi^*\left(\frac{\hat{p}^2}{2m_{\mathrm{e}}}\right)\psi\,\mathrm{d}\tau}{\int\psi^*\psi\,\mathrm{d}\tau}\qquad\left[N^2 = \frac{1}{\int\psi^*\psi\,\mathrm{d}\tau}\right]$$

$$= \frac{\frac{-\hbar^2}{2m_{\mathrm{e}}}\int\psi^*\frac{\mathrm{d}^2}{\mathrm{d}x^2}(\mathrm{e}^{ikx}\cos\chi + \mathrm{e}^{-ikx}\sin\chi)\,\mathrm{d}\tau}{\int\psi^*\psi\,\mathrm{d}\tau}$$

$$= \frac{\frac{-\hbar^2}{2m_{\mathrm{e}}}\int\psi^*(-k^2)\times(\mathrm{e}^{ikx}\cos\chi + \mathrm{e}^{-ikx}\sin\chi)\,\mathrm{d}\tau}{\int\psi^*\psi\,\mathrm{d}\tau} = \frac{\hbar^2k^2\int\psi^*\psi\,\mathrm{d}\tau}{2m_{\mathrm{e}}\int\psi^*\psi\,\mathrm{d}\tau} = \boxed{\frac{\hbar^2k^2}{2m_{\mathrm{e}}}}$$

P1.21

$$\langle r\rangle = N^2\int\psi^* r\psi\,\mathrm{d}\tau,\qquad \langle r^2\rangle = N^2\int\psi^* r^2\psi\,\mathrm{d}\tau$$

(i) The normalized wavefunction (see P1.12) is

$$\psi = N\left(2 - \frac{r}{a_0}\right)\mathrm{e}^{-r/2a_0}\quad\text{where } N = \left(\frac{1}{\pi a_0^3}\right)^{1/2}$$

$$\langle r\rangle = \frac{1}{\pi a_0^3}\int_0^\infty r\left(2 - \frac{r}{a_0}\right)^2 r^2\mathrm{e}^{-2r/a_0}\,\mathrm{d}r\times 4\pi\qquad\left[\int_0^\pi\sin\theta\,\mathrm{d}\theta\int_0^{2\pi}\mathrm{d}\phi = 4\pi\right]$$

$$= \frac{4}{a_0^3}\int_0^\infty\left(4r^3 - \frac{4r^4}{a_0} + \frac{r^5}{a_0^2}\right)\mathrm{e}^{-2r/a_0}\,\mathrm{d}r$$

$$= \frac{4}{a_0^3}\left\{4\times 3!(a_0/2)^4 - \frac{4\times 4!(a_0/2)^5}{a_0} + \frac{5!(a_0/2)^6}{a_0^2}\right\} = \boxed{3a_0/2}\qquad\left[\int_0^\infty x^n\mathrm{e}^{-ax}\,\mathrm{d}x = \frac{n!}{a^{n+1}}\right]$$

$$\langle r^2\rangle = \frac{4}{a_0^3}\int_0^\infty\left(4r^4 - \frac{4r^5}{a_0} + \frac{r^6}{a_0^2}\right)\mathrm{e}^{-2r/a_0}\,\mathrm{d}r = \frac{4}{a_0^3}a_0^5\left(\frac{4\times 4!}{2^5} - \frac{4\times 5!}{2^6} + \frac{6!}{2^7}\right)$$

$$= \boxed{9a_0^2/2}$$

(ii) The normalized wave function (see P1.12) is $\psi = Nr\sin\theta\cos\phi\,\mathrm{e}^{-r/2a_0}$ where $N = \left(\frac{1}{32\pi a_0^5}\right)^{1/2}$

$$\langle r\rangle = \frac{1}{32\pi a_0^5}\int_0^\infty r^5\mathrm{e}^{-r/a_0}\,\mathrm{d}r\times\frac{4\pi}{3} = \frac{1}{24a_0^5}\times 5!a_0^6 = \boxed{5a_0}$$

$$\langle r^2\rangle = \frac{1}{24a_0^5}\int_0^\infty r^6\mathrm{e}^{-r/a_0}\,\mathrm{d}r = \frac{1}{24a_0^5}\times 6!a_0^7 = \boxed{30a_0^2}$$

P1.20 For the hermitian operator $\hat{\Omega}$: $\left\langle \hat{\Omega}^2 \right\rangle - \int \psi^* \hat{\Omega}^2 \psi d\tau - \int \psi^* \hat{\Omega}\hat{\Omega} \psi d\tau = \left\{ \int \left(\hat{\Omega}\psi \right)^* \hat{\Omega}\psi d\tau \right\}^*$

The integrand on the far right is a function times its complex conjugate, which must always be a real, positive number. When this type of integrand is integrated over real space, the result is always a real, positive number. Thus, the expectation value of the square of a hermitian operator is always positive.

P1.25
$$\rho(\lambda) = \frac{8\pi hc}{\lambda^5}\left(\frac{1}{e^{hc/\lambda kT} - 1}\right)$$

(a) A plot of Planck's spectral energy density at 298 K is shown in Figure 1.1.

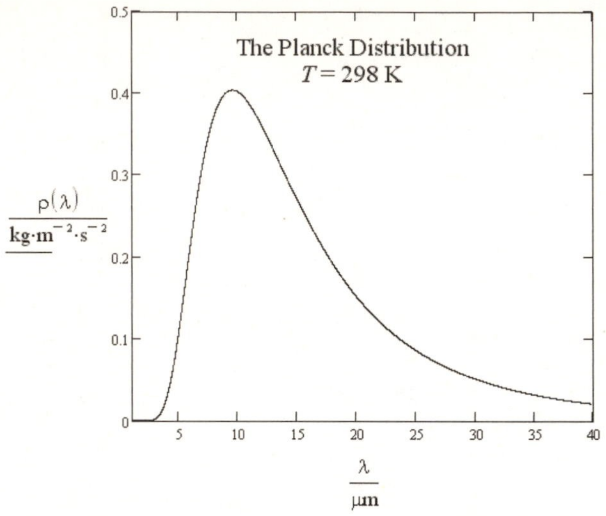

Figure 1.1

(b)
$$\lim_{\lambda\to 0}\rho(\lambda) = \lim_{\lambda\to 0}\left\{\frac{8\pi hc}{\lambda^5}\left(\frac{1}{e^{hc/\lambda kT} - 1}\right)\right\} = \lim_{\lambda\to 0}\left\{\frac{8\pi hc}{\lambda^5}\left(\frac{1}{e^{hc/\lambda kT}}\right)\right\}$$

In the limit as $\lambda \to 0$, $e^{hc/\lambda kT} \to \infty$ more rapidly than $\lambda^5 \to 0$. Hence, $\rho \to 0$ as $\lambda \to 0$. To see this in a series of mathematical manipulations, let $a = hc/kT$ and examine the denominator of the above expression in the limit as $\lambda \to 0$.

$$\lim_{\lambda\to 0}\rho(\lambda) = 8\pi hc \lim_{\lambda\to 0}\left\{\frac{1}{\lambda^5 e^{a/\lambda}}\right\} = 8\pi hc \lim_{\lambda\to 0}\left\{\frac{1}{e^{\ln\lambda^5} e^{a/\lambda}}\right\} = 8\pi hc \lim_{\lambda\to 0}\left\{\frac{1}{e^{\lambda \ln\lambda^5} e^{a}}\right\}^{1/\lambda}$$

$$= 8\pi hc \lim_{\lambda\to 0}\left\{\frac{1}{e^{\ln\lambda^{5\lambda}} e^{a}}\right\}^{1/\lambda} = 8\pi hc \lim_{\lambda\to 0}\left\{\frac{1}{e^{\lim\limits_{\lambda\to 0}\left(\ln\lambda^{5\lambda}\right)} e^{a}}\right\}^{1/\lambda} = 8\pi hc \lim_{\lambda\to 0}\left\{\frac{1}{e^{\lim\limits_{\lambda\to 0}\left(\ln\lambda^{0}\right)} e^{a}}\right\}^{1/\lambda}$$

$$= 8\pi hc \lim_{\lambda\to 0}\left\{\frac{1}{e^{\ln 1} e^{a}}\right\}^{1/\lambda} = 8\pi hc \lim_{\lambda\to 0}\left\{\frac{1}{e^{0} e^{a}}\right\}^{1/\lambda} = 8\pi hc \lim_{\lambda\to 0}\left\{\frac{1}{e^{a}}\right\}^{1/\lambda} = 8\pi hc \lim_{\lambda\to 0}\left\{\frac{1}{e^{a/\lambda}}\right\}$$

$$= \lim_{\lambda\to 0}\left\{\frac{8\pi hc}{e^{hc/\lambda kT}}\right\} = 0$$

(c) As λ increases, $hc/\lambda kT$ decreases, and at very long wavelength $hc/\lambda kT \ll 1$. Hence we can expand the exponential in a power series. Let $x = hc/\lambda kT$. Then

$$e^x = 1 + x + \frac{1}{2!}x^2 + \frac{1}{3!}x^3 + \cdots$$

$$\rho = \frac{8\pi hc}{\lambda^5}\left[\frac{1}{1 + x + \frac{1}{2!}x^2 + \frac{1}{3!}x^3 + \cdots - 1}\right]$$

$$\lim_{\lambda\to\infty}\rho = \frac{8\pi hc}{\lambda^5}\left[\frac{1}{1 + x - 1}\right] = \frac{8\pi hc}{\lambda^5}\left(\frac{1}{hc/\lambda kT}\right) = \frac{8\pi kT}{\lambda^4}$$

This is the Rayleigh–Jeans law.

Solutions to applications

P1.27

$$\lambda_{\max} = \frac{hc}{5kT}\ [\text{P1.26}] = \frac{(6.626\times10^{-34}\ \text{J s})\times(2.998\times10^{8}\ \text{m s}^{-1})}{5\times(1.381\times10^{-23}\ \text{J K}^{-1})\times(5800\ \text{K})}$$

$$= 4.96\times10^{-7}\ \text{m} = \boxed{49\bar{6}\ \text{nm, blue-green}}\ \text{(See text Fig. F.11.)}$$

P1.29 The superpositions of cosine functions of the form cos(nx) can be chosen with n equal to any integer between 1 and m. For convenience, x can be examined in the range between $-\pi/2$ and $\pi/2$ The normalization constant for each function is determined by integrating the function squared over the range of x [1.4]. Using Mathcad to perform the integration, we find

$$\int_{-\frac{\pi}{2}}^{\frac{\pi}{2}} (\cos(n\cdot x))^2\,\mathrm{d}x \rightarrow \frac{1}{2}\cdot\frac{2\cdot\cos\left(\frac{1}{2}\cdot\pi\cdot n\right)\cdot\sin\left(\frac{1}{2}\cdot\pi\cdot n\right)+\pi\cdot n}{n}$$

When n is an even integer, $\sin(\pi n/2) = 0$, and when n is an odd integer, $\cos(\pi n/2) = 0$. Consequently, when n is an integer, the above integral equals $\pi/2$ and we select $(2/\pi)^{1/2}$ as the normalization constant for the function cos(nx). The normalized function is $\phi(n, x)$. The superposition, $\psi(m, x)$, is the sum of these cosine functions from $n = 1$ to $n = m$. Since the cosine functions are orthogonal, $\psi(m, x)$ has a normalization constant equal to $(1/m)^{1/2}$. The Mathcad probability density plots of Figure 1.2 show $\psi^2(m, x)$ against x for $m = 1$, 3, and 10.

$$\phi(n,x) := \left(\frac{2}{\pi}\right)^{\frac{1}{2}}\cdot\cos(n,x) \qquad \psi(m,x) := \left(\frac{1}{m}\right)^{\frac{1}{2}}\sum_{n=1}^{m}\phi(n,x) \qquad x := \frac{-\pi}{2}, \frac{-\pi}{2}+0.001\ldots\frac{\pi}{2}$$

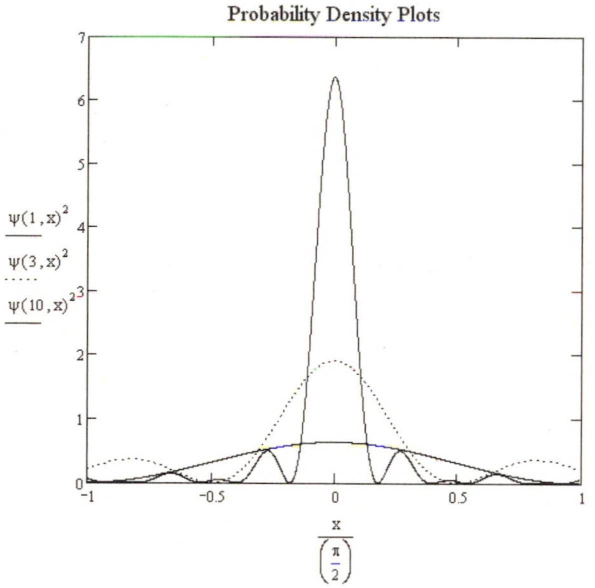

Figure 1.2

Examination of the probability density plots reveals that, when the superposition has few terms, the particle position is ill-defined. There is a great uncertainty in knowledge of position. When many terms are added to the superposition, the uncertainty narrows to a small region around $x = 0$. A plot with m greater than 10 will further confirm this conclusion.

Each function in the superposition has been assigned a weight equal to the normalization constant $(1/m)^{1/2}$. This means that each cosine function in the superposition has an identical probability contribution to the expectation value for momentum (see *Justification* 1.4). Each cosine function contributes with a probability equal to $1/m$. Furthermore, each cosine function represents a particle momentum that is proportional to the argument n (see Example 1.5). The Figure 1.3 Mathcad plot of momentum probability against momentum, as represented by n, is an interesting contrast to the plot of probability density against position.

Variables needed for the Mathcad plot: $n := 1 \ldots 12$ $\quad \text{Prob}(n,m) := \text{if}\left(n \le m, \frac{1}{m}, 0\right)$

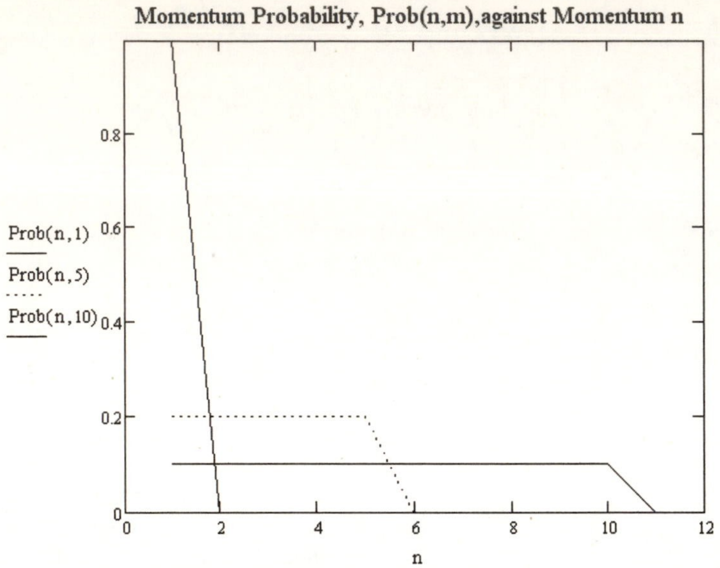

Figure 1.3

The momentum probability plot shows momentum probabilities for superpositions of m = 1, 5, and 10 terms. When there are many terms in the superposition, the range of possible momentum is very broad even though the range of observed positions becomes narrow. Position and momentum are complementary variables. As location becomes more precise with the superposition of many functions, precise knowledge of momentum decreases. This illustrates the Heisenberg uncertainty principle (1.19a and 1.23].

The plot of probability density against position clearly indicates that the superposition is symmetrical around the point $x = 0$. Consequently, the expectation position for all superpositions is $x = 0$. The expectation value for position is independent of the number of terms in the superposition.

The square root of the expectation value of x^2 is called the root-mean-square value of x, x_{rms}. The Figure 1.4 Mathcad plot of x_{rms} against m indicates that this expectation value depends upon the number of terms in the superposition. However, it does appear to very slowly converge to a very small value (zero?) when the superposition contains many functions.

$$x_{rms}(m) := \left(\int_{-\frac{\pi}{2}}^{\frac{\pi}{2}} x^2 \cdot \psi(m,x)^2 \, dx \right)^{\frac{1}{2}} \qquad m := 1 \ldots 50$$

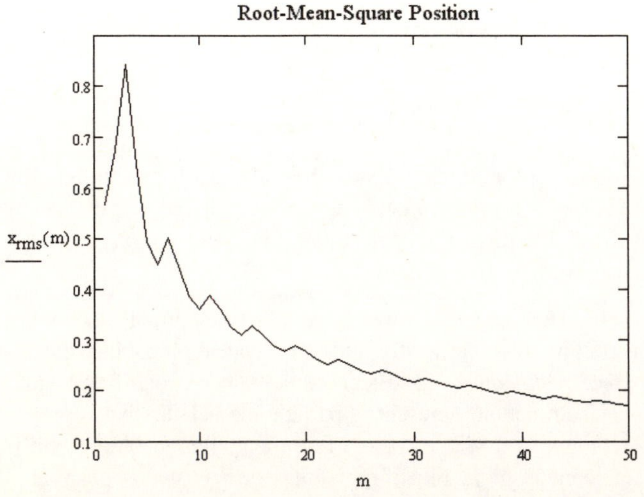

Figure 1.4

2 Nanosystems 1: motion in one dimension

Answers to discussion questions

D2.1 In quantum mechanics, particles are said to have wave characteristics. The fact of the existence of the particle then requires that the wavelengths of the waves representing it be such that the wave does not experience destructive interference upon reflection by a barrier or in its motion around a closed loop. This requirement restricts the wavelength to values $\lambda = 2/n \times L$, where L is the length of the path and n is a positive integer. Then using the relations $\lambda = h/p$ and $E = p^2/2m$, the energy is quantized at $E = n^2h^2/8mL^2$. This derivation applies specifically to the particle in a box. The derivation is similar for the particle on a ring (see Section 3.3); the same principles apply.

D2.3 The lowest energy level possible for a confined quantum mechanical system is the zero-point energy, and zero-point energy is not zero energy. The system must have at least that minimum amount of energy even at absolute zero. The physical reason is that if the particle is confined, its position is not completely uncertain, and therefore its momentum and hence its kinetic energy cannot be exactly zero. We have already seen zero-point energy in the particle in a box and the harmonic oscillator.

D2.5 Macroscopic synthesis and material development always contain elements of molecular randomness. Crystal structures are never perfect. A product of organic synthesis is never absolutely free of impurities, although impurities may be at a level that is lower than measurement techniques make possible. Alloys are grainy and slightly non-homogeneous within any particular grain. Furthermore, the random distribution of atomic/molecular positions and orientations within and between macroscopic objects causes the conversion of energy to non-useful heat during manufacturing processes. Production efficiencies are difficult to improve. Nanometer technology on the 1 nm to 100 nm scale may resolve many of these problems. Self-organization and production processes by nanoparticles and nanomachines may be able to exclude impurities and greatly improve homogeneity by effective examination and selection of each atom/molecule during nanosynthesis and nanoproduction processes. Higher efficiencies of energy usage may be achievable as nanomachines produce idealized materials at the smaller sizes and pass their products to larger nanomachines for production of larger scale materials.

The directed, non-random use of atoms and molecules by nanotechniques holds the promise of the production of smaller transistors and wires for the electronics and computer industries. Unusual material strengths, optical properties, magnetic properties, and catalytic properties may be achievable. Higher efficiencies of photoelectronic conversion would be a boon to mankind. There is hope that science will devise nanoparticles that destroy pathogens and repair tissues. See Impact 2.1 for discussion of examination of atom positions on a macroscopic surface by scanning probe microscopy and for the current nanotechnological method for positioning atoms on a surface. See Impact 3.1 for discussion of nanoquantum dots that have unusual optical and magnetic properties.

D2.7 The particle in a box can serve as a model for many kinds of bound particles. Perhaps most relevant to chemistry, it is a model for π-electrons in linear systems of conjugated double bonds. See Example 2.1, in which it is the basis for estimating the absorption wavelength for an electronic transition in the linear polyene β-carotene. More crudely yet more fundamentally, it can provide order-of-magnitude excitation energies for bound particles such as an electron confined to an atom-sized box around a nucleus or even for a nucleon confined to a nucleus-sized box. The harmonic oscillator serves as the first approximation for describing molecular vibrations. The excitation energies for stretching and bending bonds are manifested in infrared spectroscopy.

Solutions to exercises

E2.1(a) If the wavefunction is an eigenfunction of an operator, the corresponding eigenvalue is the value of the corresponding observable (Section 1.7, Postulate IV). Applying the linear momentum operator $\hat{p}=\frac{\hbar}{\mathrm{i}}\frac{\mathrm{d}}{\mathrm{d}x}$ (eqn. 1.10) to the wavefunction yields

$$\hat{p}\psi=\frac{\hbar}{\mathrm{i}}\frac{\mathrm{d}}{\mathrm{d}x}\psi=\frac{\hbar}{\mathrm{i}}\frac{\mathrm{d}}{\mathrm{d}x}\mathrm{e}^{ikx}=\hbar k\mathrm{e}^{ikx},$$

so the wavefunction is an eigenfunction of the linear momentum; thus, the value of the linear momentum is the eigenvalue

$$\hbar k=1.0546\times10^{-34}\ \mathrm{J\ s}\times3\ \mathrm{m}^{-1}=\boxed{3\times10^{-34}\ \mathrm{kg\ m\ s}^{-1}}$$

Similarly, applying the kinetic energy operator $\hat{E}_{\mathrm{k}}=-\frac{\hbar^2}{2m}\frac{\mathrm{d}^2}{\mathrm{d}x^2}$ (eqn. 1.11) to the wavefunction yields0

$$\hat{E}_{\mathrm{k}}\psi=-\frac{\hbar^2}{2m}\frac{\mathrm{d}^2}{\mathrm{d}x^2}\psi=-\frac{\hbar^2}{2m}\frac{\mathrm{d}^2}{\mathrm{d}x^2}\mathrm{e}^{ikx}=\frac{\hbar^2k^2}{2m}\mathrm{e}^{ikx},$$

so the wavefunction is an eigenfunction of this operator as well; thus, its value is the eigenvalue

$$\frac{\hbar^2k^2}{2m}=\frac{(1.0546\times10^{-34}\ \mathrm{J\ s}\times3\ \mathrm{m}^{-1})^2}{2\times9.11\times10^{-31}\ \mathrm{kg}}=\boxed{5\times10^{-38}\ \mathrm{J}}$$

E2.2(a) (a) An electron accelerated through a potential difference of 1.0 V acquires a kinetic energy of 1.0 eV. That is the definition of the electronvolt. The wavefunction for such a particle is (eqn. 2.2 with $B=0$ because the particle is moving toward positive x)

$$\psi_k=\boxed{A\mathrm{e}^{ikx}}$$

The index k is given by the relationship

$$E_{\mathrm{k}}=\frac{\hbar^2k^2}{2m}=1.0\ \mathrm{eV}$$

Thus $$k=\frac{(2mE_{\mathrm{k}})^{1/2}}{\hbar}=\frac{(2\times9.11\times10^{-31}\ \mathrm{kg}\times1.0\ \mathrm{eV}\times1.602\times10^{-19}\ \mathrm{J\ eV}^{-1})^{1/2}}{1.0546\times10^{-34}\ \mathrm{J\ s}}$$

$$=\boxed{5.1\times10^{9}\ \mathrm{m}^{-1}}$$

(b) Here $E_{\mathrm{k}}=10$ keV, so

$$k=\frac{(2mE_{\mathrm{k}})^{1/2}}{\hbar}=\frac{(2\times9.11\times10^{-31}\ \mathrm{kg}\times10\times10^{3}\ \mathrm{eV}\times1.602\times10^{-19}\ \mathrm{J\ eV}^{-1})^{1/2}}{1.0546\times10^{-34}\ \mathrm{J\ s}}$$

$$=\boxed{5.1\times10^{11}\ \mathrm{m}^{-1}}$$

E2.3(a) $$E=\frac{n^2h^2}{8m_{\mathrm{e}}L^2}\quad[2.6\mathrm{a}]$$

$$\frac{h^2}{8m_{\mathrm{e}}L^2}=\frac{(6.626\times10^{-34}\ \mathrm{J\ s})^2}{8(9.11\times10^{-31}\ \mathrm{kg})\times(1.0\times10^{-9}\ \mathrm{m})^2}=6.0\bar{2}\times10^{-20}\ \mathrm{J}$$

The conversion factors required are

$$1\ \text{eV} = 1.602\times10^{-19}\ \text{J};\quad 1\ \text{cm}^{-1} = 1.986\times10^{-23}\ \text{J};\quad 1\ \text{eV} = 96.485\ \text{kJ mol}^{-1}$$

(a) $$E_2 - E_1 = (4-1)\frac{h^2}{8m_eL^2} = \frac{3h^2}{8m_eL^2} = (3)\times(6.0\bar{2}\times10^{-20}\ \text{J})$$

$$= \boxed{1.8\times10^{-19}\ \text{J}},\ \boxed{1.1\ \text{eV}},\ \boxed{9.1\times10^3\ \text{cm}^{-1}},\ \boxed{1.1\times10^2\ \text{kJ mol}^{-1}}$$

(b) $$E_6 - E_5 = (36-25)\frac{h^2}{8m_eL^2} = \frac{11h^2}{8m_eL^2} = (11)\times(6.0\bar{2}\times10^{-20}\ \text{J})$$

$$= \boxed{6.6\times10^{-19}\ \text{J}},\ \boxed{4.1\ \text{eV}},\ \boxed{3.3\times10^4\ \text{cm}^{-1}},\ \boxed{4.0\times10^2\ \text{kJ mol}^{-1}}$$

COMMENT. The energy level separations increase as n increases.

E2.4(a) The wavefunctions are

$$\psi_n = \left(\frac{2}{L}\right)^{1/2}\sin\left(\frac{n\pi x}{L}\right)\quad [2.6\text{b}]$$

The required probability is

$$P_n = \int_{0.49L}^{0.51L}\psi_n^2\,\text{d}x \approx \psi_n^2\Delta x$$

where $\Delta x = 0.02L$ and the function is evaluated at $x = L/2$.

(a) $$P_1 = \left(\frac{2}{L}\right)\sin^2\left(\frac{\pi}{2}\right)\times0.02L = \boxed{0.04}$$

(b) $$P_2 = \left(\frac{2}{L}\right)\sin^2\left(\frac{2\pi}{2}\right)\times0.02L = \left(\frac{2}{L}\right)\sin^2\pi\times0.02L = \boxed{0}$$

E2.5(a) The wavefunction for a particle in the state $n = 1$ in a square-well potential is

$$\psi_1 = \left(\frac{2}{L}\right)^{1/2}\sin\left(\frac{\pi x}{L}\right)\quad [2.6\text{b}]$$

The expectation value of the momentum operator, $\hat{p} = \frac{\hbar}{\text{i}}\frac{\text{d}}{\text{d}x}$, is

$$\langle p\rangle = \int_0^L \psi_1{}^*\hat{p}\psi_1\,\text{d}x = \frac{2\hbar}{\text{i}L}\int_0^L \sin\left(\frac{\pi x}{L}\right)\frac{\text{d}}{\text{d}x}\sin\left(\frac{\pi x}{L}\right)\text{d}x$$

$$= \frac{2\pi\hbar}{\text{i}L^2}\int_0^L \sin\left(\frac{\pi x}{L}\right)\cos\left(\frac{\pi x}{L}\right)\text{d}x = \boxed{0}$$

The expectation value of the momentum squared operator, $\hat{p}^2 = -\hbar^2\frac{\text{d}^2}{\text{d}x^2}$, is

$$\langle p^2\rangle = -\frac{2\hbar^2}{L}\int_0^L \sin\left(\frac{\pi x}{L}\right)\frac{\text{d}^2}{\text{d}x^2}\sin\left(\frac{\pi x}{L}\right)\text{d}x = \left(\frac{2\hbar^2}{L}\right)\times\left(\frac{\pi}{L}\right)^2\int_0^L \sin^2\left(\frac{\pi x}{L}\right)\text{d}x$$

Use $\int(\sin^2 ax)\text{d}x = \frac{x}{2} - \frac{1}{4a}\sin 2ax$ with $a = \pi/L$. Thus

$$\langle p^2\rangle = \left(\frac{2\hbar^2}{L}\right)\times\left(\frac{\pi}{L}\right)^2\left(\frac{x}{2} - \frac{L}{4\pi}\sin\left(\frac{2\pi x}{L}\right)\right)\Bigg|_0^L = \left(\frac{2\hbar^2}{L}\right)\times\left(\frac{\pi}{L}\right)^2\times\left(\frac{L}{2}\right) = \boxed{\frac{h^2}{4L^2}}$$

COMMENT. The expectation value of $\hat{p}$ is zero because on average the particle moves to the left as often as the right.

E2.6(a) The zero-point energy is the ground-state energy; that is, with $n = 1$,

$$E = \frac{n^2h^2}{8m_eL^2}\,[2.6a] = \frac{h^2}{8m_eL^2}$$

Set this equal to the rest energy m_ec^2 and solve for L:

$$m_ec^2 = \frac{h^2}{8m_eL^2} \quad \text{so} \quad L = \frac{h}{8^{1/2}m_ec}$$

In absolute units, the length is

$$L = \frac{6.63\times10^{-34}\ \text{J s}}{8^{1/2}\times(9.11\times10^{-31}\ \text{kg})\times(3.00\times10^{8}\ \text{m s}^{-1})} = 8.58\times10^{-13}\ \text{m} = 0.858\ \text{pm}$$

In terms of the Compton wavelength of an electron, $\lambda_C = \dfrac{h}{m_ec}$, $\boxed{L = \dfrac{\lambda_C}{8^{1/2}}}$

E2.7(a)

$$\psi_3 = \left(\frac{2}{L}\right)^{1/2}\sin\left(\frac{3\pi x}{L}\right)\ [2.6b]$$

$$P(x) \propto \psi_3^2 \propto \sin^2\left(\frac{3\pi x}{L}\right)$$

The maxima and minima in $P(x)$ correspond to $\dfrac{dP(x)}{dx} = 0$.

$$\frac{dP(x)}{dx} \propto \frac{d\psi^2}{dx} \propto \sin\left(\frac{3\pi x}{L}\right)\cos\left(\frac{3\pi x}{L}\right) \propto \sin\left(\frac{6\pi x}{L}\right) \quad [2\sin\alpha\cos\alpha = \sin 2\alpha]$$

$\sin\theta = 0$ when θ equals an integer times π:

$$\frac{6\pi x}{L} = n'\pi \text{ for } n' = 0, 1, 2, \ldots, \text{ which corresponds to } x = \frac{n'L}{6},\ n' \le 6.$$

$n' = 0, 2, 4$, and 6 correspond to minima in ψ^2, leaving $n' = 1, 3$, and 5 for the maxima; that is,

$$\boxed{x = \frac{L}{6}, \frac{L}{2} \text{ and } \frac{5L}{6}}$$

COMMENT. Maxima in ψ^2 correspond to maxima *and* minima in ψ itself, so one can also solve this exercise by finding all points where $\dfrac{d\psi}{dx} = 0$.

E2.8(a) The transmission probability (eqn. 2.17) depends on the energy of the tunneling particle relative to the barrier height ($\varepsilon = E/V = 1.5$ eV/(2.0 eV) = 0.75), on the width of the barrier (L = 100 pm), and on the decay parameter of the wavefunction inside the barrier (κ), where

$$\kappa = \frac{\{2m(V-E)\}^{1/2}}{\hbar} = \frac{\{2\times9.11\times10^{-31}\ \text{kg}\times(2.0-1.5)\ \text{eV}\times1.602\times10^{-19}\ \text{J eV}^{-1}\}^{1/2}}{1.0546\times10^{-34}\ \text{J s}}$$
$$= 3.\bar{6}\times10^{9}\ \text{m}^{-1}$$

We note that $\kappa L = 3.\bar{6}\times10^{9}\ \text{m}^{-1}\times100\times10^{-12}\ \text{m} = 0.3\bar{6}$ is not large compared to 1, so we must use eqn. 2.17(a) for the transmission probability.

$$T = \left\{1+\frac{(e^{\kappa L}-e^{-\kappa L})^2}{16\varepsilon(1-\varepsilon)}\right\}^{-1} = \left\{1+\frac{(e^{0.3\bar{6}}-e^{-0.3\bar{6}})^2}{16\times0.75\times(1-0.75)}\right\}^{-1} = \boxed{0.8}$$

E2.9(a) The zero-point energy of a harmonic oscillator is (eqn. 2.23)

$$E_0 = \frac{1}{2}\hbar\omega = \frac{\hbar}{2}\left(\frac{k}{m}\right)^{1/2} \ [2.21] = \frac{1.0546\times10^{-34}\ \text{J s}}{2}\times\left(\frac{155\ \text{N m}^{-1}}{1.67\times10^{-27}\ \text{kg}}\right)^{1/2}$$

$$= \boxed{1.61\times10^{-20}\ \text{J}}$$

E2.10(a) The difference in adjacent energy levels is

$$\Delta E = E_{\upsilon+1} - E_{\upsilon} = \hbar\omega\ [2.22] = \hbar\left(\frac{k}{m}\right)^{1/2}\ [2.21]$$

Hence $k = m\left(\frac{\Delta E}{\hbar}\right)^2 = (1.33\times10^{-25}\,\text{kg})\times\left(\frac{4.82\times10^{-21}\ \text{J}}{1.055\times10^{-34}\ \text{J s}}\right)^2 = 278\ \text{kg s}^{-2} = \boxed{278\ \text{N m}^{-1}}$

E2.11(a) The requirement for a transition to occur is that $\Delta E(\text{system}) = E(\text{photon})$,

where $\Delta E(\text{system}) = \hbar\omega$ [2.22]

and $E(\text{photon}) = h\nu = \frac{hc}{\lambda}$

Therefore, $\frac{hc}{\lambda} = \frac{h\omega}{2\pi} = \left(\frac{h}{2\pi}\right)\times\left(\frac{k}{m}\right)^{1/2}$

$$\lambda = 2\pi c\left(\frac{m}{k}\right)^{1/2} = (2\pi)\times(2.998\times10^8\ \text{m s}^{-1})\times\left(\frac{1.673\times10^{-27}\ \text{kg}}{855\ \text{N m}^{-1}}\right)^{1/2}$$

$$= 2.63\times10^{-6}\,\text{m} = \boxed{2.63\ \mu\text{m}}$$

E2.12(a) $\lambda = 2\pi c\left(\frac{m}{k}\right)^{1/2}$ [E2.11a]

Since $\lambda \propto m^{1/2}$, $\lambda_{\text{new}} = 2^{1/2}\lambda_{\text{old}} = (2^{1/2})\times(2.63\ \mu\text{m}) = \boxed{3.72\ \mu\text{m}}$

E2.13(a) The Schrödinger equation for the linear harmonic oscillator is

$$-\frac{\hbar^2}{2m}\frac{\text{d}^2\psi}{\text{d}x^2} + \frac{1}{2}kx^2\psi = E\psi\ [2.20]$$

The ground-state wavefunction to be tested is

$$\psi_0 = N_0\text{e}^{-x^2/2\alpha^2}\ [2.25\text{a}]$$

with $\alpha = \left(\frac{\hbar^2}{mk}\right)^{1/4}$ [2.24]; so $k = \frac{\hbar^2}{m\alpha^4}$ (a)

Performing the operations,

$$\frac{\text{d}\psi_0}{\text{d}x} = \left(-\frac{1}{\alpha^2}x\right)\psi_0$$

$$\frac{\text{d}^2\psi_0}{\text{d}x^2} = \left(-\frac{1}{\alpha^2}x\right)\times\left(-\frac{1}{\alpha^2}x\right)\times\psi_0 + \left(-\frac{1}{\alpha^2}\psi_0\right) = \frac{x^2}{\alpha^4}\psi_0 - \frac{1}{\alpha^2}\psi_0 = \left(\frac{x^2}{\alpha^4} - \frac{1}{\alpha^2}\right)\psi_0$$

Substituting into the Schrödinger equation,

$$-\frac{\hbar^2}{2m}\left(\frac{x^2}{\alpha^4}-\frac{1}{\alpha^2}\right)\psi_0+\frac{1}{2}kx^2\psi_0=E_0\psi_0$$

which implies

$$E_0=\frac{-\hbar^2}{2m}\left(\frac{x^2}{\alpha^4}-\frac{1}{\alpha^2}\right)+\frac{1}{2}kx^2 \qquad \text{(b)}$$

But E_0 is a constant, independent of x; therefore the terms that contain x must drop out, which is possible only if

$$-\frac{\hbar^2}{2m\alpha^4}+\frac{1}{2}k=0$$

which is consistent with $k=\dfrac{\hbar^2}{m\alpha^4}$ as in (a). What is left in (b) is

$$E_0=\frac{\hbar^2}{2m\alpha^2}=\frac{1}{2}\hbar\omega \quad \left[\text{using } \omega=\left(\frac{k}{m}\right)^{1/2} \text{ and } k=\frac{\hbar^2}{m\alpha^4}\right]$$

Therefore, ψ_0 is a solution of the Schrödinger equation with energy $\frac{1}{2}\hbar\omega$.

E2.14(a) The harmonic oscillator wavefunctions have the form

$$\psi_v(x)=N_vH_v(y)\exp\left(-\frac{1}{2}y^2\right) \text{with } y=\frac{x}{\alpha} \text{ and } \alpha=\left(\frac{\hbar^2}{mk}\right)^{1/4} \quad [2.24]$$

The exponential function approaches zero only as x approaches $\pm\infty$, so the nodes of the wavefunction are the nodes of the Hermite polynomials.

$$H_4(y)=16y^4-48y^2+12=0 \ [\text{Table 2.1}]$$

Dividing through by 4 and letting $z=y^2$, we have a quadratic equation $4z^2-12z+3=0$.

So $$z=\frac{-b\pm\sqrt{b^2-4ac}}{2a}=\frac{12\pm\sqrt{12^2-4\times4\times3}}{2\times4}=\frac{3\pm\sqrt{6}}{2}$$

and $$y=\pm\sqrt{\frac{3\pm\sqrt{6}}{2}} \quad \text{and} \quad x=\alpha\left(\pm\sqrt{\frac{3\pm\sqrt{6}}{2}}\right)$$

Evaluating the result numerically yields z = 0.275 or 2.72, so y = ±0.525 or ±1.65. Therefore, $x=\boxed{\pm0.525\alpha \text{ or } \pm1.65\alpha}$.

COMMENT. Numerical values could also be obtained graphically by plotting $H_4(y)$.

E2.15(a) The harmonic oscillator wavefunctions have the form

$$\psi_v(x)=N_vH_v(y)\exp\left(-\frac{1}{2}y^2\right) \text{ with } y=\frac{x}{\alpha} \text{ and } \alpha=\left(\frac{\hbar^2}{mk}\right)^{1/4} \quad [2.24]$$

so the wavefunction in question is

$$\psi_2(x)=N_2(4y^2-2)e^{-y^2/2}$$

Normalization requires

$$1=\int_{-\infty}^{\infty}\psi_2{}^*\psi_2\mathrm{d}x=\int_{-\infty}^{\infty}\psi_2{}^2\alpha\mathrm{d}y=\int_{-\infty}^{\infty}\left(N_2(4y^2-2)e^{-y^2/2}\right)^2\alpha\mathrm{d}y$$

$$=4N_2{}^2\alpha\int_{-\infty}^{\infty}(4y^4-4y^2+1)e^{-y^2}\,\mathrm{d}y$$

To evaluate the integral, note that the integrand is even, so the integral is twice the integral from zero to infinity of the same function. Use

$$\int_0^\infty y^{2n}e^{-ay^2}dy = \frac{1\cdot 3\cdots(2n-1)}{2^{n+1}a^n}\sqrt{\frac{\pi}{a}} \quad \text{and} \quad \int_0^\infty e^{-ay^2}dy = \frac{1}{2}\sqrt{\frac{\pi}{a}}$$

Thus $$1 = 8N_2{}^2\alpha\left(4\times\frac{3\sqrt{\pi}}{2^3} - 4\times\frac{\sqrt{\pi}}{2^2} + \frac{\sqrt{\pi}}{2}\right) = 8N_2{}^2\alpha\sqrt{\pi}$$

and $$N_2 = \frac{1}{(8\alpha)^{1/2}\pi^{1/4}}$$

Confirming orthogonality amounts to demonstrating that

$$\int_{-\infty}^{\infty}\psi_4{}^*\psi_2 dx = 0 \qquad \text{where} \quad \psi_4(x) = N_4(16y^4 - 48y^2 + 12)e^{-y^2/2}$$

The integral in question is

$$\int_{-\infty}^{\infty} N_4(16y^4 - 48y^2 + 12)e^{-y^2/2}\, N_2(4y^2 - 2)e^{-y^2/2}\,\alpha dy$$

$$= 8N_4N_2\alpha\int_{-\infty}^{\infty}(4y^4 - 12y^2 + 3)(2y^2 - 1)e^{-y^2}\,dy$$

$$= 8N_4N_2\alpha\int_{-\infty}^{\infty}(8y^6 - 28y^4 + 18y^2 - 3)e^{-y^2}\,dy$$

$$= 8N_4N_2\alpha\left(8\times\frac{15}{2^4} - 28\times\frac{3}{2^3} + 18\times\frac{1}{2^2} - 3\times\frac{1}{2}\right)\sqrt{\pi}$$

The terms in parentheses, $\frac{15}{2} - \frac{21}{2} + \frac{9}{2} - \frac{3}{2}$, do indeed add up to zero, making the integral vanish, as required.

E2.16(a) The effective mass is

$$\mu = \frac{m_A m_B}{m_A + m_B} = \frac{m_{Cl}m_{Cl}}{m_{Cl} + m_{Cl}} = \frac{m_{Cl}}{2} = \frac{34.9688\ \text{u}\times 1.66054\times 10^{-27}\ \text{kg u}^{-1}}{2}$$
$$= 2.90335\times 10^{-26}\ \text{kg}$$

The difference in adjacent energy levels, which is equal to the energy of the photon, is

$$\Delta E = \hbar\omega\ [2.22] = h\nu$$

so, in terms of wavenumbers,

$$\hbar\left(\frac{k}{m}\right)^{1/2}[2.21] = hc\tilde{\nu} \quad \text{and} \quad \tilde{\nu} = \frac{h}{hc}\left(\frac{k}{m}\right)^{1/2} = \frac{1}{2\pi c}\left(\frac{k}{m}\right)^{1/2}$$

$$\tilde{\nu} = \frac{1}{2\pi\times 2.998\times 10^{10}\ \text{cm s}^{-1}}\left(\frac{329\ \text{N m}^{-1}}{2.90335\times 10^{-26}\ \text{kg}}\right)^{1/2} = \boxed{565\ \text{cm}^{-1}}$$

E2.17(a) *Justification* 2.2 analyzes the classical turning points of the harmonic oscillator. In terms of the dimensionless variable y, the turning points are $y_{tp} = \pm(2v + 1)^{1/2}$. The probability of extension beyond the classical turning point is

$$P = \int_{x_{tp}}^{\infty}\psi_v{}^2 dx = \alpha N_v{}^2\int_{y_{tp}}^{\infty}\{H_v(y)\}^2 e^{-y^2}dy$$

For $\quad v = 1, H_1(y) = 2y$ and $N_1 = \left(\frac{1}{2\alpha\pi^{1/2}}\right)^{1/2}$

$$P = 4\alpha N_1^2 \int_{3^{1/2}}^{\infty} y^2 e^{-y^2} dy$$

Use integration by parts:

$$\int u dv = uv - \int v du$$

where $\quad u = y, \quad dv = ye^{-y^2} dy$

so $\quad du = dy, \quad v = -\frac{1}{2} e^{-y^2}$

and $\quad P = -2\alpha N_1^2 \left(ye^{-y^2} \Big|_{3^{1/2}}^{\infty} - \int_{3^{1/2}}^{\infty} e^{-y^2} dy \right)$

$$= \pi^{-1/2} \left(3^{1/2} e^{-3} + \int_{3^{1/2}}^{\infty} e^{-y^2} dy \right)$$

The remaining integral can be expressed in terms of the error function:

$$\operatorname{erf} z = 1 - \frac{2}{\pi^{1/2}} \int_z^{\infty} e^{-y^2} dy$$

so $\quad \int_{3^{1/2}}^{\infty} e^{-y^2} dy = \frac{\pi^{1/2}(1 - \operatorname{erf} 3^{1/2})}{2}$

Finally, using $\operatorname{erf} 3^{1/2} = 0.986$,

$$P = \pi^{-1/2} \left(3^{1/2} e^{-3} + \frac{\pi^{1/2}(1 - \operatorname{erf} 3^{1/2})}{2} \right) = \boxed{0.056}$$

COMMENT. This is the probability of an extension greater than the positive classical turning point. There is an equal probability of a compression smaller than the negative classical turning point, so the total probability of finding the oscillator in a classically forbidden region is $\boxed{0.112}$.

COMMENT. Note that molecular parameters such as m and k do not enter into the calculation.

E2.18(a) Mean kinetic and potential energies are related by

$$2\langle E_k \rangle = b\langle V \rangle, \quad \text{where } V = ax^b. \text{ [2.30]}$$

So for $V \propto x^b$, we have $2\langle E_k \rangle = 3\langle V \rangle$, or $\boxed{\langle E_k \rangle = 3\langle V \rangle/2}$.

E2.19(a) The perturbation is

$$\hat{H}^{(1)} = V(x) = -\varepsilon \sin\left(\frac{2\pi x}{L}\right)$$

The first-order correction to the ground-state energy, E_1, is [2.34]

$$E_1^{(1)} = \int_0^L \psi_1^{(0)} * \hat{H}^{(1)} \psi_1^{(0)} dx = -\int_0^L \left(\frac{2}{L}\right)^{1/2} \sin\left(\frac{\pi x}{L}\right) \varepsilon \sin\left(\frac{2\pi x}{L}\right) \left(\frac{2}{L}\right)^{1/2} \sin\left(\frac{\pi x}{L}\right) dx$$

$$E_1^{(1)} = -\frac{2\varepsilon}{L} \int_0^L \sin^2\left(\frac{\pi x}{L}\right) \sin\left(\frac{2\pi x}{L}\right) dx$$

Notice that the $\sin^2$ term is symmetric about the midpoint of the region of integration while the sin term is antisymmetric. Therefore, the integral vanishes, and $E_1^{(1)} = \boxed{0}$.

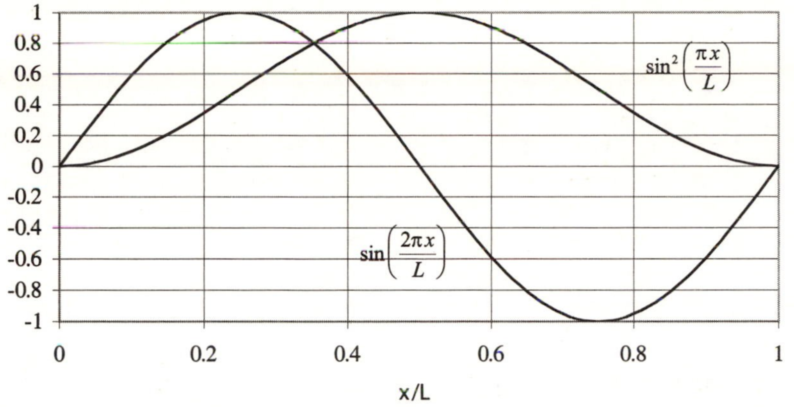

Figure 2.1

E2.20(a) The perturbation is

$$\hat{H}^{(1)} = V(x) = -\varepsilon\cos\left(\frac{2\pi x}{L}\right)\left(\frac{2\pi x}{L}\right)$$

The first-order correction to the ground-state energy, E_1, is [2.34]

$$E_1^{(1)} = \int_0^L \psi_1^{(0)} * \hat{H}^{(1)}\psi_1^{(0)}\mathrm{d}x = -\int_0^L \left(\frac{2}{L}\right)^{1/2} \sin\left(\frac{\pi x}{L}\right)\varepsilon\cos\left(\frac{2\pi x}{L}\right)\left(\frac{2}{L}\right)^{1/2} \sin\left(\frac{\pi x}{L}\right)\mathrm{d}x$$

$$E_1^{(1)} = -\frac{2\varepsilon}{L}\int_0^L \sin^2\left(\frac{\pi x}{L}\right)\cos\left(\frac{2\pi x}{L}\right)\mathrm{d}x$$

Use the trigonometric identity

$$\sin^2\left(\frac{\pi x}{L}\right) = \frac{1}{2} - \frac{1}{2}\cos\left(\frac{2\pi x}{L}\right)$$

So $$E_1^{(1)} = -\frac{\varepsilon}{L}\left\{\int_0^L \cos\left(\frac{2\pi x}{L}\right)\mathrm{d}x - \int_0^L \cos^2\left(\frac{2\pi x}{L}\right)\mathrm{d}x\right\}$$

Use $$\int \cos^2 ax\,\mathrm{d}x = \frac{x}{2} + \frac{\sin 2ax}{4a}$$

for the second integral, so

$$E_1^{(1)} = -\frac{\varepsilon}{L}\left\{\frac{L}{2\pi}\sin\left(\frac{2\pi x}{L}\right) - \frac{x}{2} - \frac{L}{8\pi}\sin\left(\frac{4\pi x}{L}\right)\right\}\Bigg|_0^L = \boxed{\frac{\varepsilon}{2}}$$

E2.21(a) The functional form of a linear slope that rises from 0 at $x = 0$ to ε at $x = L$ is

$$\hat{H}^{(1)} = V(x) = \frac{\varepsilon x}{L}$$

The first-order correction to the ground-state energy, E_1, is [2.34]

$$E_1^{(1)} = \int_0^L \psi_1^{(0)} * H^{(1)} \psi_1^{(0)} dx = \int_0^L \left(\frac{2}{L}\right)^{1/2} \sin\left(\frac{\pi x}{L}\right) \frac{\varepsilon x}{L} \left(\frac{2}{L}\right)^{1/2} \sin\left(\frac{\pi x}{L}\right) dx$$

$$E_1^{(1)} = \frac{2\varepsilon}{L^2} \int_0^L x \sin^2\left(\frac{\pi x}{L}\right) dx$$

Use $\int x \sin^2 ax\, dx = \frac{x^2}{4} - \frac{x \sin 2ax}{4a} - \frac{\cos 2ax}{8a^2}$

So $E_1^{(1)} = \frac{2\varepsilon}{L^2} \int_0^L x \sin^2\left(\frac{\pi x}{L}\right) dx = \frac{2\varepsilon}{L^2} \left(\frac{x^2}{4} - \frac{xL}{4\pi} \sin\left(\frac{2\pi x}{L}\right) - \frac{L^2}{8\pi^2} \cos\left(\frac{2\pi x}{L}\right) \right)\Bigg|_0^L = \boxed{\frac{\varepsilon}{2}}$

E2.22(a) The perturbation is

$$\hat{H}^{(1)} = V(x) = mgx$$

The first-order correction to the ground-state energy, E_0, is [2.34]

$$E_0^{(1)} = \int_{-\infty}^{\infty} \psi_0^{(0)} * \hat{H}^{(1)} \psi_0^{(0)} dx = \int_{-\infty}^{\infty} N_0 e^{-x^2/2\alpha^2} mgx N_0 e^{-x^2/2\alpha^2} dx$$

$$= mgN_0^2 \int_{-\infty}^{\infty} x e^{-x^2/\alpha^2} dx = \boxed{0}$$

COMMENT. The expression integrates to zero because the integrand is antisymmetric about the midpoint of the region of integration (i.e., an odd function of *x*). The result should not be terribly surprising, as the perturbation affects stretching and compressing the bond to an equal but opposite degree.

E2.23(a) The second-order correction to the ground-state energy, E_0, is [2.35]

$$E_0^{(2)} = \sum_{v=1}^{\infty} \frac{\left| \int_{-\infty}^{\infty} \psi_v^{(0)} * \hat{H}^{(1)} \psi_0^{(0)}\, dx \right|^2}{E_0^{(0)} - E_v^{(0)}}$$

where $\hat{H}^{(1)} = mgx$, $\psi_v^{(0)} = N_v H_v(y) e^{-y^2/2}$ and $E_v = (v + \frac{1}{2})\hbar\omega$.

The denominator in the sum is

$$E_0^{(0)} - E_v^{(0)} = \frac{1}{2}\hbar\omega - \left(v + \frac{1}{2}\right)\hbar\omega = -v\hbar\omega$$

The integral in the sum is

$$\int_{-\infty}^{\infty} \psi_v^{(0)} * \hat{H}^{(1)} \psi_0^{(0)} dx = \int_{-\infty}^{\infty} N_v H_v(y) e^{-y^2/2} mgx N_0 e^{-y^2/2} dx$$

$$= mgN_v N_0 \alpha \int_{-\infty}^{\infty} H_v(y) e^{-y^2/2} y e^{-y^2/2} dx$$

where we have used $H_0(y) = 1$ and $y = x/\alpha$. Now notice that $H_1(y) = 2y$, so the integral can be expressed as

$$\int_{-\infty}^{\infty} \psi_v^{(0)} * \hat{H}^{(1)} \psi_0^{(0)} dx = \frac{mgN_0\alpha}{2N_1} \int_{-\infty}^{\infty} N_v H_v(y) e^{-y^2/2} N_1 H_1(y) e^{-y^2/2} dx$$

$$= \frac{mgN_0\alpha}{2N_1} \int_{-\infty}^{\infty} \psi_v^{(0)} * \psi_1^{(0)} dx$$

Because the harmonic-oscillator wavefunctions are orthogonal, this integral vanishes unless $v = 1$. In that case, the remaining integral is the normalization integral.

$$E_0^{(2)} = \sum_{v=1}^{\infty} \frac{\left|\frac{mgN_0\alpha}{2N_1}\int_{-\infty}^{\infty}\psi_v^{(0)}*\psi_1^{(0)}\mathrm{d}x\right|^2}{E_0^{(0)} - E_v^{(0)}} = \frac{\left|\frac{mgN_0\alpha}{2N_1}\int_{-\infty}^{\infty}\psi_1^{(0)}*\psi_1^{(0)}\mathrm{d}x\right|^2}{E_0^{(0)} - E_1^{(0)}} = \frac{\left(\frac{mgN_0\alpha}{2N_1}\right)^2}{-\hbar\omega}$$

Using $N_0 = \left(\frac{1}{\alpha\pi^{1/2}}\right)^{1/2}$, $N_1 = \left(\frac{1}{2\alpha\pi^{1/2}}\right)^{1/2}$, $\alpha = \left(\frac{\hbar^2}{mk}\right)^{1/4}$, and $k = mw^2$, this result simplifies to

$$E_0^{(2)} = -\frac{m^2g^2N_0^{\ 2}\alpha^2}{4N_1^{\ 2}\mathrm{h}\omega} = \boxed{-\frac{mg^2}{2\omega^2}}$$

Solutions to problems

Solutions to numerical problems

P2.1 $E = \frac{n^2h^2}{8mL^2}$ [2.6a], $E_2 - E_1 = \frac{3h^2}{8mL^2}$

We take $m(O_2) = (32.000)\times(1.6605\times10^{-27}\ \text{kg})$, and find

$$E_2 - E_1 = \frac{(3)\times(6.626\times10^{-34}\ \text{J s})^2}{(8)\times(32.00)\times(1.6605\times10^{-27}\ \text{kg})\times(5.0\times10^{-2}\ \text{m})^2} = \boxed{1.24\times10^{-39}\ \text{J}}$$

We set $E = \frac{n^2h^2}{8mL^2} = \frac{1}{2}kT$ and solve for n.

From above, $\frac{h^2}{8mL^2} = \frac{E_2 - E_1}{3} = 4.13\times10^{-40}$ J; then

$$n^2\times(4.13\times10^{-40}\ \text{J}) = \left(\frac{1}{2}\right)\times(1.381\times10^{-23}\ \text{J K}^{-1})\times(300\ \text{K}) = 2.07\times10^{-21}\ \text{J}$$

We find $n = \left(\frac{2.07\times10^{-21}\ \text{J}}{4.13\times10^{-40}\ \text{J}}\right)^{1/2} = \boxed{2.2\times10^9}$

At this level,

$$E_n - E_{n-1} = \{n^2 - (n-1)^2\}\times\frac{h^2}{8mL^2} = (2n-1)\times\frac{h^2}{8mL^2} \approx (2n)\times\frac{h^2}{8mL^2}$$

$$= (4.4\times10^9)\times(4.13\times10^{-40}\ \text{J}) \approx \boxed{1.8\times10^{-30}\ \text{J}}\quad [\text{or } 1.1\ \mu\text{J mol}^{-1}]$$

P2.3 $\omega = \left(\frac{k}{\mu}\right)^{1/2}$ [2.21 with μ in place of m]

The relationship between the oscillator frequency and the wavenumber of absorbed photons comes from the fact that absorption occurs when E_{photon} equals the difference between adjacent energy levels:

$$E_{v+1} - E_v = \hbar\omega\ [2.22] = E_{\text{photon}} = h\nu = hc\tilde{\nu} \qquad \text{so} \qquad \omega = \frac{hc\tilde{\nu}}{\hbar} = 2\pi c\tilde{\nu}$$

Therefore, $k = \omega^2\mu = 4\pi^2c^2\tilde{\nu}^2\mu = \frac{4\pi^2c^2\tilde{\nu}^2m_1m_2}{m_1 + m_2}$

We draw up the following table using isotope masses from a handbook.

	$^{1}H^{35}Cl$	$^{1}H^{81}Br$	$^{1}H^{127}I$	$^{12}C^{16}O$	$^{14}N^{16}O$
$\tilde{\nu}$ / m^{-1}	299000	265000	231000	217000	190400
$10^{27}m_1$/kg	1.6735	1.6735	1.6735	19.926	23.253
$10^{27}m_2$/kg	58.066	134.36	210.72	26.560	26.560
k/(N m^{-1})	516	412	314	1902	1595

Therefore, the order of stiffness is $\boxed{\text{HI} < \text{HBr} < \text{HCl} < \text{NO} < \text{CO}}$.

P2.5 The probability is given by

$$P_n = \int |\psi_n^{\,2}| \mathrm{d}x \quad \text{[Section 1.6, Postulate II]}$$

where $|\psi_n^{\,2}| = \psi_n^{\,2} = \dfrac{2}{L}\sin^2\left(\dfrac{n\pi x}{L}\right)$ [2.11]

Use the integral

$$\int \sin^2 ax\,\mathrm{d}x = \frac{x}{2} - \frac{\sin 2ax}{4a}$$

(a) $$P_n = \frac{2}{L}\int_0^{L/2}\sin^2\left(\frac{n\pi x}{L}\right)\mathrm{d}x = \frac{2}{L}\left(\frac{x}{2} - \frac{L}{4\pi n}\sin\left(\frac{2\pi nx}{L}\right)\right)\Bigg|_0^{L/2} = \boxed{\frac{1}{2}}$$

COMMENT. The sine term vanishes at both limits of integration for all n.

(b) $$P_n = \frac{2}{L}\left\{\frac{x}{2} - \frac{L}{4\pi n}\sin\left(\frac{2\pi nx}{L}\right)\right\}\Bigg|_0^{L/4} = \frac{2}{L}\left\{\frac{L}{8} - \frac{L}{4\pi n}\sin\left(\frac{n\pi}{2}\right)\right\} = \frac{1}{4} - \frac{1}{2\pi n}\sin\left(\frac{n\pi}{2}\right)$$

For even n, the sine term vanishes. The sine term is 1 for n = 1, 5, 9, and –1 for n = 3, 7, 11, and so on. Thus

$$\boxed{P_n = \begin{cases} \frac{1}{4} & \text{for even } n \\ \frac{1}{4} - \frac{1}{2\pi n} & \text{for } n = 1,\ 5,\ 9, \text{ etc.} \\ \frac{1}{4} + \frac{1}{2\pi n} & \text{for } n = 3,\ 7,\ 11, \text{ etc.} \end{cases}}$$

(c) Here the interval of integration is small, so the integral is equal to the integrand evaluated at the midpoint of the interval times the size of the interval:

$$P_n = \frac{2}{L}\int_{\frac{1}{2}L-\delta x}^{\frac{1}{2}L+\delta x}\sin^2\left(\frac{n\pi x}{L}\right)\mathrm{d}x = \frac{2}{L}\sin^2\left(\frac{\pi n}{2}\right)\times 2\delta x = \frac{4\delta x}{L}\sin^2\left(\frac{\pi n}{2}\right)$$

Again, the sine term vanishes for even n; for odd n, squaring the sine term gives 1. So

$$\boxed{P_n = \begin{cases} 0 & \text{for even } n \\ \dfrac{4\delta x}{L} & \text{for odd } n \end{cases}}$$

P2.7 Treat the gravitational potential energy as a perturbation in the energy operator:

$$H^{(1)} = mgx$$

The first-order correction to the ground-state energy, E_1, is [2.34]

$$E_1^{(1)} = \int_0^L \psi_1^{(0)*}H^{(1)}\psi_1^{(0)}\mathrm{d}x = \int_0^L\left(\frac{2}{L}\right)^{1/2}\sin\left(\frac{\pi x}{L}\right)mgx\left(\frac{2}{L}\right)^{1/2}\sin\left(\frac{\pi x}{L}\right)\mathrm{d}x$$

$$E_1^{(1)} = \frac{2mg}{L}\int_0^L x\sin^2\left(\frac{\pi x}{L}\right)\mathrm{d}x$$

$$E_1^{(1)} = \frac{2mg}{L}\left(\frac{x^2}{4} - \frac{xL}{2\pi}\cos\left(\frac{\pi x}{L}\right)\sin\left(\frac{\pi x}{L}\right) - \frac{L^2}{4\pi^2}\cos^2\left(\frac{\pi x}{L}\right)\right)\Bigg|_0^L$$

$$E_1^{(1)} = \boxed{\tfrac{1}{2}mgL}$$

Not surprisingly, this amounts to the energy perturbation evaluated at the midpoint of the box. For $m = m_e$, $E_1^{(1)}/L = 4.47 \times 10^{-30}\ \mathrm{J\ m^{-1}}$.

P2.9 The first-order correction to the ground-state energy, E_1, is [2.34]

$$E_1^{(1)} = \int_{-\infty}^{\infty} \psi_0^{(0)*}H^{(1)}\psi_0^{(0)}\mathrm{d}x$$

where $\hat{H}^{(1)} = ax^3 + bx^4$, $\quad \psi_0^{(0)} = \left(\frac{1}{\alpha\pi^{1/2}}\right)^{1/2} \mathrm{e}^{-x^2/2\alpha^2}$, $\alpha = \left(\frac{\hbar^2}{mk}\right)^{1/4}$ [2.24]

(a) If the anharmonic perturbation is present for all values of x, then

$$E_1^{(1)} = \frac{1}{\alpha\pi^{1/2}}\int_{-\infty}^{\infty} \mathrm{e}^{-x^2/\alpha^2}(ax^3 + bx^4)\mathrm{d}x$$

Use the integrals

$$\int_0^{\infty} x^3\mathrm{e}^{-cx^2}\,\mathrm{d}x = \frac{1}{2c^2} \quad \text{and} \quad \int_0^{\infty} x^4\mathrm{e}^{-cx^2}\,\mathrm{d}x = \frac{3}{8c^2}\left(\frac{\pi}{c}\right)^{1/2}$$

and remember that $\int_{-\infty}^{\infty} f(x)\mathrm{d}x = 0$ if $f(x)$ is odd and $\int_{-\infty}^{\infty} f(x)\mathrm{d}x = 2\int_0^{\infty} f(x)\mathrm{d}x$ if $f(x)$ is even.

Thus $$E_1^{(1)} = \frac{1}{\alpha\pi^{1/2}}\left(0 + \frac{3b\alpha^5\pi^{1/2}}{4}\right) = \boxed{\frac{3b\alpha^4}{4}}$$

(b) If the anharmonic perturbation is present only during bond expansion, then the perturbation vanishes for $x < 0$.

$$E_1^{(1)} = \frac{1}{\alpha\pi^{1/2}}\int_0^{\infty} \mathrm{e}^{-x^2/\alpha^2}(ax^3 + bx^4)\mathrm{d}x = \frac{1}{\alpha\pi^{1/2}}\left(\frac{a\alpha^4}{2} + \frac{3b\alpha^5\pi^{1/2}}{8}\right) = \boxed{\frac{a\alpha^3}{2\pi^{1/2}} + \frac{3b\alpha^4}{8}}$$

(c) If the anharmonic perturbation is present only during bond compression, then the perturbation vanishes for $x > 0$. Use the fact that

$$\int_0^{\infty} f(x)\mathrm{d}x = -\int_{-\infty}^{0} f(x)\mathrm{d}x \text{ if } f(x) \text{ is odd and } \int_0^{\infty} f(x)\mathrm{d}x = \int_{-\infty}^{0} f(x)\mathrm{d}x \text{ if } f(x) \text{ is even.}$$

$$E_1^{(1)} = \frac{1}{\alpha\pi^{1/2}}\int_{-\infty}^{0} \mathrm{e}^{-x^2/\alpha^2}(ax^3 + bx^4)\mathrm{d}x = \frac{1}{\alpha\pi^{1/2}}\left(-\frac{a\alpha^4}{2} + \frac{3b\alpha^5\pi^{1/2}}{8}\right) = \boxed{-\frac{a\alpha^3}{2\pi^{1/2}} + \frac{3b\alpha^4}{8}}$$

Solutions to theoretical problems

P2.11 The normalization integral is

$$1 = \int_0^L \psi^*\psi\mathrm{d}x = \int_0^L (N\mathrm{e}^{-ikx})(N\mathrm{e}^{ikx})\mathrm{d}x = N^2\int_0^L \mathrm{d}x = N^2L$$

So the normalization constant is

$$N = \boxed{\left(\frac{1}{L}\right)^{1/2}}$$

P2.13 The text defines the transmission probability and expresses it as the ratio of $|A'|^2/|A|^2$, where the coefficients A and A' are introduced in eqns. 2.12 and 2.15. Eqn. 2.16 lists four equations for the six unknown coefficients of the full wavefunction. Once we realize that we can set B' to zero, these equations in five unknowns are

(a) $A + B = C + D$

(b) $Ce^{\kappa L} + De^{-\kappa L} = A'e^{ikL}$

(c) $ikA - ikB = \kappa C - \kappa D$

(d) $\kappa Ce^{\kappa L} - \kappa De^{-\kappa L} = ikA'e^{ikL}$

We need A' in terms of A alone, which means we must eliminate B, C, and D. Notice that B appears only in eqns. (a) and (c). Solving these equations for B and setting the results equal to each other yields

$$B = C + D - A = A - \frac{\kappa C}{ik} + \frac{\kappa D}{ik}$$

Solve this equation for C:

$$C = \frac{2A + D\left(\frac{\kappa}{ik} - 1\right)}{\frac{\kappa}{ik} + 1} = \frac{2Aik + D(\kappa - ik)}{\kappa + ik}$$

Now note that the desired A' appears only in (b) and (d). Solve these for A' and set them equal:

$$A' = e^{-ikL}(Ce^{\kappa L} + De^{-\kappa L}) = \frac{\kappa e^{-ikL}}{ik}(Ce^{\kappa L} - De^{-\kappa L})$$

Solve the resulting equation for C, and set it equal to the previously obtained expression for C:

$$C = \frac{\left(\frac{\kappa}{ik} - 1\right)De^{-2\kappa L}}{\frac{\kappa}{ik} - 1} = \frac{(\kappa + ik)De^{-2\kappa L}}{\kappa - ik} = \frac{2Aik + D(\kappa - ik)}{\kappa + ik}$$

Solve this resulting equation for D in terms of A:

$$\frac{(\kappa + ik)^2 e^{-2\kappa L} - (\kappa - ik)^2}{(\kappa - ik)(\kappa + ik)} D = \frac{2Aik}{\kappa + ik}$$

so $$D = \frac{2Aik(\kappa - ik)}{(\kappa + ik)^2 e^{-2\kappa L} - (\kappa - ik)^2}$$

Substituting this expression back into an expression for C yields

$$C = \frac{2Aik(\kappa + ik)e^{-2\kappa L}}{(\kappa + ik)^2 e^{-2\kappa L} - (\kappa - ik)^2}$$

Substituting for C and D in the expression for A' yields

$$A' = e^{-ikL}(Ce^{\kappa L} + De^{-\kappa L}) = \frac{2Aike^{-ikL}}{(\kappa + ik)^2 e^{-2\kappa L} - (\kappa - ik)^2}[(\kappa + ik)e^{-\kappa L} + (\kappa - ik)e^{-\kappa L}],$$

so $$\frac{A'}{A}=\frac{4ik\kappa e^{-\kappa L}e^{-ikL}}{(\kappa+ik)^2e^{-2\kappa L}-(\kappa-ik)^2}=\frac{4ik\kappa e^{-ikL}}{(\kappa+ik)^2e^{-\kappa L}-(\kappa-ik)^2e^{\kappa L}}$$

The transmission coefficient is

$$T=\frac{|A'|^2}{|A|^2}=\left(\frac{4ik\kappa e^{-ikL}}{(\kappa+ik)^2e^{-\kappa L}-(\kappa-ik)^2e^{\kappa L}}\right)\left(\frac{-4ik\kappa e^{ikL}}{(\kappa-ik)^2e^{-\kappa L}-(\kappa+ik)^2e^{\kappa L}}\right)$$

The denominator is worth expanding separately in several steps. It is

$$\begin{aligned}&(\kappa+ik)^2(\kappa-ik)^2e^{-2\kappa L}-(\kappa-ik)^4-(\kappa+ik)^4+(\kappa-ik)^2(\kappa+ik)^2e^{2\kappa L}\\&=(\kappa^2+k^2)^2(e^{2\kappa L}+e^{-2\kappa L})-(\kappa^2-2i\kappa k-k^2)^2-(\kappa^2+2i\kappa k-k^2)^2\\&=(\kappa^4+2\kappa^2k^2+k^4)(e^{2\kappa L}+e^{-2\kappa L})-(2\kappa^4-12\kappa^2k^2+2k^2)\end{aligned}$$

If the $12\kappa^2k^2$ term were $-4\kappa^2k^2$ instead, we could collect terms still further (completing the square), but of course we must also account for the difference between those quantities, making the denominator

$$(\kappa^4+2\kappa^2k^2+k^4)(e^{2\kappa L}-2+e^{-2\kappa L})+16\kappa^2k^2=(\kappa^2+k^2)^2(e^{\kappa L}-e^{-\kappa L})^2+16\kappa^2k^2$$

So the coefficient is

$$T=\frac{16k^2\kappa^2}{(\kappa^2+k^2)^2(e^{\kappa L}-e^{-\kappa L})^2+16\kappa^2k^2}$$

We are almost there. To get to eqn. 2.17(a), we invert the expression:

$$T=\left(\frac{(\kappa^2+k^2)^2(e^{\kappa L}-e^{-\kappa L})^2+16\kappa^2k^2}{16k^2\kappa^2}\right)^{-1}=\left(\frac{(\kappa^2+k^2)^2(e^{\kappa L}-e^{-\kappa L})^2}{16k^2\kappa^2}+1\right)^{-1}$$

Finally, we try to express $\frac{(\kappa^2+k^2)^2}{k^2\kappa^2}$ in terms of a ratio of energies, $\varepsilon=E/V$. Eqns. 2.12 and 2.14 define k and κ. The factors involving 2, $\hbar$, and the mass cancel, leaving $\kappa\propto(V-E)^{1/2}$ and $k\propto E^{1/2}$, so

$$\frac{(\kappa^2+k^2)^2}{k^2\kappa^2}=\frac{[E+(V-E)]^2}{E(V-E)}=\frac{V^2}{E(V-E)}=\frac{1}{\varepsilon(1-\varepsilon)}$$

which makes the transmission coefficient

$$T=\boxed{\left(\frac{(e^{\kappa L}-e^{-\kappa L})^2}{16\varepsilon(1-\varepsilon)}+1\right)^{-1}}$$

If $\kappa L\gg 1$, then the negative exponential is negligible compared to the positive, and the 1 inside the parentheses is negligible compared to the exponential:

$$T\approx\left(\frac{e^{2\kappa L}}{16\varepsilon(1-\varepsilon)}\right)^{-1}=\frac{16\varepsilon(1-\varepsilon)}{e^{2\kappa L}}=\boxed{16\varepsilon(1-\varepsilon)e^{-2\kappa L}}$$

P2.15 (a) The wavefunctions in each region (see Figure 2.2) are (eqns. 2.12, 2.14, and 2.15)

$$\psi_1(x)=e^{ik_1x}+B_1e^{-ik_1x}$$
$$\psi_2(x)=A_2e^{k_2x}+B_2e^{-k_2x}$$
$$\psi_3(x)=A_3e^{ik_3x}$$

With the above choice of $A_1 = 1$, the transmission probability is simply $T = |A_3|^2$. The wavefunction coefficients are determined by the criteria that both the wavefunctions and their first derivatives with respect to x be continuous at potential boundaries.

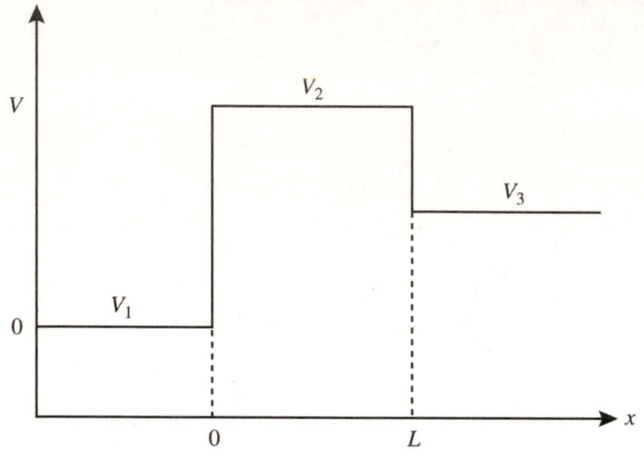

Figure 2.2

$$\psi_1(0) = \psi_2(0); \quad \psi_2(L) = \psi_3(L)$$

$$\frac{d\psi_1(0)}{dx} = \frac{d\psi_2(0)}{dx}; \quad \frac{d\psi_2(L)}{dx} = \frac{d\psi_3(L)}{dx}$$

These criteria establish the algebraic relationships:

$$1 + B_1 - A_2 - B_2 = 0$$

$$(-ik_1 - k_2)A_2 + (-ik_1 + k_2)B_2 + 2ik_1 = 0$$

$$A_2 e^{k_2 L} + B_2 e^{-k_2 L} - A_3 e^{ik_3 L} = 0$$

$$A_2 k_2 e^{k_2 L} - B_2 k_2 e^{-k_2 L} - iA_3 k_3 e^{ik_3 L} = 0$$

Solving the simultaneous equations for A_3 gives

$$A_3 = \frac{4k_1 k_2\, e^{ik_3 L}}{(ia + b)e^{k_2 L} - (ia - b)e^{-k_2 L}}$$

where $a = k_2^2 - k_1 k_3$ and $b = k_1 k_2 + k_2 k_3$.

Since $\sinh(z) = (e^z - e^{-z})/2$ or $e^z = 2\sinh(z) + e^{-z}$, substitute $e^{k_2 L} = 2\sinh(k_2 L) + e^{-k_2 L}$, giving

$$A_3 = \frac{2k_1 k_2 e^{ik_3 L}}{(ia + b)\sinh(k_2 L) + be^{-k_2 L}}$$

$$\boxed{T = |A_3|^2 = A_3 \times A_3^* = \frac{4k_1^2 k_2^2}{(a^2 + b^2)\sinh^2(k_2 L) + b^2}}$$

where $a^2 + b^2 = (k_1^2 + k_2^2)(k_2^2 + k_3^2)$ and $b^2 = k_2^2(k_1 + k_3)^2$

(b) In the special case for which $V_1 = V_3 = 0$, eqns. 2.12 and 2.15 require that $k_1 = k_3$. Additionally,

$$\left(\frac{k_1}{k_2}\right)^2 = \frac{E}{V_2 - E} = \frac{\varepsilon}{1 - \varepsilon} \text{ where } \varepsilon = E / V_2$$

$$a^2+b^2=\left(k_1^2+k_2^2\right)^2=k_2^4\left\{1+\left(\frac{k_1}{k_2}\right)^2\right\}^2$$

$$b^2=4k_1^2k_2^2$$

$$\frac{a^2+b^2}{b^2}=\frac{k_2^2\left\{1+\left(\frac{k_1}{k_2}\right)^2\right\}^2}{4k_1^2}=\frac{1}{4\varepsilon(1-\varepsilon)}$$

$$T=\frac{b^2}{b^2+(a^2+b^2)\sinh^2(k_2L)}=\frac{1}{1+\left(\dfrac{a^2+b^2}{b^2}\right)\sinh^2(k_2L)}$$

$$T=\left\{1+\frac{\sinh^2(k_2L)}{4\varepsilon(1-\varepsilon)}\right\}^{-1}=\left\{1+\frac{(e^{k_2L}-e^{-k_2L})^2}{16\varepsilon(1-\varepsilon)}\right\}^{-1}$$

This is eqn. 2.17(a). In the high, wide barrier limit, $k_2L >> 1$. This implies both that e^{-k_2L} is negligibly small compared to e^{k_2L} and that 1 is negligibly small compared to $e^{2k_2L}/\{16\varepsilon(1-\varepsilon)\}$. The previous equation simplifies to eqn. 2.17(b).

$$T=16\varepsilon(1-\varepsilon)e^{-2k_2L}$$

P2.17 The Schrödinger equation is $-\dfrac{\hbar^2}{2m}\dfrac{d^2\psi}{dx^2}+\dfrac{1}{2}kx^2\psi=E\psi$

We write the trial solution $\psi=e^{-\kappa x^2}$, so $\dfrac{d\psi}{dx}=-2\kappa xe^{-\kappa x^2}$

and $\dfrac{d^2\psi}{dx^2}=-2\kappa e^{-\kappa x^2}+4\kappa^2x^2e^{-\kappa x^2}=-2\kappa\psi+4\kappa^2x^2\psi$

Insert the trial solution into the Schrödinger equation.

$$\left(\frac{\hbar^2\kappa}{m}\right)\psi-\left(\frac{2\hbar^2\kappa^2}{m}\right)x^2\psi+\frac{1}{2}kx^2\psi=E\psi$$

so $$\left[\left(\frac{\hbar^2\kappa}{m}\right)-E\right]\psi+\left(\tfrac{1}{2}k-\frac{2\hbar^2\kappa^2}{m}\right)x^2\psi=0$$

This equation is satisfied if

$$E=\frac{\hbar^2\kappa}{m}\quad\text{and}\quad 2\hbar^2\kappa^2=\tfrac{1}{2}mk\text{, so }\kappa=\boxed{\frac{1}{2}\left(\frac{mk}{\hbar^2}\right)^{1/2}}$$

The corresponding energy is

$$E=\tfrac{1}{2}\hbar\left(\frac{k}{m}\right)^{1/2}=\tfrac{1}{2}\hbar\omega\quad\text{with }\omega=\left(\frac{k}{m}\right)^{1/2}$$

P2.19 $$\langle x^n\rangle=\alpha^n\langle y^n\rangle=\alpha^n\int_{-\infty}^{+\infty}\psi y^n\psi\,dx=\alpha^{n+1}\int_{-\infty}^{+\infty}\psi^2y^n\,dy\quad[x=\alpha y]$$

$$\langle x^3\rangle\propto\int_{-\infty}^{+\infty}\psi^2y^3\,dy=\boxed{0}\text{ by symmetry}\quad[y^3\text{ is an odd function of }y]$$

$$\langle x^4\rangle=\alpha^5\int_{-\infty}^{+\infty}\psi y^4\psi\,dy$$

$$y^4\psi = y^4 NH_v e^{-y^2/2}$$

$$y^4H_v = y^3\left(\tfrac{1}{2}H_{v+1} + vH_{v-1}\right) = y^2\left[\tfrac{1}{2}\left(\tfrac{1}{2}H_{v+2} + (v+1)H_v\right) + v\left(\tfrac{1}{2}H_v + (v-1)H_{v-2}\right)\right]$$

$$= y^2\left[\tfrac{1}{4}H_{v+2} + \left(v+\tfrac{1}{2}\right)H_v + v(v-1)H_{v-2}\right]$$

$$= y\left[\tfrac{1}{4}\left(\tfrac{1}{2}H_{v+3} + (v+2)H_{v+1}\right) + \left(v+\tfrac{1}{2}\right)\times\left(\tfrac{1}{2}H_{v+1} + vH_{v-1}\right)\right.$$

$$\left. + v(v-1)\times\left(\tfrac{1}{2}H_{v-1} + (v-2)H_{v-3}\right)\right]$$

$$= y\left(\tfrac{1}{8}H_{v+3} + \tfrac{3}{4}(v+1)H_{v+1} + \tfrac{3}{2}v^2H_{v-1} + v(v-1)\times(v-2)H_{v-3}\right)$$

Only yH_{v+1} and yH_{v-1} lead to H_v and contribute to the expectation value (since H_v is orthogonal to all except H_v) [Table 2.1]; hence

$$y^4H_v = \tfrac{3}{4}y\{(v+1)H_{v+1} + 2v^2H_{v-1}\} + \cdots$$

$$= \tfrac{3}{4}\left[(v+1)\left(\tfrac{1}{2}H_{v+2} + (v+1)H_v\right) + 2v^2\left(\tfrac{1}{2}H_v + (v-1)H_{v-2}\right)\right] + \cdots$$

$$= \tfrac{3}{4}\{(v+1)^2H_v + v^2H_v\} + \cdots$$

$$= \tfrac{3}{4}(2v^2 + 2v + 1)H_v + \cdots$$

Therefore

$$\int_{-\infty}^{+\infty}\psi y^4\psi\,dy = \tfrac{3}{4}(2v^2+2v+1)N^2\int_{-\infty}^{+\infty}H_v^2e^{-y^2}\,dy = \frac{3}{4\alpha}(2v^2+2v+1)$$

so

$$\langle x^4\rangle = (\alpha^5)\times\left(\frac{3}{4\alpha}\right)\times(2v^2+2v+1) = \boxed{\frac{3}{4}(2v^2+2v+1)\alpha^4}$$

P2.21 (a) $$\langle x\rangle = \int_0^L\left(\frac{2}{L}\right)^{1/2}\sin\left(\frac{n\pi x}{L}\right)x\left(\frac{2}{L}\right)^{1/2}\sin\left(\frac{n\pi x}{L}\right)dx$$

$$\langle x\rangle = \left(\frac{2}{L}\right)\int_0^L x\sin^2 ax\,dx \quad \left[a = \frac{n\pi}{L}\right]$$

$$= \left(\frac{2}{L}\right)\times\left(\frac{x^2}{4} - \frac{x\sin 2ax}{4a} - \frac{\cos 2ax}{8a^2}\right)\Bigg|_0^L \quad \text{[Trig functions vanish or cancel at limits.]}$$

$$= \left(\frac{2}{L}\right)\times\left(\frac{L^2}{4}\right) = \frac{L}{2} \quad \text{[by symmetry also].}$$

$$\langle x^2\rangle = \frac{2}{L}\int_0^L x^2\sin^2 ax\,dx = \left(\frac{2}{L}\right)\times\left[\frac{x^3}{6} - \left(\frac{x^2}{4a} - \frac{1}{8a^3}\right)\sin 2ax - \frac{x\cos 2ax}{4a^2}\right]\Bigg|_0^L$$

$$= \left(\frac{2}{L}\right)\times\left(\frac{L^3}{6} - \frac{L^3}{4n^2\pi^2}\right) = L^2\left(\frac{1}{3} - \frac{1}{2n^2\pi^2}\right)$$

$$\Delta x = \left[L^2\left(\frac{1}{3} - \frac{1}{2n^2\pi^2}\right) - \frac{L^2}{4}\right]^{1/2} = \boxed{L\left(\frac{1}{12} - \frac{1}{2\pi^2n^2}\right)^{1/2}}$$

$$\langle p\rangle = 0 \text{ [by symmetry]}$$

and $$\langle p^2\rangle = \frac{n^2h^2}{4L^2} \quad \left[\text{from } E = \frac{p^2}{2m}\right]$$

so $\Delta p = \left(\frac{n^2h^2}{4L^2}\right)^{1/2} = \boxed{\frac{nh}{2L}}$

$$\Delta p\Delta x = \frac{nh}{2L}\times L\left(\frac{1}{12}-\frac{1}{2\pi^2 n^2}\right)^{1/2} = \boxed{\frac{nh}{2\sqrt{3}}\left(1-\frac{1}{24\pi^2 n^2}\right)^{1/2}} > \frac{\hbar}{2}$$

(b) $\langle x\rangle = \alpha^2\int_{-\infty}^{+\infty}\psi^2 y\,\mathrm{d}y[x=\alpha y] = 0$ [by symmetry, y is an odd function].

$$\langle x^2\rangle = \frac{2}{k}\left\langle\frac{1}{2}kx^2\right\rangle = \frac{2}{k}\langle V\rangle$$

But $\langle V\rangle = \langle E_{\mathrm{k}}\rangle = \frac{E}{2} = \frac{1}{2}\left(v+\frac{1}{2}\right)\hbar\omega$ [2.29b&c],

so $\langle x^2\rangle = \left(v+\frac{1}{2}\right)\times\left(\frac{\hbar\omega}{k}\right) = \left(v+\frac{1}{2}\right)\times\left(\frac{\hbar}{\omega m}\right)$

and $\Delta x = \boxed{\left\{\left(v+\frac{1}{2}\right)\frac{\hbar}{\omega m}\right\}^{1/2}}$

$\langle p\rangle = 0$ [by symmetry, or by noting that the integrand is an odd function of x]

$$\langle p^2\rangle = 2m\langle T\rangle = (2m)\times\left(\frac{1}{2}\right)\times\left(v+\frac{1}{2}\right)\times\hbar\omega = \left(v+\frac{1}{2}\right)\hbar\omega m \text{ [Problem 2.18]}$$

$$\Delta p = \boxed{\left\{\left(v+\frac{1}{2}\right)\hbar\omega m\right\}^{1/2}}$$

$$\Delta p\Delta x = \left\{\left(v+\frac{1}{2}\right)\hbar\omega m\right\}^{1/2}\left\{\left(v+\frac{1}{2}\right)\frac{\hbar}{\omega m}\right\}^{1/2} = \boxed{\left(v+\frac{1}{2}\right)\hbar} \geq \frac{\hbar}{2}$$

Both results are consistent with the uncertainty principle, $\Delta p\Delta x \geq \frac{\hbar}{2}$. Note that for the ground-state harmonic oscillator ($v = 0$), the equality holds.

P2.23 Call the integral in question I:

$$I \equiv \int\psi_{v'}x\psi_v\,\mathrm{d}x = \alpha^2\int\psi_{v'}y\psi_v\,\mathrm{d}y \quad [x=\alpha y]$$

$$y\psi_v = N_v\left(\frac{1}{2}H_{v+1}+vH_{v-1}\right)\mathrm{e}^{-y^2/2} \text{ [Table 2.1]}$$

Hence $I = \alpha^2 N_v N_{v'}\int\left(\frac{1}{2}H_{v'}H_{v+1}+vH_{v'}H_{v-1}\right)\mathrm{e}^{-y^2}\,\mathrm{d}y = 0$ unless $v' = v\pm 1$ [Table 2.1].

For $v' = v + 1$

$$I = \frac{1}{2}\alpha^2 N_v N_{v+1}\int H_{v+1}^2\mathrm{e}^{-y^2}\,\mathrm{d}y = \frac{1}{2}\alpha^2 N_v N_{v+1}\pi^{1/2}2^{v+1}(v+1)! = \boxed{\alpha\left(\frac{v+1}{2}\right)^{1/2}}$$

For $v' = v - 1$

$$I = v\alpha^2 N_v N_{v-1}\int H_{v-1}^2\mathrm{e}^{-y^2}\,\mathrm{d}y = v\alpha^2 N_v N_{v-1}\pi^{1/2}2^{v-1}(v-1)! = \boxed{\alpha\left(\frac{v}{2}\right)^{1/2}}$$

No other values of v' result in a nonzero value for I; hence, no other transitions are allowed.

P2.25 To address this time-dependent problem, we need a time-dependent wavefunction, made up from solutions of the time-dependent Schrödinger equation

$$\hat{H}\Psi(x,t) = \mathrm{i}\hbar\frac{\partial\Psi(x,t)}{\partial t} \quad \text{[Table 1.1]}$$

If $\psi(x)$ is an eigenfunction of the energy operator with energy eigenvalue E, then

$$\Psi(x,t) = \psi(x)\mathrm{e}^{-\mathrm{i}Et/\eta}$$

is a solution of the time-dependent Schrödinger equation (provided the energy operator is not itself time dependent). To verify this, evaluate both sides of the time-dependent Schrödinger equation. On the left we have

$$\hat{H}\Psi(x,t) = \hat{H}\psi(x)\mathrm{e}^{-\mathrm{i}Et/\hbar} = E\psi(x)\mathrm{e}^{-\mathrm{i}Et/\hbar} = E\Psi(x,t)$$

On the right we have

$$\mathrm{i}\eta\frac{\partial\Psi(x,t)}{\partial t} = \mathrm{i}\eta\,\psi(x)\frac{\partial}{\partial t}\mathrm{e}^{-\mathrm{i}Et/\eta} = -\mathrm{i}^2 E\psi(x)\mathrm{e}^{-\mathrm{i}Et/\eta} = E\Psi(x,t)$$

the same as on the left. Our wavepacket is an arbitrary superposition of time-evolving harmonic oscillator wavefunctions,

$$\Psi(x,t) = \sum_{v=0} c_v\psi_v(x)\mathrm{e}^{-\mathrm{i}E_vt/\eta}$$

where $\psi_v(x)$ are time-independent harmonic-oscillator wavefunctions and

$$E_v = \left(v+\frac{1}{2}\right)\eta\omega \quad [2.21]$$

Hence, the wavepacket is

$$\Psi(x,t) = \mathrm{e}^{-\mathrm{i}\omega t/2}\sum_{v=0} c_v\psi_v(x)\mathrm{e}^{-\mathrm{i}v\omega t}$$

The angular frequency ω is related to the period T by $T = 2\pi/\omega$, so we can evaluate the wavepacket at any whole number of periods after t, that is, at a time $t + nT$, where n is any integer (*not* a quantum number.) Note that

$$t + nT = t + 2\pi n/\omega$$

so $$\Psi(x,t+nT) = \mathrm{e}^{-\mathrm{i}\omega t/2}\mathrm{e}^{-\mathrm{i}\omega nT/2}\sum_{v=0} c_v\psi_v(x)\mathrm{e}^{-\mathrm{i}v\omega t}\mathrm{e}^{-\mathrm{i}v\omega nT} = \mathrm{e}^{-\mathrm{i}\omega t/2}\mathrm{e}^{-\mathrm{i}\pi n}\sum_{v=0} c_v\psi_v(x)\mathrm{e}^{-\mathrm{i}v\omega t}\mathrm{e}^{-2\pi\mathrm{i}vn}$$

Noting that the exponential of ($2\pi\mathrm{i} \times$ any integer) = 1, we note that the last factor inside the sum is 1 for every state. Also, since $\mathrm{e}^{-\mathrm{i}n\pi} = (-1)^n$, we have

$$\Psi(x,t+nT) = (-1)^n\Psi(x,t)$$

At any whole number of periods after time t, the wavefunction is either the same as at time t or -1 times its value at time t. In any event, $|\Psi|^2$ returns to its original value in each period, so the wavepacket returns to the same spatial distribution in each period.

Solutions to applications

P2.27 (a) Treat the π electrons as particles free to move the length of the conjugated system. The energy levels are given by

$$E_n = \frac{h^2n^2}{8mL^2} \quad \text{[2.6a]}$$

and we are looking for the energy difference between $n = 6$ and $n = 7$:

$$\Delta E = \frac{h^2(7^2 - 6^2)}{8mL^2}$$

Since there are 12 atoms on the conjugated backbone, the length of the box is 11 times the bond length.

$$L = 11(140\times10^{-12}\ \text{m}) = 1.54\times10^{-9}\ \text{m}$$

so $$\Delta E = \frac{(6.626\times10^{-34}\ \text{J s})^2(49-36)}{8(9.11\times10^{-31}\ \text{kg})(1.54\times10^{-9}\ \text{m})^2} = \boxed{3.30\times10^{-19}\ \text{J}}$$

(b) The relationship between energy and frequency is

$$\Delta E = h\nu \qquad \text{so} \qquad \nu = \frac{\Delta E}{h} = \frac{3.30\times10^{-19}\ \text{J}}{6.626\times10^{-34}\ \text{J s}} = \boxed{4.98\times10^{14}\ \text{s}^{-1}}$$

The wavelength and frequency are related by $c = \lambda\nu$, so

$$\lambda = \frac{c}{\nu} = \frac{2.998\times10^{8}\ \text{m s}^{-1}}{4.98\times10^{14}\ \text{s}^{-1}} = \boxed{6.02\times10^{-7}\ \text{m} = 602\ \text{nm}}$$

(c) Look at the terms in the energy expression that change with the number of conjugated atoms, N. The energy (and frequency) are inversely proportional to L^2 and directly proportional to $(n + 1)^2 - n^2 = 2n + 1$, where n is the quantum number of the highest occupied state. Since n is proportional to N (equal to $N/2$) and L is approximately proportional to N (strictly to $N - 1$), the energy and frequency are approximately proportional to N^{-1}. So *the absorption spectrum of a linear polyene shifts to* $\boxed{\textit{lower}}$ *frequency as the number of conjugated atoms* $\boxed{\textit{increases}}$.

P2.29 In effect, we are looking for the vibrational frequency of an O atom bound, with a force constant equal to that of free CO, to an infinitely massive and immobile protein complex. The angular frequency is

$$\omega = \left(\frac{k}{m}\right)^{1/2}$$

where m is the mass of the O atom

$$m = (16.0\,\text{u})(1.66\times10^{-27}\ \text{kg u}^{-1}) = 2.66\times10^{-26}\ \text{kg}$$

and k is the same force constant as in Problem 2.3, namely, 1902 N m^{-1}:

$$\omega = \left(\frac{1902\ \text{N m}^{-1}}{2.66\times10^{-26}\ \text{kg}}\right)^{1/2} = \boxed{2.68\times10^{14}\ \text{s}^{-1}}$$

P2.31 First, let $f = n/N$; therefore, f is the fraction of the totally stretched chain represented by the end-to-end distance.

$$F = -\frac{kT}{2l}\ln\left(\frac{N+n}{N-n}\right) = -\frac{kT}{2l}\ln\left(\frac{N(1+f)}{N(1-f)}\right) = -\frac{kT}{2l}\ln\left(\frac{1+f}{1-f}\right)$$

$$= -\frac{kT}{2l}[\ln(1+f) - \ln(1-f)]$$

When $n << N$, then $f << 1$, and the natural log can be expanded: $\ln(1+f) \approx f$ and $\ln(1-f) \approx -f$. Therefore

$$F \approx -\frac{kT}{2l}[f - (-f)] = -\frac{fkT}{l} = -\frac{nkT}{Nl} = -\frac{kT}{Nl^2}x$$

In the last step, we note that the distance x between ends is equal to nl, so $n = x/l$. This is a Hooke's law force with force constant kT / Nl^2.

The root mean square displacement is $\langle x^2 \rangle^{1/2}$.

In part (b) of P2.21, $\langle x^2 \rangle$ for a harmonic oscillator was evaluated and was found to be

$$\langle x^2 \rangle = \left(v + \frac{1}{2}\right) \times \left(\frac{\hbar\omega}{k_{\text{force}}}\right) = \left(v + \frac{1}{2}\right) \times \left(\frac{\hbar^2}{mk_{\text{force}}}\right)^{1/2}$$

Therefore, putting in the appropriate values for the ground state ($v = 0$) of this model

$$\langle x^2 \rangle = \frac{1}{2} \times \left(\frac{\hbar^2}{m} \times \frac{Nl^2}{kT}\right)^{1/2} = \frac{\hbar l}{2} \times \left(\frac{N}{mkT}\right)^{1/2}$$

and $$\langle x^2 \rangle^{1/2} = \boxed{\left(\frac{hl}{2}\right)^{1/2} \times \left(\frac{N}{mkT}\right)^{1/4}}$$

3 Nanosystems 2: motion in several dimensions

Answers to discussion questions

D3.1 In quantum mechanics, particles are said to have characteristics of waves. The fact that the particle exists then requires that the waves representing it not experience destructive interference upon reflection by a barrier or in its motion around a closed loop. This requirement restricts the wavelength of a particle on a ring to wavelengths such that a whole number of wavelengths must fit on the circumference of the ring. That means

$$n\lambda = 2\pi r,\ \text{where } n = 0, 1, 2, \ldots$$

See Section 3.3(a).

D3.3 Fermions are particles with half-integral spin, 1/2, 3/2, 5/2, ..., whereas bosons have integral spin, 0, 1, 2, ... All fundamental particles that make up matter have spin 1/2 and are fermions, but composite particles can be either fermions or bosons.
Fermions: electrons, protons, neutrons, ^{3}He, ...
Bosons: photons, deuterons

COMMENT. In the standard model of particle physics, non-composite bosons are carriers of force. Photons are an example: they are not composite particles, but they are not constituents of matter. In the standard model, protons and neutrons are composite particles, but they are still fermions by virtue of their half-integral spin.

D3.5 Rotational motion on a ring and on a sphere share features such as the form of the energy (square of the angular momentum over twice the moment of inertia) and the lack of zero-point energy because the ground state does not restrict the angular position of the particle. Degeneracy is possible in both cases. In the case of the ring, the axis of rotation is specified, but it is not in the case of the sphere. This distinction gives rise to the fact that angular momenta about different perpendicular axes *cannot* be simultaneously specified: $\hat{l}_x$, $\hat{l}_y$, and $\hat{l}_z$ are complementary in the sense described in Section 1.10.

D3.7 Spin angular momentum and orbital angular momentum have in common the fact that the magnitude of angular momentum and the *z* component of angular momentum are both quantized. Both spin and orbital angular momenta give rise to magnetic effects. The spin angular momentum is an intrinsic angular momentum rather than one due to rotational motion in the classical sense.

Solutions to exercises

E3.1(a)

$$E_{n_1,n_2} = \frac{(n_1^2 + n_2^2)h^2}{8m_e L^2} \quad [3.3\text{b, with } L_1 = L_2]$$

$$\frac{h^2}{8m_e L^2} = \frac{(6.626\times10^{-34}\ \text{J s})^2}{8(9.11\times10^{-31}\ \text{kg})\times(1.0\times10^{-9}\ \text{m})^2} = 6.0\bar{2}\times10^{-20}\ \text{J}$$

The conversion factors required are

$$1\text{ eV} = 1.602\times10^{-19}\text{ J};\quad 1\text{ cm}^{-1} = 1.986\times10^{-23}\text{ J};\quad 1\text{ eV} = 96.485\text{ kJ mol}^{-1}$$

(a) $$E_{2,2}-E_{1,1}=\frac{(2^2+2^2-1^2-1^2)h^2}{8m_eL^2}=\frac{6h^2}{8m_eL^2}=6\times6.0\bar{2}\times10^{-20}\text{ J}$$

$$=\boxed{3.6\times10^{-19}\text{ J}}=\boxed{2.3\text{ eV}}=\boxed{1.8\times10^4\text{ cm}^{-1}}=\boxed{2.2\times10^2\text{ kJ mol}^{-1}}$$

(b) $$E_{6,6}-E_{5,5}=\frac{(6^2+6^2-5^2-5^2)h^2}{8m_eL^2}=\frac{22h^2}{8m_eL^2}=22\times6.0\bar{2}\times10^{-20}\text{ J}$$

$$=\boxed{1.3\times10^{-18}\text{ J}}=\boxed{8.3\text{ eV}}=\boxed{6.7\times10^4\text{ cm}^{-1}}=\boxed{8.0\times10^2\text{ kJ mol}^{-1}}$$

E3.2(a) (a) Radiation stimulates a transition if the energy of the photon is equal to the difference between energy levels. The photon energy is $h\nu$, so

$$\nu=\frac{\Delta E}{h}=\frac{3.6\times10^{-19}\text{ J}}{6.626\times10^{-34}\text{ J s}}=\boxed{5.5\times10^{14}\text{ s}^{-1}}$$

The wavelength is the reciprocal of the wavenumber, so

$$\lambda=\frac{1}{1.8\times10^4\text{ cm}^{-1}}=5.5\times10^{-5}\text{ cm}=\boxed{5.5\times10^{-7}\text{ m}}=\boxed{550\text{ nm}}$$

(b) $$\nu=\frac{\Delta E}{h}=\frac{1.3\times10^{-18}\text{ J}}{6.626\times10^{-34}\text{ J s}}=\boxed{2.0\times10^{15}\text{ s}^{-1}}$$

and $$\lambda=\frac{1}{6.7\times10^4\text{ cm}^{-1}}=1.5\times10^{-5}\text{ cm}=\boxed{1.5\times10^{-7}\text{ m}}=\boxed{150\text{ nm}}$$

E3.3(a) (a) The energy is

$$E_{n_1,n_2}=\left(\frac{n_1^2}{L_1^2}+\frac{n_2^2}{L_2^2}\right)\frac{h^2}{8m_e}\quad[3.3\text{b}]$$

so the zero-point energy is the ground-state energy:

$$E_{1,1}=\left(\frac{1^2}{(1.0\times10^{-9}\text{ m})^2}+\frac{1^2}{(2.0\times10^{-9}\text{ m})^2}\right)\frac{(6.626\times10^{-34}\text{ J s})^2}{8\times9.11\times10^{-31}\text{kg}}=7.5\times10^{-20}\text{ J}$$

The temperature at which this energy equals the typical thermal energy kT is

$$T=\frac{E_{1,1}}{k}=\frac{7.5\times10^{-20}\text{ J}}{1.381\times10^{-23}\text{ J K}^{-1}}=\boxed{5.5\times10^3\text{ K}}$$

(b) The first excited energy level is

$$E_{1,2}=\left(\frac{1^2}{(1.0\times10^{-9}\text{ m})^2}+\frac{2^2}{(2.0\times10^{-9}\text{ m})^2}\right)\frac{(6.626\times10^{-34}\text{ J s})^2}{8\times9.11\times10^{-31}\text{ kg}}=1.2\bar{0}\times10^{-19}\text{ J}$$

so the first excitation energy is

$$E_{1,2}-E_{1,1}=(1.2\bar{0}\times10^{-19}-7.5\times10^{-20})\text{ J}=4.\bar{5}\times10^{-20}\text{ J}$$

The temperature at which this energy equals the typical thermal energy kT is

$$T=\frac{4.\bar{5}\times10^{-20}\text{ J}}{1.381\times10^{-23}\text{ J K}^{-1}}=\boxed{3.\bar{3}\times10^3\text{ K}}$$

E3.4(a) The wavefunctions are

$$\psi_{n_1,n_2} = \frac{2}{L}\sin\left(\frac{n_1\pi x}{L}\right)\sin\left(\frac{n_2\pi y}{L}\right) \quad [3.3\text{a, with } L_1 = L_2]$$

The required probability is

$$P_{n_1,n_2} = \int_{0.49L}^{0.51L}\int_{0.49L}^{0.51L}\psi^2_{n_1,n_2}\,\mathrm{d}x\mathrm{d}y \approx \psi^2_{n_1,n_2}\Delta x\Delta y$$

where $\Delta x = \Delta y = 0.02L$, and the function is evaluated at $x = y = L/2$.

(a) $$P_{1,1} = \left(\frac{2}{L}\right)^2 \sin^2\left(\frac{\pi}{2}\right)\sin^2\left(\frac{\pi}{2}\right)\times 0.02L\times 0.02L = \boxed{0.0016}$$

(b) $$P_{2,2} = \left(\frac{2}{L}\right)^2 \sin^2\left(\frac{2\pi}{2}\right)\sin^2\left(\frac{2\pi}{2}\right)\times 0.02L\times 0.02L = \boxed{0}$$

E3.5(a) See the comment at the end of the solution to Exercise 2.7(a): Maxima in the probability distribution, which is equal to ψ^2, correspond to maxima *and* minima in ψ itself, so we look for points where $\frac{\partial\psi}{\partial x} = \frac{\partial\psi}{\partial y} = 0$.

$$\psi_{2,3} = \frac{2}{L}\sin\left(\frac{2\pi x}{L}\right)\sin\left(\frac{3\pi y}{L}\right) \quad [3.3\text{a, with } L_1 = L_2]$$

$$\frac{\partial\psi_{2,3}}{\partial x} = \frac{4\pi}{L^2}\cos\left(\frac{2\pi x}{L}\right)\sin\left(\frac{3\pi y}{L}\right) = 0$$

$\cos\theta = 0$ when θ equals an odd integer times $\pi/2$:

$$\frac{2\pi x}{L} = \frac{2n'+1}{2}\pi \text{ for } n' = 0, 1, 2, \ldots \text{ which corresponds to } x = \frac{L}{4} \text{ and } \frac{3L}{4}$$

Note: we disregard positions where the sine factor vanishes because they correspond to positions where the wavefunction vanishes as well.

$$\frac{\partial\psi_{2,3}}{\partial y} = \frac{6\pi}{L^2}\sin\left(\frac{2\pi x}{L}\right)\cos\left(\frac{3\pi y}{L}\right) = 0$$

$$\frac{3\pi y}{L} = \frac{2n''+1}{2}\pi \text{ for } n'' = 0, 1, 2, \ldots, \text{ which corresponds to } y = \frac{L}{6}, \frac{3L}{6}, \text{ and } \frac{5L}{6}$$

Thus, the positions of maximum probability, expressed as ordered pairs (x/L, y/L) are

$$\boxed{(1/4, 1/6), (1/4, 1/2), (1/4, 5/6), (3/4, 1/6), (3/4, 1/2), \text{ and } (3/4, 5/6)}$$

E3.6(a) Nodes of the wavefunction are sets of points where the wavefunction vanishes.

$$\psi_{2,3} = \frac{2}{L}\sin\left(\frac{2\pi x}{L}\right)\sin\left(\frac{3\pi y}{L}\right) \quad [3.3\text{a, with } L_1 = L_2]$$

$\sin\theta = 0$ when θ equals an integer times π.

$$\frac{2\pi x}{L} = n'\pi \text{ for } n' = 0, 1, 2, \ldots \text{ which corresponds to } \boxed{\frac{x}{L} = 0, \frac{1}{2}, 1}$$

$$\frac{3\pi y}{L} = n'\pi \text{ for } n' = 0, 1, 2, \ldots \text{ which corresponds to } \boxed{\frac{y}{L} = 0, \frac{1}{3}, \frac{2}{3}, 1}$$

Of course, if one factor of the wavefunction vanishes, the whole wavefunction vanishes. Thus, the wavefunction vanishes not at single points but on the vertical and horizontal lines defined above.

COMMENT. Some of these lines correspond to edges of the square box. The term *node* is often used to describe places other than the boundaries where the wavefunction vanishes. These are $x = 1/2$ and $y = 1/3, 2/3$.

E3.7(a) See Exercise 2.5(a) for the one-dimensional analog to this question. The wavefunction is

$$\psi_{1,1} = \frac{2}{L}\sin\left(\frac{\pi x}{L}\right)\sin\left(\frac{\pi y}{L}\right) \quad [3.3\text{a, with } L_1 = L_2]$$

The expectation value of the x component of the momentum operator, $\hat{p}_x = \frac{\hbar}{\mathrm{i}}\frac{\partial}{\partial x}$, is

$$\langle p_x \rangle = \int_0^L \psi_{1,1}{}^* \hat{p}_x \psi_{1,1}\,\mathrm{d}x\mathrm{d}y = \frac{4\hbar}{\mathrm{i}L^2}\int_0^L\int_0^L \sin\left(\frac{\pi x}{L}\right)\sin\left(\frac{\pi y}{L}\right)\frac{\partial}{\partial x}\sin\left(\frac{\pi x}{L}\right)\sin\left(\frac{\pi y}{L}\right)\mathrm{d}x\mathrm{d}y$$

$$= \frac{4\pi\hbar}{\mathrm{i}L^3}\int_0^L \sin\left(\frac{\pi x}{L}\right)\cos\left(\frac{\pi x}{L}\right)\mathrm{d}x\int_0^L \sin^2\left(\frac{\pi y}{L}\right)\mathrm{d}y = \boxed{0}$$

(The dx integral vanishes.) To find the expectation value of the momentum squared operator, $\hat{p}^2 = -\hbar^2\left(\frac{\partial^2}{\partial x^2} + \frac{\partial^2}{\partial y^2}\right)$, we need

$$\hat{p}^2\psi_{1,1} = -\frac{2\hbar^2}{L}\left(\frac{\partial^2}{\partial x^2} + \frac{\partial^2}{\partial y^2}\right)\sin\left(\frac{\pi x}{L}\right)\sin\left(\frac{\pi y}{L}\right)$$

$$= \frac{2\pi^2\hbar^2}{L^3}\sin\left(\frac{\pi x}{L}\right)\sin\left(\frac{\pi y}{L}\right) + \frac{2\pi^2\hbar^2}{L^3}\sin\left(\frac{\pi x}{L}\right)\sin\left(\frac{\pi y}{L}\right)$$

$$= \frac{2\pi^2\hbar^2}{L^2}\psi_{1,1} = \frac{h^2}{2L^2}\psi_{1,1}$$

$$\langle p^2 \rangle = \int_0^L\int_0^L \psi_{1,1}\hat{p}^2\psi_{1,1}\mathrm{d}x\mathrm{d}y = \frac{h^2}{2L^2}\int_0^L\int_0^L \psi_{1,1}{}^2\mathrm{d}x\mathrm{d}y = \boxed{\frac{h^2}{2L^2}}$$

E3.8(a) The zero-point energy is the ground-state energy; that is, with $n_1 = n_2 = 1$,

$$E_{1,1} = \frac{(n_1{}^2 + n_2{}^2)h^2}{8m_\mathrm{e}L^2} = \frac{(1^2+1^2)h^2}{8m_\mathrm{e}L^2} = \frac{h^2}{4m_\mathrm{e}L^2} \quad [3.3\text{b, with } L_1 = L_2]$$

Set this equal to the rest energy $m_\mathrm{e}c^2$ and solve for L:

$$m_\mathrm{e}c^2 = \frac{h^2}{4m_\mathrm{e}L^2} \quad \text{so} \quad L = \frac{h}{2m_\mathrm{e}c}$$

In absolute units, the length is

$$L = \frac{6.63\times10^{-34}\ \mathrm{J\,s}}{2\times(9.11\times10^{-31}\ \mathrm{kg})\times(3.00\times10^{8}\ \mathrm{m\,s^{-1}})} = 1.21\times10^{-12}\ \mathrm{m} = 1.21\ \mathrm{pm}$$

In terms of the Compton wavelength of an electron, $\lambda_\mathrm{C} = \frac{h}{m_\mathrm{e}c}$, $L = \frac{\lambda_\mathrm{C}}{2}$.

E3.9(a) The energy levels are given by

$$E_{n_1,n_2} = \left(\frac{n_1{}^2}{L_1{}^2} + \frac{n_2{}^2}{L_2{}^2}\right)\frac{h^2}{8m}\ [3.3\text{b}] = \left(\frac{n_1{}^2}{1} + \frac{n_2{}^2}{4}\right)\frac{h^2}{8mL^2}$$

$$E_{2,2} = \left(\frac{2^2}{1} + \frac{2^2}{4}\right)\frac{h^2}{8mL^2} = \frac{5h^2}{8mL^2}$$

We are looking for another state that has the same energy. By inspection we note that the first term in parentheses in $E_{2,2}$ works out to be 4 and the second 1; we can arrange for those values to be reversed:

$$E_{1,4} = \left(\frac{1^2}{1} + \frac{4^2}{4}\right)\frac{h^2}{8mL^2} = \frac{5h^2}{8mL^2}$$

So in this box, the state $\boxed{n_1 = 1,\ n_2 = 4}$ is degenerate to the state $n_1 = 2$, $n_2 = 2$. The question notes that degeneracy frequently accompanies symmetry, and it suggests that one might be surprised to find degeneracy in a box with unequal lengths. Symmetry is a matter of degree, as we will see in Chapter 7. This box is less symmetric than a square box, but it is more symmetric than boxes whose sides have a non-integer or irrational ratio. Similarly, there is much more degeneracy associated with a square box than with this one. Every state of a square box except those with $n_1 = n_2$ is degenerate (with the state that has n_1 and n_2 reversed). Only a few states in this box are degenerate. A box with incommensurable sides, say, L and $2^{1/2}L$, would have no degenerate levels.

E3.10(a)

$$E_{n_1,n_2n_3} = \frac{(n_1^2 + n_2^2 + n_3^2)h^2}{8m_eL^2} \quad [3.5\text{b, with } L_1 = L_2 = L_3]$$

$$E_{1,1,1} = \frac{3h^2}{8mL^2}, \qquad 3E_{1,1,1} = \frac{9h^2}{8mL^2}$$

Hence, we require the values of n_1, n_2 and n_3 that make

$$n_1^2 + n_2^2 + n_3^2 = 9$$

Therefore, $(n_1, n_2, n_3) = (1, 2, 2)$, $(2, 1, 2)$, and $(2, 2, 1)$ and the degeneracy is $\boxed{3}$.

E3.11(a)

$$E_{n_1,n_2n_3} = \frac{(n_1^2 + n_2^2 + n_3^2)h^2}{8mL^2} = \frac{K}{L^2} \quad [3.5\text{b, with } L_1 = L_2 = L_3]$$

$$\frac{\Delta E}{E} = \frac{\dfrac{K}{(0.9L)^2} - \dfrac{K}{L^2}}{\dfrac{K}{L^2}} = \frac{1}{0.81} - 1 = \boxed{0.23}, \quad \text{or} \quad \boxed{23 \text{ percent}}$$

E3.12(a) Set the particle in a cubic box energy equal to $3kT/2$.

$$E = \frac{(n_1^2 + n_2^2 + n_3^2)h^2}{8mL^2}\ [3.5\text{b}] = \frac{n^2h^2}{8mL^2} = \frac{3}{2}kT$$

$$= \left(\frac{3}{2}\right) \times (1.381\times10^{-23}\ \text{J K}^{-1}) \times (300\ \text{K}) = 6.21\bar{4}\times10^{-21}\ \text{J}$$

So $\quad n^2 = \dfrac{8mL^2}{h^2}E$

If $L^3 = 2.00\ \text{m}^3$, then $L^2 = (2.00\ \text{m}^3)^{2/3} = 1.59\ \text{m}^2$,

$$\frac{h^2}{8mL^2} = \frac{(6.626\times10^{-34}\ \text{J s})^2}{(8)\times\left(\dfrac{0.03200\ \text{kg mol}^{-1}}{6.022\times10^{23}\ \text{mol}^{-1}}\right)\times1.59\ \text{m}^2} = 6.51\times10^{-43}\ \text{J}$$

and $\quad n^2 = \dfrac{6.21\bar{4}\times10^{-21}\ \text{J}}{6.51\times10^{-43}\ \text{J}} = 9.55\times10^{21}; \quad n = \boxed{9.77\times10^{10}}$

$$\Delta E = E_{n+1} - E_n = \frac{(n+1)^2h^2}{8mL^2} - \frac{n^2h^2}{8mL^2} = (2n+1)\times\left(\frac{h^2}{8mL^2}\right)$$

$$\Delta E = E_{9.77\times10^{10}+1} - E_{9.77\times10^{10}} = [(2)\times(9.77\times10^{10})+1]\times\left(\frac{h^2}{8mL^2}\right) = \frac{19.5\overline{5}\times10^{10}h^2}{8mL^2}$$

$$= (19.5\overline{5}\times10^{10})\times(6.51\times10^{-43}\ \text{J}) = \boxed{1.27\times10^{-31}\ \text{J}}$$

The de Broglie wavelength is obtained from

$$\lambda = \frac{h}{p} = \frac{h}{mv}\quad [1.3]$$

The velocity is obtained from

$$E_{\text{K}} = \tfrac{1}{2}mv^2 = \tfrac{3}{2}kT = 6.21\overline{4}\times10^{-21}\ \text{J}$$

so
$$v^2 = \frac{6.21\overline{4}\times10^{-21}\ \text{J}}{\left(\frac{1}{2}\right)\times\left(\frac{0.03200\ \text{kg mol}^{-1}}{6.022\times10^{23}\ \text{mol}^{-1}}\right)} = 2.34\times10^{5}\ \text{m}^2\ \text{s}^{-2};\qquad v = 484\ \text{m s}^{-1}$$

$$\lambda = \frac{6.626\times10^{-34}\ \text{J s}}{(5.31\times10^{-26}\ \text{kg})\times(484\ \text{m s}^{-1})} = 2.58\times10^{-11}\ \text{m} = \boxed{25.8\ \text{pm}}$$

The conclusion to be drawn from all these calculations is that the translational motion of the oxygen molecule can be described classically. The energy of the molecule is essentially continuous,

$$\frac{\Delta E}{E} <<< 1$$

E3.13(a) Orthogonality requires that, if $m \neq n$, then

$$\int \psi_m^* \psi_n \, \text{d}\tau = 0$$

Performing the integration:

$$\int \psi_m^* \psi_n \, \text{d}\tau = \int_0^{2\pi} N\text{e}^{-\text{i}m\phi} N\text{e}^{\text{i}n\phi}\,\text{d}\phi = N^2 \int_0^{2\pi} \text{e}^{\text{i}(n-m)\phi}\,\text{d}\phi$$

If $m \neq n$, then

$$\int \psi_m^* \psi_n \, \text{d}\tau = \frac{N^2}{\text{i}(n-m)}\text{e}^{\text{i}(n-m)\phi}\bigg|_0^{2\pi} = \frac{N^2}{\text{i}(n-m)}(1-1) = 0$$

Therefore, they are orthogonal.

E3.14(a) Magnitude of angular momentum $= \{l(l+1)\}^{1/2}\hbar$ [3.22a]

Projection on arbitrary axis $= m_l\hbar$ [3.22b]

Thus,

Magnitude $= (2^{1/2})\times\hbar = \boxed{1.49\times10^{-34}\ \text{J s}}$

Possible projections $\boxed{0 \text{ or } \pm\hbar}$

E3.15(a) The diagrams are drawn by forming a vector of length $\{j(j+1)\}^{1/2}$, with $j = s$ or l as appropriate, and with a projection m_j on the z-axis (see Figure 3.1). Each vector represents the edge of a cone around the z-axis (that for $m_j = 0$ represents the side view of a disk perpendicular to z).

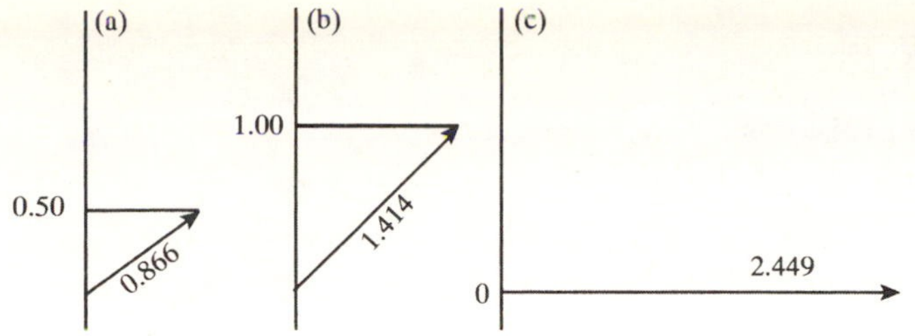

Figure 3.1

Let θ be the angle of the vector with the positive z-axis. The cosine of that angle is the ratio of $m_j / \{j(j+1)\}^{1/2}$

(a) $\cos\theta = \dfrac{0.5}{3^{1/2}/2} = 0.577$ so $\theta = \boxed{0.955 \text{ radians or } 54.7°}$

(b) $\cos\theta = \dfrac{1}{2^{1/2}}$ so $\theta = \boxed{\pi/4 \text{ radians or } 45°}$

(c) $\cos\theta = 0$ so $\theta = \boxed{\pi/2 \text{ radians or } 90°}$

E3.16(a) The rotational energy depends only on the quantum number l (eqn. 3.21), but there are distinct states for every allowed value of m_l, which can range from $-l$ to l in integer steps. For $l = 3$, possible values of $m_l = -3, -2, -1, 0, 1, 2, 3$. There are 7 such values, so the degeneracy is $\boxed{7}$.

E3.17(a) The energy levels are

$$E_{m_l} = \frac{m_l^2\hbar^2}{2I} \quad [3.10a]$$

We are looking for the difference between the ground state ($m_l = 0$) and the first excited level ($m_l = \pm 1$); that difference is equal to the energy of the photon:

$$E_{\text{photon}} = \Delta E = E_1 - E_0 = \frac{\hbar^2}{2I} = \frac{(1.0546\times10^{-34}\ \text{J s})^2}{2\times2.73\times10^{-47}\ \text{kg m}^2} = \boxed{2.04\times10^{-22}\ \text{J}}$$

$$E_{\text{photon}} = \frac{hc}{\lambda}$$

so $$\lambda = \frac{hc}{E_{\text{photon}}} = \frac{6.626\times10^{-34}\ \text{J s}\times2.998\times10^{8}\ \text{m s}^{-1}}{2.04\times10^{-22}\ \text{J}} = \boxed{9.75\times10^{-4}\ \text{m}} = \boxed{0.975\ \text{mm}}$$

Such a photon would be in the $\boxed{\text{microwave}}$ region.

E3.18(a) We can find the speed of a point on the equator if we know the rotational frequency, ω: the speed is $r\omega$, where r is the radius. Classically, angular momentum is equal to moment of inertia, I, times ω. For a sphere of mass m_e, the moment of inertia is $I = 2m_er^2/5$ [Section 3.4 or any general physics text]. An electron has spin angular momentum of $0.866\hbar$ (Section 3.5). Putting it all together,

$$0.866\hbar = \frac{2m_er^2}{5}\omega \quad \text{so} \quad \omega = \frac{5\times0.866\hbar}{2m_er^2}$$

The speed is

$$r\omega = \frac{5\times0.866\hbar}{2m_er} = \frac{5\times0.866\times1.0546\times10^{-34}\ \text{J s}}{2\times9.11\times10^{-31}\ \text{kg}\times2.82\times10^{-15}\ \text{m}} = \boxed{8.89\times10^{10}\ \text{m s}^{-1}}$$

COMMENT. This speed is about 300 times the speed of light! Clearly this quasi-classical, non-relativistic model does not capture this aspect of the electron.

Solutions to problems

Solutions to numerical problems

P3.1 The energy of the particle in a cubic box is

$$E = \frac{(n_1^2 + n_2^2 + n_3^2)h^2}{8mL^2} \quad [3.5b]$$

so the difference between the two lowest energy levels is

$$\Delta E = \frac{(1^2+1^2+2^2)h^2}{8mL^2} - \frac{(1^2+1^2+1^2)h^2}{8mL^2} = \frac{3h^2}{8mL^2}$$

The mass is $m(O_2) = (32.00)\times(1.6605\times10^{-27}\text{ kg}) = 5.31\times10^{-26}\text{ kg}$, so

$$\Delta E = \frac{3h^2}{8mL^2} = \frac{3\times(6.626\times10^{-34}\text{ J s})^2}{5.31\times10^{-26}\text{ kg}\times(5.0\times10^{-2}\text{ m})^2} = \boxed{8.86\times10^{-39}\text{ J}}$$

Set the energy equal to $3kT/2$:

$$E = \frac{3n^2h^2}{8mL^2} = \frac{3kT}{2}$$

$$n = \left(\frac{4kTmL^2}{h^2}\right)^{1/2}$$

$$= \left(\frac{4\times1.381\times10^{-23}\text{ J K}^{-1}\times300\text{ K}\times5.31\times10^{-26}\text{ kg}\times(5.0\times10^{-2}\text{ m})^2}{(6.626\times10^{-34}\text{ J s})^2}\right)^{1/2}$$

$$= 2.2\bar{4}\times10^9$$

The degenerate energy level immediately below this level has one of the quantum numbers equal to $n-1$ and the other two still equal to n.

$$\Delta E = E_{n,n,n} - E_{n,n,n-1} = \frac{3n^2h^2}{8mL^2} - \frac{\{2n^2+(n-1)^2\}h^2}{8mL^2} = \frac{(2n-1)h^2}{8mL^2}$$

$$= \frac{(2\times2.2\bar{4}\times10^9 - 1)\times(6.626\times10^{-34}\text{ J s})^2}{8\times5.31\times10^{-26}\text{ kg}\times(5.0\times10^{-2}\text{ m})^2} = \boxed{1.11\times10^{-30}\text{ J}}$$

COMMENT. The problem is similar to E3.12(a). The separation in energy levels is much smaller than the energy of the occupied level. The energy of the molecule is essentially continuous.

P3.3 In C_{60}, each carbon atom is bound to three others, so each makes three sigma bonds. That leaves one p orbital and one electron per carbon atom to go into the π system. Thus, C_{60} has 60 π electrons, which would doubly occupy the lowest 30 particle-in-a-cube "orbitals." The energy levels are

$$E_{n_1,n_2n_3} = \frac{(n_1^2 + n_2^2 + n_3^2)h^2}{8m_eL^2} \quad [3.5b, \text{ with } L_1 = L_2 = L_3]$$

Levels in which all three quantum numbers are the same are non-degenerate. The degeneracy of levels that have two quantum numbers in common is three, for the different quantum number can be in any of three places. The degeneracy of levels that have three different quantum numbers is six, the number of possible different ordered triplets. The energy levels are shown in Figure 3.2, labeled by one corresponding set of quantum numbers.

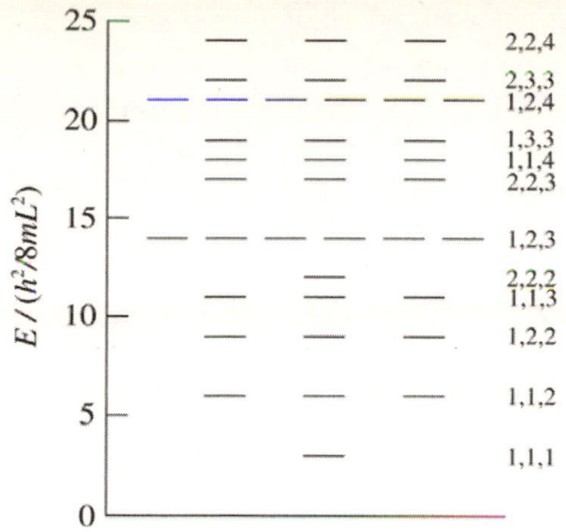

Figure 3.2

Filling the orbitals from the bottom makes the level marked (1,2,4) the highest occupied level. Its energy is $21h^2/8mL^2$. The next level has an energy of $22h^2/8mL^2$ (the level marked (2,3,3)). So the excitation energy is the difference between these levels, namely,

$$\Delta E = \frac{h^2}{8m_eL^2} = \frac{(6.626\times10^{-34}\ \text{J s})^2}{8\times9.11\times10^{-31}\ \text{kg}\times(0.7\times10^{-9}\ \text{m})^2} = 1.2\times10^{-19}\ \text{J}$$

$$\Delta E = E_{\text{photon}} = \frac{hc}{\lambda}$$

so $$\lambda = \frac{hc}{\Delta E} = \frac{6.626\times10^{-34}\ \text{J s}\times2.998\times10^{8}\ \text{m s}^{-1}}{1.2\times10^{-19}\ \text{J}} = \boxed{1.6\times10^{-6}\ \text{m}} = \boxed{1.6\ \mu\text{m}}$$

P3.5 The excitation energy is the difference between the two lowest energy levels, and that energy difference is equal to the energy of the photon.

$$\Delta E = \frac{(F_{1,1}{}^2 - F_{1,0}{}^2)h^2}{8m_ea^2} = E_{\text{photon}} = \frac{hc}{\lambda}$$

so $$a = \left(\frac{\left(F_{1,1}{}^2 - F_{1,0}{}^2\right)h\lambda}{8m_ec}\right)^{1/2} = \left(\frac{(1.430^2 - 1^2)\times6.626\times10^{-34}\ \text{J s}\times1500\times10^{-9}\ \text{m}}{8\times9.11\times10^{-31}\ \text{kg}\times2.998\times10^{8}\ \text{m s}^{-1}}\right)^{1/2}$$

$$= \boxed{6.9\times10^{-10}\ \text{m}} = \boxed{0.69\ \text{nm}}$$

P3.7 (a) The normalization integral is

$$1 = \int_0^{2\pi} N\left(\psi_{-1}{}^* - 3^{1/2}\text{i}\psi_{+1}{}^*\right)N\left(\psi_{-1} + 3^{1/2}\text{i}\psi_{+1}\right)\text{d}\varphi$$
$$= N^2\int_0^{2\pi}\left(|\psi_{-1}|^2 + 3|\psi_{+1}|^2 + 3^{1/2}\text{i}\psi_{-1}{}^*\psi_{+1} - 3^{1/2}\text{i}\psi_{+1}{}^*\psi_{-1}\right)\text{d}\varphi = 4N^2$$

In the last step, we used the fact that the particle-on-a-ring eigenfunctions are themselves normalized and orthogonal. The overall normalization constant, then, is $\boxed{N = 1/2}$, and the normalized wavefunction can be written as a superposition of particle-on-a-ring eigenfunctions:

$$\psi = \frac{\psi_{-1} + 3^{1/2}\text{i}\psi_{+1}}{2}$$

(b) The particle-on-a-ring wavefunctions are eigenfunctions of the angular momentum, so any measurement of angular momentum will result in a corresponding eigenvalue (Section 1.8, Postulate V). The measurements will be $m_l\hbar$ [3.9]. In particular

$\boxed{-\hbar}$, the eigenvalue of ψ_{-1}, with probability $|(1/2)|^2 = \boxed{1/4}$

and $\boxed{+\hbar}$, the eigenvalue of ψ_{+1}, with probability $|(3^{1/2}\mathrm{i}/2)|^2 = \boxed{3/4}$

(c) The particle-on-a-ring wavefunctions are also eigenfunctions of the energy, so any energy measurement will result in a corresponding eigenvalue. However, both energy eigenstates that make up the superposition have the same energy eigenvalue, so any measurement will yield that value, namely,

$$E_{m_l} = \frac{m_l^2\hbar^2}{2I} = \boxed{\frac{\hbar^2}{2m_p r^2}} \quad [3.10a]$$

(d) Expectation values are weighted averages of eigenvalues; the weights are the squares of the magnitude of the coefficients in the superposition wavefunction:

$$\langle\Omega\rangle = \sum_k |c_k|^2 \omega_k \quad [1.17]$$

So the expectation value of angular momentum is

$$\langle l_z\rangle = \left|\frac{1}{2}\right|^2 \times(-\hbar) + \left|\frac{3^{1/2}\mathrm{i}}{2}\right|^2 \times\hbar = \boxed{\frac{\hbar}{2}}$$

(Note that none of the angular momentum measurements will yield the expectation value.) The expectation value of the energy can be computed in the same way. We have already noted, though, that *every* energy measurement will yield the same value; therefore, the expectation value is also equal to that value.

$$\langle H\rangle = \boxed{\frac{\hbar^2}{2m_p r^2}}$$

P3.9 $$E_{m_l} = \frac{m_l^2\hbar^2}{2I}[3.10a] = \frac{m_l^2\hbar^2}{2mr^2}$$

The ground-state energy is 0, when $m_l = 0$. The first excited state energy is

$$E_1 = \frac{\hbar^2}{2mr^2} = \frac{(1.055\times10^{-34}\ \mathrm{J\ s})^2}{(2)\times(1.008)\times(1.6605\times10^{-27}\ \mathrm{kg})\times(160\times10^{-12}\ \mathrm{m})^2} = 1.30\times10^{-22}\ \mathrm{J}$$

This is also the excitation energy (the difference between the first two levels) and the energy of the photon that can excite rotation. Thus

$$\Delta E = E_{\text{photon}} = \frac{hc}{\lambda}$$

so $$\lambda = \frac{hc}{\Delta E} = \frac{6.626\times10^{-34}\ \mathrm{J\ s}\times2.998\times10^8\ \mathrm{m\ s^{-1}}}{1.30\times10^{-22}\ \mathrm{J}} = \boxed{1.53\times10^{-3}\ \mathrm{m}} = \boxed{1.53\ \mathrm{mm}}$$

The minimum angular momentum is $\boxed{\pm\hbar}$.

P3.11 This is an example of rotational motion in two dimensions, so we use results for a particle in a ring.

(a) The expectation value of angular momentum is the weighted average of possible results of angular momentum measurements.

$$\langle l_z\rangle = \tfrac{1}{4}\times3\hbar + \tfrac{3}{4}\times(-3\hbar) = \boxed{-\tfrac{3}{2}\hbar}$$

(b) The particle-on-a-ring wavefunctions are eigenfunctions of angular momentum and of energy. The angular momentum eigenvalues are $3\hbar$ and $-3\hbar$. The corresponding eigenfunctions are ψ_3 and ψ_{-3}, respectively, where

$$\psi_{m_l}(\varphi) = \frac{\mathrm{e}^{\mathrm{i}m_l\varphi}}{(2\pi)^{1/2}} \quad [3.10b]$$

The wavefunction of this molecule is a superposition of ψ_3 and ψ_{-3} (Section 1.8, Postulate V), so

$$\psi(\varphi) = c_1\psi_3 + c_2\psi_{-3}$$

The probabilities of finding the molecule in one of the eigenfunction states is the square of the magnitude of its coefficient in the superposition:

$$|c_1|^2 = 1/4 \qquad \text{and} \qquad |c_2|^2 = 3/4$$

so $\quad c_1 = 1/2 \qquad$ and $\qquad c_2 = 3^{1/2}/2$

(We can take the coefficients to be real and positive.)

$$\psi(\varphi) = \boxed{\frac{e^{3i\varphi}}{2(2\pi)^{1/2}} + \left(\frac{3}{2\pi}\right)^{1/2}\frac{e^{-3i\varphi}}{2}}$$

(c) The energy eigenvalues for the eigenfunctions are

$$E_{m_l} = \frac{m_l^{\,2}\hbar^2}{2I} = \frac{m_l^{\,2}\hbar^2}{2\mu R^2}$$

Both functions in the superposition have the same energy, and that is why energy measurements always give the same value, namely,

$$E_{m_l} = \frac{3^2(1.0546\times10^{-34}\text{ J s})^2}{2\times2.000\times10^{-26}\text{ kg}\times(250.0\times10^{-12}\text{ m})^2} = \boxed{4.004\times10^{-23}\text{ J}}$$

P3.13 The question asks about the most probable angles with respect to an arbitrary axis. We may take that axis to be the z-axis, in which case we are looking for the most probable values of the angle θ. The most probable values of θ are maxima in the probability, which correspond to positive maxima and negative minima in the wavefunction. (*Note*: We can speak as if the wavefunctions were real because the θ dependence of the spherical harmonics is real.) So we look for θ such that $\dfrac{\partial Y}{\partial\theta} = 0$.

For $l = 1$, there are two forms of dependence on θ:

$m_l = 0$: $\quad Y_{1,0} \propto \cos\theta$, which has maxima at $\boxed{\theta = 0 \;\&\; \pi}$ (i.e., along the z-axis)

$m_l = \pm1$: $\quad Y_{1,\pm1} \propto \sin\theta$, which has a maximum at $\boxed{\theta = \pi/2}$ (perpendicular to the z-axis)

For $l = 2$, there are three forms of dependence on θ:

$m_l = 0$: $\quad Y_{2,0} \propto (3\cos^2\theta - 1)$, so

$$\frac{\partial Y_{2,0}}{\partial\theta} \propto -3\times2\cos\theta\sin\theta = 0$$

The derivative vanishes at $\boxed{\theta = 0,\ \pi/2,\ \&\pi}$; these are the most probable values of θ.

$m_l = \pm1$: $\quad Y_{2,\pm1} \propto \cos\theta\sin\theta$, so

$$\frac{\partial Y_{2,\pm1}}{\partial\theta} \propto \cos^2\theta - \sin^2\theta = 0$$

The derivative vanishes where $\cos\theta = \pm\sin\theta$, namely, at $\boxed{\theta = \pi/4 \;\&\; 3\pi/4}$.

$m_l = \pm2$: $\quad Y_{2,\pm2} \propto \sin^2\theta$, so

$$\frac{\partial Y_{2,\pm2}}{\partial\theta} \propto 2\cos\theta\sin\theta = 0$$

The derivative vanishes at $\boxed{\theta = 0,\ \pi/2,\ \&\ \pi}$.

For $l = 3$, there are four forms of dependence on θ:

$m_l = 0$: $\quad Y_{3,0} \propto (5\cos^3\theta - 3\cos\theta)$, so

$$\frac{\partial Y_{3,0}}{\partial\theta} \propto -15\cos^2\theta\sin\theta + 3\sin\theta = 0 = -3\sin\theta(5\cos^2\theta - 1)$$

The derivative vanishes when $\sin\theta = 0$ and when $\cos\theta = \pm(1/5)^{1/2}$. Thus

$$\boxed{\theta = 0,\ \pi,\ 63.4°,\ \&\ 116.6°}$$

$m_l = \pm 1$: $\quad Y_{3,\pm 1} \propto (5\cos^2\theta - 1)$, so

$$\frac{\partial Y_{3,\pm 1}}{\partial\theta} \propto -10\cos\theta\sin\theta = 0$$

The derivative vanishes at $\boxed{\theta = 0,\ \pi/2,\ \&\ \pi}$.

$m_l = \pm 2$: $\quad Y_{3,\pm 2} \propto \sin^2\theta\cos\theta$, so

$$\frac{\partial Y_{3,\pm 2}}{\partial\theta} \propto 2\sin\theta\cos^2\theta - \sin^3\theta = 0 = \sin\theta(2\cos^2\theta - \sin^2\theta)$$

The derivative vanishes when $\sin\theta = 0$ and when $2\cos^2\theta = \sin^2\theta$. The latter condition amounts to $\tan^2\theta = 2$. Thus $\boxed{\theta = 0,\ \pi,\ 54.7°,\ \&\ 125.3°}$.

$m_l = \pm 3$: $\quad Y_{3,\pm 3} \propto \sin^3\theta$, so

$$\frac{\partial Y_{3,\pm 3}}{\partial\theta} \propto 3\sin^2\theta\cos\theta = 0$$

The derivative vanishes at $\boxed{\theta = 0,\ \pi/2,\ \&\ \pi}$.

Solutions to theoretical problems

P3.15 The wavefunctions in question are solutions of the Schrödinger equation. The hamiltonian inside the box is

$$\hat{H} = -\frac{\hbar^2}{2m}\left(\frac{\partial^2}{\partial x^2} + \frac{\partial^2}{\partial y^2} + \frac{\partial^2}{\partial z^2}\right)$$

so the equation to be solved is

$$-\frac{\hbar^2}{2m}\left(\frac{\partial^2\psi(x,y,z)}{\partial x^2} + \frac{\partial^2\psi(x,y,z)}{\partial y^2} + \frac{\partial^2\psi(x,y,z)}{\partial z^2}\right) = E\psi(x,y,z)$$

We look for solutions that factor into functions of one variable. That is, let

$$\psi(x,y,z) = X(x) \times Y(y) \times Z(z)$$

and see under what conditions (if any) the differential equation can be solved. In substituting our trial separated solution into the differential equation, note that each partial derivative acts only on one factor:

$$-\frac{\hbar^2}{2m}\left(Y(y)Z(z)\frac{\mathrm{d}^2X(x)}{\mathrm{d}x^2} + X(x)Z(z)\frac{\mathrm{d}^2Y(y)}{\mathrm{d}y^2} + X(x)Y(y)\frac{\mathrm{d}^2Z(z)}{\mathrm{d}z^2}\right) = EX(x)Y(y)Z(z)$$

Divide both sides of the equation by the trial wavefunction and by $-\hbar^2/2m$:

$$\frac{1}{X(x)}\frac{\mathrm{d}^2X(x)}{\mathrm{d}x^2} + \frac{1}{Y(y)}\frac{\mathrm{d}^2Y(y)}{\mathrm{d}y^2} + \frac{1}{Z(z)}\frac{\mathrm{d}^2Z(z)}{\mathrm{d}z^2} = -\frac{2mE}{\hbar^2}$$

The right side of the equation is a constant, while the left side is a sum of three terms, each of which depends only on one of the three independent variables. Because the three terms depend on independent variables, the only way that they can sum to a constant value is if *each* is equal to a constant value. Thus,

$$\frac{1}{X(x)}\frac{\mathrm{d}^2X(x)}{\mathrm{d}x^2}=-\frac{2mE_X}{\hbar^2},\quad \frac{1}{Y(y)}\frac{\mathrm{d}^2Y(y)}{\mathrm{d}y^2}=-\frac{2mE_Y}{\hbar^2},\ \text{and}\ \frac{1}{Z(z)}\frac{\mathrm{d}^2Z(z)}{\mathrm{d}z^2}=-\frac{2mE_Z}{\hbar^2}$$

where $E_X + E_Y + E_Z = E$. Each of these equations rearranges to an ordinary differential equation of the same form as eqn. 2.1(b), the Schrödinger equation for a one-dimensional particle in a box. Thus, each factor in the three-dimensional wavefunction has the form given in eqn. 2.6(b):

$$X(x)=\left(\frac{2}{L_1}\right)^{1/2}\sin\left(\frac{n_1\pi x}{L_1}\right),\ Y(y)=\left(\frac{2}{L_2}\right)^{1/2}\sin\left(\frac{n_2\pi y}{L_2}\right),\ \text{and}\ Z(z)=\left(\frac{2}{L_3}\right)^{1/2}\sin\left(\frac{n_3\pi z}{L_3}\right)$$

yielding the three-dimensional wavefunction given in eqn. 3.5(a).

P3.17 The Schrödinger equation is

$$-\frac{\hbar^2}{2m}\left(\frac{\partial^2\psi}{\partial x^2}+\frac{\partial^2\psi}{\partial y^2}\right)=E\psi\quad[3.1]$$

To verify that

$$\psi(x,y)=\frac{2}{(L_1L_2)^{1/2}}\sin\frac{n_1\pi x}{L_1}\sin\frac{n_2\pi y}{L_2}\quad[3.3\text{a}]$$

is actually a solution, we take the required partial derivatives.

$$\frac{\partial\psi}{\partial x}=\frac{2}{(L_1L_2)^{1/2}}\times\frac{n_1\pi}{L_1}\cos\frac{n_1\pi x}{L_1}\sin\frac{n_2\pi y}{L_2}$$

so
$$\frac{\partial^2\psi}{\partial x^2}=-\frac{2}{(L_1L_2)^{1/2}}\times\left(\frac{n_1\pi}{L_1}\right)^2\sin\frac{n_1\pi x}{L_1}\sin\frac{n_2\pi y}{L_2}=-\left(\frac{n_1\pi}{L_1}\right)^2\psi$$

Similarly
$$\frac{\partial^2\psi}{\partial y^2}=-\frac{2}{(L_1L_2)^{1/2}}\times\left(\frac{n_2\pi}{L_2}\right)^2\sin\frac{n_1\pi x}{L_1}\sin\frac{n_2\pi y}{L_2}=-\left(\frac{n_2\pi}{L_2}\right)^2\psi$$

Putting these derivatives into the Schrödinger equation yields

$$\frac{\hbar^2}{2m}\left\{\left(\frac{n_1\pi}{L_1}\right)^2\psi+\left(\frac{n_2\pi}{L_2}\right)^2\psi\right\}=E\psi$$

This expression is a true equation if

$$\frac{\hbar^2}{2m}\left\{\left(\frac{n_1\pi}{L_1}\right)^2+\left(\frac{n_2\pi}{L_2}\right)^2\right\}=E$$

With a little rearrangement, we arrive at eqn. 3.3(b) for the energy:

$$\frac{(h/2\pi)^2}{2m}\times\pi^2\left(\frac{n_1^2}{L_1^2}+\frac{n_2^2}{L_2^2}\right)=\boxed{\frac{h^2}{8m}\left(\frac{n_1^2}{L_1^2}+\frac{n_2^2}{L_2^2}\right)=E}$$

P3.19 (a) By hypothesis, the angular momentum is $+\hbar$, so the corresponding wavefunction is

$$\psi_{+1}(\varphi)=\frac{\mathrm{e}^{+i\varphi}}{(2\pi)^{1/2}}$$

There is $\boxed{\text{no uncertainty in } \Delta l_z}$ because the wavefunction is an eigenfunction of $\hat{l}_z$. Therefore, the uncertainty product $\Delta l_z \, \Delta \sin\varphi$ must also be zero. Because the question asks for all pieces in the uncertainty expression, we evaluate $\Delta \sin\varphi$:

$$\langle \sin\varphi \rangle = \int_0^{2\pi} \frac{e^{-i\varphi}}{(2\pi)^{1/2}} \sin\varphi \frac{e^{+i\varphi}}{(2\pi)^{1/2}} d\varphi = \frac{1}{2\pi}\int_0^{2\pi} \sin\varphi d\varphi = \frac{1}{2\pi} \times \frac{-\cos\varphi}{2}\bigg|_0^{2\pi} = 0$$

$$\langle \sin^2\varphi \rangle = \int_0^{2\pi} \frac{e^{-i\varphi}}{(2\pi)^{1/2}} \sin^2\varphi \frac{e^{+i\varphi}}{(2\pi)^{1/2}} d\varphi = \frac{1}{2\pi}\int_0^{2\pi} \sin^2\varphi d\varphi$$

$$= \frac{1}{2\pi} \times \left(\frac{\varphi}{2} - \frac{\sin 2\varphi}{4}\right)\bigg|_0^{2\pi} = \frac{1}{2}$$

So $\quad \Delta \sin\varphi = \left(\langle \sin^2\varphi \rangle - \langle \sin\varphi \rangle^2\right)^{1/2} = \left(\frac{1}{2} - 0\right)^{1/2} = \boxed{\frac{1}{2^{1/2}}}$

The only way the uncertainty relationship can hold is if the expectation value of $\sin\varphi$ also vanishes:

$$\langle \cos\varphi \rangle = \int_0^{2\pi} \frac{e^{-i\varphi}}{(2\pi)^{1/2}} \cos\varphi \frac{e^{+i\varphi}}{(2\pi)^{1/2}} d\varphi = \frac{1}{2\pi}\int_0^{2\pi} \cos\varphi d\varphi = \frac{1}{2\pi} \times \frac{\sin\varphi}{2}\bigg|_0^{2\pi} = 0$$

so $\quad \Delta l_z \Delta \sin\varphi \geq \frac{1}{2}\hbar \left|\langle \cos\varphi \rangle\right| \qquad$ because $0 \geq 0$.

(b) A wavefunction proportional to $\cos\varphi$ is a mixture of the two particle-on-a-ring wavefunctions that have $m_l = +1$ and $m_l = -1$ because $e^{-i\varphi} + e^{+i\varphi} = 2\cos\varphi$.

$$\psi(\varphi) = N\frac{e^{+i\varphi} + e^{-i\varphi}}{(2\pi)^{1/2}}$$

To compute expectation values, we need the normalization constant:

$$1 = \int_0^{2\pi} N\frac{e^{+i\varphi} + e^{-i\varphi}}{(2\pi)^{1/2}} N\frac{e^{-i\varphi} + e^{+i\varphi}}{(2\pi)^{1/2}} d\phi = \frac{N^2}{2\pi}\int_0^{2\pi} (2 + e^{-2i\varphi} + e^{+2i\varphi}) d\phi$$

$$= \frac{N^2}{2\pi}\left(2\varphi - \frac{e^{-2i\varphi}}{2i} + \frac{e^{+2i\varphi}}{2i}\right)\bigg|_0^{2\pi} = 2N^2$$

So $N = 1/2^{1/2} = c_{-1} = c_{+1}$, where the c's are coefficients in the superposition. Because particle-on-a-ring wavefunctions are eigenfunctions of angular momentum, we can use

$$\langle \Omega \rangle = \sum_k |c_k|^2 \omega_k \quad [1.17]$$

So the expectation value of angular momentum is

$$\langle l_z \rangle = \frac{1}{2} \times (-\hbar) + \frac{1}{2} \times \hbar = 0$$

Both particle-on-a-ring wavefunctions have the same eigenvalue of angular momentum squared, namely $\hbar^2$, so this is the expectation value of that operator.

$$\Delta l_z = \left(\langle l_z^2 \rangle - \langle l_z \rangle^2\right)^{1/2} = (\hbar^2 - 0)^{1/2} = \boxed{\hbar}$$

Before we look at the expectation values of trigonometric functions, note that the wavefunction can be written as

$$\psi(\varphi) = \frac{1}{2^{1/2}} \times \frac{e^{+i\varphi} + e^{-i\varphi}}{(2\pi)^{1/2}} = \frac{\cos\varphi}{\pi^{1/2}}$$

So $\langle \sin\varphi \rangle = \int_0^{2\pi} \frac{\cos\varphi}{\pi^{1/2}} \sin\varphi \frac{\cos\varphi}{\pi^{1/2}} \mathrm{d}\varphi = \frac{1}{\pi}\int_0^{2\pi} \cos^2\varphi \sin\varphi \mathrm{d}\varphi = \frac{1}{\pi} \times \left. \frac{-\cos^3\varphi}{3} \right|_0^{2\pi} = 0$

$$\langle \sin^2\varphi \rangle = \int_0^{2\pi} \frac{\cos\varphi}{\pi^{1/2}} \sin^2\varphi \frac{\cos\varphi}{\pi^{1/2}} \mathrm{d}\varphi = \frac{1}{\pi}\int_0^{2\pi} \cos^2\varphi \sin^2\varphi \mathrm{d}\varphi$$

$$= \frac{1}{\pi} \times \left. \left(\frac{\varphi}{8} - \frac{\sin 4\varphi}{32} \right) \right|_0^{2\pi} = \frac{1}{4}$$

$$\Delta \sin\varphi = (\langle \sin^2\varphi \rangle - \langle \sin\varphi \rangle^2)^{1/2} = \left(\frac{1}{4} - 0 \right)^{1/2} = \boxed{\frac{1}{2}}$$

$$\langle \cos\varphi \rangle = \int_0^{2\pi} \frac{\cos\varphi}{\pi^{1/2}} \cos\varphi \frac{\cos\varphi}{\pi^{1/2}} \mathrm{d}\varphi = \frac{1}{\pi}\int_0^{2\pi} \cos^3\varphi \mathrm{d}\varphi$$

$$= \frac{1}{\pi} \times \left. \left(\sin\varphi - \frac{\sin^3\varphi}{3} \right) \right|_0^{2\pi} = 0$$

Finally, $\Delta l_z \Delta \sin\varphi \geq \frac{1}{2}\hbar |\langle \cos\varphi \rangle|$ because $\frac{1}{2}\hbar \geq \frac{1}{2}\hbar \times 0 = 0$.

The uncertainty principle is trivially satisfied in both cases because the minimum uncertainty is zero. The cases are somewhat different, though, in that there is no uncertainty in the angular momentum in the first case, and there is uncertainty in the second case.

P3.21 The elliptical ring to which the particle is confined is defined by the set of all points that obey a certain equation. In Cartesian coordinates, that equation is

$$\frac{x^2}{a^2} + \frac{y^2}{b^2} = 1$$

as you may remember from analytical geometry. An ellipse is similar to a circle, and an appropriate change of variable can transform the ellipse of this problem into a circle. That change of variable is most conveniently described in terms of new Cartesian coordinates (X, Y) where

$$X = x \quad \text{and} \quad Y = ay/b$$

In this new coordinate system, the equation for the ellipse becomes

$$\frac{x^2}{a^2} + \frac{y^2}{b^2} = 1 \quad \Rightarrow \quad \frac{X^2}{a^2} + \frac{Y^2}{a^2} = 1 \quad \Rightarrow \quad X^2 + Y^2 = a^2$$

which we recognize as the equation of a circle of radius a centered at the origin of our (X,Y) system. The text found the eigenfunctions and eigenvalues for a particle on a circular ring by transforming from Cartesian coordinates to plane polar coordinates. Consider plane polar coordinates (R,Φ) related in the usual way to (X,Y):

$$X = R\cos\Phi \quad \text{and} \quad Y = R\sin\Phi$$

In this coordinate system, we can simply quote the results obtained in the text. The energy levels are

$$E = \frac{m_l^2\hbar^2}{2I} \quad [3.10a]$$

where the moment of inertia is the mass of the particle times the radius of the circular ring

$$I = ma^2$$

The eigenfunctions are

$$\psi = \frac{e^{im_l\Phi}}{(2\pi)^{1/2}} \quad [3.10b]$$

It is customary to express results in terms of the original coordinate system, so express Φ in terms first of X and Y, and then substitute the original coordinates:

$$\frac{Y}{X} = \tan\Phi \quad \text{so} \quad \Phi = \tan^{-1}\frac{Y}{X} = \tan^{-1}\frac{ay}{bx}$$

P3.23 The Schrödinger equation is

$$-\frac{\hbar^2}{2m}\nabla^2\psi = E\psi \ \text{[3.18b]}$$

$$\nabla^2\psi = \frac{1}{r}\frac{\partial^2(r\psi)}{\partial r^2} + \frac{1}{r^2}\Lambda^2\psi \ \text{[Table 3.1]}$$

since r = constant, the first term is eliminated and the Schrödinger equation may be rewritten

$$-\frac{\hbar^2}{2mr^2}\Lambda^2\psi = E\psi \ \text{ or } \ -\frac{\hbar^2}{2I}\Lambda^2\psi = E\psi \ [I = mr^2] \ \text{ or } \ \Lambda^2\psi = -\frac{2IE\psi}{\hbar^2}$$

$$\text{where} \quad \Lambda^2 = \frac{1}{\sin^2\theta}\frac{\partial^2}{\partial\phi^2} + \frac{1}{\sin\theta}\frac{\partial}{\partial\theta}\sin\theta\frac{\partial}{\partial\theta}$$

Now use the specified $\psi = Y_{l,m_l}$ from Table 3.2, and see if they satisfy this equation.

(a) Because $Y_{0,0}$ is a constant, all derivatives with respect to angles are zero, so $\Lambda^2 Y_{0,0} = \boxed{0}$, implying that $E = \boxed{0}$ and angular momentum = $\boxed{0}$ [from $\{l(l+1)\}^{1/2}\hbar$, eqn. 3.22a].

(b) $$\Lambda^2 Y_{2,-1} = \frac{1}{\sin^2\theta}\frac{\partial^2 Y_{2,-1}}{\partial\phi^2} + \frac{1}{\sin\theta}\frac{\partial}{\partial\theta}\sin\theta\frac{\partial Y_{2,-1}}{\partial\theta} \qquad \text{where } Y_{2,-1} = N\cos\theta\sin\theta\, e^{-i\phi}$$

$$\frac{\partial Y_{2,-1}}{\partial\theta} = Ne^{-i\phi}(\cos^2\theta - \sin^2\theta)$$

$$\begin{aligned}\frac{1}{\sin\theta}\frac{\partial}{\partial\theta}\sin\theta\frac{\partial Y_{2,-1}}{\partial\theta} &= \frac{1}{\sin\theta}\frac{\partial}{\partial\theta}\sin\theta Ne^{-i\phi}(\cos^2\theta-\sin^2\theta)\\ &= \frac{Ne^{-i\phi}}{\sin\theta}\left(\sin\theta(-4\cos\theta\sin\theta) + \cos\theta(\cos^2\theta - \sin^2\theta)\right)\\ &= Ne^{-i\phi}\left(-6\cos\theta\sin\theta + \frac{\cos\theta}{\sin\theta}\right) \quad [\cos^3\theta = \cos\theta(1-\sin^2\theta)]\end{aligned}$$

$$\frac{1}{\sin^2\theta}\frac{\partial^2 Y_{2,-1}}{\partial\phi^2} = \frac{-N\cos\theta\sin\theta\, e^{-i\phi}}{\sin^2\theta} = \frac{-N\cos\theta\, e^{-i\phi}}{\sin\theta}$$

so $$\Lambda^2 Y_{2,-1} = Ne^{-i\phi}(-6\cos\theta\sin\theta) = -6Y_{2,-1} = -2(2+1)Y_{2,-1} \ \text{[i.e., } l = 2]$$

and hence

$$-6Y_{2,-1} = -\frac{2IE}{\eta^2}Y_{2,-1}, \ \text{implying that} \ \boxed{E = \frac{3\hbar^2}{I}}$$

and the angular momentum is $\{2(2+1)\}^{1/2}\hbar = \boxed{6^{1/2}\hbar}$.

(c) $$\Lambda^2 Y_{3,3} = \frac{1}{\sin^2\theta}\frac{\partial^2 Y_{3,3}}{\partial\phi^2} + \frac{1}{\sin\theta}\frac{\partial}{\partial\theta}\sin\theta\frac{\partial Y_{3,3}}{\partial\theta} \ \text{where} \quad Y_{3,3} = N\sin^3\theta\, e^{3i\phi}$$

$$\frac{\partial Y_{3,3}}{\partial\theta} = 3N\sin^2\theta\cos\theta\, e^{3i\phi}$$

$$\frac{1}{\sin\theta}\frac{\partial}{\partial\theta}\sin\theta\frac{\partial Y_{3,3}}{\partial\theta} = \frac{1}{\sin\theta}\frac{\partial}{\partial\theta}3N\sin^3\theta\cos\theta\,\mathrm{e}^{3\mathrm{i}\phi}$$

$$= \frac{3N\mathrm{e}^{3\mathrm{i}\phi}}{\sin\theta}(3\sin^2\theta\cos^2\theta - \sin^4\theta) = 3N\mathrm{e}^{3\mathrm{i}\phi}\sin\theta(3\cos^2\theta - \sin^2\theta)$$

$$= 3N\mathrm{e}^{3\mathrm{i}\phi}\sin\theta(3 - 4\sin^2\theta) \quad [\cos^3\theta = \cos\theta(1-\sin^2\theta)]$$

$$\frac{1}{\sin^2\theta}\frac{\partial^2 Y_{3,3}}{\partial\phi^2} = \frac{-9N\sin^3\theta\,\mathrm{e}^{3\mathrm{i}\phi}}{\sin^2\theta} = -9N\sin\theta\,\mathrm{e}^{3\mathrm{i}\phi}$$

so $\quad \Lambda^2 Y_{3,3} = -12N\sin^3\theta\,\mathrm{e}^{3\mathrm{i}\phi} = -12Y_{3,3} = -3(3+1)Y_{3,3}$ [i.e., $l = 3$]

and hence

$$-12Y_{3,3} = -\frac{2IE}{\eta^2}Y_{3,3}, \text{ implying that } \boxed{E = \frac{6\hbar^2}{I}}$$

and the angular momentum is $\{3(3+1)\}^{1/2}\hbar = \boxed{2\sqrt{3}\hbar}$.

P3.25 $$\int_0^{\pi}\int_0^{2\pi} Y_{3,3}^* Y_{3,3}\sin\theta\,\mathrm{d}\theta\,\mathrm{d}\phi = \left(\frac{1}{64}\right)\times\left(\frac{35}{\pi}\right)\int_0^{\pi}\sin^6\theta\sin\theta\,\mathrm{d}\theta\int_0^{2\pi}\mathrm{d}\phi \text{ [Table 3.2]}$$

$$= \left(\frac{1}{64}\right)\times\left(\frac{35}{\pi}\right)\times(2\pi)\int_{-1}^{1}(1-\cos^2\theta)^3\,\mathrm{d}\cos\theta \quad [\sin\theta\,\mathrm{d}\theta = \mathrm{d}\cos\theta,\ \sin^2\theta = 1-\cos^2\theta]$$

$$= \frac{35}{32}\int_{-1}^{1}(1 - 3x^2 + 3x^4 - x^6)\,\mathrm{d}x \quad [x = \cos\theta]$$

$$= \frac{35}{32}\left(x - x^3 + \tfrac{3}{5}x^5 - \tfrac{1}{7}x^7\right)\Big|_{-1}^{1} = \frac{35}{32}\times\frac{32}{35} = \boxed{1}$$

P3.27 $$\hat{\mathbf{l}} = \hat{\mathbf{r}}\times\hat{\mathbf{p}} = \begin{vmatrix} \mathbf{i} & \mathbf{j} & \mathbf{k} \\ \hat{x} & \hat{y} & \hat{z} \\ \hat{p}_x & \hat{p}_y & \hat{p}_z \end{vmatrix} \text{ [see any book treating the vector product of vectors]}$$

$$= \mathbf{i}(\hat{y}\hat{p}_z - \hat{z}\hat{p}_y) + \mathbf{j}(\hat{z}\hat{p}_x - \hat{x}\hat{p}_z) + \mathbf{k}(\hat{x}\hat{p}_y - \hat{y}\hat{p}_x)$$

Therefore,

$$\hat{l}_x = (\hat{y}\hat{p}_z - \hat{z}\hat{p}_y) = \boxed{\frac{\hbar}{\mathrm{i}}\left(y\frac{\partial}{\partial z} - z\frac{\partial}{\partial y}\right)}$$

$$\hat{l}_y = (\hat{z}\hat{p}_x - \hat{x}\hat{p}_z) = \boxed{\frac{\hbar}{\mathrm{i}}\left(z\frac{\partial}{\partial x} - x\frac{\partial}{\partial z}\right)}$$

$$\hat{l}_z = (\hat{x}\hat{p}_y - \hat{y}\hat{p}_x) = \boxed{\frac{\hbar}{\mathrm{i}}\left(x\frac{\partial}{\partial y} - y\frac{\partial}{\partial x}\right)}$$

We have used $\hat{p}_x = \frac{\hbar}{\mathrm{i}}\frac{\partial}{\partial x}$, etc. The commutator of $\hat{l}_x$ and $\hat{l}_y$ is $[\hat{l}_x, \hat{l}_y] = (\hat{l}_x\hat{l}_y - \hat{l}_y\hat{l}_x)$. Recall that the operators always imply operation on a function. We form

$$\hat{l}_x\hat{l}_y f = -\hbar^2\left(y\frac{\partial}{\partial z} - z\frac{\partial}{\partial y}\right)\left(z\frac{\partial}{\partial x} - x\frac{\partial}{\partial z}\right)f$$

$$= -\hbar^2\left(yz\frac{\partial^2 f}{\partial z\partial x} + y\frac{\partial f}{\partial x} - yx\frac{\partial^2 f}{\partial z^2} - z^2\frac{\partial^2 f}{\partial y\partial x} + zx\frac{\partial^2 f}{\partial z\partial y}\right)$$

and $$\hat{l}_y\hat{l}_x f = -\hbar^2\left(z\frac{\partial}{\partial x}-x\frac{\partial}{\partial z}\right)\left(y\frac{\partial}{\partial z}-z\frac{\partial}{\partial y}\right)f$$

$$= -\hbar^2\left(zy\frac{\partial^2 f}{\partial x\partial z}-z^2\frac{\partial^2 f}{\partial x\partial y}-xy\frac{\partial^2 f}{\partial z^2}+xz\frac{\partial^2 f}{\partial z\partial y}+x\frac{\partial f}{\partial y}\right)$$

Since multiplication and differentiation is each commutative, the results of the operation $\hat{l}_x\hat{l}_y$ and $\hat{l}_y\hat{l}_x$ differ only in one term. For $\hat{l}_y\hat{l}_x f$, $x\frac{\partial f}{\partial y}$ replaces $y\frac{\partial f}{\partial x}$. Hence, the commutator of the operations, $(\hat{l}_x\hat{l}_y-\hat{l}_y\hat{l}_x)$, is $-\hbar^2\left(y\frac{\partial}{\partial x}-x\frac{\partial}{\partial y}\right)$ or $\boxed{-\frac{\hbar}{\mathrm{i}}\hat{l}_z}$.

COMMENT. We also would find

$$(\hat{l}_y\hat{l}_z-\hat{l}_z\hat{l}_y) = -\frac{\hbar}{\mathrm{i}}\hat{l}_x \quad \text{and} \quad (\hat{l}_z\hat{l}_x-\hat{l}_x\hat{l}_z) = -\frac{\hbar}{\mathrm{i}}\hat{l}_y$$

P3.29 We are to show that $[\hat{l}^2,\hat{l}_z]=0$. Start by expanding the first operator:

$$[\hat{l}^2,\hat{l}_z]=[\hat{l}_x^2+\hat{l}_y^2+\hat{l}_z^2,\hat{l}_z]=[\hat{l}_x^2,\hat{l}_z]+[\hat{l}_y^2,\hat{l}_z]+[\hat{l}_z^2,\hat{l}_z]$$

The three commutators are

$$[\hat{l}_z^2,\hat{l}_z]=\hat{l}_z^2\hat{l}_z-\hat{l}_z\hat{l}_z^2=\hat{l}_z^3-\hat{l}_z^3=0$$

$$[\hat{l}_x^2,\hat{l}_z]=\hat{l}_x^2\hat{l}_z-\hat{l}_z\hat{l}_x^2=\hat{l}_x^2\hat{l}_z-\hat{l}_x\hat{l}_z\hat{l}_x+\hat{l}_x\hat{l}_z\hat{l}_x-\hat{l}_z\hat{l}_x^2$$

$$=\hat{l}_x(\hat{l}_x\hat{l}_z-\hat{l}_z\hat{l}_x)+(\hat{l}_x\hat{l}_z-\hat{l}_z\hat{l}_x)\hat{l}_x=\hat{l}_x[\hat{l}_x,\hat{l}_z]+[\hat{l}_x,\hat{l}_z]\hat{l}_x$$

$$=\hat{l}_x(-\mathrm{i}\hbar\hat{l}_y)+(-\mathrm{i}\hbar\hat{l}_y)\hat{l}_x=-\mathrm{i}\hbar(\hat{l}_x\hat{l}_y+\hat{l}_y\hat{l}_x) \quad [3.24]$$

$$[\hat{l}_y^2,\hat{l}_z]=\hat{l}_y^2\hat{l}_z-\hat{l}_z\hat{l}_y^2=\hat{l}_y^2\hat{l}_z-\hat{l}_y\hat{l}_z\hat{l}_y+\hat{l}_y\hat{l}_z\hat{l}_y-\hat{l}_z\hat{l}_y^2$$

$$=\hat{l}_y(\hat{l}_y\hat{l}_z-\hat{l}_z\hat{l}_y)+(\hat{l}_y\hat{l}_z-\hat{l}_z\hat{l}_y)\hat{l}_y=\hat{l}_y[\hat{l}_y,\hat{l}_z]+[\hat{l}_y,\hat{l}_z]\hat{l}_y$$

$$=\hat{l}_y(\mathrm{i}\hbar\hat{l}_x)+(\mathrm{i}\hbar\hat{l}_x)\hat{l}_y=\mathrm{i}\hbar(\hat{l}_y\hat{l}_x+\hat{l}_x\hat{l}_y) \quad [3.24]$$

Therefore, $$[\hat{l}^2,\hat{l}_z]=-\mathrm{i}\hbar(\hat{l}_x\hat{l}_y+\hat{l}_y\hat{l}_x)+\mathrm{i}\hbar(\hat{l}_x\hat{l}_y+\hat{l}_y\hat{l}_x)+0=0$$

We may also conclude that $[\hat{l}^2,\hat{l}_x]=0$ and $[\hat{l}^2,\hat{l}_y]=0$ because $\hat{l}_x$, $\hat{l}_y$, and $\hat{l}_z$ occur symmetrically in $\hat{l}^2$.

Solutions to applications

P3.31 P3.30 states that there are 22 electrons in the conjugated systems, so they would doubly occupy the lowest 11 particle-in-a-square "orbitals." The energy levels are

$$E_{n_1,n_2}=\frac{(n_1^2+n_2^2)h^2}{8m_eL^2} \quad [3.3\text{b, with } L_1=L_2]$$

Levels in which both quantum numbers are the same are non-degenerate; levels whose two quantum numbers are different are doubly degenerate. The energy levels are shown in Figure 3.3(a), labeled by one corresponding set of quantum numbers.

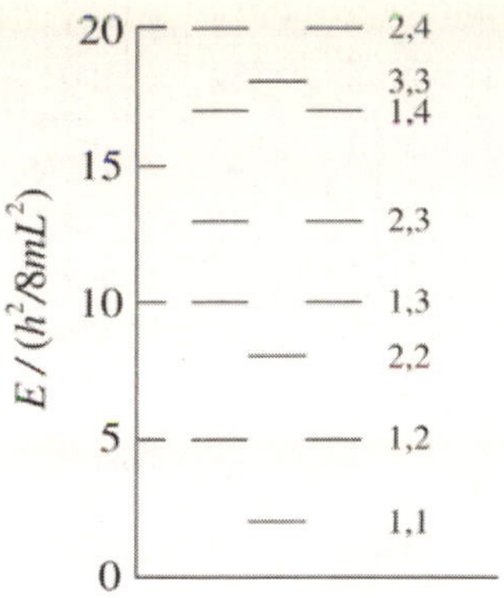

Figure 3.3(a)

Filling the orbitals from the bottom makes the (3,3) level the highest occupied level. So the wavefunction to be plotted is

$$\psi_{3,3}(x,y)=\frac{2}{L}\sin\left(\frac{3\pi x}{L}\right)\sin\left(\frac{3\pi y}{L}\right) \quad [3.3a, \text{ with } L_1=L_2]$$

and the corresponding probability density is

$$\psi^2_{3,3}(x,y)=\frac{4}{L^2}\sin^2\left(\frac{3\pi x}{L}\right)\sin^2\left(\frac{3\pi y}{L}\right)$$

To generalize our plot to any size L, we will neglect the normalization constant and redefine the coordinates with $\chi = x/L$ and $\eta = y/L$. The coordinate origin is $(\chi,\eta) = (0,0)$ and the ranges are $0 \le \chi \le 1$ and $0 \le \eta \le 1$. The function $f(\chi,\eta) = \sin(n_1\pi\chi)\sin(n_2\pi\eta)$ will represent our wavefunction. Mathcad™ contour and surface plots are set up by defining a grid of 101×101 coordinate points, assigning values to n_1 and n_2, and calculating the value of f at each point:

$$\mathbf{N} := 100 \qquad \mathbf{i} := 0...\,\mathrm{N} \qquad \mathbf{j} := 0...\,\mathrm{N} \qquad \chi_i := \frac{i}{N} \qquad \eta_j := \frac{j}{N}$$

$$n_1 := 3 \qquad n_2 := 3$$

$$f_{i,j} := \sin(n_1\cdot\pi\cdot\chi_i)\cdot\sin(n_2\cdot\pi\cdot\eta_j)$$

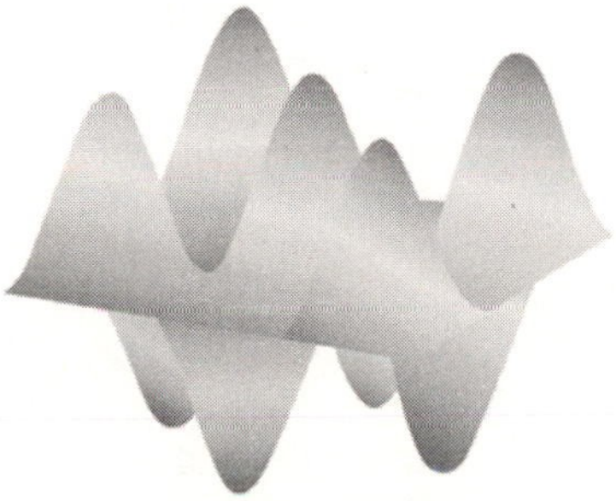

Figure 3.3(b)

The Mathcad surface plotting feature easily produces the wavefunction graph of Figure 3.3(b). Figure 3.3(c) is a surface plot of the HOMO probability density, which is simply f squared. A portion of the porphine molecule has been sketched onto the Figure 3.3(c) probability density and the nitrogen atoms of the porphine ring should be viewed as being near the ring's center.

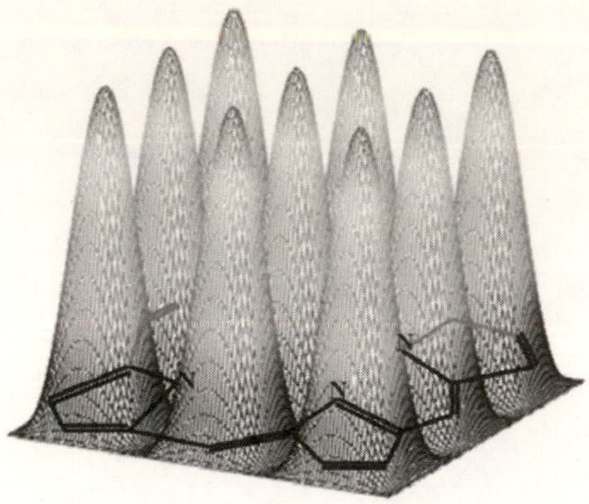

Figure 3.3(c)

Figure 3.3(c) shows a remarkable resemblance between the probability density of the particle-in-a-square and the conjugated π-system of porphine suggested by the structure. It is even possible to imagine the center spike in the probability density as being representative of the lone pair electrons on the nitrogen atoms. However useful this model is for conceptualizing the electron density in porphine, far more accurate computations, to be studied in Chapter 6, are needed to determine the actual distribution of electron density.

P3.33 (a) In the sphere, the Schrödinger equation is

$$-\frac{\hbar^2}{2m}\left(\frac{\partial^2}{\partial r^2}+\frac{2}{r}\frac{\partial}{\partial r}+\frac{1}{r^2}\Lambda^2\right)\psi = E\psi \quad \text{[Table 3.1]}$$

where Λ^2 is an operator that contains derivatives with respect to θ and ϕ only.

Let $\quad \psi(r, \theta, \phi) = R(r)Y(\theta, \phi)$.

Substituting into the Schrödinger equation gives

$$-\frac{\hbar^2}{2m}\left(Y\frac{\partial^2 R}{\partial r^2}+\frac{2Y}{r}\frac{\partial R}{\partial r}+\frac{R}{r^2}\Lambda^2 Y\right) = ERY$$

Divide both sides by RY:

$$-\frac{\hbar^2}{2m}\left(\frac{1}{R}\frac{\partial^2 R}{\partial r^2}+\frac{2}{Rr}\frac{\partial R}{\partial r}+\frac{1}{Yr^2}\Lambda^2 Y\right) = E$$

The first two terms in parentheses depend only on r, but the last one depends on both r and angles; however, multiplying both sides of the equation by r^2 will effect the desired separation:

$$-\frac{\hbar^2}{2m}\left(\frac{r^2}{R}\frac{\partial^2 R}{\partial r^2}+\frac{2r}{R}\frac{\partial R}{\partial r}+\frac{1}{Y}\Lambda^2 Y\right) = Er^2$$

Put all the terms involving angles on the right side and the terms involving distance on the left:

$$-\frac{\hbar^2}{2m}\left(\frac{r^2}{R}\frac{\partial^2 R}{\partial r^2}+\frac{2r}{R}\frac{\partial R}{\partial r}\right) - Er^2 = \frac{\hbar^2}{2mY}\Lambda^2 Y$$

Note that the right side depends only on θ and ϕ, while the left side depends on r. The only way that the two sides can be equal to each other for all r, θ, and ϕ is if they are both equal to a constant. Call that constant $-\frac{\hbar^2 l(l+1)}{2m}$ (with l as yet undefined), and we have, from the right side of the equation,

$$\frac{\hbar^2}{2mY}\Lambda^2 Y = -\frac{\hbar^2 l(l+1)}{2m} \quad \text{so} \quad \Lambda^2 Y = -l(l+1)Y$$

From the left side of the equation, we have

$$-\frac{\hbar^2}{2m}\left(\frac{r^2}{R}\frac{\partial^2 R}{\partial r^2}+\frac{2r}{R}\frac{\partial R}{\partial r}\right)-Er^2=-\frac{\hbar^2 l(l+1)}{2m}$$

After multiplying both sides by R/r^2 and rearranging, we get the desired radial equation:

$$-\frac{\hbar^2}{2m}\left(\frac{\partial^2 R}{\partial r^2}+\frac{2}{r}\frac{\partial R}{\partial r}\right)+\frac{\hbar^2 l(l+1)}{2mr^2}R=ER$$

Thus, the assumption that the wavefunction can be written as a product of functions is a valid one, for we can find separate differential equations for the assumed factors. That is what it means for a partial differential equation to be separable.

(b) The radial equation with $l = 0$ can be rearranged to read

$$\frac{\partial^2 R}{\partial r^2}+\frac{2}{r}\frac{\partial R}{\partial r}=-\frac{2mER}{\hbar^2}$$

Form the following derivatives of the proposed solution:

$$\frac{\partial R}{\partial r}=(2\pi a)^{-1/2}\left[\frac{\cos(n\pi r/a)}{r}\left(\frac{n\pi}{a}\right)-\frac{\sin(n\pi r/a)}{r^2}\right]$$

$$\frac{\partial^2 R}{\partial r^2}=(2\pi a)^{-1/2}\left[-\frac{\sin(n\pi r/a)}{r}\left(\frac{n\pi}{a}\right)^2-\frac{2\cos(n\pi r/a)}{r^2}\left(\frac{n\pi}{a}\right)+\frac{2\sin(n\pi r/a)}{r^3}\right]$$

Substituting into the left side of the rearranged radial equation yields

$$(2\pi a)^{-1/2}\left[-\frac{\sin(n\pi r/a)}{r}\left(\frac{n\pi}{a}\right)^2-\frac{2\cos(n\pi r/a)}{r^2}\left(\frac{n\pi}{a}\right)+\frac{2\sin(n\pi r/a)}{r^3}\right]$$
$$+(2\pi a)^{-1/2}\left[\frac{2\cos(n\pi r/a)}{r^2}\left(\frac{n\pi}{a}\right)-\frac{2\sin(n\pi r/a)}{r^3}\right]$$
$$=-(2\pi a)^{-1/2}\frac{\sin(n\pi r/a)}{r}\left(\frac{n\pi}{a}\right)^2=-\left(\frac{n\pi}{a}\right)^2 R$$

Acting on the proposed solution by taking the prescribed derivatives yields the function back-multiplied by a constant, so the proposed solution is in fact a solution.

(c) Comparing this result to the right side of the rearranged radial equation gives an equation for the energy

$$\left(\frac{n\pi}{a}\right)^2=\frac{2mE}{\hbar^2} \quad \text{so} \quad E=\left(\frac{n\pi}{a}\right)^2\frac{\hbar^2}{2m}=\frac{n^2\pi^2}{2ma^2}\left(\frac{h}{2\pi}\right)^2=\boxed{\frac{n^2h^2}{8ma^2}}$$

PART 2 Atoms, molecules, and assemblies

4 Atomic structure and atomic spectra

Answers to discussion questions

D4.1 When hydrogen atoms are excited into higher energy levels, they emit energy at a series of discrete wavenumbers that show up as a series of lines in the spectrum. The lines in the Lyman series, for which the emission of energy corresponds to the final state $n_1 = 1$, are in the ultraviolet region of the electromagnetic spectrum. The Balmer series, for which $n_1 = 2$, is in the visible region. The Paschen series, $n_1 = 3$, is in the infrared. The Brackett series, $n_1 = 4$, is also in the infrared. The Pfund series, $n_1 = 5$, is in the far infrared. The Humphrey series, $n_1 = 6$, is in the very far infrared.

D4.3

(1) The principal quantum number, n, determines the energy of a hydrogenic atomic orbital through eqn. 4.8.

(2) The azimuthal quantum number, l, determines the magnitude of the angular momentum of a hydrogenic atomic orbital through the formula $\{l(l+1)\}^{1/2}\hbar$.

(3) The magnetic quantum number, m_l, determines the z-component of the angular momentum of a hydrogenic orbital through the formula $m_l\hbar$.

(4) The spin quantum number, s, determines the magnitude of the spin angular momentum through the formula $\{s(s+1)\}^{1/2}\hbar$. For hydrogenic atomic orbitals, s can only be $1/2$.

(5) The spin quantum number, m_s, determines the z-component of the spin angular momentum through the formula $m_s\hbar$. For hydrogenic atomic orbitals, m_s can only be $\pm 1/2$.

D4.5 The selection rules are

$$\Delta n = \pm 1, \pm 2, \ldots \qquad \Delta l = \pm 1 \qquad \Delta m_l = 0, \pm 1$$

In a spectroscopic transition the atom emits or absorbs a photon. Photons have a spin angular momentum of 1. Therefore, as a result of the transition, the angular momentum of the electromagnetic field has changed by $\pm 1\hbar$. The principle of the conservation of angular momentum then requires that the angular momentum of the atom has undergone an equal and opposite change in angular momentum. Hence, the selection rule on $\Delta l = \pm 1$. The principal quantum number n can change by any amount since n does not directly relate to angular momentum. The selection rule on Δm_l is harder to account for on the basis of these simple considerations alone. One has to evaluate the transition dipole moment between the wavefunctions representing the initial and final states involved in the transition. See *Justification* 4.4 for an example of this procedure.

The selection rules are strictly valid (1) only for hydrogenic atoms in which the states can be described by pure and exact wavefunctions and (2) if there are no other external perturbations (interactions) that could lead to a mixing of the states.

D4.7 See Fig. 4.8 of the text. The periodic table shows the manner in which the chemical and physical properties of an element, composed of atoms of one kind only, varies as a function of the atomic number of the element. This variation is periodic in the atomic numbers of the elements. The configuration of the electrons in an atom is what determines the chemical and physical properties of the atom (element); hence the periodicity of the properties is necessarily a result of the electron configurations.

The order of occupation of atomic orbitals, forming the electron configuration of the atom, mirrors the structure of the periodic table.

The s and p blocks form the main groups of the periodic table. The similar electron configurations for the elements in a main group are the reason for the similar properties of these elements. Each new period corresponds to the occupation of a shell with a higher principal quantum number. This correspondence explains the different lengths of the periods. Period 1 consists of only two elements, H and He, in which the single 1s orbital of the $n = 1$ shell is being filled with its two electrons. Period 2 consists of the eight elements Li through Ne, in which the one 2s and three 2p orbitals are being filled with eight more electrons. In Period 3 (Na through Ar), the 3s and 3p orbitals are being occupied by eight additional electrons. In Period 4, not only are the electrons of the 4s and 4p orbitals being added but so are the 10 electrons of the 3d orbitals. Hence there are 18 elements in Period 4. Period 5 elements add another 18 electrons as the 5s, 4d, and 5p orbitals are filled. In Period 6, a total of 32 electrons are added, because 14 electrons are also being added to the seven 4f orbitals.

D4.9 In the crudest form of the orbital approximation, the many-electron wavefunctions for atoms are represented as a simple product of one-electron wavefunctions. At a somewhat more sophisticated level, the many electron wavefunctions are written as linear combinations of such simple product functions that explicitly satisfy the Pauli exclusion principle. Relatively good one-electron functions are generated by the Hartree–Fock self-consistent field method described in Section 4.4(f). If we place no restrictions on the form of the one-electron functions, we reach the Hartree–Fock limit that gives us the best value of the calculated energy within the orbital approximation. The orbital approximation is based on the disregard of significant portions of the electron–electron interaction terms in the many-electron hamiltonian, so we cannot expect that it will be quantitatively accurate. By abandoning the orbital approximation, we could in principle obtain essentially exact energies; however, there are significant conceptual advantages to retaining the orbital approximation. Increased accuracy can be obtained by reintroducing the neglected electron–electron interaction terms and including their effects on the energies of the atom by a form of perturbation theory similar to that described in *Further Information* 2.1. For a more complete discussion consult the reference in Footnote 3.

D4.11 An electron has a magnetic moment and magnetic field due to its orbital angular momentum. It also has a magnetic moment and magnetic field due to its spin angular momentum. There is an interaction energy between magnetic moments and magnetic fields. The energy between the spin magnetic moment and the magnetic field generated by the orbital motion is called spin–orbit coupling. The energy of interaction is proportional to the scalar product of the two vectors representing the spin and orbital angular momenta and hence depends upon the orientation of the two vectors. See text Fig. 4.26. The total angular momentum of an electron in an atom is the vector sum of the orbital and spin angular momenta as expressed in eqn. 4.36. The spin–orbit coupling results in a splitting of the energy levels associated with atomic terms as shown on the top of text Fig. 4.28. This splitting shows up in atomic spectra as a fine structure as illustrated on the bottom of Fig. 4.28.

Solutions to exercises

E4.1(a) Eqn. 4.1 implies that the shortest wavelength corresponds to $n_2 = \infty$, the longest to $n_2 = 2$. Solve eqn. 4.1 for λ.

$$\lambda = \frac{(1/n_1^2 - 1/n_2^2)^{-1}}{R_H}$$

Shortest: $$\lambda = \frac{(1/1^2 - 1/\infty^2)^{-1}}{109677\text{cm}^{-1}} = \boxed{9.118\times10^{-6}\ \text{cm}}$$

Longest: $$\lambda = \frac{(1/1^2 - 1/2^2)^{-1}}{109677\text{cm}^{-1}} = \boxed{1.216\times10^{-5}\ \text{cm}}$$

E4.2(a) For atoms A, eqn. 4.8 may be rewritten in terms of the Rydberg constant R as

$$E_n = \frac{Z^2\mu_A hcR}{m_e n^2} \approx \frac{Z^2 hcR}{n^2}$$

where to within 0.01% the ratio μ_A/m_e is unity. Eqn. 4.1 can then be rewritten as

$$\tilde{\nu} = Z^2R\left(\frac{1}{n_1^2} - \frac{1}{n_2^2}\right) \qquad \lambda = \frac{1}{\tilde{\nu}} \qquad \nu = \frac{c}{\lambda}$$

$$\tilde{\nu} = 4\times 109737 \text{ cm}^{-1}\left(\frac{1}{1^2} - \frac{1}{2^2}\right) = \boxed{3.292\times 10^5 \text{ cm}^{-1}} \qquad \lambda = \boxed{3.038\times 10^{-6}\text{ cm}}$$

$$\nu = \frac{2.9978\times 10^{10} \text{ cm s}^{-1}}{3.0376\times 10^{-1} \text{ cm}} = \boxed{9.869\times 10^{15} \text{ s}^{-1}}$$

E4.3(a) Identify l and use angular momentum $= \{l(l+1)\}^{1/2}\hbar$.

(a) $l = 0$, so angular momentum $= 0$
(b) $l = 1$, so angular momentum $= \sqrt{2}h$
(c) $l = 3$, so angular momentum $= 2\sqrt{3}h$

The total number of nodes is equal to $n-1$ and the number of angular nodes is equal to l; hence the number of radial nodes is equal to $n-l-1$. We can draw up the following table.

	2s	3p	5f
n, l	2,0	3,1	5,3
Angular nodes	0	1	3
Radial nodes	1	1	1

E4.4(a) The ionization energy for one-electron atoms and ions (A) is given by the negative of eqn. 4.8 with $n = 1$.

$$I_A = \frac{Z^2\mu_A e^4}{32\pi^2\varepsilon_0^2\hbar^2 n^2}$$

This may be rewritten in terms of the Rydberg constant as

$$I_A = Z^2hcR_A = Z^2hc\frac{\mu_A}{m_e}R = Z^2\frac{\mu_A}{\mu_H}I_H$$

$I_H = 2.1788$ aJ. Reduced masses are calculated from eqn. 4.4, the ionization energies from the above equation.

μ_H	μ_{He^+}	μ_{He^+}/μ_H	I_{He^+}
5.4828×10^{-4} u	5.4850×10^{-4} u	1.00040	8.7188 aJ

E4.5(a) This is essentially the photoelectric effect [eqn. 1.2 of Section 1.2] with the ionization energy of the ejected electron being the work function Φ.

$$h\nu = \frac{1}{2}m_e v^2 + I$$

$$I = h\nu - \frac{1}{2}m_e v^2 = (6.626\times10^{-34}\,\mathrm{J\,Hz^{-1}})\times\left(\frac{2.998\times10^8\,\mathrm{m\,s^{-1}}}{58.4\times10^{-9}\,\mathrm{m}}\right)$$

$$-(\tfrac{1}{2})\times(9.109\times10^{-31}\,\mathrm{kg})\times(1.59\times10^6\,\mathrm{m\,s^{-1}})^2\times$$

$$= 2.25\times10^{-18}\,\mathrm{J}, \text{ corresponding to } \boxed{14.0\,\mathrm{eV}}$$

E4.6(a) The L shell corresponds to $n = 2$. For $n = 2$, $l = 1$ (p subshell) and 0 (s subshell). The degeneracies of these subshells are $2l + 1$. Hence the total degeneracy of the L shell is $\boxed{4}$.

E4.7(a) The energies are $E = -\frac{hcR_H}{n^2}$ [4.8 with 4.11], and the orbital degeneracy g of an energy level of principal quantum number n is

$$g = \sum_{l=0}^{n-1}(2l-1) = 1+3+5+\cdots+2n-1 = \frac{(1+2n-1)n}{2} = n^2$$

(a) $E = -hcR_H$ implies that $n = 1$, so $\boxed{g = 1}$ [the 1s orbital].

(b) $E = -\frac{hcR_H}{4}$ implies that $n = 2$, so $\boxed{g = 4}$ (2s orbital and three 2p orbitals).

(c) $E = -\frac{hcR_H}{16}$ implies that $n = 4$, so $\boxed{g = 16}$ (the 4s orbital, the three 4p orbitals, the five 4d orbitals, and the seven 4*f* orbitals).

E4.8(a)

$$R_{1,0} = N\mathrm{e}^{-r/a_0}$$

$$\int_0^\infty R^2 r^2 \mathrm{d}r = 1 = \int_0^\infty N^2 r^2 e^{-2r/a_0}\,\mathrm{d}r = N^2 \times \frac{2!}{\left(\frac{2}{a_0}\right)^3} = 1 \quad \left[\int_0^\infty x^n \mathrm{e}^{-ax}\mathrm{d}x = \frac{n!}{a^{n+1}}\right]$$

$$N^2 = \frac{4}{a_0^3}, \quad \boxed{N = \frac{2}{a_0^{3/2}}}$$

Thus,

$$R_{1,0} = 2\left(\frac{1}{a_0}\right)^{3/2}\mathrm{e}^{-r/a_0}$$

which agrees with Table 4.1.

E4.9(a)

$$R_{2,0} \propto \left(2 - \frac{\rho}{2}\right)\mathrm{e}^{-\rho/4} \quad \text{with } \rho = \frac{2r}{a_0} \text{ [Table 4.1]}$$

$$\frac{\mathrm{d}R}{\mathrm{d}r} = \frac{2}{a_0}\frac{\mathrm{d}R}{\mathrm{d}\rho} = \frac{2}{a_0}\left(-\frac{1}{2} - \frac{1}{2} + \frac{1}{8}\rho\right)\mathrm{e}^{-\rho/4} = 0 \quad \text{when } \rho = 8$$

Hence, the wavefunction has an extremum at $r = \boxed{4a_0}$. Because $2 - \frac{\rho}{2} < 0$, $\psi < 0$ and the extremum is a minimum (more formally: $\frac{\mathrm{d}^2\psi}{\mathrm{d}r^2} > 0$ at $\rho = 8$).

The second extremum is at $\boxed{r = 0.}$ It is not a minimum and in fact is a physical maximum, though not one that can be obtained by differentiation. To see that it is maximum, substitute $\rho = 0$ into $R_{2,0}$.

E4.10(a) The 2s radial function is proportional to $(2-\rho)e^{-\rho/2}$. This factor goes to zero when $\rho = 2$. Since $\rho = (2Z/na)r$, $r = na\rho/2Z$. If we assume $a = a_0$, then

$$r = 4a_0/2 = 2a_0 = 2\times 5.292\times 10^{-11}\,\text{m} = \boxed{1.058\times 10^{-10}\,\text{m}}$$

E4.11(a) We will solve this exercise by straightforward integration.

$$\psi_{1,0,0} = R_{1,0}Y_{0,0} = \left(\frac{Z^3}{\pi a_0^3}\right)^{1/2} e^{-r/a_0} \text{ [Tables 3.2 and 4.1]}$$

The potential energy operator is

$$V = -\frac{Ze^2}{4\pi\varepsilon_0}\times\left(\frac{1}{r}\right) = -k\left(\frac{1}{r}\right)$$

$$\langle V\rangle = -k\left\langle\frac{1}{r}\right\rangle\left[k = \frac{e^2}{4\pi\varepsilon_0}\right] = -k\int_0^\infty\int_0^\pi\int_0^{2\pi}\left(\frac{1}{\pi a_0^3}\right)e^{-r/a_0}\left(\frac{1}{r}\right)e^{-r/a_0}r^2\text{d}r\sin\theta\,\text{d}\theta\,\text{d}\phi$$

$$= -k\times(4\pi)\times\left(\frac{1}{\pi a_0^3}\right)\int_0^\infty re^{-2r/a_0}\text{d}r = -k\times\left(\frac{4}{a_0^3}\right)\times\left(\frac{a_0^2}{4}\right) = -k\left(\frac{1}{a_0}\right)$$

$$\left[\text{We have used }\int_0^\pi\sin\theta\,\text{d}\theta = 2,\ \int_0^{2\pi}\text{d}\theta = 2\pi,\text{ and }\int_0^\infty x^n e^{-ax}\text{d}x = \frac{n!}{a^{n+1}}.\right]$$

Hence,

$$\langle V\rangle = -\frac{Ze^2}{4\pi\varepsilon_0 a_0} = \boxed{2E_{1s}}$$

The kinetic energy operator is $-\frac{\hbar^2}{2\mu}\nabla^2$ [4.4 and *Further Information* 4.1]; hence

$$\langle E_K\rangle \equiv \langle T\rangle = \int\psi_{1s}^*\left(-\frac{\hbar^2}{2\mu}\right)\nabla^2\psi_{1s}\,\text{d}\tau$$

$$\nabla^2\psi_{1s} = \frac{1}{r}\frac{\partial^2(r\psi_{1s})}{\partial r^2} + \frac{1}{r^2}\Lambda^2\psi_{1s} \text{ [Problem 3.23]}$$

$$= \left(\frac{1}{\pi a_0^3}\right)^{1/2}\times\left(\frac{1}{r}\right)\times\left(\frac{\text{d}^2}{\text{d}r^2}\right)re^{-r/a_0}$$

$$[\Lambda^2\psi_{1s} = 0,\ \psi_{1s}\text{ contains no angular variables}]$$

$$= \left(\frac{Z^3}{\pi a_0^3}\right)^{1/2}\left[-\left(\frac{2}{a_0 r}\right)+\left(\frac{1}{a_0^2}\right)\right]e^{-r/a_0}$$

$$\langle T\rangle = -\left(\frac{\hbar^2}{2\mu}\right)\times\left(\frac{1}{\pi a_0^3}\right)\int_0^\infty\left[-\left(\frac{2}{a_0 r}\right)+\left(\frac{1}{a_0^2}\right)\right]e^{-2r/a_0}r^2\,\text{d}r\times\int_0^\pi\sin\theta\,\text{d}\theta\int_0^{2\pi}\text{d}\phi$$

$$= -\left(\frac{2\hbar^2}{\mu a_0^3}\right)\int_0^\infty\left[-\left(\frac{2r}{a_0}\right)+\left(\frac{r^2}{a_0^2}\right)\right]e^{-2r/a_0}\,\text{d}r = -\left(\frac{2\hbar^2}{\mu a_0^3}\right)\times\left(-\frac{a_0}{4}\right) = \frac{\hbar^2}{2\mu a_0^2}$$

$$= \boxed{-E_{1s}}$$

Hence, $\langle T\rangle + \langle V\rangle = -E_{1s} + 2E_{1s} = E_{1s}$.

We determined the ionization energy of the He^+ ion in the solution to Exercise 4.4(a) with the result $I(He^+) = 8.7188$ aJ. Using the relationships demonstrated above, we obtain for the potential and kinetic energies of the He^+ ion

$$V(\text{He}^+) = 2)E_{1s}(\text{He}^+) = -2I(\text{He}^+) = -2\times 8.7188\text{ aJ} = \boxed{-17.4376\text{ aJ}}$$

$$T(\text{He}^+) = -E_{1s}(\text{He}^+) = I(\text{He}^+) = \boxed{8.7188\text{ aJ}}$$

COMMENT. E_{1s} may also be written as

$$E_{1s} = -\frac{\mu e^4}{32\pi^2\varepsilon_0^2\hbar^2}$$

Question. Are the three different expressions for E_{1s} given in this exercise all equivalent?

E4.12(a) See Example 4.4. The mean radius is given by $\langle r\rangle = \int_0^\infty r^3 R_{n,l}^2 \mathrm{d}r$. After substituting eqn. 4.10 for $R_{n,l}$, and integrating, the general expression for the mean radius of a hydrogenic orbital with quantum numbers l and n is obtained (see eqn. 10.19 of Atkins and DePaula, *Physical Chemistry*, 8th edition).

$$\langle r_{n,l}\rangle = n^2\left\{1+\frac{1}{2}\left(1-\frac{l(l+1)}{n^2}\right)\right\}\frac{a_0}{Z}$$

For the 2s orbital $n = 2$ and $l = 0$; hence $\boxed{\langle r_{2s}\rangle = \frac{6a_0}{Z}}$.

See Example 4.5, which illustrates the calculation of the most probable radius. It is obtained by solving $\mathrm{d}P(r)/\mathrm{d}r = 0$ with $P(r) = r^2R(r)^2$. For the 2s orbital, $P(r) \propto r^2\left(2-\frac{Z}{a}r\right)^2 \mathrm{e}^{-\frac{Z}{a}r}$. Differentiating this expression with respect to r and solving for those values of r that make the derivative equal to zero is most readily performed with mathematical software. Here we use MathCad®.

$$r^2\cdot\left(2-Z\cdot\frac{r}{a}\right)^2\cdot e^{-Z\cdot\frac{r}{a}}$$

$$2\cdot r\cdot\left(2-Z\cdot\frac{r}{a}\right)^2\cdot\exp\left(-Z\cdot\frac{r}{a}\right)-2\cdot r^2\cdot\left(2-Z\cdot\frac{r}{a}\right)\cdot\exp\left(-Z\cdot\frac{r}{a}\right)\cdot\frac{Z}{a}-r^2\cdot\left(2-Z\cdot\frac{r}{a}\right)^2\cdot\frac{Z}{a}\cdot\exp\left(-Z\cdot\frac{r}{a}\right)=0 \text{ solve}, r \rightarrow \begin{bmatrix} 0 \\ 2\cdot\frac{a}{Z} \\ 2\cdot\left(\frac{3}{2}+\frac{1}{2}\cdot\sqrt{5}\right)\cdot\frac{a}{Z} \\ 2\cdot\left(\frac{3}{2}-\frac{1}{2}\cdot\sqrt{5}\right)\cdot\frac{a}{Z} \end{bmatrix}$$

The first and second of these roots correspond to minima in the probability. The third and fourth roots correspond to maxima, with the third being the largest, and hence it is the most probable value of r. This may be rewritten as $\boxed{r^* = (3+\sqrt{5})\frac{a}{Z}}$.

E4.13(a) The radial distribution function is defined as

$$P = 4\pi r^2\psi^2 \quad \text{so} \quad P_{3s} = 4\pi r^2(Y_{0,0}R_{3,0})^2$$

$$\boxed{P_{3s} = 4\pi r^2\left(\frac{1}{4\pi}\right)\times\left(\frac{1}{243}\right)\times\left(\frac{Z}{\alpha_0}\right)^3\times(6-6\rho+\rho^2)^2\mathrm{e}^{-\rho}}$$

where $\rho \equiv \frac{2Zr}{na_0} = \frac{2Zr}{3a_0}$. But we want to find the most likely radius, so it would help to simplify the function by expressing it in terms of either r or ρ but not both. To find the most likely radius, we could set the derivative of P_{3s} equal to zero; therefore, we can collect all multiplicative constants together (including the factors of a_0/Z needed to turn the initial r^2 into ρ^2) because they will eventually be divided into zero.

$$P_{3s} = C^2\rho^2(6-6\rho+\rho^2)^2 e^{-\rho}$$

Note that not all the extrema of P are maxima; some are minima. But all the extrema of $(P_{3s})^{1/2}$ correspond to maxima of P_{3s}. So let us find the extrema of $(P_{3s})^{1/2}$.

$$\frac{d(P_{3s})^{1/2}}{d\rho} = 0 = \frac{d}{d\rho}C\rho(6-6\rho+\rho^2)e^{-\rho/2}$$
$$= C[\rho(6-6\rho+\rho^2)\times(-\tfrac{1}{2})+(6-12\rho+3\rho]e^{-\rho/2}$$

$$0 = C\left(6-15\rho+6\rho^2-\tfrac{1}{2}\rho^3\right)e^{-\rho/2} \quad \text{so} \quad 12-30\rho+12\rho^2-\rho^3=0$$

Numerical solution of this cubic equation yields

$$\rho = 0.49,\ 2.79,\ \text{and } 8.72$$

corresponding to

$$r = \boxed{0.74a_0/Z, 4.19a_0/Z, \text{ and } 13.08a_0/Z}$$

COMMENT. If numerical methods are to be used to find the roots of the equation that locates the extrema, then graphical/numerical methods might as well be used to locate the maxima directly. That is, the student may simply have a spreadsheet compute P_{3s} and examine or manipulate the spreadsheet to locate the maxima.

E4.14(a) See Fig. 4.14 of the text. The number of angular nodes is the value of the quantum number l, which for p orbitals is 1. Hence, each of the three p orbitals has one angular node. To locate the angular nodes look for the value of θ that makes the wavefunction zero.

p_z orbital: see eqn. 4.17(a) and Fig. 4.14. The nodal plane is the $\boxed{xy \text{ plane}}$ and $\boxed{\theta = \pi/2}$ is an angular node.

p_x orbital: see eqns. 4.18(a) and 4.18(b) and Fig. 4.14. The nodal plane is the $\boxed{yz}$ plane and $\boxed{\theta = 0}$ is an angular node.

p_y orbital: see eqns. 4.18(a) and 4.18(c) and Fig. 4.14. The nodal plane is the $\boxed{xz}$ plane and $\boxed{\theta = 0}$ is an angular node.

E4.15(a) The selection rules for a many-electron atom are given by the set [4.39]. For a single-electron transition these amount to Δn = any integer; $\Delta l = \pm 1$. Hence

(a) $3s \rightarrow 1s$; $\Delta l = 0$, $\boxed{\text{forbidden}}$

(b) $3p \rightarrow 2s$; $\Delta l = -1$, $\boxed{\text{allowed}}$

(c) $5d \rightarrow 2p$; $\Delta l = -1$, $\boxed{\text{allowed}}$

E4.16(a) A source approaching an observer appears to be emitting light of frequency

$$\nu_{\text{approaching}} = \frac{\nu}{1-\frac{s}{c}} \quad [4.26b]$$

Since $\nu \propto \dfrac{1}{\lambda}$, $\lambda_{\text{obs}} = \left(1-\dfrac{s}{c}\right)\lambda$

$s = 60\ \text{km h}^{-1} = 16.\bar{7}\ \text{m s}^{-1}$. Hence

$$\lambda_{\text{obs}} = \left(1-\frac{16.\bar{7}\ \text{m s}^{-1}}{2.998\times10^8\ \text{ms}^{-1}}\right)\times(680\ \text{nm}) = \boxed{0.999999944\times680\ \text{nm}}$$

For all practical purposes, there is no shift at all at this speed.

E4.17(a) $\delta\tilde{\nu} \approx \dfrac{5.31\,\text{cm}^{-1}}{\tau/\text{ps}}$ [4.28b], implying that $\tau \approx \dfrac{5.31\text{ ps}}{\delta\tilde{\nu}/\text{cm}^{-1}}$

(a) $\tau \approx \dfrac{5.31\,\text{ps}}{0.20} = \boxed{27\,\text{ps}}$

(b) $\tau \approx \dfrac{5.31\,\text{ps}}{2.0} = \boxed{2.7\,\text{ps}}$

E4.18(a) $\delta\tilde{\nu} \approx \dfrac{5.31\,\text{cm}^{-1}}{\tau/\text{ps}}$ [4.28b]

(a) $\tau \approx 1.0\times 10^{-13}\text{ s} = 0.10\text{ ps}$, implying that $\delta\tilde{\nu} \approx \boxed{53\,\text{cm}^{-1}}$

(b) $\tau \approx (100)\times(1.0\times 10^{-13}\text{ s}) = 10\text{ ps}$, implying that $\delta\tilde{\nu} \approx \boxed{0.53\,\text{cm}^{-1}}$

E4.19(a)

$$\Psi(1,2,\ldots,12) = \frac{1}{(12!)^{1/2}}\begin{vmatrix}
1s\alpha(1) & 1s\beta(1) & 2s\alpha(1) & 2s\beta(1) & 2p_1\alpha(1) & 2p_1\beta(1) & 2p_0\alpha(1) & 2p_0\beta(1) & 2p_{-1}\alpha(1) & 2p_{-1}\beta(1) & 3s\alpha(1) & 3s\beta(1) \\
1s\alpha(2) & 1s\beta(2) & 2s\alpha(2) & 2s\beta(2) & 2p_1\alpha(2) & 2p_1\beta(2) & 2p_0\alpha(2) & 2p_0\beta(2) & 2p_{-1}\alpha(2) & 2p_{-1}\beta(2) & 3s\alpha(2) & 3s\beta(2) \\
1s\alpha(3) & 1s\beta(3) & 2s\alpha(3) & 2s\beta(3) & 2p_1\alpha(3) & 2p_1\beta(3) & 2p_0\alpha(3) & 2p_0\beta(3) & 2p_{-1}\alpha(3) & 2p_{-1}\beta(3) & 3s\alpha(3) & 3s\beta(3) \\
1s\alpha(4) & 1s\beta(4) & 2s\alpha(4) & 2s\beta(4) & 2p_1\alpha(4) & 2p_1\beta(4) & 2p_0\alpha(4) & 2p_0\beta(4) & 2p_{-1}\alpha(4) & 2p_{-1}\beta(4) & 3s\alpha(4) & 3s\beta(4) \\
1s\alpha(5) & 1s\beta(5) & 2s\alpha(5) & 2s\beta(5) & 2p_1\alpha(5) & 2p_1\beta(5) & 2p_0\alpha(5) & 2p_0\beta(5) & 2p_{-1}\alpha(5) & 2p_{-1}\beta(5) & 3s\alpha(5) & 3s\beta(5) \\
1s\alpha(6) & 1s\beta(6) & 2s\alpha(6) & 2s\beta(6) & 2p_1\alpha(6) & 2p_1\beta(6) & 2p_0\alpha(6) & 2p_0\beta(6) & 2p_{-1}\alpha(6) & 2p_{-1}\beta(6) & 3s\alpha(6) & 3s\beta(6) \\
1s\alpha(7) & 1s\beta(7) & 2s\alpha(7) & 2s\beta(7) & 2p_1\alpha(7) & 2p_1\beta(7) & 2p_0\alpha(7) & 2p_0\beta(7) & 2p_{-1}\alpha(7) & 2p_{-1}\beta(7) & 3s\alpha(7) & 3s\beta(7) \\
1s\alpha(8) & 1s\beta(8) & 2s\alpha(8) & 2s\beta(8) & 2p_1\alpha(8) & 2p_1\beta(8) & 2p_0\alpha(8) & 2p_0\beta(8) & 2p_{-1}\alpha(8) & 2p_{-1}\beta(8) & 3s\alpha(8) & 3s\beta(8) \\
1s\alpha(9) & 1s\beta(9) & 2s\alpha(9) & 2s\beta(9) & 2p_1\alpha(9) & 2p_1\beta(9) & 2p_0\alpha(9) & 2p_0\beta(9) & 2p_{-1}\alpha(9) & 2p_{-1}\beta(9) & 3s\alpha(9) & 3s\beta(9) \\
1s\alpha(10) & 1s\beta(10) & 2s\alpha(10) & 2s\beta(10) & 2p_1\alpha(10) & 2p_1\beta(10) & 2p_0\alpha(10) & 2p_0\beta(10) & 2p_{-1}\alpha(10) & 2p_{-1}\beta(10) & 3s\alpha(10) & 3s\beta(10) \\
1s\alpha(11) & 1s\beta(11) & 2s\alpha(11) & 2s\beta(11) & 2p_1\alpha(11) & 2p_1\beta(11) & 2p_0\alpha(11) & 2p_0\beta(11) & 2p_{-1}\alpha(11) & 2p_{-1}\beta(11) & 3s\alpha(11) & 3s\beta(11) \\
1s\alpha(12) & 1s\beta(12) & 2s\alpha(12) & 2s\beta(12) & 2p_1\alpha(12) & 2p_1\beta(12) & 2p_0\alpha(12) & 2p_0\beta(12) & 2p_{-1}\alpha(12) & 2p_{-1}\beta(12) & 3s\alpha(12) & 3s\beta(12)
\end{vmatrix}$$

E4.20(a) The ground state configuration of the carbon atom is $1s^2 2s^2 2p^2$. The normal valence of carbon is 4 and each of the electrons with principal quantum number 2 can participate in bonding. In the free carbon atom, the three 2p orbitals are degenerate; hence we have no knowledge about in which two of the three 2p orbitals the electrons actually reside. We do know that the two 2p electrons are unpaired and that the total spin is 1, but whether they are both $m_s = +1/2$ or both $m_s = -1/2$ is unknown. The table below summarizes one possible set of quantum numbers.

Electron	n	l	m_l	s	m_s
1	2	0	0	1/2	+1/2
2	2	0	0	1/2	−1/2
3	2	1	1	1/2	+1/2
4	2	1	0	1/2	+1/2

E4.21(a) Sc: $[\text{Ar}]4s^2 3d^1$

Ti: $[\text{Ar}]4s^2 3d^2$

V: $[\text{Ar}]4s^2 3d^3$

Cr: $[\text{Ar}]4s^2 3d^4$ or $[\text{Ar}]4s^1 3d^5$ (most probable)

Mn: $[\text{Ar}]4s^2 3d^5$

Fe: $[\text{Ar}]4s^2 3d^6$

Co: $[\mathrm{Ar}]4s^2 3d^7$

Ni: $[\mathrm{Ar}]4s^2 3d^8$

Cu: $[\mathrm{Ar}]4s^2 3d^9$ or $[\mathrm{Ar}]4s^1 3d^{10}$ (most probable)

Zn: $[\mathrm{Ar}]4s^2 3d^{10}$

E4.22(a) (a) Pd^{2+}: $\boxed{[\mathrm{Kr}]4d^8}$

(b) All subshells except 4d are filled and hence have no net spin. Applying Hund's rule to $4d^8$ shows that there are two unpaired spins. The paired spins do not contribute to the net spin; hence we consider only $s_1 = \frac{1}{2}$ and $s_2 = \frac{1}{2}$. The Clebsch–Gordan series [4.35] produces

$$S = s_1 + s_2, \ldots, |s_1 - s_2|, \quad \text{hence } \boxed{S = 1, 0}$$

$$M_S = -S, -S+1, \ldots, S$$

For $S = 1$, $\boxed{M_S = -1, 0, +1}$

$S = 0$, $\boxed{M_S = 0}$

E4.23(a) We use the Clebsch–Gordan series [4.36] in the form

$$j = l + s,\ l + s - 1, \ldots, |l - s| \qquad \text{[lowercase for a single electron]}$$

(a) $l = 2,\quad s = \frac{1}{2};\quad \text{so } j = \boxed{\frac{5}{2}, \frac{3}{2}}$

(b) $l = 3,\quad s = \frac{1}{2};\quad \text{so } j = \boxed{\frac{7}{2}, \frac{5}{2}}$

E4.24(a) The Clebsch–Gordan series for $\boxed{l = 3 \text{ or } 2}$ and $s = \frac{1}{2}$ leads to $j = \frac{5}{2}$.

The Clebsch–Gordan series for $\boxed{l = 1 \text{ or } 0}$ and $s = \frac{1}{2}$ leads to $j = \frac{1}{2}$.

E4.25(a) Use the Clebsch–Gordan series in the form

$$J = j_1 + j_2, j_1 + j_2 - 1, \ldots, |j_1 - j_2|$$

Then, with $j_1 = 1$ and $j_2 = 2$,

$$J = \boxed{3, 2, 1}$$

E4.26(a) The letter P indicates that $L = 1$, the superscript 3 is the value of $2S + 1$, so $S = 1$ and the subscript 2 is the value of J. Hence, $\boxed{L = 1,\ S = 1,\ J = 2}$.

E4.27(a) Use the Clebsch–Gordan series in the form

$$S' = s_1 + s_2, s_1 + s_2 - 1, \ldots, |s_1 - s_2|$$

and

$$S = S' + s_1, S' + s_1 - 1, \ldots, |S' - s_1|$$

in succession. The multiplicity is $2S + 1$.

(a) $S = \frac{1}{2} + \frac{1}{2}, \frac{1}{2} - \frac{1}{2} = \boxed{1, 0}$ with multiplicities $\boxed{3, 1}$ respectively.

(b) $S' = 1, 0$; then $S = \boxed{\frac{3}{2}, \frac{1}{2}}$ [from 1], $\boxed{\text{and } \frac{1}{2}}$ [from 0], with multiplicities $\boxed{4, 2, 2}$.

E4.28(a) These electrons are not equivalent (different subshells), hence all the terms that arise from the vector model and the Clebsch–Gordan series are allowed (Example 4.6).

$$L = l_1 + l_2, \ldots, |l_1 - l_2| \; [4.34] = 2 \text{ only}$$

$$S = s_1 + s_2, \ldots, |s_1 - s_2| = 1, 0$$

The allowed terms are then $^3\mathrm{D}$ and $^1\mathrm{D}$. The possible values of J are given by

$$J = L + S, \ldots, |L - S| \, [4.36] = 3, 2, 1 \text{ for } {}^3\mathrm{D} \text{ and } 2 \text{ for } {}^1\mathrm{D}$$

The allowed complete term symbols are then

$$\boxed{{}^3\mathrm{D}_3, \ {}^3\mathrm{D}_2, \ {}^3\mathrm{D}_1, \ {}^1\mathrm{D}_2}$$

The $\boxed{^3\text{D set of terms are the lower in energy}}$ [Hund's rule].

COMMENT. Hund's rule in the form given in the text does not allow the energies of the triplet terms to be distinguished. Experimental evidence indicates that $^3\mathrm{D}_1$ is lowest.

E4.29(a) Use the Clebsch–Gordan series in the form

$$J = L + S, L + S - 1, \ldots, |L - S|$$

The number of states (M_J values) is $2J + 1$ in each case.

(a) $L = 0, S = 1$; hence $\boxed{J = 1}$ ($^3\mathrm{S}_1$) and there are $\boxed{\text{3 states}}$.

(b) $L = 2, S = \frac{1}{2}$; hence $\boxed{J = 5/2, 3/2 \text{ respectively}}$ $\left({}^2\mathrm{D}_{5/2}, {}^2\mathrm{D}_{3/2}\right)$ with $\boxed{\text{6, and 2 states respectively}}$.

(c) $L = 1, S = 0$; hence $\boxed{J = 1}$ ($^1\mathrm{P}_1$) with $\boxed{\text{3 states}}$.

E4.30(a) Closed shells and subshells do not contribute to either L or S and thus are ignored in what follows.

(a) Na: $[\mathrm{Ne}]3\mathrm{s}^1: S = \frac{1}{2}, \ L = 0; \ J = \frac{1}{2}$, so the only term is $\boxed{^2\mathrm{S}_{1/2}}$

(b) K: $[\mathrm{Ar}]3\mathrm{d}^1: S = \frac{1}{2}, \ L = 2; \ J = \frac{5}{2}, \frac{3}{2}$, so the terms are $\boxed{^2\mathrm{D}_{5/2} \text{ and } {}^2\mathrm{D}_{3/2}}$

E4.31(a) See eqn. 4.39 for the selection rules.

(a) $\boxed{\text{allowed}}$

(b) $\boxed{\text{forbidden}}$

(c) $\boxed{\text{allowed}}$

Solutions to problems

Solutions to numerical problems

P4.1 All lines in the hydrogen spectrum fit the Rydberg formula

$$\frac{1}{\lambda} = R_\mathrm{H}\left(\frac{1}{n_1^2} - \frac{1}{n_2^2}\right) \quad \left[4.1, \text{ with } \tilde{\nu} = \frac{1}{\lambda}\right] \quad R_\mathrm{H} = 109677 \,\mathrm{cm}^{-1}$$

Find n_1 from the value of $\lambda_{\max}$, which arises from the transition $n_1+1 \rightarrow n_1$

$$\frac{1}{\lambda_{\max} R_{\mathrm{H}}}=\frac{1}{n_1^2}-\frac{1}{(n_1+1)^2}=\frac{2n_1+1}{n_1^2(n_1+1)^2}$$

$$\lambda_{\max} R_{\mathrm{H}}=\frac{n_1^2(n_1+1)^2}{2n_1+1}=(12368\times10^{-9}\ \mathrm{m})\times(109677\times10^2\ \mathrm{m}^{-1})=135.65$$

Since $n_1=1, 2, 3,$ and 4 have already been accounted for, try $n_1=5, 6,\ldots$. With $n_1=6$ we get $\frac{n_1^2(n_1+1)^2}{2n_1+1}=136$. Hence, the Humphreys series is $\boxed{n_2 \rightarrow 6}$ and the transitions are given by

$$\frac{1}{\lambda}=(109677\ \mathrm{cm}^{-1})\times\left(\frac{1}{36}-\frac{1}{n_2^2}\right), \quad n_2=7,8,\ldots$$

and occur at 12372 nm, 7503 nm, 5908 nm, 5129 nm, …, 3908 nm (at $n_2=15$), converging to 3282 nm as $n_2 \rightarrow \infty$, in agreement with the quoted experimental result.

P4.3 A Lyman series corresponds to $n_1=1$; hence,

$$\tilde{\nu}=R_{\mathrm{Li}^{2+}}\left(1-\frac{1}{n^2}\right), \quad n=2,3,\ldots \quad \left[\tilde{\nu}=\frac{1}{\lambda}\right]$$

Therefore, if the formula is appropriate, we expect to find that $\tilde{\nu}\left(1-\frac{1}{n^2}\right)^{-1}$ is a constant ($R_{\mathrm{Li}^{2+}}$). We therefore draw up the following table.

n	2	3	4
$\tilde{\nu}/\mathrm{cm}^{-1}$	740747	877924	925933
$\tilde{\nu}\left(1-\frac{1}{n^2}\right)^{-1}/\mathrm{cm}^{-1}$	987663	987665	987662

Hence, the formula does describe the transitions, and $\boxed{R_{\mathrm{Li}^{2+}}=987663\ \mathrm{cm}^{-1}}$. The Balmer transitions lie at

$$\tilde{\nu}=R_{\mathrm{Li}^{2+}}\left(\frac{1}{4}-\frac{1}{n^2}\right) \quad n=3,4,\ldots$$

$$=(987663\ \mathrm{cm}^{-1})\times\left(\frac{1}{4}-\frac{1}{n^2}\right)=\boxed{137175\ \mathrm{cm}^{-1}}, \boxed{185187\ \mathrm{cm}^{-1}},\ldots$$

The ionization energy of the ground-state ion is given by

$$\tilde{\nu}=R_{\mathrm{Li}^{2+}}\left(1-\frac{1}{n^2}\right), \quad n \rightarrow \infty$$

and hence corresponds to

$$\tilde{\nu}=987663\ \mathrm{cm}^{-1}, \quad \text{or} \quad \boxed{122.5\ \mathrm{eV}}$$

P4.5 The $7p$ configuration has just one electron outside a closed subshell. That electron has $l=1$, $s=1/2$, and $j=1/2$ or $3/2$, so the atom has $L=1$, $S=1/2$, and $J=1/2$ or $3/2$. The term symbols are $\boxed{{}^2\mathrm{P}_{1/2}\ \text{and}\ {}^2\mathrm{P}_{3/2}}$, of which the former has the lower energy. The $6d$ configuration also has just one electron outside a closed subshell; that electron has $l=2$, $s=1/2$, and $j=3/2$ or $5/2$, so the atom has $L=2$, $S=1/2$, and $J=3/2$ or $5/2$. The term symbols are $\boxed{{}^2\mathrm{D}_{3/2}\ \text{and}\ {}^2\mathrm{D}_{5/2}}$, of which the former has the lower energy. According to the simple treatment of spin–orbit coupling, the energy is given by

$$E_{l,s,j}=\tfrac{1}{2}\lambda h^2[j(j+1)-l(l+1)-s(s+1)]$$

where λ is the spin–orbit coupling constant. So

$$E(^2\mathrm{P}_{1/2}) = \tfrac{1}{2}\lambda\hbar^2[\tfrac{1}{2}(1/2+1) - 1(1+1) - \tfrac{1}{2}(1/2+1)] = -\lambda\hbar^2$$

and $$E(^2\mathrm{D}_{3/2}) = \tfrac{1}{2}\lambda\hbar^2[\tfrac{3}{2}(3/2+1) - 2(2+1) - \tfrac{1}{2}(1/2+1)] = -\tfrac{3}{2}\lambda\hbar^2$$

This approach would predict the ground state to be $\boxed{^2\mathrm{D}_{3/2}}$.

COMMENT. The computational study cited finds the $^2\mathrm{P}_{1/2}$ level to be lowest, but the authors caution that the error of similar calculations on Y and Lu is comparable to the computed difference between levels.

P4.7 $$R_\mathrm{H} = k\mu_\mathrm{H}, \quad R_\mathrm{D} = k\mu_\mathrm{D}, \quad R = k\mu \text{ [4.12]}$$

where R corresponds to an infinitely heavy nucleus, with $\mu = m_\mathrm{e}$.

Since $$\mu = \frac{m_\mathrm{e}m_\mathrm{N}}{m_\mathrm{e}+m_\mathrm{N}} [\mathrm{N} = \mathrm{p \text{ or } d}]$$

$$R_\mathrm{H} = k\mu_\mathrm{H} = \frac{km_\mathrm{e}}{1+\frac{m_\mathrm{e}}{m_\mathrm{p}}} = \frac{R}{1+\frac{m_\mathrm{e}}{m_\mathrm{p}}}$$

Likewise, $R_\mathrm{D} = \dfrac{R}{1+\frac{m_\mathrm{e}}{m_\mathrm{d}}}$, where m_p is the mass of the proton and m_d the mass of the deuteron. The two lines in question lie at

$$\frac{1}{\lambda_\mathrm{H}} = R_\mathrm{H}\left(1-\frac{1}{4}\right) = \tfrac{3}{4}R_\mathrm{H} \qquad \frac{1}{\lambda_\mathrm{D}} = R_\mathrm{D}(1-\tfrac{1}{4}) = \tfrac{3}{4}R_\mathrm{D}$$

and hence

$$\frac{R_\mathrm{H}}{R_\mathrm{D}} = \frac{\lambda_\mathrm{D}}{\lambda_\mathrm{H}} = \frac{\tilde{\nu}_\mathrm{H}}{\tilde{\nu}_\mathrm{D}}$$

Then, since

$$\frac{R_\mathrm{H}}{R_\mathrm{D}} = \frac{1+\frac{m_\mathrm{e}}{m_\mathrm{d}}}{1+\frac{m_\mathrm{e}}{m_\mathrm{p}}}, \quad m_\mathrm{d} = \frac{m_\mathrm{e}}{\left(1+\frac{m_\mathrm{e}}{m_\mathrm{p}}\right)\frac{R_\mathrm{H}}{R_\mathrm{D}} - 1}$$

we can calculate m_d from

$$m_\mathrm{d} = \frac{m_\mathrm{e}}{\left(1+\frac{m_\mathrm{e}}{m_\mathrm{p}}\right)\frac{\lambda_\mathrm{D}}{\lambda_\mathrm{H}} - 1} = \frac{m_\mathrm{e}}{\left(1+\frac{m_\mathrm{e}}{m_\mathrm{p}}\right)\frac{\tilde{\nu}_\mathrm{H}}{\tilde{\nu}_\mathrm{D}} - 1}$$

$$= \frac{9.10939\times10^{-31}\,\mathrm{kg}}{\left(1+\frac{9.1039\times10^{-31}\,\mathrm{kg}}{1.67262\times10^{-27}\,\mathrm{kg}}\right)\times\left(\frac{82259.098\,\mathrm{cm}^{-1}}{82281.476\,\mathrm{cm}^{-1}}\right)^{-1}} = \boxed{3.3429\times10^{-27}\,\mathrm{kg}}$$

Since $I = Rhc$,

$$\frac{I_\mathrm{D}}{I_\mathrm{H}} = \frac{R_\mathrm{D}}{R_\mathrm{H}} = \frac{\tilde{\nu}_\mathrm{D}}{\tilde{\nu}_\mathrm{H}} = \frac{82281.476\ \mathrm{cm}^{-1}}{82259.098\ \mathrm{cm}^{-1}} = \boxed{1.000272}$$

P4.9 (a) The splitting of adjacent energy levels is related to the difference in wavenumber of the spectral lines as follows:

$$hc\Delta\tilde{\nu} = \Delta E = \mu_\mathrm{B}\mathcal{B}, \quad \text{so} \quad \Delta\tilde{\nu} = \frac{\mu_\mathrm{B}\mathcal{B}}{hc} = \frac{(9.274\times10^{-24}\ \mathrm{J\,T^{-1}})(2\ \mathrm{T})}{(6.626\times10^{-34}\ \mathrm{J\,s})(2.998\times10^{10}\ \mathrm{cm\,s^{-1}})}$$

$$\Delta\tilde{\nu} = \boxed{0.9\ \mathrm{cm}^{-1}}$$

(b) Transitions induced by absorbing visible light have wavenumbers in the tens of thousands of reciprocal centimeters, so normal Zeeman splitting is $\boxed{\text{small}}$ compared to the difference in energy of the states involved in the transition. Take a wavenumber from the middle of the visible spectrum as typical:

$$\tilde{\nu}=\frac{1}{\lambda}=\frac{1}{600\ \text{nm}}\left(\frac{10^9\ \text{nm m}^{-1}}{10^2\ \text{cm m}^{-1}}\right)=1.7\times10^4\ \text{cm}^{-1}$$

Or take the Balmer series as an example, as suggested in the problem; the Balmer wavenumbers are (eqn. 4.1)

$$\tilde{\nu}=R_{\text{H}}\left(\frac{1}{2^2}-\frac{1}{n^3}\right)$$

The smallest Balmer wavenumber is

$$\tilde{\nu}=(109677\ \text{cm}^{-1})\times(1/4-1/9)=15233\ \text{cm}^{-1}$$

and the upper limit is

$$\tilde{\nu}=(109677\ \text{cm}^{-1})\times(1/4-0)=27419\ \text{cm}^{-1}$$

P4.11 (a) $^2\text{S}_{1/2}\rightarrow{}^2\text{P}_{3/2}$ (or $^2\text{P}_{1/2}$)

(b) See the solution to Exercise 4.2(a).

For atoms A, eqn. 4.8 may be rewritten in terms of the Rydberg constant R as

$$E_n=\frac{Z^2\mu_{\text{A}}hcR}{m_{\text{e}}n^2}\approx\frac{Z^2hcR}{n^2}$$

where to within 0.01% the ratio $\mu_{\text{A}}/m_{\text{e}}$ is unity. Eqn. 4.1 can then be rewritten as

$$\tilde{\nu}=Z^2R\left(\frac{1}{n_1^2}-\frac{1}{n_2^2}\right)\qquad\lambda=\frac{1}{\tilde{\nu}}\qquad\nu=\frac{c}{\lambda}$$

$$\tilde{\nu}=4\times109737\ \text{cm}^{-1}\left(\frac{1}{1^2}-\frac{1}{4^2}\right)=\boxed{4.115\times10^5\ \text{cm}^{-1}}\qquad\lambda=\boxed{2.430\times10^{-6}\ \text{cm}}$$

$$\nu=\frac{2.9978\times10^{10}\ \text{cm s}^{-1}}{2.430\times10^{-6}\ \text{cm}}=\boxed{1.234\times10^{16}\ \text{s}^{-1}}$$

(c) See the solution to Exercise 4.12(a). The formula for the mean radius is

$$\langle r_{n,l}\rangle=n^2\left\{1+\frac{1}{2}\left(1-\frac{l(l+1)}{n^2}\right)\right\}\frac{a_0}{Z}$$

$$\langle r_{4,1}\rangle=\boxed{\frac{23}{2}a_0}$$

Hence, since $\langle r_{1,0}\rangle=\boxed{\frac{3}{4}a_0}$, the mean radius has increased by $\boxed{\frac{43}{4}a_0}$.

P4.13 The ground state wavefunction for a hydrogen-like atom is

$$\psi_{1,0,0}=R_{1,0}Y_{0,0}=\left(\frac{Z^3}{\pi a_0^3}\right)^{1/2}\text{e}^{-Zr/a_0}\ \text{[Tables 3.2 and 4.1]}$$

This may be rewritten as $\psi_{1,0,0}=R_{1,0}Y_{0,0}=\left(\frac{b^3}{\pi}\right)^{1/2}\text{e}^{-br}$ where $b=Z/a_0$.

(a) For the probability of being in a sphere of radius R centered on the nucleus, allow for the r dependence of ψ:

$$P = \int_{\text{volume}} \psi^2 \mathrm{d}\tau = (b^3/\pi)\int_0^{2\pi}\mathrm{d}\phi\int_0^{\pi}\sin\theta\mathrm{d}\theta\int_0^R r^2\mathrm{d}r(\mathrm{e}^{-2br})$$

$$= (b^3/\pi)(2\pi)(2)\int_0^R r^2\mathrm{e}^{-2br}\mathrm{d}r = 1-[1+2br+2(br)^2]\mathrm{e}^{-2br}$$

For $Z = 1$, $b = (1/53\text{ pm})$, and for $R = 53$, pm $= 1/b$.

$$P = 1-[1+2+2]\mathrm{e}^{-2} = 1-5\mathrm{e}^{-2} = \boxed{0.323}$$

(b) We find the 90% boundary surface by solving

$$0.90 = 1-[1+2br+2(br)^2]\mathrm{e}^{-2br}$$

That is, solve

$$[1+2br+2(br)^2]\mathrm{e}^{-2br} = 0.10$$

Using mathematical software, for example, MathCad®, this solves to $bR = 2.66$, so $R = \boxed{141\text{ pm}}$.

Solutions to theoretical problems

P4.15 (a) Consider $\psi_{2p_z} = \psi_{2,1,0}$, which extends along the z-axis. The most probable point along the z-axis is where the radial function has its maximum value (for ψ^2 is also a maximum at that point). From Table 4.1 we know that

$$R_{21} \propto \rho\mathrm{e}^{-\rho/2}$$

$$\text{so } \frac{\mathrm{d}R}{\mathrm{d}\rho} = \left(1-\tfrac{1}{2}\rho\right)\mathrm{e}^{-\rho/2} = 0 \text{ when } \rho = 2.$$

Therefore, $r^* = \dfrac{2a_0}{Z}$, and the point of maximum probability lies at $z = \pm\dfrac{2a_0}{Z} = \boxed{\pm 106\,\text{pm}}$.

(b) The most probable radius occurs when the radial distribution function is a maximum. At this point the derivative of the function w/r/t either r or ρ equals zero.

$$\left(\frac{\mathrm{d}R_{31}}{\mathrm{d}\rho}\right)_{\max} = 0 = \left(\frac{\mathrm{d}((4-\rho)\rho\mathrm{e}^{-\rho/2})}{\mathrm{d}\rho}\right)_{\max} [\text{Table 4.1}] = \left(4-4\rho+\frac{\rho^2}{2}\right)\mathrm{e}^{-\rho/2}$$

The function is a maximum when the polynomial equals zero. The quadratic equation gives the roots $\rho = 4+2\sqrt{2} = 6.89$ and $\rho = 4-2\sqrt{2} = 1.17$. Since $\rho = (2Z/na_0)r$ and $n = 3$, these correspond to $r = 10.3\times a_0/Z$ and $r = 1.76\times a_0/Z$. However, $\left|\dfrac{R_{31}(\rho_1)}{R_{31}(\rho_2)}\right| = \left|\dfrac{R_{31}(1.17)}{R_{31}(10.3)}\right| = 4.90$. So we conclude that the function is a maximum at $\rho = \pm 1.17$, which corresponds to $\boxed{r = \pm 1.76\, a_0/Z}$. See the solutions to Exercises 4.12(b) and 4.13(b) for the most probable radii of $2p$ and $3p$ electrons. The results are $4a_0/Z$ and $12a_0/Z$, respectively, both much larger than the points of maximum probability.

COMMENT. Since the radial portion of any 2p or any 3p function is the same, the same result would have been obtained for all of them. The direction of the most probable point would, however, be different.

P4.17 (a) We must show that $\int|\psi_{3p_x}|^2\,d\tau=1$. The integrations are most easily performed in spherical coordinates (Fig. 3.12).

$$\int|\psi_{3p_x}|^2\,d\tau=\int_0^{2\pi}\int_0^{\pi}\int_0^{\infty}\left|\psi_{3p_x}\right|^2 r^2\sin(\theta)\,dr\,d\theta\,d\phi$$

$$=\int_0^{2\pi}\int_0^{\pi}\int_0^{\infty}\left|R_{31}(\rho)\left\{\frac{Y_{1-1}-Y_{11}}{\sqrt{2}}\right\}\right|^2 r^2\sin(\theta)\,dr\,d\theta\,d\phi \text{ [Table 4.1, eqn 4.18]}$$

where $\rho=2r/a_0, r=\rho a_0/2, dr=(a_0/2)d\rho$.

$$=\frac{1}{2}\int_0^{2\pi}\int_0^{\pi}\int_0^{\infty}\left(\frac{a_0}{2}\right)^3\left|\left[\left(\frac{1}{27(6)^{1/2}}\right)\left(\frac{1}{a_0}\right)^{3/2}\left(4-\frac{1}{3}\rho\right)\rho e^{-\rho/6}\right]\right.$$

$$\left.\times\left[\left(\frac{3}{8\pi}\right)^{1/2}2\sin(\theta)\cos(\phi)\right]\right|^2\rho^2\sin(\theta)\,d\rho\,d\theta\,d\phi$$

$$=\frac{1}{46656\pi}\int_0^{2\pi}\int_0^{\pi}\int_0^{\infty}\left|\left(4-\frac{1}{3}\rho\right)\rho e^{-\rho/6}\sin(\theta)\cos(\phi)\right|^2\rho^2\sin(\theta)\,d\rho\,d\theta\,d\phi$$

$$=\frac{1}{46656\pi}\underbrace{\int_0^{2\pi}\cos^2(\phi)\,d\phi}_{\pi}\underbrace{\int_0^{\pi}\sin^3(\theta)\,d\theta}_{4/3}\underbrace{\int_0^{\infty}\left(4-\frac{1}{3}\rho\right)^2\rho^4 e^{-\rho/3}\,d\rho}_{34992}$$

$=1$ Thus, ψ_{3p_x} is normalized to 1.

We must also show that $\int\psi_{3p_x}\psi_{3d_{xy}}\,d\tau=0$.

Using Tables 3.2 and 4.1, we find that

$$\psi_{3p_x}=\frac{1}{54(2\pi)^{1/2}}\left(\frac{1}{a_0}\right)^{3/2}\left(4-\frac{1}{3}\rho\right)\rho e^{-\rho/6}\sin(\theta)\cos(\phi)$$

$$\psi_{3d_{xy}}=R_{32}\left\{\frac{Y_{22}-Y_{2-2}}{\sqrt{2}i}\right\}$$

$$=\frac{1}{32(2\pi)^{1/2}}\left(\frac{1}{a_0}\right)^{3/2}\rho^2 e^{-\rho/6}\sin^2(\theta)\sin(2\phi)$$

where $\rho=2r/a_0, r=\rho a_0/2, dr=(a_0/2)d\rho$.

$$\int\psi_{3p_x}\psi_{3d_{xy}}\,d\tau=\text{constant}\times\int_0^{\infty}\rho^5 e^{-\rho/3}d\rho\underbrace{\int_0^{2\pi}\cos(\phi)\sin(2\phi)d\phi}_{0}\int_0^{\pi}\sin^4(\theta)d\theta$$

Since the integral equals zero, ψ_{3p_x} and $\psi_{3d_{xy}}$ are orthogonal.

(b) Radial nodes are determined by finding the ρ values $(\rho=2r/a_0)$ for which the radial wavefunction equals zero. These values are the roots of the polynomial portion of the wavefunction. For the 3s orbital, $6-6\rho+\rho^2=0$, when $\boxed{\rho_{\text{node}}=3+\sqrt{3} \text{ and } \rho_{\text{node}}=3-\sqrt{3}}$.

The 3*s* orbital has these two spherically symmetrical nodes. There is no node at $\rho=0$ so we conclude that there is a finite probability of finding a 3*s* electron at the nucleus.

For the 3_{Px} orbital, $(4-\rho)(\rho)=0$ when $\boxed{\rho_{\text{node}}=0 \text{ and } \rho_{\text{node}}=4}$. There is a zero probability of finding a $3p_x$ electron at the nucleus.

For the $3d_{xy}$ orbital $\boxed{\rho_{\text{node}}=0}$ is the only radial node.

(c) $\langle r\rangle_{3s} = \int |R_{10}Y_{00}|^2 r\,d\tau = \int |R_{10}Y_{00}|^2 r^3 \sin(\theta)\,dr\,d\theta\,d\phi$

$$= \int_0^\infty R_{10}^2 r^3 dr \underbrace{\int_0^{2\pi}\int_0^{\pi} |Y_{00}|^2 \sin(\theta)\,d\theta\,d\phi}_{1} = \frac{a_0}{3888}\underbrace{\int_0^\infty (6-2\rho+\rho^2/9)^2 \rho^3 e^{-\rho/3} d\rho}_{52488}$$

$$\boxed{\langle r\rangle_{3s} = \frac{27a_0}{2}}$$

(d)

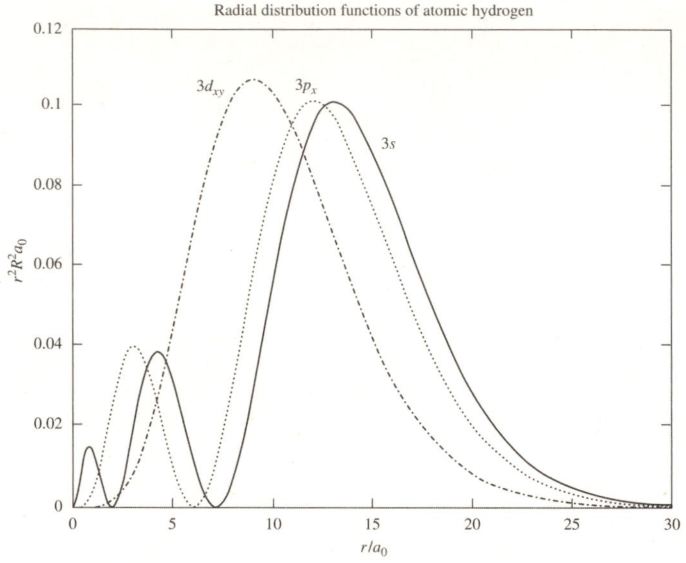

Figure 4.1

The plot, Figure 4.1, shows that the 3s orbital has larger values of the radial distribution function for $r < a_0$. This penetration of inner core electrons of multi-electron atoms means that a 3s electron experiences a larger effective nuclear charge and, consequently, has a lower energy than either a 3p or $3d_{xy}$ electron. This reasoning also leads us to conclude that a $3p_x$ electron has less energy than a $3d_{xy}$ electron.

$$E_{3s} < E_{3p_x} < E_{3d_{xy}}$$

(e) Polar plots with $\theta = 90°$, Figure 4.2

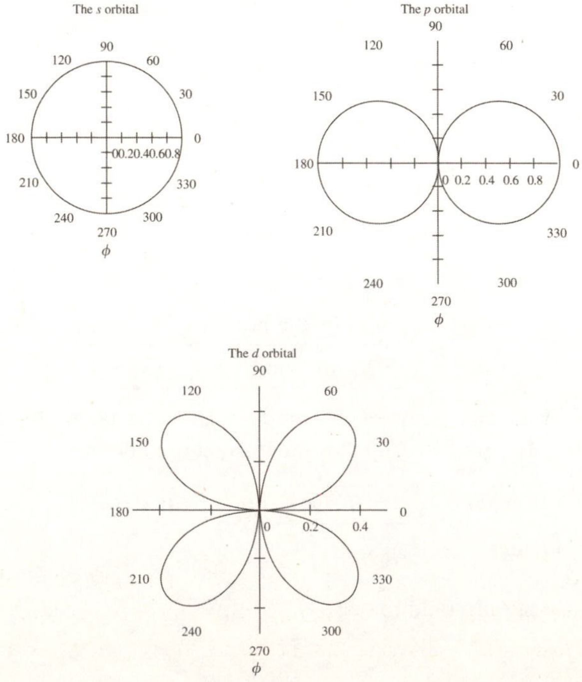

Figure 4.2

Boundary surface plots, Figure 4.3

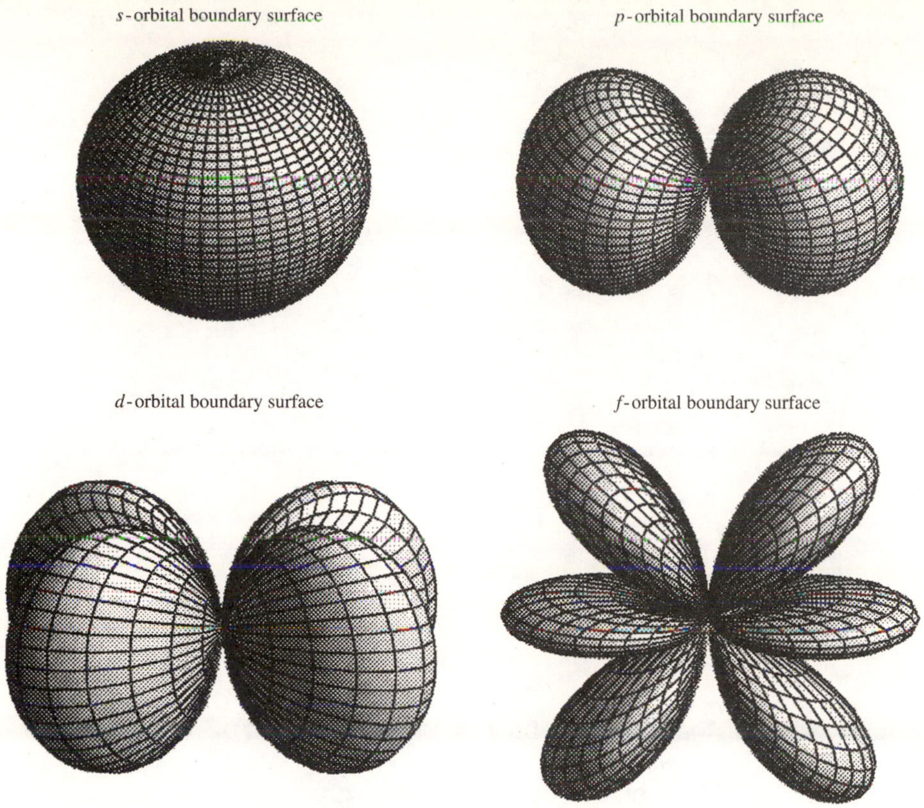

Figure 4.3

P4.19 $$\psi_{1s} = \left(\frac{1}{\pi a_0^3}\right)^{1/2} \mathrm{e}^{-r/a_0} \,[4.14]$$

The probability of the electron being within a sphere of radius r' is

$$\int_0^{r'}\int_0^{\pi}\int_0^{2\pi} \psi_{1s}^2 r^2 \,\mathrm{d}r \,\sin\theta \,\mathrm{d}\theta \,\mathrm{d}\phi$$

We set this equal to 0.90 and solve for r'. The integral over θ and ϕ gives a factor of 4π; thus

$$0.90 = \frac{4}{a_0^3}\int_0^{r'} r^2 \mathrm{e}^{-2r/a_0}\,\mathrm{d}r$$

$\int_0^{r'} r^2 \mathrm{e}^{-2r/a_0}\mathrm{d}r$ is integrated by parts to yield

$$-\frac{a_0 r^2 \mathrm{e}^{-2r/a_0}}{2}\Bigg|_0^{r'} + a_0\left[-\frac{a_0 r \mathrm{e}^{-2r/a_0}}{2}\Bigg|_0^{r'} + \frac{a_0}{2}\left(-\frac{a_0 \mathrm{e}^{-2r/a_0}}{2}\right)\Bigg|_0^{r'}\right]$$

$$= -\frac{a_0 (r')^2 \mathrm{e}^{-2r'/a_0}}{2} - \frac{a_0^2 r'}{2}\mathrm{e}^{-2r'/a_0} - \frac{a_0^3}{4}\mathrm{e}^{-2r'/a_0} + \frac{a_0^3}{4}$$

Multiplying by $\frac{4}{a_0^3}$ and factoring e^{-2r'/a_0},

$$0.90 = \left[-2\left(\frac{r'}{a_0}\right)^2 - 2\left(\frac{r'}{a_0}\right) - 1\right]\mathrm{e}^{-2r'/a_0} + 1 \text{ or } 2\left(\frac{r'}{a_0}\right)^2 + 2\left(\frac{r'}{a_0}\right) + 1 = 0.10\mathrm{e}^{2r'/a_0}$$

It is easiest to solve this numerically. It is seen that $\boxed{r' = 2.66\,a_0}$ satisfies the above equation. Mathematical software has powerful features for handling this type of problem. Plots are very convenient to both make and use. Solve blocks can be used as functions. Both features are demonstrated below using Mathcad®.

Let $z = r/a_0$. The probability, Prob(z), that a 1s electron is within a sphere of radius z is

$$\text{Prob}(z) := 4 \cdot \int_0^z x^2 \cdot e^{-2x} dx$$

Variables needed for plot: $N := 800$ $\quad i := 0 \ldots N$ $\quad z_{max} := 5$ $\quad z_i := \frac{z_{max} \cdot i}{N}$

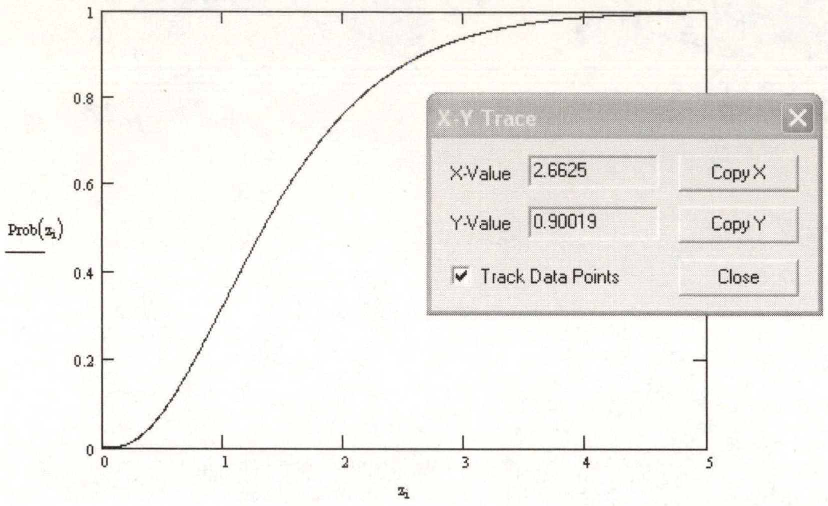

The plot indicates that the probability of finding the electron in a sphere of radius is sigmoidal. The trace feature of Mathcad is used to find that with $z = 2.66$ ($r = 2.66\, a_0$) there is a 90.0% probability of finding the electron in the sphere.

The following Mathcad® document develops a function for calculating the radius for any desired probability. The probability is presented to the function as an argument

$z := 2$ Estimate of z needed for computation within following Given/Find solve block for the function z(Probability).

Given

$$\text{Probability} = 4 \cdot \int_0^z x^2 \cdot e^{-2x} dx$$

z(Probability) := Find(z)

z(.9) = 2.661

P4.21

$$\langle r^m \rangle_{nl} = \int r^m |\psi_{nl}|^2 \, d\tau = \int_0^\infty \int_0^{2\pi} \int_0^{\pi} r^{m+2} |R_{nl} Y_{l0}|^2 \sin(\theta) d\theta d\phi dr$$

$$= \int_0^\infty r^{m+2} |R_{nl}|^2 \, dr \int_0^{2\pi} \int_0^{\pi} |Y_{l0}|^2 \sin(\theta) d\theta d\phi = \int_0^\infty r^{m+2} |R_{nl}|^2 \, dr$$

With $r = (na_0/2Z)\rho$ and $m = -1$ the expectation value is

$$\langle r^{-1} \rangle_{nl} = \left(\frac{na_0}{2Z}\right)^2 \int_0^\infty \rho |R_{nl}|^2 \, d\rho$$

(a) $$\langle r^{-1} \rangle_{1s} = \left(\frac{a_0}{2Z}\right)^2 \left\{ 2\left(\frac{Z}{a_0}\right)^{3/2} \right\}^2 \int_0^\infty \rho \, e^{-\rho} d\rho \qquad \text{[Table 4.1]}$$

$$= \boxed{\frac{Z}{a_0}} \qquad \text{because} \quad \int_0^\infty \rho \, e^{-\rho} d\rho = 1$$

(b) $$\langle r^{-1}\rangle_{2s} = \left(\frac{a_0}{Z}\right)^2 \left\{\frac{1}{8^{1/2}}\left(\frac{Z}{a_0}\right)^{3/2}\right\}^2 \int_0^\infty \rho(2-\rho)^2\,\mathrm{e}^{-\rho}\mathrm{d}\rho \qquad \text{[Table 4.1]}$$

$$= \frac{Z}{8a_0}(2) \qquad \text{because} \qquad \int_0^\infty \rho(2-\rho)^2\,\mathrm{e}^{-\rho}\mathrm{d}\rho = 2$$

$$\langle r^{-1}\rangle_{2s} = \boxed{\frac{Z}{4a_0}}$$

(c) $$\langle r^{-1}\rangle_{2p} = \left(\frac{a_0}{Z}\right)^2 \left\{\frac{1}{24^{1/2}}\left(\frac{Z}{a_0}\right)^{3/2}\right\}^2 \int_0^\infty \rho^3\,\mathrm{e}^{-\rho}\mathrm{d}\rho \qquad \text{[Table 4.1]}$$

$$= \frac{Z}{24a_0}(6) \qquad \text{because} \qquad \int_0^\infty \rho^3\,\mathrm{e}^{-\rho}\mathrm{d}\rho = 6$$

$$\langle r^{-1}\rangle_{2p} = \boxed{\frac{Z}{4a_0}}$$

The general formula for a hydrogenic orbital is $\langle r^{-1}\rangle_{nl} = \dfrac{Z}{n^2 a_0}$.

P4.23 *Justification* 4.4 noted that the transition dipole moment, μ_{fi}, had to be non-zero for a transition to be allowed. The Justification examined conditions that allowed the z component of this quantity to be non-zero; now examine the x and y components.

$$\mu_{x,\mathrm{fi}} = -e\int \psi_{\mathrm{f}}^* x \psi_i \mathrm{d}\tau \quad \text{and} \quad \mu_{y,\mathrm{fi}} = -e\int \psi_{\mathrm{f}}^* y \psi_i \mathrm{d}\tau$$

As in the Justification, express the relevant Cartesian variables in terms of the spherical harmonics, $Y_{l,m}$. Start by expressing them in spherical polar coordinates:

$$x = r\sin\theta\cos\phi \quad \text{and} \quad y = r\sin\theta\sin\phi$$

Note that $Y_{1,1}$ and $Y_{1,-1}$ have factors of $\sin\theta$. They also contain complex exponentials that can be related to the sine and cosine of ϕ through the identities

$$\cos\phi = 1/2(\mathrm{e}^{\mathrm{i}\phi} + \mathrm{e}^{-\mathrm{i}\phi}) \quad \text{and} \quad \sin\phi = 1/2i(\mathrm{e}^{\mathrm{i}\phi} - \mathrm{e}^{-\mathrm{i}\phi})$$

These relations motivate us to try linear combinations $Y_{1,1} - Y_{1,-1}$ and $Y_{1,1} - Y_{1,-1}$ (from Table 3.2; note that c here corresponds to the normalization constant in the table):

$$Y_{1,1} + Y_{1,-1} = -c\sin\theta(\mathrm{e}^{\mathrm{i}\phi} + \mathrm{e}^{-\mathrm{i}\phi}) = -2c\sin\theta\cos\phi = -2cx/r$$

so $$x = -(Y_{1,1} + Y_{1,-1})r/2c$$

$$Y_{1,1} - Y_{1,-1} = c\sin\theta(\mathrm{e}^{\mathrm{i}\phi} - \mathrm{e}^{-\mathrm{i}\phi}) = 2ic\sin\theta\sin\phi = 2icy/r$$

so $$y = (Y_{1,1} - Y_{1,-1})r/2ic$$

Now we can express the integrals in terms of radial wavefunctions $R_{n,l}$ and spherical harmonics Y_{l,m_l}:

$$\mu_{x,\mathrm{fi}} = \frac{e}{2c}\int_0^\infty R_{n_\mathrm{f},l_\mathrm{f}} r R_{n_\mathrm{i},l_\mathrm{i}} r^2\,dr \int_0^\pi\int_0^{2\pi} Y^*_{l_\mathrm{f},m_{l_\mathrm{f}}}(Y_{1,1} + Y_{1,-1})Y_{l_\mathrm{i},m_{l_\mathrm{i}}} \sin\theta\,\mathrm{d}\theta\,\mathrm{d}\phi$$

The angular integral can be broken into two, one of which contains $Y_{1,1}$ and the other $Y_{1,-1}$. A "triple integral" over spherical harmonics of the form

$$\int_0^\pi\int_0^{2\pi} Y^*_{l_f,m_{l_f}} Y_{1,1} Y_{l_i,m_{l_i}} \sin\theta\,\mathrm{d}\theta\,\mathrm{d}\phi$$

vanishes unless $l_f = l_i \pm 1$ and $m_{l_f} = m_{l_i} \pm 1$. The integral that contains $Y_{1,-1}$ introduces no further constraints; it vanishes unless $l_f = l_i \pm 1$ and $m_{l_f} = m_{l_i} \pm 1$. Similarly, the y component introduces no further constraints, for it involves the same spherical harmonics as the x component. The whole set of selection rules, then, is that transitions are allowed only if

$$\boxed{\Delta l = \pm 1 \text{ and } \Delta m_l = 0 \text{ or } \pm 1}$$

P4.25 (a) The Slater wavefunction [4.32] is

$$\psi(1,2,3,\ldots,N) = \frac{1}{(N!)^{1/2}} \begin{vmatrix} \psi_a(1)\alpha(1) & \psi_a(2)\alpha(2) & \psi_a(3)\alpha(3) & \cdots & \psi_a(N)\alpha(N) \\ \psi_a(1)\beta(1) & \psi_a(2)\beta(2) & \psi_a(3)\beta(3) & \cdots & \psi_a(N)\beta(N) \\ \psi_b(1)\alpha(1) & \psi_b(2)\alpha(2) & \psi_b(3)\alpha(3) & \cdots & \psi_b(N)\alpha(N) \\ \vdots & \vdots & \vdots & \vdots & \vdots \\ \psi_z(1)\beta(1) & \psi_z(2)\beta(2) & \psi_z(3)\beta(3) & \cdots & \psi_z(N)\beta(N) \end{vmatrix}$$

Interchanging any two columns or rows leaves the function unchanged except for a change in sign. For example, interchanging the first and second columns of the above determinant gives

$$\psi(1,2,3,\ldots,N) = \frac{-1}{(N!)^{1/2}} \begin{vmatrix} \psi_a(2)\alpha(2) & \psi_a(1)\alpha(1) & \psi_a(3)\alpha(3) & \cdots & \psi_a(N)\alpha(N) \\ \psi_a(2)\beta(2) & \psi_a(1)\beta(1) & \psi_a(3)\beta(3) & \cdots & \psi_a(N)\beta(N) \\ \psi_b(2)\alpha(2) & \psi_b(1)\alpha(1) & \psi_b(3)\alpha(3) & \cdots & \psi_b(N)\alpha(N) \\ \vdots & \vdots & \vdots & \vdots & \vdots \\ \psi_z(2)\beta(2) & \psi_z(1)\beta(1) & \psi_z(3)\beta(3) & \cdots & \psi_z(N)\beta(N) \end{vmatrix}$$

$$= -\psi(2,1,3,\ldots,N)$$

This demonstrates that a Slater determinant is antisymmetric under particle exchange.

(b) The possibility that two electrons occupy the same orbital with the same spin can be explored by making any two rows of the Slater determinant identical, thereby providing identical orbital and spin functions to two rows. Rows 1 and 2 are identical in the Slater wavefunction below. Interchanging these two rows causes the sign to change without in any way changing the determinant.

$$\psi(1,2,3,\ldots,N) = \frac{1}{(N!)^{1/2}} \begin{vmatrix} \psi_a(1)\alpha(1) & \psi_a(2)\alpha(2) & \psi_a(3)\alpha(3) & \cdots & \psi_a(N)\alpha(N) \\ \psi_a(1)\alpha(1) & \psi_a(2)\alpha(2) & \psi_a(3)\alpha(3) & \cdots & \psi_a(N)\alpha(N) \\ \psi_b(1)\alpha(1) & \psi_b(2)\alpha(2) & \psi_b(3)\alpha(3) & \cdots & \psi_b(N)\alpha(N) \\ \vdots & \vdots & \vdots & \vdots & \vdots \\ \psi_z(1)\beta(1) & \psi_z(2)\beta(2) & \psi_z(3)\beta(3) & \cdots & \psi_z(N)\beta(N) \end{vmatrix}$$

$$= -\psi(2,1,3,\ldots,N) = -\psi(1,2,3,\ldots,N)$$

Only the null function satisfies a relationship in which it is the negative of itself so we conclude that, since the null function is inconsistent with existence, the Slater determinant satisfies the Pauli Exclusion Principle [Section 4.4 b]. No two electrons can occupy the same orbital with the same spin.

Solutions to applications

P4.27 The wavenumber of a spectroscopic transition is related to the difference in the relevant energy levels. For a one-electron atom or ion, the relationship is

$$hc\tilde{\nu} = \Delta E = \frac{Z^2\mu_{\text{He}}e^4}{32\pi^2\varepsilon_0^2\hbar^2 n_1^2} - \frac{Z^2\mu_{\text{He}}e^4}{32\pi^2\varepsilon_0^2\hbar^2 n_2^2} = \frac{Z^2\mu_{\text{He}}e^4}{32\pi^2\varepsilon_0^2\hbar^2}\left(\frac{1}{n_2^2} - \frac{1}{n_1^2}\right)$$

Solving for $\tilde{\nu}$, using the definition $\hbar = h/2\pi$ and the fact that $Z = 2$ for He, yields

$$\tilde{\nu} = \frac{\mu_{\mathrm{He}} e^4}{2\varepsilon_0^2 h^3 c}\left(\frac{1}{n_2^2} - \frac{1}{n_1^2}\right)$$

Note that the wavenumbers are proportional to the reduced mass, which is very close to the mass of the electron for both isotopes. In order to distinguish between them, we need to carry lots of significant figures in the calculation.

$$\tilde{\nu} = \frac{\mu_{\mathrm{He}}(1.60218\times10^{-19}\,\mathrm{C})^4}{2(8.85419\times10^{-12}\,\mathrm{J^{-1}C^2m^{-1}})^2\times(6.62607\times10^{-34}\,\mathrm{Js})^3\times(2.99792\times10^{10}\,\mathrm{cm\,s^{-1}})}$$

$$\times\left(\frac{1}{n_2^2} - \frac{1}{n^2}\right)$$

$$\tilde{\nu}/\mathrm{cm^{-1}} = 4.81870\times10^{35}(\mu_{\mathrm{He}}/kg)\left(\frac{1}{n_2^2} - \frac{1}{n_1^2}\right)$$

The reduced masses for the $^4\mathrm{He}$ and $^3\mathrm{He}$ nuclei are

$$\mu = \frac{m_{\mathrm{e}} m_{\mathrm{nuc}}}{m_{\mathrm{e}} + m_{\mathrm{nuc}}}$$

where $m_{\mathrm{nuc}} = 4.00260\mathrm{u}$ for $^4\mathrm{He}$ and 3.01603 u for $^3\mathrm{He}$, or, in kg

$$^4\mathrm{He}\ m_{\mathrm{nuc}} = (4.00260\mathrm{u})\times(1.66054\times10^{-27}\,\mathrm{kg\,u^{-1}}) = 6.64648\times10^{-27}\,\mathrm{kg}$$

$$^3\mathrm{He}\ m_{\mathrm{nuc}} = (3.01603\mathrm{u})\times(1.66054\times10^{-27}\,\mathrm{kg\,u^{-1}}) = 5.00824\times10^{-27}\,\mathrm{kg}$$

The reduced masses are

$$^4\mathrm{He}\ \mu = \frac{(9.10939\times10^{-31}\,\mathrm{kg})\times(6.64648\times10^{-27}\,\mathrm{kg})}{(9.10939\times10^{-31} + 6.64648\times10^{-27})\mathrm{kg}} = 9.10814\times10^{-31}\,\mathrm{kg}$$

$$^3\mathrm{He}\ \mu = \frac{(9.10939\times10^{-31}\,\mathrm{kg})\times(5.00824\times10^{-27}\,\mathrm{kg})}{(9.10939\times10^{-31} + 5.00824\times10^{-27})\mathrm{kg}} = 9.10773\times10^{-31}\,\mathrm{kg}$$

Finally, the wavenumbers for $n = 3 \rightarrow n = 2$ are

$$^4\mathrm{He}\ \tilde{\nu} = (4.81870\times10^{35})\times(9.10814\times10^{-31})\times(1/4 - 1/9)\mathrm{cm^{-1}} = \boxed{60957.4\,\mathrm{cm^{-1}}}$$

$$^3\mathrm{He}\ \tilde{\nu} = (4.81870\times10^{35})\times(9.10773\times10^{-31})\times(1/4 - 1/9)\mathrm{cm^{-1}} = \boxed{60954.7\mathrm{cm^{-1}}}$$

The wavenumbers for $n = 2 \rightarrow n = 1$ are

$$^4\mathrm{He}\ \tilde{\nu} = (4.81870\times10^{35})\times(9.10814\times10^{-31})\times(1/1 - 1/4)\mathrm{cm^{-1}} = \boxed{329170\,\mathrm{cm^{-1}}}$$

$$^3\mathrm{He}\ \tilde{\nu} = (4.81870\times10^{35})\times(9.10773\times10^{-31})\times(1/1 - 1/4)\mathrm{cm^{-1}} = \boxed{329155\,\mathrm{cm^{-1}}}$$

P4.29 (a) Compute the ratios ν_{star}/ν for all three lines. We are given wavelength data, so we can use

$$\frac{\nu_{\mathrm{star}}}{\nu} = \frac{\lambda}{\lambda_{\mathrm{star}}}.$$

The ratios are

$$\frac{438.392\text{ nm}}{438.882\text{ nm}} = 0.998884, \quad \frac{440.510\text{ nm}}{441.000\text{ nm}} = 0.998889, \text{ and } \frac{441.510\text{ nm}}{442.020\text{ nm}} = 0.998846$$

The frequencies of the stellar lines are all less than those of the stationary lines, so we infer that the star is $\boxed{\text{receding}}$ from earth. The Doppler effect follows:

$$\nu_{\text{receding}} = \nu f \quad \text{where } f = \left(\frac{1-s/c}{1+s/c}\right)^{1/2}, \text{ so}$$

$$f^2(1+s/c) = (1-s/c), \quad (f^2+1)s/c = 1-f^2, \quad s = \frac{1-f^2}{1+f^2}c$$

Our average value of f is 0.998873. (*Note*: The uncertainty is actually greater than the significant figures here imply, and a more careful analysis would treat uncertainty explicitly.) So the speed of recession with respect to the earth is

$$s = \left(\frac{1-0.997747}{1+0.997747}\right)c = \boxed{1.128\times10^{-3}\text{ c}} = \boxed{3.381\times10^{5}\text{ ms}^{-1}}$$

(b) One could compute the star's radial velocity with respect to the sun if one knew the earth's speed with respect to the sun along the sun–star vector at the time of the spectral observation. This could be estimated from quantities available through astronomical observation: the earth's orbital velocity times the cosine of the angle between that velocity vector and the earth–star vector at the time of the spectral observation. (The earth–star direction, which is observable by earth-based astronomers, is practically identical to the sun–star direction, which is technically the direction needed.) Alternatively, repeat the experiment half a year later. At that time, the earth's motion with respect to the sun is approximately equal in magnitude and opposite in direction compared to the original experiment. Averaging f values over the two experiments would yield f values in which the earth's motion is effectively averaged out.

P4.31 Electronic configurations of neutral, 4th-period transition atoms in the ground state are summarized in the following table along with observed positive oxidation states. The most common positive oxidation states are indicated with white boxing.

Group	3	4	5	6	7	8	9	10	11	12
Oxidation State	Sc	Ti	V	Cr	Mn	Fe	Co	Ni	Cu	Zn
0	$3d4s^2$	$3d^24s^2$	$3d^34s^2$	$3d^54s$	$3d^54s^2$	$3d^64s^2$	$3d^74s^2$	$3d^84s^2$	$3d^{10}4s$	$3d^{10}4s^2$
+1			☺	☺	☺		☺	☺	☺	
+2		☺	☺	☺	☺	☺	☺	☺	☺	☺
+3	☺	☺	☺	☺	☺	☺	☺	☺		
+4		☺	☺	☺	☺	☺	☺	☺		
+5			☺	☺	☺	☺				
+6				☺	☺	☺				
+7					☺					

Toward the middle of the first transition series (Cr, Mn, and Fe), elements exhibit the widest ranges of oxidation states. This phenomenon is related to the availability of both electrons and orbitals favorable for bonding. Elements to the left (Sc and Ti) of the series have few electrons, and a relatively low effective nuclear charge leaves d orbitals at high energies that are relatively unsuitable for bonding. To the far right (Cu and Zn), effective nuclear charge may be higher but there are few, if any, orbitals available for bonding. Consequently, it is more difficult to produce a range of compounds that promote a wide range of oxidation states for elements at either end of the series. At the middle and right of the series, the +2 oxidation state is very commonly observed because normal reactions can provide the requisite ionization energies for the removal of 4s electrons. The readily available +2 and +3 oxidation states of Mn, Fe, and the +1 and +2 oxidation states of Cu make these cations useful in electron transfer processes occurring in chains of specialized proteins within biological cells. The special size and charge of the Zn^{2+} cation makes it useful for the function of some enzymes. The tendency of Fe^{2+} and Cu^{+} to bind oxygen proves very useful in hemoglobin and electron transport (respiratory) chain, respectively.

5 The chemical bond

Answers to discussion questions

D5.1 Our comparison of the two theories will focus on the manner of construction of the trial wavefunctions for the hydrogen molecule in the simplest versions of both theories. In the valence bond method, the trial function is a linear combination of two simple product wavefunctions, in which one electron resides totally in an atomic orbital on atom A and the other totally in an orbital on atom B. See eqns. 5.1 and 5.2, as well as Fig. 5.2 in the main text. There is no contribution to the wavefunction from products in which both electrons reside on either atom A or B. So the valence bond approach undervalues any ionic contribution to the trial function by totally neglecting it. It is a totally covalent function. The molecular orbital function for the hydrogen molecule is a product of two functions of the form of eqn. 5.7, one for each electron:

$$\psi = [A(1) \pm B(1)][A(2) \pm B(2)] = A(1)A(2) + B(1)B(2) + A(1)B(2) + B(1)A(2)$$

This function gives as much weight to the ionic forms as to the covalent forms. So the molecular orbital approach greatly overvalues the ionic contributions. At these crude levels of approximation, the valence bond method gives dissociation energies closer to the experimental values. However, the molecular orbital approach, albeit in more sophisticated versions, is the method of choice for obtaining quantitative results on both diatomic and polyatomic molecules. See Sections 5.6–5.9.

D5.3 The ion has 24 valence electrons as shown in the Lewis structure. The hybridizations are (from left to right) sp^2 for the first O atom and for the N and sp^3 for the next two O atoms. The first bond is a double bond whose σ component arises from the overlap of sp^2 orbitals and whose π component comes from overlap of unhybridized p orbitals. The next bond is a σ bond involving Nsp^2 and Osp^3 orbitals. The O–O bond is a σ bond involving sp^3 orbitals.

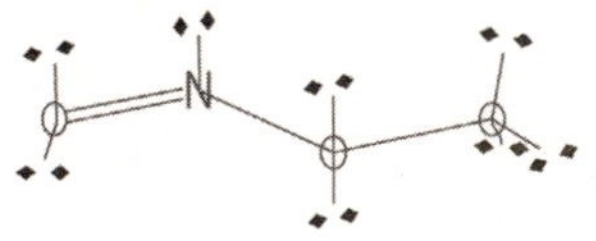

D5.5 Both the Pauling and Mulliken methods for measuring the attracting power of atoms for electrons seem to make good chemical sense. If we look at eqn. 5.22 (the Pauling scale), we see that if D(A–B) were equal to $\frac{1}{2}[D(\text{A–A}) + D(\text{B–B})]$, the calculated electronegativity difference would be zero, as expected for completely non-polar bonds. Hence, any increased strength of the A–B bond over the average of the A–A and B–B bonds can reasonably be thought of as being due to the polarity of the A–B bond, which in turn is due to the difference in electronegativity of the atoms involved. Therefore, this difference in bond strengths can be used as a measure of electronegativity difference. To obtain numerical values for individual atoms, a reference state (atom) for electronegativity must be established. The value for fluorine is arbitrarily set at 4.0.

The Mulliken scale may be more intuitive than the Pauling scale because we are used to thinking of ionization energies and electron affinities as measures of the electron attracting powers of atoms. The choice of factor $\frac{1}{2}$, however, is arbitrary, though reasonable, but no more arbitrary than the specific form of eqn. 5.22 that defines the Pauling scale.

D5.7 The Hückel method parameterizes rather than calculates the energy integrals that arise in molecular orbital theory. In the simplest version of the method, the overlap integral is also parameterized. The energy integrals, α and β, are always considered to be adjustable parameters; their numerical values emerge only at the end of the calculation by comparison to experimental energies. The simple form of the method has three other rather drastic approximations, listed in Section 5.6 of the text, which eliminate many terms from the secular determinant and make it easier to solve. Ease of solution was important in the early days of quantum chemistry before the advent of computers. Without the use of these approximations, calculations on polyatomic molecules would have been difficult to accomplish.

The simple Hückel method is usually applied only to the calculation of π-electron energies in conjugated organic systems. It is based on the assumption of the separability of the σ- and π-electron systems in the molecule. This is a very crude approximation and works best when the energy level pattern is determined largely by the symmetry of the molecule. (See Chapter 7.)

D5.9 See *Impact* I5.1 for a more detailed discussion. The ground electronic configurations of the valence electrons are found in Figs. 5.29, 5.31, and 5.34 of the text.

N_2	$1\sigma_g^2 1\sigma_u^2 1\pi_u^4 2\sigma_g^2$	$b = 3$	$2S + 1 = 0$
O_2	$1\sigma_g^2 1\sigma_u^2 2\sigma_g^2 1\pi_u^4 1\pi_g^2$	$b = 2$	$2S + 1 = 3$
NO	$1\sigma^2 2\sigma^2 3\sigma^2 1\pi^4 2\pi^1$	$b = 2^1/_2$	$2S + 1 = 2$

The following figures show HOMOs of each configuration. Shaded versus unshaded atomic-orbital (AO) lobes represent opposite signs of the wave funcstions. A relatively large AO represents the major contribution to the MO.

N_2 2σMO

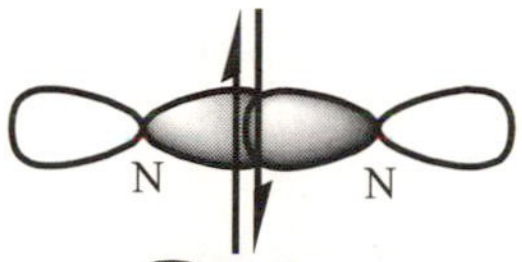

O_2 $1\pi_g$ MO, doubly degenerate

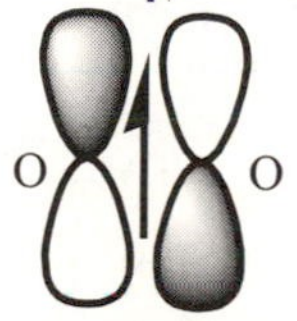

NO 2π

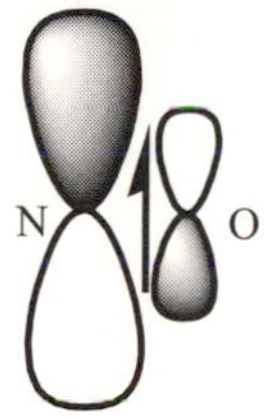

Dinitrogen with a bond order of three and paired electrons in relatively low-energy MOs is very unreactive. Special biological or industrial processes are needed to channel energy for promotion of 2σ electrons into high-energy, reactive states. The high-energy $1\pi_g$ LUMO is not expected to form stable complexes with electron donors.

Molecular nitrogen is very stable in most biological organisms, and as a result the task of converting plentiful atmospheric N_2 to the fixed forms of nitrogen that can be incorporated into proteins is a difficult one. The fact that N_2 possesses no unpaired electrons is itself an obstacle to facile reactivity, and the great strength (large dissociation energy) of the N_2 bond is another obstacle. Molecular orbital theory explains both these obstacles by assigning N_2 a configuration that gives rise to a high bond order (triple bond) with all electrons paired. (See Fig. 5.31 of the text.)

Dioxygen is kinetically stable because of a bond order equal to two and a high effective nuclear charge that causes the MOs to have relatively low energy. But two electrons are in the high-energy $1\pi_g$ HOMO level, which is doubly degenerate. These two electrons are unpaired and can contribute to bonding of dioxygen with other species such as the atomic radicals Fe(II) of hemoglobin and Cu(II) of the electron transport chain. When sufficient though not excessively large energy is available, biological processes can channel an electron into this HOMO to produce the reactive superoxide anion of bond order $1^1/_2$. As a result, O_2 is very reactive in biological systems in ways that promote function (such as respiration) and in ways that disrupt it (damaging cells).

Although the bond order of nitric oxide is $2^1/_2$, the nitrogen nucleus has a smaller effective nuclear change than an oxygen atom would have. Thus, compared to the HOMO of dioxygen, the one electron of the 2π HOMO is a high-energy, reactive radical. Additionally, the HOMO, being anti-bonding and predominantly centered on the nitrogen atom, is expected to bond through the nitrogen. Oxidation can result from the loss of the radical electron to form the nitrosyl ion, NO^+, which has a bond order equal to 2. Even though it has a rather high bond order, NO is readily converted to the damagingly reactive peroxynitrite ion ($ONOO^-$) by reaction with O_2^-—without breaking the NO linkage.

Solutions to exercises

E5.1(a) The valence bond description of P_2 is similar to that of N_2, a triple bond. The three bonds are a σ from the overlap of sp hybrid orbitals and two π bonds from the overlap of unhybridized 3p orbitals.

In the tetrahedral P_4 molecule there are six single P–P bonds of roughly 200 kJ mol^{-1} bond enthalpy each. So the total bonding enthalpy is roughly 1200 kJ mol^{-1}. In the transformation

$$P_4 \rightarrow 2P_2$$

there is a loss of about 800 kJ mol^{-1} in σ-bond enthalpy. This loss is not likely to be made up by the formation of 4 P–P π bonds. Period 3 atoms, such as P, are too large to get close enough to each other to form strong π bonds.

E5.2(a) All of the carbon atoms are sp^2 hybridized. The C–H σ bonds are formed by the overlap of Csp^2 orbitals with H1s orbitals. The C–C σ bonds are formed by the overlap of Csp^2 orbitals. The C–C π bonds are formed by the overlap of C2p orbitals. This description predicts double bonds between carbon atoms 1 and 2 and 3 and 4 but (unlike a simple molecular-orbital description of the bonding) no double-bond character between carbon atoms 2 and 3.

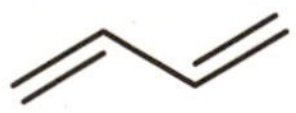

E5.3(a) $h_1 = s + p_x + p_y + p_z \qquad h_2 = s - p_x - p_y + p_z$

We need to evaluate

$$\int h_1 h_2 \, d\tau = \int (s + p_x + p_y + p_z)(s - p_x - p_y + p_z) \, d\tau$$

We assume that the atomic orbitals are normalized atomic orbitals that are mutually orthogonal. We expand the integrand, noting that all cross terms integrate to zero. The remaining terms integrate to one, yielding

$$\int h_1 h_2 \, d\tau = \int s^2 d\tau - \int p_x^2 d\tau - \int p_y^2 d\tau + \int p_z^2 d\tau = 1 - 1 - 1 + 1 = 0$$

E5.4(a) Normalization requires

$$\int \psi^* \psi d\tau = 1$$

where $\psi = Nh = N(s + 2^{1/2}p)$. Solve for the normalization constant N:

$$1 = N^2 \int (s + 2^{1/2}p)^* (s + 2^{1/2}p)d\tau = N^2 \int (|s|^2 + 2^{1/2}p^* s + 2^{1/2}s^* p + 2|p|^2)d\tau = 3N^2$$

In the last step, we used the fact that the s and p orbitals are normalized to see that the squared terms integrate to 1 and the fact that they are orthogonal to see that the cross terms integrate to 0. Thus

$$\boxed{N = 3^{-1/2}} \quad \text{and} \quad \boxed{\psi = 3^{-1/2}(s + 2^{1/2}p)}$$

E5.5(a) Refer to Fig. 5.31 of the text. Place two of the valence electrons in each orbital starting with the lowest-energy orbital, until all valence electrons are used up. Apply Hund's rule to the filling of degenerate orbitals.

(a) Li_2 (2 electrons) $\boxed{1\sigma_g^2,\ b = 1}$

(b) Be_2 (4 electrons) $\boxed{1\sigma_g^2 1\sigma_u^2,\ b = 0}$

(c) C_2 (8 electrons) $\boxed{1\sigma_g^2 1\sigma_u^2 1\pi_u^4,\ b = 2}$

E5.6(a) Note that CO and CN^- are isoelectronic with N_2, so refer to Fig. 5.31 of the text for them; note, however, that the σ and π orbitals no longer have u or g symmetry, so they are simply labeled consecutively. For NO, refer to Fig. 5.34 of the text.

(a) CO (10 electrons) $\boxed{1\sigma^2 2\sigma^2 1\pi^4 3\sigma^2,\ b = 3}$

(b) NO (11 electrons) $\boxed{1\sigma^2 2\sigma^2 3\sigma^2 1\pi^4 2\pi^1,\ b = 2.5}$

(c) CN^- (10 electrons) $\boxed{1\sigma^2 2\sigma^2 1\pi^4 3\sigma^2,\ b = 3}$

E5.7(a) B_2 (6 electrons): $1\sigma_g^2 1\sigma_u^2 1\pi_u^2$ $b = 1$

C_2 (8 electrons): $1\sigma_g^2 1\sigma_u^2 1\pi_u^4$ $b = 2$

The bond orders of B_2 and C_2 are, respectively, 1 and 2, so $\boxed{C_2}$ should have the greater bond dissociation enthalpy. The experimental values are approximately 4eV and 6eV, respectively.

E5.8(a) Decide whether the added electron increases or decreases the bond order. The simplest procedure is to decide whether the added electron would occupy a bonding or antibonding orbital. We can draw up the following table, which denotes the relevant orbital. An asterisk (*) in the table denotes an orbital whose character is primarily antibonding.

	N_2	NO	O_2	C_2	F_2	CN
AB^-	$1\pi_g$*	2π*	$1\pi_g$*	$2\sigma_g$	$2\sigma_u$*	3σ
Change in bond order	−1/2	−1/2	−1/2	+1/2	−1/2	+1/2

Therefore, $\boxed{C_2 \text{ and CN}}$ would be stabilized (have lower energy) by anion formation.

E5.9(a) We can use a version of Fig. 5.29 of the text in which the energy levels of F are lower than those of Xe, as in Figure 5.1.

For XeF we insert 15 valence electrons. The bond order is increased when XeF^+ is formed from XeF because an electron is removed from an antibonding orbital. Therefore, we predict that XeF^+ has a shorter bond length than XeF.

E5.10(a) Form the electron configurations and find the bond order:

NO^- (12 electrons) $1\sigma^2 2\sigma^2 3\sigma^2 1\pi^4 1\pi^2$ $b = 2$

O_2^+ (11 electrons) $1\sigma_g^2 1\sigma_u^2 2\sigma_g^2 1\pi_u^4 1\pi_g^1$ $b = 2.5$

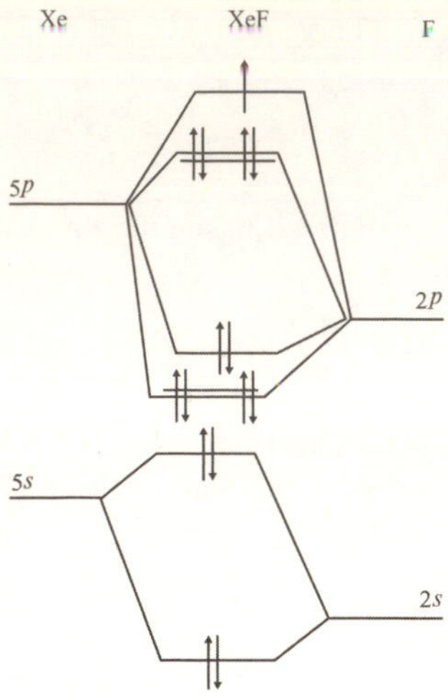

Figure 5.1

Based on the electron configurations, we would expect $\boxed{O_2^+}$ to have the stronger and therefore the shorter bond. Keep in mind, though, that we are comparing molecules with different—albeit not terribly different—nuclei.

E5.11(a) O_2^+ (11 electrons) $1\sigma_g^2 1\sigma_u^2 2\sigma_g^2 1\pi_u^4 1\pi_g^1$

O_2 (12 electrons) $1\sigma_g^2 1\sigma_u^2 2\sigma_g^2 1\pi_u^4 1\pi_g^2$

O_2^- (13 electrons) $1\sigma_g^2 1\sigma_u^2 2\sigma_g^2 1\pi_u^4 1\pi_g^3$

O_2^{2-} (14 electrons) $1\sigma_g^2 1\sigma_u^2 2\sigma_g^2 1\pi_u^4 1\pi_g^4$

In each case, the HOMO is an antibonding $\boxed{1\pi_g}$ orbital.

E5.12(a) Refer to eqn. 2.24, which defines the harmonic oscillator wavefunctions. The parity of the wavefunction is the parity of the Hermite polynomial portion; those polynomials are given in Table 2.1. Examine the inversion through the center of these functions, i.e., replace x with $-x$. For the first four levels,

$$v = 0,\ \psi \rightarrow \psi, \text{ so } \boxed{\text{g}}$$

$$v = 1,\ \psi \rightarrow -\psi, \text{ so } \boxed{\text{u}}$$

$$v = 2,\ \psi \rightarrow \psi, \text{ so } \boxed{\text{g}}$$

$$v = 3,\ \psi \rightarrow -\psi, \text{ so } \boxed{\text{u}}$$

If v is even, ψ_v is $\boxed{\text{g}}$. If v is odd, ψ_v is $\boxed{\text{u}}$.

E5.13(a)

$$\int \psi^2\,d\tau = N^2 \int (\psi_A + \lambda\psi_B)^2\,d\tau = N^2 \int (\psi_A^2 + \lambda^2\psi_B^2 + 2\lambda\psi_A\psi_B)\,d\tau = 1$$

$$= N^2(1 + \lambda^2 + 2\lambda S) \quad \left[\int \psi_A\psi_B\,d\tau = S\right]$$

Hence $\boxed{N = \left(\dfrac{1}{1 + 2\lambda S + \lambda^2}\right)^{1/2}}$.

E5.14(a) Let $\psi_1 = N(0.145A + 0.844B)$ and $\psi_2 = aA + bB$.

First, let us normalize ψ_1:

$$\int \psi_1{}^*\psi_1\,d\tau = 1 = N^2 \int (0.145A + 0.844B)^*(0.145A + 0.844B)\,d\tau$$

$$0.0210 \int |A|^2\,d\tau + 0.712 \int |B|^2\,d\tau + 0.122 \int B^*A\,d\tau + 0.122 \int A^*B\,d\tau = \frac{1}{N^2}$$

The first two integrals are 1 due to normalization and the latter two are the overlap integral, S = 0.250. So

$$0.210+0.712+2\times 0.122\times 0.250=\frac{1}{N^2}=0.795$$

so $\boxed{N=1.12}$ which makes $\boxed{\psi_1=0.163A+0.947B}$.

Orthogonality of the two molecular orbitals requires

$$\int \psi_1 * \psi_2 \mathrm{d}\tau = 0 = N\int (0.145A+0.844B)*(aA+bB)\mathrm{d}\tau$$

Dividing by N yields

$$0.145a\int |A|^2\,\mathrm{d}\tau + 0.844b\int |B|^2\,\mathrm{d}\tau + 0.844a\int B*A\mathrm{d}\tau + 0.145b\int A*B\mathrm{d}\tau = 0,$$

$$0.145\mathrm{a} + 0.844\mathrm{b} + 0.844a\times 0.250 + 0.145b\times 0.250 = 0 = 0.356a + 0.880b$$

Normalization of ψ_2 requires

$$\int \psi_2 * \psi_2 \mathrm{d}\tau = 1 = \int (aA+bB)*(aA+bB)\mathrm{d}\tau,$$

$$a^2\int |A|^2\,\mathrm{d}\tau + b^2\int |B|^2\,\mathrm{d}\tau + ab\int B*A\mathrm{d}\tau + ab\int A*B\mathrm{d}\tau = 1$$

So $a^2 + b^2 + 2ab\times 0.250 = 1 = a^2 + b^2 + 0.500ab.$

We have two equations in the two unknown coefficients a and b. Solve the first equation for a in terms of b:

$$a = -0.880b/0.356 = -2.47b \text{ s}$$

Substitute this result into the second (quadratic) equation:

$$1 = (-2.47b)^2 + b^2 + 0.500(-2.47b)b = 5.88b^2$$

So $\boxed{b=0.412}$, $\boxed{a=-1.02}$, and $\boxed{\psi_2=-1.02A+0.412B}$.

E5.15(a) Energy is conserved, so when the photon is absorbed, its energy is transferred to the electron. Part of it overcomes the binding energy (ionization energy) and the remainder is manifest as the now freed electron's kinetic energy.

$$E_{\text{photon}} = I + E_{\text{kinetic}}$$

so
$$E_{\text{kinetic}} = E_{\text{photon}} - I = \frac{hc}{\lambda} - I = \frac{(6.626\times 10^{-34}\text{ J s})\times(2.998\times 10^{8}\text{ m s}^{-1})}{(100\times 10^{-9}\text{ m})\times(1.602\times 10^{-19}\text{ J eV}^{-1})} - 12.0\text{ eV}$$
$$= 0.4\bar{0}\text{ eV} = 6.\bar{4}\times 10^{-20}\text{ J}$$

The speed is obtained from the kinetic energy:

$$E_{\text{kinetic}} = \frac{mv^2}{2} \quad\text{so}\quad v = \sqrt{\frac{2E_{\text{kinetic}}}{m}} = \sqrt{\frac{2\times 6.\bar{4}\times 10^{-20}\text{ J}}{9.11\times 10^{-31}\text{ kg}}} = \boxed{4\times 10^{5}\text{ m s}^{-1}}$$

E5.16(a) Draw up the following table using data from Table 5.4.

Element	Li	Be	B	C	N	O	F	Ne
χ_M	1.28	1.99	1.83	2.67	3.08	3.22	4.43	4.60
$(\chi_M)^{1/2}$	1.13	1.41	1.35	1.63	1.75	1.79	2.10	2.14
χ_P (from table)	0.98	1.57	2.04	2.55	3.04	3.44	3.98	
χ_P (from formula)	0.16	0.53	0.46	0.84	1.00	1.05	1.47	1.53

A plot of the Pauling electronegativities (actual and from the formula) versus the square root of the Mulliken electronegativities shows that the formula does a poor job (see Figure 5.2). The formula consistently underestimates the Pauling electronegativity, and it underestimates the rise in electronegativity across the period.

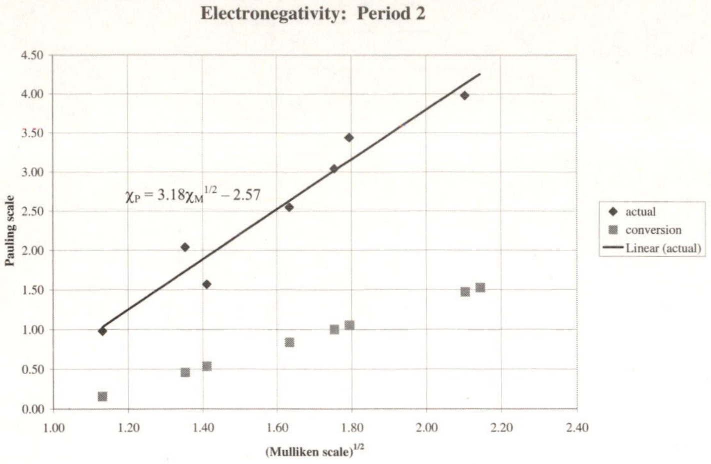

Figure 5.2

E5.17(a) The molecular orbitals of the fragments and the molecular orbitals they form are shown in Figure 5.3.

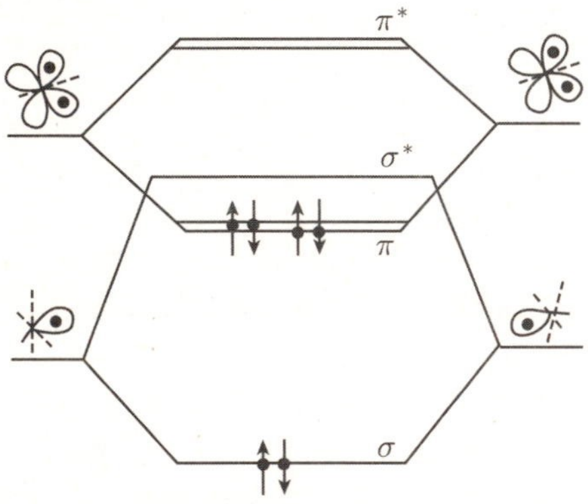

Figure 5.3

COMMENT. Note that the π-bonding orbital must be lower in energy than the σ-antibonding orbital for π-bonding to exist in ethene.

Question. Would the ethene molecule exist if the order of the energies of the π and σ^* orbitals were reversed?

E5.18(a) In setting up the secular determinant we use the Hückel approximations outlined in Section 5.6.

$$\text{(a)}\quad \begin{vmatrix} \alpha-E & \beta & 0 \\ \beta & \alpha-E & \beta \\ 0 & \beta & \alpha-E \end{vmatrix} = 0 \qquad \text{(b)}\quad \begin{vmatrix} \alpha-E & \beta & \beta \\ \beta & \alpha-E & \beta \\ \beta & \beta & \alpha-E \end{vmatrix} = 0$$

The atomic orbital basis is $1s_A$, $1s_B$, $1s_C$ in each case; in linear H_3 we ignore A, C overlap because A and C are not neighboring atoms; in triangular H_3 we include it because they are neighboring atoms.

E5.19(a) We use the molecular orbital energy level diagram in Fig. 5.38 of the text. As usual, we fill the orbitals starting with the lowest-energy orbital, obeying the Pauli principle and Hund's rule. We then write

(a) $C_6H_6^-$ (7 electrons): $\boxed{a_{2u}^2 e_{1g}^4 e_{2u}^1}$

$$E_\pi = 2(\alpha + 2\beta) + 4(\alpha + \beta) + (\alpha - \beta) = \boxed{7\alpha + 7\beta}$$

(b) $C_6H_6^+$ (5 electrons): $\boxed{a_{2u}^2 e_{1g}^3}$

$$E_\pi = 2(\alpha + 2\beta) + 3(\alpha + \beta) = \boxed{5\alpha + 7\beta}$$

E5.20(a) The π-bond formation energy is the difference between the π-electron binding energy and the Coulomb energies α:

$$E_{\rm bf} = E_\pi - N\alpha \, [5.46]$$

The delocalization energy is the difference between E_π and the energy of isolated π bonds:

$$E_{\rm delocal} = E_\pi - N(\alpha + \beta)$$

(a) $E_\pi = 7\alpha + 7\beta$ [E5.19(a)]

so $E_{\rm bf} = 7\alpha + 7\beta - 7\alpha = \boxed{7\beta}$

and $E_{\rm delocal} = 7\alpha + 7\beta - 7(\alpha + \beta) = \boxed{0}$

COMMENT. With an odd number of π electrons, we do not have a whole number of π bonds in the formula for delocalization energy. In effect, we compare the π-electron binding energy to the energy of 3.5 isolated π bonds—whatever that means. The result is that the benzene anion has none of the "extra" stabilization we associate with aromaticity.

(b) $E_\pi = 5\alpha + 7\beta$ [E5.19(a)]

so $E_{\rm bf} = 5\alpha + 7\beta - 5\alpha = \boxed{7\beta}$

and $E_{\rm delocal} = 5\alpha + 7\beta - 5(\alpha + \beta) = \boxed{2\beta}$

E5.21(a) The structures are numbered to match the row and column numbers shown in the determinants:

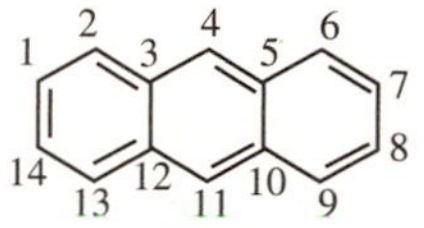

anthracene

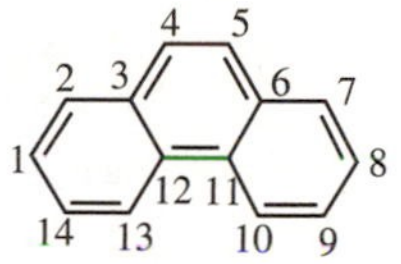

phenanthrene

(a) The secular determinant of anthracene in the Hückel approximation is

	1	2	3	4	5	6	7	8	9	10	11	12	13	14
1	$\alpha-E$	β	0	0	0	0	0	0	0	0	0	0	0	β
2	β	$\alpha-E$	β	0	0	0	0	0	0	0	0	0	0	0
3	0	β	$\alpha-E$	β	0	0	0	0	0	0	0	β	0	0
4	0	0	β	$\alpha-E$	β	0	0	0	0	0	0	0	0	0
5	0	0	0	β	$\alpha-E$	β	0	0	0	β	0	0	0	0
6	0	0	0	0	β	$\alpha-E$	β	0	0	0	0	0	0	0
7	0	0	0	0	0	β	$\alpha-E$	β	0	0	0	0	0	0
8	0	0	0	0	0	0	β	$\alpha-E$	β	0	0	0	0	0
9	0	0	0	0	0	0	0	β	$\alpha-E$	β	0	0	0	0
10	0	0	0	0	β	0	0	0	β	$\alpha-E$	β	0	0	0
11	0	0	0	0	0	0	0	0	0	β	$\alpha-E$	β	0	0
12	0	0	β	0	0	0	0	0	0	0	β	$\alpha-E$	β	0
13	0	0	0	0	0	0	0	0	0	0	0	β	$\alpha-E$	β
14	β	0	0	0	0	0	0	0	0	0	0	0	β	$\alpha-E$

(b) The secular determinant of phenanthrene in the Hückel approximation is

	1	2	3	4	5	6	7	8	9	10	11	12	13	14
1	$\alpha-E$	β	0	0	0	0	0	0	0	0	0	0	0	β
2	β	$\alpha-E$	β	0	0	0	0	0	0	0	0	0	0	0
3	0	β	$\alpha-E$	β	0	0	0	0	0	0	0	β	0	0
4	0	0	β	$\alpha-E$	β	0	0	0	0	0	0	0	0	0
5	0	0	0	β	$\alpha-E$	β	0	0	0	0	0	0	0	0
6	0	0	0	0	β	$\alpha-E$	β	0	0	0	β	0	0	0
7	0	0	0	0	0	β	$\alpha-E$	β	0	0	0	0	0	0
8	0	0	0	0	0	0	β	$\alpha-E$	β	0	0	0	0	0
9	0	0	0	0	0	0	0	β	$\alpha-E$	β	0	0	0	0
10	0	0	0	0	0	0	0	0	β	$\alpha-E$	β	0	0	0
11	0	0	0	0	0	β	0	0	0	β	$\alpha-E$	β	0	0
12	0	0	β	0	0	0	0	0	0	0	β	$\alpha-E$	β	0
13	0	0	0	0	0	0	0	0	0	0	0	β	$\alpha-E$	β
14	β	0	0	0	0	0	0	0	0	0	0	0	β	$\alpha-E$

E5.22(a) The secular determinants from E5.21(a) can be diagonalized with the assistance of general-purpose mathematical software. Alternatively, programs specifically designed for Hückel calculations (such as the Simple Huckel Molecular Orbital Theory Calculator at the University of Calgary, http://www.chem.ucalgary.ca/SHMO/ or Hückel software in *Explorations of Physical Chemistry*, 2nd Ed., by Julio de Paula, Valerie Walters, and Peter Atkins, http://ebooks.bfwpub.com/explorations.php) can be used. In both molecules, 14 π electrons fill seven orbitals.

(a) In anthracene, the energies of the filled orbitals are $\alpha + 2.41421\beta$, $\alpha + 2\beta$, $\alpha + 1.41421\beta$ (doubly degenerate), $\alpha + \beta$ (doubly degenerate), and $\alpha + 0.41421\beta$, so the total π-electron binding energy is $\boxed{14\alpha + 19.31368\beta}$.

(b) For phenanthrene, the energies of the filled orbitals are $\alpha + 2.43476\beta$, $\alpha + 1.95063\beta$, $\alpha + 1.51627\beta$, $\alpha + 1.30580\beta$, $\alpha + 1.14238\beta$, $\alpha + 0.76905\beta$, and $\alpha + 0.60523\beta$, so the total π-electron binding energy is $\boxed{14\alpha + 19.44824\beta}$.

Solutions to problems

Solutions to numerical problems

P5.1 The energy of repulsion is given by Coulomb's law:

$$E = \frac{e^2}{4\pi\varepsilon_0 R} = \frac{(1.602\times10^{-19}\ \mathrm{C})^2}{4\pi\times8.854\times10^{-12}\ \mathrm{J^{-1}\ C^2\ m^{-1}}\times74.1\times10^{-12}\ \mathrm{m}} = 3.11\times10^{-18}\ \mathrm{J}$$

The repulsion energy of a mole of H_2 is

$$E_{\mathrm{molar}} = N_{\mathrm{A}}E = 6.022\times10^{23}\ \mathrm{mol^{-1}}\times3.11\times10^{-18}\ \mathrm{J} = \boxed{1.87\times10^{6}\ \mathrm{J\ mol^{-1}} = 1.87\ \mathrm{MJ\ mol^{-1}}}$$

The gravitational energy is

$$E_{\mathrm{grav}} = \frac{-Gm_1m_2}{R} = \frac{-6.673\times10^{-11}\ \mathrm{m^3\ kg^{-1}\ s^{-2}}\times(1.672\times10^{-27}\ \mathrm{kg})^{-2}}{74.1\times10^{-12}\ \mathrm{m}} = -2.52\times10^{-54}\ \mathrm{J}$$

The gravitational energy is many, many orders of magnitude smaller than the energy of electrostatic repulsion. This is an illustration of the fact that gravitational effects tend to be negligible on the atomic scale.

P5.3 Draw up the following table.

R/a_0	0	1	2	3	4	5	6	7	8	9	10
S	1.000	0.858	0.586	0.349	0.189	0.097	0.047	0.022	0.010	0.005	0.002

The points are plotted in Figure 5.4. At large separations, the curve is dominated by exponential decay. At small separations, however, the polynomial factor makes the decay less pronounced than exponential.

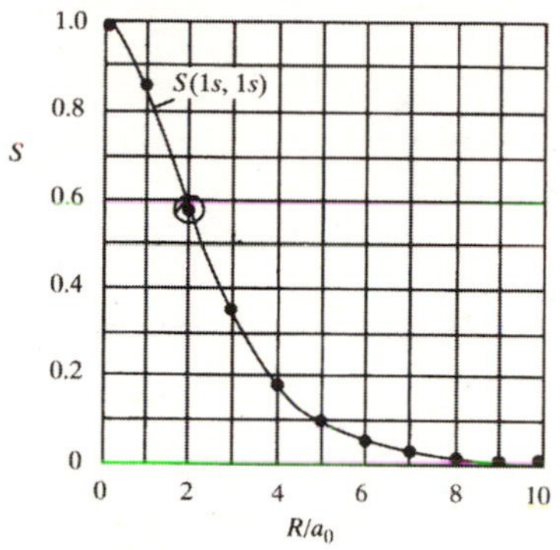

Figure 5.4

P5.5 We require the properly normalized functions

$$\psi_{\pm} = \left(\frac{1}{2(1 \pm S)}\right)^{1/2} (A \pm B) \ \ [5.7 \text{ and Example } 5.1]$$

We first calculate the overlap integral at $R = 106$ pm $= 2a_0$. (The expression for the overlap integral, S, is given in problem 5.3.)

$$S = \left(1 + 2 + \frac{1}{3}(2)^2\right) e^{-2} = 0.586$$

Then $$N_+ = \left(\frac{1}{2(1+S)}\right)^{1/2} = \left(\frac{1}{2(1+0.586)}\right)^{1/2} = 0.561$$

$$N_- = \left(\frac{1}{2(1-S)}\right)^{1/2} = \left(\frac{1}{2(1-0.586)}\right)^{1/2} = 1.09\overline{9}$$

We then calculate with $A = \left(\frac{1}{\pi a_0^3}\right)^{1/2} e^{-r_A/a_0}, \psi_{\pm} = N_{\pm}\left(\frac{1}{\pi a_0^3}\right)^{1/2} \{e^{-r_A/a_0} \pm e^{-r_B/a_0}\}$, with r_A and r_B both measured from nucleus A, that is,

$$\psi_{\pm} = N_{\pm}\left(\frac{1}{\pi a_0^3}\right)^{1/2} \{e^{-|z|/a_0} \pm e^{-|z-R|/a_0}\}$$

with z measured from A along the axis toward B. We draw up the following table with $R = 106$ pm and $a_0 = 52.9$ pm.

z /pm	−100	−80	−60	−40	−20	0	20	40	60	80	100	120	140	160	180	200
$\dfrac{\psi_+}{\left(\frac{1}{\pi a_0^3}\right)^{1/2}}$	0.096	0.14	0.20	0.30	0.44	0.64	0.49	0.42	0.42	0.47	0.59	0.49	0.33	0.23	0.16	0.11
$\dfrac{\psi_-}{\left(\frac{1}{\pi a_0^3}\right)^{1/2}}$	0.14	0.21	0.31	0.45	0.65	0.95	0.54	0.20	−0.11	−0.43	−0.81	−0.73	−0.50	−0.23	−0.23	−0.16

The points are plotted in Figure 5.5.

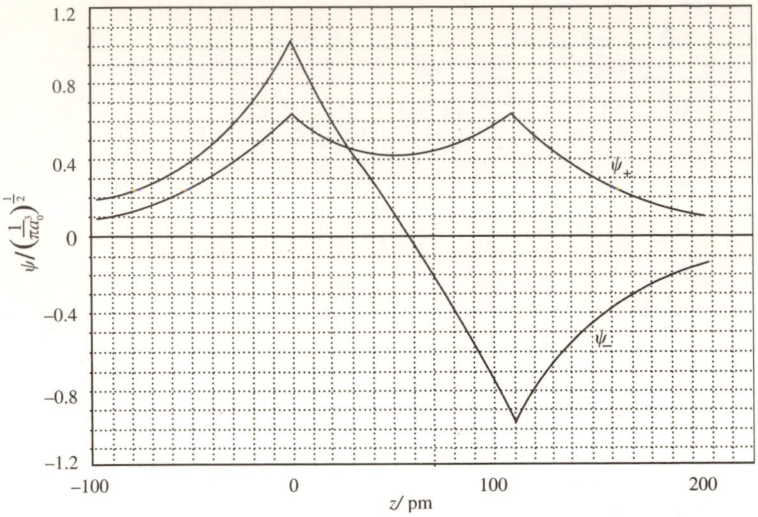

Figure 5.5

P5.7

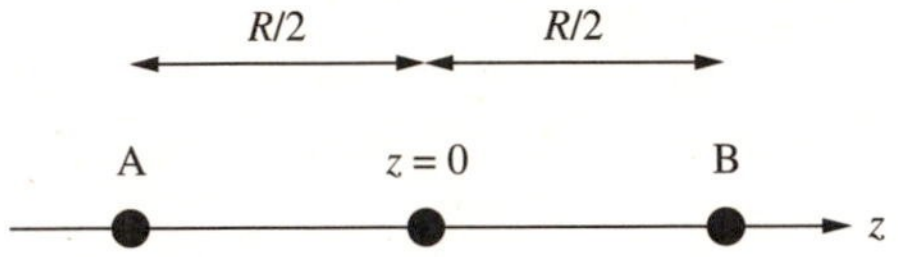

(a) With spatial dimensions in units (multiples) of a_0, the atomic orbitals of atom A and atom B may be written in the form

$$p_{z,A} = \frac{1}{4(2\pi)^{1/2}}(z + R/2)e^{-[x^2+y^2+(z+R/2)^2]^{1/2}/2}$$

$$p_{z,B} = \frac{1}{4(2\pi)^{1/2}}(z - R/2)e^{-[x^2+y^2+(z-R/2)^2]^{1/2}/2}$$

Following eqn. 5.7 and Example 5.1, we form LCAO-MOs of the form

$$\psi_{\sigma^*} = \frac{p_{z,A} + p_{z,B}}{\{2(1+S)\}^{1/2}} \text{[antibonding]} \quad \text{and} \quad \psi_{\sigma} = \frac{p_{z,A} - p_{z,B}}{\{2(1-S)\}^{1/2}} \text{[bonding]}$$

where $S = \int_{-\infty}^{\infty}\int_{-\infty}^{\infty}\int_{-\infty}^{\infty} p_{z,A}p_{z,B}\,dx\,dy\,dz$ [5.17].

Computations and plots are readily prepared with mathematical software such as Mathcad (see Figure 5.6).

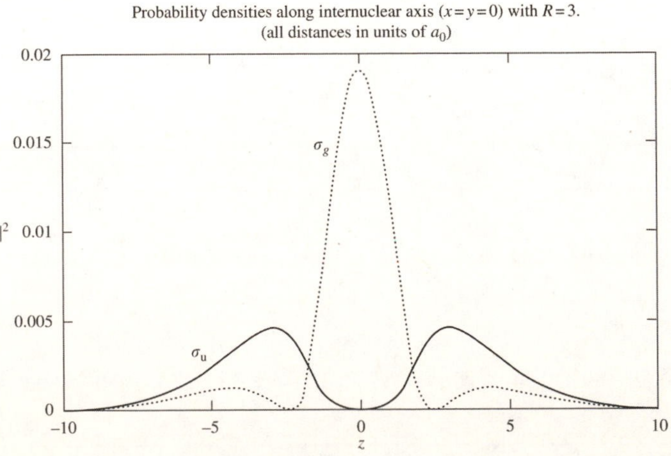

Figure 5.6

(b) With spatial dimensions in units of a_0, the atomic orbitals for the construction of π molecular orbitals are

$$p_{x,A} = \frac{1}{4(2\pi)^{1/2}} x e^{-[x^2+y^2+(z+R/2)^2]^{1/2}/2}$$

$$p_{x,B} = \frac{1}{4(2\pi)^{1/2}} x e^{-[x^2+y^2+(z-R/2)^2]^{1/2}/2}$$

See Figures 5.7 and 5.8.

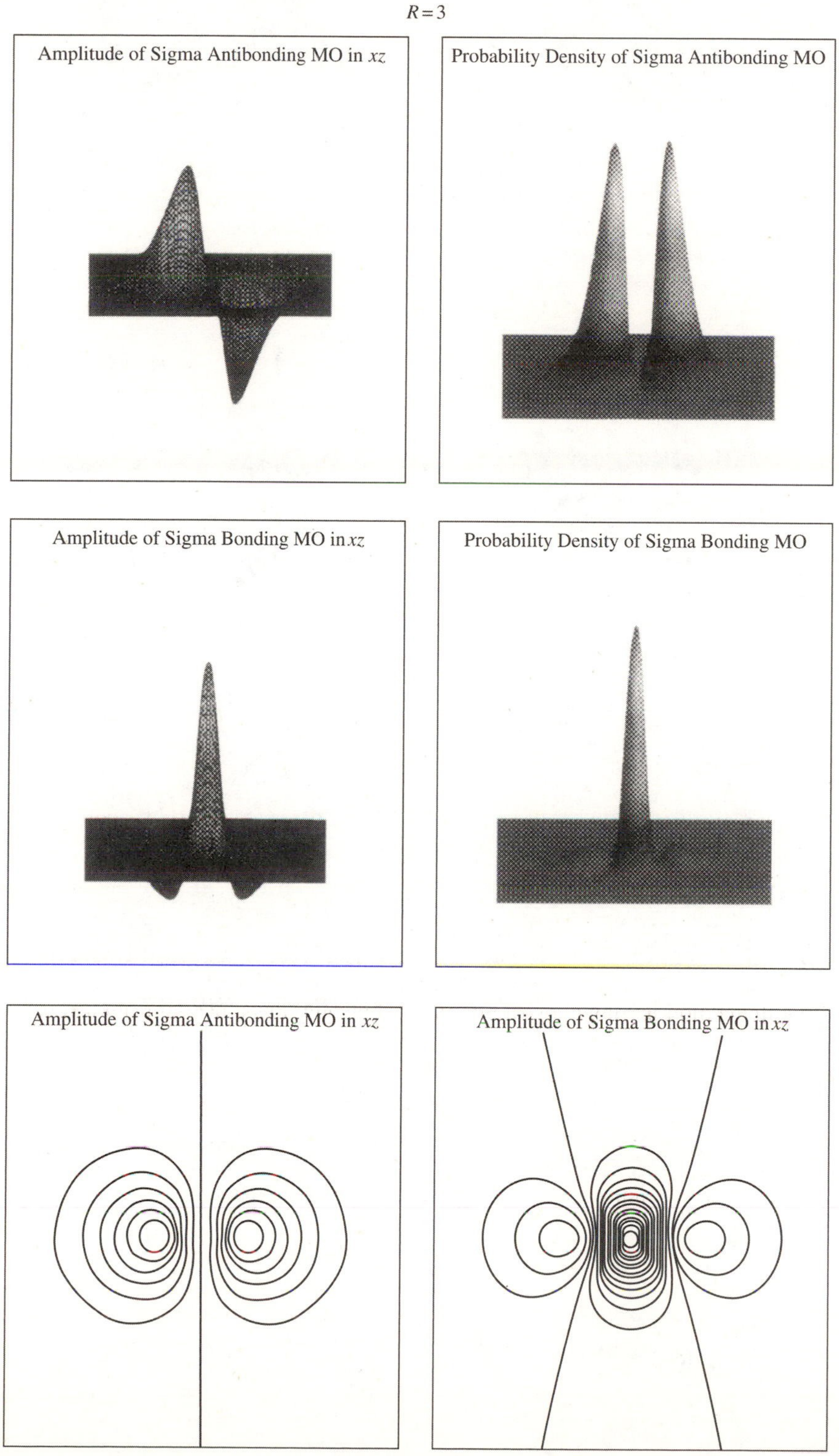

Figure 5.7

$R = 3$

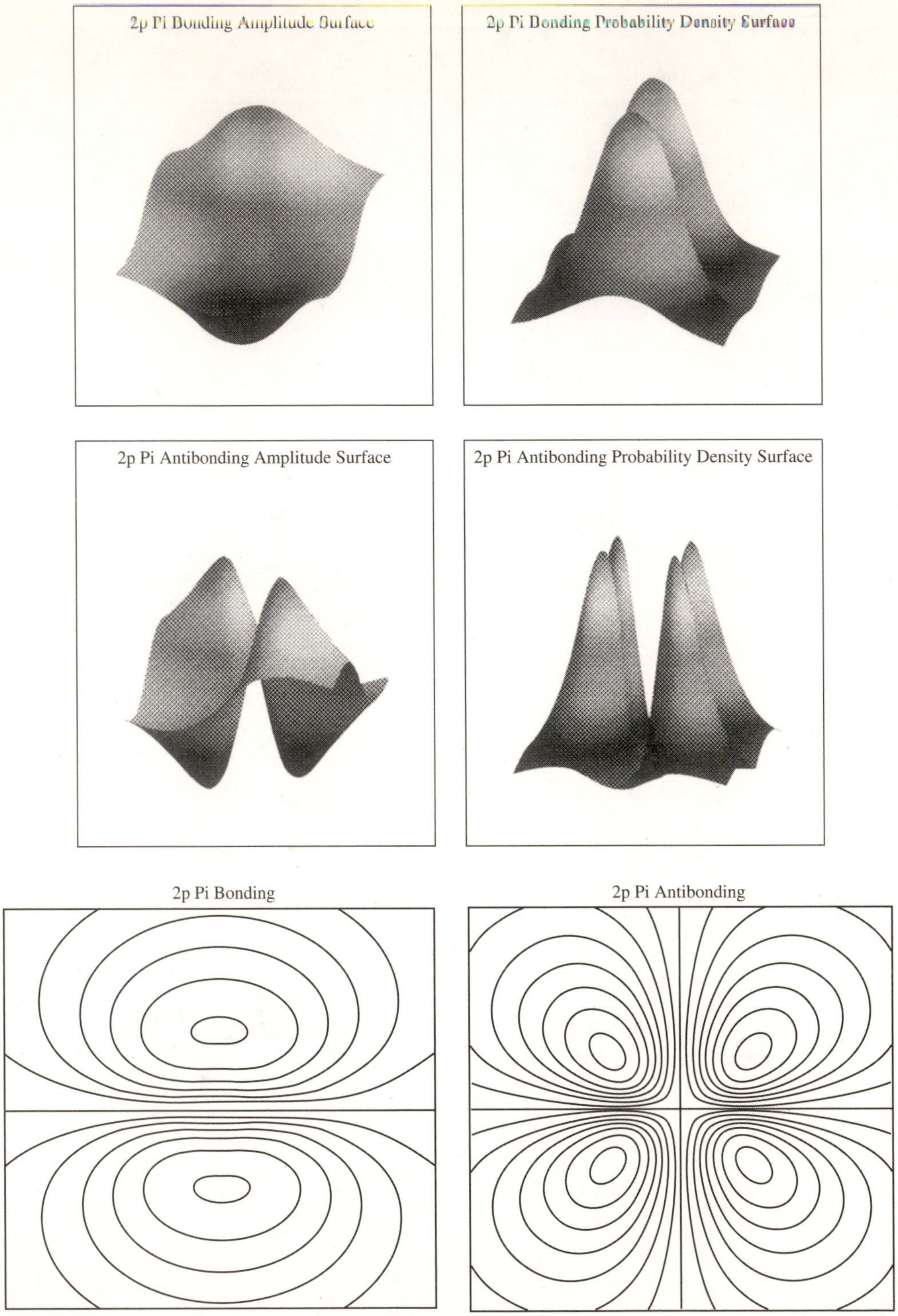

Figure 5.8

The π MOs are

$$\psi_{\pi} = \frac{\mathrm{p}_{x,\mathrm{A}} + \mathrm{p}_{x,\mathrm{B}}}{\{2(1+S)\}^{1/2}} \text{ [bonding]} \quad \text{and} \quad \psi_{\pi^*} = \frac{\mathrm{p}_{x,\mathrm{A}} - \mathrm{p}_{x,\mathrm{B}}}{\{2(1-S)\}^{1/2}} \text{ [antibonding]}$$

where $S = \int_{-\infty}^{\infty}\int_{-\infty}^{\infty}\int_{-\infty}^{\infty} \mathrm{p}_{x,\mathrm{A}}\mathrm{p}_{x,\mathrm{B}}\, \mathrm{d}x\, \mathrm{d}y\, \mathrm{d}z.$

The plots clearly show the constructive interference that makes a bonding molecular orbital. Nodal planes created by destructive interference are clearly seen in the antibonding molecular orbitals. When calculations and plots are produced for the $R = 10$ case, constructive and destructive interference is seen to be much weaker because of the weak atomic orbital overlap.

P5.9 $E_H = E_1 = -hcR_H$ [Section 4.2(a)]

Draw up the following table using the data in the question and using

$$\frac{e^2}{4\pi\varepsilon_0 R} = \frac{e^2}{4\pi\varepsilon_0 a_0} \times \frac{a_0}{R} = \frac{e^2}{4\pi\varepsilon_0 \times (4\pi\varepsilon_0 \hbar^2 / m_e e^2)} \times \frac{a_0}{R}$$

$$= \frac{m_e e^4}{16\pi^2 \varepsilon_0^2 \hbar^2} \times \frac{a_0}{R} = E_h \times \frac{a_0}{R} \quad \left[E_h \equiv \frac{m_e e^4}{16\pi^2 \varepsilon_0^2 \hbar^2} = 2hcR_H \right]$$

so that $\dfrac{e^2}{4\pi\varepsilon_0 R} \times \dfrac{1}{E_h} = \dfrac{a_0}{R}$.

R / a_0	0	1	2	3	4	∞
$\dfrac{e^2}{4\pi\varepsilon_0 R} \times \dfrac{1}{E_h}$	∞	1	0.500	0.333	0.250	0
$(V_1 + V_2) / E_h$	2.000	1.465	0.843	0.529	0.342	0
$(E - E_H) / E_h$	∞	0.212	–0.031	–0.059	–0.038	0

The points are plotted in Figure 5.9.

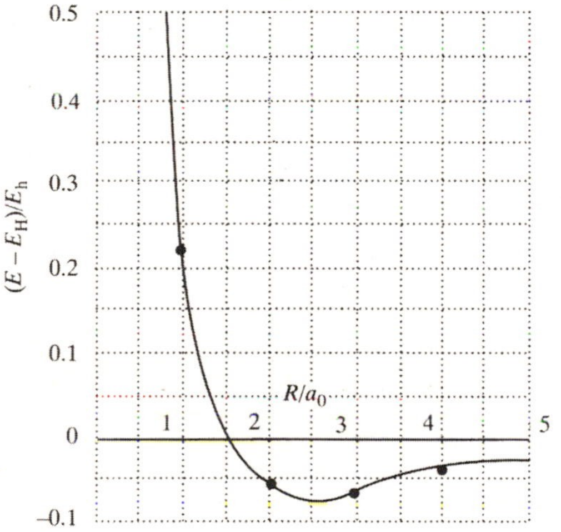

Figure 5.9

The minimum occurs at $R = 2.5a_0$, so $R = 130$ pm. At that bond length

$$E - E_H = -0.07E_h = -1.91\,\text{eV}$$

Hence, the dissociation energy is predicted to be about $\boxed{1.9\text{ eV}}$ and the equilibrium bond length about $\boxed{130\text{ pm}}$.

P5.11 (a) Start from eqn. 5.11 (whose derivation is covered in problem 5.23).

$$\langle \hat{H} \rangle = E_H - \frac{J + K}{1 + S} + \frac{j_0}{R}$$

where we define the Coulomb integral J and the resonance integral K as in eqn. 5.12 and use the relationship of the normalization constant N to the overlap integral S as defined in Example 5.1. We must write the integrals S, J, and K explicitly to evaluate them. Use

$$A = \frac{e^{-r_A/a_0}}{(\pi a_0^3)^{1/2}} \quad \text{and} \quad B = \frac{e^{-r_B/a_0}}{(\pi a_0^3)^{1/2}} \quad [5.8]$$

Define a coordinate system centered on nucleus A so that r_A is a coordinate and

$$r_B = (r_A^2 + R^2 - 2r_A R \cos\theta)^{1/2} \text{ [5.9]}$$

The volume element is

$$d\tau = r_A^2 \sin\theta \, dr_A \, d\theta \, d\phi$$

Thus, the overlap integral is

$$S = \int AB d\tau = \int_0^{2\pi}\int_0^{\pi}\int_0^{\infty} \frac{e^{-r_A/a_0}}{(\pi a_0^3)^{1/2}} \frac{e^{-r_B/a_0}}{(\pi a_0^3)^{1/2}} r_A^2 \sin\theta dr_A d\theta d\phi$$

Integration over $d\phi$ yields a factor of 2π. Numerical integration over $d\theta$ and dr_A can be done by approximating the differentials by small but finite $\delta\theta$ and δr_A and summing over a series of θ_j and $r_{A,i}$:

$$S = \frac{2}{a_0^3}\int_0^{\pi}\int_0^{\infty} e^{-r_A/a_0} e^{-r_B/a_0} r_A^2 \sin\theta dr_A d\theta \approx 2\delta\theta \frac{\delta r_A}{a_0} \sum_j \sum_i e^{-r_{A,i}/a_0} e^{-r_{B,ij}/a_0} \left(\frac{r_{A,i}}{a_0}\right)^2 \sin\theta_j$$

Similar treatment of the other integrals yields

$$J \approx 2\delta\theta \frac{\delta r_A}{a_0} \times \frac{j_0}{a_0} \sum_j \sum_i \frac{e^{-2r_{A,i}/a_0}}{(r_{B,ij}/a_0)} \left(\frac{r_{A,i}}{a_0}\right)^2 \sin\theta_j$$

$$K \approx 2\delta\theta \frac{\delta r_A}{a_0} \times \frac{j_0}{a_0} \sum_j \sum_i \frac{e^{-r_{A,i}/a_0} e^{-r_{B,ij}/a_0}}{(r_{B,ij}/a_0)} \left(\frac{r_{A,i}}{a_0}\right)^2 \sin\theta_j$$

To evaluate the integrals, express the distances in units of a_0. S is dimensionless. J and K have dimensions of energy; they have units of $j_0 / a_0 = E_h$. (See problem 5.9 for E_h.) In the following table and plots, the numerical values of the integrals were computed using θ_j from 0 to π in increments that correspond to 10° and using $r_{A,i}$ from 0 to $5a_0$ in increments of $0.25a_0$.

R/a_0	S_{num}	J_{num}/E_h	K_{num}/E_h	S_{an}	J_{an}/E_h	K_{an}/E_h
1.00	0.8524	0.7227	0.7220	0.8584	0.7293	0.7358
1.25	0.7875	0.6460	0.6287	0.7939	0.6522	0.6446
1.50	0.7183	0.5781	0.5408	0.7252	0.5837	0.5578
1.75	0.6480	0.5192	0.4606	0.6553	0.5240	0.4779
2.00	0.5789	0.4684	0.3890	0.5865	0.4725	0.4060
2.25	0.5125	0.4249	0.3262	0.5204	0.4284	0.3425
2.50	0.4501	0.3876	0.2718	0.4583	0.3906	0.2873
2.75	0.3924	0.3555	0.2253	0.4009	0.3581	0.2397
3.00	0.3397	0.3279	0.1858	0.3485	0.3300	0.1991
3.25	0.2922	0.3038	0.1525	0.3013	0.3057	0.1648
3.50	0.2498	0.2829	0.1245	0.2592	0.2845	0.1359
3.75	0.2122	0.2644	0.1012	0.2219	0.2660	0.1117
4.00	0.1791	0.2482	0.0817	0.1893	0.2496	0.0916

A plot of S versus R is shown in Figure 5.10(a) and a plot of J and K versus R is shown in Figure 5.10(b). Note that the numerical results at this level of approximation agree well but not identically with the analytical ones. The disagreement becomes more noticeable at larger R, which is not surprising given that all sums were truncated at $r_A = 5a_0$. The disagreement is more pronounced in K than in S or J. The numerical results can be made arbitrarily close to the analytical by using smaller increments and extending the range of r_A.

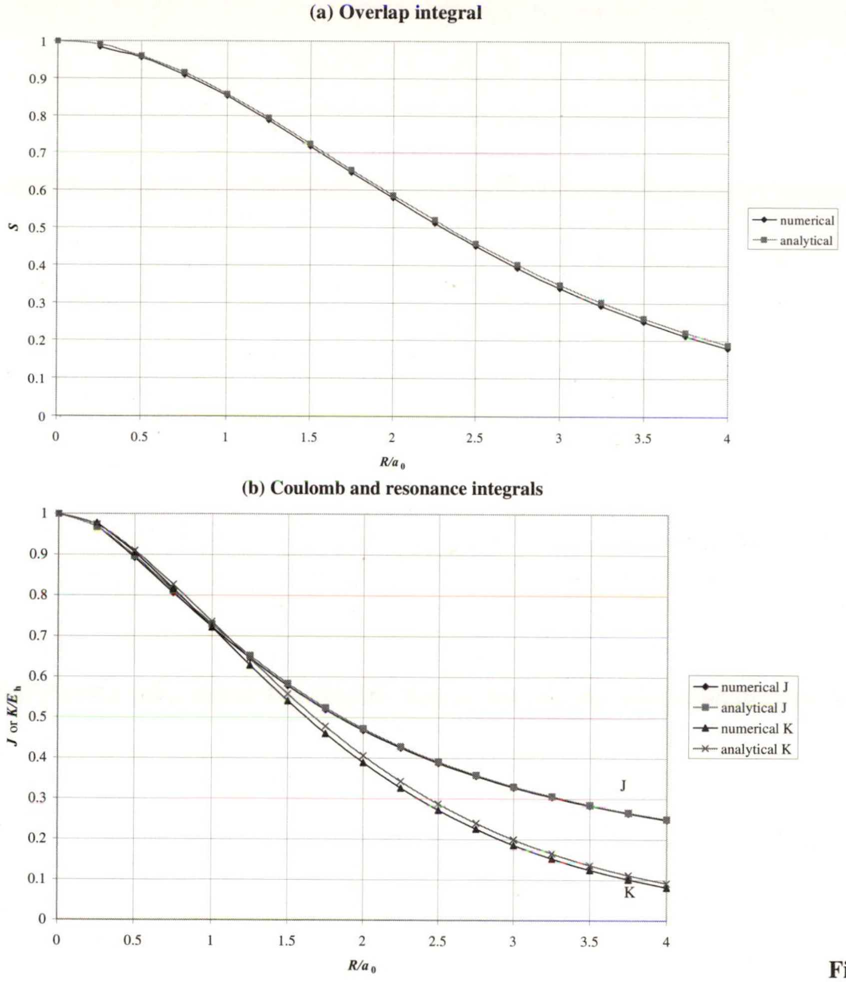

Figure 5.10

(b) Use eqn. 5.11 to compute the total energy, whether from numerical or analytical values of the integrals. Results are shown in the table below and in Figure 5.11.

R/a_0	E_{num} / E_h	E_{an} / E_h
1.00	–0.280	–0.288
1.25	–0.413	–0.423
1.50	–0.485	–0.495
1.75	–0.523	–0.534
2.00	–0.543	–0.554
2.25	–0.552	–0.563
2.50	–0.555	–0.565
2.75	–0.554	–0.563
3.00	–0.550	–0.559
3.25	–0.545	–0.554
3.50	–0.540	–0.548
3.75	–0.535	–0.542
4.00	–0.530	–0.537

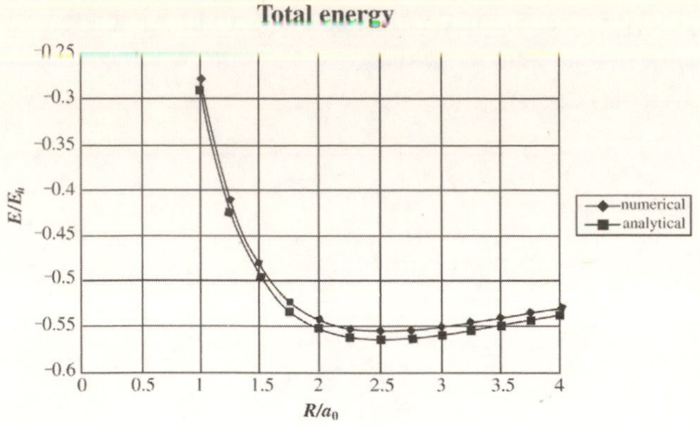

Figure 5.11

The minimum in both total energy curves lies at $R = \boxed{2.5a_0 = 1.3\times10^{-10}\ \text{m}}$. The minimum total energy is $E_{\text{min}} = \boxed{-0.555E_\text{h} = -15.1\ \text{eV}}$ (numerical) and $\boxed{-0.565E_\text{h} = -15.4\ \text{eV}}$ (analytical). The bond dissociation energy is

$$D_\text{e} = E_\text{H} - E_{\text{min}} = (-0.500 + 0.555)E_\text{h} = \boxed{0.055E_\text{h} = 1.5\ \text{eV}}\ \text{(numerical)}$$

$$D_\text{e} = (-0.500 + 0.565)E_\text{h} = \boxed{0.065E_\text{h} = 1.8\ \text{eV}}\ \text{(analytical)}$$

COMMENT. Dissociation energies are typically (relatively) small differences between larger quantities (electron-binding energies). Even small errors in the latter lead to relatively large errors in the former. In this case, the numerical binding energy differed by about 2% from the analytical value; the difference in D_e was closer to 20%.

P5.13 In the simple Hückel approximation,

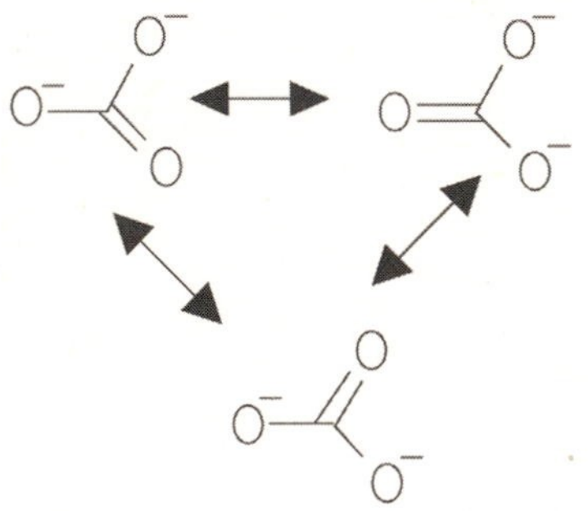

$$\begin{vmatrix} \alpha_\text{O} - E & 0 & 0 & \beta \\ 0 & \alpha_\text{O} - E & 0 & \beta \\ 0 & 0 & \alpha_\text{O} - E & \beta \\ \beta & \beta & \beta & \alpha_\text{C} - E \end{vmatrix} = 0$$

$$(E-\alpha_\text{O})^2 \times \{(E-\alpha_\text{O})\times(E-\alpha_\text{C})-3\beta^2\} = 0$$

Therefore, the factors yield

$$E - \alpha_\text{O} = 0\ \text{(twice)}$$

so $\boxed{E = \alpha_\text{O}}$ and $(E - \alpha_\text{O}) \times (E - \alpha_\text{C}) - 3\beta^2 = 0 = E^2 - E(\alpha_\text{O} + \alpha_\text{C}) + \alpha_\text{O}\alpha_\text{C} - 3\beta^2$

so $$E_{\pm}=\frac{\alpha_O+\alpha_C\pm\sqrt{(\alpha_O+\alpha_C)^2-4(\alpha_O\alpha_C-3\beta^2)}}{2}=\frac{\alpha_O+\alpha_C\pm\sqrt{(\alpha_O-\alpha_C)^2+12\beta^2}}{2}$$

$$=\boxed{\frac{1}{2}\left(\alpha_O+\alpha_C\pm(\alpha_O-\alpha_C)\sqrt{1+\frac{12\beta^2}{(\alpha_O-\alpha_C)^2}}\right)}$$

Because Coulomb integrals approximate ionization energies in magnitude, α_O is more negative than α_C. Therefore, E_+ corresponds to the lowest-energy orbital and E_- to the highest-energy orbital; the degenerate orbitals at $E = \alpha_O$ fall in between.

The π energies in the absence of resonance are derived for just one of the three structures, i.e., for a structure containing a single localized π bond.

$$\begin{vmatrix}\alpha_O-E & \beta \\ \beta & \alpha_C-E\end{vmatrix}=0$$

Expanding the determinant and solving for E gives

$$(E-\alpha_O)\times(E-\alpha_C)-\beta^2=0=E^2-E(\alpha_O+\alpha_C)+\alpha_O\alpha_C-\beta^2$$

so $$E_{\text{local}\pm}=\frac{1}{2}\left(\alpha_O+\alpha_C\pm(\alpha_O-\alpha_C)\sqrt{1+\frac{4\beta^2}{(\alpha_O-\alpha_C)^2}}\right)$$

There are two π electrons in the system, so the delocalization energy is

$$2E_+-2E_{\text{local}+}=\boxed{(\alpha_O-\alpha_C)\left(\sqrt{1+\frac{12\beta^2}{(\alpha_O-\alpha_C)^2}}-\sqrt{1+\frac{4\beta^2}{(\alpha_O-\alpha_C)^2}}\right)}$$

If $12\beta^2 \ll (\alpha_O-\alpha_C)^2$, we can use $(1+x)^{1/2}\approx 1+x/2$, so the delocalization energy is

$$\approx(\alpha_O-\alpha_C)\left(1+\frac{12\beta^2}{2(\alpha_O-\alpha_C)^2}-1-\frac{4\beta^2}{2(\alpha_O-\alpha_C)^2}\right)=\boxed{\frac{4\beta^2}{(\alpha_O-\alpha_C)}}$$

P5.15 (a) The transitions occur for photons whose energies are equal to the difference in energy between the highest occupied and lowest unoccupied orbital energies:

$$E_{\text{photon}}=E_{\text{LUMO}}-E_{\text{HOMO}}$$

If N is the number of carbon atoms in these species, then the number of π electrons is also N. These N electrons occupy the first $N/2$ orbitals, so orbital number $N/2$ is the HOMO and orbital number $1 + N/2$ is the LUMO. Writing the photon energy in terms of the wavenumber, substituting the given energy expressions with this identification of the HOMO and LUMO, gives

$$hc\tilde{\nu}=\left(\alpha+2\beta\cos\frac{(\frac{1}{2}N+1)\pi}{N+1}\right)-\left(\alpha+2\beta\cos\frac{\frac{1}{2}N\pi}{N+1}\right)$$

$$=2\beta\left(\cos\frac{(\frac{1}{2}N+1)\pi}{N+1}-\cos\frac{\frac{1}{2}N\pi}{N+1}\right).$$

Solving for β yields

$$\beta=\frac{hc\tilde{\nu}}{2\left(\cos\frac{(\frac{1}{2}N+1)\pi}{N+1}-\cos\frac{\frac{1}{2}N\pi}{N+1}\right)}$$

Draw up the following table.

Species	N	$\tilde{\nu}$ / cm^{-1}	Estimated β/eV
C_2H_4	2	61500	–3.813
C_4H_6	4	46080	–4.623
C_6H_8	6	39750	–5.538
C_8H_{10}	8	32900	–5.873

(b) The total energy of the π electron system is the sum of the energies of occupied orbitals weighted by the number of electrons that occupy them. In C_8H_{10}, each of the first four orbitals are doubly occupied, so

$$E_\pi = 2\sum_{k=1}^{4} E_k = 2\sum_{k=1}^{4}\left(\alpha + 2\beta\cos\frac{k\pi}{9}\right) = 8\alpha + 4\beta\sum_{k=1}^{4}\cos\frac{k\pi}{9} = 8\alpha + 9.518\beta$$

The delocalization energy is the difference between this quantity and that of four isolated double bonds:

$$E_{\text{deloc}} = E_\pi - 8(\alpha+\beta) = 8\alpha + 9.518\beta - 8(\alpha+\beta) = \boxed{1.518\beta}$$

Using the estimate of β from part (a) yields $E_{\text{deloc}} = \boxed{8.913\text{ eV}}$.

(c) Draw up the following table, in which the orbital energy decreases as we go down. For the purpose of comparison, we express orbital energies as $(E_k - \alpha)/\beta$. Recall that β is negative (as is α for that matter), so the orbital with the greatest value of $(E_k - \alpha)/\beta$ has the lowest energy.

	Energy	Coefficients					
Orbital	$(E_k-\alpha)/\beta$	1	2	3	4	5	6
6	–1.8019	0.2319	–0.4179	0.5211	–0.5211	0.4179	–0.2319
5	–1.2470	0.4179	–0.5211	0.2319	0.2319	–0.5211	0.4179
4	–0.4450	0.5211	–0.2319	–0.4179	0.4179	0.2319	–0.5211
3	0.4450	0.5211	0.2319	–0.4179	–0.4179	0.2319	0.5211
2	1.2470	0.4179	0.5211	0.2319	–0.2319	–0.5211	–0.4179
1	1.8019	0.2319	0.4179	0.5211	0.5211	0.4179	0.2319

The orbitals are shown schematically in Figure 5.12, with each vertical pair of lobes representing a p orbital on one of the carbons in hexatriene. Shaded lobes represent one sign of the wavefunction (say, positive) and unshaded lobes the other sign. Where adjacent atoms have atomic orbitals of the same sign, the resulting molecular orbital is bonding with respect to those atoms; where adjacent atoms have a different sign, there is a node between the atoms and the resulting molecular orbital is antibonding with respect to them. The lowest-energy orbital is totally bonding (no nodes between atoms) and the highest-energy orbital is totally antibonding (nodes between each adjacent pair). Note that the orbitals have increasing antibonding character as their energy increases. The size of each atomic p orbital is proportional to the magnitude of the coefficient of that orbital in the molecular orbital. So, for example, in orbitals 1 and 6, the largest lobes are in the center of the molecule, so electrons that occupy those orbitals are more likely to be found near the center of the molecule than on the ends. In the ground state of the molecule, there are two electrons in each of orbitals 1, 2, and 3, with the result that the probability of finding a π electron in hexatriene is uniform over the entire molecule.

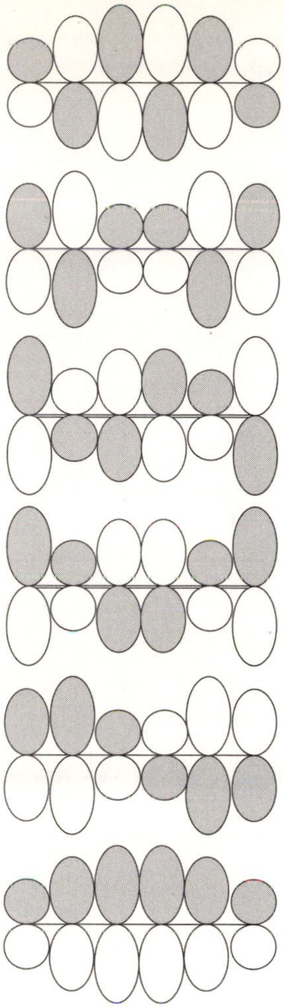

Figure 5.12

P5.17 We use the Hückel approximation, neglecting overlap integrals.

The secular determinant of ethene is $\begin{vmatrix} \alpha - E & \beta \\ \beta & \alpha - E \end{vmatrix}$

Mathematical software (such as the Simple Huckel Molecular Orbital Theory Calculator at Canada's University of Calgary, http://www.chem.ucalgary.ca/SHMO/) diagonalizes the hamiltonian matrix to

$$\boldsymbol{E} = \begin{pmatrix} \alpha + \beta & 0 \\ 0 & \alpha - \beta \end{pmatrix}$$

and the matrix that achieves the diagonalization is

$$\boldsymbol{C} = \begin{pmatrix} 0.707 & 0.707 \\ 0.707 & -0.707 \end{pmatrix}$$

The secular determinant of butadiene is $\begin{vmatrix} \alpha - E & \beta & 0 & 0 \\ \beta & \alpha - E & \beta & 0 \\ 0 & \beta & \alpha - E & \beta \\ 0 & 0 & \beta & \alpha - E \end{vmatrix}$

The hamiltonian matrix is diagonalized to

$$\boldsymbol{E}=\begin{pmatrix}\alpha+1.618\beta & 0 & 0 & 0\\ 0 & \alpha+0.618\beta & 0 & 0\\ 0 & 0 & \alpha-0.618\beta & 0\\ 0 & 0 & 0 & \alpha-1.618\beta\end{pmatrix}$$

and the matrix that achieves the diagonalization is

$$\boldsymbol{C}=\begin{pmatrix}0.372 & 0.602 & 0.602 & 0.372\\ 0.602 & 0.372 & -0.372 & -0.602\\ 0.602 & -0.372 & -0.372 & 0.602\\ 0.372 & -0.602 & 0.602 & -0.372\end{pmatrix}$$

The secular determinant of hexatriene is
$$\begin{vmatrix}\alpha-E & \beta & 0 & 0 & 0 & 0\\ \beta & \alpha-E & \beta & 0 & 0 & 0\\ 0 & \beta & \alpha-E & \beta & 0 & 0\\ 0 & 0 & \beta & \alpha-E & \beta & 0\\ 0 & 0 & 0 & \beta & \alpha-E & \beta\\ 0 & 0 & 0 & 0 & \beta & \alpha-E\end{vmatrix}$$

The hamiltonian matrix is diagonalized to

$$\boldsymbol{E}=\begin{pmatrix}\alpha+1.802\beta & 0 & 0 & 0 & 0 & 0\\ 0 & \alpha+1.247\beta & 0 & 0 & 0 & 0\\ 0 & 0 & \alpha+0.445\beta & 0 & 0 & 0\\ 0 & 0 & 0 & \alpha-0.445\beta & 0 & 0\\ 0 & 0 & 0 & 0 & \alpha-1.247\beta & 0\\ 0 & 0 & 0 & 0 & 0 & \alpha-1.802\beta\end{pmatrix}$$

and the matrix that achieves the diagonalization is

$$\boldsymbol{C}=\begin{pmatrix}0.232 & 0.418 & 0.521 & 0.521 & 0.418 & 0.232\\ 0.418 & 0.521 & 0.232 & -0.232 & -0.521 & -0.418\\ 0.521 & 0.232 & -0.418 & -0.418 & 0.232 & 0.521\\ 0.521 & -0.232 & -0.418 & 0.418 & 0.232 & -0.521\\ 0.418 & -0.521 & 0.232 & 0.232 & -0.521 & 0.418\\ 0.232 & -0.418 & 0.521 & -0.521 & 0.418 & -0.232\end{pmatrix}$$

The secular determinant of octatetraene is

$$\begin{vmatrix}\alpha-E & \beta & 0 & 0 & 0 & 0 & 0 & 0\\ \beta & \alpha-E & \beta & 0 & 0 & 0 & 0 & 0\\ 0 & \beta & \alpha-E & \beta & 0 & 0 & 0 & 0\\ 0 & 0 & \beta & \alpha-E & \beta & 0 & 0 & 0\\ 0 & 0 & 0 & \beta & \alpha-E & \beta & 0 & 0\\ 0 & 0 & 0 & 0 & \beta & \alpha-E & \beta & 0\\ 0 & 0 & 0 & 0 & 0 & \beta & \alpha-E & \beta\\ 0 & 0 & 0 & 0 & 0 & 0 & \beta & \alpha-E\end{vmatrix}$$

The hamiltonian matrix is diagonalized to

$$E=\begin{pmatrix} \alpha+1.879\beta & 0 & 0 & 0 & 0 & 0 & 0 & 0 \\ 0 & \alpha+1.532\beta & 0 & 0 & 0 & 0 & 0 & 0 \\ 0 & 0 & \alpha+\beta & 0 & 0 & 0 & 0 & 0 \\ 0 & 0 & 0 & \alpha+0.347\beta & 0 & 0 & 0 & 0 \\ 0 & 0 & 0 & 0 & \alpha-0.347\beta & 0 & 0 & 0 \\ 0 & 0 & 0 & 0 & 0 & \alpha-\beta & 0 & 0 \\ 0 & 0 & 0 & 0 & 0 & 0 & \alpha-1.532\beta & 0 \\ 0 & 0 & 0 & 0 & 0 & 0 & 0 & \alpha-1.879\beta \end{pmatrix}$$

and the matrix that achieves the diagonalization is

$$C=\begin{pmatrix} 0.161 & 0.303 & 0.408 & 0.464 & 0.464 & 0.408 & 0.303 & 0.161 \\ 0.303 & 0.464 & 0.408 & 0.161 & -0.161 & -0.408 & -0.464 & -0.303 \\ 0.408 & 0.408 & 0 & -0.408 & -0.408 & 0 & 0.408 & 0.408 \\ 0.464 & 0.161 & -0.408 & -0.303 & 0.303 & 0.408 & -0.161 & -0.464 \\ 0.464 & -0.161 & -0.408 & 0.303 & 0.303 & -0.408 & -0.161 & 0.464 \\ 0.408 & -0.408 & 0 & 0.408 & -0.408 & 0 & 0.408 & -0.408 \\ 0.303 & -0.464 & 0.408 & -0.161 & -0.161 & 0.408 & -0.464 & 0.303 \\ 0.161 & -0.303 & 0.408 & -0.464 & 0.464 & -0.408 & 0.303 & -0.161 \end{pmatrix}$$

The columns of the C matrices are coefficients of the atomic orbitals in each molecular orbital. The first column represents the lowest-energy molecular orbital. In each molecule, all the coefficients are positive, meaning that the lowest-energy MO has no nodes. Note also that the coefficients rise from one end of the molecule to the middle and then decrease to the other end; thus the lowest-energy MO extends over all the carbon atoms of the chain. Note that the number of sign changes within a given column increases as one moves from the lowest-energy MO (leftmost column) to the highest-energy (rightmost column). Each time the sign of a coefficient changes, there is a node in the wavefunction. So, for example, the second column in octatetraene has four positive and four negative coefficients but only one sign change and one node. The eighth column also has four positive and four negative coefficients, but the sign changes between every atom; there are seven nodes. The number of nodes increases with the energy of the MO.

P5.19 In butadiene, the HOMO is bonding with respect to the two bonds on the ends (the ones conventionally depicted as double bonds) and antibonding with respect to the central bond (the one conventionally depicted as a single bond). The situation is reversed in the LUMO. Thus, a HOMO $\rightarrow$ LUMO transition weakens the terminal C=C bonds and strengthens the central C–C bond. (See Figure 5.13a.)

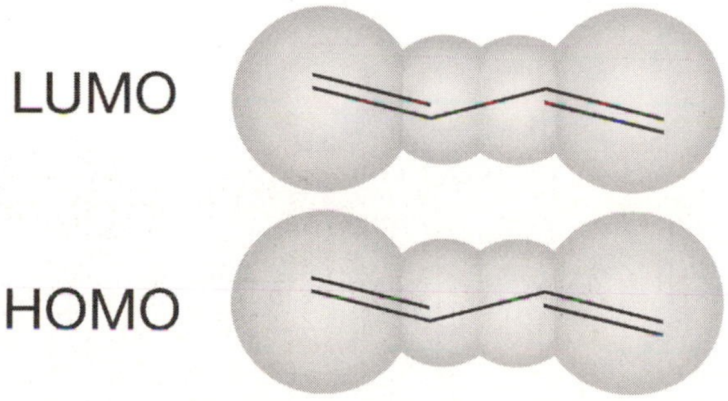

Figure 5.13(a)

In benzene, the situation is more complicated because there are two degenerate LUMOs. (See Figure 5.13b.) We will analyze these orbitals in terms of the top and bottom halves as shown in the figure. The HOMO on the left is bonding *between* the two halves but antibonding *within* each half; the HOMO on the right is completely bonding *within* each half but antibonding *between* the halves. The LUMO on the left is also bonding *between* the halves but completely antibonding *within* each half (more antibonding than the HOMO on the left); the LUMO on the right is antibonding ***between*** the halves and antibonding ***within*** each half as well. A transition from the right-hand HOMO to the right-hand LUMO or left-hand HOMO to left-hand LUMO would not affect the bonding ***between*** the halves, and it would destabilize the bonding ***within*** each half. A transition from the left-hand HOMO to the right-hand LUMO would weaken the bonding ***between*** the halves, leaving the bonding ***within*** each half unchanged. A transition from the right-hand HOMO to the left-hand LUMO would strengthen the bonding ***between*** the halves and destabilize the bonding between each atom ***within*** each half.

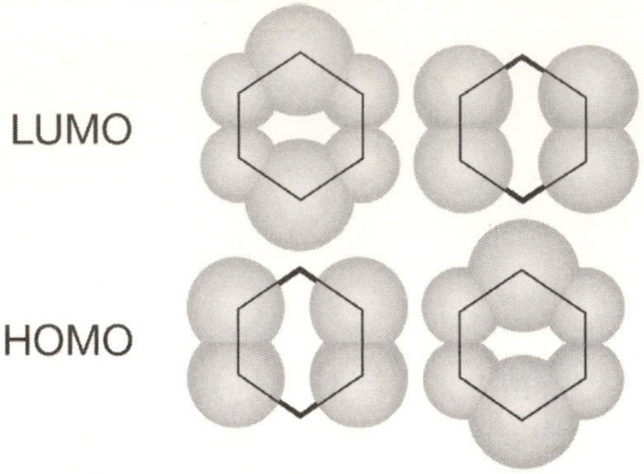

Figure 5.13(b)

Solutions to theoretical problems

P5.21 What does it mean to say that the antibonding orbital is more antibonding than the bonding orbital is bonding? It means that the energy level of the antibonding orbital is farther above the energy level of a free atom (i.e., a non-bonding atom) than the energy level of the bonding orbital is ***below*** the same free-atom level. The free-atom energy level is simply the energy of the ground state of the hydrogen atom, E_H. In symbols, then, we are asked to show that

$$E_{anti} - E_H > E_H - E_{bond}$$

which amounts to showing that

$$E_{anti} + E_{bond} > 2E_H$$

or, equivalently, that the average of the bonding and antibonding orbital energies is greater than the energy in the free atom:

$$(E_{anti} + E_{bond})/2 > E_H$$

$$E_{anti} + E_{bond} = E_H + \frac{e^2}{4\pi\varepsilon_0 R} - \frac{V_1 - V_2}{1-S} + E_H + \frac{e^2}{4\pi\varepsilon_0 R} - \frac{V_1 + V_2}{1+S}$$

$$= 2E_H + \frac{2e^2}{4\pi\varepsilon_0 R} - \frac{\{(V_1 - V_2)\times(1+S) + (1-S)\times(V_1+V_2)\}}{(1-S)\times(1+S)}$$

$$= 2E_H + \frac{2e^2}{4\pi\varepsilon_0 R} + \frac{2(SV_2 - V_1)}{1-S^2},$$

So $$\frac{E_{anti} + E_{bond}}{2} = E_H + \frac{e^2}{4\pi\varepsilon_0 R} + \frac{SV_2 - V_1}{1-S^2}$$

The nuclear repulsion term is always positive and always tends to raise the mean energy of the orbitals above E_H, although its effect decreases with increasing internuclear distance. Clearly the overlap integral decreases with increasing R, and inspection of the analytical expressions for V_1 and V_2 (J and K, respectively, in eqn. 5.12) shows that they also go to zero as R increases. Thus the last term vanishes as R increases, but it is difficult to assess analytically how it compares to the nuclear repulsion term. So we use numerical values given in problem 5.9, and we can add other points using the analytical expressions given in eqn. 5.12. Draw up the following table.

R/a_0	0.25	0.50	1	2	3	4	5	6
$\dfrac{e^2}{4\pi\varepsilon_0 R}\times\dfrac{1}{E_h}$	4.000	2.000	1	0.500	0.333	0.250	0.200	0.167
$\dfrac{SV_2-V_1}{1-S^2}\times\dfrac{1}{E_h}$	–0.188	–0.291	–0.370	–0.357	–0.297	–0.241	–0.198	–0.166
$\{(E_{\text{bond}}+E_{\text{anti}})/2 - E_{\text{H}}\}/E_{\text{h}}$	3.812	1.709	0.630	0.143	0.037	0.009	0.002	0.001

The last term indeed is complicated, appearing to go to zero at both small and large R. At small R, it appears to be negligible compared to the nuclear repulsion term. At large R, its absolute value appears to follow closely that of the nuclear repulsion term but to remain smaller. Thus, the antibonding orbital appears to be more antibonding than the bonding orbital is bonding at all values of the internuclear distance. (See Figure 5.14.)

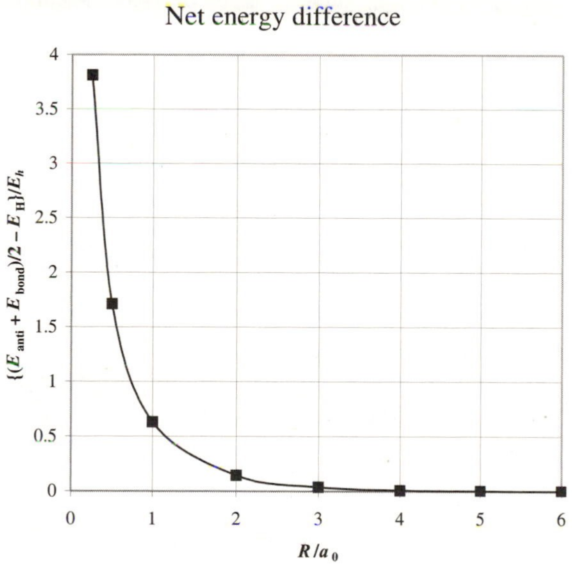

Figure 5.14

P5.23 The hamiltonian is

$$\hat{H}=-\frac{\hbar^2}{2m}\nabla^2-\frac{e^2}{4\pi\varepsilon_0}\left(\frac{1}{r_{\text{A}}}+\frac{1}{r_{\text{B}}}-\frac{1}{R}\right)=-\frac{\hbar^2}{2m}\nabla^2-j_0\left(\frac{1}{r_{\text{A}}}+\frac{1}{r_{\text{B}}}-\frac{1}{R}\right)\quad[5.6]$$

where we have introduced the abbreviation j_0 for the collection of constants $e^2/4\pi\varepsilon_0$. Its expectation value is

$$\langle H\rangle=\int\psi^*\hat{H}\psi\,\mathrm{d}\tau,$$

where $\psi_\pm=N_\pm(A\pm B)$ [5.7] and $N_\pm=\dfrac{1}{\{2(1+S)\}^{1/2}}$ [Example 5.1]

Thus, $$\langle\hat{H}\rangle=\int\psi_\pm^*\left\{-\frac{\hbar^2}{2m}\nabla^2-j_0\left(\frac{1}{r_{\text{A}}}+\frac{1}{r_{\text{B}}}-\frac{1}{R}\right)\right\}\psi_\pm\,\mathrm{d}\tau$$

Pull out the nuclear repulsion term:

$$\left\langle \hat{H} \right\rangle = \int \psi_\pm{}^* \left(-\frac{\hbar^2}{2m}\nabla^2 - \frac{j_0}{r_\mathrm{A}} - \frac{j_0}{r_\mathrm{B}} \right) \psi_\pm \mathrm{d}\tau + \frac{j_0}{R} \int \psi_\pm{}^* \psi_\pm \mathrm{d}\tau$$

$$= N_\pm{}^2 \int (A \pm B) \left(-\frac{\hbar^2}{2m}\nabla^2 - \frac{j_0}{r_\mathrm{A}} - \frac{j_0}{r_\mathrm{B}} \right) (A \pm B) \mathrm{d}\tau + \frac{j_0}{R}$$

Expand the remaining integral into four integrals, letting $\hat{H}_\mathrm{H}$ represent the hamiltonian operator of a hydrogen atom. Note that

$$\hat{H}_\mathrm{H} A = -\frac{\hbar^2}{2m}\nabla^2 A - \frac{j_0}{r_\mathrm{A}} A = E_\mathrm{H} A \quad \text{and} \quad \hat{H}_\mathrm{H} B = -\frac{\hbar^2}{2m}\nabla^2 B - \frac{j_0}{r_\mathrm{B}} B = E_\mathrm{H} B$$

So $$\left\langle \hat{H} \right\rangle = N_\pm{}^2 \int A \left(\hat{H}_\mathrm{H} - \frac{j_0}{r_\mathrm{B}} \right) A \mathrm{d}\tau \pm N_\pm{}^2 \int A \left(\hat{H}_\mathrm{H} - \frac{j_0}{r_\mathrm{A}} \right) B \mathrm{d}\tau$$

$$\pm N_\pm{}^2 \int B \left(\hat{H}_\mathrm{H} - \frac{j_0}{r_\mathrm{B}} \right) A \mathrm{d}\tau + N_\pm{}^2 \int B \left(\hat{H}_\mathrm{H} - \frac{j_0}{r_\mathrm{A}} \right) B \mathrm{d}\tau + \frac{j_0}{R}$$

and $$\left\langle \hat{H} \right\rangle = N_\pm{}^2 \int A \left(E_\mathrm{H} - \frac{j_0}{r_\mathrm{B}} \right) A \mathrm{d}\tau \pm N_\pm{}^2 \int A \left(E_\mathrm{H} - \frac{j_0}{r_\mathrm{A}} \right) B \mathrm{d}\tau$$

$$\pm N_\pm{}^2 \int B \left(E_\mathrm{H} - \frac{j_0}{r_\mathrm{B}} \right) A \mathrm{d}\tau + N_\pm{}^2 \int B \left(E_\mathrm{H} - \frac{j_0}{r_\mathrm{A}} \right) B \mathrm{d}\tau + \frac{j_0}{R}$$

Pull out the terms involving E_H:

$$\left\langle \hat{H} \right\rangle = E_\mathrm{H} N_\pm{}^2 \int (A^2 \pm 2AB + B^2) \mathrm{d}\tau$$

$$- j_0 N_\pm{}^2 \int \frac{A^2}{r_\mathrm{B}} \mathrm{d}\tau \;\mathrm{m}\; j_0 N_\pm{}^2 \int \frac{AB}{r_\mathrm{A}} \mathrm{d}\tau \;\mathrm{m}\; j_0 N_\pm{}^2 \int \frac{AB}{r_\mathrm{B}} \mathrm{d}\tau - j_0 N_\pm{}^2 \int \frac{B^2}{r_\mathrm{A}} \mathrm{d}\tau + \frac{j_0}{R}$$

What multiplies E_H is the normalization integral, which equals 1. Now examine the remaining four integrals. The first and fourth integrals are equal to each other, as are the second and third. (This can be demonstrated by interchanging the labels A and B, which are arbitrary.) Thus

$$\left\langle \hat{H} \right\rangle = E_\mathrm{H} - 2N_\pm{}^2 \left(j_0 \int \frac{A^2}{r_\mathrm{B}} \mathrm{d}\tau \pm j_0 \int \frac{AB}{r_\mathrm{A}} \mathrm{d}\tau \right) + \frac{j_0}{R}$$

$$= E_\mathrm{H} - 2N_\pm{}^2 (J \pm K) + \frac{j_0}{R} = E_\mathrm{H} - \frac{J \pm K}{1 \pm S} + \frac{j_0}{R}$$

This works out to eqns. 5.11 and 5.14:

$$\left\langle \hat{H} \right\rangle_+ = \boxed{E_\mathrm{H} - \frac{J+K}{1+S} + \frac{j_0}{R}} \quad \text{and} \quad \left\langle \hat{H} \right\rangle_- = \boxed{E_\mathrm{H} - \frac{J-K}{1-S} + \frac{j_0}{R}}$$

P5.25 (a) Expanding the determinant yields

$$\begin{vmatrix} \alpha_\mathrm{A} - E & \beta \\ \beta & \alpha_\mathrm{B} - E \end{vmatrix} = 0 = (\alpha_\mathrm{A} - E)(\alpha_\mathrm{B} - E) - \beta^2 = E^2 - (\alpha_\mathrm{A} + \alpha_\mathrm{B})E + \alpha_\mathrm{A}\alpha_\mathrm{B} - \beta^2$$

This is a quadratic equation in E where $a = 1$, $b = -(\alpha_A + \alpha_B)$, and $c = \alpha_A\alpha_B - \beta^2$. The solution is

$$E_\pm = \frac{-b \pm \sqrt{b^2 - 4ac}}{2a} = \frac{\alpha_A + \alpha_B}{2} \pm \frac{(\alpha_A{}^2 + 2\alpha_A\alpha_B + \alpha_B{}^2 - 4\alpha_A\alpha_B + 4\beta^2)^{1/2}}{2}$$

$$= \frac{\alpha_A + \alpha_B}{2} \pm \frac{[(\alpha_A - \alpha_B)^2 + 4\beta^2]^{1/2}}{2}$$

$$= \boxed{\frac{\alpha_A + \alpha_B}{2} \pm \frac{\alpha_A - \alpha_B}{2}\left(1 + \frac{4\beta^2}{(\alpha_A - \alpha_B)^2}\right)^{1/2}}$$

(b) By hypothesis, $(\alpha_A - \alpha_B)^2 >> \beta^2$, so the term in parentheses can be expanded:

$$E_\pm = \frac{\alpha_A + \alpha_B}{2} \pm \frac{\alpha_A - \alpha_B}{2}\left(1 + \frac{4\beta^2}{2(\alpha_A - \alpha_B)^2} - \dots\right),$$

so $$E_+ \approx \frac{\alpha_A + \alpha_B}{2} + \frac{\alpha_A - \alpha_B}{2} + \frac{\alpha_A - \alpha_B}{2}\left(\frac{4\beta^2}{2(\alpha_A - \alpha_B)^2}\right) = \boxed{\alpha_A + \frac{\beta^2}{\alpha_A - \alpha_B}}$$

and $$E_- \approx \frac{\alpha_A + \alpha_B}{2} - \frac{\alpha_A - \alpha_B}{2} - \frac{\alpha_A - \alpha_B}{2}\left(\frac{4\beta^2}{2(\alpha_A - \alpha_B)^2}\right) = \boxed{\alpha_B - \frac{\beta^2}{\alpha_A - \alpha_B}}$$

Thus, the energies of these molecular orbitals differ only slightly from the energies of the component atomic orbitals.

P5.27 The secular determinant for an N-carbon linear polyene (call the determinant P_N) has the form

$$\begin{array}{cccccccc} 1 & 2 & 3 & \dots & \dots & \dots & N-1 & N \end{array}$$
$$\begin{vmatrix} x & 1 & 0 & 0 & 0 & \dots & 0 & 0 \\ 1 & x & 1 & 0 & 0 & \dots & 0 & 0 \\ 0 & 1 & x & 1 & 0 & \dots & 0 & 0 \\ 0 & 0 & 1 & x & 1 & \dots & 0 & 0 \\ \vdots & \vdots & \vdots & \vdots & \vdots & \vdots & \vdots & \vdots \\ 0 & 0 & 0 & 0 & 0 & \dots & x & 1 \\ 0 & 0 & 0 & 0 & 0 & \dots & 1 & x \end{vmatrix} = P_N$$

where $x = \dfrac{\alpha - E}{\beta}$. The determinant can be expanded by cofactors; use the elements of the first row:

$$P_N = M_{11}c_{11} - M_{12}c_{12} + M_{13}c_{13} - \cdots + (-1)^{N+1}M_{1N}c_{1N}$$

In this notation, M_{1n} is the element in the first row, nth column of the determinant and c_{1n} is the cofactor of that element. The cofactor c_{1n} is a determinant whose elements are the elements of the original determinant with row 1 and column n removed. Note that the elements of the first row after M_{11} and M_{12} are zero, so $P_N = M_{11}c_{11} - M_{12}c_{12}$, $M_{11} = x$, and c_{11} is the secular determinant of an $(N-1)$-carbon polyene; that is, the c_{11} cofactor in P_N is P_{N-1}.

$M_{12} = 1$, so we are almost there:

$$P_N = xP_{N-1} - c_{12}$$

We need to examine c_{12}:

$$c_{12} = \begin{vmatrix} \cancel{x} & \boxed{\cancel{1}} & \cancel{0} & \cancel{0} & \cancel{0} & \cancel{\dots} & \cancel{0} & \cancel{0} \\ 1 & \cancel{x} & 1 & 0 & 0 & \dots & 0 & 0 \\ 0 & \cancel{1} & x & 1 & 0 & \dots & 0 & 0 \\ 0 & \cancel{0} & 1 & x & 1 & \dots & 0 & 0 \\ \vdots & \cancel{\vdots} & \vdots & \vdots & \vdots & \vdots & \vdots & \vdots \\ 0 & \cancel{0} & 0 & 0 & 0 & \dots & x & 1 \\ 0 & \cancel{0} & 0 & 0 & 0 & \dots & 1 & x \end{vmatrix}$$

Now c_{12} is not itself a secular determinant of a polyene; its first row has ones as its first two elements and no x elements. Its first column, though, has only one element, a one. Expanding this determinant by the elements and cofactors of its first column yields a one-term sum whose cofactor is illustrated here:

$$c_{12} = 1 \times \begin{vmatrix} \cancel{x} & \cancel{1} & \cancel{0} & \cancel{0} & \cancel{0} & \cancel{\dots} & \cancel{0} & \cancel{0} \\ \boxed{\cancel{1}} & \cancel{x} & \cancel{1} & \cancel{0} & \cancel{0} & \cancel{\dots} & \cancel{0} & \cancel{0} \\ \cancel{0} & \cancel{1} & x & 1 & 0 & \dots & 0 & 0 \\ \cancel{0} & \cancel{0} & 1 & x & 1 & \dots & 0 & 0 \\ \cancel{\vdots} & \cancel{\vdots} & \vdots & \vdots & \vdots & \vdots & \vdots & \vdots \\ \cancel{0} & \cancel{0} & 0 & 0 & 0 & \dots & x & 1 \\ \cancel{0} & \cancel{0} & 0 & 0 & 0 & \dots & 1 & x \end{vmatrix} = 1 \times P_{N-2} = P_{N-2}$$

That is, the cofactor involved in evaluating c_{12} is the secular determinant of an (N–2)-carbon polyene, i.e., P_{N-2} . So $\boxed{P_N = xP_{N-1} - P_{N-2}}$.

Solutions to applications

P5.29 (a)
$$\begin{vmatrix} \alpha - E & \beta & \beta \\ \beta & \alpha - E & \beta \\ \beta & \beta & \alpha - E \end{vmatrix} = 0$$

$$(\alpha - E)\begin{vmatrix} \alpha - E & \beta \\ \beta & \alpha - E \end{vmatrix} - \beta\begin{vmatrix} \beta & \beta \\ \beta & \alpha - E \end{vmatrix} + \beta\begin{vmatrix} \beta & \alpha - E \\ \beta & \beta \end{vmatrix} = 0$$

$$(\alpha - E) \times \{(\alpha - E)^2 - \beta^2\} - \beta\{\beta(\alpha - E) - \beta^2\} + \beta\{\beta^2 - (\alpha - E)\beta\} = 0$$

$$(\alpha - E) \times \{(\alpha - E)^2 - \beta^2\} - 2\beta^2\{\alpha - E - \beta\} = 0$$

$$(\alpha - E) \times (\alpha - E + \beta) \times (\alpha - E - \beta) - 2\beta^2(\alpha - E - \beta) = 0$$

$$(\alpha - E - \beta) \times \{(\alpha - E) \times (\alpha - E + \beta) - 2\beta^2\} = 0$$

$$(\alpha - E - \beta) \times \{(\alpha - E) \times (\alpha - E + 2\beta) - \beta(\alpha - E) - 2\beta^2\} = 0$$

$$(\alpha - E - \beta) \times \{(\alpha - E) \times (\alpha - E + 2\beta) - \beta(\alpha - E + 2\beta)\} = 0$$

$$(\alpha - E - \beta) \times (\alpha - E + 2\beta) \times (\alpha - E - \beta) = 0$$

Therefore, the desired roots are $E = \boxed{\alpha-\beta,\ \alpha-\beta \text{ and } \alpha+2\beta}$. The energy level diagram is shown in Figure 5.15.

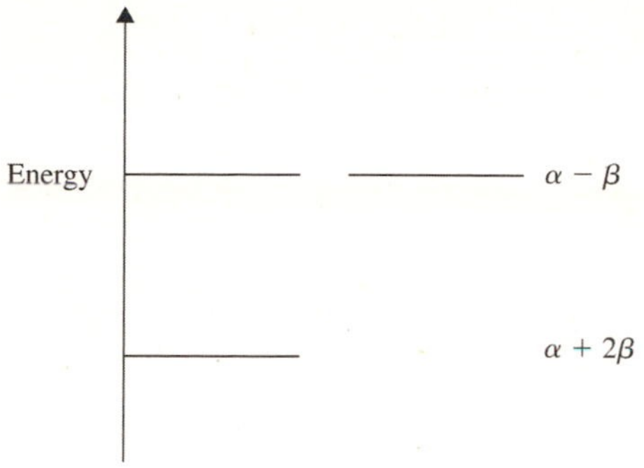

Figure 5.15

The binding energies are shown in the following table.

Species	Number of e^-	Binding energy
H_3^+	2	$2(\alpha+2\beta) = 2\alpha+4\beta$
H_3	3	$2(\alpha+2\beta)+(\alpha-\beta) = 3\alpha+3\beta$
H_3^-	4	$2(\alpha+2\beta)+2(\alpha-\beta) = 4\alpha+2\beta$

(b)

$$\begin{array}{ll} H_3^+(g) \rightarrow 2H(g) + H^+(g) & \Delta H_1 = 849 \text{ kJ mol}^{-1} \\ H^+(g) + H_2(g) \rightarrow H_3^+(g) & \Delta H_2 = ? \\ \hline H_2(g) \rightarrow 2H(g) & \Delta H_3 = [2(217.97) - 0] \text{ kJ mol}^{-1} \end{array}$$

$$\Delta H_2 = \Delta H_3 - \Delta H_1 = \{2(217.97) - 849\} \text{ kJ mol}^{-1}$$

$$\Delta H_2 = \boxed{-413 \text{ kJ mol}^{-1}}$$

This is only slightly less than the binding energy of H_2 (435.94 kJ mol^{-1}).

(c) $2\alpha + 4\beta = -\Delta H_1 = -849 \text{ kJ mol}^{-1}$

so $\beta = \dfrac{-\Delta H_1 - 2\alpha}{4}$ where $\Delta H_1 = 849 \text{ kJ mol}^{-1}$

Species	Binding energy
H_3^+	$2\alpha+4\beta = -\Delta H_1 = \boxed{-849 \text{ kJ mol}^{-1}}$
H_3	$3\alpha+3\beta = 3\left(\alpha - \dfrac{\Delta H_1 + 2\alpha}{4}\right) = 3\left(\dfrac{1}{2}\alpha - \dfrac{\Delta H_1}{4}\right) = \boxed{3(\alpha/2) - 212 \text{ kJ mol}^{-1}}$
H_3^-	$4\alpha+2\beta = 4\alpha - \dfrac{\Delta H_1 + 2\alpha}{2} = 3\alpha - \dfrac{\Delta H_1}{2} = \boxed{3\alpha - 425 \text{ kJ mol}^{-1}}$

As α is a negative quantity, all three species are expected to be stable.

P5.31 (a) The orbitals are sketched in Figure 5.16. ψ_1 is a bonding orbital, showing no nodes between adjacent atoms, and ψ_3 is antibonding with respect to all three atoms. ψ_2 is non-bonding, with neither constructive nor destructive interaction of the atomic orbitals of adjacent atoms.

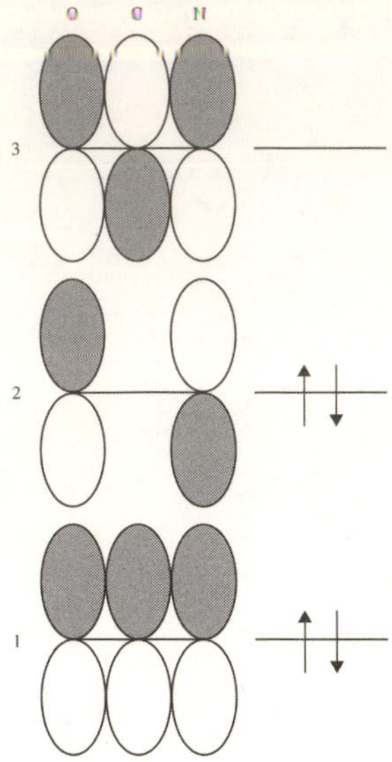

Figure 5.16

(b) This arrangement works only if the entire peptide link is coplanar. Let us call the plane defined by the O, C, and N atoms the *xy* plane; therefore, the p orbitals used to make the three MOs sketched above are p_z orbitals. If the p_z orbital of N is used in the π system, then the σ bonds it makes must be in the *xy* plane. Hence the H atom and the atom labeled $C_{\alpha 2}$ must also be in the *xy* plane. Likewise, if the p_z orbital of the C atom in the peptide link is used in the π system, then its σ bonds must also lie in the *xy* plane, putting the atom labeled $C_{\alpha 1}$ in that plane as well.

(c) The relative energies of the orbitals and their occupancy are shown in Figure 5.16. Four electrons are to be distributed. If we look at the conventional representation of the peptide link (**8** in the text), the two electrons represented by the C=O π bond are obviously part of the π system, leaving the two lone pairs on O, the C–O σ bond, and the two other σ bonds of C as part of the σ system. Turning now to the Lewis octet of electrons around the N atom, we must assign two electrons to each of the σ bonds involving N; clearly they cannot be part of the π system. That leaves the lone pair on N, which must occupy the other orbital that N contributes to the molecule, namely, the p_z orbital that is part of the π system.

(d) The orbitals are sketched in Figure 5.17. ψ_4 is a bonding orbital with respect to C and O, and ψ_6 is antibonding with respect to C and O. ψ_5 is non-bonding, involving only the N atom. There are four electrons to be placed in this system, as before, two each in a bonding and a non-bonding orbital.

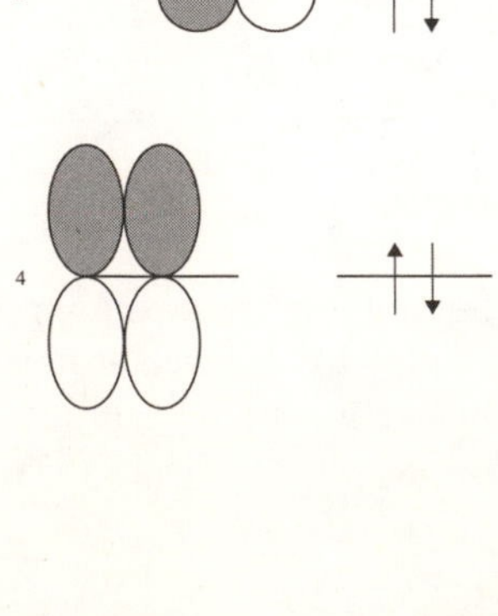

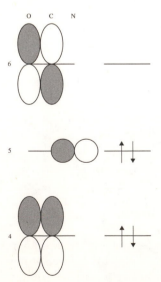

Figure 5.17

(e) This system cannot be planar. As before, the atom labeled $C_{\alpha 1}$ must be in the xy plane. Additionally, the atoms bound to N must be in a plane perpendicular to the orbital that N contributes to this system, which is itself in the xy plane; the bonding partners of N are therefore forced out of the xy plane.

(f) The bonding MO ψ_1 must have a lower energy than the bonding MO ψ_4, for ψ_1 is bonding (stabilizing) with respect to all three atoms, while ψ_4 is bonding with respect to only two of them. Likewise, the antibonding MO ψ_3 must have a higher energy than the antibonding MO ψ_6, for ψ_3 is antibonding (destabilizing) with respect to all three atoms pair-wise, while ψ_6 is antibonding only with respect to two of them. The non-bonding MOs ψ_2 and ψ_5 must have similar energies, not much different than the parameter α, for there is no significant constructive or destructive interference between adjacent atoms in either one.

(g) Because bonding orbital ψ_1 has a lower energy than ψ_4, the planar arrangement has a lower energy than the non-planar one. The total energy of the planar arrangement is

$$E_{\text{planar}} = 2E_1 + 2E_2$$

Compare this to the energy of the non-planar arrangement:

$$E_{\text{non-planar}} = 2E_4 + 2E_5 > 2E_1 + 2E_2 = E_{\text{planar}}$$

The fact that $E_3 > E_6$ is immaterial, for neither of those orbitals is occupied.

6 Computational chemistry

Answers to discussion questions

D6.1 The Fock operator, $f_1 = h_1 + V_{\text{Coulomb}} + V_{\text{Exchange}}$ [6.6], is the sum of three terms. The core hamiltonian for electron 1, h_1 [6.2], is a one-electron hamiltonian that accounts for the electron kinetic energy and the potential energy of attraction between the electron and all nuclei. The Coulomb repulsion term, V_{Coulomb}, accounts for the average potential energy due to repulsions between electron 1 and all other electrons. The exchange potential term provides an average correction to the Coulomb potential for the phenomenon of **electron spin correlation** in which electrons of the same spin tend to avoid each other. Spin correlation reduces the energy description provided by the Coulomb repulsion, and consequently the exchange potential correction has a negative value.

D6.3 The Hartree–Fock equations are

$$f_1\psi_a(1) = \varepsilon_a\psi_a(1) \qquad [6.6]$$

$$f_1\psi_a(1) = h_1\psi_a(1) + \sum_m \{2J_m(1) - K_m(1)\}\psi_a(1) \qquad [6.8]$$

$$h_1 = -\frac{\hbar^2}{2m_e}\nabla_1^2 - j_0\sum_I^{N_n}\frac{Z_I}{r_{I1}} \qquad [6.2]$$

$$J_m(1)\psi_a(1) = j_0\int \psi_a(1)\frac{1}{r_{12}}\psi_m^*(2)\,\psi_m(2)\,\mathrm{d}\tau_2 \qquad [6.7a]$$

$$K_m(1)\psi_a(1) = j_0\int \psi_m(1)\frac{1}{r_{12}}\psi_m^*(2)\,\psi_a(2)\,\mathrm{d}\tau_2 \qquad [6.7b]$$

The equations are solved iteratively. The first computational step is to select, or guess, the initial form of all the one-electron wavefunctions $\psi_m(1)$. They are substituted into the right side of eqns. 6.7a and 6.7b and the integrals are evaluated numerically. The results are used in eqn. 6.8/6.6 to enable the numerical determination of the one-electron wavefunctions and their energies. The newly found wavefunctions are then used to recalculate the integrals of eqns. 6.7a and 6.7b, and the process is repeated until successive iterations find the wavefunctions and corresponding energies to be unchanged to within a chosen criterion. Because iteration continues until self-consistency is achieved, the method is said to provide a **self-consistent field** (SCF).

D6.5 By writing one-electron wavefunctions as a linear combination of basis functions (the LCAO method), the extreme difficulty of solving the Hartree-Fock equations by the HF-SCF procedure is reduced to the somewhat simpler problem of solving the Roothaan equations for LCAO coefficients. Even so, molecular orbital calculations on even moderately large molecules can demand more computer memory and more expensive computational time than is available to the chemist because the very large number of four-center, two-electron integrals (AB|CD) that must be evaluated increases as N_b^4. Consequently, practical molecular orbital calculations require the judicious selection of a finite basis set.

Commonly used basis sets are listed in Table 6.1. They include

(a) The STO-3G minimal basis set in which each Slater-type atomic orbital (STO) is expanded in terms of three Gaussian functions. The linear coefficients have been determined by least-squares best fit to the STOs. The minimal STO basis set uses one STO to represent each of the orbitals in an elementary valence theory treatment of the molecule. This is the fewest number of functions needed to hold all electrons on an atom and maintain spherical symmetry.
(b) The 6-31G basis set in which each core atomic orbital is expanded in terms of six Gaussians. Valence AOs are split into two basis set functions, called "inner" and "outer" functions. The inner basis consists of three Gaussians while the outer basis consists of a single Gaussian. Expansion coefficients have been determined by Hartree-Fock energy minimization on atomic ground states.
(c) The 6-31G* basis set in which d-type polarization functions have been added to the 6-31G basis set. The addition of the d-type AO, written as a single Gaussian, removes the restriction of AO spherical symmetry, thereby accounting for the distortion (or polarization) of AO by adjacent atoms when bonds form in molecules.

D6.7 In ab initio methods an attempt is made to evaluate all integrals that appear in the secular determinant. Approximations are still employed, but they are mainly associated with the construction of the wavefunctions involved in the integrals. In semi-empirical methods, many of the integrals are expressed in terms of spectroscopic data or physical properties. Semi-empirical methods exist at several levels. At some levels, to simplify the calculations, many of the integrals are set equal to zero. Density functional theory (DFT) is different from the Hartree–Fock self-consistent field (HF-SCF) methods and the ab initio methods in that DFT focuses on the electron density while HF/SCF methods focus on the wavefunction. However, they both attempt to evaluate integrals from first principles. DFT and ab initio methods are both iterative self consistent methods in that the calculations are repeated until the energy and wavefunctions (HF-SCF) or energy and electron density (DFT) are unchanged to within some acceptable tolerance.

D6.9 **Density functional theory** (DFT) is not a semi-empirical method. In semi-empirical methods, many of the integrals are expressed in terms of spectroscopic data or physical properties or selectively set equal to zero to simplify calculations and simultaneously give reasonable agreement with select experimental data. DFT attempts to evaluate the integrals numerically, using as input only the values of fundamental constants and atomic numbers of the atoms present in the molecule. In fact, if the **exchange-correlation energy functional** [6.22] were known, DFT would be an exact method.

The exchange-correlation energy functional is modeled in DFT methods. One widely used approximation is the uniform electron gas model for which the exchange-correlation energy is found to be the sum of an exchange contribution and a correlation contribution. Ignoring the complicated correlation contribution for simplicity, the uniform gas approximation for the exchange-correlation energy (see Self-test 6.3 and *A brief illustration*, Section 6.9) is

$$E_{\mathrm{XC}}[\rho] = A\int \rho^{4/3}\mathrm{d}\boldsymbol{r} \qquad \text{with } A = -(9/8)(3/\pi)^{1/2}\, j_0$$

An iteration procedure similar to the HF-SCF methods uses the density functional to calculate an improved exchange-correlation energy functional which is then used to find an improved density functional, and so on.

Solutions to exercises

E6.1(a) $$\hat{H} = -\frac{\hbar^2}{2m_\mathrm{e}}\{\nabla_1^2+\nabla_2^2\} - 2j_0\left\{\frac{1}{r_{\mathrm{He1}}}+\frac{1}{r_{\mathrm{He2}}}\right\} + j_0\left\{\frac{1}{r_{12}}\right\}$$

E6.2(a) $$\hat{V} = -3j_0\left\{\frac{1}{r_{\mathrm{Li1}}}+\frac{1}{r_{\mathrm{Li2}}}+\frac{1}{r_{\mathrm{Li3}}}+\frac{1}{r_{\mathrm{Li4}}}\right\} - j_0\left\{\frac{1}{r_{\mathrm{H1}}}+\frac{1}{r_{\mathrm{H2}}}+\frac{1}{r_{\mathrm{H3}}}+\frac{1}{r_{\mathrm{H4}}}\right\} + j_0\left\{\frac{1}{r_{12}}+\frac{1}{r_{13}}+\frac{1}{r_{14}}+\frac{1}{r_{23}}+\frac{1}{r_{24}}+\frac{1}{r_{34}}\right\}$$

E6.3(a) $N_e = 2$ for HeH^+

$$\hat{H} = -\frac{\hbar^2}{2m_e}\{\nabla_1^2 + \nabla_2^2\} - 2j_0\left\{\frac{1}{r_{He1}} + \frac{1}{r_{He2}}\right\} - j_0\left\{\frac{1}{r_{H1}} + \frac{1}{r_{H2}}\right\} + j_0\left\{\frac{1}{r_{12}}\right\}$$

E6.4(a) $N_e = 2$ for HeH^+ $\Psi = \frac{1}{\sqrt{2}}\begin{vmatrix}\psi_a^{\alpha}(1) & \psi_a^{\beta}(1)\\ \psi_a^{\alpha}(2) & \psi_a^{\beta}(2)\end{vmatrix}$

E6.5(a) $N_e = 2$ for HeH^+

$$f_1\psi_a(1) = h_1\psi_a(1) + 2J_a(1)\psi_a(1) - K_a(1)\psi_a(1) = \varepsilon_a\psi_a(1) \quad [6.8] \text{ and } [6.6]$$

where

$$h_1 = -\frac{\hbar^2}{2m_e}\nabla_1^2 - j_0\left\{\frac{1}{r_{He1}} + \frac{1}{r_{H1}}\right\} \quad [6.2]$$

$$J_a(1)\psi_a(1) = j_0\int\psi_a(1)\frac{1}{r_{12}}\psi_a^*(2)\psi_a(2)\,d\tau_2 = K_a(1)\psi_a(1) \quad [6.7a] \text{ and } [6.7b]$$

Consequently,

$$\boxed{f_1\psi_a(1) = h_1\psi_a(1) + J_a(1)\psi_a(1) = \varepsilon_a\psi_a(1)}$$

E6.6(a) The complete set of equations for the $N_e = 2$ of HeH^+ is

(i) $\psi_a = c_{Hea}\chi_{He} + c_{Ha}\chi_H$ and $\psi_b = c_{Heb}\chi_{He} + c_{Hb}\chi_H$ [6.9]

(ii) $f_1\chi_{He}(1) = h_1\chi_{He}(1) + 2J_a(1)\chi_{He}(1) - K_a(1)\chi_{He}(1)$ [6.8]

$f_1\chi_H(1) = h_1\chi_H(1) + 2J_a(1)\chi_H(1) - K_a(1)\chi_H(1)$ [6.8]

where

$$h_1 = -\frac{h^2}{2m_e}\nabla_1^2 - j_0\left\{\frac{1}{r_{He1}} + \frac{1}{r_{H1}}\right\} \quad [6.2]$$

$$J_a(1)\chi_{He}(1) = j_0\int\chi_{He}(1)\frac{1}{r_{12}}\psi_a^*(2)\psi_a(2)\,d\tau_2 \quad [6.7a]$$

$$K_a(1)\chi_{He}(1) = j_0\int\psi_a(1)\frac{1}{r_{12}}\psi_a^*(2)\chi_{He}(2)\,d\tau_2 \quad [6.7b]$$

$$J_a(1)\chi_H(1) = j_0\int\chi_H(1)\frac{1}{r_{12}}\psi_a^*(2)\psi_a(2)\,d\tau_2 \quad [6.7a]$$

$$K_a(1)\chi_H(1) = j_0\int\psi_a(1)\frac{1}{r_{12}}\psi_a^*(2)\chi_H(2)\,d\tau_2 \quad [6.7b]$$

(iii) $\boldsymbol{Fc} = \boldsymbol{Sc\varepsilon}$ [6.10] and [6.11]

where

$$\boldsymbol{F} = \begin{pmatrix}\int\chi_{He}(1)f_1\chi_{He}(1)d\tau_1 & \int\chi_{He}(1)f_1\chi_H(1)d\tau_1\\ \int\chi_H(1)f_1\chi_{He}(1)d\tau_1 & \int\chi_H(1)f_1\chi_H(1)d\tau_1\end{pmatrix} \quad (F_{12} = F_{21})$$

$$\boldsymbol{c} = \begin{pmatrix}c_{Hea} & c_{Heb}\\ c_{Ha} & c_{Hb}\end{pmatrix}$$

$$S=\begin{pmatrix}\int\chi_{He}(1)\chi_{He}(1)d\tau_1 & \int\chi_{He}(1)\chi_{H}(1)d\tau_1\\ \int\chi_{H}(1)\chi_{He}(1)d\tau_1 & \int\chi_{H}(1)\chi_{H}(1)d\tau_1\end{pmatrix}=\begin{pmatrix}1 & S\\ S & 1\end{pmatrix}$$

Note: $\int\chi_{He}(1)\chi_{H}(1)d\tau_1=\int\chi_{H}(1)\chi_{He}(1)d\tau_1=S$

$$\varepsilon=\begin{pmatrix}\varepsilon_a & 0\\ 0 & \varepsilon_b\end{pmatrix}$$

E6.7(a) For the $N_e = 2$ of HeH^+,

$$F_{AA}=\int\chi_{He}(1)f_1\chi_{He}(1)d\tau_1=\int\chi_{He}(1)\{h_1\chi_{He}(1)+2J_a(1)\chi_{He}(1)-K_a(1)\chi_{He}(1)\}d\tau_1\quad\text{[6.11a] and [6.8]}$$
$$=\int\chi_{He}(1)h_1\chi_{He}(1)d\tau_1+2\int\chi_{He}(1)\{J_a(1)\chi_{He}(1)\}d\tau_1-\int\chi_{He}(1)\{K_a(1)\chi_{He}(1)\}d\tau_1$$
$$=\int\chi_{He}(1)h_1\chi_{He}(1)d\tau_1+2\int\chi_{He}(1)\left\{j_0\int\chi_{He}(1)\frac{1}{r_{12}}\psi_a^*(2)\psi_a(2)d\tau_2\right\}d\tau_1$$
$$-\int\chi_{He}(1)\left\{j_0\int\psi_a(1)\frac{1}{r_{12}}\psi_a^*(2)\chi_{He}(2)d\tau_2\right\}d\tau_1$$
$$=\int\chi_{He}(1)h_1\chi_{He}(1)d\tau_1+2j_0\int\int\chi_{He}(1)\chi_{He}(1)\frac{1}{r_{12}}\psi_a^*(2)\psi_a(2)d\tau_1d\tau_2$$
$$-j_0\int\int\chi_{He}(1)\psi_a(1)\frac{1}{r_{12}}\psi_a^*(2)\chi_{He}(2)d\tau_1d\tau_2$$
$$=E_{He}+2j_0\int\int\chi_{He}(1)\chi_{He}(1)\frac{1}{r_{12}}\{c_{Hea}\chi_{He}(2)+c_{Ha}\chi_{H}(2)\}^2d\tau_1d\tau_2$$
$$-j_0\int\int\chi_{He}(1)\{c_{Hea}\chi_{He}(1)+c_{Ha}\chi_{H}(1)\}\frac{1}{r_{12}}\{c_{Hea}\chi_{He}(2)+c_{Ha}\chi_{H}(2)\}\chi_{He}(2)d\tau_1d\tau_2$$
$$=E_{He}+2j_0\int\int\chi_{He}(1)\chi_{He}(1)\frac{1}{r_{12}}\{c_{Hea}^2\chi_{He}^2(2)+2c_{Hea}c_{Ha}\chi_{He}(2)\chi_{H}(2)+c_{Ha}^2\chi_{H}^2(2)\}d\tau_1d\tau_2$$
$$-j_0\int\int\{c_{Hea}\chi_{He}(1)\chi_{He}(1)+c_{Ha}\chi_{He}(1)\chi_{H}(1)\}\frac{1}{r_{12}}\{c_{Hea}\chi_{He}(2)\chi_{He}(2)+c_{Ha}\chi_{He}(2)\chi_{H}(2)\}d\tau_1d\tau_2$$

Multiplication of the factors and substituting the symbolism

$(AB|CD)=j_0\int\int\chi_A(1)\chi_B(1)\frac{1}{r_{12}}\chi_C(2)\chi_D(2)d\tau_1d\tau_2$ [6.14] gives

$$F_{AA}=E_{He}+\{2c_{Hea}^2(HeHe|HeHe)+4c_{Hea}c_{Ha}(HeHe|HeH)+2c_{Ha}^2(HeHe|HH)\}$$
$$-\{c_{Hea}^2(HeHe|HeHe)+c_{Hea}c_{Ha}(HeHe|HeH)+c_{Hea}c_{Ha}(HeH|HeHe)+c_{Ha}^2(HeH|HeH)\}$$

$$\boxed{F_{AA}=E_{He}+c_{Hea}^2(HeHe|HeHe)+2c_{Hea}c_{Ha}(HeHe|HeH)+c_{Ha}^2\{2(HeHe|HH)-(HeH|HeH)\}}$$

$$F_{AB}=\int\chi_{He}(1)f_1\chi_{H}(1)d\tau_1=\int\chi_{He}(1)\{h_1\chi_{H}(1)+2J_a(1)\chi_{H}(1)-K_a(1)\chi_{H}(1)\}d\tau_1\quad\text{[6.11a] and [6.8]}$$
$$=\int\chi_{He}(1)h_1\chi_{H}(1)d\tau_1+2\int\chi_{He}(1)\{J_a(1)\chi_{H}(1)\}d\tau_1-\int\chi_{He}(1)\{K_a(1)\chi_{H}(1)\}d\tau_1$$
$$=\int\chi_{He}(1)h_1\chi_{H}(1)d\tau_1+2\int\chi_{He}(1)\left\{j_0\int\chi_{H}(1)\frac{1}{r_{12}}\psi_a^*(2)\psi_a(2)d\tau_2\right\}d\tau_1$$
$$-\int\chi_{He}(1)\left\{j_0\int\psi_a(1)\frac{1}{r_{12}}\psi_a^*(2)\chi_{H}(2)d\tau_2\right\}d\tau_1$$
$$=\int\chi_{He}(1)h_1\chi_{H}(1)d\tau_1+2j_0\int\int\chi_{He}(1)\chi_{H}(1)\frac{1}{r_{12}}\psi_a^*(2)\psi_a(2)d\tau_1d\tau_2$$
$$-j_0\int\int\chi_{He}(1)\psi_a(1)\frac{1}{r_{12}}\psi_a^*(2)\chi_{H}(2)d\tau_1d\tau_2$$

$$= \int \chi_{He}(1) h_1 \chi_H(1) d\tau_1 + 2j_0 \int\int \chi_{He}(1)\chi_H(1)\frac{1}{r_{12}}\{c_{Hea}\chi_{He}(2) + c_{Ha}\chi_H(2)\}^2 d\tau_1 d\tau_2$$

$$- j_0 \int\int \chi_{He}(1)\{c_{Hea}\chi_{He}(1) + c_{Ha}\chi_H(1)\}\frac{1}{r_{12}}\{c_{Hea}\chi_{He}(2) + c_{Ha}\chi_H(2)\}\chi_H(2) d\tau_1 d\tau_2$$

$$= \int \chi_{He}(1) h_1 \chi_H(1) d\tau_1 + 2j_0 \int\int \chi_{He}(1)\chi_H(1)\frac{1}{r_{12}}\{c_{Hea}^2\chi_{He}^2(2) + 2c_{Hea}c_{Ha}\chi_{He}(2)\chi_H(2) + c_{Ha}^2\chi_H^2(2)\} d\tau_1 d\tau_2$$

$$- j_0 \int\int \{c_{Hea}\chi_{He}(1)\chi_{He}(1) + c_{Ha}\chi_{He}(1)\chi_H(1)\}\frac{1}{r_{12}}\{c_{Hea}\chi_{He}(2)\chi_H(2) + c_{Ha}\chi_H^2(2)\} d\tau_1 d\tau_2$$

Multiplication of the factors and substituting the symbolism

$(AB|CD) = j_0 \int\int \chi_A(1)\chi_B(1)\frac{1}{r_{12}}\chi_C(2)\chi_D(2) d\tau_1 d\tau_2$ [6.14] gives

$$F_{AB} = \int \chi_{He}(1) h_1 \chi_H(1) d\tau_1 + \{2c_{Hea}^2(HeH|HeHe) + 4c_{Hea}c_{Ha}(HeH|HeH) + 2c_{Ha}^2(HeH|HH)\}$$

$$- \{c_{Hea}^2(HeHe|HeH) + c_{Hea}c_{Ha}(HeHe|HH) + c_{Hea}c_{Ha}(HeH|HeH) + c_{Ha}^2(HeH|HH)\}$$

$$\boxed{F_{AB} = \int \chi_{He}(1) h_1 \chi_H(1) d\tau_1 + c_{Hea}^2(HeH|HeHe) + c_{Hea}c_{Ha}\{3(HeH|HeH) - (HeHe|HH)\} + c_{Ha}^2(HeH|HH)}$$

Notes:

(i) $E_{He} = \int \chi_{He}(1) h_1 \chi_{He}(1) d\tau_1$ [6.15] is the energy of an electron in orbital χ_{He}, taking into account its interaction with both the He and H nuclei.

(ii) In the above manipulations we have used the relationship (see E6.8a and E6.8b)

$$(AB|AA) = (AA|AB)$$

(iii) We have explicitly shown the integral signs for both the integration over all τ_2 values and the integration over all τ_1 values. For example, we have written the four-center, two-electron integral as

$$(AB|CD) = j_0 \int\int \chi_A(1)\chi_B(1)\frac{1}{r_{12}}\chi_C(2)\chi_D(2) d\tau_1 d\tau_2$$

This has the same meaning as the equation that shows a single integral sign only:

$$(AB|CD) = j_0 \int \chi_A(1)\chi_B(1)\frac{1}{r_{12}}\chi_C(2)\chi_D(2) d\tau_1 d\tau_2 \quad [6.14]$$

We explicitly show the double integration sign to help with the visual recognition of the requisite mathematics.

E6.8(a)

$$(AA|AB) = j_0 \int \chi_A(1)\chi_A(1)\frac{1}{r_{12}}\chi_A(2)\chi_B(2) d\tau_1 d\tau_2 \quad [6.14]$$

$$= j_0 \int\int \chi_A(1)\chi_A(1)\frac{1}{r_{12}}\chi_A(2)\chi_B(2) d\tau_1 d\tau_2$$

This integral is symmetric with the interchange of electron 1 and electron 2. Consequently, (AA|AB) = (AB|AA). Furthermore, the functions A and B, like all functions, commute, so the integral may also be written in the forms (AA|BA) and (BA|AA).

E6.9(a) CH_3Cl

(a) Minimal basis set: one basis function for the 1s orbital of each of the three hydrogen atoms; one basis function for each of the 1s, 2s, and three 2p orbitals for the carbon atom; one basis function for each of the 1s, 2s, three 2p, 3s, and three 3p orbitals for the chlorine atom. Thus, the minimal basis set consists of 17 basis functions.

(b) Split-valence basis set: two basis functions for the 1s orbital of each of the three hydrogen atoms; one basis function for the 1s orbital of carbon and two basis functions for each of the 2s and three 2p orbitals of the carbon atom; one basis function for the 1s, 2s, and three 2p orbitals of the chlorine atom and two basis functions for each of the 3s and three 3p orbitals of the chlorine atom. Thus, the split-valence basis set consists of 28 basis functions.

(c) Double-zeta basis set: each basis function in the minimal basis is replaced by two functions. Thus, the double-zeta basis set consists of 34 basis functions.

E6.10(a) p-type Gaussian: $\chi_p = Nx^i y^j z^k e^{-\alpha r^2}$ where $i + j + k = 1$ [6.17]. Example: $\chi_{p_x} = Nx\, e^{-\alpha r^2}$

E6.11(a) The s-type Gaussian, $\chi_s = Ne^{-\alpha r^2}$ [6.17], can be written as the product of these three one-dimensional Gaussian functions:

$$N_x e^{-\alpha x^2},\ N_y e^{-\alpha y^2},\ N_z e^{-\alpha z^2}$$

Proof:

$$\left(N_x e^{-\alpha x^2}\right)\times\left(N_y e^{-\alpha y^2}\right)\times\left(N_z e^{-\alpha z^2}\right) = N_x N_y N_z e^{-\alpha x^2-\alpha y^2-\alpha z^2} = Ne^{-\alpha\left(x^2+y^2+z^2\right)} = Ne^{-\alpha r^2} \text{ where } N = N_x N_y N_z$$

E6.12(a) Analysis of the s-type Gaussian for HeH^+ is facilitated by placing the He nucleus at the origin of the Cartesian coordinate system and placing the H nucleus at the coordinate $(x, y, z) = (d, 0, 0)$. In general, an s-type Gaussian centered on the point $(x, y, z) = (x_c, y_c, z_c)$ has the form $Ne^{-\alpha\left\{(x-x_c)^2+(y-y_c)^2+(z-z_c)^2\right\}}$ [6.17]. So with our choice for the coordinate system, the s-type Gaussians on He and H are

$$\chi_{He} = N_{He}e^{-\alpha_{He}r^2} = N_{He}e^{-\alpha_{He}\left\{x^2+y^2+z^2\right\}} \quad \text{and} \quad \chi_H = N_H e^{-\alpha_H\left\{(x-d)^2+y^2+z^2\right\}}$$

Their product is

$$\chi_{He}\chi_H = N_{He}N_H e^{-\alpha_{He}\left\{x^2+y^2+z^2\right\}} e^{-\alpha_H\left\{(x-d)^2+y^2+z^2\right\}}$$

$$= N_{He}N_H e^{-\left[\alpha_{He}\left\{x^2+y^2+z^2\right\}+\alpha_H\left\{(x-d)^2+y^2+z^2\right\}\right]}$$

and it will have the s-type Gaussian form and be in an intermediate position provided that the exponent equals $-\alpha\left\{(x-e)^2+y^2+z^2\right\}-\beta$ where α, e, and β are constants. If we can find these constants, which are independent of coördinates, we will have demonstrated that the product of two s-type Gaussians is a Gaussian.

$$\alpha_{He}\left\{x^2+y^2+z^2\right\}+\alpha_H\left\{(x-d)^2+y^2+z^2\right\}=\alpha\left\{(x-e)^2+y^2+z^2\right\}+\beta$$

$$\alpha_{He}\left\{x^2+y^2+z^2\right\}+\alpha_H\left\{x^2-2dx+d^2+y^2+z^2\right\}=\alpha\left\{x^2-2ex+e^2+y^2+z^2\right\}+\beta$$

$$(\alpha_{He}+\alpha_H-\alpha)(x^2+y^2+z^2)-2(\alpha_H d-\alpha e)x+\alpha_H d^2-\alpha e^2-\beta=0$$

This last expression can equal zero for any chosen value of the coordinate point (x, y, z) provided that the coefficients of the polynomial expression each equal zero. Thus, from the quadratic term we find that $\alpha = \alpha_{He} + \alpha_H$. The first-order term yields

$$e = \frac{\alpha_H d}{\alpha} = \frac{\alpha_H d}{\alpha_{He}+\alpha_H}$$

and the constant term yields

$$\beta = \alpha_{\mathrm{H}} d^2 - \alpha e^2 = \alpha_{\mathrm{H}} d^2 - (\alpha_{\mathrm{He}} + \alpha_{\mathrm{H}})\left(\frac{\alpha_{\mathrm{H}} d}{\alpha_{\mathrm{He}} + \alpha_{\mathrm{H}}}\right)^2 = \alpha_{\mathrm{H}} d^2 - \frac{(\alpha_{\mathrm{H}} d)^2}{\alpha_{\mathrm{He}} + \alpha_{\mathrm{H}}} = \frac{\alpha_{\mathrm{He}}\alpha_{\mathrm{H}} d^2}{\alpha_{\mathrm{He}} + \alpha_{\mathrm{H}}}$$

Thus,

$$\chi_{\mathrm{He}}\chi_{\mathrm{H}} = N_{\mathrm{He}}N_{\mathrm{H}}\mathrm{e}^{-\alpha\{(x-e)^2+y^2+z^2\}-\beta} = \left(N_{\mathrm{He}}N_{\mathrm{H}}\mathrm{e}^{-\beta}\right)\times \mathrm{e}^{-\alpha\{(x-e)^2+y^2+z^2\}}$$

where $\alpha = \alpha_{\mathrm{He}} + \alpha_{\mathrm{H}}$, $e = \dfrac{\alpha_{\mathrm{H}} d}{\alpha_{\mathrm{He}} + \alpha_{\mathrm{H}}}$, and $\beta = \dfrac{\alpha_{\mathrm{He}}\alpha_{\mathrm{H}} d^2}{\alpha_{\mathrm{He}} + \alpha_{\mathrm{H}}}$.

This demonstrates that the product is a Gaussian centered at $(x, y, z) = (e, 0, 0)$, which is positioned at a point intermediate between the two nuclei because $e = \dfrac{\alpha_{\mathrm{H}} d}{\alpha_{\mathrm{He}} + \alpha_{\mathrm{H}}} < d$.

E6.13(a) CH_3Cl

(a) The 6-31G* basis set consists of the 6-31G basis set plus polarization functions (see Table 6.1). In the 6-31G basis set, each core atomic basis function is expanded as a linear combination of six Gaussian functions. Valence basis functions are split into two basis set functions, called "inner" and "outer" functions. The inner basis consists of three Gaussians while the outer basis consists of a single Gaussian. To form the 6-31G* basis set, six d-type polarization functions are added for each atom other than hydrogen.

6-31G Basis Set for CH_3Cl

Atom	Core Orbitals Basis Functions	Valence Orbitals Basis Functions
H		1s 2
C	1s 1	2s, three 2p 4(2) = 8
Cl	1s, 2s, three 2p 5	3s, three 3p 4(2) = 8
Total number of basis functions (include 3 Hs) = 3(2) + 9 + 13 = 28		

Adding 6 polarization functions for both the carbon atom and the chlorine atom gives a 6-31G* basis set that consists of **40 basis functions**.

(b) The 6-311G** basis set is a 6-311G basis set plus both a set of five d-type polarization functions for each atom other than hydrogen and a set of three p-type polarization functions for each hydrogen atom. In the 6-311G basis set each core atomic basis function is expanded as a linear combination of six Gaussian functions. Valence basis functions are split into three basis set functions consisting of three, one, and one Gaussians.

6-311G Basis Set for CH_3Cl

Atom	Core Orbitals Basis Functions	Valence Orbitals Basis Functions
H		1s 3(1) = 3
C	1s 1	2s, three 2p 4(3) = 12
Cl	1s, 2s, three 2p 5	3s, three 3p 4(3) = 12
Total number of basis functions (include 3 Hs) = 3(3) + 13 + 17 = 39		

Adding 5 polarization functions for both the carbon atom and the chlorine atom and an addition 3 polarization functions for each hydrogen atom gives a 6-311G** basis set that consists of 58 basis functions.

(c) The 6-311++G basis set is a 6-311G basis set plus both an s-type and three p-type diffuse functions for each atom other than hydrogen and one s-type diffuse functions for each hydrogen atom. In the 6-311G basis set each core atomic basis function is expanded as a linear combination of six Gaussian functions. Valence orbitals are split into three basis set functions consisting of three, one, and one Gaussians.

Atom	Core Orbitals Basis Functions	Valence and Diffuse Orbitals Basis Functions
H		1s, one diffuse s 3(1) + 1 = 4
C	1s 1	2s, three 2p, diffuse s and three p 4(3) + 4 = 16
Cl	1s, 2s, three 2p 5	3s, three 3p, diffuse s and three p 4(3) + 4 = 16
Total number of basis functions (include 3 Hs) = 3(4) + 17 + 21 = 50		

E6.14(a) Very general equations for setting up the Roothaan equations are used in the first and second *A Brief Illustration* of Section 6.2. By replacing the index A with He and the index B with H, we find that for a linear combination of two basis functions, such as $\chi_{\rm He}$ and $\chi_{\rm H}$, the coefficients of the molecular orbitals in $\psi_a = c_{{\rm He}a}\chi_{\rm He} + c_{{\rm H}a}\chi_{\rm H}$ and $\psi_b = c_{{\rm He}b}\chi_{\rm He} + c_{{\rm H}b}\chi_{\rm H}$ are given by the equations

$$\varepsilon_a = \frac{F_{\rm HeHe} + F_{\rm HH} - S(F_{\rm HeH} - F_{\rm HHe}) - (\{F_{\rm HeHe} + F_{\rm HH} - S(F_{\rm HeH} - F_{\rm HHe})\}^2 - 4\{1 - S^2\}\{F_{\rm HeHe}F_{\rm HH} - F_{\rm HeH}F_{\rm HHe}\})^{1/2}}{2\{1 - S^2\}}$$

and

$$c_{{\rm He}a} = -\frac{F_{\rm HeH} - S\varepsilon_a}{F_{\rm HeHe} - \varepsilon_a} c_{{\rm H}a} \quad \text{and} \quad S = \int \chi_{\rm He}\chi_{\rm H}\,{\rm d}\tau$$

in conjunction with the normalization condition $c_{{\rm He}a}^2 + c_{{\rm H}a}^2 + 2c_{{\rm He}a}c_{{\rm H}a}S = 1$.

Applying the Hückel approximation $S = 0$ and using the symbols $\alpha_{\rm He} = F_{\rm HeHe}$, $\alpha_{\rm H} = F_{\rm HH}$, and $F_{\rm HeH} = F_{\rm HHe} = \beta$ simplifies the equations to

$$\varepsilon_a = \frac{\alpha_{\rm He} + \alpha_{\rm H} - (\{\alpha_{\rm He} + \alpha_{\rm H}\}^2 - 4\{\alpha_{\rm He}\alpha_{\rm H} - \beta^2\})^{1/2}}{2} = \frac{\alpha_{\rm He} + \alpha_{\rm H} - (\{\alpha_{\rm He} - \alpha_{\rm H}\}^2 + 4\beta^2)^{1/2}}{2}$$

and

$$c_{{\rm He}a} = -\frac{\beta}{\alpha_{\rm He} - \varepsilon_a} c_{{\rm H}a}$$

in conjunction with the normalization condition $c_{{\rm He}a}^2 + c_{{\rm H}a}^2 = 1$. Thus,

$$\left(\frac{\beta}{\alpha_{\rm He} - \varepsilon_a}\right)^2 c_{{\rm H}a}^2 + c_{{\rm H}a}^2 = 1 \text{ or } \boxed{c_{{\rm H}a} = \left(\frac{1}{1 + \left(\frac{\beta}{\alpha_{\rm He} - \varepsilon_a}\right)^2}\right)^{1/2}}$$

Alternatively, we may use the Hückel approximation much earlier in the derivation and thereby avoid the general equations of the *Brief illustrations*. Using this method, we first write the secular determinant and apply the Hückel approximation before expanding the determinant.

$$\begin{vmatrix} F_{\text{HeHe}}-\varepsilon & F_{\text{HeH}}-S\varepsilon \\ F_{\text{HHe}}-S\varepsilon & F_{\text{HH}}-\varepsilon \end{vmatrix} = \begin{vmatrix} \alpha_{\text{He}}-\varepsilon & \beta \\ \beta & \alpha_{\text{H}}-\varepsilon \end{vmatrix} = 0$$

$$(\alpha_{\text{He}}-\varepsilon)(\alpha_{\text{H}}-\varepsilon)-\beta^2=0$$

$$\varepsilon^2-(\alpha_{\text{He}}+\alpha_{\text{H}})\varepsilon+\alpha_{\text{He}}\alpha_{\text{H}}-\beta^2=0$$

$$\varepsilon=\frac{\alpha_{\text{He}}+\alpha_{\text{H}}\pm\left(\{\alpha_{\text{He}}+\alpha_{\text{H}}\}^2-4\{\alpha_{\text{He}}\alpha_{\text{H}}-\beta^2\}\right)^{1/2}}{2}=\frac{\alpha_{\text{He}}+\alpha_{\text{H}}\pm\left(\{\alpha_{\text{He}}-\alpha_{\text{H}}\}^2+4\beta^2\right)^{1/2}}{2}$$

Thus, $\boxed{\varepsilon_a=\dfrac{\alpha_{\text{He}}+\alpha_{\text{H}}-\left(\{\alpha_{\text{He}}-\alpha_{\text{H}}\}^2+4\beta^2\right)^{1/2}}{2}}$ and $\varepsilon_b=\dfrac{\alpha_{\text{He}}+\alpha_{\text{H}}+\left(\{\alpha_{\text{He}}-\alpha_{\text{H}}\}^2+4\beta^2\right)^{1/2}}{2}$.

The matrix form of the Roothaan equation can be used to find the wavefunction coefficients.

$$\begin{pmatrix} F_{\text{HeHe}} & F_{\text{HeH}} \\ F_{\text{HHe}} & F_{\text{HH}} \end{pmatrix}\begin{pmatrix} c_{\text{He}a} & c_{\text{He}b} \\ c_{\text{H}a} & c_{\text{H}b} \end{pmatrix}=\begin{pmatrix} S_{\text{HeHe}} & S_{\text{HeH}} \\ S_{\text{HHe}} & S_{\text{HH}} \end{pmatrix}\begin{pmatrix} c_{\text{He}a} & c_{\text{He}b} \\ c_{\text{H}a} & c_{\text{H}b} \end{pmatrix}\begin{pmatrix} \varepsilon_a & 0 \\ 0 & \varepsilon_b \end{pmatrix}$$

$$\begin{pmatrix} \alpha_{\text{He}} & \beta \\ \beta & \alpha_{\text{H}} \end{pmatrix}\begin{pmatrix} c_{\text{He}a} & c_{\text{He}b} \\ c_{\text{H}a} & c_{\text{H}b} \end{pmatrix}=\begin{pmatrix} 1 & 0 \\ 0 & 1 \end{pmatrix}\begin{pmatrix} \varepsilon_a c_{\text{He}a} & \varepsilon_b c_{\text{He}b} \\ \varepsilon_a c_{\text{H}a} & \varepsilon_b c_{\text{H}b} \end{pmatrix}=\begin{pmatrix} \varepsilon_a c_{\text{He}a} & \varepsilon_b c_{\text{He}b} \\ \varepsilon_a c_{\text{H}a} & \varepsilon_b c_{\text{H}b} \end{pmatrix}$$

$$\begin{pmatrix} \alpha_{\text{He}}c_{\text{He}a}+\beta c_{\text{H}a} & \alpha_{\text{He}}c_{\text{He}b}+\beta c_{\text{H}b} \\ \beta c_{\text{He}a}+\alpha_{\text{H}}c_{\text{H}a} & \beta c_{\text{He}b}+\alpha_{\text{H}}c_{\text{H}b} \end{pmatrix}=\begin{pmatrix} \varepsilon_a c_{\text{He}a} & \varepsilon_b c_{\text{He}b} \\ \varepsilon_a c_{\text{H}a} & \varepsilon_b c_{\text{H}b} \end{pmatrix}$$

For example, by equating the matrix$_{11}$ elements of the above equation and substituting the normalization condition $c_{\text{He}a}{}^2+c_{\text{H}a}{}^2=1$, we find the coefficient $c_{\text{H}a}$.

$$\alpha_{\text{He}}c_{\text{He}a}+\beta c_{\text{H}a}=\varepsilon_a c_{\text{He}a}$$
$$(\alpha_{\text{He}}-\varepsilon_a)c_{\text{He}a}=-\beta c_{\text{H}a}$$

$$(\alpha_{\text{He}}-\varepsilon_a)(1-c_{\text{H}a}{}^2)^{1/2}=-\beta c_{\text{H}a}$$
$$(\alpha_{\text{He}}-\varepsilon_a)^2(1-c_{\text{H}a}{}^2)=\beta^2 c_{\text{H}a}{}^2$$

$$\boxed{c_{\text{H}a}=\left(\frac{1}{1+\left(\dfrac{\beta}{\alpha_{\text{He}}-\varepsilon_a}\right)^2}\right)^{1/2}}$$

E6.15(a) (a) The semi-empirical **CNDO (complete neglect of differential overlap) method** sets all two-electron integrals of the type (AB|CD) equal to zero unless $\chi_A=\chi_B$ and $\chi_C=\chi_D$. Only integrals of the type (AA|CC) survive.

(b) The semi-empirical **INDO (intermediate neglect of differential overlap) method** sets all two-electron integrals of the type (AB|CD) equal to zero unless $\chi_A=\chi_B$ and $\chi_C=\chi_D$ except for the type (AB|AB) when the basis functions χ_A and χ_B are both centerd on the same nucleus.

E6.16(a) Twenty basis set functions generate twenty atomic orbitals. The silicon atom has 14 electrons, which are paired in the 7 lowest AOs to form the ground state. This leaves $\boxed{13}$ virtual orbitals unoccupied.

E6.17(a) $N_e=2$ for H_2

$$\boxed{\Psi_1=(1/2)^{1/2}\begin{vmatrix} \psi_a^\alpha(1) & \psi_a^\beta(1) \\ \psi_b^\alpha(2) & \psi_b^\beta(2) \end{vmatrix}}$$ [6.5*a*] where $\psi_a=(1/2)^{1/2}\{\chi_A+\chi_B\}$ and $\psi_b=(1/2)^{1/2}\{\chi_A-\chi_B\}$.

This may be expanded to show that this singly excited determinant is composed of the singlet spin-state wavefunction $\sigma_-(1,2)=(1/2)^{1/2}\{\alpha(1)\beta(2)-\beta(1)\alpha(2)\}$.

$$\begin{aligned}\Psi_1&=(1/2)^{1/2}\{\psi_a(1)\alpha(1)\psi_b(2)\beta(2)-\psi_a(1)\beta(1)\psi_b(2)\alpha(2)\}\\&=(1/2)^{1/2}\psi_a(1)\psi_b(2)\{\alpha(1)\beta(2)-\beta(1)\alpha(2)\}\\&=\psi_a(1)\psi_b(2)\sigma_-(1,2)\end{aligned}$$

E6.18(a) $N_e = 2$ for HeH^+

$$\psi_a = c_{\text{He}a}\chi_{\text{He}}+c_{\text{H}a}\chi_{\text{H}} \quad \text{and} \quad \psi_b = c_{\text{He}b}\chi_{\text{He}}+c_{\text{H}b}\chi_{\text{H}}$$

where the coefficients are determined with the Roothaan equations. The **configuration state functions (CSF)** are

$$\Psi_0=(1/2)^{1/2}\begin{vmatrix}\psi_a^\alpha(1) & \psi_a^\beta(1)\\ \psi_a^\alpha(2) & \psi_a^\beta(2)\end{vmatrix}=(1/2)^{1/2}\psi_a(1)\psi_a(2)\{\alpha(1)\beta(2)-\beta(1)\alpha(2)\}=\psi_a(1)\psi_a(2)\sigma_-(1,2)$$

$$\Psi_1=(1/2)^{1/2}\begin{vmatrix}\psi_a^\alpha(1) & \psi_a^\beta(1)\\ \psi_b^\alpha(2) & \psi_b^\beta(2)\end{vmatrix}=(1/2)^{1/2}\psi_a(1)\psi_b(2)\{\alpha(1)\beta(2)-\beta(1)\alpha(2)\}=\psi_a(1)\psi_b(2)\sigma_-(1,2)$$

Absorbing the normalization constants $(1/2)^{1/2}$ into the coefficients of the **configuration interaction (CI)** linear combination of CSFs gives the CI ground-state wavefunction:

$$\boxed{\Psi=C_0\Psi_0+C_1\Psi_1=C_0\begin{vmatrix}\psi_a^\alpha(1) & \psi_a^\beta(1)\\ \psi_a^\alpha(2) & \psi_a^\beta(2)\end{vmatrix}+C_1\begin{vmatrix}\psi_a^\alpha(1) & \psi_a^\beta(1)\\ \psi_b^\alpha(2) & \psi_b^\beta(2)\end{vmatrix}} \quad [6.18]$$

The coefficients of the CI ground state wavefunction are determined with the variation principle.

E6.19(a) $N_e = 2$ for HeH^+

The molecular orbitals are

$$\psi_a = c_{\text{He}a}\chi_{\text{He}}+c_{\text{H}a}\chi_{\text{H}} \quad \text{and} \quad \psi_b = c_{\text{He}b}\chi_{\text{He}}+c_{\text{H}b}\chi_{\text{H}} \quad [6.9]$$

while the ground and doubly excited configuration state functions are

$$\Psi_0=\psi_a(1)\psi_a(2)\sigma_-(1,2) \quad \text{and} \quad \Psi_2=\psi_b(1)\psi_b(2)\sigma_-(1,2)$$

The first-order perturbation correction [6.20] to the hamiltonian is

$$\hat{H}^{(1)}(1)=\frac{j_0}{r_{12}}-2J_a(1)+K_a(1)$$

The second-order perturbation correction to the energy [6.21] is $E_0^{(2)}=\dfrac{\left|\int\Psi_2\hat{H}^{(1)}\Psi_0\mathrm{d}\tau\right|^2}{E_0^{(0)}-E_2^{(0)}}$. It is our intent to evaluate the integral in the numerator of the $E_0^{(2)}$ expression. Since the spin functions factor from all integrations, they will not be written in the algebraic manipulations.

$$\begin{aligned}\hat{H}^{(1)}\Psi_0&=\left\{\frac{j_0}{r_{12}}-2J_a(1)+K_a(1)\right\}\{\psi_a(1)\psi_a(2)\}\\&=\frac{j_0\psi_a(1)\psi_a(2)}{r_{12}}-2J_a(1)\psi_a(1)\psi_a(2)+K_a(1)\psi_a(1)\psi_a(2)\\&=\frac{j_0\psi_a(1)\psi_a(2)}{r_{12}}-2\psi_a(2)\left(j_0\int\psi_a(1)\frac{1}{r_{12}}\psi_a^*(2)\psi_a(2)\mathrm{d}\tau_2\ [6.7\text{a}]\right)+\psi_a(2)\left(j_0\int\psi_a(1)\frac{1}{r_{12}}\psi_a^*(2)\psi_a(2)\mathrm{d}\tau_2\ [6.7\text{b}]\right)\end{aligned}$$

$$\begin{aligned}\int\Psi_2\hat{H}^{(1)}\Psi_0\mathrm{d}\tau&=\iint\psi_b(1)\psi_b(2)\hat{H}^{(1)}\Psi_0\mathrm{d}\tau_1\mathrm{d}\tau_2\\&=\iint\frac{j_0\psi_b(1)\psi_b(2)\psi_a(1)\psi_a(2)}{r_{12}}\mathrm{d}\tau_1\mathrm{d}\tau_2-2\cancel{\int\psi_b(2)\psi_a(2)\mathrm{d}\tau_2}\left(j_0\int\psi_b(1)\psi_a(1)\frac{1}{r_{12}}\psi_a^*(2)\psi_a(2)\mathrm{d}\tau_2\mathrm{d}\tau_1\right)\\&\quad+\cancel{\int\psi_b(2)\psi_a(2)\mathrm{d}\tau_2}\left(j_0\int\psi_b(1)\psi_a(1)\frac{1}{r_{12}}\int\psi_a^*(2)\psi_a(2)\mathrm{d}\tau_2\mathrm{d}\tau_1\right)\end{aligned}$$

The last two terms vanish because ψ_a and ψ_b are orthogonal. Thus,

$$\int \Psi_2 \hat{H}^{(1)} \Psi_0 \mathrm{d}\tau = \int\int \frac{j_0 \psi_b(1)\psi_b(2)\psi_a(1)\psi_a(2)}{r_{12}} \mathrm{d}\tau_1 \mathrm{d}\tau_2 = j_0 \int\int \psi_a(1)\psi_b(1)\frac{1}{r_{12}}\psi_a(2)\psi_b(2)\mathrm{d}\tau_1\mathrm{d}\tau_2$$

$$= j_0 \int\int \frac{1}{r_{12}} (c_{\mathrm{He}a}\chi_{\mathrm{He}}(1) + c_{\mathrm{H}a}\chi_{\mathrm{H}}(1))(c_{\mathrm{He}b}\chi_{\mathrm{He}}(1) + c_{\mathrm{H}b}\chi_{\mathrm{H}}(1))(c_{\mathrm{He}a}\chi_{\mathrm{He}}(2) + c_{\mathrm{H}a}\chi_{\mathrm{H}}(2))(c_{\mathrm{He}b}\chi_{\mathrm{He}}(2) + c_{\mathrm{H}b}\chi_{\mathrm{H}}(2))\mathrm{d}\tau_1\mathrm{d}\tau_2$$

$$= j_0 \int\int \frac{1}{r_{12}} \left(c_{\mathrm{He}a}c_{\mathrm{He}b}\chi_{\mathrm{He}}(1)^2 + c_{\mathrm{He}a}c_{\mathrm{H}b}\chi_{\mathrm{He}}(1)\chi_{\mathrm{H}}(1) + c_{\mathrm{H}a}c_{\mathrm{He}b}\chi_{\mathrm{He}}(1)\chi_{\mathrm{H}}(1) + c_{\mathrm{H}a}c_{\mathrm{H}b}\chi_{\mathrm{H}}(1)^2\right)$$

$$\times\left(c_{\mathrm{He}a}c_{\mathrm{He}b}\chi_{\mathrm{He}}(2)^2 + c_{\mathrm{He}a}c_{\mathrm{H}b}\chi_{\mathrm{He}}(2)\chi_{\mathrm{H}}(2) + c_{\mathrm{H}a}c_{\mathrm{He}b}\chi_{\mathrm{He}}(2)\chi_{\mathrm{H}}(2) + c_{\mathrm{H}a}c_{\mathrm{H}b}\chi_{\mathrm{H}}(2)^2\right)\mathrm{d}\tau_1\mathrm{d}\tau_2$$

Multiplying the terms of the integrand and using the definition

$$(\mathrm{AB|CD}) = j_0 \int\int \chi_{\mathrm{A}}(1)\chi_{\mathrm{B}}(1)\frac{1}{r_{12}}\chi_{\mathrm{C}}(2)\chi_{\mathrm{D}}(2)\mathrm{d}\tau_1\mathrm{d}\tau_2 \quad [6.14]$$

yields the result

$$\boxed{\begin{aligned}\int \Psi_2 \hat{H}^{(1)} \Psi_0 \mathrm{d}\tau &= c_{\mathrm{He}a}{}^2 c_{\mathrm{He}b}{}^2 (\mathrm{HeHe|HeHe}) + 2\left(c_{\mathrm{He}a}{}^2 c_{\mathrm{He}b}c_{\mathrm{H}b} + c_{\mathrm{He}a}c_{\mathrm{H}a}c_{\mathrm{He}b}{}^2\right)(\mathrm{HeH|HeHe}) \\ &+\left(c_{\mathrm{He}a}{}^2 c_{\mathrm{H}b}{}^2 + 2c_{\mathrm{He}a}c_{\mathrm{H}a}c_{\mathrm{He}b}c_{\mathrm{H}b} + c_{\mathrm{H}a}{}^2 c_{\mathrm{He}b}{}^2\right)(\mathrm{HeH|HeH}) + 2c_{\mathrm{He}a}c_{\mathrm{H}a}c_{\mathrm{He}b}c_{\mathrm{H}b}(\mathrm{HH|HeHe}) \\ &+2\left(c_{\mathrm{H}a}{}^2 c_{\mathrm{He}b}c_{\mathrm{H}b} + c_{\mathrm{He}a}c_{\mathrm{H}a}c_{\mathrm{H}b}{}^2\right)(\mathrm{HH|HHe}) + c_{\mathrm{H}a}{}^2 c_{\mathrm{H}b}{}^2(\mathrm{HH|HH})\end{aligned}}$$

where terms have been combined using the relationships (AB|CD) = (CD|AB) and (AB|CD) = (AB|DC) = (BA|CD). See E6.8(a) and E6.8(b).

E6.20(a) The **function** $f(x)$ is a mathematical procedure that assigns a number for a specified value of the independent variable x. The **functional** $G[f]$ assigns a number for a specified function $f(x)$ over a specified range of the variable x. A functional is a function of a function.

Functions: (a) $f(x) = \mathrm{d}(x^3)/\mathrm{d}x = 3x^2$

(c) $f(x) = \int x^3 \mathrm{d}x = x^4/4$ + constant

Functionals: $\boxed{\text{(b)}}$ $G[f] = \mathrm{d}(x^3)/\mathrm{d}x|_{x=1} = 3$ because the number 3 is associated with the function $f(x) = x^3$.

$\boxed{\text{(d)}}$ $G[f] = \int_1^3 x^3 \mathrm{d}x = 20$ because the number 20 is associated with a function $f(x) = x^3$.

E6.21(a) $N_e = 4$ for LiH

$$\rho(\boldsymbol{r}) = \sum_{i=1}^{N_e=4} |\psi_i(\boldsymbol{r})|^2 \ [6.23] = \boxed{2\left\{|\psi_a(\boldsymbol{r})|^2 + |\psi_b(\boldsymbol{r})|^2\right\}}$$

E6.22(a) $N_e = 2$ for HeH^+

$$\rho(\boldsymbol{r}) = \sum_{i=1}^{N_e=2} |\psi_i(\boldsymbol{r})|^2 \ [6.23] = 2|\psi_1(\boldsymbol{r}_1)|^2$$

$$\left\{h_1 + j_0 \int \frac{\rho(\boldsymbol{r}_2)}{r_{12}}\mathrm{d}\boldsymbol{r}_2 - \tfrac{3}{2}(3/\pi)^{1/3} j_0 \rho(\boldsymbol{r}_1)^{1/3}\right\}\psi_1(\boldsymbol{r}_1) = \varepsilon_1\psi_1(\boldsymbol{r}_1) \quad [6.24]$$

where $h_1 = -\dfrac{\hbar^2}{2m_e}\nabla_1^2 - j_0\left\{\dfrac{1}{r_{\mathrm{He}1}} + \dfrac{1}{r_{\mathrm{H}1}}\right\}$ [6.2]

E6.23(a) We expect that the basis set having the greatest number of well-chosen basis functions gives a result that is closer to the Hartree-Fock limit.

The minimal basis set consists of 5 functions for each carbon and oxygen atom and 1 function for each hydrogen atom. Thus, the minimal basis set of C_2H_5OH consists of 21 basis functions.

(a) The double-zeta basis set replaces each function in the minimal basis set with 2 functions. It consists of 42 basis functions for ethanol.
(b) The split-valence basis set consists of 1 function for each core atomic orbital and each valence atomic orbital is replaced by 2 basis functions. Consequently, there are 9 basis functions for each carbon and oxygen atom and 2 functions for each hydrogen. This basis set consists of 39 basis functions for ethanol.
(c) The triple-zeta basis set replaced each function in the minimal basis set with 3 functions, which means that it consists of 63 basis functions for ethanol.

Of these three basis sets, the triple-zeta basis set is expected to give a result that is closer to the Hartree-Fock limit.

Solutions to problems

Solutions to numerical problems

P6.1 For H_2, the 6-31G* basis set is equivalent to the 6-31G basis set because the star indicates that the basis set adds d-type polarization functions for each atom other than hydrogen. 1 au = 27.2114 eV. The experimental ground electronic energy of dihydrogen is calculated as $D_e + 2I$.

*Features Calculated with HF-SCF Method**
Bond Length (R) in pm and Ground-State Energy (E_0) in eV

H_2	(a) 6-31G*	(b) 6-311+G**	exp
R	73.0	73.5	74.1
E_0	−30.6626	−30.8167	−32.06
F_2	**(a) 3-31G***	**(b) 6-311+G****	**exp**
R	134.5	132.9	141.8
E_0	−5406.30	−5407.92	

*Spartan '06™

Both computational basis sets give satisfactory bond length agreement with the experimental value for H_2. However, the 6-31G* basis set is not as accurate as the larger basis set. The smaller set also yields a higher ground-state energy. Figure 6.1 shows the variation of the dihyrogen ground-state energy with the internuclear distance.

It is surprising that the 6-311+G** basis set gives a significantly shorter bond length for F_2. This might be an indication that the method should be used with caution when fluorine is present in a molecule.

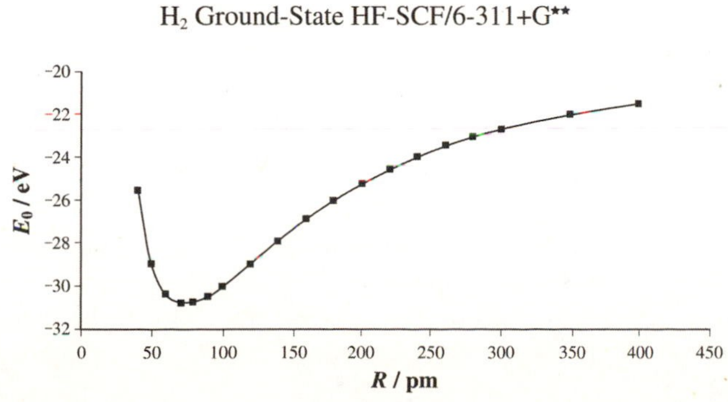

Figure 6.1

P6.3 (a) Ethanol

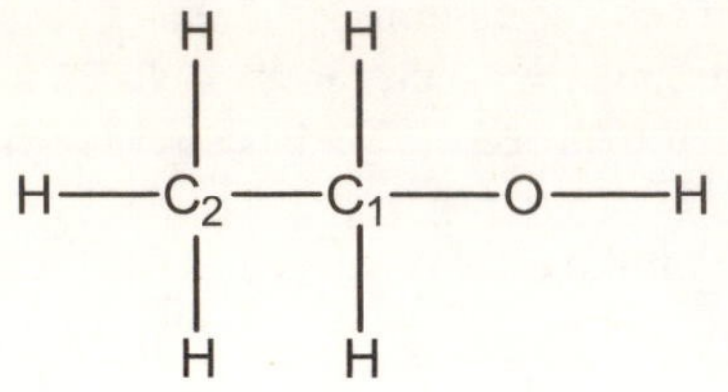

Ethanol	AM1*	PM3*	exp
$R(C_1C_2)$/pm	151.2	151.8	153.0
$R(C_1O)$	142.0	141.0	142.5
$R(C_1H)$	112.4	110.8	110
$R(C_2H)$	111.6	109.7	109
$R(OH)$	96.4	94.7	97.1
$\angle C_2C_1O$	107.34	107.81	107.8
$\Delta_f H^{\ominus}/kJ\ mol^{-1}$	–262.180	–237.881	–235.10(g)

*Spartan '06™

(b) 1,4-dichlorobenzene

Both methods give fair agreement with experiment. However, the PM3 calculation appears to provide a more reliable standard enthalpy of formation.

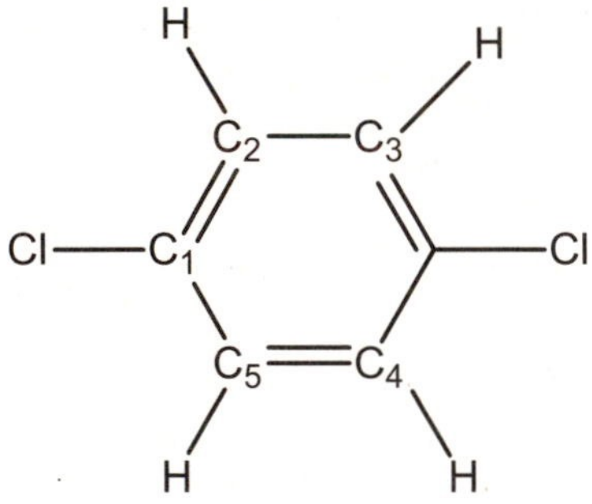

1,4-dichlorobenzene	AM1*	PM3*	exp
$R(C_1C_2)$/pm	139.9	139.4	138.8
$R(C_2C_3)$	139.3	138.9	138.8
$R(CCl)$	169.9	168.5	173.9
$\angle C_6C_1C_2$	120.58	121.09	
$\angle C_1C_2C_3$	119.71	119.45	
$\Delta_f H^{\ominus}$ / kJ mol^{-1}	33.363	42.306	24.6(g)

*Spartan '06™

Once again, both methods give fair agreement with experiment. However, this time the AM1 calculation appears to provide a more reliable standard enthalpy of formation. Parts (a) and (b) together suggest that both methods are in broad agreement with experiment and that neither of them gives consistently better agreement with experiment.

P6.5 (a) The standard enthalpy of formation ($\Delta_f H^{\ominus}$/ kJ mol^{-1}) of ethene and the first few linear polyenes is listed below.

Species	Computed*	Experimental†	% error
C_2H_4	69.580	52.46694	32.6
C_4H_6	129.834	108.8 ± 0.79	19.3
		111.9 ± 0.96	16.0
C_6H_8	188.523	168. ± 3	12.2
C_8H_{10}	246.848	295.9‡	16.6

*Semi-empirical, PM3 level, PC Spartan Pro™
†http://webbook.nist.gov/chemistry/
‡Pedley, Naylor, and Kirby, *Thermodynamic Data of Organic Compounds.*

(b) The % error, shown in the table, is defined by

$$\% \text{ error} = \left|\frac{\Delta_f H^{\ominus}(\text{calc}) - \Delta_f H^{\ominus}(\text{expt})}{\Delta_f H^{\ominus}(\text{expt})}\right| \times 100\%$$

(c) For all the molecules, the computed enthalpies of formation deviate from the experimental values by much more than the uncertainty in the experimental value. This observation serves to illustrate that molecular modeling software is not a substitute for experimentation when it comes to quantitative measures. It is also worth noting, however, that the experimental uncertainty can vary a great deal. The NIST database reports $\Delta_f H^{\ominus}$ for C_2H_4 to seven significant figures (with no explicit uncertainty). Even if the figure is not accurate to 1 part in 5000000, it is clearly a very precisely known quantity—as one should expect in such a familiar and well-studied substance. The database lists two different determinations for $\Delta_f H^{\ominus}(C_4H_6)$, and the experimental values differ by more than the uncertainty claimed for each; a critical evaluation of the experimental data is called for. The uncertainty claimed for $\Delta_f H^{\ominus}(C_6H_8)$ is greater still (but still only about 2%). Finally, it should go without saying that not all of the figures reported by the molecular modeling software are physically significant.

P6.7 (a) The table displays both the experimental and calculated* (HF-SCF/6-311G*) ^{13}C chemical shifts and computed* atomic charges on the carbon atom para to a number of substituents in substituted benzenes. Three sets of charges are shown, one derived by fitting the electrostatic potential, another by Mulliken population analysis, and the other by the method of "natural" charges.

Substituent	CH_3	H	CF_3	CN	NO_2
δ_{exp}	128.4	128.5	128.9	129.1	129.4
δ_{calc}*	134.4	133.5	132.1	138.8	141.8
Electrostatic charge* / e	−0.240	−0.135	−0.138	−0.102	−0.116
Mulliken charge* / e	−0.231	−0.217	−0.205	−0.199	−0.182
Natural charge* / e	−0.199	−0.183	−0.158	−0.148	−0.135

Spartan '06TM; HF-SCF/6-311G

The idea of atomic charges within molecules is very useful on a conceptual level but this concept is not a measureable property. Nuclei definitely define the positions of atoms, but the wave character of electrons causes them to be distributed throughout the molecule in molecular orbitals. Some of the electrons (e.g., core electrons) may be largely localized around a particular nucleus while others are more widely distributed in bonding and antibonding MOs. The case for which a bonding MO is localized around two adjacent nuclei illustrates the impossibility of defining an absolute scale of measurable atomic charges as there is no way to define which fraction of the MO should be assigned to one or the other nucleus. Consequently, there are many ways of defining atomic charges in a molecule. The most popular three (electrostatic charge, Mulliken charge, and natural charge) have been computed for this molecular series.

The three methods for calculating atomic charge agree that in going across the substituent series $–CH_3$, –H, $–CF_3$, –CN, $–NO_2$ the magnitude of the negative charge on the para-carbon decreases. The electrostatic charge shows the greatest change across the series; the natural charge shows the least change. The decrease in the magnitude of negative charge across the series is compatible with the elementary concepts of electronegativity effects at a bond and the effects of substituent releasing or withdrawing electrons to or from an aromatic benzene ring. The atomic charges show that an alkyl substituent (e.g., $–CH_3$) is electron-releasing with respect to –H while $–CF_3$, –CN, and $–NO_2$ are progressively stronger electron-withdrawing substituents. See Figure 6.2.

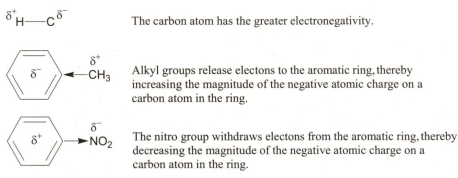

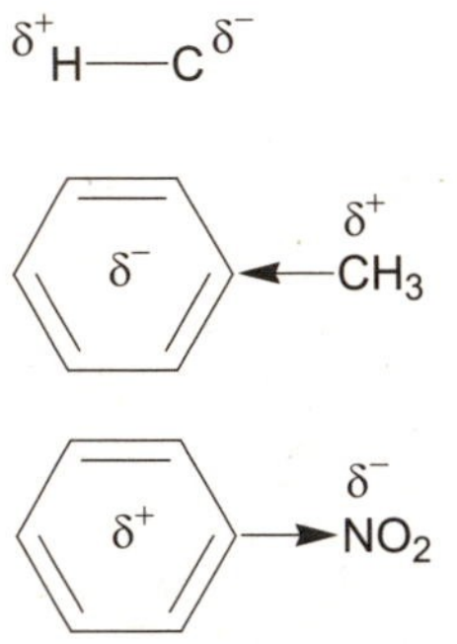

The carbon atom has the greater electronegativity.

Alkyl groups release electons to the aromatic ring, thereby increasing the magnitude of the negative atomic charge on a carbon atom in the ring.

The nitro group withdraws electons from the aromatic ring, thereby decreasing the magnitude of the negative atomic charge on a carbon atom in the ring.

Figure 6.2

(b) The very small increase across the series in the experimental chemical shift of the para-^{13}C strongly correlates with a decrease in the magnitude of the negative atomic charge across the series. This effect will be further explored in Chapter 12. Furthermore, it is evident that the uncertainty in the computed chemical shifts exceeds the actual variation in the chemical shifts across the series. The calculated shifts do indicate that the cyano and nitro groups are strong electron-withdrawing groups.

Solutions to theoretical problems

P6.9 We construct a table to determine the Russell-Saunders term symbol for each Slater determinant. The terms must be consistent with the Pauli Exclusion Principle. L is the total orbital angular momentum quantum number. S is the total spin angular momentum quantum number. $2S + 1$ is the spin multiplicity.

Part	Slater Determinant	L	S	2S + 1	Term Symbol(s)
(a)	$\lvert\psi_{1s}{}^{\alpha}\,\psi_{1s}{}^{\beta}\psi_{2s}{}^{\alpha}\rvert$	0	1/2	2	^2S
(b)	$\lvert\psi_{1s}{}^{\alpha}\,\psi_{1s}{}^{\beta}\psi_{2s}{}^{\beta}\rvert$	0	1/2	2	^2S
(c)	$\lvert\psi_{1s}{}^{\alpha}\,\psi_{1s}{}^{\beta}\psi_{2p}{}^{\alpha}\rvert$	1	1/2	2	^2P
(d)	$\lvert\psi_{1s}{}^{\alpha}\,\psi_{2p}{}^{\alpha}\psi_{2p}{}^{\beta}\rvert$	2, 0	1/2	2	^2D, ^2S
(e)	$\lvert\psi_{1s}{}^{\alpha}\,\psi_{3d}{}^{\alpha}\psi_{3d}{}^{\beta}\rvert$	4, 2, 0	1/2	2	^2G, ^2D, ^2S
(f)	$\lvert\psi_{1s}{}^{\alpha}\,\psi_{2s}{}^{\alpha}\psi_{3s}{}^{\alpha}\rvert$	0	3/2	4	^4G

The Slater determinants of parts (a), (b), (d), and (e) have the correct total angular momentum and spin multiplicity to contribute to the ground state ^{2}S of Li.

P6.11 Non-degenerate, time-independent perturbation theory is mathematically developed in *Further Information* 2.1. We outline the results for the ground-state wavefunction, ψ_0, corrected to first order.

After expanding the hamiltonian, the ground-state wavefunction, and the ground-state energy in successive powers of the perturbation parameter λ, the expansions are substituted into the stationary-state Schrödinger equation and powers of λ are collected. Because λ is an independent parameter, the successive powers in the resultant polynomial must equal zero. The first-order term gives

$$\left(H^{(0)} - E_0^{(0)}\right)\psi_0^{(1)} = \left(E_0^{(1)} - H^{(1)}\right)\psi_0^{(0)} \qquad \text{(i)}$$

where $H^{(0)}$, $E_0^{(0)}$, and $\psi_0^{(0)}$ are the unperturbed hamiltonian, ground-state energy, and ground-state wavefunction. $H^{(1)}$, $E_0^{(1)}$, and $\psi_0^{(1)}$ are the first order perturbations to the hamiltonian, the ground-state energy, and the ground-sate wavefunction. The eigenfunctions of the unperturbed hamiltonian, $\psi_n^{(0)}$, form a complete orthonormal basis set of functions such that $H_0^{(0)}\,\psi_n^{(0)} = E_n^{(0)}\,\psi_n^{(0)}$.

Let $\psi_0^{(1)} = \sum_n a_n\psi_n^{(0)}$. Substitution of the expanded first-order wavefunction correction into equation (i) gives

$$\sum_n a_n\left(E_n^0 - E_0^{(0)}\right)\psi_n^{(0)} = \left(E_n^{(1)} - H^{(1)}\right)\psi_0^{(0)}$$

Left multiply by $\psi_k^{(0)}$ with $k \neq 0$ and integrate over all space.

$$\sum_n a_n\left(E_n^0 - E_0^{(0)}\right)\int\psi_{k(\neq 0)}^{(0)}\psi_n^{(0)}\mathrm{d}\tau = E_0^{(1)}\int\psi_{k(\neq 0)}^{(0)}\psi_0^{(0)}\mathrm{d}\tau - \int\psi_{k(\neq 0)}^{(0)}H^{(1)}\psi_0^{(0)}\mathrm{d}\tau$$

$$\sum_n a_n\left(E_n^0 - E_0^{(0)}\right)\delta_{k(\neq 0)n} = -\int\psi_{k(\neq 0)}^{(0)}H^{(1)}\psi_0^{(0)}\mathrm{d}\tau$$

$$a_{k(\neq 0)}\left(E_{k(\neq 0)}^0 - E_0^{(0)}\right) = -\int\psi_{k(\neq 0)}^{(0)}H^{(1)}\psi_0^{(0)}\mathrm{d}\tau$$

$$a_{k(\neq 0)} = \frac{\int\psi_{k(\neq 0)}^{(0)}H^{(1)}\psi_0^{(0)}\mathrm{d}\tau}{E_0^{(0)} - E_{k(\neq 0)}^0}$$

Thus, the ground-state wavefunction corrected to first order in the perturbation is

$$\boxed{\psi_0 = \psi_0^{(0)} + \psi_0^{(1)} = \psi_0^{(0)} + \sum_{n(\neq 0)}\left\{\frac{\int\psi_n^{(0)}H^{(1)}\psi_0^{(0)}\mathrm{d}\tau}{E_0^{(0)} - E_n^0}\right\}\psi_n^{(0)}}$$

where the first-order correction to the hamiltonian is given by eqn. 6.20.

P6.13 (a) $N_e = 2$ for HeH^+

Very general equations are used in the text *Brief illustration* for setting up the Roothaan equations (also see E6.6a and E6.14a). By replacing the index A with He and the index B with H we find that for a linear combination of two basis functions, such as χ_{He} and χ_H, the coefficients of the molecular orbitals in

$$\psi_a = c_{Hea}\chi_{He} + c_{Ha}\chi_H \text{ and } \psi_b = c_{Heb}\chi_{He} + c_{Hb}\chi_H$$

are given by the equations

$$\varepsilon_a = \frac{F_{HeHe} + F_{HH} - S(F_{HeH} - F_{HHe}) - \left(\{F_{HeHe} + F_{HH} - S(F_{HeH} - F_{HHe})\}^2 - 4\{1 - S^2\}\{F_{HeHe}F_{HH} - F_{HeH}F_{HHe}\}\right)^{1/2}}{2\{1 - S^2\}}$$

and

$$c_{Hea} = -\frac{F_{HeH} - S\varepsilon_a}{F_{HeHe} - \varepsilon_a}c_{Ha} \quad \text{and} \quad S = \int\chi_{He}\chi_H\mathrm{d}\tau$$

in conjunction with the normalization condition $c_{\text{He}a}{}^2 + c_{\text{H}a}{}^2 + 2c_{\text{He}a}c_{\text{H}a}S = 1$.

Also,

$$\varepsilon_b = \frac{F_{\text{HeHe}} + F_{\text{HH}} - S(F_{\text{HeH}} - F_{\text{HHe}}) + \left(\{F_{\text{HeHe}} + F_{\text{HH}} - S(F_{\text{HeH}} - F_{\text{HHe}})\}^2 - 4\{1 - S^2\}\{F_{\text{HeHe}}F_{\text{HH}} - F_{\text{HeH}}F_{\text{HHe}}\}\right)^{1/2}}{2\{1 - S^2\}}$$

and

$$c_{\text{He}b} = -\frac{F_{\text{HeH}} - S\varepsilon_b}{F_{\text{HeHe}} - \varepsilon_b} c_{\text{H}b} \quad \text{and} \quad S = \int \chi_{\text{He}}\chi_{\text{H}}\,\text{d}\tau$$

in conjunction with the normalization condition $c_{\text{He}b}{}^2 + c_{\text{H}b}{}^2 + 2c_{\text{He}b}c_{\text{H}b}S = 1$.

(b) $N_e = 2$ for LiH^{2+}

Very general equations are used in the text brief illustration for setting up the Roothaan equations (also see E6.6b and E6.14b). By replacing the index A with Li and the index B with H we find that for a linear combination of two basis functions, such as χ_{Li} and χ_{H}, the coefficients of the molecular orbitals

$$\psi_a = c_{\text{Li}a}\chi_{\text{Li}} + c_{\text{H}a}\chi_{\text{H}} \text{ and } \psi_b = c_{\text{Li}b}\chi_{\text{Li}} + c_{\text{H}b}\chi_{\text{H}}$$

are given by the equations

$$\varepsilon_a = \frac{F_{\text{LiLi}} + F_{\text{HH}} - S(F_{\text{LiH}} - F_{\text{HLi}}) - \left(\{F_{\text{LiLi}} + F_{\text{HH}} - S(F_{\text{LiH}} - F_{\text{HLi}})\}^2 - 4\{1 - S^2\}\{F_{\text{LiLi}}F_{\text{HH}} - F_{\text{LiH}}F_{\text{HLi}}\}\right)^{1/2}}{2\{1 - S^2\}}$$

and

$$c_{\text{Li}a} = -\frac{F_{\text{LiH}} - S\varepsilon_a}{F_{\text{LiLi}} - \varepsilon_a} c_{\text{H}a} \quad \text{and} \quad S = \int \chi_{\text{Li}}\chi_{\text{H}}\,\text{d}\tau$$

in conjunction with the normalization condition $c_{\text{Li}a}{}^2 + c_{\text{H}a}{}^2 + 2c_{\text{Li}a}c_{\text{H}a}S = 1$.

Also,

$$\varepsilon_b = \frac{F_{\text{LiLi}} + F_{\text{HH}} - S(F_{\text{LiH}} - F_{\text{HLi}}) + \left(\{F_{\text{LiLi}} + F_{\text{HH}} - S(F_{\text{LiH}} - F_{\text{HLi}})\}^2 - 4\{1 - S^2\}\{F_{\text{LiLi}}F_{\text{HH}} - F_{\text{LiH}}F_{\text{HLi}}\}\right)^{1/2}}{2\{1 - S^2\}}$$

and

$$c_{\text{Li}b} = -\frac{F_{\text{LiH}} - S\varepsilon_b}{F_{\text{LiLi}} - \varepsilon_b} c_{\text{H}b} \quad \text{and} \quad S = \int \chi_{\text{Li}}\chi_{\text{H}}\,\text{d}\tau$$

in conjunction with the normalization condition $c_{\text{Li}b}{}^2 + c_{\text{H}b}{}^2 + 2c_{\text{Li}b}c_{\text{H}b}S = 1$.

P6.15 (a) Applying the Hückel approximation $S = 0$ and using the symbols $\alpha_{\text{He}} = F_{\text{HeHe}}$, $\alpha_{\text{H}} = F_{\text{HH}}$, and $F_{\text{HeH}} = F_{\text{HHe}} = \beta$ simplifies the secular determinant [6.12] for HeH^+ and rapidly produces an approximation for ε_a and ε_b.

$$\begin{vmatrix} F_{\text{HeHe}} - \varepsilon & F_{\text{HeH}} - S\varepsilon \\ F_{\text{HHe}} - S\varepsilon & F_{\text{HH}} - \varepsilon \end{vmatrix} = \begin{vmatrix} \alpha_{\text{He}} - \varepsilon & \beta \\ \beta & \alpha_{\text{H}} - \varepsilon \end{vmatrix} = 0$$

$$(\alpha_{\text{He}} - \varepsilon)(\alpha_{\text{H}} - \varepsilon) - \beta^2 = 0$$

$$\varepsilon^2 - (\alpha_{\text{He}} + \alpha_{\text{H}})\varepsilon + \alpha_{\text{He}}\alpha_{\text{H}} - \beta^2 = 0$$

$$\varepsilon = \frac{\alpha_{\text{He}} + \alpha_{\text{H}} \pm \left(\{\alpha_{\text{He}} + \alpha_{\text{H}}\}^2 - 4\{\alpha_{\text{He}}\alpha_{\text{H}} - \beta^2\}\right)^{1/2}}{2} = \frac{\alpha_{\text{He}} + \alpha_{\text{H}} \pm \left(\{\alpha_{\text{He}} - \alpha_{\text{H}}\}^2 + 4\beta^2\right)^{1/2}}{2}$$

Thus, $$\boxed{\varepsilon_a = \frac{\alpha_{\text{He}} + \alpha_{\text{H}} - \left(\{\alpha_{\text{He}} - \alpha_{\text{H}}\}^2 + 4\beta^2\right)^{1/2}}{2}} \text{ and } \boxed{\varepsilon_b = \frac{\alpha_{\text{He}} + \alpha_{\text{H}} + \left(\{\alpha_{\text{He}} - \alpha_{\text{H}}\}^2 + 4\beta^2\right)^{1/2}}{2}}$$

(b) Applying the Hückel approximation $S = 0$ and using the symbols $\alpha_{\text{Li}} = F_{\text{LiLi}}$, $\alpha_{\text{H}} = F_{\text{HH}}$, and $F_{\text{LiH}} = F_{\text{HLi}} = \beta$ simplifies the secular determinant [6.12] for LiH^{2+}.

$$\begin{vmatrix} F_{\text{LiLi}} - \varepsilon & F_{\text{LiH}} - S\varepsilon \\ F_{\text{HLi}} - S\varepsilon & F_{\text{HH}} - \varepsilon \end{vmatrix} = \begin{vmatrix} \alpha_{\text{Li}} - \varepsilon & \beta \\ \beta & \alpha_{\text{H}} - \varepsilon \end{vmatrix} = 0$$

$$(\alpha_{\text{Li}} - \varepsilon)(\alpha_{\text{H}} - \varepsilon) - \beta^2 = 0$$

$$\varepsilon^2 - (\alpha_{\text{Li}} + \alpha_{\text{H}})\varepsilon + \alpha_{\text{Li}}\alpha_{\text{H}} - \beta^2 = 0$$

$$\varepsilon = \frac{\alpha_{\text{Li}} + \alpha_{\text{H}} \pm \left(\{\alpha_{\text{Li}} + \alpha_{\text{H}}\}^2 - 4\{\alpha_{\text{Li}}\alpha_{\text{H}} - \beta^2\}\right)^{1/2}}{2} = \frac{\alpha_{\text{Li}} + \alpha_{\text{H}} \pm \left(\{\alpha_{\text{Li}} - \alpha_{\text{H}}\}^2 + 4\beta^2\right)^{1/2}}{2}$$

Thus, $\boxed{\varepsilon_a = \dfrac{\alpha_{\text{Li}} + \alpha_{\text{H}} - \left(\{\alpha_{\text{Li}} - \alpha_{\text{H}}\}^2 + 4\beta^2\right)^{1/2}}{2}}$ and $\boxed{\varepsilon_b = \dfrac{\alpha_{\text{Li}} + \alpha_{\text{H}} + \left(\{\alpha_{\text{Li}} - \alpha_{\text{H}}\}^2 + 4\beta^2\right)^{1/2}}{2}}$

P6.17 We begin by choosing a coordinate system centered on the N atom of ammonia, which is shown in Figure 6.3 and is compatible with the one used in E6.12(a). d is the bond length. θ is the bond angle, and the xz plane bisects the H_3NH_2 bond angle.

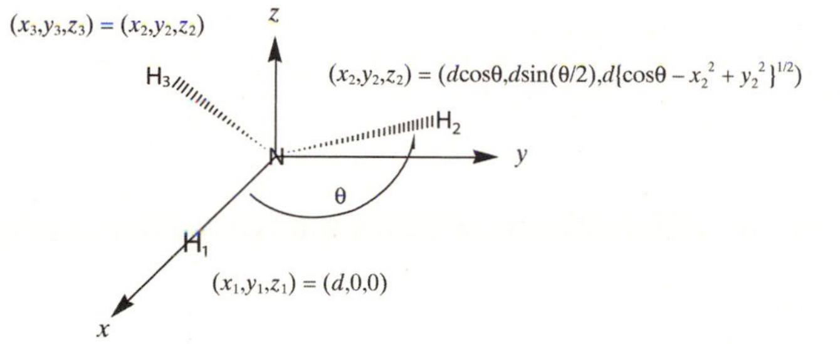

Figure 6.3

The s-type Gaussians on the nitrogen atom and the hydrogen atoms of ammonia are

$$\chi_{\text{N}} = N_{\text{N}} e^{-\alpha_{\text{N}}|r|^2},\ \chi_{\text{H}_1} = N_{\text{H}_1} e^{-\alpha_{\text{H}}|r - r_{\text{H}_1}|^2},\ \chi_{\text{H}_2} = N_{\text{H}_2} e^{-\alpha_{\text{H}}|r - r_{\text{H}_2}|^2},\ \chi_{\text{H}_3} = N_{\text{H}_3} e^{-\alpha_{\text{H}}|r - r_{\text{H}_3}|^2}$$

where $\boldsymbol{r}_{\text{H}_1}$, $\boldsymbol{r}_{\text{H}_2}$, and $\boldsymbol{r}_{\text{H}_3}$ are position vectors of the atoms as shown in the figure.

In E6.12(a) it is shown that

$$\chi_{\text{N}}\chi_{\text{H}_1} = \left(N_{\text{N}}N_{\text{H}_1}e^{-\beta_1}\right) \times e^{-\alpha_1|r - r_1|^2}$$

where $\alpha_1 = \alpha_{\text{N}} + \alpha_{\text{H}}$, $r_1 = \left(\dfrac{\alpha_{\text{H}}d}{\alpha_{\text{N}} + \alpha_{\text{H}}}, 0, 0\right)$, and $\beta_1 = \dfrac{\alpha_{\text{N}}\alpha_{\text{H}}d^2}{\alpha_{\text{N}} + \alpha_{\text{H}}}$.

Thus, the two-center $\chi_i\chi_j$ s-type Gaussian can be replaced with a one-center Gaussian at the position $\boldsymbol{r}_{\text{NH}_1} = (e_1, 0, 0)$ between N and H_1 atoms. This result also allows us to conclude that the product $\chi_{\text{H}_2}\chi_{\text{H}_3}$ can be replaced with a one-center Gaussian that is halfway between H_2 and H_3 because the hydrogen atoms have identical α values.

$$\chi_{\text{H}_2}\chi_{\text{H}_3} = \left(N_{\text{H}}{}^2 e^{-\beta_{\text{H}}}\right) \times e^{-\alpha|r - r_c|^2}$$

where $\alpha = 2\alpha_{\text{H}}$, $\beta = \dfrac{d_{\text{HH}}{}^2}{2} = 2y_2{}^2$, and $\boldsymbol{r}_c = (x_2, 0, z_2)$.

These products can be substituted in the four-center, two-electron integral (NH|HH).

$$(\text{NH}_1|\text{H}_2\text{H}_3) = j_0 \int \chi_{\text{N}}(1)\chi_{\text{H}_1}(1)\frac{1}{r_{12}}\chi_{\text{H}_2}(2)\chi_{\text{H}_3}(2)\,d\tau = j_0 \int \chi_{\text{N}}(1)\chi_{\text{H}_1}(1)\frac{1}{r_{12}}\chi_{\text{H}_2}(2)\chi_{\text{H}_3}(2)\,d\tau_1 d\tau_2$$

$$= N \int e^{-\alpha_1|r(1) - r_1|^2}\frac{1}{r_{12}}e^{-\alpha|r(2) - r_c|^2}\,d\tau_1 d\tau_2 \quad \text{where } N = j_0 N_{\text{N}} N_{\text{H}}{}^3 e^{-\beta_1 - \beta_{\text{H}}}$$

This shows that the four-center, two-electron integral reduces to an integral over two different centers at $\boldsymbol{r}_1$ and $\boldsymbol{r}_c$. The demonstration may also proceed using any arbitrarily placed coordinate system using the results of E6.12(b) where it is shown that

$$\chi_A\chi_B = Ne^{-\alpha|r-r_c|^2} \text{ where } N = N_A N_B e^{-\alpha^{-1}\alpha_A\alpha_B|r_A-r_B|^2},\ \alpha = \alpha_A + \alpha_B \text{ and } \boldsymbol{r}_c = (\alpha_A\boldsymbol{r}_A + \alpha_B\boldsymbol{r}_B)/\alpha$$

P6.19 $N_e = 2$ for H_2 and from E6.17(a) and E6.17(b),

$$\psi_a = (1/2)^{1/2}\{\chi_A + \chi_B\} \text{ and } \psi_b = (1/2)^{1/2}\{\chi_A - \chi_B\},\ \sigma_-(1,2) = (1/2)^{1/2}\{\alpha(1)\beta(2) - \beta(1)\alpha(2)\}$$

$$\Psi_0 = \psi_a(1)\psi_a(2)\sigma_-(1,2),\ \Psi_1 = \psi_a(1)\psi_b(2)\sigma_-(1,2),\ \Psi_2 = \psi_b(1)\psi_b(2)\sigma_-(1,2)$$

$$H_{22} = \int\Psi_2\left(h_1 + h_2 + \frac{j_0}{r_{12}}\right)\Psi_2\,d\tau$$

$$= \int\psi_b(1)\psi_b(2)\left(h_1 + h_2 + \frac{j_0}{r_{12}}\right)\psi_b(1)\psi_b(2)\,d\tau \quad \text{(The spin states are normalized.)}$$

$$= \int\psi_b(1)\psi_b(2)h_1\psi_b(1)\psi_b(2)\,d\tau + \int\psi_b(1)\psi_b(2)h_2\psi_b(1)\psi_b(2)\,d\tau + \int\psi_b(1)\psi_b(2)\frac{j_0}{r_{12}}\psi_b(1)\psi_b(2)\,d\tau \quad \text{(i)}$$

Evaluating the first two integrals to the right, we find

$$\int\psi_b(1)\psi_b(2)h_1\psi_b(1)\psi_b(2)\,d\tau = \int\psi_b(1)h_1\psi_b(1)\,d\tau_1\int\psi_b(2)\psi_b(2)\,d\tau_2 = \int\psi_b(1)h_1\psi_b(1)\,d\tau_1 \quad \text{(ii)}$$

Similarly, $\int\psi_b(1)\psi_b(2)h_2\psi_b(1)\psi_b(2)\,d\tau = \int\psi_b(2)h_2\psi_b(2)\,d\tau_2$, which is identical to integral (ii).

Evaluation of the third integral to the right of equation (i) proceeds as follows.

$$4\int\psi_b(1)\psi_b(2)\frac{j_0}{r_{12}}\psi_b(1)\psi_b(2)\,d\tau = 4\int\psi_b(1)^2\frac{j_0}{r_{12}}\psi_b(2)^2\,d\tau$$

$$= \int\{\chi_A(1) - \chi_B(1)\}^2\frac{j_0}{r_{12}}\{\chi_A(2) - \chi_B(2)\}^2\,d\tau$$

$$= \int\{\chi_A(1)^2 - 2\chi_A(1)\chi_B(1) + \chi_B(1)^2\}\frac{j_0}{r_{12}}\{\chi_A(2)^2 - 2\chi_A(2)\chi_B(2) + \chi_B(2)^2\}\,d\tau$$

$$= (AA|AA) - 2(AA|AB) + (AA|BB) - 2(AB|AA) + 4(AB|AB) - 2(AB|BB)$$

$$+ (BB|AA) - 2(BB|AB) + (BB|BB)$$

$$= 2(AA|AA) - 8(AA|AB) + 2(AA|BB) + 4(AB|AB) \quad \text{(iii)}$$

where we have used $(BB|BB) = (AA|AA)$, $(AB|AA) = (AA|AB) = (AB|BB) = (BB|AB)$, and $(BB|AA) = (AA|BB)$.

With (i), (ii), and (iii) we find that

$$\boxed{H_{22} = 2\int\psi_b(1)h_1\psi_b(1)\,d\tau_1 + \tfrac{1}{2}\{(AA|AA) - 4(AA|AB) + (AA|BB) + 2(AB|AB)\}}$$

Likewise,

$$\boxed{H_{00} = 2E_H + \tfrac{1}{2}\{(AA|AA) + 4(AA|AB) + (AA|BB) + 2(AB|AB)\}}$$

where $E_H = \int\psi_a(1)h_1\psi_a(1)\,d\tau_1$.

$$H_{02} = \int \Psi_0 \left(h_1 + h_2 + \frac{j_0}{r_{12}} \right) \Psi_2 \mathrm{d}\tau_1 \mathrm{d}\tau_2$$

$$= \int \psi_a(1)\psi_a(2) \left(h_1 + h_2 + \frac{j_0}{r_{12}} \right) \psi_b(1)\psi_b(2) \mathrm{d}\tau \quad \text{(The spin states are normalized.)}$$

$$= \int \psi_a(1)\psi_a(2) h_1 \psi_b(1)\psi_b(2) \mathrm{d}\tau + \int \psi_a(1)\psi_a(2) h_2 \psi_b(1)\psi_b(2) \mathrm{d}\tau + \int \psi_a(1)\psi_a(2) \frac{j_0}{r_{12}} \psi_b(1)\psi_b(2) \mathrm{d}\tau \quad \text{(iv)}$$

$$\int \psi_a(1)\psi_a(2) h_1 \psi_b(1)\psi_b(2) \mathrm{d}\tau = \int \psi_a(1) h_1 \psi_b(1) \mathrm{d}\tau_1 \int \psi_a(2)\psi_b(2) \mathrm{d}\tau_2 = 0 \quad \text{(v)}$$

because ψ_a and ψ_b are orthogonal.

Likewise, $\int \psi_a(1)\psi_a(2) h_2 \psi_b(1)\psi_b(2) \mathrm{d}\tau = 0.$ (vi)

$$4\int \psi_a(1)\psi_a(2) \frac{j_0}{r_{12}} \psi_b(1)\psi_b(2) \mathrm{d}\tau = 4\int \psi_a(1)\psi_b(1) \frac{j_0}{r_{12}} \psi_a(2)\psi_b(2) \mathrm{d}\tau$$

$$= \int \{\chi_A(1) + \chi_B(1)\}\{\chi_A(1) - \chi_B(1)\} \frac{j_0}{r_{12}} \{\chi_A(2) + \chi_B(2)\}\{\chi_A(2) - \chi_B(2)\} \mathrm{d}\tau$$

$$= \int \{\chi_A(1)^2 - \chi_B(1)^2\} \frac{j_0}{r_{12}} \{\chi_A(2)^2 - \chi_B(2)^2\} \mathrm{d}\tau$$

$$= (\mathrm{AA}|\mathrm{AA}) - (\mathrm{AA}|\mathrm{BB}) - (\mathrm{BB}|\mathrm{AA}) + (\mathrm{BB}|\mathrm{BB})$$

$$= 2(\mathrm{AA}|\mathrm{AA}) - 2(\mathrm{AA}|\mathrm{BB}) \quad \text{(vii)}$$

where we have used $(\mathrm{BB}|\mathrm{BB}) = (\mathrm{AA}|\mathrm{AA})$ and $(\mathrm{BB}|\mathrm{AA}) = (\mathrm{AA}|\mathrm{BB})$

With (iv)–(vii) we find that

$$\boxed{H_{02} = \tfrac{1}{2}\{(\mathrm{AA}|\mathrm{AA}) - (\mathrm{AA}|\mathrm{BB})\} = H_{20}} \quad \text{(by symmetry)}$$

P6.21 This solution is based on the solution to problem 9.17 in P. W. Atkins and R. S. Friedman, *Solutions Manual for Molecular Quantum Mechanics*, 3rd ed., Oxford University Press, 1997.

We are to show that

$$\int \Psi_0 H \Psi_a^p \mathrm{d}\tau = 0$$

where Ψ_0 is the HF ground-state wavefunction and Ψ_a^p is a singly excited determinant in which the virtual spinorbital ψ_p replaces the occupied spinorbital ψ_a of the ground-state.

The hamiltonian is

$$H = \sum_i^{N_e} h_i + \tfrac{1}{2} \sum_{\substack{i,j \\ i \neq j}}^{N_e} \frac{j_0}{r_{ij}} \text{ [6.1a] where } h_i = -\frac{\hbar^2}{2m_e} \nabla_i^2 - \sum_I^{N_n} \frac{Z_I j_0}{r_{Ii}}$$

The spinorbitals are orthonormal eigenfunctions of the Fock operator f_1 [6.6], so

$$\int \psi_a(1) f_1 \psi_p(1) \mathrm{d}\tau_1 = \varepsilon_p \int \psi_a(1) \psi_p(1) \mathrm{d}\tau_1 = 0 \quad \text{(i)}$$

We begin by showing that

$$\int \psi_a(1) f_1 \psi_p(1) \mathrm{d}\tau_1 = \int \psi_a(1) h_1 \psi_p(1) \mathrm{d}\tau_1 + \sum_m \{(\psi_a\psi_p|\psi_m\psi_m) - (\psi_a\psi_m|\psi_m\psi_p)\} \quad \text{(ii)}$$

where $(\psi_a\psi_b|\psi_c\psi_d) = \int \psi_a(1)\psi_b(1) \left(\frac{j_0}{r_{12}} \right) \psi_c(2)\psi_d(2) \mathrm{d}\tau_1 \mathrm{d}\tau_2$

$$\int \psi_a(1) f_1 \psi_p(1)\,\mathrm{d}\tau_1 = \int \psi_a(1)\left\{ h_1\psi_p(1) + \sum_{\substack{m \\ \text{(singly occupied orbitals)}}} \{J_m(1) - K_m(1)\}\psi_p(1)\right\}\mathrm{d}\tau_1 \quad [6.8]$$

$$= \int \psi_a(1) h_1 \psi_p(1)\,\mathrm{d}\tau_1 + \sum_m \int \psi_a(1)\left\{ \int \psi_p(1)\frac{j_0}{r_{12}}\psi_m(2)\psi_m(2)\,\mathrm{d}\tau_2\right\}\mathrm{d}\tau_1$$

$$- \sum_m \int \psi_a(1)\left\{ \int \psi_m(1)\frac{j_0}{r_{12}}\psi_m(2)\psi_p(2)\,\mathrm{d}\tau_2\right\}\mathrm{d}\tau_1$$

$$= \int \psi_a(1) h_1 \psi_p(1)\,\mathrm{d}\tau_1 + \sum_m \{(\psi_a\psi_p|\psi_m\psi_m) - (\psi_a\psi_m|\psi_m\psi_p)\}$$

We also need to see that

$$\sum_m^{N_e} \int \Psi_0 h_m \Psi_a^p\,\mathrm{d}\tau = N_e \int \Psi_0 h_1 \Psi_a^p\,\mathrm{d}\tau \quad \text{(iii)}$$

The reason is that h_m is a one-electron operator and electrons in the determinant wavefunction are indistinguishable, so $\int \Psi_0 h_1 \Psi_a^p\,\mathrm{d}\tau = \int \Psi_0 h_2 \Psi_a^p\,\mathrm{d}\tau = \int \Psi_0 h_3 \Psi_a^p\,\mathrm{d}\tau = \cdots = \int \Psi_0 h_{N_e} \Psi_a^p\,\mathrm{d}\tau$.

Thus,

$$\sum_m^{N_e} \int \Psi_0 h_m \Psi_a^p\,\mathrm{d}\tau = \int \Psi_0 h_1 \Psi_a^p\,\mathrm{d}\tau + \int \Psi_0 h_2 \Psi_a^p\,\mathrm{d}\tau + \int \Psi_0 h_3 \Psi_a^p\,\mathrm{d}\tau + \cdots + \int \Psi_0 h_{N_e} \Psi_a^p\,\mathrm{d}\tau = N_e \int \Psi_0 h_1 \Psi_a^p\,\mathrm{d}\tau$$

To find a simplifying relationship for $\int \Psi_0 h_1 \Psi_a^p\,\mathrm{d}\tau$, we must find the number of possible orbital permutations given by the Slater determinants. With ψ_a occupied by electron 1 in Ψ_0 and ψ_p occupied by electron 1 in Ψ_a^p, there are $(N_e - 1)!$ ways of permuting electrons 2, 3, ..., N_e among the other $N_e - 1$ spinorbitals in each of the determinants Ψ_0 and Ψ_a^p. Each of the $(N_e - 1)!$ permutations in Ψ_0 makes a nonzero contribution to the determinant expansion if it matches identically one of the $(N_e - 1)!$ permutations in Ψ_a^p. Thus, there are a total of $(N_e - 1)!$ terms that contribute to the summation and since the spinorbitals are normalized, each of the $(N_e - 1)!$ terms gives an equal contribution of $\int \psi_a(1) h_1 \psi_p(1)\,\mathrm{d}\tau$.

$$\int \Psi_0 h_1 \Psi_a^p\,\mathrm{d}\tau = \left(\frac{1}{N_e!}\right)\int \begin{vmatrix} \psi_a(1) & \psi_b(1) & \cdots & \psi_z(1) \\ \psi_a(2) & \psi_b(2) & \cdots & \vdots \\ \vdots & \vdots & \cdots & \vdots \\ \psi_a(N_e) & \psi_b(N_e) & \cdots & \psi_z(N_e) \end{vmatrix} h_1 \begin{vmatrix} \psi_p(1) & \psi_b(1) & \cdots & \psi_z(1) \\ \psi_p(2) & \psi_b(2) & \cdots & \vdots \\ \vdots & \vdots & \cdots & \vdots \\ \psi_p(N_e) & \psi_b(N_e) & \cdots & \psi_z(N_e) \end{vmatrix}\mathrm{d}\tau$$

$$= \frac{(N_e - 1)!}{N_e!}\int \psi_a(1) h_1 \psi_p(1)\,\mathrm{d}\tau = \frac{1}{N_e}\int \psi_a(1) h_1 \psi_p(1)\,\mathrm{d}\tau \quad \text{(iv)}$$

One additional relationship must be demonstrated before these equations can be used to prove the Brillouin theorem. It is

$$\tfrac{1}{2}\sum_{\substack{i,j \\ i\neq j}}^{N_e} \int \Psi_0 \frac{j_0}{r_{ij}} \Psi_a^p\,\mathrm{d}\tau = \sum_m \{(\psi_a\psi_p|\psi_m\psi_m) - (\psi_a\psi_m|\psi_m\psi_p)\} \quad \text{(v)}$$

To demonstrate the relationship, first expand the sum.

$$\tfrac{1}{2}\sum_{\substack{i,j \\ i\neq j}}^{N_e} \int \Psi_0 \frac{j_0}{r_{ij}} \Psi_a^p\,\mathrm{d}\tau = \tfrac{1}{2} j_0 \int \Psi_0 \{r_{12}^{-1} + r_{21}^{-1} + r_{13}^{-1} + r_{31}^{-1} \cdots\}\Psi_a^p\,\mathrm{d}\tau$$

Since $r_{12} = r_{21}$, $r_{13} = r_{31}$, and so on,

$$\tfrac{1}{2}\sum_{\substack{i,j \\ i\neq j}}^{N_e} \int \Psi_0 \frac{j_0}{r_{ij}} \Psi_a^p\,\mathrm{d}\tau = j_0 \int \Psi_0 \{r_{12}^{-1} + r_{13}^{-1} \cdots + r_{23}^{-1} + r_{24}^{-1} + \cdots + r_{(N_e-1)N_e}^{-1}\}\Psi_a^p\,\mathrm{d}\tau$$

The electrons in the determinants are indistinguishable, so each term in the summation gives an identical result. There are $\frac{1}{2}N_e(N_e-1)$ equal terms so the previous equation simplifies to

$$\frac{1}{2}\sum_{\substack{i,j\\i\neq j}}^{N_e}\int\Psi_0\frac{j_0}{r_{ij}}\Psi_a^p\,\mathrm{d}\tau=\tfrac{1}{2}N_e(N_e-1)\,j_0\int\Psi_0 r_{12}^{-1}\Psi_a^p\,\mathrm{d}\tau$$

The operator r_{12}^{-1} within the above integral is a two electron operator leaving $N_e - 2$ electrons that can be permutated in $(N_e - 2)!$ ways in the determinant wavefunctions with each term in the resultant sum being equal.

$$\frac{1}{2}\sum_{\substack{i,j\\i\neq j}}^{N_e}\int\Psi_0\frac{j_0}{r_{ij}}\Psi_a^p\,\mathrm{d}\tau=\frac{\frac{1}{2}N_e(N_e-1)}{N_e!}\,j_0\int\begin{vmatrix}\psi_a(1)&\psi_b(1)&\cdots&\psi_z(1)\\\psi_a(2)&\psi_b(2)&\cdots&\vdots\\\vdots&\vdots&\cdots&\vdots\\\psi_a(N_e)&\psi_b(N_e)&\cdots&\psi_z(N_e)\end{vmatrix}r_{12}^{-1}\begin{vmatrix}\psi_p(1)&\psi_b(1)&\cdots&\psi_z(1)\\\psi_p(2)&\psi_b(2)&\cdots&\vdots\\\vdots&\vdots&\cdots&\vdots\\\psi_p(N_e)&\psi_b(N_e)&\cdots&\psi_z(N_e)\end{vmatrix}\mathrm{d}\tau$$

$$=\frac{\frac{1}{2}N_e(N_e-1)(N_e-2)!}{N_e!}\,j_0\sum_{\substack{m\\m\neq a,p}}^{N_e}\left\{\begin{aligned}&\int\psi_a(1)\psi_m(2)r_{12}^{-1}\psi_p(1)\psi_m(2)\,\mathrm{d}\tau-\int\psi_a(1)\psi_m(2)r_{12}^{-1}\psi_p(2)\psi_m(1)\,\mathrm{d}\tau\\&+\int\psi_m(1)\psi_a(2)r_{12}^{-1}\psi_m(1)\psi_p(2)\,\mathrm{d}\tau-\int\psi_m(1)\psi_a(2)r_{12}^{-1}\psi_m(2)\psi_p(1)\,\mathrm{d}\tau\end{aligned}\right\}$$

The above minus signs appear because permutations in Ψ_a^p differ by a sign if two electrons are interchanged. Because electrons 1 and 2 are indistinguishable, the first and third terms in the sum are equal as are the second and fourth terms. Furthermore, the restriction that $m \neq a, p$ can be removed because, if included in the sum, the subtractions will cause their elimination.

$$\frac{1}{2}\sum_{\substack{i,j\\i\neq j}}^{N_e}\int\Psi_0\frac{j_0}{r_{ij}}\Psi_a^p\,\mathrm{d}\tau$$

$$=\frac{2(\frac{1}{2})N_e(N_e-1)(N_e-2)!}{N_e!}\,j_0\sum_{m}^{N_e}\left\{\int\psi_a(1)\psi_m(2)r_{12}^{-1}\psi_p(1)\psi_m(2)\,\mathrm{d}\tau-\int\psi_a(1)\psi_m(2)r_{12}^{-1}\psi_p(2)\psi_m(1)\,\mathrm{d}\tau\right\}$$

$$=j_0\sum_{m}^{N_e}\left\{\int\psi_a(1)\psi_m(2)r_{12}^{-1}\psi_p(1)\psi_m(2)\,\mathrm{d}\tau-\int\psi_a(1)\psi_m(2)r_{12}^{-1}\psi_p(2)\psi_m(1)\,\mathrm{d}\tau\right\}$$

$$=\sum_{m}^{N_e}\left\{(\psi_a\psi_p|\psi_m\psi_m)-(\psi_a\psi_m|\psi_m\psi_p)\right\},$$ which completes the demonstration of (v).

Equations (i)–(v) are used to show that $\int\Psi_0 H\Psi_a^p\,\mathrm{d}\tau=0$.

$$\int\Psi_0H\Psi_a^p\,\mathrm{d}\tau=\int\Psi_0\left\{\sum_i^{N_e}h_i+\frac{1}{2}\sum_{\substack{i,j\\i\neq j}}^{N_e}\frac{j_0}{r_{ij}}\right\}\Psi_a^p\,\mathrm{d}\tau=\sum_i^{N_e}\int\Psi_0h_i\Psi_a^p\,\mathrm{d}\tau+\frac{1}{2}\sum_{\substack{i,j\\i\neq j}}^{N_e}\int\Psi_0\left(\frac{j_0}{r_{ij}}\right)\Psi_a^p\,\mathrm{d}\tau$$

$$=N_e\int\Psi_0h_1\Psi_a^p\,\mathrm{d}\tau+\sum_m^{N_e}\left\{(\psi_a\psi_p|\psi_m\psi_m)-(\psi_a\psi_m|\psi_m\psi_p)\right\}\qquad\text{by (iii) and (v)}$$

$$=\int\psi_a(1)h_1\psi_p(1)\,\mathrm{d}\tau+\sum_m^{N_e}\left\{(\psi_a\psi_p|\psi_m\psi_m)-(\psi_a\psi_m|\psi_m\psi_p)\right\}\qquad\text{by (iv)}$$

$$=\int\psi_a(1)f_1\psi_p(1)\,\mathrm{d}\tau_1=0\qquad\text{by (ii) and (i)}$$

This proves the Brillouin theorem: $\int\Psi_0H\Psi_a^p\,\mathrm{d}\tau=0$.

P6.23 $$E_{XC}[\rho] = \int C\rho^{5/3}\mathrm{d}\boldsymbol{r}$$

When the density changes form $\rho[\boldsymbol{r}]$ to $\rho[\boldsymbol{r}] + \delta\rho[\boldsymbol{r}]$ at each point, the functional changes form $E_{XC}[\rho]$ to $E_{XC}[\rho+\delta\rho]$:

$$E_{XC}[\rho+\delta\rho] = \int C(\rho+\delta\rho)^{5/3}\,\mathrm{d}\boldsymbol{r}$$

The integrand can be expanded in a Taylor series:

$$(\rho+\delta\rho)^{5/3} = \rho^{5/3} + \tfrac{5}{3}\rho^{2/3}\delta\rho + \tfrac{1}{2}\left(\tfrac{5}{3}\right)\left(\tfrac{2}{3}\right)\rho^{-1/3}\delta\rho^2 + \cdots$$

Discarding terms of order $\delta\rho^2$ and higher we obtain

$$E_{XC}[\rho+\delta\rho] = \int C\left(\rho^{5/3} + \tfrac{5}{3}\rho^{2/3}\delta\rho\right)\boldsymbol{r} = E_{XC}[\rho] + \int \tfrac{5}{3}C\rho^{2/3}\delta\rho\boldsymbol{r}$$

Therefore, the differential δE_{XC} of the functional (the difference $E_{XC}[\rho+\delta\rho] - E_{XC}[\rho]$ that depends linearly on $\delta\rho$) is

$$\delta E_{XC}[\rho] = \int \tfrac{5}{3}C\rho^{2/3}\delta\rho\mathrm{d}\boldsymbol{r}$$

Comparison with $\delta E_{XC}[\rho] = \int V_{XC}(\boldsymbol{r})\delta\rho\mathrm{d}\boldsymbol{r}$ [6.25] yields

$$\boxed{V_{XC}(\boldsymbol{r}) = \tfrac{5}{3}C\rho(\boldsymbol{r})^{2/3}}$$

Solutions to applications

P6.25 (a) The hydrocarbons in questions form a homologous series. They are straight-chain alkanes of the formula C_nH_{2n+2}, or R–H where $R = C_nH_{2n+1}$. Draw up the following table.

n	1	2	3	4	5
π	0.5	1.0	1.5	2.0	2.5

The relationship here is evident by inspection: $\pi = n/2$, so we predict for the seven-carbon hydrocarbon in question

$$\pi = 7/2 = \boxed{3.5}.$$

(b) The plot, shown in Figure 6.4, is consistent with a linear relationship, for $R^2 = 0.997$ is close to unity. The best linear fit is

$$\log K = -1.95 - 1.49\pi$$

so $\boxed{\text{slope} = -1.49}$ and $\boxed{\text{intercept} = -1.95}$.

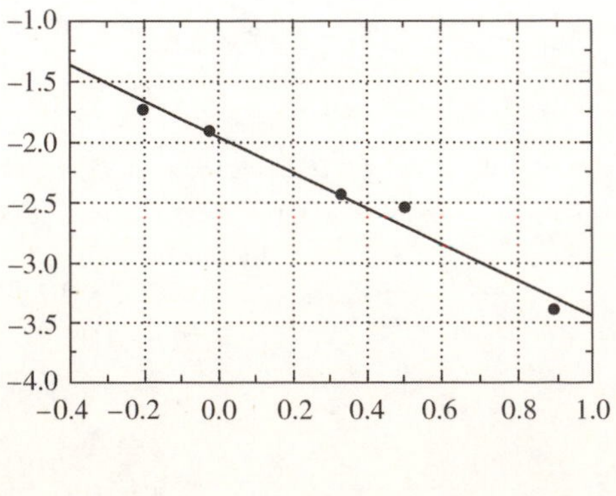

Figure 6.4

(c) If we know π for the substituent R = H, then we can use the linear SAR just derived. Our best estimate of π can be obtained by considering the zero-carbon "alkane" H_2, whose radical H ought to have a hydrophobicity constant $\pi = 0/2 = 0$. This value yields

$$\log K = -1.95 - 1.49(0) = -1.95 \qquad \text{so} \qquad K = 10^{-1.95} = \boxed{1.12 \times 10^{-2}}$$

Note: The assumption that R = H is part of the homologous series of straight-chain alkanes is a reasonable but questionable one.

7 Molecular symmetry

Answers to discussion questions

D7.1 The point group to which a molecule belongs is determined by the symmetry elements it possesses. Therefore the first step is to examine a model (which can be a mental picture) of the molecule for all its symmetry elements. All possible symmetry elements are described in Section 7.1. We list all that apply to the molecule of interest and then follow the assignment procedure summarized by the flow diagram in Fig. 7.7 of the text.

D7.3 Within the context of quantum theory and molecular symmetry a **group** is a collection of transformations (R, R', and so on) that satisfy these criteria:

(1) One of the transformations is the identity, E.
(2) For every transformation R, the inverse transformation R^{-1} is included in the collection so that the combination RR^{-1} is equivalent to the identity: $RR^{-1} = E$.
(3) The combination RR' is equivalent to a single member of the collection of transformations.
(4) The combination $R(R'R'') = (RR')R''$.

D7.5 A molecule may be chiral and therefore optically active only if it does not posses an axis of improper rotation, S_n. An improper rotation is a rotation followed by a reflection and this combination of operations always converts a right-handed object into a left-handed object and vice versa; hence an S_n axis guarantees that a molecule cannot exist in chiral forms. When discussing optical activity, it is helpful to remember that

(a) the presence of both a C_n and a σ_h is equivalent to an S_n.
(b) $i = S_2$.
(c) $\sigma = S_1$. Thus, a molecule cannot be optically active if it possesses a center of symmetry or a mirror plane.

D7.7 See Sections 7.4(a) and 7.4(b).

D7.9 The letters and subscripts of a symmetry species provide information about the symmetry behavior of the species. An A or a B is used to denote a one-dimensional representation; A is used if the character under the principal rotation is +1 (symmetric behavior), and B is used if the character is –1 (antisymmetric behavior). Subscripts are used to distinguish the irreducible representations if there is more than one of the same type: A_1 is reserved for the representation with the character 1 under all operations; E denotes a two-dimensional irreducible representation; T indicates a three-dimensional irreducible representation. These labels are called **Mulliken symbols**. For groups with an inversion center, a subscript g (gerade) indicates symmetric behavior under the inversion operation; a subscript u (ungerade) indicates antisymmetric behavior. A horizontal mirror is assigned $'$ or $''$ superscripts if the behavior is symmetric or antisymmetric, respectively, under the σ_h operation.

Solutions to exercises

E7.1(a) Chloromethane belongs to the point group C_{3v}. The elements other than the $\boxed{\text{identity } E}$ are a $\boxed{C_3 \text{ axis}}$ and $\boxed{\text{three vertical mirror planes } \sigma_v}$. The symmetry axis passes through the C–Cl nuclei.

The mirror planes are defined by the three C1CH planes. The C_3 principal axis and one of the σ_v mirror planes are shown in Figure 7.1.

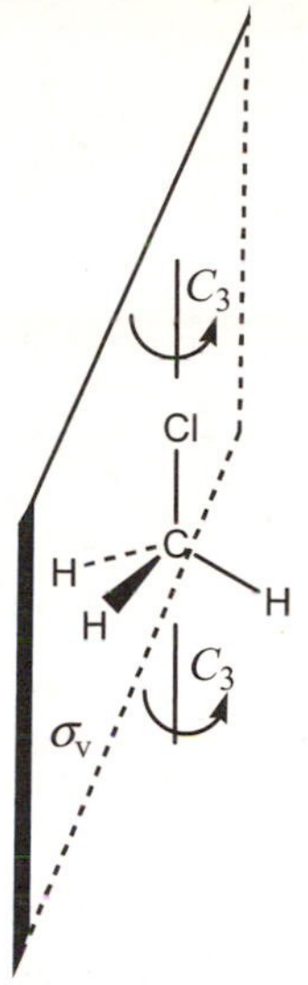

Figure 7.1

E7.2(a) Naphthalene belongs to the point group D_{2h} and it has the symmetry elements shown in Figure 7.2. There are $3C_2$ axes, a center of inversion, and $3\sigma_h$ mirror planes.

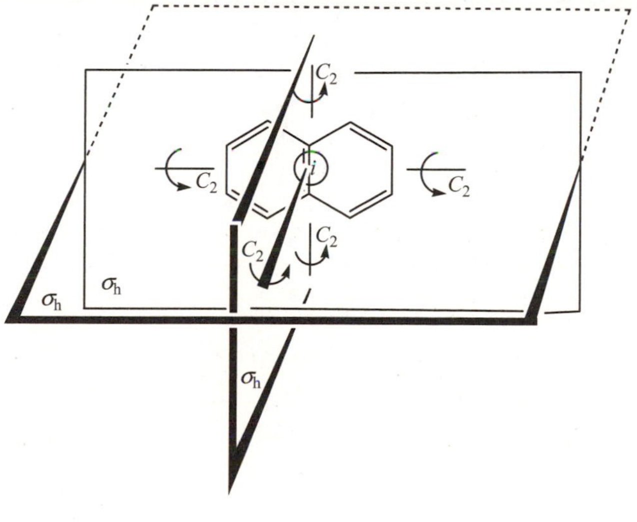

Figure 7.2

E7.3(a) List the symmetry elements of the objects (the principal ones, not necessarily all the implied ones); then use the remarks in Section 7.2, and Figure 7.3 below. Also refer to Figs. 7.7 and 7.8 of the text.

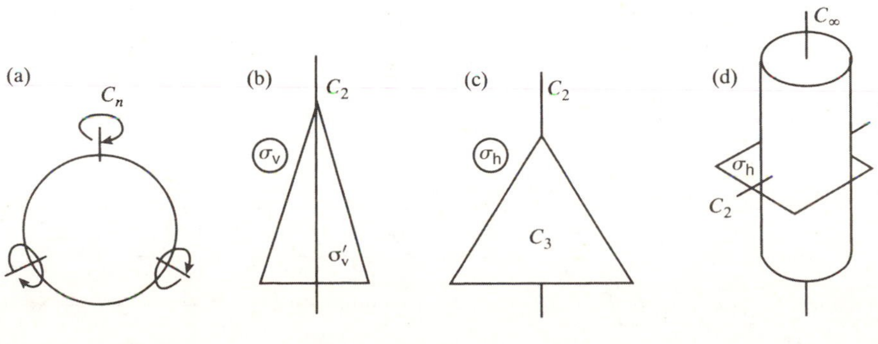

Figure 7.3

(a) Sphere: an infinite number of symmetry axes; therefore $\boxed{R_3}$

(b) Isosceles triangles; E, C_2, σ_v, and σ'_v; therefore $\boxed{C_{2v}}$

(c) Equilateral triangle: $\underbrace{\underbrace{E,\ C_3,\ C_2}_{D_3},\ \sigma_h}_{\boxed{D_{3h}}}$

(d) Cylinder: $E,\ C_\infty,\ C_2,\ \sigma_h$; therefore $\boxed{D_{\infty h}}$

E7.4(a) (a) NO_2: $E,\ C_2,\ \sigma_v,\ \sigma'_v$; $\boxed{C_{2v}}$

(b) N_2O: $E,\ C_\infty,\ C_2,\ \sigma_v$; $\boxed{C_{\infty v}}$

(c) $CHCl_3$: $E,\ C_3,\ 3\sigma_v$; $\boxed{C_{3v}}$

(d) $CH_2{=}CH_2$: $E, C_2, 2C'_2, \sigma_h$; $\boxed{D_{2h}}$

E7.5(a) (a) *cis*-CHCl=CHCl; $E,\ C_2,\ \sigma_v,\ \sigma'_v$; $\boxed{C_{2v}}$

(b) *trans*-CHCl=CHCl; $E,\ C_2,\ \sigma_h,\ i$; $\boxed{C_{2h}}$

E7.6(a) Only molecules belonging to the groups C_n, C_{nv}, and C_s may be polar [Section 7.3a]; hence, of the molecules listed, only $\boxed{\text{(a) pyridine}}$ and $\boxed{\text{(b) nitroethane}}$ are polar.

E7.7(a) The parent of the dichloronaphthalene isomers is shown to the right.

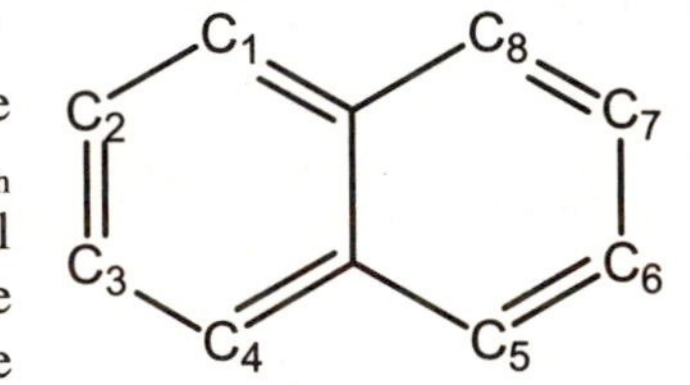

Care must be taken when determining possible isomers because naphthalene is a flat molecule that belongs to the point group D_{2h} (see E7.2a). It has an inversion center, mirror planes, and rotational axes that cause superficially distinct visual images to actually be the same molecule viewed from different angles. For example, the structures in Figure 7.4 are all 1,3-dichloronaphthalene. By drawing figures that avoid the redundancy caused by the symmetry elements, you will find a total of ten dichloronaphthalene isomers.

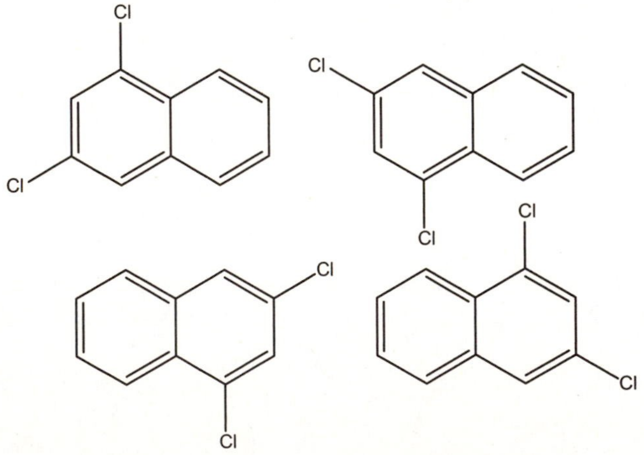

Figure 7.4

The names and point groups of the ten isomers are summarized in the following table.

Isomers and Point Groups of m,n-*Dichloronaphthalene*

m,n	1,2	1,3	1,4	1,5	1,6	1,7	1,8	2,3	2,6	2,7
Point group	C_s	C_s	C_{2v}	C_{2h}	C_s	C_s	C_{2v}	C_{2v}	C_{2h}	C_{2v}

E7.8(a) Since the p_z orbitals are perpendicular to the molecular plane, we recognize that the set of p_z orbitals on each atom of BF_3 experiences the σ_h change.

$$(p_B, p_{F1}, p_{F2}, p_{F3})\boldsymbol{D}(\sigma_h) = (-p_B, -p_{F1}, -p_{F2}, -p_{F3}).$$

Consequently, we find by inspection that

$$\boxed{\boldsymbol{D}(\sigma_h) = \begin{pmatrix} -1 & 0 & 0 & 0 \\ 0 & -1 & 0 & 0 \\ 0 & 0 & -1 & 0 \\ 0 & 0 & 0 & -1 \end{pmatrix}}$$

E7.9(a) The matrix representations of the operations σ_h and C_3 are deduced in E7.8(a) and E7.8(b). According to the precepts of group theory, the successive application of these operations yields another member of the D_{3h} group to which BF_3 belongs, and, in fact, by definition the operation $\sigma_h C_3$ should yield the S_3 symmetry operation. The matrix representation of S_3 can be found by matrix multiplication of the component operations.

$$\boldsymbol{D}(\sigma_h)\boldsymbol{D}(C_3) = \begin{pmatrix} -1 & 0 & 0 & 0 \\ 0 & -1 & 0 & 0 \\ 0 & 0 & -1 & 0 \\ 0 & 0 & 0 & -1 \end{pmatrix}\begin{pmatrix} 1 & 0 & 0 & 0 \\ 0 & 0 & 1 & 0 \\ 0 & 0 & 0 & 1 \\ 0 & 1 & 0 & 0 \end{pmatrix} = \boxed{\begin{pmatrix} -1 & 0 & 0 & 0 \\ 0 & 0 & -1 & 0 \\ 0 & 0 & 0 & -1 \\ 0 & -1 & 0 & 0 \end{pmatrix} = \boldsymbol{D}(S_3)}$$

The result may be checked by matrix operation on the p_z orbital vector where the effort should yield

$$(p_B, p_{F1}, p_{F2}, p_{F3})\boldsymbol{D}(S_3) = (-p_B, -p_{F3}, -p_{F1}, -p_{F2})$$

$$(p_B, p_{F1}, p_{F2}, p_{F3})\begin{pmatrix} -1 & 0 & 0 & 0 \\ 0 & 0 & -1 & 0 \\ 0 & 0 & 0 & -1 \\ 0 & -1 & 0 & 0 \end{pmatrix} = (-p_B, -p_{F3}, -p_{F1}, -p_{F2}),$$ which is the expected result.

E7.10(a) Consider the equilateral triangle $P_1P_2P_3$, which belongs to the D_{3h} point group (text Fig. 7.8). The three C_2 axes and the three σ_v mirror planes of this triangle are shown in Figure 7.5.

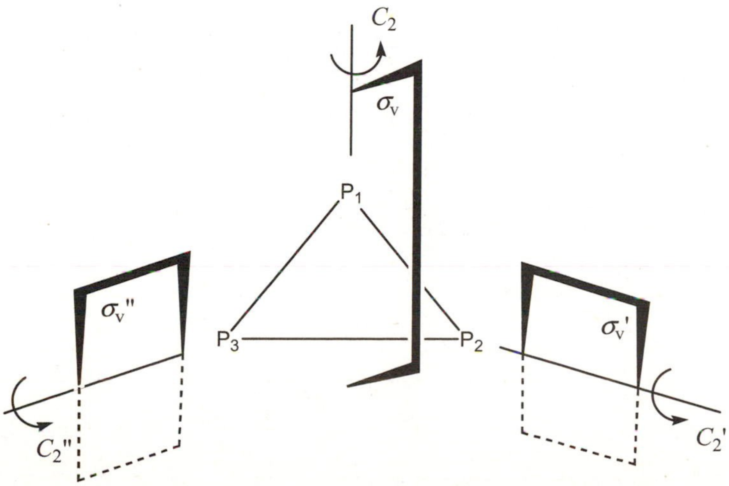

Figure 7.5

The C_2 and C_2' axes belong to the same class if there is a member S of the group such that $C_2' = S^{-1}C_2S$ [7.5] where S^{-1} is the inverse of S. We will work with $S = \sigma_v''$, an operator for which $S^{-1} = \sigma_v''$. By comparison of the action of $S^{-1}C_2S$ on the vector (P_1,P_2,P_3) with the action of C_2' on the same vector, we can determine whether or not the equality of eqn. 7.5 holds. If it does, C_2 and C_2' axes belong to the same class.

$$\begin{aligned} S^{-1}C_2S(P_1,P_2,P_3) &= \sigma_v''C_2\sigma_v''(P_1,P_2,P_3) \\ &= \sigma_v''C_2(P_2,P_1,P_3) \\ &= \sigma_v''(P_2,P_3,P_1) \\ &= (P_3,P_2,P_1) \qquad \text{(i)} \end{aligned}$$

$$C_2'(P_1,P_2,P_3) = (P_3,P_2,P_1) \qquad \text{(ii)}$$

Eqns. (i) and (ii) indicate that $C_2' = S^{-1}C_2S$ where $S = \sigma_v''$, and we conclude that C_2 and C_2' belong to the same class.

By either using the same argument or seeing the necessities of symmetry, we find that C_2 and C_2'' also belong to the same class. Consequently, C_2, C_2' and C_2'' all belong to the same class.

E7.11(a) The p_x orbital spans B_1 of the C_{2v} point group while z and p_z span A_1. Following the first *A Brief Illustration* of Section 7.5(a), we write a table of the characters of each function and multiply the rows.

C_{2v}	E	C_2	σ_v	σ_v'
p_x	1	−1	1	−1
z	1	1	1	1
p_z	1	1	1	1
p_xzp_z	1	−1	1	−1

The characters of the product p_xzp_z are those of B_1 alone, so the integrand does not span A_1. It follows that the integral must be zero.

E7.12(a) For a C_{3v} molecule, x and y span E while z spans A_1. Thus, the x and y components of the dipole moment [7.10] have transition integrands that span $A_2 \times E \times A_1$ for the $A_1 \rightarrow A_2$ transition. By inspection of the C_{3v} character table, we find the **decomposition of the direct product** to be $A_2 \times E \times A_1 = E$. The integrand spans E alone. Since it does not span A_1, the x and y components of the transition integral must be zero. The transition integrand for the z component spans $A_2 \times A_1 \times A_1 = A_2$ for the $A_1 \rightarrow A_2$ transition. Consequently, the z component of the transition integral must also equal zero, and we conclude that the transition is forbidden.

Should these considerations prove confusing, write a table with columns headed by the three components of the electric dipole moment operator, μ.

Component of μ	x	Y	z
A_1	1 1 1	1 1 1	1 1 1
$\Gamma(\mu)$	2 −1 0	2 −1 0	1 1 1
A_2	1 1 −1	1 1 −1	1 1 −1
$A_1\,\Gamma(\mu)\,A_2$	2 −1 0	2 −1 0	1 1 −1
	E	E	A_2

Since A_1 is not present in any product, the transition dipole moment must be zero.

E7.13(a) We first determine how x and y individually transform under the operations of the C_{4v} group. Using these results we determine how the product xy transforms. The transform of xy is the product of the transforms of x and y.

Under each operation the functions transform as follows.

	E	C_2	C_4	σ_v	σ_d
x	X	$-x$	y	x	$-y$
y	Y	$-y$	$-x$	$-y$	$-x$
xy	xy	xy	$-xy$	$-xy$	Xy
χ	1	1	−1	−1	1

From the C_{4v} character table, we see that this set of characters belongs to B_2.

E7.14(a) In each molecule we must look for an improper rotation axis, perhaps in a disguised form $(S_1 = \sigma,\ S_2 = \mathrm{i})$ (Section 7.3b). If present, the molecule cannot be chiral. D_{2h} contains $\boxed{i}$, and C_{3h} contains $\boxed{\sigma_h}$; therefore, molecules belonging to these point groups cannot be chiral and cannot be optically active.

E7.15(a) Because the largest character is 3 in the column headed E in the O_h character table, we know that the maximum orbital degeneracy is $\boxed{3}$. (See Section 7.4c).

E7.16(a) Benzene belongs to the D_{6h} point group. Because the largest character is 2 in the column headed E in the D_{6h} character table, we know that the maximum orbital degeneracy is $\boxed{2}$. (See Section 7.4c).

E7.17(a) Recall that $\mathrm{p}_x \propto x$, $\mathrm{p}_y \propto y$, $\mathrm{p}_z \propto z$, $\mathrm{d}_{xy} \propto xy$, $\mathrm{d}_{xz} \propto xz$, $\mathrm{d}_{yz} \propto yz$, $\mathrm{d}_{z^2} \propto z^2$, $\mathrm{d}_{x^2-y^2} \propto x^2 - y^2$ (Section 4.2g). Additionally, when the functions f_1 and f_2 of an overlap integral are bases for irreducible representations of a group, the integral must vanish if they are different symmetry species; if they are the same symmetry species, then the integral may be nonzero (Section 7.5a). Since the combination $\mathrm{p}_x(\mathrm{A}) - \mathrm{p}_x(\mathrm{B})$ of the two O atoms (with x perpendicular to the plane) spans A_2, the orbital on N must span A_2 for a non-zero overlap. Now refer to the C_{2v} character table. The s orbital spans A_1 and the p orbitals of the central N atom span $A_1(\mathrm{p}_z)$, $B_1(\mathrm{p}_x)$, and $B_2(\mathrm{p}_y)$. Therefore, $\boxed{\text{no orbitals}}$ span A_2, and hence $\mathrm{p}_x(\mathrm{A}) - \mathrm{p}_x(\mathrm{B})$ is a non-bonding combination. If d orbitals are available, as they are in S of the SO_2 molecule, we could form a molecular orbital with $\boxed{\mathrm{d}_{xy}}$, which is a basis for A_2.

E7.18(a) The electric dipole moment operator transforms as $x(B_1)$, $y(B_2)$, and $z(A_1)$ (C_{2v} character table).

Transitions are allowed if $\int \psi_f^* \mu \psi_i \, d\tau$ is non-zero (Example 7.6) and hence are forbidden unless $\Gamma_f \times \Gamma(\mu) \times \Gamma_i$ contains A_1. Since $\Gamma_i = A_1$, $\Gamma_f \times \Gamma(\mu) = A_1$. Since $B_1 \times B_1 = A_1$, $B_2 \times B_2 = A_1$, and $A_1 \times A_1 = A_1$, x-polarized light may cause a transition to a B_1 term, y-polarized light to a B_2 term, and z-polarized light to an A_1 term.

E7.19(a)

C_{4v}, $h = 8$	E	C_2	$2C_4$	$2\sigma_v$	$2\sigma_d$
A_1	1	1	1	1	1
A_2	1	1	1	−1	−1
B_1	1	1	−1	1	−1
B_2	1	1	−1	−1	1
E	2	−2	0	0	0

$$n(\Gamma)=\frac{1}{h}\sum_{R}\chi^{(\Gamma)}(R)\chi(R) \quad [7.7] \quad \text{where } \chi(R)=(5,1,1,3,1)$$

$$n(A_1)=\tfrac{1}{8}\{1(1\times5)+1(1\times1)+2(1\times1)+2(1\times3)+2(1\times1)\}=2$$

$$n(A_2)=\tfrac{1}{8}\{1(1\times5)+1(1\times1)+2(1\times1)+2(-1\times3)+2(-1\times1)\}=0$$

$$n(B_1)=\tfrac{1}{8}\{1(1\times5)+1(1\times1)+2(-1\times1)+2(1\times3)+2(-1\times1)\}=1$$

$$n(B_2)=\tfrac{1}{8}\{1(1\times5)+1(1\times1)+2(-1\times1)+2(-1\times3)+2(1\times1)\}=0$$

$$n(E)=\tfrac{1}{8}\{1(2\times5)+1(-2\times1)+2(0\times1)+2(0\times3)+2(0\times1)\}=1$$

Thus, this set of basis functions spans $\boxed{2A_1+B_1+E}$.

E7.20(a) (a) The point group of benzene is D_{6h}. In D_{6h}, μ spans $E_{1u}(x, y)$ and $A_{2u}(z)$, and the ground term is A_{1g}. Then, using $A_{2u}\times A_{1g}=A_{2u}$, $E_{1u}\times A_{1g}=E_{1u}$, $A_{2u}\times A_{2u}=A_{1g}$, and $E_{1u}\times E_{1u}=A_{1g}+A_{2g}+E_{2g}$, we conclude that the upper term is $\boxed{\text{either } E_{1u} \text{ or } A_{2u}}$.

(b) Naphthalene belongs to D_{2h}. In D_{2h} itself, the components span $B_{3u}(x)$, $B_{2u}(y)$, and $B_{1u}(z)$, and the ground term is A_g. Hence, since $A_g\times\Gamma=\Gamma$ in this group, the upper terms are $\boxed{B_{3u}\,(x\text{-polarized})}$, $\boxed{B_{2u}\,(y\text{-polarized})}$, and $\boxed{B_{1u}\,(z\text{-polarized})}$.

E7.21(a) We consider the integral

$$I=\int_{-a}^{a} f_1 f_2\,\mathrm{d}\theta=\int_{-a}^{a}\sin\theta\cos\theta\,\mathrm{d}\theta$$

and hence draw up the following table for the effect of operations in the group C_s (see Figure 7.6).

	E	σ_h
$f_1=\sin\theta$	$\sin\theta$	$-\sin\theta$
$f_2=\cos\theta$	$\cos\theta$	$\cos\theta$

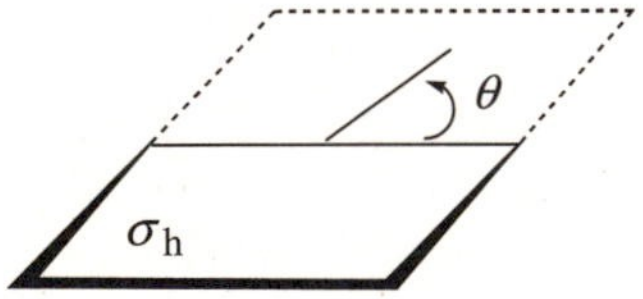

Figure 7.6

In terms of characters,

	E	σ_h	Symmetry Species
f_1	1	–1	A″
f_2	1	1	A′
f_1f_2	1	–1	A″

Since the product does not span the totally symmetric species A′, the integral is necessarily zero.

Solutions to problems

Solutions to numerical problems

P7.1 (a) Staggered CH_3CH_3: E, C_3, C_2, i, $3\sigma_d$, S_6; $\boxed{D_{3d}}$ [see Fig. 7.6b of the text]

(b) Chair C_6H_{12}: E, C_3, C_2, i, $3\sigma_d$, S_6; $\boxed{D_{3d}}$

Boat C_6H_{12}: E, C_2, σ_v, σ_v'; $\boxed{C_{2v}}$

(c) B_2H_6: E, C_2, $2C_2'$, i, σ_h, $2\sigma_v'$; $\boxed{D_{2h}}$

(d) $[Co(en)_3]^{3+}$: E, $2C_3$, $3C_2$; $\boxed{D_3}$

(e) Crown S_8: E, C_4, C_2, $4C_2'$, $4\sigma_d$, $2S_8$; $\boxed{D_{4d}}$

Only boat C_6H_{12} may be polar since all the others are D point groups. Only $[Co(en)_3]^{3+}$ belongs to a group without an improper rotation axis ($S_1 = \sigma$) and hence is chiral.

P7.3 The operations are illustrated in Figure 7.7. Note that $R^2 = E$ for all the operations of the groups, that $ER = RE = R$ always, and that $RR' = R'R$ for this group. Since $C_2\sigma_h = i, \sigma_h i = C_2$, and $iC_2 = \sigma_h$, we can draw up the following group multiplication table.

	E	C_2	σ_h	i
E	E	C_2	σ_h	i
C_2	C_2	E	i	σ_h
σ_h	σ_h	i	E	C_2
I	i	σ_h	C_2	E

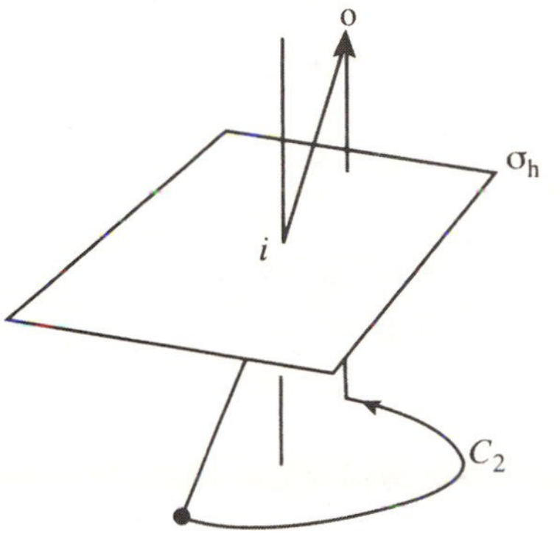

Figure 7.7

The $\boxed{\textit{trans}\text{-CHCl=CHCl}}$ molecule belongs to the group C_{2h}.

COMMENT. Note that the multiplication table for C_{2h} can be put into a one-to-one correspondence with the multiplication table of D_2 obtained in Exercise 7.13(b). We say that they both belong to the same abstract group and are isomorphous.

Question. Can you find another abstract group of order 4 and obtain its multiplication table? There is only one other.

P7.5 Refer to Fig.7.3 of the text. Place orbitals h_1 and h_2 on the H atoms and s, p_x, p_y, and p_z on the O atom. The z-axis is the C_2 axis; x lies perpendicular to σ_v' and y lies perpendicular to σ_v. Then draw up the following table of the effect of the operations on the basis.

	E	C_2	σ_v	σ_v'
h_1	h_1	h_2	h_2	h_1
h_2	h_2	h_1	h_1	h_2
s	s	s	s	s
p_x	p_x	$-p_x$	p_x	$-p_x$
p_y	p_y	$-p_y$	$-p_y$	p_y
p_z	p_z	p_z	p_z	p_z

Express the columns headed by each operation R in the form

$$(\text{new}) = D(R)(\text{original})$$

where $D(R)$ is the 6×6 representative of the operation R. We use the methods set out in Section 7.4(a).

(i) E: (h_1, h_2, s, p_x, p_y, p_z) ← (h_1, h_2, s, p_x, p_y, p_z) is reproduced by the 6 × 6 unit matrix.

(ii) C_2: (h_2, h_1, s, $-p_x$, $-p_y$, p_z) ← (h_1, h_2, s, p_x, p_y, p_z) is reproduced by

$$\boldsymbol{D}(C_2)=\begin{bmatrix}0&1&0&0&0&0\\1&0&0&0&0&0\\0&0&1&0&0&0\\0&0&0&-1&0&0\\0&0&0&0&-1&0\\0&0&0&0&0&1\end{bmatrix}$$

(iii) σ_v: (h_2, h_1, s, p_x, $-p_y$, p_z) ← (h_1, h_2, s, p_x, p_y, p_z) is reproduced by

$$\boldsymbol{D}(\sigma_v)=\begin{bmatrix}0&1&0&0&0&0\\1&0&0&0&0&0\\0&0&1&0&0&0\\0&0&0&1&0&0\\0&0&0&0&-1&0\\0&0&0&0&0&1\end{bmatrix}$$

(iv) σ_v' : (h_1, h_2, s, $-p_x$, p_y, p_z) ← (h_1, h_2, s, p_x, p_y, p_z) is reproduced by

$$\boldsymbol{D}(\sigma_v')=\begin{bmatrix}1&0&0&0&0&0\\0&1&0&0&0&0\\0&0&1&0&0&0\\0&0&0&-1&0&0\\0&0&0&0&1&0\\0&0&0&0&0&1\end{bmatrix}$$

(a) To confirm the correct representation of $C_2\sigma_v=\sigma_v'$, we write

$$\boldsymbol{D}(C_2)\boldsymbol{D}(\sigma_v)=\begin{bmatrix}0&1&0&0&0&0\\1&0&0&0&0&0\\0&0&1&0&0&0\\0&0&0&-1&0&0\\0&0&0&0&-1&0\\0&0&0&0&0&1\end{bmatrix}\begin{bmatrix}0&1&0&0&0&0\\1&0&0&0&0&0\\0&0&1&0&0&0\\0&0&0&1&0&0\\0&0&0&0&-1&0\\0&0&0&0&0&1\end{bmatrix}$$

$$=\begin{bmatrix}1&0&0&0&0&0\\0&1&0&0&0&0\\0&0&1&0&0&0\\0&0&0&-1&0&0\\0&0&0&0&1&0\\0&0&0&0&0&1\end{bmatrix}=\boldsymbol{D}(\sigma_v')$$

(b) Similarly, to confirm the correct representation of $\sigma_v \sigma_v' = C_2$, we write

$$D(\sigma_v)D(\sigma_v') = \begin{bmatrix} 0 & 1 & 0 & 0 & 0 & 0 \\ 1 & 0 & 0 & 0 & 0 & 0 \\ 0 & 0 & 1 & 0 & 0 & 0 \\ 0 & 0 & 0 & 1 & 0 & 0 \\ 0 & 0 & 0 & 0 & -1 & 0 \\ 0 & 0 & 0 & 0 & 0 & 1 \end{bmatrix} \begin{bmatrix} 1 & 0 & 0 & 0 & 0 & 0 \\ 0 & 1 & 0 & 0 & 0 & 0 \\ 0 & 0 & 1 & 0 & 0 & 0 \\ 0 & 0 & 0 & -1 & 0 & 0 \\ 0 & 0 & 0 & 0 & 1 & 0 \\ 0 & 0 & 0 & 0 & 0 & 1 \end{bmatrix}$$

$$= \begin{bmatrix} 0 & 1 & 0 & 0 & 0 & 0 \\ 1 & 0 & 0 & 0 & 0 & 0 \\ 0 & 0 & 1 & 0 & 0 & 0 \\ 0 & 0 & 0 & -1 & 0 & 0 \\ 0 & 0 & 0 & 0 & -1 & 0 \\ 0 & 0 & 0 & 0 & 0 & 1 \end{bmatrix} = D(C_2)$$

(a) The characters of the representatives are the sums of their diagonal elements:

E	C_2	σ_v	σ_v'
6	0	2	4

(b) The characters are not those of any one irreducible representation, so the representation is reducible.

(c) The sum of the characters of the specified sum is

	E	C_2	σ_v	σ_v'
$3A_1$	3	3	3	3
B_1	1	−1	1	−1
$2B_2$	2	−2	−2	2
$3A_1 + B_1 + 2B_2$	6	0	2	4

which is the same as the original. Therefore the representation is $3A_1 + B_1 + 2B_2$.

P7.7 The basis set consists of four 1s hydrogen orbitals, which is written as $\boldsymbol{f}$ = (A, B, C, D) when positioned as shown in Figure 7.8. When reflected in the mirror plane σ_{dAB} of the figure, the basis vector becomes (B,A,C,D). The order of the T_d group is 24, so there are 24 matrices to find. In addition to E there are $8C_3$ operations; clockwise and counterclockwise C_3 axes along the C—H_A bond, shown in the figure, are labeled C_{3A}^+ and C_{3A}^-. Likewise, the clockwise and counterclockwise C_3 axes along the C—H_B bond, shown in the figure, are labeled C_{3B}^+ and C_{3B}^-. There are $3C_2$ operations; the one that bisects the H_A—C—H_B angle is labeled C_{2AB}. There are $6S_4$. The clockwise and counterclockwise S_4 axes that bisect the H_A—C—H_B angle are labeled S_{4AB}^+ and S_{4AB}^-. Finally, there are $6\sigma_d$ mirror planes; the one, shown in the figure, that bisects the H_A—C—H_B angle is labeled σ_{dAB}.

$$E\boldsymbol{f} = (\mathrm{A,B,C,D}) = \boldsymbol{f}\begin{pmatrix} 1 & 0 & 0 & 0 \\ 0 & 1 & 0 & 0 \\ 0 & 0 & 1 & 0 \\ 0 & 0 & 0 & 1 \end{pmatrix} = \boldsymbol{f}\boldsymbol{D}(E);\ \chi = 4$$

$$C_{3A}^{+}\boldsymbol{f}=(\text{A,C,D,B})=\boldsymbol{f}\begin{pmatrix}1&0&0&0\\0&0&0&1\\0&1&0&0\\0&0&1&0\end{pmatrix}=\boldsymbol{f}\boldsymbol{D}(C_{3A}^{+});\ \chi=1$$

$$C_{3A}^{-}\boldsymbol{f}=(\text{A,D,B,C})=\boldsymbol{f}\begin{pmatrix}1&0&0&0\\0&0&1&0\\0&0&0&1\\0&1&0&0\end{pmatrix}=\boldsymbol{f}\boldsymbol{D}(C_{3A}^{-});\ \chi=1$$

$$C_{3B}^{+}\boldsymbol{f}=(\text{D,B,A,C})=\boldsymbol{f}\begin{pmatrix}0&0&1&0\\0&1&0&0\\0&0&0&1\\1&0&0&0\end{pmatrix}=\boldsymbol{f}\boldsymbol{D}(C_{3B}^{+});\ \chi=1$$

$$C_{3B}^{-}\boldsymbol{f}=(\text{C,B,D,A})=\boldsymbol{f}\begin{pmatrix}0&0&0&1\\0&1&0&0\\1&0&0&0\\0&0&1&0\end{pmatrix}=\boldsymbol{f}\boldsymbol{D}(C_{3B}^{-});\ \chi=1$$

$$C_{3C}^{+}\boldsymbol{f}=(\text{B,D,C,A})=\boldsymbol{f}\begin{pmatrix}0&0&0&1\\1&0&0&0\\0&0&1&0\\0&1&0&0\end{pmatrix}=\boldsymbol{f}\boldsymbol{D}(C_{3C}^{+});\ \chi=1$$

$$C_{3C}^{-}\boldsymbol{f}=(\text{D,A,C,B})=\boldsymbol{f}\begin{pmatrix}0&1&0&0\\0&0&0&1\\0&0&1&0\\1&0&0&0\end{pmatrix}=\boldsymbol{f}\boldsymbol{D}(C_{3C}^{-});\ \chi=1$$

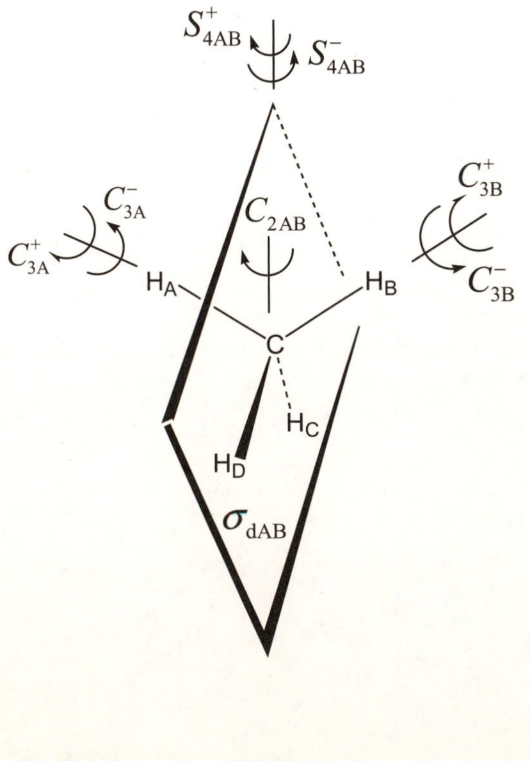

Figure 7.8

$$C_{3D}^{+}\boldsymbol{f} = (\mathrm{C,A,B,D}) = \boldsymbol{f}\begin{pmatrix} 0 & 1 & 0 & 0 \\ 0 & 0 & 1 & 0 \\ 1 & 0 & 0 & 0 \\ 0 & 0 & 0 & 1 \end{pmatrix} = \boldsymbol{fD}\left(C_{3D}^{+}\right);\ \chi = 1$$

$$C_{3D}^{-}\boldsymbol{f} = (\mathrm{B,C,A,D}) = \boldsymbol{f}\begin{pmatrix} 0 & 0 & 1 & 0 \\ 1 & 0 & 0 & 0 \\ 0 & 1 & 0 & 0 \\ 0 & 0 & 0 & 1 \end{pmatrix} = \boldsymbol{fD}\left(C_{3D}^{-}\right);\ \chi = 1$$

$$C_{2AB}\boldsymbol{f} = (\mathrm{B,A,D,C}) = \boldsymbol{f}\begin{pmatrix} 0 & 1 & 0 & 0 \\ 1 & 0 & 0 & 0 \\ 0 & 0 & 0 & 1 \\ 0 & 0 & 1 & 0 \end{pmatrix} = \boldsymbol{fD}\left(C_{2AB}\right);\ \chi = 0$$

$$C_{2BC}\boldsymbol{f} = (\mathrm{D,C,B,A}) = \boldsymbol{f}\begin{pmatrix} 0 & 0 & 0 & 1 \\ 0 & 0 & 1 & 0 \\ 0 & 1 & 0 & 0 \\ 1 & 0 & 0 & 0 \end{pmatrix} = \boldsymbol{fD}\left(C_{2BC}\right);\ \chi = 0$$

$$C_{2AC}\boldsymbol{f} = (\mathrm{C,D,A,B}) = \boldsymbol{f}\begin{pmatrix} 0 & 0 & 1 & 0 \\ 0 & 0 & 0 & 1 \\ 1 & 0 & 0 & 0 \\ 0 & 1 & 0 & 0 \end{pmatrix} = \boldsymbol{fD}\left(C_{2AC}\right);\ \chi = 0$$

$$S_{4AB}^{+}\boldsymbol{f} = (\mathrm{D,C,A,B}) = \boldsymbol{f}\begin{pmatrix} 0 & 0 & 1 & 0 \\ 0 & 0 & 0 & 1 \\ 0 & 1 & 0 & 0 \\ 1 & 0 & 0 & 0 \end{pmatrix} = \boldsymbol{fD}\left(S_{4AB}^{+}\right);\ \chi = 0$$

$$S_{4AB}^{-}\boldsymbol{f} = (\mathrm{C,D,B,A}) = \boldsymbol{f}\begin{pmatrix} 0 & 0 & 0 & 1 \\ 0 & 0 & 1 & 0 \\ 1 & 0 & 0 & 0 \\ 0 & 1 & 0 & 0 \end{pmatrix} = \boldsymbol{fD}\left(S_{4AB}^{-}\right);\ \chi = 0$$

$$S_{4BC}^{+}\boldsymbol{f} = (\mathrm{B,D,A,C}) = \boldsymbol{f}\begin{pmatrix} 0 & 0 & 1 & 0 \\ 1 & 0 & 0 & 0 \\ 0 & 0 & 0 & 1 \\ 0 & 1 & 0 & 0 \end{pmatrix} = \boldsymbol{fD}\left(S_{4BC}^{+}\right);\ \chi = 0$$

$$S_{4BC}^{-}\boldsymbol{f} = (\mathrm{C,A,D,B}) = \boldsymbol{f}\begin{pmatrix} 0 & 1 & 0 & 0 \\ 0 & 0 & 0 & 1 \\ 1 & 0 & 0 & 0 \\ 0 & 0 & 1 & 0 \end{pmatrix} = \boldsymbol{fD}\left(S_{4BC}^{-}\right);\ \chi = 0$$

$$S_{4AC}^{+}\boldsymbol{f} = (\mathrm{B,C,D,A}) = \boldsymbol{f}\begin{pmatrix} 0 & 0 & 0 & 1 \\ 1 & 0 & 0 & 0 \\ 0 & 1 & 0 & 0 \\ 0 & 0 & 1 & 0 \end{pmatrix} = \boldsymbol{fD}\left(S_{4AC}^{+}\right);\ \chi = 0$$

$$S_{4AC}^{-}\boldsymbol{f} = (\text{D,A,B,C}) = \boldsymbol{f}\begin{pmatrix} 0 & 1 & 0 & 0 \\ 0 & 0 & 1 & 0 \\ 0 & 0 & 0 & 1 \\ 1 & 0 & 0 & 0 \end{pmatrix} = \boldsymbol{fD}(S_{4AC}^{-});\ \chi = 0$$

$$\sigma_{dAB}\boldsymbol{f} = (\text{B,A,C,D}) = \boldsymbol{f}\begin{pmatrix} 0 & 1 & 0 & 0 \\ 1 & 0 & 0 & 0 \\ 0 & 0 & 1 & 0 \\ 0 & 0 & 0 & 1 \end{pmatrix} = \boldsymbol{fD}(\sigma_{dAB});\ \chi = 2$$

$$\sigma_{dAC}\boldsymbol{f} = (\text{C,B,A,D}) = \boldsymbol{f}\begin{pmatrix} 0 & 0 & 1 & 0 \\ 0 & 1 & 0 & 0 \\ 1 & 0 & 0 & 0 \\ 0 & 0 & 0 & 1 \end{pmatrix} = \boldsymbol{fD}(\sigma_{dAC});\ \chi = 2$$

$$\sigma_{dAD}\boldsymbol{f} = (\text{D,B,C,A}) = \boldsymbol{f}\begin{pmatrix} 0 & 0 & 0 & 1 \\ 0 & 1 & 0 & 0 \\ 0 & 0 & 1 & 0 \\ 1 & 0 & 0 & 0 \end{pmatrix} = \boldsymbol{fD}(\sigma_{dAD});\ \chi = 2$$

$$\sigma_{dBC}\boldsymbol{f} = (\text{A,C,B,D}) = \boldsymbol{f}\begin{pmatrix} 1 & 0 & 0 & 0 \\ 0 & 0 & 1 & 0 \\ 0 & 1 & 0 & 0 \\ 0 & 0 & 0 & 1 \end{pmatrix} = \boldsymbol{fD}(\sigma_{dBC});\ \chi = 2$$

$$\sigma_{dBD}\boldsymbol{f} = (\text{A,D,C,B}) = \boldsymbol{f}\begin{pmatrix} 1 & 0 & 0 & 0 \\ 0 & 0 & 0 & 1 \\ 0 & 0 & 1 & 0 \\ 0 & 1 & 0 & 0 \end{pmatrix} = \boldsymbol{fD}(\sigma_{dBD});\ \chi = 2$$

$$\sigma_{dCD}\boldsymbol{f} = (\text{A,B,D,C}) = \boldsymbol{f}\begin{pmatrix} 1 & 0 & 0 & 0 \\ 0 & 1 & 0 & 0 \\ 0 & 0 & 0 & 1 \\ 0 & 0 & 1 & 0 \end{pmatrix} = \boldsymbol{fD}(\sigma_{dCD});\ \chi = 2$$

Exercise: Find the representations for the 1*s* orbital basis at the corners of a regular trigonal bipyramid.

P7.9 *Representation 1*

$$\boldsymbol{D}(C_3)\boldsymbol{D}(C_2) = 1\times 1 = 1 = \boldsymbol{D}(C_6)$$

and from the C_{6v} character table, this is either A_1 or A_2. Hence, either $\boldsymbol{D}(\sigma_v) = \boldsymbol{D}(\sigma_d) = \boxed{+1 \text{ or } -1}$, respectively.

Representation 2

$$\boldsymbol{D}(C_3)\boldsymbol{D}(C_2) = 1\times(-1) = -1 = \boldsymbol{D}(C_6)$$

and from the C_{6v} character table this is either B_1 or B_2. Hence, either $\boldsymbol{D}(\sigma_v) = -\boldsymbol{D}(\sigma_d) = \boxed{1}$ or $\boldsymbol{D}(\sigma_v) = -\boldsymbol{D}(\sigma_d) = \boxed{-1}$, respectively.

P7.11 A quick rule for determining the character without first having to set up the matrix representation is to count 1 each time a basis function is left unchanged by the operation because only these functions give a non-zero entry on the diagonal of the matrix representative. In some cases there is a sign change, $(\ldots -f \ldots) \leftarrow (\ldots f \ldots)$; then –1 occurs on the diagonal, so count –1. The character of the identity is always equal to the dimension of the basis since each function contributes 1 to the trace.

E: All four orbitals are left unchanged; hence $\chi = 4$

C_3: One orbital is left unchanged; hence $\chi = 1$

C_2: No orbitals are left unchanged; hence $\chi = 0$

σ_d: Two orbitals are left unchanged; hence $\chi = 2$

S_4: No orbitals are left unchanged; hence $\chi = 0$

The character set 4, 1, 0, 2, 0 spans $\boxed{A_1 + T_2}$. Inspection of the character table of the group T_d shows that an s orbital spans A_1 and that the three p orbitals on the C atom span T_2. Hence, the $\boxed{\text{s and p}}$ orbitals of the C atom may form molecular orbitals with the four H1s orbitals. In T_d, the d orbitals of the central atom span E + T_2 (character table, final column), so only the T_2 set $\boxed{(d_{xy}, d_{yz}, d_{zx})}$ may contribute to molecular orbital formation with the H orbitals.

P7.13 The most distinctive symmetry operation is the $\boxed{S_4}$ axis through the central atom and aromatic nitrogens on both ligands. That axis is also a $\boxed{C_2}$ axis. The group is $\boxed{S_4}$.

P7.15 (a) C_{2v}. The functions x^2, y^2, and z^2 are invariant under all operations of the group, so $z(5z^2 - 3r^2)$ transforms as $z(A_1)$, $y(5y^2 - 3r^2)$ as $y(B_2)$, $x(5x^2 - 3r^2)$ as $x(B_1)$, and likewise for $z(x^2 - y^2)$, $y(x^2 - z^2)$, and $x(z^2 - y^2)$. The function xyz transforms as $B_1 \times B_2 \times A_1 = A_2$. Therefore, in group C_{2v}, $f \rightarrow \boxed{2A_1 + A_2 + 2B_1 + 2B_2}$.

(b) C_{3v}. In C_{3v}, z transforms as A_1, and hence so does z^3. From the C_{3v} character table, $(x^2 - y^2, xy)$ is a basis for E, so $(xyz,\ z(x^2 - y^2))$ is a basis for $A_1 \times E = E$. The linear combinations $y(5y^2 - 3r^2) + 5y(x^2 - z^2) \propto y$ and $x(5x^2 - 3r^2) + 5x(z^2 - y^2) \propto x$ are a basis for E. Likewise, the two linear combinations orthogonal to these are another basis for E. Hence, in the group C_{3v}, $f \rightarrow \boxed{A_1 + 3E}$.

(c) T_d. Make the inspired guess that the f orbitals are a basis of dimension 3 + 3 + 1, suggesting the decomposition T + T + A. Is the A representation A_1 or A_2? We see from the character table that the effect of S_4 discriminates between A_1 and A_2. Under S_4, $x \rightarrow y$, $y \rightarrow -x$, $z \rightarrow -z$, so $xyz \rightarrow xyz$. The character is $\chi = 1$, so xyz spans A_1. Likewise, $(x^3, y^3, z^3) \rightarrow (y^3, -x^3, -z^3)$ and $\chi = 0 + 0 - 1 = -1$. Hence, this trio spans T_2. Finally,

$$\{x(z^2 - y^2),\ y(z^2 - x^2),\ z(x^2 - y^2)\} \rightarrow \{y(z^2 - x^2),\ -x(z^2 - y^2),\ -z(y^2 - x^2)\}$$

resulting in $\chi = 1$, indicating T_1. Therefore, in T_d, $f \rightarrow \boxed{A_1 + T_1 + T_2}$.

(d) O_h. Anticipate an A + T + T decomposition as in the other cubic group. Since x, y, and z all have odd parity, all the irreducible representatives will be u. Under S_4, $xyz \rightarrow xyz$ (as in (c)), and so the representation is $\chi = 1$ (see the character table). Under S_4, $(x^3, y^3, z^3) \rightarrow (y^3, -x^3, -z^3)$, as before, and $\chi = -1$, indicating T_{1u}. In the same way, the remaining three functions span T_{2u}. Hence, in O_h, $f \rightarrow \boxed{A_{2u} + T_{1u} + T_{2u}}$.

(The shapes of the orbitals are shown in D. F. Shriver and P. W. Atkins, *Inorganic Chemistry*, 3rd Ed., Oxford University Press and W. H. Freeman and Company, 1999.)

The f orbitals will cluster into sets according to their irreducible representations.

Thus, (a) $f \rightarrow A_1 + T_1 + T_2$ in T_d symmetry, and there is one non-degenerate orbital and two sets of triply degenerate orbitals. (b) $f \rightarrow A_{2u} + T_{1u} + T_{2u}$, and the pattern of splitting (but not the order of energies) is the same.

P7.17 We begin by drawing up the following table for the C_{2v} group.

	N2s	N2p_x	N2p_y	N2p_z	O2p_x	O2p_y	O2p_z	O′2p_x	O′2p_y	O′2p_z	χ
E	N2s	N2p_x	N2p_y	N2p_z	O2p_x	O2p_y	O2p_z	O′2p_x	O′2p_y	O′2p_z	10
C_2	N2s	–N2p_x	–N2p_y	N2p_z	–O′2p_x	–O′2p_y	O′2p_z	–O2p_x	–O2p_y	O2p_z	0
σ_v	N2s	N2p_x	–N2p_y	N2p_z	O′2p_x	–O′2p_y	O′2p_z	O2p_x	–O2p_y	O2p_z	2
σ_v'	N2s	–N2p_x	N2p_y	N2p_z	–O2p_x	O2p_y	O2p_z	–O′2p_x	O′2p_y	O′2p_z	4

The character set (10, 0, 2, 4) decomposes into $\boxed{4A_1+2B_1+3B_2+A_2}$. We then form symmetry-adapted linear combinations as described in Section 7.5(e).

$\psi(A_1) = N2s$	(column 1)	$\psi(B_1) = O2p_x + O'2p_x$	(column 5)
$\psi(A_1) = N2p_z$	(column 4)	$\psi(B_2) = N2p_y$	(column 3)
$\psi(A_1) = O2p_z + O'2p_z$	(column 7)	$\psi(B_2) = O2p_y + O'2p_y$	(column 6)
$\psi(A_1) = -O2p_y + O'2p_y$	(column 9)	$\psi(B_2) = O2p_z - O'2p_z$	(column 7)
$\psi(B_1) = N2p_x$	(column 2)	$\psi(A_2) = O2p_x - O'2p_x$	(column 5)

(The other columns yield the same combinations.)

P7.19 Consider phenanthrene with carbon atoms as labeled in the structure below.

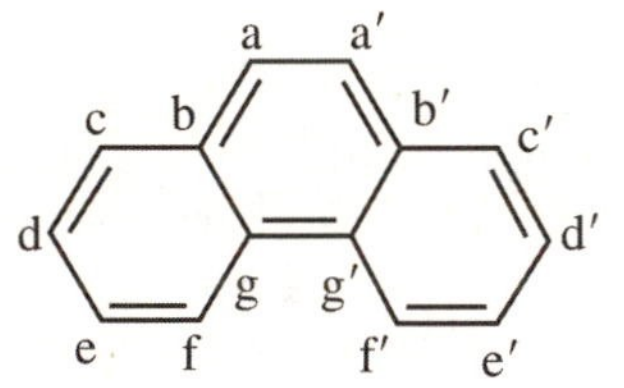

(a) The 2p orbitals involved in the π system are the basis we are interested in. To find the irreducible representations spanned by this basis, consider how each basis is transformed under the symmetry operations of the C_{2v} group. To find the character of an operation in this basis, sum the coefficients of the basis terms that are unchanged by the operation.

	a	a′	b	b′	c	C′	d	d′	e	e′	f	f′	g	g′	χ
E	a	a′	b	b′	c	C′	d	d′	e	e′	f	f′	g	g′	14
C_2	–a′	–a	–b′	–b	–c′	–c	–d′	–d	–e′	–e	–f′	–f	–g′	–g	0
σ_v	a′	a	B′	b	c′	c	d′	d	e′	e	f′	f	g′	g	0
σ_4'	–a	–a′	–b	–b′	–c	–c′	–d	–d′	–e	–e′	–f	–f′	–g	–g′	–14

To find the irreducible representations that these orbitals span, multiply the characters in the representation of the orbitals by the characters of the irreducible representations, sum those products, and divide the sum by the order h of the group (as in Section 7.5a). The table below illustrates the procedure, beginning at left with the C_{2v} character table.

	E	C_2	σ_v	σ_v'	Product	E	C_2	σ_v	σ_v'	Sum/h
A_1	1	1	1	1		14	0	0	–14	0
A_2	1	1	–1	–1		14	0	0	14	7
B_1	1	–1	1	–1		14	0	0	14	7
B_2	1	–1	–1	1		14	0	0	–14	0

The orbitals span $\boxed{7A_2 + 7B_1}$.

To find symmetry-adapted linear combinations (SALCs), follow the procedure described in Section 7.5(c). Refer to the table above that displays the transformations of the original basis orbitals. To find SALCs of a given symmetry species, take a column of the table, multiply each entry by the character of the species' irreducible representation, sum the terms in the column, and divide by the order of the group. For example, the characters of species A_1 are 1, 1, 1, 1, so the columns to be summed are identical to the columns in the table above. Each column sums to zero, so we conclude that there are no SALCs of A_1 symmetry. (No surprise here: the orbitals span only A_2 and B_1.) An A_2 SALC is obtained by multiplying the characters 1, 1, –1, –1 by the first column:

$$\frac{1}{4}(a - a' - a' + a) = \frac{1}{2}(a - a')$$

The A_2 combination from the second column is the same. There are seven distinct A_2 combinations in all: $\boxed{\frac{1}{2}(a - a'),\ \frac{1}{2}(b - b'), \ldots, \frac{1}{2}(g - g')}$.

The B_1 combination from the first column is $\frac{1}{4}(a + a' + a' + a) = \frac{1}{2}(a + a')$.

The B_1 combination from the second column is the same. There are seven distinct B_1 combinations in all: $\boxed{\frac{1}{2}(a + a'), \frac{1}{2}(b + b'), \ldots, \frac{1}{2}(g + g')}$. There are no B_2 combinations, as the columns sum to zero.

(b) The structure is labeled to match the row and column numbers shown in the determinant. The Hückel secular determinant of phenanthrene is

	a	b	c	d	e	f	g	g′	f′	e′	d′	c′	b′	a′
a	$\alpha-E$	β	0	0	0	0	0	0	0	0	0	0	0	β
b	β	$\alpha-E$	β	0	0	0	β	0	0	0	0	0	0	0
c	0	β	$\alpha-E$	β	0	0	0	0	0	0	0	0	0	0
d	0	0	β	$\alpha-E$	β	0	0	0	0	0	0	0	0	0
e	0	0	0	β	$\alpha-E$	β	0	0	0	0	0	0	0	0
f	0	0	0	0	β	$\alpha-E$	β	0	0	0	0	0	0	0
g	0	β	0	0	0	β	$\alpha-E$	β	0	0	0	0	0	0
g′	0	0	0	0	0	0	β	$\alpha-E$	β	0	0	0	β	0
f′	0	0	0	0	0	0	0	β	$\alpha-E$	β	0	0	0	0
e′	0	0	0	0	0	0	0	0	β	$\alpha-E$	β	0	0	0
d′	0	0	0	0	0	0	0	0	0	β	$\alpha-E$	β	0	0
c′	0	0	0	0	0	0	0	0	0	0	β	$\alpha-E$	β	0
b′	0	0	0	0	0	0	0	β	0	0	0	β	$\alpha-E$	β
a′	β	0	0	0	0	0	0	0	0	0	0	0	β	$\alpha-E$

This determinant has the same eigenvalues as in exercise 5.22a(b).

(c) The ground state of the molecule has A_1 symmetry because its wavefunction is the product of doubly occupied orbitals, and the product of any two orbitals of the same symmetry has A_1 character. If a transition is to be allowed, the transition dipole must be non-zero, which in turn can happen only if the representation of the product $\Psi_f^* \mu \Psi_i$ includes the totally symmetric species A_1. Consider first transitions to another A_1 wavefunction, in which case we need the product $A_1 \mu A_1$. Now $A_1 A_1 = A_1$, and the only character that returns A_1 when multiplied by A_1 is A_1 itself. The *z* component of the dipole operator belongs to species A_1, so *z*-polarized $A_1 \leftarrow A_1$ transitions are allowed. (*Note:* Transitions from the A_1 ground state to an A_1 excited state are transitions from an orbital occupied in the ground state to an excited-state orbital of the same symmetry.)

The other possibility is a transition from an orbital of one symmetry (A_2 or B_1) to the other; in that case, the excited state wavefunction will have symmetry of $A_2B_1 - B_2$ from the two singly occupied orbitals in the excited state. The symmetry of the transition dipole, then, is $A_1\mu B_2 = \mu B_2$, and the only species that yields A_1 when multiplied by B_2 is B_2 itself. The y component of the dipole operator belongs to species B_2, so these transitions are also allowed (y-polarized).

P7.21 The p_z orbitals of the fluorine atoms in XeF_4 are shown in Figure 7.9. (The orbital has a positive wavefunction sign in shaded lobes and a negative wavefunction sign in unshaded lobes.) These orbitals may form π molecular orbitals with the xenon atom provided that a non-zero overlap integral exists.

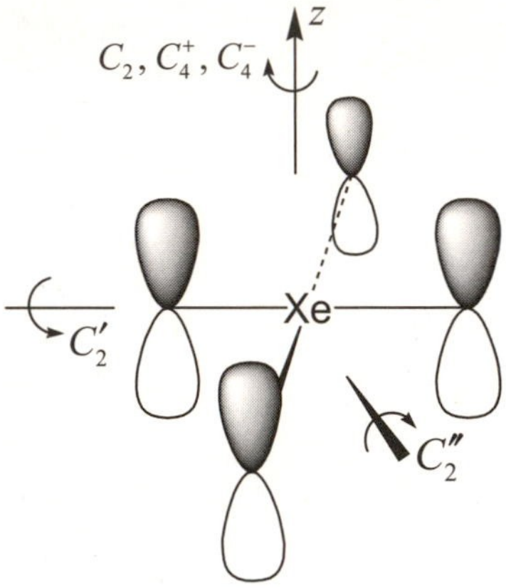

Figure 7.9

To find the symmetry species spanned by the fluorine p_z orbitals, we use a quick rule for determining the character of the basis set under each symmetry operation of the group: count 1 each time a basis function is left unchanged by the operation because only these functions give a non-zero entry on the diagonal of the matrix representative. In some cases there is a sign change, $(\ldots -f \ldots) \leftarrow (\ldots f \ldots)$; then –1 occurs on the diagonal, so count –1. The character of the identity is always equal to the dimension of the basis since each function contributes 1 to the trace. Although XeF_4 belongs to the D_{4h} group, we will use the D_4 subgroup for convenience. Here is a summary of the characters exhibited by the fluorine orbital under each symmetry operation.

D_4	E	C_2	$2C_4$	$2C_2'$	$2C_2''$
Fluorine p_z orbitals	4	0	0	–2	0

Inspection of the D_4 character table shows that the fluorine p_z orbitals span $A_2 + B_2 + E$. Further inspection of the D_4 character table reveals that p_z belongs to A_2, d_{xy} belongs to B_2, and both the (p_x,p_y) set and the (d_{xz},d_{yz}) set belong to E. Consequently, only these orbitals of the central atom may possibly have non-zero overlap with the fluorine p_z orbitals. We must now use the procedure of Section 7.5(c) to find the four **symmetry-adapted linear combinations (SALC)** of the fluorine p_z orbitals. Using clockwise labeling of the Fp_z orbitals, we write a table that summarizes the effect of each operation on each orbital. (The symbol A is used to represent the orbital p_z(A) at fluorine atom A, and so on.)

D_4	A	B	C	D
E	A	B	C	D
C_2	C	D	A	B
C_4^+	D	A	B	C
C_4^-	B	C	D	A
C_2'(A-C)	–A	–D	–C	–B
C_2'(B-D)	–C	–B	–A	–D
C_2''(A-Xe-B)	–B	–A	–D	–C
C_2''(B-Xe-C)	–D	–C	–B	–A

To generate the B_2 combination, we take the characters for B_2 (1,1,–1, –1, –1, –1,1,1) and multiply column 3 and sum the terms. Then we divide by the order of the group (8):

$$p_1(B_2) = {}^1/_8\{C + A - B - D + C + A - D - B\} = {}^1/_4\{p_z(A) - p_z(B) + p_z(C) - p_z(D)\}$$

To generate the A_2 combination, we take the characters for A_2 and multiply column 1:

$$p_2(A_2) = {}^1/_4\{p_z(A) + p_z(A) + p_z(C) + p_z(D)\}$$

To generate the two E combinations, we take the characters for E and multiply column 1 for one of them and multiply column 2 for the other:

$$p_3(E) = {}^1/_4\{p_z(A) - p_z(C)\}$$

$$p_4(E) = {}^1/_4\{p_z(B) - p_z(D)\}$$

Other column multiplications yield these same combinations or zero.

Figure 7.10(a) shows that the $p_z(A_2)$ orbital of the central atom does have nonzero overlap with p_2. Figure 7.10(b) shows that $d_{xz}(E)$ has a non-zero overlap with p_3. However, Figure 7.10(c) shows that, because of the balance between constructive and destructive interference, there is zero overlap between $d_{xy}(B_2)$ and p_1. Thus, there is no orbital of the central atom that forms a non-zero overlap with the p_1 combination, so p_1 is nonbonding.

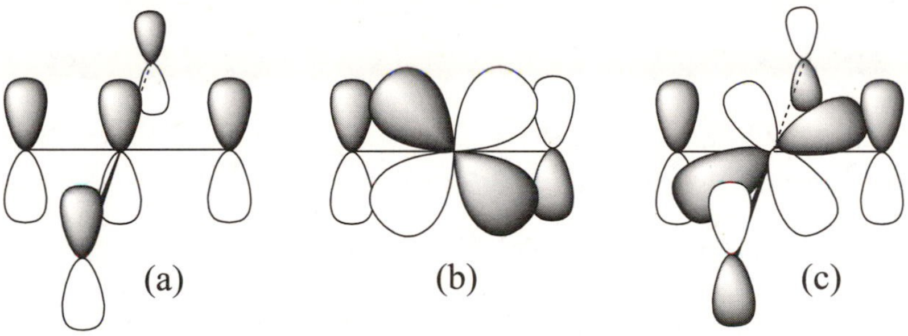

Figure 7.10

Solutions to applications

P7.23 (a) Following the flow chart in Fig. 12.7 of the text, note that the molecule is not linear (at least not in the mathematical sense); there is only one C_n axis (a C_2), and there is a σ_h. The point group, then, is $\boxed{C_{2h}}$.

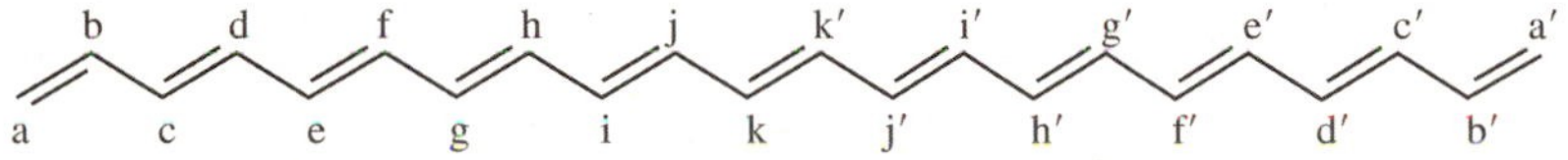

(b) The $2p_z$ orbitals are transformed under the symmetry operations of the C_{2h} group as follows.

	a	a′	b	b′	c	c′	…	j	j′	k	k′	χ
E	a	a′	b	b′	c	c′	…	j	j′	k	k′	22
C_2	a′	a	b	b	c′	c	…	j′	j	k′	k	0
i	–a′	–a	– b′	–b	– c′	–c	…	– j′	– j′	– k′	–k	0
σ_h	–a	– a′	–b	– b′	–c	– c′	…	–j	– j′	–k	– k′	–22

To find the irreproducible representations that these orbitals span, we multiply the characters of orbitals by the characters of the irreproducible representations, sum those products, and divide the sum by the order h of the group (as in Section 12.5a). The table below illustrates the procedure, beginning at left with the C_{2h} character table.

	E	C_2	i	σ_h	product	E	C_2	i	σ_h	sum/h
A_g	1	1	1	1		22	0	0	−22	0
A_u	1	1	−1	−1		22	0	0	22	11
B_g	1	−1	1	−1		22	0	0	22	11
B_u	1	−1	−1	1		22	0	0	−22	0

The orbitals span $\boxed{11A_u + 11B_g}$.

To find symmetry-adapted linear combinations (SALCs), follow the procedure described in Section 12.5(c). Refer to the table above that displays the transformations of the original basis orbitals. To find SALCs of a given symmetry species, take a column of the table, multiply each entry by the character of the species' irreducible representation, sum the terms in the column, and divide by the order of the group. For example, the characters of species A_u are 1, 1, 1, 1, so the columns to be summed are identical to the columns in the table above. Each column sums to zero, so we conclude that there are no SALCs of A_g symmetry. (No surprise: the orbitals span only A_u and B_g). An A_u SALC is obtained by multiplying the characters 1, 1, –1, –1 by the first column:

$$\frac{1}{4}(a+a'+a'+a)=\frac{1}{2}(a+a')$$

The A_u combination from the second column is the same. There are 11 distinct A_u combinations in all: $\boxed{\frac{1}{2}(a+a'), \frac{1}{2}(b+b'), \ldots \frac{1}{2}(k+k')}$.

The B_g combination from the first column is: $\frac{1}{4}(a-a'-a'+a)=\frac{1}{2}(a-a')$.

The B_g combination from the second column is the same. There are 11 distinct B_g combinations in all: $\boxed{\frac{1}{2}(a-a'), \frac{1}{2}(b-b'), \ldots \frac{1}{2}(k-k')}$. There are no B_u combinations, as the columns sum to zero.

(c) The structure is labeled to match the row and column numbers shown in the determinant. The Hückel secular determinant is

	a	b	c	…	i	j	k	k′	j′	i′	…	c′	b′	a′
a	$\alpha-E$	β	0	…	0	0	0	0	0	0	…	0	0	0
b	β	$\alpha-E$	β	…	0	0	0	0	0	0	…	0	0	0
c	0	β	$\alpha-E$	…	0	0	0	0	0	0	…	0	0	0
…	…	…	…	…	…	…	…	…	…	…	…	…	…	…
i	0	0	0	…	$\alpha-E$	β	0	0	0	0	…	0	0	0
j	0	0	0	…	β	$\alpha-E$	β	0	0	0	…	0	0	0
k	0	0	0	…	0	β	$\alpha-E$	β	0	0	…	0	0	0
k′	0	0	0	…	0	0	β	$\alpha-E$	β	0	…	0	0	0
j′	0	0	0	…	0	0	0	β	$\alpha-E$	β	…	0	0	0
i′	0	0	0	…	0	0		0	0	β	$\alpha-E$	0	0	0
…	…	…	…	…	…	…	…	…	…	…	…	…	…	…
c′	0	0	0	…	0	0	0	0	0	0	…	$\alpha-E$	β	0
b′	0	0	0	…	0	0	0	0	0	0	…	β	$\alpha-E$	β
a′	0	0	0	…	0	0	0	0	0	0	…	0	β	$\alpha-E$

The energies of the filled orbitals are $\alpha + 1.98137\beta$, $\alpha + 1.92583\beta$, $\alpha + 1.83442\beta$, $\alpha + 1.70884\beta$, $\alpha + 1.55142\beta$, $\alpha + 1.36511\beta$, $\alpha + 1.15336\beta$, $\alpha + 0.92013\beta$, $\alpha + 0.66976\beta$, $\alpha + 0.40691\beta$, and $\alpha + 0.13648\beta$. The π energy is 27.30729β.

(d) The ground state of the molecule has A_g symmetry because its wavefunction is the product of doubly occupied orbitals, and the product of any two orbitals of the same symmetry has A_g character. If a transition is to be allowed, the transition dipole must be non-zero, which in turn can happen only if the representation of the product $\Psi_f^* \boldsymbol{\mu} \Psi_i$ includes the totally symmetric species A_g. Consider first transitions to another A_g wavefunction, in which case we need the product $A_g\mu A_g$. Now $A_gA_g = A_g$, and the only character that returns A_g when multiplied by A_g is A_g itself. No component of the dipole operator belongs to species A_g, so no $A_g \leftarrow A_g$ transitions are allowed. (*Note:* Such transitions are transitions from an orbital occupied in the ground state to an excited-state orbital of the same symmetry.) The other possibility is a transition from an orbital of one symmetry (A_u or B_g) to the other; in that case, the excited-state wavefunction will have symmetry of $A_uB_g = B_u$ from the two singly occupied orbitals in the excited state. The symmetry of the transition dipole, then, is $A_g\mu B_u = \mu B_u$, and the only species that yields A_g when multiplied by B_u is B_u itself. The *x* and *y* components of the dipole operator belong to species B_u, so these transitions are allowed.

8 Molecular assemblies

Answers to discussion questions

D8.1 Molecules with a permanent separation of electric charge have a permanent dipole moment. In molecules containing atoms of differing electronegativity, the bonding electrons may be displaced in such a way as to produce a net separation of charge in the molecule. Separation of charge may also arise from a difference in atomic radii of the bonded atoms. The separation of charges in the bonds is usually, though not always, in the direction of the more electronegative atom but depends on the precise bonding situation in the molecule as described in Section 8.2. A heteronuclear diatomic molecule necessarily has a dipole moment if there is a difference in electronegativity between the atoms, but the situation in polyatomic molecules is more complex. A polyatomic molecule has a permanent dipole moment only if it fulfills certain symmetry requirements as discussed in Section 7.3(a) of the text.

An external electric field can distort the electron density in both polar and nonpolar molecules, which results in an induced dipole moment that is proportional to the field. The constant of proportionality is called the polarizability.

D8.3 See Fig. 8.4 of the text for typical charge arrays corresponding to electric multipoles.

The interaction potential between a point charge q_2 (monopole, $n = 1$) and any of the multipoles ($m = 2$ or 3 or ...) is given by eqn. 8.5 as $V \propto \dfrac{1}{r^m}$ where r is the separation distance between q_2 and the multipole. This is a steeper potential energy decrease with r than that observed for the interaction between two point charges: $V \propto \dfrac{1}{r}$ [8.1a]. The steeper decline originates in the case for which $r >> l$, where l is the separation of charge within the multipole because, as r becomes relatively large, the array of charges in the multipole appears to blend together into neutrality causing lower order interaction terms to cancel. For example, the dipole terms within the monopole-quadrupole ($m = 3$) interaction potential cancel leaving only a $1/r^3$ term. We use the linear quadrupole charge arrangement shown in Figure 8.1 to show this cancellation of lower order terms. Since we are interested in the case $x = l/r << 1$, the following Taylor series expansions are useful substitutions:

$$(1+x)^{-1} = 1 - x + x^2 - x^3 + \cdots \qquad \text{and} \qquad (1-x)^{-1} = 1 + x + x^2 + x^3 + \cdots$$

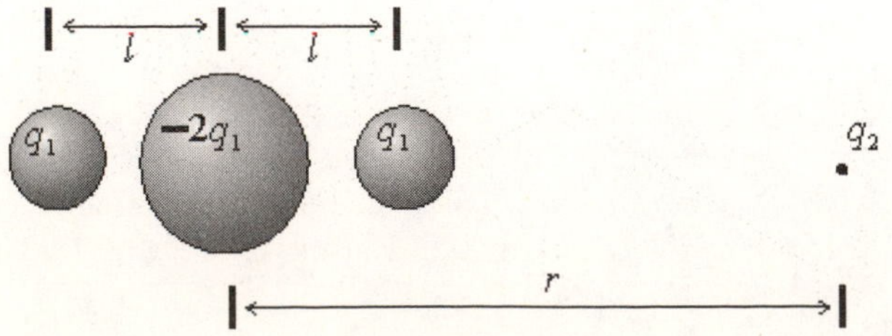

Figure 8.1

Begin by adding the terms for the Coulomb potential interaction [8.1a] between the charge array of the quadrupole and the monopole q_2, substitute $x = l/r$, and perform Taylor series expansions on the functions of x.

$$
\begin{aligned}
4\pi\varepsilon_0 V &= \frac{q_1 q_2}{r+l} - \frac{2q_1 q_2}{r} + \frac{q_1 q_2}{r-l} \\
&= \frac{q_1 q_2}{r}\left\{\frac{1}{1+x} - 2 + \frac{1}{1-x}\right\} \\
&= \frac{q_1 q_2}{r}\left\{\cancel{1} \cancel{-x} + x^2 - x^3 + x^4 + \cdots \cancel{-2} \cancel{+1} \cancel{+x} + x^2 + x^3 + x^4 \cdots\right\} \\
&= \frac{2x^2 q_1 q_2}{r}\left\{1 + x^2 + x^4 + \cdots\right\}
\end{aligned}
$$

The higher order terms within the polynomial are negligibly small compared to 1 in the case for which $x = l/r << 1$, thereby leaving the simple expression

$$V = \frac{2x^2 q_1 q_2}{4\pi\varepsilon_0 r} = \frac{l^2 q_1 q_2}{2\pi\varepsilon_0 r^3} \quad \text{or} \quad V \propto \frac{1}{r^3}$$

D8.5 A **hydrogen bond** ($\cdots$) is an attractive interaction between two species that arises from a link of the form $A-H\ldots B$, where A and B are highly electronegative elements (usually nitrogen, oxygen, or fluorine) and B possesses a lone pair of electrons. It is a contact-like attraction that requires AH to touch B. Experimental evidences supports a linear or near-linear structural arrangement and a bond strength of about 20 kJ mol^{-1}. The hydrogen bond strength is considerably weaker than a covalent bond but it is larger than and dominates other intermolecular attractions such as dipole-dipole attractions. Its formation can be understood in terms of either the (a) electrostatic interaction model or (b) molecular orbital calculations (Chapter 6).

(a) A and B, being highly electronegative, are viewed as having partial negative charges (δ^-) in the electrostatic interaction model of the hydrogen bond. Hydrogen, being less electronegative than A, is viewed as having a partial positive (δ^+). The linear structure maximizes the electrostatic attraction between H and B:

δ^- δ^+ δ^-
A——H··············:B

This model is conceptually very useful. However, it is impossible to exactly calculate the interaction strength with this model because the partial atomic charges cannot be precisely defined. There is no way to define which fraction of the electrons of the AB covalent bond should be assigned to one or the other nucleus. (See numerical problem 6.7.)

(b) Ab initio quantum calculations are needed in order to explore questions about the linear structure, the role of the lone pair, the shape of the potential energy surface, and the extent to which the hydrogen bond has covalent sigma bond character. Yes, the hydrogen bond appears to have some sigma bond character. This was initially suggested by Linus Pauling in the 1930s, and more recent experiments with Compton scattering of X-rays and NMR techniques indicate that the covalent character may provide as much as 20% of the hydrogen bond strength. A three-center molecular orbital model provides a degree of insight. A linear combination of an appropriate sigma orbital on A, the 1*s* hydrogen orbital, and an appropriate orbital for the lone pair on B yields a total of three molecular orbitals. One of the MOs is bonding, one is almost nonbonding, and the third is antibonding. Both bonding MO and the almost nonbonding orbital are occupied by two electrons (the sigma-bonding electrons of A–H and the lone pair of B). The antibonding MO is empty. Thus, depending on the precise location of the almost nonbonding orbital, the nonbonding orbital may lower the total energy and account for the hydrogen bond.

D8.7 The formation of micelles is favored by the interaction between hydrocarbon tails and is opposed by charge repulsion of the polar groups that are placed close together at the micelle surface. As salt concentration is increased, the repulsion of head groups is reduced because their charges are partly shielded by the ions of the salt. This favors micelle formation causing the micelles to be larger and the critical micelle concentration to be smaller.

D8.9 The van der Waals equation of state (eqn. 8.24) provides an accounting of both the repulsive and attractive interactions between molecules by employing the van der Waals constants a and b to account for the attractive and repulsive interaction, respectively. These interactions are negligibly small in the limit of zero pressure because the average intermolecular distance is very large compared to the size of a molecule in this limit. Consequently, the perfect gas law accurately describes p, V_m, T behavior in the limit of zero pressure.

$$p = \frac{RT}{V_m - b} - \frac{a}{V_m^2} \quad [8.24b]$$

The van der Waals constants a and b do not exactly reflect the fact that interaction potential energy is a function of distance from a molecule. The potential is a periodic function in solids, in liquids, and in gases at small molar volumes because molecular closeness causes the potential to depend on many nearest-neighbor molecules (covalent compound), ions (ionic compound), and atoms (element). The potential is not a constant. However, the van der Waals constants do represent an average value of the repulsive and attractive interactions, which is an improvement over the perfect gas equation at higher pressures and lower temperatures. The van der Waals constant b treats the repulsive interaction as if molecules were hard spheres for which the repulsion is zero until there is contact between two colliding spheres with an ensuing infinite repulsion potential.

Fig. 8.30 of the text illustrates the physically implausible van der Waals loops that are produced by the van der Waals equation in regions of gas-liquid equilibrium. This error is resolved by using a **Maxwell construction** to replace the loops with horizontal straight lines so that the loops have equal areas above and below the lines. One end of the isotherm line provides a molar volume estimate for the gas phase; the other end gives an estimate for the molar volume of the liquid phase.

D8.11 The **radial distribution function, *g*(*r*)**, is defined so that $g(r)r^2\mathrm{d}r$ is the probability that a molecule will be found in the range dr at a distance r from the molecule. Figure 8.2(a) illustrates a sketch of a radial distribution function for a solid. $g(r)$ spikes at intervals that depend on the crystal structure and the identifiable spikes continue to the edge of the crystal because of the long-range order of the perfect atomic array. Figure 8.2(b) illustrates a radial distribution function for a liquid. The distribution shows several broad peaks, which indicate some short-range local order around each molecule, but there is considerable variance in the positions of even nearest-neighbor molecules. This happens because molecules are no longer at fixed lattice sites. They have more freedom to move and slip past one another. All r values beyond the closest contact distance become possible, causing $g(r)$ to exhibit broadened short-range peaks and to never decline to zero. Molecular mobility also destroys long-range order so that no spike in seen after several molecular diameters.

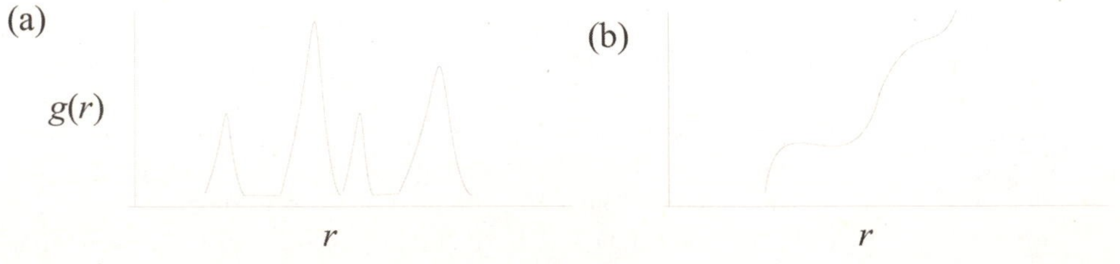

Figure 8.2

D8.13 Kevlar is a polyaromatic amide. Phenyl groups provide aromaticity and a planar, rigid structure. The amide group is expected to be like the peptide bond that connects amino acid residues within protein molecules. This group is also planar because resonance produces partial double-bond character between the carbon and nitrogen atoms. There is a substantial energy barrier preventing free rotation about the CN bond. The two bulky phenyl groups on the ends of an amide group are trans because steric hindrance makes the cis conformation unfavorable.

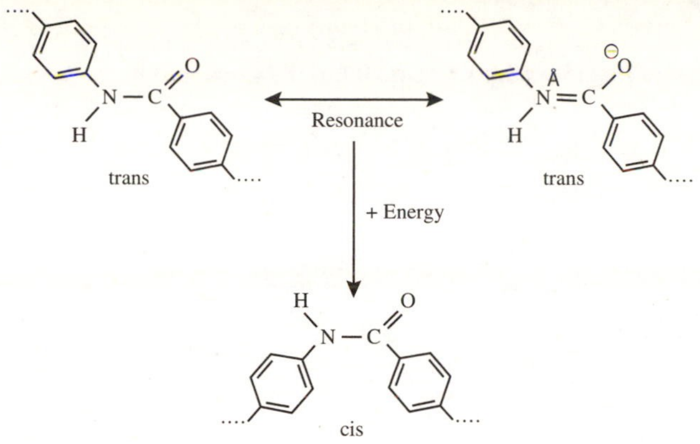

The flatness of the Kevlar polymeric molecule makes it possible to process the material so that many molecules with parallel alignment form highly ordered, untangled crystal bundles. The alignment makes possible both considerable van der Waals attractions between adjacent molecules and strong hydrogen bonding between the polar amide groups on adjacent molecules. These bonding forces create the high thermal stability and mechanical strength observed in Kevlar.

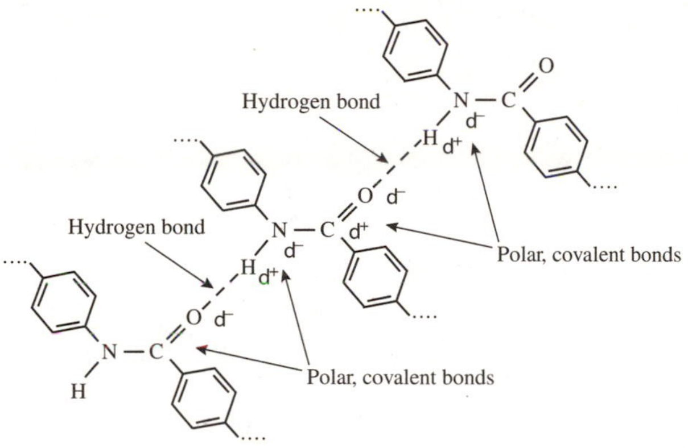

Kevlar is able to absorb great quantities of energy, such as the kinetic energy of a speeding bullet, through hydrogen bond breakage and the transition to the cis conformation.

Solutions to exercises

E8.1(a) A molecule with a center of symmetry may not be polar, but molecules belonging to the groups C_n and C_{nv} may be polar (Section 7.3). The C atom of CIF_3 is approximately sp^3 hybridized, which causes the molecule to belong to the C_{3v} point group. The highly electronegative F atoms cause the C–F bonds to be very polar and the average dipole of the three very polar C–F bonds is unbalanced by the less polar C–I bond. Therefore, $\boxed{CIF_3}$ is polar.

Ozone is a bent molecule that belongs to the C_{2v} point group. Lewis resonance structures for the molecule show a central atom with a double bond to an end oxygen atom and a single bond to oxygen at the other end. The central oxygen has a formal charge of +1 while the double-bonded end oxygen has a formal charge of zero. The single-bonded end oxygen has a –1 formal charge. The average position of the –1 formal charge of the two resonance structures predicts a small negative charge that lies halfway between the extremities and a fraction of a bond length away from the central oxygen, which is expected to have a small positive charge. Consequently, $\boxed{O_3}$ is polar.

Hydrogen peroxide belongs to the C_2 point group with each carbon atom approximately sp^3 hybridized. The H–O–O bond angles are 96.87°. The H–O–O plane of one hydrogen atom is at a 93.85° angle with the O–O–H plane of the other hydrogen atom. Consequently, the dipole moments of the two polar O–H bonds do not cancel and the $\boxed{H_2O_2}$ molecule is polar.

E8.2(a) $\mu = (\mu_1^2 + \mu_2^2 + 2\mu_1\mu_2 \cos\theta)^{1/2}$ [8.2a]

$$= [(1.5)^2 + (0.80)^2 + (2)\times(1.5)\times(0.80)\times(\cos 109.5°)]^{1/2}\,\mathrm{D} = \boxed{1.4\ \mathrm{D}}$$

E8.3(a) The dipole moment is the vector sum (see Figure 8.3).

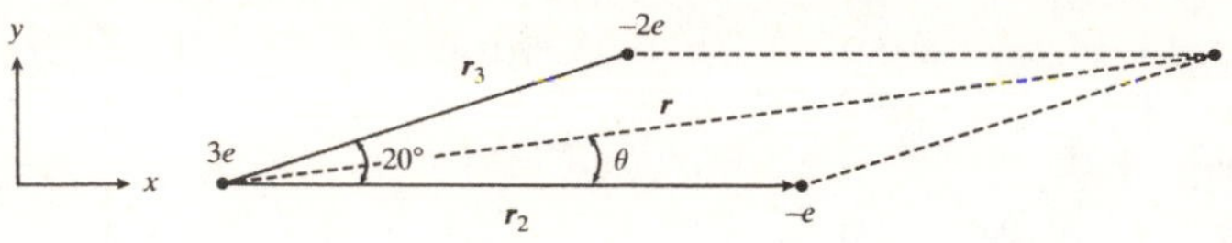

Figure 8.3

$$\mu = \sum_i q_i r_i = 3e(0) - er_2 - 2er_3$$

$$r_2 = \mathbf{i}x_2, \qquad r_3 = \mathbf{i}x_3 + \mathbf{j}\ {}_3$$

$$x_2 = +0.32\,\mathrm{nm}$$

$$x_3 = r_3 \cos 20° = (+0.23\,\mathrm{nm})\times(0.940) = 0.21\overline{6}\,\mathrm{nm}$$

$$ {}_3 = r_3 \sin 20° = (+0.23\,\mathrm{nm})\times(0.342) = 0.078\overline{7}\,\mathrm{nm}$$

The components of the vector sum are the sums of the components. That is (with all distances in nm),

$$\mu_x = -ex_2 - 2ex_3 = -(e)\times\{(0.32) + (2)\times(0.21\overline{6})\} = -(e)\times(0.752\,\mathrm{nm})$$

$$\mu\ = -2e\ {}_3 = -(e)\times(2)\times(0.078\overline{7}) = -(e)\times(0.1574\,\mathrm{nm})$$

$$\mu = (\mu_x^2 + \mu^2)^{1/2} = (e)\times(0.76\overline{8}\,\mathrm{nm}) = (1.602\times10^{-19}\,\mathrm{C})\times(0.76\overline{8}\times10^{-9}\,\mathrm{m})$$

$$= 1.2\overline{3}\times10^{-28}\,\mathrm{C\,m} = \boxed{37\,\mathrm{D}}$$

The angle that μ makes with the x-axis is given by

$$\cos\theta = \frac{|\mu_x|}{\mu} = \frac{0.752}{0.768}; \quad \boxed{\theta = 11.7°}$$

E8.4(a) The O–H bond length of a water molecule is 95.85 pm and the Li^+ cation is 100 pm from the dipole center. Because these lengths are comparable, a calculation based on the assumption that the water dipole acts like a point dipole with a dipole length much shorter than the dipole-ion distance is unlikely to provide an accurate value of the dipole-ion interaction energy. However, such a calculation does provide an "order-of-magnitude" estimate. The minimum value of the dipole-ion interaction occurs with the dipole pointing toward the cation.

$$ {}_{\mathrm{min}} \sim -\frac{\mu_{\mathrm{H_2O}} q_{\mathrm{Li^+}}}{4\pi\varepsilon_0 r^2}\ [8.4] = -\frac{\mu_{\mathrm{H_2O}} e}{4\pi\varepsilon_0 r^2}$$

$$\sim -\frac{(1.85\,\mathrm{D})\times(3.336\times10^{-30}\,\mathrm{C\,m\,D^{-1}})\times(1.602\times10^{-19}\,\mathrm{C})}{(1.113\times10^{-10}\,\mathrm{J^{-1}\,C^2\,m^{-1}})\times(100\times10^{-12}\,\mathrm{m})^2}$$

$$\sim -8.88\times10^{-19}\,\mathrm{J}$$

The interaction potential becomes a maximum upon flipping the dipole. This effectively changes the sign of the dipole in the previous calculation:

$$ {}_{\mathrm{max}} \sim 8.88\times10^{-19}\,\mathrm{J}$$

The work w required to flip the dipole is the difference $V_{max} - V_{min}$.

$$w \sim V_{max} - V_{min} = 1.78\times10^{-18}\ \text{J}$$

$$w_m = wN_A \sim \boxed{1.07\times10^3\ \text{kJ mol}^{-1}}$$

E8.5(a) The two linear quadrupoles are shown with a collinear configuration in Figure 8.4.

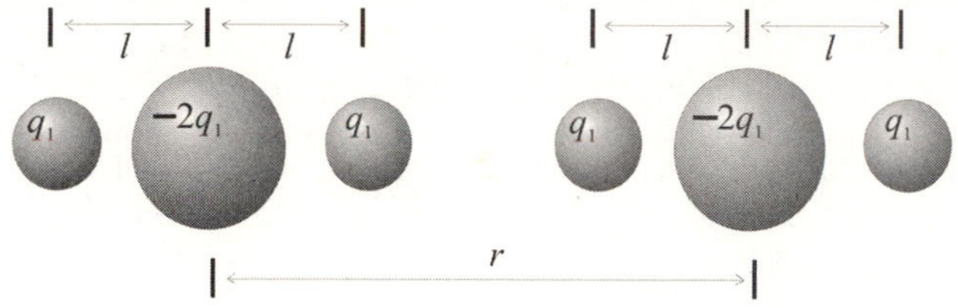

Figure 8.4

The total potential energy of the interaction between the quadrupoles is

$$4\pi\varepsilon_0 V = \frac{q_1^2}{r} - \frac{2q_1^2}{r+l} + \frac{q_1^2}{r+2l} - \frac{2q_1^2}{r-l} + \frac{4q_1^2}{r} - \frac{2q_1^2}{r+l} + \frac{q_1^2}{r-2l} - \frac{2q_1^2}{r-l} + \frac{q_1^2}{r}$$

$$\frac{4\pi\varepsilon_0 rV}{q_1^2} = 6 - \frac{4}{1+x} - \frac{4}{1-x} + \frac{1}{1+2x} + \frac{1}{1-2x} \quad \text{where} \quad x = \frac{l}{r}$$

With the point quadrupole condition that $x << 1$, the last four terms in the above expression can be expanded with the Taylor series:

$$(1+z)^{-1} = 1 - z + z^2 - z^3 + z^4 - \cdots \quad \text{and} \quad (1-z)^{-1} = 1 + z + z^2 + z^3 + z^4 + \cdots$$

where z is either $2x$ or x.

$$\frac{4\pi\varepsilon_0 rV}{q_1^2} = 6 - 4\{1 - x + x^2 - x^3 + x^4 - \cdots + 1 + x + x^2 + x^3 + x^4 + \cdots\}$$

$$+\{1-(2x)+(2x)^2-(2x)^3+(2x)^4-\cdots\}+\{1+(2x)+(2x)^2+(2x)^3+(2x)^4+\cdots\}$$

$$= -8x^2 - 8x^4 + 8x^2 + 32x^4 + \cdots$$

$$= 24x^4 + \text{ higher order terms}$$

In the limit of small x values, the higher order terms are negligibly small, thereby leaving

$$V = \frac{6x^4 q_1^2}{\pi\varepsilon_0 r} = \boxed{\frac{6l^4 q_1^2}{\pi\varepsilon_0 r^5}}$$

Thus, $V \propto \frac{1}{r^5}$ for the quadrupole-quadrupole interaction is in agreement with eqn. 8.5.

E8.6(a) The induced dipole moment is

$$\mu^*_{H_2O} = \alpha\mathcal{E}\ [8.10] = 4\pi\varepsilon_0\alpha'_{H_2O}\mathcal{E}\ [8.11]$$

$$= (4\pi)\times(8.854\times10^{-12}\ \text{J}^{-1}\ \text{C}^2\ \text{m}^{-1})\times(1.48\times10^{-30}\ \text{m}^3)\times(1.0\times10^5\ \text{V m}^{-1})$$

$$= 1.6\times10^{-35}\ \text{C m} \quad [1\ \text{J} = 1\ \text{C V}]$$

which corresponds to $\boxed{4.9\ \mu\text{D}}$

E8.7(a)

$$V_{\text{London}} = -\frac{3(\alpha'_{\text{He}} I_{\text{He}})^2}{2(I_{\text{He}}+I_{\text{He}})r^6}\ [8.14] = -\frac{3(\alpha'_{\text{He}})^2 I_{\text{He}}}{4r^6}$$

$$= -\frac{3\times(0.20\times10^{-30}\ \text{m}^3)^2\times(2372.3\ \text{kJ mol}^{-1})}{4\times(1.0\times10^{-9}\ \text{m})^6}$$

$$= \boxed{0.071\ \text{J mol}^{-1}}$$

E8.8(a) Using the partial charge presented in Table 8.2, we estimate the partial charge on each hydrogen atom of a water molecule to be $q_{\text{H}} = \delta e$ where $\delta = 0.42$. The electroneutrality of an H_2O molecule implies that the estimated partial charge on the oxygen atom is $q_{\text{O}} = -2\delta e$. With a hydrogen bond length of 170 pm, the point charge model of the hydrogen bond in a vacuum estimates the potential of interaction to be

$$V = \frac{q_{\text{H}}q_{\text{O}}}{4\pi\varepsilon_0 r} = -\frac{2(\delta e)^2}{4\pi\varepsilon_0 r}\ [8.1a]$$

$$= -\frac{2(0.42\times1.60\times10^{-19}\ \text{C})^2}{4\pi(8.85\times10^{-12}\ \text{J}^{-1}\ \text{C}^2\ \text{m}^{-1})\times(170\times10^{-12}\ \text{m})} = -4.8\times10^{-19}\ \text{J}$$

The molar energy required to break these bonds is

$$E_{\text{m}} = -N_{\text{A}}V = -(6.022\times10^{23}\ \text{mol}^{-1})\times(-4.8\times10^{-19}\ \text{J}) = \boxed{28\bar{9}\ \text{kJ mol}^{-1}}$$

E8.9(a)

$$n_{\text{N}_2} = 3.15\ \text{g}\times\left(\frac{1\ \text{mol N}_2}{28.01\ \text{g}}\right) = 0.112\ \text{mol N}_2$$

$$V_{\text{m}} = \frac{V}{n_{\text{N}_2}} = \frac{2.05\ \text{dm}^3}{0.112\ \text{mol N}_2} = 0.0183\ \text{m}^3\ \text{mol}^{-1} = 1.83\times10^4\ \text{cm}^3\ \text{mol}^{-1}$$

Assuming that the dinitrogen virial equation of state has $C/V_{\text{m}}^2 << B/V_{\text{m}}$,

$$p_{\text{virial}} = \frac{RT}{V_{\text{m}}}\left(1+\frac{B}{V_m}\right)[8.22\text{b}] = \frac{(8.3145\ \text{J mol}^{-1}\ \text{K}^{-1})(273\ \text{K})}{0.0183\ \text{m}^3\ \text{mol}^{-1}}\left(1-\frac{10.5\ \text{cm}^3\ \text{mol}^{-1}}{1.83\times10^4\ \text{cm}^3\ \text{mol}^{-1}}\right)\ [\text{Table 8.5}]$$

$$= 1.24\times10^5\ \text{Pa} = \boxed{1.24\ \text{bar}}$$

According to the perfect (ideal) gas equation of state,

$$p_{\text{perfect}} = \frac{RT}{V_{\text{m}}} = \frac{(8.3145\ \text{J mol}^{-1}\ \text{K}^{-1})(273\ \text{K})}{0.0183\ \text{m}^3\ \text{mol}^{-1}} = 1.24\times10^5\ \text{Pa} = \boxed{1.24\ \text{bar}}$$

The fact that $p_{\text{virial}} = p_{\text{perfect}}$ for nitrogen under these conditions results from the low pressure, high temperature conditions for which the molar volume is large enough to cause $B/V_{\text{m}} << 1$. The second coefficient of the virial equation of state is negligibly small under these conditions and the perfect gas equation of state provides accuracy.

E8.10(a) The second virial coefficient equals zero at the Boyle temperature, so the perfect gas law will accurately describe the pressure.

$$n_{\text{CO}_2} = 4.56\ \text{g}\times\left(\frac{1\ \text{mol CO}_2}{43.99\ \text{g}}\right) = 0.104\ \text{mol CO}_2$$

$$V_{\text{m}} = \frac{V}{n_{\text{CO}_2}} = \frac{2.25\ \text{dm}^3}{0.104\ \text{mol CO}_2} = 0.0216\ \text{m}^3\ \text{mol}^{-1}$$

$$p_{\text{perfect}} = \frac{RT_{\text{B}}}{V_{\text{m}}} = \frac{(8.3145\text{ J mol}^{-1}\text{ K}^{-1})(714.8\text{ K})}{0.0216\text{ m}^3\text{ mol}^{-1}}\ [\text{Table 8.6}] = 2.75\times10^5\text{ Pa} = \boxed{2.75\text{ bar}}$$

E8.11(a) (a) The amount of xenon is $n = \dfrac{131\text{g}}{131\text{g mol}^{-1}} = 1.00\text{mol}$.

$$V_{\text{m}} = \frac{V}{n} = \frac{1.0\text{ dm}^3}{1.00\text{ mol}} = 1.0\text{ dm}^3\text{ mol}^{-1}$$

$$p_{\text{perfect}} = \frac{RT}{V_{\text{m}}} = \frac{(8.3145\text{ J K}^{-1}\text{mol}^{-1})\times(298.15\text{K})}{1.0\times10^{-3}\text{ m}^3\text{ mol}^{-1}} = 25\times10^5\text{ Pa} = \boxed{25\text{bar}}$$

(b) For xenon, Table 8.7 gives $a = 4.137\text{dm}^6\text{ atm mol}^{-2}$ and $b = 5.16\times10^{-2}\text{ dm}^3\text{ mol}^{-1}$.

$$p_{\text{van der Waals}} = \frac{RT}{V_{\text{m}}-b} - \frac{a}{V_{\text{m}}^2}$$

$$= \frac{(8.3145\text{ J K}^{-1}\text{mol}^{-1})\times(298.15\text{K})}{(1.0-0.0516)\times10^{-3}\text{ m}^3\text{ mol}^{-1}} - \frac{4.137\times10^{-6}\text{ atm m}^6\text{ mol}^{-2}}{(1.0\times10^{-3}\text{ m}^3\text{ mol}^{-1})^2}\left(\frac{1.013\times10^5\text{ Pa}}{1\text{ atm}}\right)$$

$$= 22\times10^5\text{ Pa} = \boxed{22\text{ bar}}$$

E8.12(a) The conversions needed are as follows:

$1\text{ atm} = 1.013\times10^5\text{ Pa}$ $\quad$ $1\text{ Pa} = 1\text{ kg m}^{-1}\text{ s}^{-2}$ $\quad$ $1\text{ dm}^6 = 10^{-6}\text{ m}^6$ $\quad$ $1\text{ dm}^3 = 10^{-3}\text{ m}^3$

Therefore,

$a = 0.751\text{ atm dm}^6\text{ mol}^{-2}$ becomes, after substitution of the conversions,

$a = \boxed{7.61\times10^{-2}\text{ kg m}^5\text{ s}^{-2}\text{ mol}^{-2}}$, and

$b = 0.0226\text{ dm}^3\text{ mol}^{-1}$ becomes

$b = \boxed{2.26\times10^{-5}\text{ m}^3\text{ mol}^{-1}}$.

E8.13(a) The definition of Z is $Z = \dfrac{pV_{\text{m}}}{RT} = \dfrac{V_{\text{m}}}{V_{\text{m}}^{\circ}}$ [8.20 and 8.21].

V_{m} is the actual molar volume, V_{m}° is the perfect gas molar volume. $V_{\text{m}}^{\circ} = \dfrac{RT}{p}$. Since V_{m} is 8 percent smaller than that of a perfect gas, $V_{\text{m}} = 0.92\ V_{\text{m}}^{\circ}$, and

(a) $$Z = \frac{0.92\ V_{\text{m}}^{\circ}}{V_{\text{m}}^{\circ}} = \boxed{0.92}$$

(b) $$V_{\text{m}} = \frac{ZRT}{p} = \frac{(0.92)\times(8.206\times10^{-2}\text{ dm}^3\text{ atm K}^{-1}\text{mol}^{-1})\times(250\text{K})}{12\text{atm}} = \boxed{1.6\text{dm}^3\text{ mol}^{-1}}$$

Since $V_{\text{m}} < V_{\text{m}}^{\circ}$ $\boxed{\text{attractive}}$ forces dominate.

E8.14(a) The amount of gas is first determined from its mass. Then, the van der Waals equation is used to determine its pressure at the working temperature. The initial conditions of 300 K and 100 atm are in a sense superfluous information.

$$n = \frac{92.4\text{ kg}}{28.02\times10^{-3}\text{kg mol}^{-1}} = 3.30\times10^3\text{mol}$$

$$V = 1.000\text{ m}^3 = 1.000\times10^3\text{ dm}^3$$

$$p=\frac{nRT}{V-nb}-\frac{an^2}{V^2}[8.24a]=\frac{(3.30\times10^3\text{ mol})\times(0.08206\text{ dm}^3\text{ atm K}^{-1}\text{mol}^{-1})\times(500\text{ K})}{(1.000\times10^3\text{dm}^3)-(3.30\times10^3\text{mol})\times(0.0391\text{ dm}^3\text{ mol}^{-1})}-\frac{(1.39\text{ dm}^6\text{ atm mol}^{-2})\times(3.30\times10^3\text{ mol})^2}{(1.000\times10^3\text{ dm}^3)^2}$$

$$=(155-15)\text{ atm}=\boxed{140\text{ atm}}$$

E8.15(a) The Boyle temperature, T_B, is the temperature at which $B=0$. To express T_B in terms of a and b, the van der Waals equation must be recast in the form of the virial equation.

$$p=\frac{RT}{V_m-b}-\frac{a}{V_m^2}\ [8.24b]$$

Factoring out $\frac{RT}{V_m}$ yields $p=\frac{RT}{V_m}\left\{\frac{1}{1-b/V_m}-\frac{a}{RTV_m}\right\}$.

So long as $b/V_m<1$, the first term inside the brackets can be expanded using $(1-x)^{-1}=1+x+x^2+\cdots$, which gives

$$p=\frac{RT}{V_m}\left\{1+\left(b-\frac{a}{RT}\right)\times\left(\frac{1}{V_m}\right)+\cdots\right\}$$

We can now identify the second virial coefficient as $B=b-\frac{a}{RT}$.

Since at the Boyle temperature $B=0$, $T_B=\frac{a}{bR}$.

From Table 8.7, $a=6.260\text{ dm}^6\text{ atm mol}^{-2}$, $b=5.42\times10^{-2}\text{ dm}^3\text{ mol}^{-1}$. Therefore,

$$T_B=\frac{6.260\text{ dm}^6\text{ atm mol}^{-2}}{(5.42\times10^{-2}\text{ dm}^3\text{ mol}^{-1})\times(8.206\times10^{-2}\text{ dm}^3\text{ atm K}^{-1}\text{ mol}^{-1})}=\boxed{1.41\times10^3\text{ K}}$$

E8.16(a) Figure 8.5 shows how the collision of two hard-sphere molecules establishes an excluded volume. The closest distance between two molecules of radius r and volume $V_{molecule}={}^4/_3\pi r^3$ is $2r$, so the volume excluded is ${}^4/_3\pi(2r)^3$, or $8V_{molecule}$. The volume excluded per molecule, a quantity that provides an estimate of the van der Waals coefficient b, is one-half this volume, or $4V_{molecule}$.

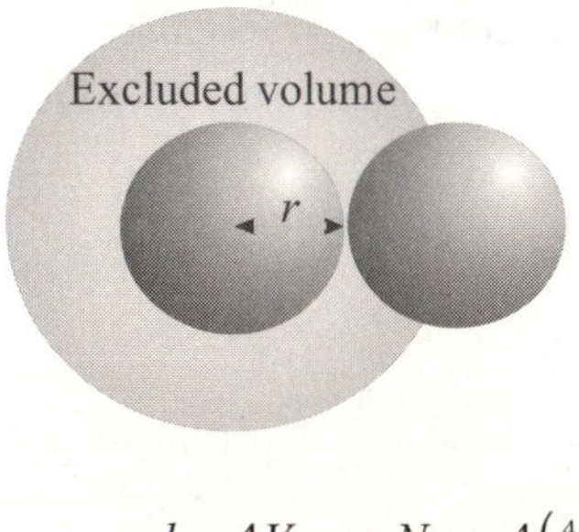

Figure 8.5

$$b=4V_{molecule}N_A=4\left(\tfrac{4}{3}\pi r^3\right)N_A \quad \text{where } b_{Cl_2}=5.42\times10^{-5}\text{ m}^3\text{ mol}^{-1}\text{ [Table 8.7]}$$

$$r=\left(\frac{3b}{16\pi N_A}\right)^{1/3}=\left(\frac{3\left(5.42\times10^{-5}\text{ m}^3\text{ mol}^{-1}\right)}{16\pi\left(6.022\times10^{23}\text{ mol}^{-1}\right)}\right)^{1/3}=\boxed{175\text{ pm}}$$

E8.17(a) The van der Waals equation [8.24b] is solved for b, which yields

$$b=V_m-\frac{RT}{\left(p+\frac{a}{V_m^2}\right)}$$

Substituting the data,

$$b = 5.00\times10^{-4}\ \text{m}^3\ \text{mol}^{-1} - \frac{(8.3145\ \text{J K}^{-1}\ \text{mol}^{-1})\times(273\ \text{K})}{\left\{(3.0\times10^6\ \text{Pa}) + \left(\dfrac{0.50\ \text{m}^6\ \text{Pa mol}^{-2}}{(5.00\times10^{-4}\ \text{m}^3\ \text{mol}^{-1})^2}\right)\right\}}$$

$$= \boxed{0.46\times10^{-4}\ \text{m}^3\ \text{mol}^{-1}}$$

E8.18(a) $$Z = \frac{pV_\text{m}}{RT}\ [8.21] = \frac{(3.0\times10^6\ \text{Pa})\times(5.00\times10^{-4}\ \text{m}^3)}{(8.3145\ \text{J K}^{-1}\ \text{mol}^{-1})\times(273\ \text{K})} = \boxed{0.66}$$

COMMENT. The definition of Z involves the actual pressure, volume, and temperature and does not depend on the equation of state used to relate these variables.

Solutions to problems

Solutions to numerical problems

P8.1 The positive end of the dipole, which bisects the hydrogen atom positions, will lie closer to the (negative) anion. The electric field generated by a dipole is

$$\mathcal{E} = \left(\frac{\mu}{4\pi\varepsilon_0}\right)\times\left(\frac{2}{r^3}\right)\ [8.7]$$

$$= \frac{(2)\times(1.85)\times(3.34\times10^{-30}\ \text{C m})}{(4\pi)\times(8.854\times10^{-12}\ \text{J}^{-1}\ \text{C}^2\ \text{m}^{-1})\times r^3} = \frac{1.11\times10^{-19}\ \text{V m}^{-1}}{(r/\text{m})^3} = \frac{1.11\times10^{8}\ \text{V m}^{-1}}{(r/\text{nm})^3}$$

(a) $\mathcal{E} = \boxed{1.1\times10^8\ \text{V m}^{-1}}$ when $r = 1.0\ \text{nm}$

(b) $\mathcal{E} = \dfrac{1.11\times10^8\ \text{V m}^{-1}}{0.3^3} = \boxed{4\times10^9\ \text{V m}^{-1}}$ for $r = 0.3\ \text{nm}$

(c) $\mathcal{E} = \dfrac{1.11\times10^8\ \text{V m}^{-1}}{30^3} = \boxed{4\ \text{kV m}^{-1}}$ for $r = 30\ \text{nm}$.

P8.3 The point charge model can be used to estimate the magnitude of the electric dipole moment of hydrogen peroxide as a function of φ (defined in Figure 8.6(b) as a view down the z-axis of the O–O bond). Each hydrogen atom has a partial charge of δ; each oxygen atom has a partial charge of $-\delta$. The dipole moment magnitude is

$$\mu = (\mu\cdot\mu)^{1/2} = \left(\mu_x^{\,2}+\mu_y^{\,2}+\mu_z^{\,2}\right)^{1/2} \quad \text{where} \quad \mu_x = \sum_\text{J} q_\text{J}x_\text{J} = \delta\times\left\{x_{\text{H}_1} - x_{\text{O}_1} - x_{\text{O}_2} + x_{\text{H}_2}\right\}\ [8.3\text{a}], \text{ etc.}$$

We will use the Cartesian coordinate system defined in Figure 8.6(a). The bond lengths are $l_\text{OH} = 97$ pm and $l_\text{OO} = 149$ pm. We also use the ratio $l_\text{ratio} = l_\text{OO}\ /\ l_\text{OH} = 1.54$ and calculate μ in units of δl_OH so that it is unnecessary to estimate the magnitude of δ (though an estimate is found in Table 8.2). The O–O–H bond angle may be estimated as 90°, but we will use the experimental value of 100°. The computations of μ_x, μ_y, and μ_z require the coordinates of each atom; those of H_1 and the oxygen atoms are shown in Figure 8.6(a).

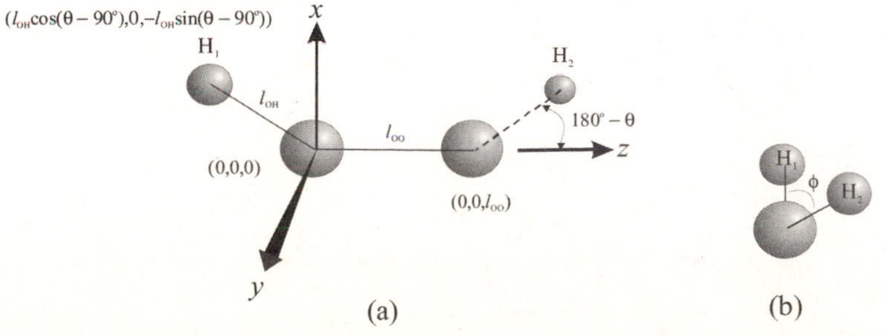

Figure 8.6

The coordinates of H_2 can be determined by analogy to the relationships between Cartesian coordinates and spherical polar coordinates. They are

$$x = l_{OH}\sin(180° - \theta)\cos\phi$$

$$y = l_{OH}\sin(180° - \theta)\sin\phi$$

$$z = l_{OO} + l_{OH}\cos(180° - \theta)$$

Substitution of variables into eqn. 8.3(a) yields

$$(\mu/\delta l_{OH})^2 = (\mu_x/\delta l_{OH})^2 + (\mu_y/\delta l_{OH})^2 + (\mu_z/\delta l_{OH})^2$$

$$= \{\cos(10°) + \sin(80°)\cos\phi\}^2 + \{\sin(80°)\sin\phi\}^2 + \{-\sin(10°) - l_{ratio} + l_{ratio} + \cos(80°)\}^2$$

$$= \{\cos(10°) + \sin(80°)\cos\phi\}^2 + \{\sin(80°)\sin\phi\}^2 + \{-\sin(10°) + \cos(80°)\}^2$$

We now draw a table to calculate $(\mu/\delta l_{OH})^2$ in φ increments of 15° and subsequently calculate $\mu/\delta l_{OH}$ values at each φ. Figure 8.7 is a plot of the variation. As expected, there the dipole is a maximum of almost twice the single O–H bond dipole when the hydrogen atoms are eclipsed, and it is zero when they have a gauche conformation.

ϕ / deg	ϕ / radians	sq(μ / δl)	μ / δl
0	0	3.879385	1.969616
15	0.261799	3.813292	1.952765
30	0.523599	3.619516	1.902502
45	0.785398	3.311262	1.819687
60	1.047198	2.909539	1.705737

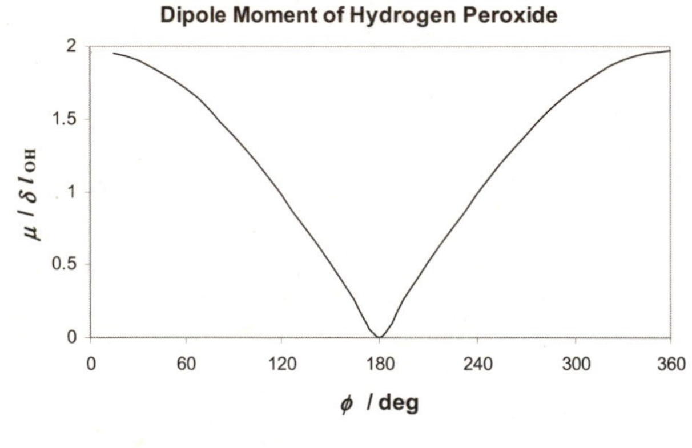

Figure 8.7

P8.5 The energy of the dipole is $-\mu_1\mathcal{E}$ in the electric field $\mathcal{E}$. To flip it over requires a change in energy of $2\mu_1\mathcal{E}$. This will occur when the energy of interaction of the dipole with the induced dipole of the Ar atom equals $2\mu_1\mathcal{E}$. The magnitude of the dipole–induced dipole interaction is

$$V = \frac{\mu_1^2\alpha_2'}{4\pi\varepsilon_0 r^6}\ [8.13] = 2\mu_1\mathcal{E}\ \text{[after flipping over]}$$

$$r^6 = \frac{\mu_1\alpha_2'}{8\pi\varepsilon_0\mathcal{E}} = \frac{(6.17\times10^{-30}\ \text{C m})\times(1.66\times10^{-30}\ \text{m}^3)}{(8\pi)\times(8.854\times10^{-12}\ \text{J}^{-1}\ \text{C}^2\ \text{m}^{-1})\times(1.0\times10^3\ \text{V m}^{-1})} = 4.6\times10^{-53}\ \text{m}^6$$

$$r = 1.9\times10^{-9}\ \text{m} = \boxed{1.9\ \text{nm}}$$

COMMENT. This distance is about 11 times the radius of the Ar atom.

P8.7 The induced dipole moment μ^* is given by

$$\mu^* = \alpha\mathcal{E}\ [8.10] = 4\pi\varepsilon_0\alpha'\mathcal{E}\ [8.11] = \frac{4\pi\varepsilon_0\alpha' e}{4\pi\varepsilon_0 r^2} = \frac{\alpha' e}{r^2}$$

Consequently, the dipole-proton distance needed to induce a particular dipole is

$$r = \left(\frac{\alpha' e}{\mu^*}\right)^{1/2}$$

$$= \left(\frac{(1.48\times10^{-30}\ \text{m}^3)\times(1.602\times10^{-19}\ \text{C})}{(1.85\ \text{D})\times(3.336\times10^{-30}\ \text{C m D}^{-1})}\right)^{1/2} = \boxed{196\ \text{pm}}$$

P8.9 Let the partial charge on the carbon atom equal δe and the N-to-C distance equal l. Then

$$\mu = \delta el\ [8.3a] \quad \text{or} \quad \delta = \frac{\mu}{el}$$

$$\delta = \frac{(1.77\ \text{D})\times(3.3356\times10^{-30}\ \text{C m D}^{-1})}{(1.602\times10^{-19}\ \text{C})\times(299\times10^{-12}\ \text{m})} = \boxed{0.123}$$

P8.11 By the law of cosines, $r_{\text{O}\cdot\cdot\text{H}}^2 = r_{\text{O-H}}^2 + r_{\text{O-O}}^2 - 2r_{\text{O-H}}r_{\text{O-O}}\cos\theta$. Therefore,

$$r_{\text{O}\cdot\cdot\text{H}} = f(\theta) = \left(r_{\text{O-H}}^2 + r_{\text{O-O}}^2 - 2r_{\text{O-H}}r_{\text{O-O}}\cos\theta\right)^{1/2}$$

$$V_\text{m} = \frac{N_\text{A}e^2}{4\pi\varepsilon_0}\left\{\frac{\delta_\text{O}\delta_\text{H}}{r_\text{O-H}} + \frac{\delta_\text{O}\delta_\text{H}}{r_{\text{O}\cdot\cdot\text{H}}} + \frac{\delta_\text{O-O}^2}{r_\text{O-O}}\right\} = \frac{N_\text{A}e^2}{4\pi\varepsilon_0}\left\{\delta_\text{O}\delta_\text{H}\left(\frac{1}{r_\text{O-H}} + \frac{1}{f(\theta)}\right) + \frac{\delta_\text{O-O}^2}{r_\text{O-O}}\right\}$$

$$= \frac{N_\text{A}e^2}{4\pi\varepsilon_0\times(10^{-12}\ \text{m})}\left\{\delta_\text{O}\delta_\text{H}\left(\frac{1}{r_\text{O-H}/\text{pm}} + \frac{1}{f(\theta)/\text{pm}}\right) + \frac{\delta_\text{O-O}^2}{r_\text{O-O}/\text{pm}}\right\}$$

$$= (139\ \text{MJ mol}^{-1})\times\left\{\delta_\text{O}\delta_\text{H}\left(\frac{1}{r_\text{O-H}/\text{pm}} + \frac{1}{f(\theta)/\text{pm}}\right) + \frac{\delta_\text{O-O}^2}{r_\text{O-O}/\text{pm}}\right\}$$

With $\delta_\text{O} = -0.83$, $\delta_\text{H} = 0.45$, $r_\text{O-H} = 95.7$ pm, and $r_\text{O-O} = 200$ pm, we draw up a tabular computation of $f(\theta)$ and $V_\text{m}(\theta)$ over the range $0 \le \theta \le 2\pi$ and plot $V_\text{m}(\theta)$ in Figure 8.8. As expected, the potential is a minimum when $\theta = 0$ because at that angle the hydrogen lies directly between the two oxygen atoms, which repel.

‾ / deg	‾ / radian	f(‾)	V / kJ/mol
0	0	104.30	−561
15	0.261799	110.38	−534
30	0.523599	126.52	−474
45	0.785398	148.63	−413
60	1.047198	173.26	−363
75	1.308997	198.12	−326
90	1.570796	221.72	−298
105	1.832596	243.04	−277
120	2.094395	261.34	−262
135	2.356194	276.09	−252
150	2.617994	286.90	−245
165	2.879793	293.49	−241
180	3.141593	295.70	−239

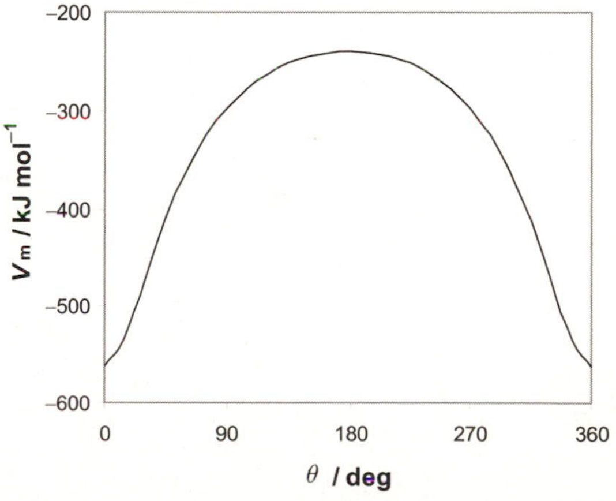

Figure 8.8

P8.13 (a) $$p = \frac{nRT}{V} = \frac{(10.0\,\text{mol})\times(0.08206\ \text{dm}^3\ \text{atm}\,\text{K}^{-1}\ \text{mol}^{-1})\times(300\,\text{K})}{4.860\,\text{dm}^3} = \boxed{50.7\,\text{atm}}$$

(b) $$p = \frac{nRT}{V-nb} - a\left(\frac{n}{V}\right)^2 \quad [8.24a]$$

$$= \frac{(10.0\,\text{mol})\times(0.08206\ \text{dm}^3\,\text{atm}\,\text{K}^{-1}\text{mol}^{-1})\times(300\,\text{K})}{(4.860\,\text{dm}^3)-(10.0\,\text{mol})\times(0.06380\,\text{dm}^3\,\text{mol}^{-1})} - (5.489\,\text{dm}^6\text{atm}\,\text{mol}^{-2})\times\left(\frac{10.0\,\text{mol}}{4.860\,\text{dm}^3}\right)^2$$

$$= \boxed{35.1\ \text{atm}}$$

The compression factor is calculated from its definition [8.20 and 8.21] after inserting $V_\text{m} = \frac{V}{n}$.

To complete the calculation of Z, a value for the pressure, p, is required. The implication in the definition [8.21] is that p is the actual pressure as determined experimentally. This pressure is neither the perfect gas pressure nor the van der Waals pressure. However, on the assumption that the van der Waals equation provides a value for the pressure close to the experimental value, we can calculate the compression factor as follows:

$$Z = \frac{pV}{nRT} = \frac{(35.1\,\text{atm})\times(4.860\,\text{dm}^3)}{(10.0\,\text{mol})\times(0.08206\,\text{dm}^3\ \text{atm}\,\text{K}^{-1}\ \text{mol}^{-1})\times(300\,\text{K})} = \boxed{0.693}$$

COMMENT. If the perfect gas pressure had been used, Z would have been 1, the perfect gas value.

P8.15 From the definition of Z [8.21] and the virial equation [8.22b], Z may be expressed in virial form as

$$Z = 1 + B\left(\frac{1}{V_\text{m}}\right) + C\left(\frac{1}{V_\text{m}}\right)^2 + \cdots$$

Since $V_\text{m} = \frac{RT}{p}$ [assumption of perfect gas], $\frac{1}{V_\text{m}} = \frac{p}{RT}$; hence, on substitution and dropping terms beyond the second power of $\left(\frac{1}{V_\text{m}}\right)$,

$$Z = 1 + B\left(\frac{p}{RT}\right) + C\left(\frac{p}{RT}\right)^2$$

$$Z = 1 + (-21.7\times10^{-3}\ \text{dm}^3\ \text{mol}^{-1})\times\left(\frac{100\,\text{atm}}{(0.0821\,\text{dm}^3\ \text{atm}\,\text{K}^{-1}\text{mol}^{-1})\times(273\,\text{K})}\right)$$

$$+ (1.200\times10^{-3}\ \text{dm}^6\ \text{mol}^{-2})\times\left(\frac{100\,\text{atm}}{(0.0821\,\text{dm}^3\ \text{atm}\,\text{K}^{-1}\ \text{mol}^{-1})\times(273\,\text{K})}\right)^2$$

$$Z = 1 - (0.0968) + (0.0239) = \boxed{0.927}$$

$$V_\text{m} = (0.927)\times\left(\frac{RT}{p}\right) = (0.927)\times\left(\frac{(0.0821\,\text{dm}^3\ \text{atm}\,\text{K}^{-1}\ \text{mol}^{-1})\times(273\,\text{K})}{100\,\text{atm}}\right) = \boxed{0.208\ \text{dm}^3}$$

Question. What is the value of Z obtained from the next approximation using the value of V_m just calculated? Which value of Z is likely to be more accurate?

P8.17 The virial equation is

$$pV_\text{m} = RT\left(1 + \frac{B}{V_\text{m}} + \frac{C}{V_\text{m}^2} + \cdots\right) \quad \text{or} \quad \frac{pV_\text{m}}{RT} = 1 + \frac{B}{V_\text{m}} + \frac{C}{V_\text{m}^2} + \cdots$$

(a) If we assume that the series may be truncated after the B term, then a plot of $\frac{pV_m}{RT}$ versus $\frac{1}{V_m}$ will be linear with slope B and a y-intercept equal to 1. Transforming the data gives

p/MPa	V_m / (dm^3 mol^{-1})	pV_m/RT	V_m^{-1} / (mol dm^{-3})
0.4000	6.2208	0.9976	0.1608
0.5000	4.9736	0.9970	0.2011
0.6000	4.1423	0.9964	0.2414
0.8000	3.1031	0.9952	0.3223
1.000	2.4795	0.9941	0.4033
1.500	1.6483	0.9912	0.6067
2.000	1.2328	0.9885	0.8112
2.500	0.98357	0.9858	1.017
3.000	0.81746	0.9832	1.223
4.000	0.60998	0.9782	1.639

A plot of the data in the third column against the data in the fourth column is shown in Figure 8.9. The data fit a straight line reasonably well, and the y-intercept is very close to 1. The regression yields $B = \boxed{-1.32\times10^{-2}\ \text{dm}^3\ \text{mol}^{-1}}$.

(b) A quadratic function fits the data somewhat better (Figure 8.10) with a slightly better correlation coefficient and a y-intercept closer to 1. This fit implies that truncation of the virial series after the term with C is more accurate than after just the B term. The regression then yields

$$B = \boxed{-1.51\times10^{-2}\ \text{dm}^3\ \text{mol}^{-1}} \quad \text{and} \quad C = \boxed{1.07\times10^{-3}\ \text{dm}^6\ \text{mol}^{-2}}$$

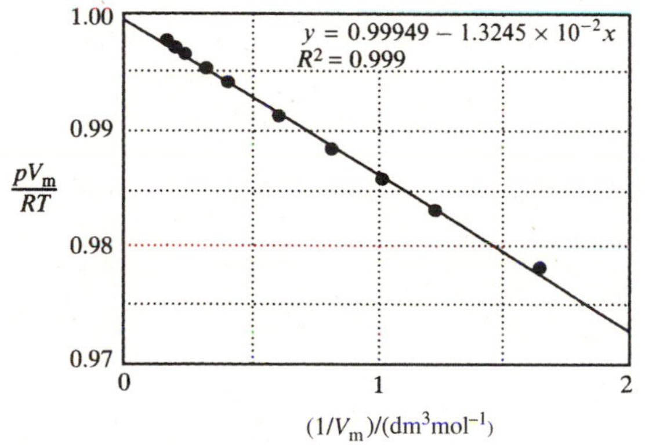

Figure 8.9

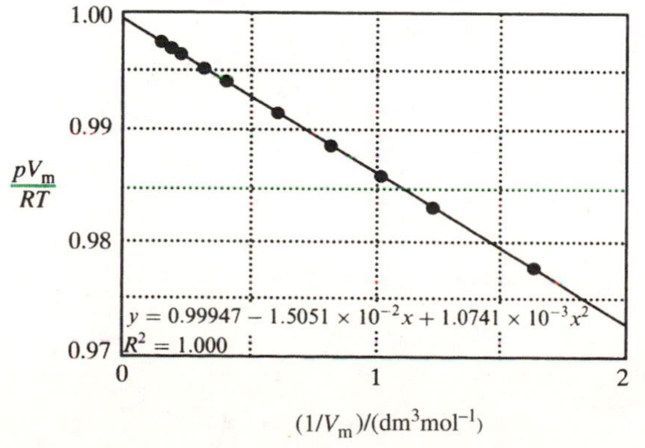

Figure 8.10

P8.19 (a) The classical perturbation energy for interaction between the vector dipole $\vec{\mu}$ and an applied vector electric field $\vec{\mathcal{E}}$ is given by the dot (scalar) product expression

$$H_{\text{perturbation}} = -\vec{\mu}\cdot\vec{\mathcal{E}} = -\{\mu_x\mathcal{E}_x + \mu_y\mathcal{E}_y + \mu_z\mathcal{E}_z\}$$

The dipole moment components of a one-dimensional (x) oscillator are (μ_x,0, 0). For a perpendicular field the components are something like (0, $\mathcal{E}_y$, 0). The dot product of these vectors is clearly zero, meaning that a perpendicular field cannot perturb the one-dimensional oscillator. Eqn. 8.10, $\vec{\mu}^* = \alpha\vec{\mathcal{E}}$, addresses the question as to whether or not the applied field can induce a dipole. The induced response is in the direction of the applied field and the response is the same irrespective of the direction of the field. The constant of proportionality can be called the isotropic polarizability. In this problem, however, we ask whether $\mathcal{E}_y$ can elicit the response μ_x^*. An affirmative answer requires an anisotropic material and polarizability; label it α_{xy}, for which $\mu_x^* = \alpha_{xy}\mathcal{E}_y$. For a one-dimensional harmonic oscillator, $\boxed{\alpha_{xy}\text{ must equal zero}}$ because there is no possible coupling of electron motion in the perpendicular directions.

(b) Eqn.8.12 is used to calculate the polarizability when the applied field is parallel to the one-dimensional oscillator, which is chosen to be the x direction in part (a).

$$\alpha = 2\sum_{n\neq 0}\frac{|\mu_{x,0n}|^2}{E_n^{(0)} - E_0^{(0)}} \sim 2\frac{|\mu_{x,01}|^2}{E_1^{(0)} - E_0^{(0)}}$$

$$\sim \frac{2}{\hbar\omega}\left\{e\left(\frac{\hbar}{2m\omega}\right)^{1/2}\right\}^2 \quad [2.21] = \frac{2e^2}{2m\omega^2} = \frac{e^2}{m(k/m)} \quad [2.21]$$

$$\sim \boxed{\frac{e^2}{k}}$$

The mass independence of the polarizability arises from the fact that the static (zero-frequency) polarizability is a response to the stationary electric field and does not depend on the inertial properties of the oscillator (the rate at which it responds to a changing force).

P8.21 Consider a single molecule surrounded by $N - 1(\approx \text{N})$ others in a container of volume V. The number of molecules in a spherical shell of thickness dR is $4\pi R^2 \times \frac{N}{V}\text{d}R$. Molecules cannot approach more closely than the molecular diameter d so $R \geq d$, and therefore the pair interaction energy is

$$u = \int_d^\infty 4\pi R^2 \times \left(\frac{N}{V}\right)\times\left(\frac{-C_6}{R^6}\right)\text{d}R = \frac{-4\pi NC_6}{V}\int_d^\infty \frac{\text{d}R}{R^4} = \left(\frac{4\pi NC_6}{3V}\right)\times\left(\frac{1}{\infty^3} - \frac{1}{d^3}\right)$$

$$= \frac{-4\pi NC_6}{3Vd^3}$$

The mutual pairwise interaction energy of all N molecules is $U = \frac{1}{2}Nu$ (the $\frac{1}{2}$ appears because each pair must be counted only once, i.e., A with B but not A with B and B with A). Therefore,

$$U = \boxed{\frac{-2\pi N^2 C_6}{3Vd^3}}$$

For a van der Waals gas, $\frac{n^2 a}{V^2} = \left(\frac{\partial U}{\partial V}\right)_T = \frac{2\pi N^2 C_6}{3V^2 d^3}$,

and therefore $a = \boxed{\frac{2\pi N_A^2 C_6}{3d^3}}$ $[N = nN_A]$.

P8.23 $$p=\frac{RT}{V_m-b}-\frac{a}{V_m^2}\quad[8.24b]$$

Factoring out $\frac{RT}{V_m}$ yields $p=\frac{RT}{V_m}\left\{\frac{1}{1-b/V_m}-\frac{a}{RTV_m}\right\}$

$$Z=\frac{pV_m}{RT}=\frac{1}{(1-b/V_m)}-\frac{a}{RTV_m}$$

which upon the Taylor series expansion $\left(1-\frac{b}{V_m}\right)^{-1}=1+\frac{b}{V_m}+\left(\frac{b}{V_m}\right)^2+\cdots$ yields

$$Z=1+\left(b-\frac{a}{RT}\right)\times\left(\frac{1}{V_m}\right)+b^2\left(\frac{1}{V_m}\right)^2+\cdots$$

We note that all terms beyond the second are necessarily positive, so only if

$$\frac{a}{RTV_m}>\frac{b}{V_m}+\left(\frac{b}{V_m}\right)^2+\cdots$$

can Z be less than one. If we ignore terms beyond $\frac{b}{V_m}$, the conditions are simply stated as

$$Z<1\quad\text{when}\ \frac{a}{RT}>b\qquad Z>1\quad\text{when}\ \frac{a}{RT}<b$$

Thus, $Z < 1$ when attractive forces predominate, and $Z > 1$ when size effects (short-range repulsions) predominate.

P8.25 For a real gas we may use the virial expansion in terms of p [8.22a]:

$$p=\frac{nRT}{V}(1+B'p+\cdots)=\rho\frac{RT}{M}(1+B'p+\cdots)$$

which rearranges to $\frac{p}{\rho}=\frac{RT}{M}+\frac{RTB'}{M}p+\cdots$

Therefore, if terms that are second order and higher in p are negligibly small, $\frac{p}{\rho}=\frac{RT}{M}+\frac{RTB'}{M}p$. This has the linear form $\frac{p}{\rho}=\text{intercept}+\text{slope}\times p$ where the intercept equals $\frac{RT}{M}$ and the slope equals $\frac{RTB'}{M}$.

Thus, the slope is proportional to B'.

P8.27 Refer to Figure 8.11.

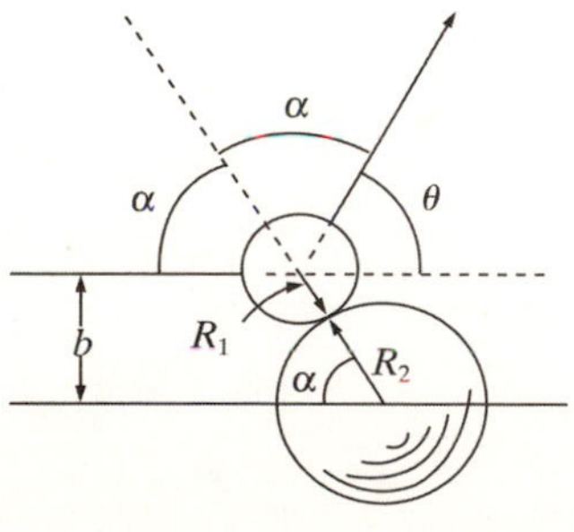

Figure 8.11

The scattering angle is $\theta = \pi - 2\alpha$ if specular reflection occurs in the collision (angle of impact equal to angle of departure from the surface). For $b \le R_1 + R_2$, $\sin\alpha = \dfrac{b}{R_1 + R_2}$.

$$\theta = \begin{cases} \pi - 2\arcsin\left(\dfrac{b}{R_1+R_2}\right) & b \le R_1 + R_2 \\ 0 & b > R_1 + R_2 \end{cases}$$

The function is plotted in the Figure 8.12.

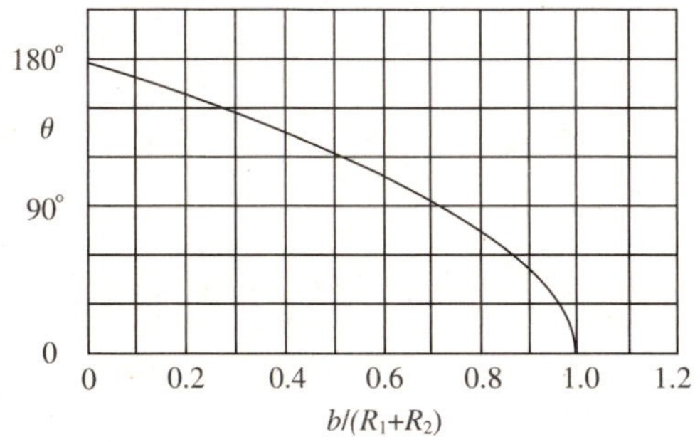

Figure 8.12

P8.29 The number of molecules in a volume element dτ is $\dfrac{N\,\mathrm{d}\tau}{V} = \mathrm{N}\,\mathrm{d}\tau$. The energy of interaction of these molecules with one at a distance R is $V(R)\mathrm{N}\,\mathrm{d}\tau$. The total interaction energy, taking into account the entire sample volume, is therefore

$$u = \int V(R)\mathrm{N}\,\mathrm{d}\tau = \mathrm{N}\int V(R)\,\mathrm{d}\tau \quad [V(R)\text{ is the interaction energy, not the volume}]$$

The total interaction energy of a sample of N molecules is ½Nu (the ½ is included to avoid double counting), so the cohesive energy density is

$$\mathrm{U} = -\frac{U}{V} = \frac{-\frac{1}{2}Nu}{V} = -\frac{1}{2}\mathrm{N}u = -\frac{1}{2}\mathrm{N}^2\int V(R)\,\mathrm{d}\tau$$

For $V(R) = -\dfrac{C_6}{R^6}$ and $\mathrm{d}\tau = 4\pi R^2\,\mathrm{d}R$,

$$-\frac{U}{V} = 2\pi\mathrm{N}^2 C_6\int_d^\infty \frac{\mathrm{d}R}{R^4} = \frac{2\pi}{3}\times\frac{\mathrm{N}^2 C_6}{d^3}$$

However, $\mathrm{N} = \dfrac{N_A\rho}{M}$, where M is the molar mass; therefore

$$\mathrm{U} = \boxed{\left(\frac{2\pi}{3}\right)\times\left(\frac{N_A\rho}{M}\right)^2\times\left(\frac{C_6}{d^3}\right)}$$

Solutions to applications

P8.31 The interaction is a dipole-induced-dipole interaction. The energy is given by eqn. 8.13:

$$V = -\frac{\mu_1^2\alpha_2'}{4\pi\varepsilon_0 r^6} = -\frac{[(2.7\,\mathrm{D})(3.336\times10^{-30}\,\mathrm{C\,m\,D^{-1}})]^2(1.04\times10^{-29}\,\mathrm{m^3})}{4\pi(8.854\times10^{-12}\,\mathrm{J^{-1}\,C^2\,m^{-1}})(4.0\times10^{-9}\,\mathrm{m})^6}$$

$$V = \boxed{-1.8\times10^{-27}\,\mathrm{J} = -1.1\times10^{-3}\,\mathrm{J\,mol^{-1}}}.$$

COMMENT. This value seems exceedingly small. The distance suggested in the problem may be too large compared to typical values. A distance of 0.40 nm, yields $V = -1.1\ \mathrm{kJ\ mol^{-1}}$.

P8.33 (a) The left side molecule in Figure 8.13 is methyladenine. Please note that we have taken the liberty of placing the methyl group in the position that would be occupied by a sugar in RNA and DNA. The wavefunction, structure, and atomic electrostatic charges (shown in the figure) calculation was performed with Spartan '06™ using a Hartree-Fock procedure with a 6-31G* basis set. The atomic electrostatic charge (ESP) numerical method generates charges that reproduce the electrostatic field from the entire wavefunction. The right side molecule in Figure 8.13 is methylthymine.

(b) The two molecules will hydrogen bond into a stable dimer in an orientation for which hydrogen bonding is linear, maximized, and steric hindrance is avoided. We expect hydrogen bonds of the type N–H·····O and N–H·····N with the N and O atoms having large negative electrostatic charges and the H atoms having large positive charges. These atoms are evident in the figure.

(c) Figure 8.13 shows one of three arrangements of hydrogen bonding between the two molecules. Another can be drawn by rotating methylthymine over the top of methyladenine and a third involves rotation to the bottom. The dashed lines show the alignments of two strong hydrogen bonds between the molecules.

(d) The A-to-T base pairing shown in Figure 8.13 has the largest charges in the most favorable positions for strong hydrogen bonding. Also the N-to-O distance of one hydrogen bond equals the N-to-N distance of the other, a favorable feature in RNA and DNA polymers where this pairing and alignment is observed naturally.

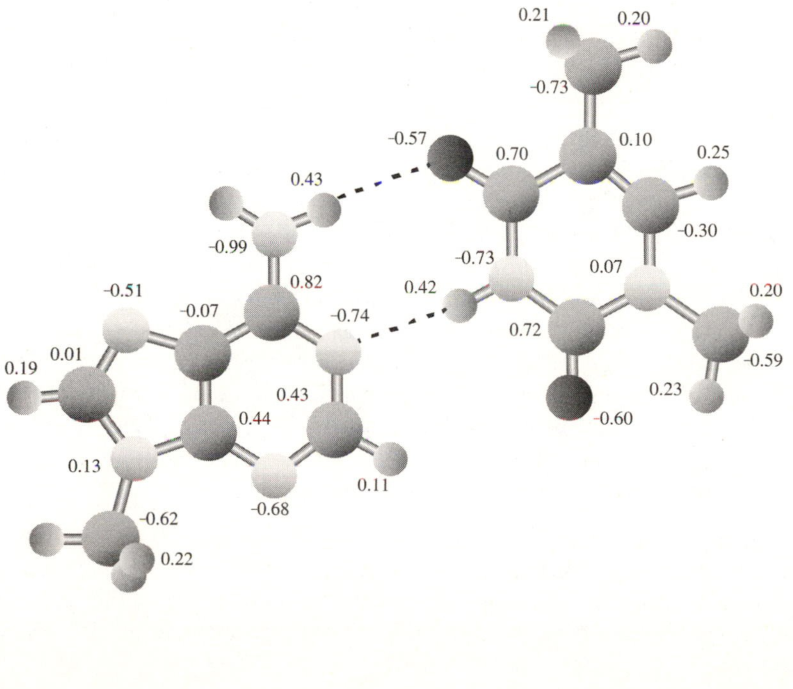

Figure 8.13

(e) The favorable orientation for hydrogen bonding between methylguanine (left side molecule) and methylcytosine (right side molecule) is shown in Figure 8.14. Large counter charges align within distances that result in three strong hydrogen bonds.

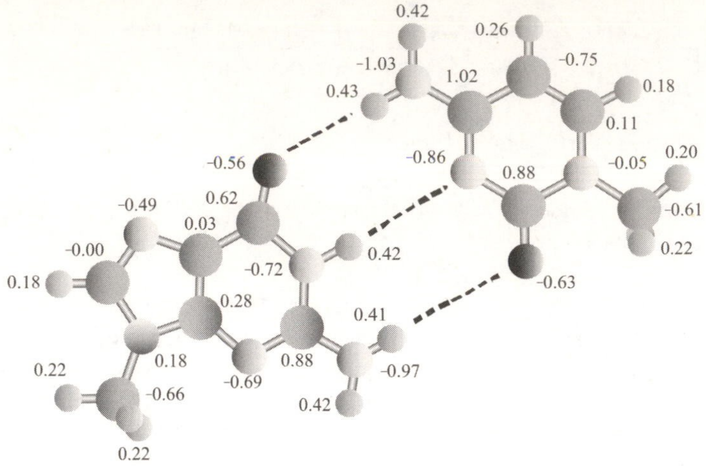

Figure 8.14

P8.35 Molecular mechanics computations with the AMBER force field using the HyperChem package are reported below. The value of the total potential energy will vary between different force fields, as will the shape of the potential energy surface. The local energy minimum at $\phi = -179.6°$ and $\psi = -4.1°$ is found to have a potential energy equal to 28.64 kJ mol^{-1} when R = H. This value is used as a reference to calculate energy differences (ΔE) on the potential energy surface. ΔE values give the relative stability of different conformations with higher values indicating energetically unstable conformations. Similarly, ΔE values were calculated with respect to the local energy minimum at $\phi = -152.3°$ and $\psi = 163.2°$ when R = CH_3.

	Initial		Optimized			
	ϕ / °	ψ / °	ϕ / °	ψ / °	E/(kJ mol^{-1})	ΔE/(kJ mol^{-1})
(a) R = H	75	–65	–176.0	8.3	28.765	0.126
	180	180	180	180	32.154	3.515
	65	35	–179.6	–4.1	28.639	0.000
(b) R = CH_3	75	–65	54.5	19.7	46.338	7.531
	180	180	–152.3	163.2	38.807	0.000
	65	35	52.9	24.1	46.250	7.443

The computations were set up by using the software's "model build" command—that is, initially setting default values for bond lengths and angles except for the specified initial values of ϕ and ψ. Care must be taken to build the proper chirality at the central carbon when R = CH_3. Then the constraints were removed, and the entire structure was allowed to relax to a minimum energy. Not all the initial conformations relaxed to the same final conformation. The different final conformations appear to represent local energy minima. It ought not to be surprising that there are several such minima in even a short peptide chain that contains several nearly free internal rotations. It is instructive to compare the all-trans ($\phi = \psi = 180°$) initial conformations in the R = H and R = CH_3 cases. In the former, neither angle changes, but the resulting structure is not the lowest energy structure. In the latter, the methyl group appears to push the planes of the peptide link away from each other (albeit not far) due to steric effects; however, the resulting energy is lower than that of the other conformations examined.

Two of the initial conformations of each molecule converge to the same energy minimum. These energy wells are rather broad, and the exact angle at which the computation stops within the minimum depends on details of convergence criteria used in the iterative methodology of the software as well as details of the force field. Both sets of computations also found a second local energy minimum.

An alternative method for studying the energy dependence on ϕ and ψ involves a method like that specified above but with the AMBER computation performed at fixed values of both angles. Figure 8.15 summarizes a set of computations with $-180° < \phi < 180°$ and $\psi = 90°$. To characterize the energy surface, one would carry out similar calculations for several values of ψ.

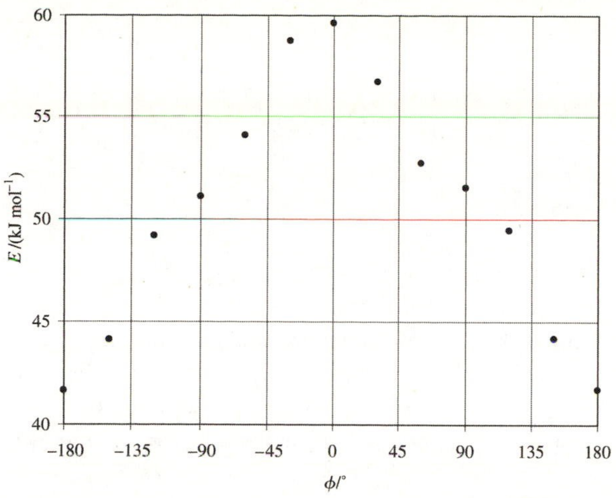

Figure 8.15

9 Solids

Answers to discussion questions

D9.1 A **space lattice** is the three-dimensional structural pattern formed by lattice points representing the locations of motifs that may be atoms, molecules, or groups of atoms, molecules, or ions within a crystal. All points of the space lattice have identical environments and they define the crystal structure. The **unit cell** is an imaginary parallelepiped from which the entire crystal structure can be generated, without leaving gaps, using translations of the unit cell alone. Each unit cell is defined in terms of lattice points. The smallest possible unit cell is called the **primitive unit cell**. **Non-primitive unit cells** may exhibit lattice points within the cell, at the cell center, on cell faces, or on cell edges.

D9.3 We can use the Debye–Scherrer powder diffraction method, follow the procedure in Section 9.3(b)–(d), and in particular look for systematic absences in the diffraction patterns. We can proceed through the following sequence:

(1) Measure distances of the lines in the diffraction pattern from the center.
(2) From the known radius of the camera, convert the distances to angles.
(3) Calculate $\sin^2\theta$.
(4) Find the common factor $A = \lambda^2/4a^2$ in $\sin^2\theta = (\lambda^2/4a^2)(h^2+k^2+l^2)$.
(5) Index the lines using $\sin^2\theta / A = h^2+k^2+l^2$.
(6) Look for the systematic absences in (*hkl*). See Fig. 9.22 of the text. Body-centered cubic diffraction lines corresponding to $h+k+l$ that are odd will be absent. For face-centered cubic, only lines for which *h*, *k*, and *l* are either all even or all odd will be present; others will be absent.
(7) Solve $A = \lambda^2/4a^2$ for *a*.

D9.5 There are two large-angle meridional reflection spots in the x-ray diffraction pattern obtained from parallel fibres of B-DNA (Fig. 9.26 of the text). They are reflections off the relatively close planes (340 pm) of nucleotide bases that run perpendicular to the axis of the molecule (see Fig. 9.28 of the text). Between these meridional spots is an X-shaped distribution of reflection spots that originates from planes of longer periodicity (3400 pm; see Fig. 9.27 of the text). The pitch *p* of the helix, the distance *d* between parallel, reflective planes, and the angle α from the meridional of the spots of the X-pattern are shown in Figure 9.1. The x-rays approach the helix on a path almost perpendicular to the molecular axis. Regions of high electron density act like reflective planes with each coil of the helix providing one plane that reflects at a clockwise angle and another that reflects at a counterclockwise angle. A set of parallel planes that reflect the x-rays at the angle α is shown in Figure 9.1; one of the set of parallel planes that reflect the x-rays at the angle $-\alpha$ is also shown. Reflections from each set of parallel planes constructively interfere to produce one of the legs of the crossed spot patterns seen at the center of text Fig. 9.26. Constructive interference requires that $d = p\cos\alpha$. The sequence of spots outward along a leg corresponds to the first-, second-, ... order diffraction (n = 1, 2, ... where by Bragg's law, $n\lambda = 2d\sin\theta$ [9.2a]). Examination of Figure 9.1 reveals that α increases as the pitch increases.

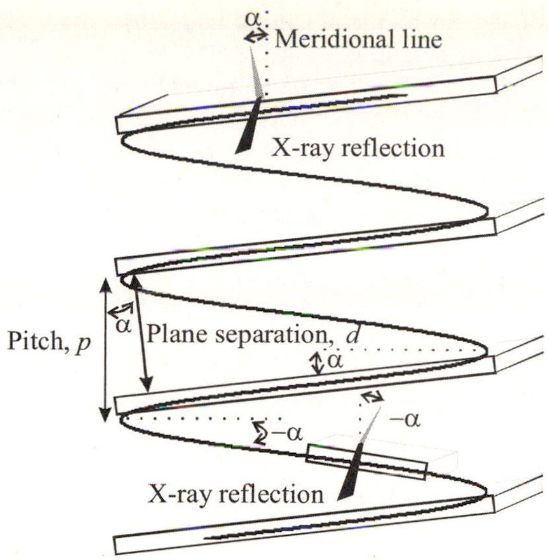

Figure 9.1

D9.7 The **structure factor** F_{hkl} is the sum over all j atoms of terms each of which has a scattering factor f_j:

$$F_{hkl} = \sum_j f_j e^{i\phi_{hkl}(j)} \quad \text{where} \quad \phi_{hkl}(j) = 2\pi\left(hx_j + ky_j + lz_j\right) \quad [9.4]$$

The importance of the structure factor to the x-ray crystallographic method of structure determination is its relationship to the electron density distribution, $\rho(r)$, within the crystal:

$$\rho(\boldsymbol{r}) = \frac{1}{V}\sum_{hkl} F_{hkl} e^{-2\pi i(hx+ky+lz)} \quad [9.5]$$

Eqn. 9.5 reveals that if the structure factors for the lattice planes can be measured, then the electron density distribution can be calculated by performing the indicated sum. Therein lies the **phase problem**. Measurement detectors yield only the intensity of scattered radiation, which is proportional to $|F_{hkl}|^2$, and give no direct information about F_{hkl}. To see this, consider the structure factor form $F_{hkl} = |F_{hkl}| e^{i\alpha_{hkl}}$ where α_{hkl} is the phase of the hkl reflection plane. Then

$$|F_{hkl}|^2 = \left\{|F_{hkl}| e^{i\alpha_{hkl}}\right\}^* \left\{|F_{hkl}| e^{i\alpha_{hkl}}\right\} = |F_{hkl}| \times |F_{hkl}| e^{-i\alpha_{hkl}} e^{i\alpha_{hkl}} = |F_{hkl}| \times |F_{hkl}|$$

and we see that all information about the phase is lost in an intensity measurement. It seems impossible to perform the sum of eqn. 9.5 since we do not have the important factor $e^{i\alpha_{hkl}}$. Crystallographers have developed numerous methods to resolve the phase problem. In the **Patterson synthesis**, x-ray diffraction spot intensities are used to acquire separation and relative orientations of atom pairs. Another method uses the dominance of heavy atom scattering to deduce phase. **Heavy atom replacement** may be necessary for this type of application. **Direct methods** dominate modern x-ray diffraction analysis. These methods use statistical techniques and the considerable computational capacity of the modern computer to compute the probabilities that the phases have a particular value.

D9.9 In a face-centered cubic close-packed lattice, there is an octahedral hole in the centre. The rock-salt structure can be thought of as being derived from an fcc structure of Cl^- ions in which Na^+ ions have filled the octahedral holes.

The cesium chloride structure can be considered to be derived from the ccp structure by having Cl^- ions occupy all the primitive lattice points and octahedral sites, with all tetrahedral sites occupied by Cs^+ ions. This is exceedingly difficult to visualize and describe without carefully constructed figures or models. Refer to S.-M. Ho and B. E. Douglas, *J. Chem. Educ.* **46**, 208 (1969) for the appropriate diagrams.

D9.11 The Davydov splitting in the exciton bands of a crystal can be understood by considering the allowed transitions of interacting dipoles of neighboring molecules. As shown in Fig. 9.58 of the text, the parallel alignment of transition dipoles is energetically unfavorable, and the exciton absorption is shifted to higher energy (blue-shifted) than in the isolated molecules. However, a head-to-tail alignment is energetically favorable, and the transition occurs at a lower frequency (red-shifted) than in the isolated molecules. Transition dipoles are rarely in a single parallel or head-to-tail alignment. As shown in Fig. 9.59 of the text, transition moments may often be either energetically favored or energetically unfavorable. These crystals exhibit both the blue-shifted and the red-shifted band and the separation of the bands is the **Davydov or exciton splitting**.

Solutions to exercises

E9.1(a) The $\boxed{\text{center of each edge}}$ of the body-centered cubic unit cell is equivalent to the point (0,½,0).

E9.2(a) The monoclinic P unit cell is defined by the small spheres of Figure 9.2. The separation between (100) planes is $d = a\cos(\beta - 90°)$.

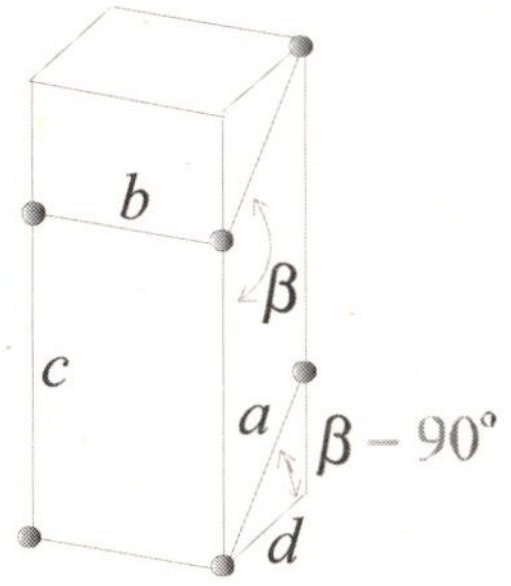

Figure 9.2

$$\begin{aligned} V &= bcd = bc\{a\cos(\beta - 90°)\} \\ &= abc\{\cos(\beta)\cos(90°) + \sin(\beta)\sin(90°)\} \quad \text{(standard trigonometric addition formula)} \\ &= abc\{\cos(\beta)\times 0 + \sin(\beta)\times 1\} \\ &\boxed{V = abc\sin\beta} \end{aligned}$$

E9.3(a) When the axis intersections are ($3a$, $2b$, c) and ($2a, \infty b, \infty c$), the Miller indices of the planes are {236} and {100}. Simply take the inverse of the intersection multiple and clear fractions to find these planes.

E9.4(a) For the cubic unit cell, $d_{hkl} = \dfrac{a}{(h^2 + k^2 + l^2)^{1/2}}$ [9.1b].

$$d_{112} = \frac{a}{(1^2 + 1^2 + 2^2)^{1/2}} = \frac{562\text{ pm}}{(6)^{1/2}} = \boxed{229\text{ pm}}$$

$$d_{110} = \frac{a}{(1^2 + 1^2 + 0^2)^{1/2}} = \frac{562\text{ pm}}{(2)^{1/2}} = \boxed{397\text{ pm}}$$

$$d_{224} = \frac{a}{(2^2 + 2^2 + 4^2)^{1/2}} = \frac{562\text{ pm}}{(24)^{1/2}} = \boxed{115\text{ pm}}$$

E9.5(a) Refer to Fig. 9.22 of the text. Systematic absences correspond to $h+k+l=\text{odd}$. Hence the first three lines are from planes (110), (200), and (211).

$$\sin\theta_{hkl} = \frac{\lambda}{2d_{hkl}}\ [9.2\text{b}] \qquad d_{hkl} = \frac{a}{(h^2+k^2+l^2)^{1/2}}\ [9.1\text{b}]\text{, then}$$

$$\sin\theta_{hkl} = (h^2+k^2+l^2)^{1/2}\times\left(\frac{\lambda}{2a}\right)$$

In a bcc unit cell, the body diagonal of the cube is $4R$ where R is the atomic radius. The relationship of the side of the unit cell to R is therefore (using the Pythagorean theorem twice)

$$(4R)^2 = a^2 + 2a^2 = 3a^2 \quad \text{or} \quad a = \frac{4R}{3^{1/2}}$$

This can be seen from Figure 9.3.

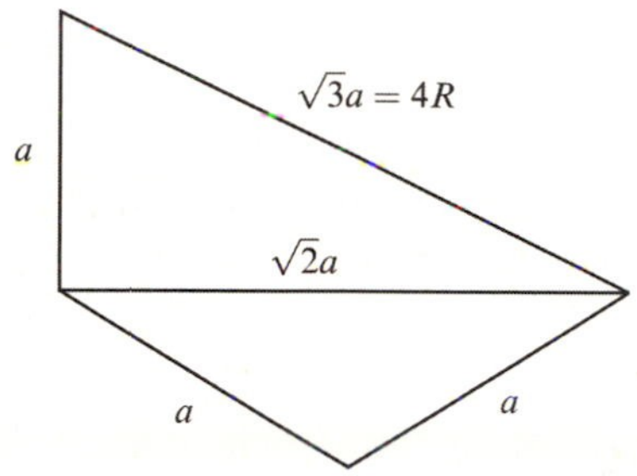

Figure 9.3

$$a = \frac{4\times 126\ \text{pm}}{3^{1/2}} = 291\ \text{pm}$$

$$\frac{\lambda}{2a} = \frac{72\ \text{pm}}{(2)\times(291\ \text{pm})} = 0.12\overline{4}$$

$$\sin\theta_{110} = \sqrt{2}\times(0.12\overline{4}) = 0.17\overline{5} \qquad \theta_{110} = \boxed{10.1°}$$

$$\sin\theta_{200} = (2)\times(0.12\overline{4}) = 0.24\overline{8} \qquad \theta_{200} = \boxed{14.4°}$$

$$\sin\theta_{211} = \sqrt{6}\times(0.12\overline{4}) = 0.30\overline{4} \qquad \theta_{211} = \boxed{17.7°}$$

E9.6(a) $$\theta = \arcsin\frac{\lambda}{2d} \quad [9.2\text{b},\ \arcsin \equiv \sin^{-1}]$$

$$\Delta\theta = \arcsin\frac{\lambda_2}{2d} - \arcsin\frac{\lambda_1}{2d} = \arcsin\left(\frac{154.433\ \text{pm}}{(2)\times(77.8\ \text{pm})}\right) - \arcsin\left(\frac{154.051\ \text{pm}}{(2)\times(77.8\ \text{pm})}\right)$$

$$= \boxed{1.07°} = 0.0187\ \text{rad}$$

E9.7(a) *Justification* 9.1 demonstrates that the scattering factor in the forward direction equals the number of electrons in the atom or simple ion. Consequently, $\boxed{f_{\text{Br}^-} = 36}$.

E9.8(a) $$\rho = \frac{\text{mass of unit cell}}{\text{volume of unit cell}} = \frac{m}{V}$$

$$m = nM = \frac{N}{N_\text{A}}M \ [N \text{ is the number of formula units per unit cell}]$$

Then, $\rho = \dfrac{NM}{VN_\text{A}}$

and $$N=\frac{\rho V N_A}{M}$$

$$=\frac{(3.9\times10^6\ \mathrm{g\,m^{-3}})\times(634)\times(784)\times(516\times10^{-36}\ \mathrm{m^3})\times(6.022\times10^{23}\ \mathrm{mol^{-1}})}{154.77\ \mathrm{g\,mol^{-1}}}=3.9$$

Therefore, $\boxed{N=4}$ and the true calculated density (in the absence of defects) is

$$\rho=\frac{(4)\times(154.77\ \mathrm{g\,mol^{-1}})}{(634)\times(784)\times(516\times10^{-30}\ \mathrm{cm^3})\times(6.022\times10^{23})\ \mathrm{mol^{-1}}}=\boxed{4.01\ \mathrm{g\,cm^{-3}}}$$

E9.9(a) $$d_{hkl}=\left[\left(\frac{h}{a}\right)^2+\left(\frac{k}{b}\right)^2+\left(\frac{l}{c}\right)^2\right]^{-1/2}\quad[9.1c]$$

$$d_{321}=\left[\left(\frac{3}{812}\right)^2+\left(\frac{2}{947}\right)^2+\left(\frac{1}{637}\right)^2\right]^{-1/2}\ \mathrm{pm}=\boxed{220\ \mathrm{pm}}$$

E9.10(a) $$\theta_{hkl}=\arcsin\frac{\lambda}{2d_{hkl}}[9.2b]=\arcsin\left\{\frac{\lambda}{2}\left[\left(\frac{h}{a}\right)^2+\left(\frac{k}{b}\right)^2+\left(\frac{l}{c}\right)^2\right]^{1/2}\right\}[9.1c]$$

$$\theta_{100}=\arcsin\left\{\frac{154}{2}\times\left[\left(\frac{1}{542}\right)^2+\left(\frac{0}{917}\right)^2+\left(\frac{0}{645}\right)^2\right]^{1/2}\right\}=\boxed{8.16^\circ}$$

$$\theta_{010}=\arcsin\left\{\frac{154}{2}\times\left[\left(\frac{0}{542}\right)^2+\left(\frac{1}{917}\right)^2+\left(\frac{0}{645}\right)^2\right]^{1/2}\right\}=\boxed{4.82^\circ}$$

$$\theta_{111}=\arcsin\left\{\frac{154}{2}\times\left[\left(\frac{1}{542}\right)^2+\left(\frac{1}{917}\right)^2+\left(\frac{1}{645}\right)^2\right]^{1/2}\right\}=\boxed{11.7\bar{5}^\circ}$$

E9.11(a) From the discussion of systematic absences (Section 9.3 and Fig. 9.22 of the text) we can conclude that the unit cell is $\boxed{\text{face-centered cubic}}$.

E9.12(a) $$F_{hkl}=\sum_j f_j e^{2\pi i(hx_j+ky_j+lz_j)}\quad[9.4]\ \text{with}\ f_j=\tfrac{1}{8}f\ \text{(each atom is shared by eight cells).}$$

Therefore, $F_{hkl}=\frac{1}{8}f\{1+e^{2\pi ih}+e^{2\pi ik}+e^{2\pi il}+e^{2\pi i(h+k)}+e^{2\pi i(h+l)}+e^{2\pi i(k+l)}+e^{2\pi i(h+k+l)}\}$.

However all the exponential terms equal 1 since h, k, and l are all integers and

$$e^{i\theta}=\cos\theta+i\sin\theta\ [\theta=2\pi h,2\pi k,\ldots]=\cos\theta=1$$

Therefore, $F_{hkl}=\boxed{f}$.

E9.13(a) The structure factor is given by

$$F_{hkl}=\sum_i f_i e^{i\phi_i}\quad\text{where}\quad \phi_i=2\pi(hx_i+ky_i+lz_i)\ [9.4]$$

All eight of the vertices of the cube are shared by eight cubes, so each vertex has a scattering factor of $f/8$ and their total contribution to the structure factor equals f (see E9.12a). The two side-centered (C) ions of the unit cell are shared by two cells, and in this problem each has a scattering factor of $2f$. Their coordinates are (½a, ½a, 0) and (½a, ½a, a). Thus, the contribution of the face-centered ions to the structure factor is

$$\sum_{i\text{(face-centered)}} f_i e^{i\phi_i}=\tfrac{1}{2}(2f)e^{2\pi i(h/2+k/2)}+\tfrac{1}{2}(2f)e^{2\pi i(h/2+k/2+l)}=fe^{\pi i(h+k)}\left\{1+e^{2\pi il}\right\}$$

But, according to Euler's identity ($e^{i\theta} = \cos\theta + i \sin\theta$),

$$e^{\pi i(h+k)} = \cos(\pi\{h+k\}) + i\sin(\pi\{h+k\}) = \cos(\pi\{h+k\}) = (-1)^{h+k}$$

and

$$e^{2\pi i l} = \cos(2\pi l) + i\sin(2\pi l) = \cos(2\pi l) = 1$$

Thus, $\sum\limits_{i(\text{face-centered})} f_i e^{i\phi_i} = f(-1)^{h+k}\{1+1\} = 2(-1)^{h+k} f$

and the structure factor becomes

$$F_{hkl} = \sum_{i(\text{vertices})} f_i e^{i\phi_i} + \sum_{i(\text{face-centered})} f_i e^{i\phi_i} = f + 2(-1)^{h+k} f = f\left\{1 + 2(-1)^{h+k}\right\}$$

$$F_{hkl} = \boxed{3f \text{ for } h+k \text{ even and } -f \text{ for } h+k \text{ odd}}$$

E9.14(a) The electron density is given by

$$\rho(r) = \frac{1}{V}\sum_{hkl} F_{hkl} e^{-2\pi i(hx+ky+lz)} \quad [9.5]$$

The component along the x direction is

$$\rho(x) = \frac{1}{V}\sum_{h} F_h e^{-2\pi i hx}$$

Using the data of this problem, we sum from $h = -9$ to $+9$ and use the relationship $F_h = F_{|h|}$. The following Mathcad computation of $\rho(0.5)$ uses unit volume. Figure 9.4 shows high electron density at the center of a unit cell edge.

$$\mathbf{F} := \begin{pmatrix} 10 \\ -10 \\ 8 \\ -8 \\ 6 \\ -6 \\ 4 \\ -4 \\ 2 \\ -2 \end{pmatrix}$$

$$\rho(x) := \sum_{h=-9}^{9} (F_{|h|} \cdot \cos(2 \cdot \pi \cdot h \cdot x))$$

$$\rho(.5) = 110$$

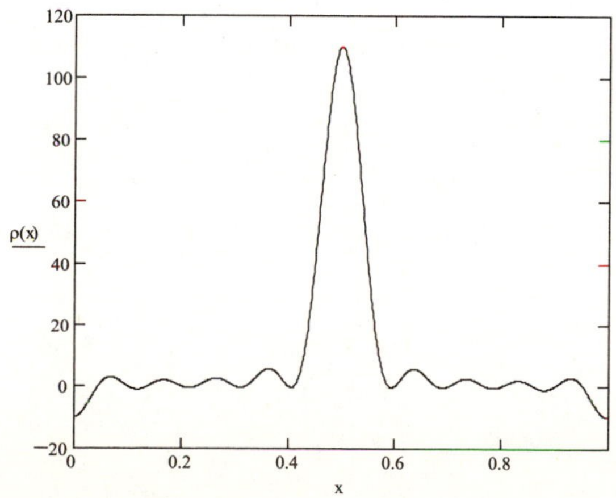

Figure 9.4

E9.15(a) Using the information of E9.14(a), the Mathcad computation of $P(1.0)$ is performed with eqn. 9.6.

$$F := \begin{pmatrix} 10 \\ -10 \\ 8 \\ -8 \\ 6 \\ -6 \\ 4 \\ -4 \\ 2 \\ -2 \end{pmatrix} \qquad P(x) := \sum_{h=-9}^{9} \left[(F_{|h|})^2 \cdot e^{-2\cdot\pi\cdot h\cdot x} \right] \qquad P(1.0) = 780$$

The Patterson synthesis $P(x)$ of Figure 9.5 shows that atoms represented by this data are separated by 1 a unit along the x-axis.

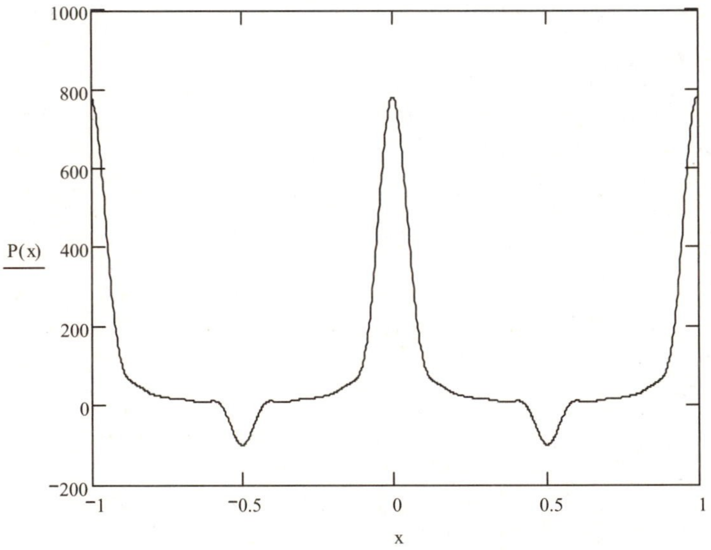

Figure 9.5

E9.16(a) Draw points corresponding to the vectors joining each pair of atoms. Heavier atoms give more intense contribution than light atoms. Remember that there are two vectors joining any pair of atoms ($\overrightarrow{AB}$ and $\overleftarrow{AB}$); don't forget the AA zero vectors for the center point of the diagram. See Figure 9.6.

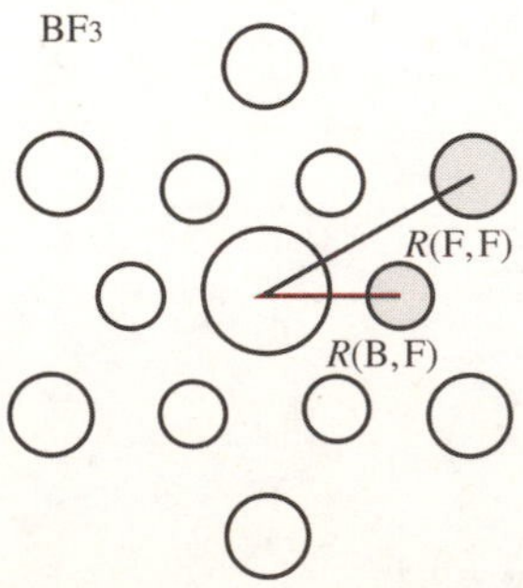

Figure 9.6

E9.17(a) $\lambda = \dfrac{h}{p} = \dfrac{h}{mv}$

Hence, $v = \dfrac{h}{m\lambda} = \dfrac{6.626\times10^{-34}\ \text{J s}}{(1.675\times10^{-27}\ \text{kg})\times(65\times10^{-12}\ \text{m})} = \boxed{6.1\ \text{km s}^{-1}}$

E9.18(a) Combine $E = \frac{1}{2}kT$ and $E = \frac{1}{2}mv^2 = \dfrac{h^2}{2m\lambda^2}$ to obtain

$$\lambda = \frac{h}{(mkT)^{1/2}} = \frac{6.626\times10^{-34}\ \text{J s}}{[(1.675\times10^{-27}\ \text{kg})\times(1.381\times10^{-23}\ \text{J K}^{-1})\times(350\ \text{K})]^{1/2}} = \boxed{233\ \text{pm}}$$

E9.19(a) The hatched area in Figure 9.7 is $h\times2R = 3^{1/2}R\times2R = 2\sqrt{3}R^2$ where $h = 2R\cos30°$. The net number of cylinders in a hatched area is 1, and the area of the cylinder's base is πR^2. The volume of the prism (of which the hatched area is the base) is $2\sqrt{3}R^2L$, and the volume occupied by the cylinders is $\pi R^2 L$. Hence, the packing fraction is

$$f = \frac{\pi R^2 L}{2\sqrt{3}R^2 L} = \frac{\pi}{2\sqrt{3}} = \boxed{0.9069}$$

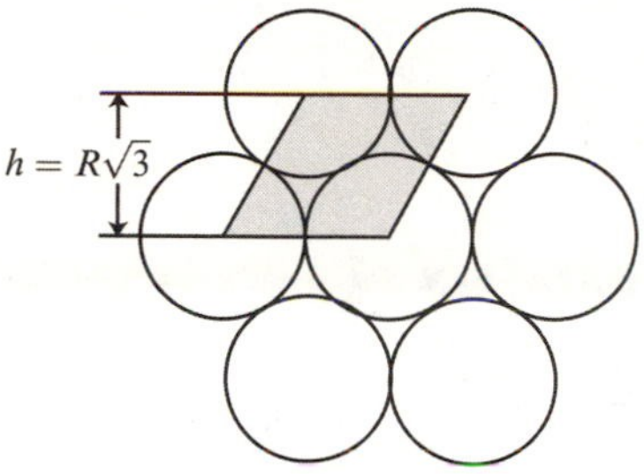

Figure 9.7

E9.20(a) $f = \dfrac{NV_a}{V_c}$ where N is the number of atoms in each unit cell, V_a their individual volumes, and V_c the volume of the unit cell itself. Refer to Figure 9.8 for a view of the primitive unit cell, the bcc unit cell, and the fcc unit cell.

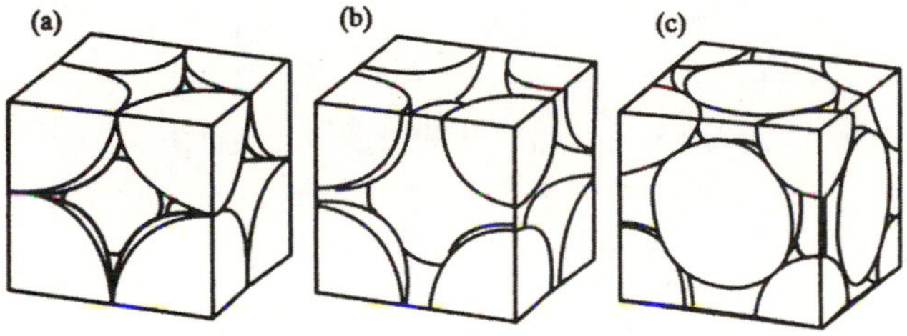

Figure 9.8

(a) For a primitive unit cell, $N = 1$, $V_a = \dfrac{4}{3}\pi R^3$, and $V_c = (2R)^3$.

$$f = \frac{\left(\frac{4}{3}\pi R^3\right)}{(2R)^3} = \frac{\pi}{6} = \boxed{0.5236}$$

(b) For the bcc unit cell, $N = 2$, $V_a = \dfrac{4}{3}\pi R^3$, and $V_c = \left(\dfrac{4R}{\sqrt{3}}\right)^3$ [body diagonal of a unit cube is $4R$].

$$f = \frac{2\times\frac{4}{3}\pi R^3}{\left(\frac{4R}{\sqrt{3}}\right)^3} = \frac{\pi\sqrt{3}}{8} = \boxed{0.6802}$$

(c) For the fcc unit cell, $N = 4$, $V_a = \frac{4}{3}\pi R^3$, and $V_c = (2\sqrt{2}R)^3$.

$$f = \frac{4 \times \frac{4}{3}\pi R^3}{(2\sqrt{2}R)^3} = \frac{\pi}{3\sqrt{2}} = \boxed{0.7405}$$

E9.21(a) (a) For sixfold coordination, see Figure 9.9.

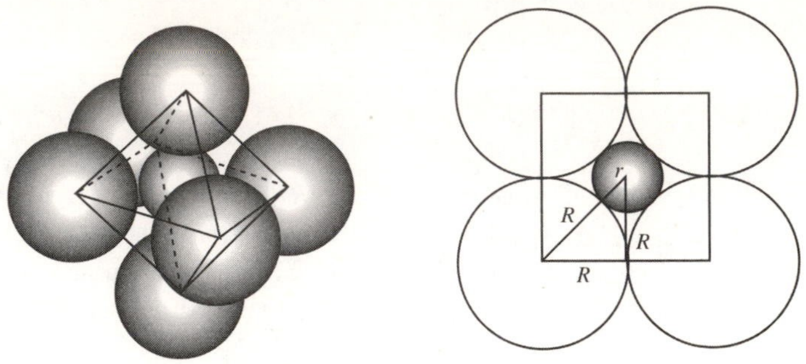

Figure 9.9

We assume that the larger spheres of radius R touch each other and that they also touch the smaller interior sphere. Hence, by the Pythagorean theorem,

$$(R+r)^2 = 2(R)^2 \quad \text{or} \quad \left(1+\frac{r}{R}\right)^2 = 2$$

Thus, $\frac{r}{R} = 0.414$ and $r = 0.414R = (0.414)\times(181\text{ pm}) = \boxed{74.9\text{ pm}}$.

(b) For eightfold coordination, see Figure 9.10.

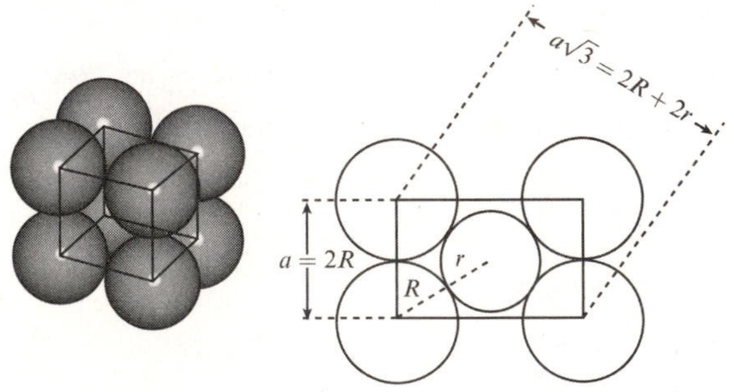

Figure 9.10

The body diagonal of a cube is $a\sqrt{3}$. Hence,

$$a\sqrt{3} = 2R + 2r \quad \text{or} \quad \sqrt{3}R = R + r \quad [a = 2R]$$

$$\frac{r}{R} = 0.732 \quad \text{and} \quad r = 0.732R = (0.732)\times(181\text{ pm}) = \boxed{132\text{ pm}}$$

E9.22(a) The volume change is a result of two partially counteracting factors: (1) different packing fraction (f) and (2) different radii.

$$\frac{V(\text{bcc})}{V(\text{hcp})} = \frac{f(\text{hcp})}{f(\text{bcc})} \times \frac{v(\text{bcc})}{v(\text{hcp})}$$

$f(\text{hcp}) = 0.7405$, $f(\text{bcc}) = 0.6802$ [*Justification* 9.3 and E9.20a]

$$\frac{V(\text{bcc})}{V(\text{hcp})} = \frac{0.7405}{0.6802} \times \frac{(142.5)^3}{(145.8)^3} = 1.016$$

Hence, there is an $\boxed{\text{expansion}}$ of 1.6%.

E9.23(a) The lattice enthalpy is the difference in enthalpy between an ionic solid and the corresponding isolated ions. In this exercise, it is the enthalpy corresponding to the process

$$\mathrm{CaO(s)} \rightarrow \mathrm{Ca^{2+}(g)} + \mathrm{O^{2-}(g)}$$

The standard lattice enthalpy can be computed from the standard enthalpies given in the exercise by considering the formation of CaO(s) from its elements as occurring through the following steps: sublimation of Ca(s), removal of two electrons from Ca(g), atomization of O_2(g), two-electron attachment to O(g), and formation of CaO(s) lattice from gaseous ions. The formation reaction of CaO(s) is

$$\mathrm{Ca(s)} + \tfrac{1}{2}\,\mathrm{O_2(g)} \rightarrow \mathrm{CaO(s)}$$

$$\Delta_f H^\ominus(\mathrm{CaO,s}) = \Delta_{sub} H^\ominus(\mathrm{Ca,s}) + \Delta_{ion} H^\ominus(\mathrm{Ca,g}) + \tfrac{1}{2}\Delta_{bond\ diss} H^\ominus(\mathrm{O_2,g}) + \Delta_{eg} H^\ominus(\mathrm{O,g}) + \Delta_{eg} H^\ominus(\mathrm{O^-,g}) - \Delta_L H^\ominus(\mathrm{CaO,s})$$

So the lattice enthalpy is

$$\Delta_L H^\ominus(\mathrm{CaO,s}) = \Delta_{sub} H^\ominus(\mathrm{Ca,s}) + \Delta_{ion} H^\ominus(\mathrm{Ca,g}) + \tfrac{1}{2}\Delta_{bond\ diss} H^\ominus(\mathrm{O_2,g}) + \Delta_{eg} H^\ominus(\mathrm{O,g}) + \Delta_{eg} H^\ominus(\mathrm{O^-,g}) - \Delta_f H^\ominus(\mathrm{CaO,s})$$

$$\Delta_L H^\ominus(\mathrm{CaO,s}) = [178 + 1735 + \tfrac{1}{2}(497) - 141 + 844 + 635]\ \mathrm{kJ\ mol^{-1}} = \boxed{3500.\ \mathrm{kJ\ mol^{-1}}}$$

E9.24(a) Young's modulus: $E = \dfrac{\text{normal stress}}{\text{normal strain}}\,[9.13a] = 1.2\ \mathrm{GPa} = 1.2\times10^{9}\ \mathrm{kg\ m^{-1}\ s^{-2}}$

We solve this relation for the normal strain after calculating the normal stress from the data provided.

normal stress = force per unit area = F/A

normal strain = relative elongation = $\Delta L/L$

$$\Delta L/L = \frac{F/A}{E} = \frac{mg/A}{E} = \frac{mg}{AE} = \frac{mg}{\pi(d/2)^2 E}$$

$$= \frac{1.5\ \mathrm{kg}\times 9.81\ \mathrm{m\ s^{-2}}}{\pi(2.0\times10^{-3}\ \mathrm{m}/2)^2 \times 1.2\times10^{9}\ \mathrm{kg\ m^{-1}\ s^{-2}}}$$

$= \boxed{0.0039}$ or about 0.4% elongation

E9.25(a) Poisson's ratio: $\nu_p = \dfrac{\text{transverse strain}}{\text{normal strain}}\,[9.15] = 0.45$

We note that the transverse strain is usually a contraction and that it is usually evenly distributed in both transverse directions. That is, if $(\Delta L/L)_z$ is the normal strain, then the transverse strains, $(\Delta L/L)_x$ and $(\Delta L/L)_y$, are equal. In this case of a 2.0% uniaxial stress,

$$\left(\frac{\Delta L}{L}\right)_z = +0.020 \qquad \left(\frac{\Delta L}{L}\right)_x = -0.0090 = \left(\frac{\Delta L}{L}\right)_y$$

$$\text{New volume} = (1-0.0090)^2 \times (1+0.020) \times 1.0\ \mathrm{cm^3} = 1.0017\ \mathrm{cm^3}$$

The change in volume is $\boxed{1.7\times10^{-3}\ \mathrm{cm^3}}$.

E9.26(a) Answer: n-type; the dopant, arsenic, belongs to Group 15, whereas germanium belongs to Group 14.

E9.27(a) $$E_g = h\nu_{min} = \frac{hc}{\lambda_{max}} = \frac{(6.626\times10^{-34}\,\text{J s})(2.998\times10^{8}\,\text{m s}^{-1})}{350.\times10^{-9}\,\text{m}}\left(\frac{1\,\text{eV}}{1.602\times10^{-19}\,\text{J}}\right) = \boxed{3.54\,\text{eV}}$$

E9.28(a) $$m = g_e\{S(S+1)\}^{1/2}\mu_B \quad [9.27, \text{ with } S \text{ in place of } s]$$

Therefore, because $m = 3.81\,\mu_B$ and $g_e \approx 2$,

$$S(S+1) = (\tfrac{1}{4})\times(3.81)^2 = 3.63\text{, implying that } S = 1.47$$

Because $S \approx \frac{3}{2}$, there must be $\boxed{\text{three unpaired spins}}$.

E9.29(a) $$\chi_m = \chi V_m[9.23] = \chi M/\rho = (-7.2\times10^{-7})\times(78.11\,\text{g mol}^{-1})/(0.879\,\text{g cm}^{-3})$$

$$= \boxed{-6.4\times10^{-5}\,\text{cm}^3\,\text{mol}^{-1}} = \boxed{-6.4\times10^{-11}\,\text{m}^3\,\text{mol}^{-1}}$$

E9.30(a) The molar susceptibility is given by

$$\chi_m = \frac{N_A g_e^2\mu_0\mu_B^2\,S(S+1)}{3kT}\,[9.28] \qquad \text{so} \qquad S(S+1) = \frac{3kT\chi_m}{N_A g_e^2\mu_0\mu_B^2}$$

$$S(S+1) = \frac{3(1.381\times10^{-23}\,\text{J K}^{-1})\times(294.53\,\text{K})\times(0.1463\times10^{-6}\,\text{m}^3\,\text{mol}^{-1})}{(6.022\times10^{23}\,\text{mol}^{-1})\times(2.0023)^2\times(4\pi\times10^{-7}\,\text{T}^2\,\text{J}^{-1}\,\text{m}^3)\times(9.27\times10^{-24}\,\text{J T}^{-1})^2}$$
$$= 6.84$$

so

$$S^2 + S - 6.841 = 0 \quad \text{and} \quad S = \frac{-1+\sqrt{1+4(6.841)}}{2} = 2.163$$

corresponding to $\boxed{4.326}$ effective unpaired spins. The theoretical number is $\boxed{5}$, corresponding to the $3d^5$ electronic configuration of Mn^{2+}.

COMMENT. The discrepancy between the two values is accounted for by an antiferromagnetic interaction between the spins, which alters χ_m from the form of eqn 9.28.

E9.31(a) $$\chi_m = (6.3001\times10^{-6})\times\left(\frac{S(S+1)}{T/\text{K}}\text{m}^3\,\text{mol}^{-1}\right) \quad \text{[See *A Brief Illustration* in Section 9.11(b).]}$$

Since Cu(II) is a d^9 species, it has one unpaired spin, so $S = s = \frac{1}{2}$. Therefore,

$$\chi_m = \frac{(6.3001\times10^{-6})\times(\frac{1}{2})\times(\frac{3}{2})}{298}\,\text{m}^3\,\text{mol}^{-1} = \boxed{+1.6\times10^{-8}\,\text{m}^3\,\text{mol}^{-1}}$$

E9.32(a) The relationship between critical temperature and critical magnetic field is given by

$$H_c(T) = H_c(0)\left(1-\frac{T^2}{T_c^2}\right) \quad [9.29]$$

Solving for T gives the critical temperature for a given magnetic field.

$$T = T_c\left(1-\frac{H_c(T)}{H_c(0)}\right)^{1/2} = (7.19\,\text{K})\times\left(1-\frac{20\,\text{kA m}^{-1}}{63.9\,\text{kA m}^{-1}}\right)^{1/2} = \boxed{6.0\,\text{K}}$$

Solutions to problems

Solutions to numerical problems

P9.1 $$\lambda = 2d_{hkl}\sin\theta_{hkl} = \frac{2a\sin\theta_{hkl}}{(h^2+k^2+l^2)^{1/2}} \text{ [9.2b and 9.1b]} = 2a\sin 6.0° = 0.209a$$

In a NaCl unit cell (Figure 9.11) the number of formula units is 4 (each corner ion is shared by 8 cells, each edge ion by 4, and each face ion by 2).

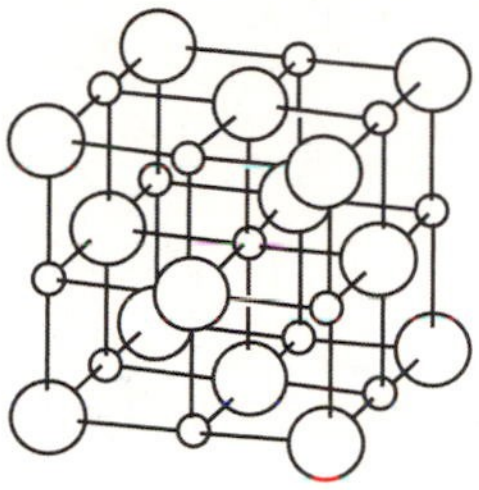

Figure 9.11

Therefore,

$$\rho = \frac{NM}{VN_A} = \frac{4M}{a^3N_A}, \quad \text{implying that} \quad a = \left(\frac{4M}{\rho N_A}\right)^{1/3} \quad \text{[Exercise 9.8a]}$$

$$a = \left(\frac{(4)\times(58.44\,\text{g mol}^{-1})}{(2.17\times10^6\,\text{g m}^{-3})\times(6.022\times10^{23}\,\text{mol}^{-1})}\right)^{1/3} = 563.\bar{5}\,\text{pm}$$

and hence $\lambda = (0.209)\times(563.\bar{5}\,\text{pm}) = \boxed{118\,\text{pm}}$.

P9.3 For the three given reflections

$\sin 19.076° = 0.32682 \qquad \sin 22.171° = 0.37737 \qquad \sin 32.256° = 0.53370$

For cubic lattices $\sin\theta_{hkl} = \dfrac{\lambda(h^2+k^2+l^2)^{1/2}}{2a}$ [9.2b with 9.1b]

First, consider the possibility of simple cubic; the first three reflections are (100), (110), and (200). (See Fig. 9.22 of the text.)

$$\frac{\sin\theta_{100}}{\sin\theta_{110}} = \frac{1}{\sqrt{2}} \neq \frac{0.32682}{0.37737} \quad \text{[not simple cubic]}$$

Consider next the possibility of body-centered cubic; the first three reflections are (110), (200), and (211).

$$\frac{\sin\theta_{110}}{\sin\theta_{200}} = \frac{\sqrt{2}}{\sqrt{4}} = \frac{1}{\sqrt{2}} \neq \frac{0.32682}{0.37737} \quad \text{(not bcc)}$$

Consider finally face-centered cubic; the first three reflections are (111), (200), and (220).

$$\frac{\sin\theta_{111}}{\sin\theta_{200}} = \frac{\sqrt{3}}{\sqrt{4}} = 0.86603$$

which compares very favorably to $\dfrac{0.32682}{0.37737} = 0.86605$. Therefore, the lattice is $\boxed{\text{face-centered cubic}}$.

This conclusion may easily be confirmed in the same manner using the second and third reflection.

$$a = \frac{\lambda}{2\sin\theta_{111}}(h^2+k^2+l^2)^{1/2} = \left(\frac{154.18\,\text{pm}}{(2)\times(0.32682)}\right)\times\sqrt{3} = \boxed{408.55\,\text{pm}}$$

$$\rho = \frac{NM}{N_A V} \text{ [Exercise 9.8a]} = \frac{(4)\times(107.87\,\text{g mol}^{-1})}{(6.0221\times10^{23}\,\text{mol}^{-1})\times(4.0855\times10^{-8}\,\text{cm})^3}$$
$$= \boxed{10.507\,\text{g cm}^{-3}}$$

This compares favorably to the value listed in the *Data Section*.

P9.5 $\theta_{111}(100\,\text{K}) = 22.0403°,\quad \theta_{111}(300\,\text{K}) = 21.9664°$

$$\sin\theta_{111}(100\,\text{K}) = 0.37526,\quad \sin\theta_{111}(300\,\text{K}) = 0.37406$$

$$\frac{\sin\theta(300\,\text{K})}{\sin\theta(100\,\text{K})} = 0.99681 = \frac{a(100\,\text{K})}{a(300\,\text{K})} \text{ [see Problem 9.4]}$$

$$a(300\,\text{K}) = \frac{\lambda\sqrt{3}}{2\sin\theta_{111}} = \frac{(154.0562\,\text{pm})\times\sqrt{3}}{(2)\times(0.37406)} = 356.67\,\text{pm}$$

$$a(100\,\text{K}) = (0.99681)\times(356.67\,\text{pm}) = 355.53\,\text{pm}$$

$$\frac{\delta a}{a} = \frac{356.67-355.53}{355.53} = 3.206\times10^{-3}$$

$$\frac{\delta V}{V} = \frac{356.67^3-355.53^3}{355.53^3} = 9.650\times10^{-3}$$

$$\alpha_{\text{volume}} = \frac{1}{V}\frac{\delta V}{\delta T} = \frac{9.560\times10^{-3}}{200\,\text{K}} = \boxed{4.8\times10^{-5}\,\text{K}^{-1}}$$

$$\alpha_{\text{linear}} = \frac{1}{a}\frac{\delta a}{\delta T} = \frac{3.206\times10^{-3}}{200\,\text{K}} = \boxed{1.6\times10^{-5}\,\text{K}^{-1}}$$

P9.7 $V = abc\sin\beta$

and the information given tells us that $a = 1.377b$, $c = 1.436b$, and $\beta = 122°49'$; hence

$$V = (1.377)\times(1.436b^3)\sin 122°49' = 1.662b^3$$

Since $\rho = \dfrac{NM}{VN_A} = \dfrac{2M}{1.662b^3 N_A}$, we find that

$$b = \left(\frac{2M}{1.662\rho N_A}\right)^{1/3}$$
$$= \left(\frac{(2)\times(128.18\,\text{g mol}^{-1})}{(1.662)\times(1.152\times10^6\,\text{g m}^{-3})\times(6.022\times10^{23}\,\text{mol}^{-1})}\right)^{1/3} = 605.8\,\text{pm}$$

Therefore, $a = \boxed{834\,\text{pm}}$, $b = \boxed{606\,\text{pm}}$, $c = \boxed{870\,\text{pm}}$.

P9.9 As shown in E9.2(a), the volume of a primitive monoclinic unit cell is

$$V = abc\,\sin\beta = (1.0427\,\text{nm})\times(0.8876\,\text{nm})\times(1.3777\,\text{nm})\times\sin(93.254°)$$
$$= 1.2730\,\text{nm}^3$$

The mass per unit cell is

$$m = \rho V = (2.024\,\text{g cm}^{-3})\times(1.2730\,\text{nm}^3)\times(10^{-7}\,\text{cm nm}^{-1})^3 = 2.577\times10^{-21}\,\text{g}$$

The monomer is $CuC_7H_{13}N_5O_8S$, so its molar mass is

$$M = [63.546 + 7(12.011) + 13(1.008) + 5(14.007) + 8(15.999) + 32.066] \text{ g mol}^{-1}$$
$$= 390.82 \text{ g mol}^{-1}$$

The number of monomer units, then, is the mass of the unit cell divided by the mass of the monomer.

$$N = \frac{mN_A}{M} = \frac{(2.577\times10^{-21}\text{ g})\times(6.022\times10^{23}\text{ mol}^{-1})}{390.82\text{ g mol}^{-1}} = 3.97 \quad \text{or} \quad \boxed{4}$$

P9.11 The problem asks for an estimate of $\Delta_f H^\ominus(\text{CaCl,s})$. A Born–Haber cycle would envision formation of CaCl(s) from its elements as sublimation of Ca(s), ionization of Ca(g), atomization of $Cl_2(g)$, electron gain of Cl(g), and formation of CaCl(s) from gaseous ions. Therefore,

$$\Delta_f H^\ominus(\text{CaCl,s}) = \Delta_{\text{sub}}H^\ominus(\text{Ca,s}) + \Delta_{\text{ion}}H^\ominus(\text{Ca,g}) + \Delta_f H^\ominus(\text{Cl,g}) + \Delta_{\text{eg}}H^\ominus(\text{Cl,g}) - \Delta_L H^\ominus(\text{CaCl,s})$$

Before we can estimate the lattice enthalpy of CaCl, we select a lattice with the aid of the radius-ratio rule. The ionic radius for Cl^- is 181 pm (Table 9.3); use the ionic radius of K^+ (138) for Ca^+

$$\gamma = \frac{138\text{ pm}}{181\text{ pm}} = 0.762$$

suggesting the CsCl structure (A = 1.763, Table 9.4). We can interpret the Born–Mayer equation (eqn. 9.12) as giving the negative of the lattice enthalpy.

$$\Delta_L H^\ominus \approx \frac{A|z_1z_2|N_Ae^2}{4\pi\varepsilon_0 d}\left(1-\frac{d^*}{d}\right) \quad \text{where } d^* \text{ is taken to be 34.5 pm (common choice)}$$

The distance d is $d = (138{+}181)\text{ pm} = 319\text{ pm}$.

$$\Delta_L H^\ominus \approx \frac{(1.763)\times|(1)(-1)|\times(6.022\times10^{23}\text{ mol}^{-1})\times(1.602\times10^{-19}\text{ C})^2}{4\pi(8.854\times10^{-12}\text{ J}^{-1}\text{C}^2\text{ m}^{-1})\times(319\times10^{-12}\text{ m})}\left(1-\frac{34.5\text{ pm}}{319\text{ pm}}\right)$$
$$\approx 6.85\times10^5\text{ J mol}^{-1} = 685\text{ kJ mol}^{-1}$$

The enthalpy of formation, then, is

$$\Delta_f H^\ominus(\text{CaCl,s}) \approx [176 + 589.7 + 121.7 - 348.7 - 685]\text{ kJ mol}^{-1} = \boxed{-146\text{ kJ mol}^{-1}}$$

Although formation of CaCl(s) from its elements is exothermic, formation of $CaCl_2(s)$ is still more favored energetically. Consider the disproportionation reaction 2 CaCl(s) → Ca(s) + $CaCl_2$(s) for which

$$\Delta H^\ominus = \Delta_f H^\ominus(\text{Ca,s}) + \Delta_f H^\ominus(\text{CaCl}_2\text{,s}) - 2\Delta_f H^\ominus(\text{CaCl,s})$$
$$\approx [0 - 795.8 - 2(-146)]\text{kJ mol}^{-1}$$
$$\approx -504\text{ kJ mol}^{-1}$$

and the thermodynamic instability of CaCl(s) toward disproportionation to Ca(s) and $CaCl_2$(s) becomes apparent.

Note: Using the tabulated ionic radius of Ca (i.e., that of Ca^{2+}) would be less valid than using the atomic radius of a neighboring monovalent ion, for the problem asks about a hypothetical compound of monovalent calcium. Predictions with the smaller Ca^{2+} radius (100 pm) differ from those listed above, but the conclusion remains the same: the expected structure changes to rock salt, the lattice enthalpy to 758 kJ mol^{-1}, $\Delta_f H^\ominus$(CaCl,s) to −219 kJ mol^{-1}, and the disproportionation enthalpy to −358 kJ mol^{-1}.

P9.13 The molar magnetic susceptibility is given by

$$\chi_m = \frac{N_A g_e^2 \mu_0 \mu_B^2 S(S+1)}{3kT} = (6.3001\times10^{-6})\times\frac{S(S+1)}{T/K}\,\mathrm{m^3\,mol^{-1}}$$ [*A brief illustration* in Section 9.11(b)]

For $S=2$, $\chi_m = \frac{(6.3001\times10^{-6})\times(2)\times(2+1)}{298}\,\mathrm{m^3mol^{-1}} = \boxed{0.127\times10^{-6}\,\mathrm{m^3\,mol^{-1}}}$.

For $S=3$, $\chi_m = \frac{(6.3001\times10^{-6})\times(3)\times(3+1)}{298}\,\mathrm{m^3mol^{-1}} = \boxed{0.254\times10^{-6}\,\mathrm{m^3\,mol^{-1}}}$.

For $S=4$, $\chi_m = \frac{(6.3001\times10^{-6})\times(4)\times(4+1)}{298}\,\mathrm{m^3mol^{-1}} = \boxed{0.423\times10^{-6}\,\mathrm{m^3\,mol^{-1}}}$.

Instead of a single value of S we use an average weighted by the Boltzmann factor.

$$\exp\left(\frac{-50\times10^3\ \mathrm{J\,mol^{-1}}}{(8.3145\,\mathrm{J\,mol^{-1}\,K^{-1}})\times(298\,\mathrm{K})}\right) = 1.7\times10^{-9}$$

Thus the $S=2$ and $S=4$ forms are present in negligible quantities compared to the $S=3$ form. The compound's susceptibility, then, is that of the $S=3$ form, namely, $\boxed{0.254\ \mathrm{cm^3\ mol^{-1}}}$.

Solutions to theoretical problems

P9.15 Consider for simplicity the two-dimensional lattice and planes shown in Figure 9.12.

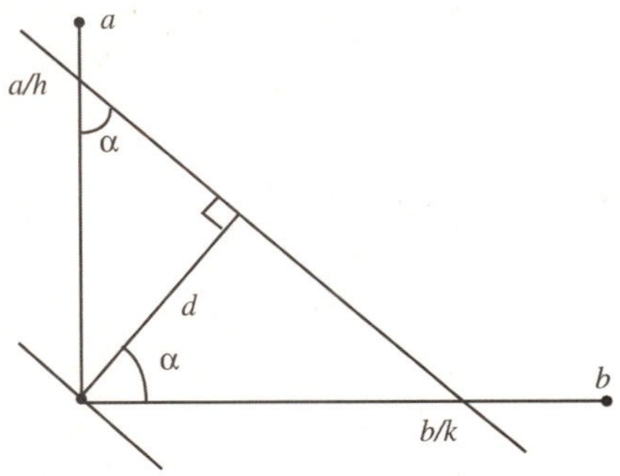

Figure 9.12

The (hk) planes cut the a and b axes at a/h and b/k, and we have

$$\sin\alpha = \frac{d}{(a/h)} = \frac{hd}{a}, \qquad \cos\alpha = \frac{d}{(b/k)} = \frac{kd}{b}$$

Then, since $\sin^2\alpha + \cos^2\alpha = 1$, we can write

$$\left(\frac{hd}{a}\right)^2 + \left(\frac{kd}{b}\right)^2 = 1$$

and therefore

$$\frac{1}{d^2} = \left(\frac{h}{a}\right)^2 + \left(\frac{k}{b}\right)^2$$

The same argument extends by analogy (or further trigonometry) to three dimensions, to give

$$\boxed{\frac{1}{d^2} = \left(\frac{h}{a}\right)^2 + \left(\frac{k}{b}\right)^2 + \left(\frac{l}{c}\right)^2} \quad [9.1c]$$

P9.17 The scattering factor is given by

$$f = 4\pi\int_0^\infty \rho(r)\frac{\sin kr}{kr}r^2\mathrm{d}r \ [9.3] \quad \text{where} \quad k = \frac{4\pi}{\lambda}\sin\theta$$

and the integral may be divided into segments such as

$$f = 4\pi\int_0^R \rho(r)\frac{\sin kr}{kr}r^2\mathrm{d}r + 4\pi\int_R^\infty \rho(r)\frac{\sin kr}{kr}r^2\mathrm{d}r$$

In the case for which $\rho = 3Z/4\pi R^3$ when $0 \le r \le R$ and $\rho = 0$ when $r > R$, the second integral vanishes, leaving

$$f = \frac{3Z}{R^3}\int_0^R \frac{\sin kr}{kr}r^2\mathrm{d}r = \frac{3Z}{R^3k}\int_0^R \sin(kr)\,r\mathrm{d}r$$

$$= \frac{3Z}{R^3k}\left[\frac{\sin kr}{k^2} - \frac{r\cos kr}{k}\right]_{r=0}^{r=R}$$

$$= \frac{3Z}{(kR)^3}\left\{\frac{\sin kR}{kR} - \cos kR\right\}$$

This shows that the scattering factor is proportional to the atomic number Z, which is illustrated in a plot of f against $\sin(\theta)/\lambda$ in Figure 9.13. As expected, $f = Z$ in the forward direction.

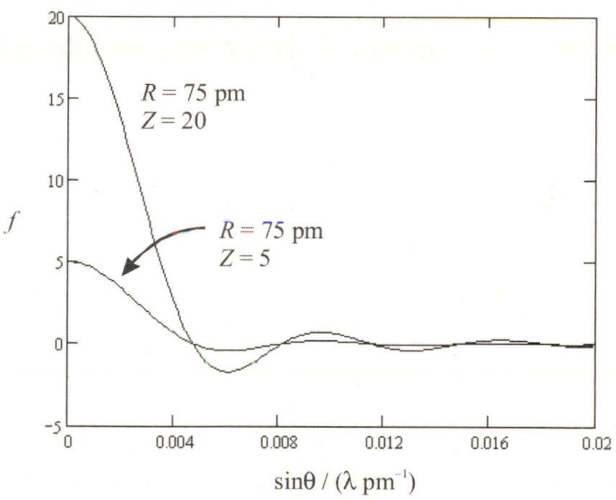

Figure 9.13

Figure 9.14 is a plot of the scattering factor for several R values at constant Z. As R increases, the scattering factor shifts to the forward direction.

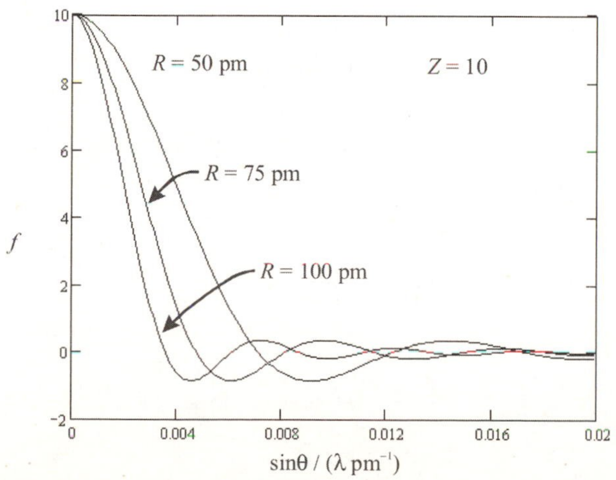

Figure 9.14

P9.19 Replacing an exponentially decreasing 1s wavefunction with a Gaussian function effectively replaces the cusp of the exponential decay near $r = 0$ with a broader, flatter function while simultaneously decreasing the long tail of the exponential. Physically, this is analogous to contraction of the orbital by an increase in the nuclear charge Z. In Problem 9.18 an increase in Z was found to cause a shift of the scattering factor away from the forward direction. To confirm this hypothesis, we first fit the $1s(Z = 1)$ exponential $\exp(\rho/2)$ with the $\exp(-br^2)$ Gaussian by choosing the constant b so as to minimize the sum of the squared errors (SSE) in the range $0 \le r \le 2a$. Doing this in a Mathcad worksheet yields the following b value.

$$N = 40 \qquad j := 0 \ldots N \qquad r_{max} := 2\cdot a \qquad r_j := \frac{j\cdot r_{max}}{N}$$

$$SSE(b) := \sum_j \left[e^{\frac{-\rho(r_j,1)}{2}} - e^{-b(r_j)^2} \right]^2 \qquad \text{Estimate of b:} \qquad b := 1\cdot pm^{-2}$$

Given $SSE(b) = 0$ **b**:= Minerr(b)

$$b = 0.00035\ pm^{-2}$$

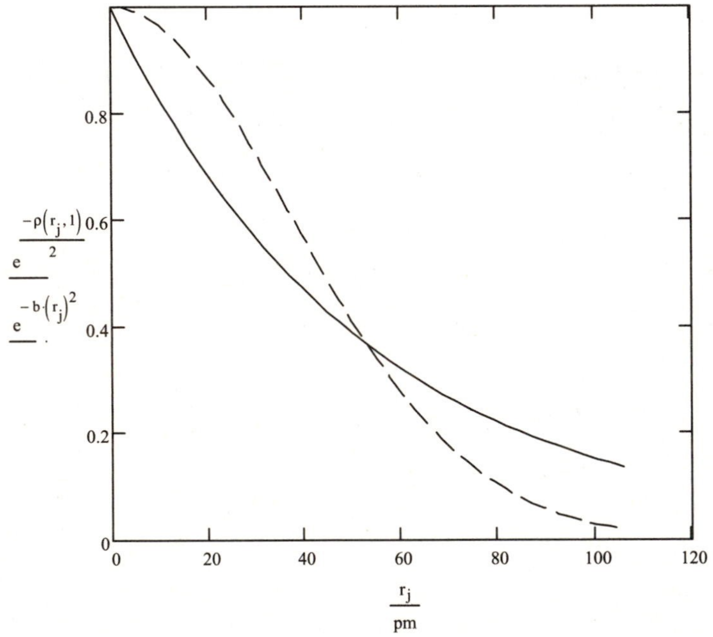

The absence of both cusp and long tail in the Gaussian is clearly apparent in the worksheet plot.

For the one-electron hydrogen-like atom, $4\pi\rho(r) = R_{1s}(r)^2$ [4.15b] or $4\pi\rho(r) = g(r)^2$, and the scattering factor is given by

$$f(r,Z) = 4\pi \int_0^\infty \rho(r,Z)\frac{\sin(4\pi xr)}{4\pi xr} r^2 dr \ [9.3] \quad \text{where} \quad x = \frac{\sin\theta}{\lambda}$$

The next section of the worksheet normalizes the Gaussian $g(r)$ and defines the Gaussian scattering factor f_g.

$$N := \left[\frac{1}{\int_{0\cdot pm}^{50\cdot a} e^{-b\cdot(r)^2}\cdot e^{-b\cdot(r)^2}\cdot r^2 dr} \right]^{\frac{1}{2}} \qquad N = 0.00649\ pm^{\frac{-3}{2}}$$

$$g(r) = N\cdot e^{-b\cdot(r)^2} \qquad f_g(x) := \int_{0\cdot pm}^{100\cdot a} \frac{g(r)^2\cdot \sin(4\cdot\pi\cdot x\cdot r)\cdot r}{4\cdot\pi\cdot x} dr$$

The worksheet setup for the 1s radial wavefunction (Table 4.1) and its scattering factor f follows.

$$\text{pm} := 10^{-12}\cdot \text{m} \qquad a := 52.91772108\ \text{pm}$$

$$\rho(r,Z) := \frac{2\cdot Z\cdot r}{a} \qquad R_{1s}(r,Z) := 2\cdot\left(\frac{Z}{a}\right)^{\frac{3}{2}}\cdot e^{\frac{-\rho(r,Z)}{2}}$$

$$f(x,Z) := \int_{0\cdot \text{pm}}^{100\cdot a} \frac{R_{1s}(r,Z)^2\cdot \sin(4\cdot\pi\cdot x\cdot r)\cdot r}{4\cdot\pi\cdot x}\,dr$$

The subsequent plots of $f(x,1)$ and $f_g(x)$ confirm the hypothesis that replacement of the 1s wavefunction by a Gaussian function causes the scattering factor to shift away from the forward direction.

$$N := 200 \qquad i := 0 \ldots N \qquad x_{max} := 0.01\cdot \text{pm}^{-1} \qquad x_i := \frac{i\cdot x_{max}}{N}$$

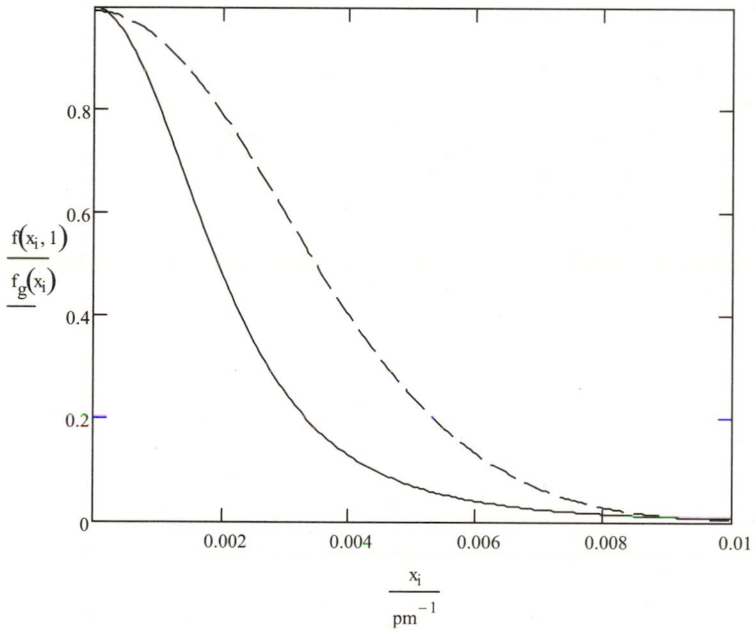

P9.21 Close-packed rods of length L and elliptical cross-section with semi-major axis a and semi-minor axis b have a rod volume of πabL. The parallelepiped unit cell is shown in Figure 9.15(a) in a cross-section of the packed rods. Examination of the figure reveals that each unit cell contains one rod. Thus, the packing fraction f is given by

$$f = \frac{\pi abL}{LA_p} = \frac{\pi ab}{A_p} \quad \text{where } A_p \text{ is the cross-section area of the unit cell}$$

Figure 9.15(b) defines the parameters needed for the determination of A_p. Examination of the center positions of a pair of stacked ellipses reveals that they have the relative coordinates (0, 0) and (0, $2b$). The adjacent ellipse column is centered b higher, and consequently the vertical contact point between the adjacent ellipses is necessarily at $b/2$. The horizontal component of the contact point, x, is calculated with the formula for an ellipse using $y = b/2$.

$$\left(\frac{x}{a}\right)^2 + \left(\frac{y}{b}\right)^2 = 1$$

$$\left(\frac{x}{a}\right)^2 + \left(\frac{b/2}{b}\right)^2 = 1 \quad \text{or} \quad x = \tfrac{1}{2}\sqrt{3}a$$

The parallelepiped area has a base of $2b$ and its height h is

$$h = 2x = \sqrt{3}a$$

Thus,

$$A_{\rm p} = 2bh = 2\sqrt{3}ab$$

and

$$f = \frac{\pi ab}{2\sqrt{3}ab} = \frac{\pi}{2\sqrt{3}} = \boxed{0.907}$$

The above formula for the packing fraction shows that it is independent of the ellipse eccentricity.

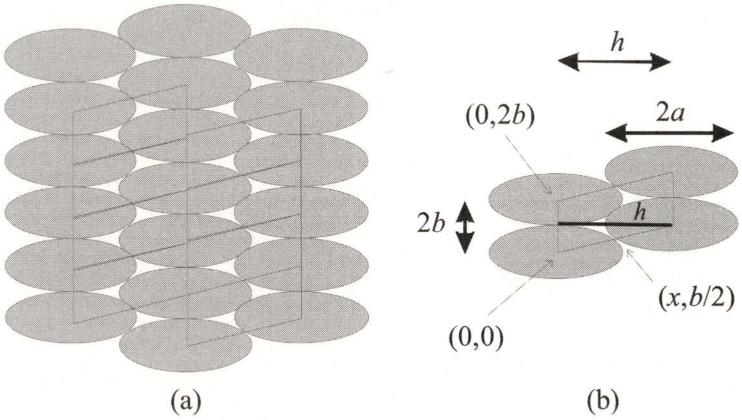

Figure 9.15

P9.23 $$F_{hkl} = \sum_i f_i e^{2\pi i(hx_i + ky_i + lz_i)} \quad [9.4]$$

For each A atom use $\frac{1}{8}f_{\rm A}$ (each A atom shared by eight cells), but use $f_{\rm B}$ for the central atom (since it contributes solely to the cell).

$$\begin{aligned} F_{hkl} &= \tfrac{1}{8}f_{\rm A}\{1 + e^{2\pi ih} + e^{2\pi ik} + e^{2\pi il} + e^{2\pi i(h+k)} + e^{2\pi i(h+l)} + e^{2\pi i(k+l)} + e^{2\pi i(h+k+l)}\} \\ &\quad + f_{\rm B}e^{\pi i(h+k+l)} \\ &= f_{\rm A} + (-1)^{(h+k+l)} f_{\rm B} \quad [h, k, l \text{ are all integers}, e^{i\pi} = -1] \end{aligned}$$

(a) $f_{\rm A} = f,\ \ f_{\rm B} = 0;\ \ F_{hkl} = f$ $\boxed{\text{no systematic absences}}$

(b) $f_{\rm B} = \frac{1}{2}f_{\rm A};\ \ F_{hkl} = f_{\rm A}\left[1 + \frac{1}{2}(-1)^{(h+k+l)}\right]$

Therefore, when $h+k+l$ is odd, $F_{hkl} = f_{\rm A}\left(1 - \frac{1}{2}\right) = \frac{1}{2}f_{\rm A}$, and when $h+k+l$ is even, $F_{hkl} = \frac{3}{2}f_{\rm A}$.

That is, there is an $\boxed{\text{alternation of intensity}}$ ($I \propto F^2$) according to $\boxed{\text{whether } h+k+l \text{ is odd or even}}$.

(c) $f_{\rm A} = f_{\rm B} = f;\ \ F_{hkl} = f\{1 + (-1)^{h+k+l}\} = 0 \quad \text{if } h+k+l \text{ is odd}$

Thus, $\boxed{\text{all } h+k+l \text{ odd lines are missing}}$.

P9.25 The distance between adjacent ions on the two diagonal lines is d, and consequently the distance between adjacent ions on the horizontal and vertical lines is $c = d\cos(45°) = d/2^{1/2}$. These four lines with their constant, close distance between adjacent ions determine the Madelung constant when there are very large numbers of ions in a line. The textbook figure shows that distances between ions on different lines quickly grow to a large value, and consequently the total interaction between ions on different lines contributes very negligibly to the lattice energy per mole of ions. Thus, the Madelung constant of this ion arrangement is twice the Madelung constant for a line of alternating

$+e$ cations and $-e$ anions separated by d plus twice the Madelung constant for a line of alternating $+e$ cations and $-e$ anions separated by c (see the one-dimensional model discussion in Section 9.6b of the textbook).

$$E_P = 2\times(-2\ln 2)\times\left(\frac{N_A e^2}{4\pi\varepsilon_0 d}\right) + 2\times(-2\ln 2)\times\left(\frac{N_A e^2}{4\pi\varepsilon_0 c}\right)$$

$$= 2\times(-2\ln 2)\times\left(\frac{N_A e^2}{4\pi\varepsilon_0 d}\right) + 2\times(-2\ln 2)\times\left(\frac{2^{1/2} N_A e^2}{4\pi\varepsilon_0 d}\right)$$

$$= -2\times(2\ln 2)\times\left(1+2^{1/2}\right)\times\left(\frac{N_A e^2}{4\pi\varepsilon_0 d}\right)$$

Comparing the above equation with eqn. 9.10 yields the Madelung constant.

$$A = 2\times(2\ln 2)\times\left(1+2^{1/2}\right) = \boxed{6.694}$$

P9.27 In *Justification* 9.4 it is shown that the bulk modulus at equilibrium volume is

$$K = \frac{V}{9a^2r^4}\left(\frac{\partial^2 U}{\partial r^2}\right)_T$$

where $r = r_e$ is the interatomic separation, a is a constant such that $V = ar^3$, the internal energy U at $T = 0$ is the potential multiplied by N, and $-|\varepsilon|$ is the minimum value of the potential.

(a) General Mie potential: $V_{\text{potential}} = \frac{C_n}{r^n} - \frac{C_m}{r^m}$ where $n > m$

$$\frac{\partial V_{\text{potential}}}{\partial r} = -\frac{nC_n}{r^{n+1}} + \frac{mC_m}{r^{m+1}}$$

$$\frac{\partial^2 V_{\text{potential}}}{\partial r^2} = \frac{n(n+1)C_n}{r^{n+2}} - \frac{m(m+1)C_m}{r^{m+2}}$$

$$K = \frac{VN}{9a^2r^4}\left\{\frac{n(n+1)C_n}{r^{n+2}} - \frac{m(m+1)C_m}{r^{m+2}}\right\}$$

The Mie potential is very general, so we will simplify the discussion of K by examining the special case for which $n = 2m$. In this case, it can be shown that

$$C_n = |\varepsilon| r_e^{2m},\ C_m = 2C_n / r_e^{m},\ \text{and}\ (r_e / r_0)^m = 2$$

Furthermore, r_0 is the separation at which the potential equals zero. Substitution of the special case values into the K equation gives

$$K = \frac{VN|\varepsilon|r^{2m}}{9a^2r^4}\left\{\frac{2m(2m+1)}{r^{2m+2}} - \frac{2m(m+1)}{r^{2m+2}}\right\}$$

$$= \frac{2m^2VN|\varepsilon|}{9a^2r^6}$$

Finally, we recognize that $ar^3 = V$ and obtain

$$K = \frac{2m^2N|\varepsilon|}{9V} = \boxed{\frac{2m^2|\varepsilon|}{9V_{\text{molecule}}} \text{ or } \frac{2m^2N_A|\varepsilon|}{9V_m}} \text{ for } n = 2m \text{ special case}$$

(b) Exp-6 potential: $V_{\text{potential}} = -A|\varepsilon|\{e^{-r/r_0} - (r_0/r)^6\}$

where $V(0.7525\, r_e) = V(1.2269\, r_0) = 0$, $V(r_e) = -|\varepsilon|$, $V(r_0) = 0.6321A|\varepsilon|$, $A = 7.0121$, and $r_e = 1.6305\, r_0$.

$$\frac{\partial V_{\text{potential}}}{\partial r} = -\frac{A|\varepsilon|}{r_0}\{-e^{-r/r_0} + 6(r_0/r)^7\}$$

$$\frac{\partial^2 V_{\text{potential}}}{\partial r^2} = -\frac{A|\varepsilon|}{r_0^2}\{e^{-r/r_0} - 42(r_0/r)^8\}$$

$$K = \frac{VN}{9a^2r^4}\left\{-\frac{A|\varepsilon|}{r_0^2}\{e^{-r/r_0} - 42(r_0/r)^8\}\right\}$$

It follows that the bulk modulus at the separation $r = r_e = 1.6305r_0$, or $r_0 = 0.6133r$, is

$$K = \frac{VN}{9a^2(1.6305r_0)^4}\left\{-\frac{A|\varepsilon|}{r_0^2}\left\{e^{-1.6305} - \frac{42}{1.6305^8}\right\}\right\}$$
$$= \frac{0.07110|\varepsilon|VN}{a^2r_0^6} = \frac{1.3359|\varepsilon|VN}{a^2r^6}$$

Finally, we recognize that $ar^3 = V$, and obtain

$$K = \frac{1.3359N|\varepsilon|}{V} = \boxed{\frac{1.3359|\varepsilon|}{V_{\text{molecule}}} \text{ or } \frac{1.3359N_A|\varepsilon|}{V_m}}.$$

COMMENT. The exp-6 potential used above has the disadvantage that it is non-zero at $r = r_0$. The exp-6 potential that is developed in problem 8.22, $V_{\text{potential}} = -A|\varepsilon|\{e^{1-r/r_0} - (r_0/r)^6\}$, does equal zero at $r = r_0$. Its properties are $V(r_0) = 0$, $V(r_e) = -|\varepsilon|$, $A = 1.8531$, and $r_e = 1.3598\, r_0$. This form of the potential yields

$$K = \frac{1.1022N|\varepsilon|}{V} = \frac{1.1022|\varepsilon|}{V_{\text{molecule}}} \text{ or } \frac{1.1022N_A|\varepsilon|}{V_m}$$

P9.29 Permitted states at the low-energy edge of the band must have a relatively long characteristic wavelength while the permitted states at the high-energy edge of the band must have a relatively short characteristic wavelength. There are few wavefunctions that have these characteristics, so the density of states is lowest at the edges. This is analogous to the MO picture that shows a few bonding MOs that lack nodes and a few antibonding MOs that have the maximum number of nodes.

Another insightful view is provided by consideration of the spatially periodic potential that the electron experiences within a crystal. The periodicity demands that the electron wavefunction be a periodic function of the position vector $\vec{r}$. We can approximate it with a Bloch wave: $\psi \propto e^{i\vec{k}\cdot\vec{r}}$ where $\vec{k} = k_x\hat{i} + k_y\hat{j} + k_z\hat{k}$ is called the wave number vector. This is a bold "free" electron approximation and, in the spirit of searching for a conceptual explanation, not an accurate solution. Suppose that the wavefunction satisfies a Hamiltonian in which the potential can be neglected: $\hat{H} = -(\hbar^2/2m)\nabla^2$. The eigenvalues of the Bloch wave are $E = \hbar^2|\vec{k}|^2/2m$. The Bloch wave is periodic when the components of the wave number vector are multiples of a basic repeating unit. Writing the repeating unit as $2\pi/L$ where L is a length that depends on the structure of the unit cell, we find $k_x = 2n_x\pi/L$ where $n_x = 0, \pm1, \pm2, \ldots$. Similar equations can be written for k_y and k_z, and with substitution the eigenvalues become $E = (1/2m)(2\pi\hbar/L)^2(n_x^2 + n_y^2 + n_z^2)$. This equation suggests that the density of states for energy level E can be visually evaluated by looking at a plot of permitted n_x, n_y, n_z values as shown in the graph of Figure 9.16. The number of n_x, n_y, n_z values within a thin, spherical shell around the origin equals the density of states which have energy E. Three shells, labeled 1, 2, and 3, are shown in the graph. All have the same width but their energies increase with their distance from the origin. It is obvious that the low-energy shell 1 has a much lower density of states than the intermediate-energy shell 2. The sphere of shell 3 has been cut into the shape determined by the periodic potential pattern of the crystal, and because of this phenomenon, it also has a lower density of states than the intermediate-energy shell 2. The general concept is that the low-energy and high-energy edges of a band have lower density of states than that of the band center.

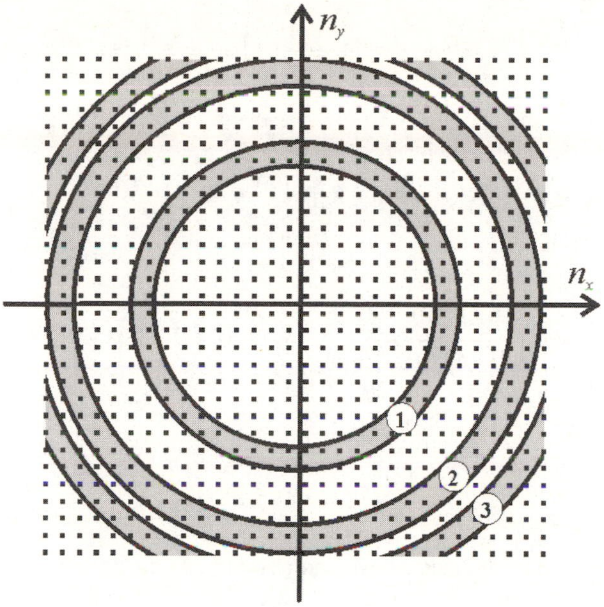

Figure 9.16

P9.31 $P(E)=\dfrac{1}{e^{(E-\mu)/kT}+1}$ [9.20] where E is measured from in minimum value of $E_{min}=0$.

In the case for which $E<\mu$,

$$\lim_{T\to 0}(E-\mu)/kT=\lim_{T\to 0}\left(-|E-\mu|/kT\right)=-\infty$$

Thus,

$$\lim_{T\to 0}P(E)=\frac{1}{e^{-\infty}+1}=\frac{1}{0+1}=1$$

Conclusion: $\boxed{\lim_{T\to 0}P(E)=1 \text{ when } E<\mu}$

In the case for which $E>\mu$,

$$\lim_{T\to 0}(E-\mu)/kT=\lim_{T\to 0}\left(|E-\mu|/kT\right)=+\infty$$

Thus,

$$\lim_{T\to 0}P(E)=\frac{1}{e^{+\infty}+1}=\frac{1}{\infty+1}=0$$

Conclusion: $\boxed{\lim_{T\to 0}P(E)=0 \text{ when } E>\mu}$

Since $N_e=\int_0^{\infty}\rho(E)P(E)\,dE=\int_0^{\mu}\rho(E)P(E)\,dE+\int_{\mu}^{\infty}\rho(E)P(E)\,dE$, we find μ at $T=0$ by substitution of $P=1$ in the first integral and $P=0$ in the second integral.

$$N_e=\int_0^{\mu(0)}\rho(E)\times 1\,dE+\int_{\mu(0)}^{\infty}\rho(E)\times 0\,dE=\int_0^{\mu(0)}\rho(E)\,dE$$

$$=\int_0^{\mu(0)}\rho(E)dE=C\int_0^{\mu(0)}E^{1/2}dE \quad \text{where} \quad C=4\pi V\left(2m_e/h^2\right)^{3/2}$$

$$=\left.\frac{2CE^{3/2}}{3}\right|_{E=0}^{E=\mu(0)}=\frac{2C\mu(0)^{3/2}}{3}$$

Solving for $\mu(0)$ yields

$$\mu(0) = \left(\frac{3N_e}{2C}\right)^{2/3} = \boxed{(3\mathcal{N}/8\pi)^{2/3}\left(h^2/2m_e\right)} \quad \text{where} \quad \mathcal{N} = N_e/V$$

Because a sodium atom contributes a single valence electron and sodium has a density of about 0.97 g cm^{-3},

$$\begin{aligned}\mathcal{N} &= N_e/V = N_A/V_m = N_A \times \text{mass density / molar mass} \\ &= \left(6.022\times10^{23}\ \text{mol}^{-1}\right)\times\left(0.97\ \text{g cm}^{-3}\right)/\left(23\ \text{g mol}^{-1}\right) = 2.5\times10^{28}\ \text{m}^{-3}\end{aligned}$$

and

$$\mu(0) = \left(\frac{3\left(2.5\times10^{28}\ \text{m}^{-3}\right)}{8\pi}\right)^{2/3}\frac{\left(6.63\times10^{-34}\ \text{J s}\right)^2}{2\left(9.11\times10^{-31}\ \text{kg}\right)} = 5.0\times10^{-19}\ \text{J} = \boxed{3.1\ \text{eV}}$$

P9.33 If a substance responds non-linearly to an electric field E, then it induces a dipole moment:

$$\mu = \alpha E + \beta E^2$$

If the electric field is oscillating at two frequencies, we can write the electric field as

$$E = E_1 \cos \omega_1 t + E_2 \cos \omega_2 t$$

and the non-linear response as

$$\beta E^2 = \beta(E_1 \cos \omega_1 t + E_2 \cos \omega_2 t)^2$$

$$\beta E^2 = \beta\left(E_1^2 \cos^2 \omega_1 t + E_2^2 \cos^2 \omega_2 t + 2E_1E_2 \cos \omega_1 t \cos \omega_2 t\right)$$

Application of trigonometric identities allows a product of cosines to be rewritten as a sum:

$$\cos A \cos B = \tfrac{1}{2}\cos(A-B) + \tfrac{1}{2}\cos(A+B)$$

Using this result (a special case of which applies to the $\cos^2$ terms), yields

$$\beta E^2 = \tfrac{1}{2}\beta\left[E_1^2(1+\cos 2\omega_1 t) + E^2(1+\cos 2\omega_2 t) + 2E_1E_2(\cos(\omega_1+\omega_2)t + \cos(\omega_1-\omega_2)t\right]$$

This expression includes responses at twice the original frequencies as well as at the sum and difference frequencies.

P9.35 (a) $\hat{H}\psi_{+\text{or}-} = \tilde{\nu}_{+\text{or}-}\psi_{+\text{or}-}$ and $\left(\hat{H} - \tilde{\nu}_{+\text{or}-}\right)\psi_{+\text{or}-} = 0$

$$\begin{pmatrix}\tilde{\nu}_{\text{mon}} - \tilde{\nu}_{+\text{or}-} & \beta \\ \beta & \tilde{\nu}_{\text{mon}} - \tilde{\nu}_{+\text{or}-}\end{pmatrix}\psi_{+\text{or}-} = 0 \quad \text{where} \quad \beta = \frac{\mu_{\text{mon}}^2}{4\pi\varepsilon_0 hcr^3}\left(1-3\cos^2\theta\right)$$

$$\begin{pmatrix}x_{+\text{or}-} & 1 \\ 1 & x_{+\text{or}-}\end{pmatrix}\psi_{+\text{or}-} = 0 \quad \text{where} \quad x_{+\text{or}-} = \left(\tilde{\nu}_{\text{mon}} - \tilde{\nu}_{+\text{or}-}\right)/\beta$$

$$\begin{vmatrix}x_{+\text{or}-} & 1 \\ 1 & x_{+\text{or}-}\end{vmatrix} = x_{+\text{or}-}^2 - 1 = 0$$

$$x_{+\text{or}-} = \left(\tilde{\nu}_{\text{mon}} - \tilde{\nu}_{+\text{or}-}\right)/\beta = \pm 1 \quad \text{and} \quad \tilde{\nu}_{+\text{or}-} = \tilde{\nu}_{\text{mon}} \pm \beta$$

$$\tilde{\nu}_+ = \tilde{\nu}_{\text{mon}} - \beta \quad \text{and} \quad \tilde{\nu}_- = \tilde{\nu}_{\text{mon}} + \beta$$

$\tilde{\nu}_+$ and $\tilde{\nu}_-$ are plotted in Figure 9.17 as a function of θ using $\mu_{\text{mon}} = 4.00$ D, $\tilde{\nu}_{\text{mon}} = 25000\ \text{cm}^{-1}$, and $r = 0.5$ nm.

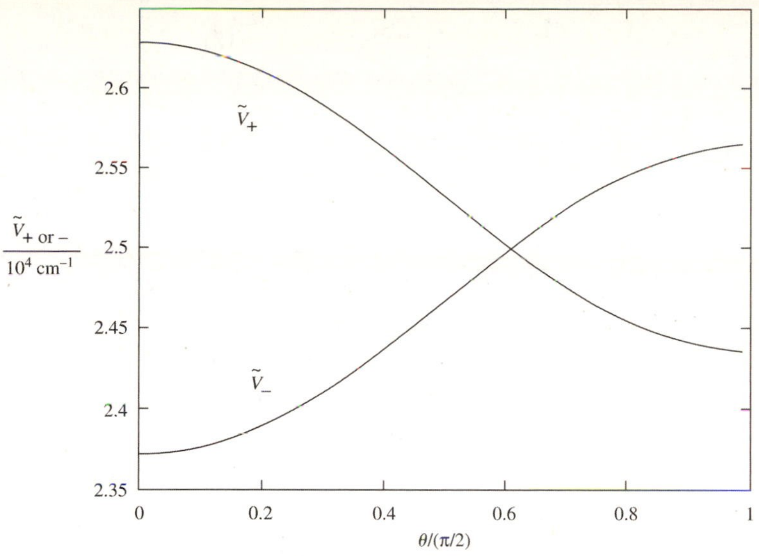

Figure 9.17

The ratio of μ_+^2/μ_-^2 (and the relative intensities of the dimer transitions) doesn't depend on β or θ because $\mu_+ = 0$. To see this, we use the coefficients of the normalized wavefunctions for ψ_+ and ψ_- and the overlap integral $S = \langle \psi_1 | \psi_2 \rangle$.

$$\begin{pmatrix} x_{+\text{or}-} & 1 \\ 1 & x_{+\text{or}-} \end{pmatrix} \begin{pmatrix} c_{+\text{or}-,1} \\ c_{+\text{or}-,2} \end{pmatrix} = 0 \qquad \text{where} \qquad x_{+\text{or}-} = \pm 1$$

$$x_{+\text{or}-} c_{+\text{or}-,1} + c_{+\text{or}-,2} = 0$$

$$c_{+\text{or}-,2} = -x_{+\text{or}-} c_{+\text{or}-,1} \qquad \text{(i)}$$

The coefficients must also satisfy the normalization condition.

$$\begin{aligned} \langle \psi_{+\text{or}-} | \psi_{+\text{or}-} \rangle &= \langle c_{+\text{or}-,1}\psi_1 + c_{+\text{or}-,2}\psi_2 | c_{+\text{or}-,1}\psi_1 + c_{+\text{or}-,2}\psi_2 \rangle \\ &= c_{+\text{or}-,1}^2 + c_{+\text{or}-,2}^2 + 2c_{+\text{or}-,1}c_{+\text{or}-,2}S \\ &= c_{+\text{or}-,1}^2 + c_{+\text{or}-,1}^2 - 2x_{+\text{or}-}c_{+\text{or}-,1}^2 S = 1 \end{aligned} \qquad \text{(ii)}$$

Thus,

$$c_{+,1} = \frac{1}{\{2(1-S)\}^{1/2}} \qquad c_{+,2} = -c_{+,1}$$

and

$$c_{-,1} = \frac{1}{\{2(1+S)\}^{1/2}} \qquad c_{-,2} = c_{-,1}$$

$$\frac{\mu_+^2}{\mu_-^2} = \left(\frac{\mu_+}{\mu_-}\right)^2 = \left(\frac{(c_{+,1}+c_{+,2})\mu_{\text{mon}}}{(c_{-,1}+c_{-,2})\mu_{\text{mon}}}\right)^2 \text{ [see Problem 9.34]} = \left(\frac{c_{+,1}-c_{+,1}}{c_{-,1}+c_{-,1}}\right)^2 = 0$$

(b) The secular determinant for N monomers has the dimension $N \times N$.

$$\begin{vmatrix} \tilde{\nu}_{\text{mon}} - \tilde{\nu}_{\text{dimer}} & V & 0 & \cdots \\ V & \tilde{\nu}_{\text{mon}} - \tilde{\nu}_{\text{dimer}} & V & \cdots \\ 0 & V & \tilde{\nu}_{\text{mon}} - \tilde{\nu}_{\text{dimer}} & \cdots \\ \vdots & \vdots & \vdots & \cdots \end{vmatrix} = 0$$

$$\tilde{\nu}_{\text{dimer}} = \tilde{\nu}_{\text{mon}} + 2V\cos\left(\frac{k\pi}{N+1}\right) \qquad k = 1, 2, 3, \ldots, N \qquad [9.18]$$

$$V = \beta(0) = \frac{\mu_{\text{mon}}^2}{4\pi\varepsilon_0 hcr^3}\left(1 - 3\cos^2 0\right) = \frac{-\mu_{\text{mon}}^2}{2\pi\varepsilon_0 hcr^3}$$

The plot in Figure 9.18 shows the dimer transitions for $\theta = 0$ and $N = 15$. The shape of the transition distribution changes slightly with N and transition energies are symmetrically distributed around the monomer transition. The lowest energy transition changes only slightly with N giving a value that goes to

$$25000\ \text{cm}^{-1} + 2V = 25000\ \text{cm}^{-1} + 2 \times (-1289\ \text{cm}^{-1}) = 22422\ \text{cm}^{-1} \text{ as } N \to \infty$$

Since the model considers only nearest-neighbor interactions, the transition dipole moment of the lowest energy transition doesn't depend on the size of the chain.

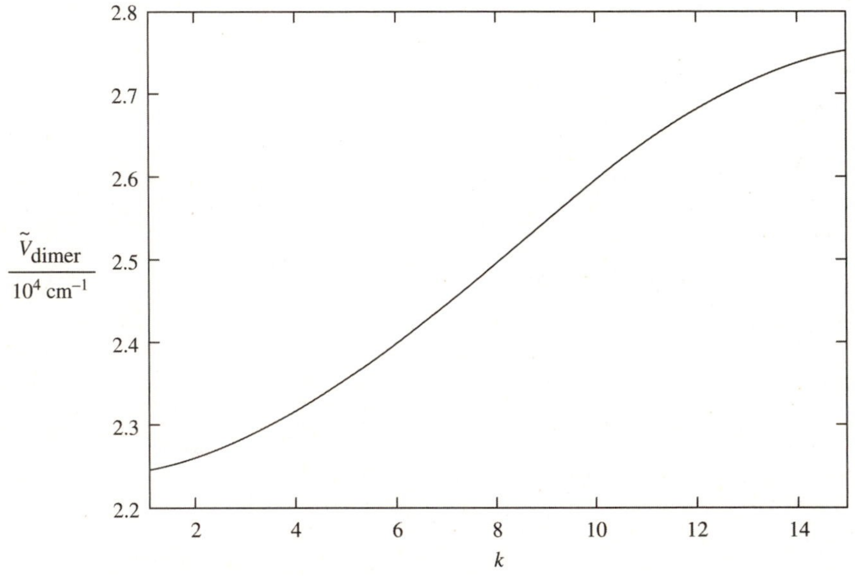

Figure 9.18

P9.37

$$N_2O_4(g) \underset{}{\overset{K}{\rightleftharpoons}} 2\ NO_2(g)$$

$(1-\alpha)n$	$2\alpha n$	amounts
$\dfrac{1-\alpha}{1+\alpha}$	$\dfrac{2\alpha}{1+\alpha}$	mole fractions
$\left(\dfrac{1-\alpha}{1+\alpha}\right)p$	$\left(\dfrac{2\alpha}{1+\alpha}\right)p$	partial pressures [$p \equiv p/p^{\ominus}$ here]

$$K = \frac{(2\alpha/1+\alpha)^2 p}{(1-\alpha/1+\alpha)} = \frac{4\alpha^2}{1-\alpha^2}p$$

Now solve for α.

$$\alpha^2 = \frac{K}{4p+K}, \qquad \alpha = \left(\frac{K}{4p+K}\right)^{1/2}$$

The degree of dimerization is $d = 1 - \alpha = 1 - \left(\dfrac{K}{4p+K}\right)^{1/2} = \boxed{1 - \left(\dfrac{1}{4(p/K)+1}\right)^{1/2}}$.

The susceptibility varies in proportion to $\alpha = 1 - d$. As pressure increases, α decreases and the susceptibility $\boxed{\text{decreases}}$.

To determine the effect of temperature we need $\Delta_r H \approx \Delta_r H^{\ominus}$ for the reaction above.

$$\Delta_r H^{\ominus} = 2\times(33.18\ \text{kJ mol}^{-1}) - 9.16\ \text{kJ mol}^{-1} = +57.2\ \text{kJ mol}^{-1}$$

A positive $\Delta_r H^{\ominus}$ indicates that $NO_2(g)$ is favored as the temperature increases; hence, the susceptibility $\boxed{\text{increases}}$ with temperature.

Solutions for applications

P9.39 The density of a face-centered cubic crystal is $4m/V$ where m is the mass of the unit hung on each lattice point and V is the volume of the unit cell. (The 4 comes from the fact that each of the cell's 8 vertices is shared by 8 cells, and each of the cell's 6 faces is shared by 2 cells.)

So $$\rho = \frac{4m}{a^3} = \frac{4M}{N_A a^3} \text{ and } M = \tfrac{1}{4}\rho N_A a^3$$

$$M = \tfrac{1}{4}(1.287\ \text{g cm}^{-3})\times(6.022\times10^{23}\ \text{mol}^{-1})\times(12.3\times10^{-7}\ \text{cm})^3$$
$$= \boxed{3.61\times10^5\ \text{g mol}^{-1}}$$

P9.41 Tans and coworkers (S. J. Tans et al., *Nature*, **393**, 49 (1998)) have draped a semiconducting carbon nanotube (CNT) over metal (gold in Figure 9.19) electrodes that are 400 nm apart atop a silicon surface coated with silicon dioxide. A bias voltage between the electrodes provides the source and drain of the molecular field-effect transistor (FET). The silicon serves as a gate electrode and the thin silicon oxide layer (at least 100 nm thick) insulates the gate from the CNT circuit. By adjusting the magnitude of an electric field applied to the gate, current flow across the CNT may be turned on and off.

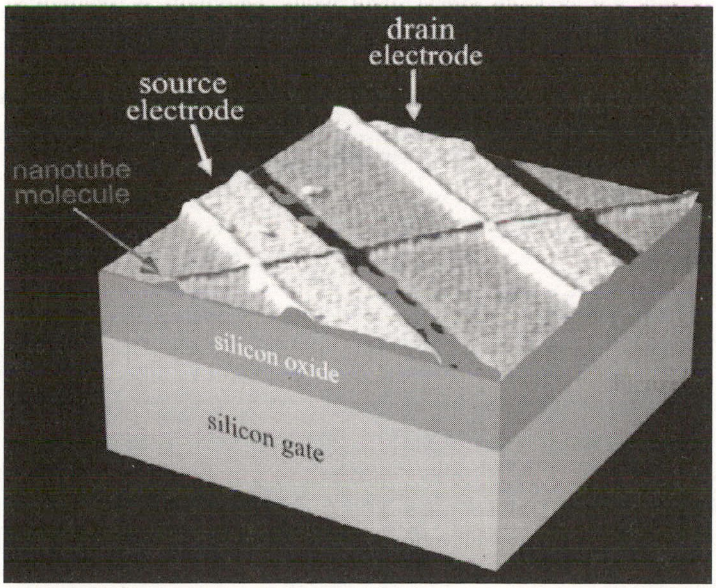

Figure 9.19

Wind and coworkers (S. J. Wind et al., *Applied Physics Letters*, **80**, *20* (May 20), 3817 (2002)) have designed a CNT-FET of improved current carrying capability (Figure 9.20). The gate electrode is above the conduction channel and separated from the channel by a thin oxide dielectric. In this manner the CNT-to-air contact is eliminated, an arrangement that prevents the circuit from acting like a p-type transistor. This arrangement also reduces the gate oxide thickness to about 15 nm, allowing for much smaller gate voltages and a steeper subthreshold slope, which is a measure of how well a transistor turns on or off.

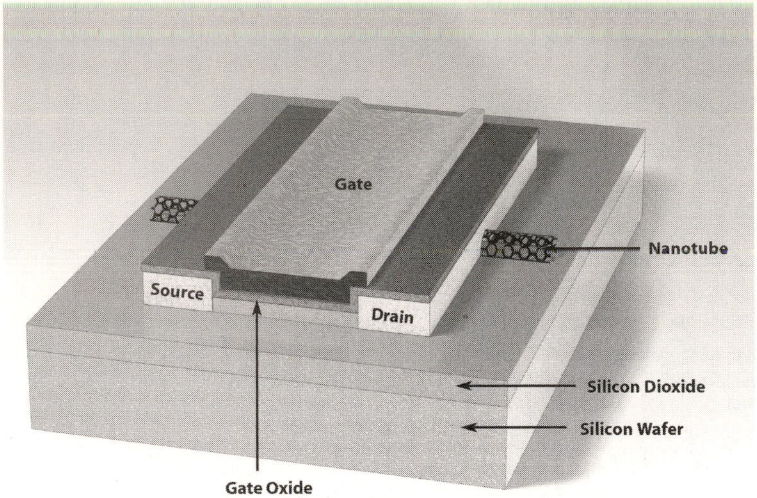

Figure 9.20

A single-electron transistor (SET) has been prepared by Cees Dekker and coworkers (*Science*, **293**, 76 (2001)) with a CNT. The SET is prepared by putting two bends in a CNT with the tip of an AFM (Figure 9.21). Bending causes two buckles that, at a distance of 20 nm, serves as a conductance barrier. When an appropriate voltage is applied to the gate below the barrier, electrons tunnel one at a time across the barrier.

Figure 9.21

Weitz et al. (*Phys. Stat. Sol. (b)*, **243**, *13*, 3394 (2006)) report on the construction of a single-wall CNT using a silane-based organic self-assembled monolayer (SAM) as a gate dielectric on top of a highly doped silicon wafer (Figure 9.22). The organic SAM is made of 18-phenoxyoctadecyltrichlorosilane. This ultrathin layer ensures strong gate coupling and therefore low operation voltages. Single-electron transistors (SETs) were obtained from individual metallic SWCNTs. Field-effect transistors made from individual semiconducting SWCNTs operate with gate-source voltages of –2 V, and they show good saturation, small hysteresis (200 mV), and a low subthreshold swing (290 mV/dec).

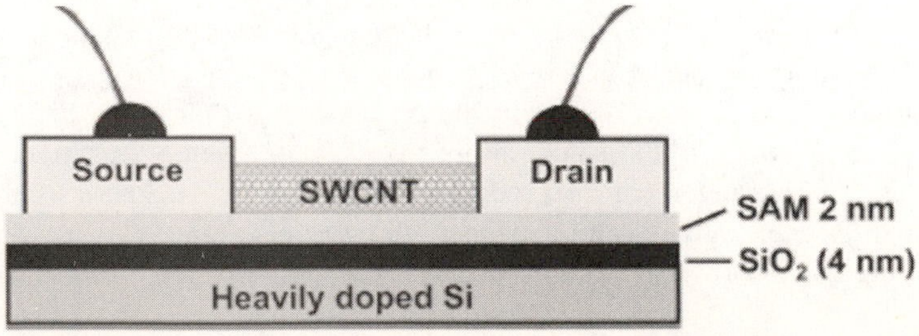

Figure 9.22

John Rodgers and researchers at the University of Illinois have reported a technique for producing near-perfect alignment of CNT transistors (Figure 9.23). The array is prepared by patterning thin strips of an iron catalyst on quartz crystals and then growing nanometer-wide CNTs along the strips using conventional carbon vapor deposition. The quartz crystal aligns the nanotubes. Transistor development then includes depositing source, drain, and gate electrodes using conventional photolithography. Transistors made with about 2,000 nanotubes can carry currents of 1 ampere, which is several orders of magnitude larger than the current possible with single nanotubes. The research group also developed a technique for transferring the nanotube arrays onto any substrate, including silicon, plastic, and glass. See Coskun Kocabas, Seong Jun Kang, Taner Ozel, Moonsub Shim, and John A. Rogers, *J. Phys. Chem. C*, **2007**, *111*, 17879, Improved Synthesis of Aligned Arrays of Single-Walled Carbon Nanotubes and Their Implementation in Thin Film Type Transistors.

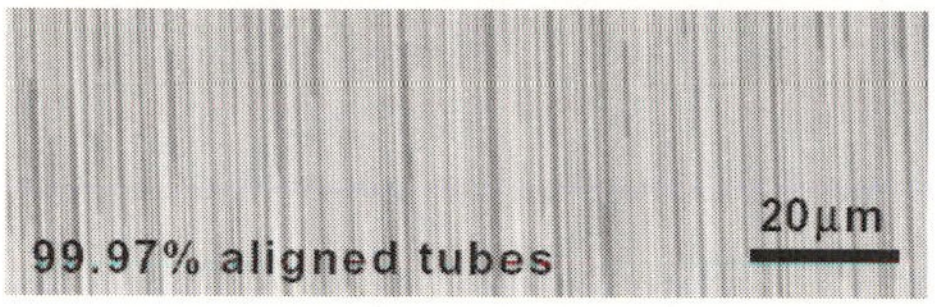

Figure 9.23

PART 3 Molecular spectroscopy

10 Rotational and vibrational spectra

Answers to discussion questions

D10.1 Symmetric rotor: the energy depends on J and K^2 [eqn. 10.10], hence each level except the $K = 0$ level is doubly degenerate. In addition, states of a given J have a component of their angular momentum along an external, laboratory-fixed axis, characterized by the quantum number M_J, that can take on $2J + 1$ values. The quantum number M_J does not affect the energy; consequently all $2J + 1$ orientations of the molecule have the same energy. It follows that a symmetric rotor level is $2(2J + 1)$-fold degenerate for $K \neq 0$ and $2J + 1$ degenerate for $K = 0$.

Linear rotor: A linear rotor has K fixed at 0, but there are still $2J + 1$ values of M_J, so the degeneracy is $2J + 1$.

Spherical rotor: A spherical rotor can be regarded as a version of a symmetric rotor in which $A = B$, and consequently, the energy is independent of the $2J + 1$ values that K can assume. Hence there is simultaneous degeneracy of $2J + 1$ in both K and M_J, resulting in a total degeneracy of $(2J + 1)^2$.

Asymmetric rotor: In an asymmetric rotor there is no longer a preferred direction in the molecule that carries out a simple rotation about J. Thus for each value of J , there is only a degeneracy of $2J + 1$, as in a linear rotor.

If a decrease in rigidity affects the symmetry of the molecule, the rotational degeneracy could be affected also. For example, if a wobble resulted in a spherical rotor being transformed into an asymmetric rotor, the rotational degeneracy of the molecule would change.

D10.3 In addition to the quantum numbers J and K, the rotational states of molecules are characterized by a quantum number M_J, which can take on $2J + 1$ values $0, \pm 1, \pm 2, \ldots, \pm J$. In the absence of external fields, the energy of the molecule does not depend on M_J, and for linear rotors, for example, there is a degeneracy of $2J + 1$ associated with the quantum number M_J in the state J. This degeneracy is partly removed when an electric field is applied to a polar molecule, such as HCl. This splitting of the energies of the M_J states is called the Stark effect and is proportional to the permanent dipole moment of the molecule. The observation of the Stark effect can therefore be used to measure molecular dipole moments in microwave spectroscopy. The Stark effect is also of importance in the technology of microwave spectrometers because an oscillating electric field will modulate the position of spectral lines and therefore the intensity of absorption or emission at a particular frequency. Spectrometers using detectors that make use of this oscillation of the intensity are called Stark modulation spectrometers.

D10.5

(1) *Rotational Raman spectroscopy*. The gross selection rule is that the molecule must be anisotropically polarizable, which is to say that its polarizability, α, depends on the direction of the electric field relative to the molecule. Non-spherical rotors satisfy this condition. Therefore, linear and symmetric rotors are rotationally Raman active.

(2) *Vibrational Raman spectroscopy*. The gross selection rule is that the polarizability of the molecule must change as the molecule vibrates. All diatomic molecules satisfy this condition as the molecules swell and contract during a vibration, the control of the nuclei over the electrons varies, and the molecular polarizability changes. Hence both homonuclear and heteronuclear diatomics are vibrationally Raman active. In polyatomic molecules it is usually quite difficult to judge by inspection whether or not the molecule is anisotropically polarizable; hence group

theoretical methods are relied on for judging the Raman activity of the various normal modes of vibration. The procedure is discussed in Section 10.15(b) and demonstrated in *A Brief Illustration*.

D10.7 $^1H^{12}C\equiv{}^{12}C^1H$ is a linear molecule. It has hydrogen nuclei with half integral spin and carbon nuclei with zero spin. Rotation by 180 degrees produces an equivalent form. This rotation is equivalent to the interchange of identical hydrogen nuclei. Since hydrogen is a fermion, the overall rotational wave function must be antisymmetric with respect to exchange of identical particles. The overall statistics are Fermi statistics. This leads to the result that symmetric hydrogen nuclear spin functions must be paired with antisymmetrical spatial wave functions for rotation and vice versa. Consequently $^1H^{12}C\equiv{}^{12}C^1H$ will exhibit ortho and para forms just as for H_2. The (odd J)/(even J) statistical weight ratio is 3/1.

$^2H^{12}C\equiv{}^{12}C^2H$ is a linear molecule. It has deuterium nuclei with integral spin and carbon nuclei with zero spin. Rotation by 180 degrees produces an equivalent form. This rotation is equivalent to the interchange of identical deuterium nuclei. Since deuterium is a boson, the overall rotational wave function must be symmetric with respect to exchange of identical particles. The overall statistics are Bose statistics. This leads to the result that symmetric deuterium nuclear spin functions must be paired with symmetrical spatial wave functions for rotation and also leads to the result that antisymmetric deuterium nuclear spin functions must be paired with antisymmetrical spatial wave functions for rotation. Consequently $^2H^{12}C\equiv{}^{12}C^2H$ will exhibit ortho and para forms. The (odd J)/(even J) statistical weight ratio is 3/6.

From the point of view of the effect of nuclear statistics on the occupation of rotational energy levels, the situation in $^1H^{13}C\equiv{}^{13}C^1H$ is similar to the case of $^2H^{12}C\equiv{}^{12}C^2H$. $^1H^{13}C\equiv{}^{13}C^1H$ will exhibit ortho and para forms. The (odd J)/(even J) statistical weight ratio is 6/10.

D10.9 The rotational constants in vibrationally excited states are smaller than the constants in the vibrational ground state and continue to get smaller as the vibrational level increases. Any anharmonicity in the vibration causes a slight extension of the bond length in the excited state. This results in an increase in the moment of inertia and a consequent decrease in the rotational constant. Problem 10.13 gives the equation that describes how the rotational constant changes with increasing vibrational level. That equation is $\tilde{B}_v=\tilde{B}_e-a(v+\frac{1}{2})$, where $\tilde{B}_e$ is a constant and $\tilde{B}_v$ is the rotational constant in level v.

D10.11 The exclusion rule applies to the benzene molecule because it has a center of symmetry. Consequently, none of the normal modes of vibration of benzene can be both infrared and Raman active. If we wish to characterize all the normal modes, we must obtain both kinds of spectra. See the solutions to Exercises 10.25(a) and 10.25(b) for specific illustrations of which modes are IR active and which are Raman active.

Solutions to exercises

E10.1(a) Ozone is an asymmetric rotor similar to H_2O; hence we follow the method of Example 10.1 and use $I=\sum_i m_i x_i^2$ [10.1] to calculate the moment of inertia, I_C, about the C_2 axis.

$$I=m_O\left(x_O^2+0+x_O^2\right)=2m_Ox_O^2=2m_O(R\sin\phi)^2$$

where the bond angle is denoted 2ϕ and the bond length is denoted R. Substitution of the data gives

$$I=2\times15.9949\text{u}\times1.66054\times10^{-27}\,\text{kg/u}\times(1.28\times10^{-10}\,\text{m}\times\sin58.5°)^2$$

$$=\boxed{6.33\times10^{-46}\,\text{kg m}^2}$$

The corresponding rotational constant is

$$A = \frac{\hbar}{4\pi c I_c} = \frac{1.05447\times10^{-34}\ \text{J s}}{4\pi\times2.998\times10^{8}\ \text{m s}^{-1}\times6.33\times10^{-46}\ \text{kg m}^2} = 44.21\ \text{m}^{-1} = \boxed{0.4421\ \text{cm}^{-1}}$$

E10.2(a) In order to conform to the symbols used in the first symmetric rotor figure of Table 10.1, we will use the molecular formula BA_4. $I_{\parallel}$ is along a bond and $I_{\perp}$ is perpendicular to both $I_{\parallel}$ and a molecular face that does not contain $I_{\parallel}$ (see the spherical rotor of text Figure 10.4). For our molecule, $R' = R$, $m_C = m_A$, and the equations of Table 10.1 simplify to

$$I_{\parallel}/m_A R^2 = 2(1-\cos\theta)$$

$$I_{\perp}/m_A R^2 = 1-\cos\theta + (m_A + m_B)\times(1+2\cos\theta)/m$$

$$+\left\{(3m_A + m_B) + 6m_A\left[\tfrac{1}{3}(1+2\cos\theta)\right]^{1/2}\right\}/m \text{ where } m = 4m_A + m_B$$

The $I_{\parallel}/m_A R^2$ moment of inertia ratio does not depend on specific atomic masses, so the plot of this ratio against θ, shown in Figure 10.1, describes the angular dependence of all molecules having the formula BA_4. However, the $I_{\perp}/m_A R^2$ moment of inertia ratio does have a specific atomic mass dependency, so we will plot its angular dependence for CH_4, an important fuel and powerful greenhouse gas. The computational equation is

$$I_{\perp}/m_A R^2 = 1-\cos\theta + 13(1+2\cos\theta)/16 + \left\{15 + 6\left[\tfrac{1}{3}(1+2\cos\theta)\right]^{1/2}\right\}/16$$

and its plot is also found in Figure 10.1. As θ increases, atoms move away from the axis of $I_{\parallel}$, which causes the moment of inertia around this axis to increase. Atoms move toward the axis of $I_{\perp}$ as θ increases, thereby decreasing the moment of inertia around this axis.

COMMENT. Figure 10.1 suggests that $I_{\parallel}(\theta_{\text{tetra}}) = I_{\perp}(\theta_{\text{tetra}})$ for all tetrahedral molecules where θ_{tetra} is the tetrahedral angle (approximately 109.471°). Can you provide an analytic proof that this is true? Hint: The exact value of θ_{tetra} is $\pi/2 + \sin^{-1}(1/3)$ in radians where $\sin^{-1}()$ is the arcsin( ) function.

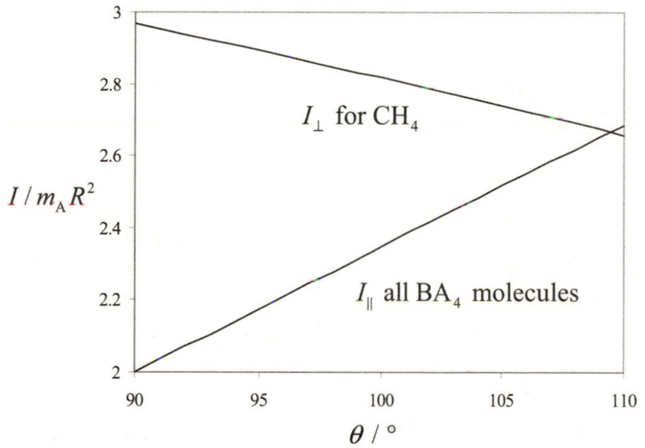

Figure 10.1

E10.3(a) (a) asymmetric (b) oblate symmetric (c) spherical (d) prolate symmetric

E10.4(a) NO is a linear rotor, and we assume that there is little centrifugal distortion; hence

$$\tilde{F}(J) = \tilde{B}J(J+1)\ [10.12]$$

with $\tilde{B} = \dfrac{\hbar}{4\pi c I}$, $I = m_{\text{eff}}R^2$ [Table 10.1], and

$$m_{\text{eff}} = \frac{m_N m_O}{m_N + m_O}\ \text{[nuclide masses from the } \textit{Data Section}\text{]}$$

$$= \left(\frac{(14.003\,\text{u})\times(15.995\,\text{u})}{(14.003\,\text{u})+(15.995\,\text{u})}\right)\times(1.6605\times10^{-27}\ \text{kg u}^{-1}) = 1.240\times10^{-26}\ \text{kg}$$

Then $I=(1.240\times10^{-26}\,\text{kg})\times(1.15\times10^{-10}\,\text{m})^2=1.64\overline{0}\times10^{-46}\,\text{kg m}^2$

and $\tilde{B}=\dfrac{1.0546\times10^{-34}\,\text{Js}}{(4\pi)\times(2.998\times10^8\,\text{m s}^{-1})\times(1.64\overline{0}\times10^{-46}\,\text{kg m}^2)}=170.\overline{7}\,\text{m}^{-1}=1.70\overline{7}\,\text{cm}^{-1}$

The wavenumber of the $J=3\leftarrow2$ transition is

$$\tilde{\nu}=2\tilde{B}(J+1)[10.17\text{a}]=6\tilde{B}[J=2]=(6)\times(1.70\overline{7}\,\text{cm}^{-1})=10.2\overline{4}\,\text{cm}^{-1}$$

The frequency is

$$\nu=\tilde{\nu}c=(10.2\overline{4}\,\text{cm}^{-1})\times\left(\frac{10^2\,\text{m}^{-1}}{1\,\text{cm}^{-1}}\right)\times(2.998\times10^8\,\text{ms}^{-1})=\boxed{3.07\times10^{11}\,\text{Hz}}$$

When centrifugal distortion is taken into account, the frequency decreases, as can be seen by considering eqn. 10.17(b).

Question. What is the percentage change in these calculated values if centrifugal distortion is included?

E10.5(a) The wavenumber of the transition is related to the rotational constant by

$$hc\tilde{\nu}=\Delta E=hc\tilde{B}[J(J+1)-(J-1)J]=2hc\tilde{B}J\ [10.6,\ 10.8]$$

where J refer to the upper state $(J=3)$. The rotational constant is related to molecular structure by

$$\tilde{B}=\frac{\hbar}{4\pi cI}\ [10.5]$$

where I is moment of inertia. Putting these expressions together yields

$$\tilde{\nu}=2\tilde{B}J=\frac{\hbar J}{2\pi cI}\quad\text{so}\quad I=\frac{\hbar J}{2\pi c\tilde{\nu}}=\frac{(1.0546\times10^{-34}\,\text{J s})\times(3)}{2\pi(2.998\times10^{10}\,\text{cm s}^{-1})\times(63.56\,\text{cm}^{-1})}$$
$$=2.642\times10^{-47}\,\text{kg m}^2$$

The moment of inertia is related to the bond length by

$$I=m_{\text{eff}}R^2\ \text{so}\ R=\sqrt{\frac{I}{m_{\text{eff}}}}$$

$$m_{\text{eff}}^{-1}=m_{\text{H}}^{-1}+m_{\text{Cl}}^{-1}=\frac{(1.0078\,\text{u})^{-1}+(34.9688\,\text{u})^{-1}}{1.66054\times10^{-27}\,\text{kg u}^{-1}}=6.1477\times10^{26}\,\text{kg}^{-1}$$

and $R=\sqrt{(6.1477\times10^{26}\,\text{kg}^{-1})\times(2.642\times10^{-47}\,\text{kg m}^2)}=1.274\times10^{-10}\,\text{m}=\boxed{127.4\,\text{pm}}$

E10.6(a) If the spacing of lines is constant, the effects of centrifugal distortion are negligible. Hence we can use for the wavenumbers of the transitions

$$\tilde{F}(J)-\tilde{F}(J-1)=2\tilde{B}J\ [10.8]$$

Since $J=1,2,3,\ldots$, the spacing of the lines is $2\tilde{B}$.

$$12.604\,\text{cm}^{-1}=2\tilde{B}$$

$$\tilde{B}=6.302\,\text{cm}^{-1}=6.302\times10^2\,\text{m}^{-1}$$

$$I=\frac{\hbar}{4\pi c\tilde{B}}=m_{\text{eff}}R^2$$

$$\frac{\hbar}{4\pi c}=\frac{1.0546\times10^{-34}\,\text{Js}}{(4\pi)\times(2.9979\times10^8\,\text{ms}^{-1})}=2.7993\times10^{-44}\,\text{kg m}$$

$$I = \frac{2.7993\times10^{-44}\,\text{kg m}}{6.302\times10^{2}\text{m}^{-1}} = \boxed{4.442\times10^{-47}\,\text{kg m}^2}$$

$$m_{\text{eff}} = \frac{m_{\text{Al}}m_{\text{H}}}{m_{\text{Al}}+m_{\text{H}}}$$

$$= \left(\frac{(26.98)\times(1.008)}{(26.98)+(1.008)}\right)\text{u}\times\left(1.6605\times10^{-27}\text{kg u}^{-1}\right) = 1.61\overline{3}6\times10^{-27}\,\text{kg}$$

$$R = \left(\frac{I}{m_{\text{eff}}}\right)^{1/2} = \left(\frac{4.442\times10^{-47}\,\text{kg m}^2}{1.6136\times10^{-27}\,\text{kg}}\right)^{1/2} = 1.659\times10^{-10}\,\text{m} = \boxed{165.9\text{ pm}}$$

E10.7(a) The determination of two unknowns requires data from two independent experiments and the equation that relates the unknowns to the experimental data. In this exercise, two independently determined values of B for two isotopically different HCN molecules are used to obtain the moments of inertia of the molecules, and from these, by use of the equation for the moment of inertia of linear triatomic rotors (Table 10.1), the interatomic distances R_{HC} and R_{CN} are calculated.

Rotational constants that are usually expressed in wavenumber (cm^{-1}) are sometimes expressed in frequency units (Hz). The conversion between the two is

$$B/\text{Hz} = c\times\tilde{B}/\text{cm}^{-1}\quad [c \text{ in cm s}^{-1}]$$

Thus, $B(\text{in Hz}) = \dfrac{\hbar}{4\pi I}$ and $I = \dfrac{\hbar}{4\pi B}$

Let $^1\text{H}=\text{H}$, $^2\text{H}=\text{D}$, $R_{\text{HC}}=R_{\text{DC}}=R$, $R_{\text{CN}}=R'$. Then

$$I(\text{HCN}) = \frac{1.05457\times10^{-34}\,\text{J s}}{(4\pi)\times\left(4.4316\times10^{10}\text{s}^{-1}\right)} = 1.8937\times10^{-46}\,\text{kg m}^2$$

$$I(\text{DCN}) = \frac{1.05457\times10^{-34}\,\text{Js}}{(4\pi)\times(3.6208\times10^{10}\text{s}^{-1})} = 2.3178\times10^{-46}\,\text{kg m}^2$$

and from Table 10.1 with isotope masses from the *Data Section*,

$$I(\text{HCN}) = m_{\text{H}}\text{R}^2 + m_N R'^2 - \frac{\left(m_{\text{H}}R - m_{\text{N}}R'\right)^2}{m_{\text{H}}+m_C+m_N}$$

$$I(\text{HCN}) = \left[\left(1.0078R^2\right)+\left(14.0031R'^2\right)-\left(\frac{\left(1.0078R-14.0031R'\right)^2}{1.0078+12.0000+14.0031}\right)\right]\text{u}$$

Multiplying through by $m/u = \left(m_{\text{H}}+m_{\text{C}}+m_{\text{N}}\right)/\text{u} = 27.0109$,

$$27.0109\times I(\text{HCN}) = \{27.0109\times(1.0078R^2+14.0031R'^2)-\left(1.0078R-14.0031R'\right)^2\}\text{u}$$

or $$\left(\frac{27.0109}{1.66054\times10^{-27}\,\text{kg}}\right)\times\left(1.8937\times10^{-46}\,\text{kg m}^2\right) = 3.0804\times10^{-18}\,\text{m}^2$$

$$= \{27.0109\times\left(1.0078R^2+14.0031R'^2\right)-(1.0078R-14.0031R')^2\} \qquad \text{(a)}$$

In a similar manner we find for DCN

$$\left(\frac{28.0172}{1.66054\times10^{-27}\,\text{kg}}\right)\times\left(2.3178\times10^{-46}\,\text{kg m}^2\right) = 3.9107\times10^{-18}\,\text{m}^2$$

$$= \left\{28.0172\times\left(2.0141R^2+14.0031R'^2\right)-(2.0141R-14.0031R')^2\right\} \qquad \text{(b)}$$

Thus there are two simultaneous quadratic equations (a) and (b) to solve for R and R'. These equations are most easily solved by readily available computer programs or by successive approximations. The results are

$$R = 1.065\times10^{-10}\,\text{m} = \boxed{106.5\text{ pm}} \quad \text{and} \quad R' = 1.156\times10^{-10}\,\text{m} = \boxed{115.6\text{ pm}}$$

These values are easily verified by direct substitution into the equations and agree well with the accepted values $R_{HC} = 1.064 \times 10^{-10}$ m and $R_{CN} = 1.156 \times 10^{-10}$ m.

E10.8(a) The Stokes lines appear at

$$\tilde{\nu}(J+2 \leftarrow J) = \tilde{\nu}_i - 2\tilde{B}(2J+3)[10.20a] \quad \text{with } J = 0,\ \tilde{\nu} = \tilde{\nu}_i - 6\tilde{B}$$

Since $\tilde{B} = 1.9987\,\text{cm}^{-1}$ (Table 10.2), the Stokes line appears at

$$\tilde{\nu} = (20487) - (6)\times(1.9987\,\text{cm}^{-1}) = \boxed{20475\,\text{cm}^{-1}}$$

E10.9(a) The separation of lines is $4\tilde{B}$ [10.20a and 10.20b], so $\tilde{B} = 0.2438\,\text{cm}^{-1}$. Then we use

$$R = \left(\frac{\hbar}{4\pi m_{\text{eff}} c\tilde{B}}\right)^{1/2}$$

with $m_{\text{eff}} = \frac{1}{2}m(^{35}\text{Cl}) = (\frac{1}{2})\times(34.9688\text{ u}) = 17.4844\text{ u}$

Therefore

$$R = \left(\frac{1.05457\times10^{-34}\,\text{J s}}{(4\pi)\times(17.4844)\times(1.6605\times10^{-27}\,\text{kg})\times(2.9979\times10^{10}\,\text{cm s}^{-1})\times(0.2438\,\text{cm}^{-1})}\right)^{1/2}$$

$$= 1.989\times10^{-10}\,\text{m} = \boxed{198.9\,\text{pm}}$$

E10.10(a) Polar molecules show a pure rotational absorption spectrum. Therefore, select the polar molecules based on their well-known structures. Alternatively, determine the point groups of the molecules and use the rule that only molecules belonging to C_n, C_{nv}, and C_s may be polar, and in the case of C_n and C_{nv}, the dipole must lie along the rotation axis. Hence the polar molecules are

(b) HCl (d) CH_3Cl (e) CH_2Cl_2

Their point group symmetries are

(b) $C_{\infty v}$ (d) C_{3v} (e) C_{2h} (*trans*), C_{2v} (*cis*)

COMMENT. Note that the *cis* form of CH_2Cl_2 is polar, but the *trans* form is not.

E10.11(a) We select the molecules with an anisotropic polarizability. A practical rule to apply is that spherical rotors do not have anisotropic polarizabilities. Therefore $\boxed{\text{(c) CH}_4\text{ is inactive}}$. All others are active.

E10.12(a) For diatomic molecules, we can use eqn. 10.21 to determine the statistical ratio of weights of populations. For chlorine-35 $I = 3/2$, and hence

$$\text{Ratio of (odd J/even J) weights of populations is } \frac{I+1}{I} = \boxed{\frac{5}{3}}$$

E10.13(a)

$$\omega = 2\pi\nu = \left(\frac{k}{m}\right)^{1/2}$$

$$k = 4\pi^2\nu^2 m = 4\pi^2\times(2.0\,\text{s}^{-1})^2\times(0.100\,\text{kg}) = 16\,\text{kg s}^{-2} = \boxed{16\,\text{N m}^{-1}}$$

E10.14(a) $\omega=\left(\dfrac{k}{m_{\text{eff}}}\right)^{1/2}$ [10.32]

The fractional difference is

$$\frac{\omega'-\omega}{\omega}=\frac{\left(\dfrac{k}{m'_{\text{eff}}}\right)^{1/2}-\left(\dfrac{k}{m_{\text{eff}}}\right)^{1/2}}{\left(\dfrac{k}{m_{\text{eff}}}\right)^{1/2}}=\frac{\left(\dfrac{1}{m'_{\text{eff}}}\right)^{1/2}-\left(\dfrac{1}{m_{\text{eff}}}\right)^{1/2}}{\left(\dfrac{1}{m_{\text{eff}}}\right)^{1/2}}=\left(\frac{m_{\text{eff}}}{m'_{\text{eff}}}\right)^{1/2}-1$$

$$=\left(\frac{m(^{23}\text{Na})m(^{35}\text{Cl})\{m(^{23}\text{Na})+m(^{37}\text{Cl})\}}{\{m(^{23}\text{Na})+m(^{35}\text{Cl})\}m(^{23}\text{Na})m(^{37}\text{Cl})}\right)^{1/2}-1$$

$$=\left(\frac{m(^{35}\text{Cl})}{m(^{37}\text{Cl})}\times\frac{m(^{23}\text{Na})+m(^{37}\text{Cl})}{m(^{23}\text{Na})+m(^{35}\text{Cl})}\right)^{1/2}-1$$

$$=\left(\frac{34.9688}{36.9651}\times\frac{22.9898+36.9651}{22.9898+34.9688}\right)^{1/2}-1=-0.01077$$

Hence, the difference is $\boxed{1.077\text{ percent}}$.

E10.15(a) $\omega=\left(\dfrac{k}{m_{\text{eff}}}\right)^{1/2}$ [10.32]; $\omega=2\pi\nu=2\pi\left(\dfrac{c}{\lambda}\right)=2\pi c\tilde{\nu}$

Therefore, $k=m_{\text{eff}}\omega^2=4\pi^2m_{\text{eff}}c^2\tilde{\nu}^2, m_{\text{eff}}=\tfrac{1}{2}m(^{35}\text{Cl})$

$$=(4\pi^2)\times\left(\frac{34.9688}{2}\right)\times(1.66054\times10^{-27}\text{ kg})\times[(2.997924\times10^{10}\text{ cm s}^{-1})\times(564.9\text{ cm}^{-1})]^2$$

$$=\boxed{328.7\text{ N m}^{-1}}$$

E10.16(a) $\omega=\left(\dfrac{k}{m_{\text{eff}}}\right)^{1/2}$ [10.32], so $k=m_{\text{eff}}\omega^2=4\pi^2m_{\text{eff}}c^2\tilde{\nu}^2$

$$m_{\text{eff}}=\frac{m_1m_2}{m_1+m_2}\text{[10.31]}$$

$$m_{\text{eff}}\left(\text{H}^{19}\text{F}\right)=\frac{(1.0078)\times(18.9984)}{(1.0078)+(18.9984)}\text{u}=0.9570\text{u}$$

$$m_{\text{eff}}\left(\text{H}^{35}\text{Cl}\right)=\frac{(1.0078)\times(34.9688)}{(1.0078)+(34.9688)}\text{u}=0.9796\text{u}$$

$$m_{\text{eff}}\left(\text{H}^{81}\text{Br}\right)=\frac{(1.0078)\times(80.9163)}{(1.0078)+(80.9163)}\text{u}=0.9954\text{u}$$

$$m_{\text{eff}}(\text{H}^{127}\text{I})=\frac{(1.0078)\times(126.9045)}{(1.0078)+(126.9045)}\text{u}=0.9999\text{u}$$

We draw up the following table.

	HF	HCl	HBr	HI
$\tilde{\nu}$ / cm^{-1}	4141.3	2988.9	2649.7	2309.5
m_{eff} / u	0.9570	0.9697	0.9954	0.9999
k / (N m^{-1})	$\boxed{967.0}$	$\boxed{515.6}$	$\boxed{411.8}$	$\boxed{314.2}$

Note the order of stiffness HF > HCl > HBr > HI.

Question. Which ratio, $\dfrac{k}{B(\text{A}-\text{B})}$ or $\dfrac{\tilde{\nu}}{B(\text{A}-\text{B})}$, where $B(\text{A}-\text{B})$ are the bond energies, is the more nearly constant across the series of hydrogen halides? Why?

E10.17(a) Data on three transitions are provided. Only two are necessary to obtain the value of $\tilde{\nu}$ and x_e. The third datum can then be used to check the accuracy of the calculated values.

$$\Delta\tilde{G}(v=1\leftarrow 0)=\tilde{\nu}-2\tilde{\nu}x_e=1556.22\,\text{cm}^{-1}[10.38]$$
$$\Delta\tilde{G}(v=2\leftarrow 0)=2\tilde{\nu}-6\tilde{\nu}x_e=3088.28\,\text{cm}^{-1}[10.39]$$

Multiply the first equation by 3, then subtract the second

$$\tilde{\nu}=(3)\times(1556.22\text{ cm}^{-1})-(3088.28\text{ cm}^{-1})=\boxed{1580.38\text{ cm}^{-1}}$$

Then from the first equation

$$x_e=\frac{\tilde{\nu}-1556.22\,\text{cm}^{-1}}{2\tilde{\nu}}=\frac{(1580.38-1556.22)\,\text{cm}^{-1}}{(2)\times(1580.38\,\text{cm}^{-1})}=\boxed{7.644\times10^{-3}}$$

x_e data are usually reported as $x_e\tilde{\nu}$, which is

$$x_e\tilde{\nu}=12.08\,\text{cm}^{-1}$$

$$\Delta\tilde{G}(v=3\leftarrow 0)=3\tilde{\nu}-12\tilde{\nu}x_e$$
$$=(3)\times(1580.38\text{ cm}^{-1})-(12)\times(12.08\text{ cm}^{-1})=4596.18\,\text{cm}^{-1}$$

which is very close to the experimental value.

E10.18(a) The R branch obeys the relation

$$\tilde{\nu}_R(J)=\tilde{\nu}+2\tilde{B}(J+1)\,[10.42c]$$

Hence, $\tilde{\nu}_R(2)=\tilde{\nu}+6\tilde{B}=(2648.98\text{ cm}^{-1})+(6)\times(8.465\text{ cm}^{-1})$ [Table 10.2] $=\boxed{2699.77\text{ cm}^{-1}}$.

E10.19(a) See *A Brief Illustration*. Select the molecules in which a vibration gives rise to a change in dipole moment. It is helpful to write down the structural formulas of the compounds. The infrared active compounds are

(b) HCl (c) CO_2 (d) H_2O

COMMENT. A more powerful method for determining infrared activity based on symmetry considerations is described in Section 10.15. Also see Exercises 10.23–10.24.

E10.20(a) The number of normal modes of vibration is given by (Section 10.15)

$$N_{\text{vib}}=\begin{cases}3N-5 \text{ for linear molecules}\\ 3N-6 \text{ for nonlinear molecules}\end{cases}$$

where N is the number of atoms in the molecule. Hence, since none of these molecules are linear,

(a) 3 (b) 6 (c) 12

COMMENT. Even for moderately sized molecules, the number of normal modes of vibration is large and they are usually difficult to visualize.

E10.21(a) This molecule is linear; hence the number of vibrational modes is $3N-5$. $N = 44$ in this case; therefore, the number of vibrational modes is $\boxed{127}$.

E10.22(a) $G_q(v) = \left(v + \frac{1}{2}\right)\tilde{\nu}_q \qquad \tilde{\nu}_q = \frac{1}{2\pi c}\left(\frac{k_q}{m_q}\right)^{1/2}$ [10.46]

The lowest energy term is $\tilde{\nu}_2$, corresponding to the normal mode for bending. For this mode the oxygen atom may be considered to remain stationary and the effective mass is approximately $m_q = \frac{2m_H m_O}{2m_H + m_O}$. For the other modes, the effective mass expression is more complicated and is beyond the scope of this text. However we know that in the ground vibrational state all normal modes have $\upsilon = 0$. Thus, since H_2O has the three normal modes shown in text Fig. 10.41, the ground vibrational term is the sum of eqn. 10.46 normal mode terms: $G_{\text{ground}} = G_1(0) + G_2(0) + G_3(0) = \boxed{\frac{1}{2}(\tilde{\nu}_1 + \tilde{\nu}_2 + \tilde{\nu}_3)}$.

E10.23(a) See Figs. 10.41(H_2O, bent) and 10.40(CO_2, linear) of the text as well as *A Brief Illustration.* Decide which modes correspond to (i) a changing electric dipole moment and (ii) a changing polarizability, and take note of the exclusion rule (Sections 10.13a–10.15).

(a) Nonlinear: All modes are both infrared and Raman active.
(b) Linear: The symmetric stretch is infrared inactive but Raman active.

The antisymmetric stretch is infrared active and (by the exclusion rule) Raman inactive. The two bending modes are infrared active and therefore Raman inactive.

E10.24(a) The uniform expansion is depicted in Figure 10.2.

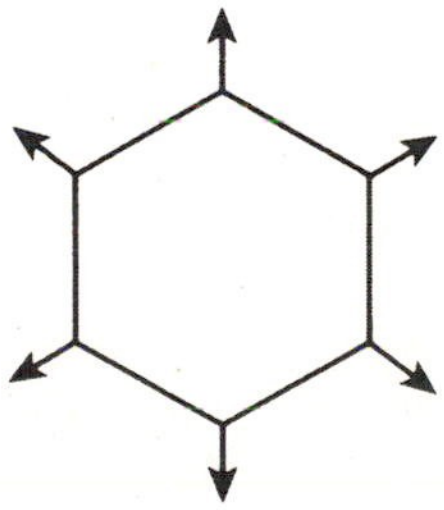

Figure 10.2

Benzene is centrosymmetric, so the exclusion rule applies (Section 10.14). The mode is infrared inactive (symmetric breathing leaves the molecular dipole moment unchanged at zero), and therefore the mode may be (and is) $\boxed{\text{Raman active}}$. In group theoretical terms, the breathing mode has symmetry A_{1g} in D_{6h}, which is the point group for benzene, and quadratic forms $x^2 + y^2$ and z^2 have this symmetry (see the character table for C_{6h}, a subgroup of D_{6h}). Hence, the mode is Raman active.

E10.25(a) Use the character table for the group C_{2v} (and see Example 10.3). The rotations span $A_2 + B_1 + B_2$. The translations span $A_1 + B_1 + B_2$. Hence the normal modes of vibration span the difference, $\boxed{4A_1 + A_2 + 2B_1 + 2B_2}$.

COMMENT. A_1, B_1 and B_2 are infrared active; all modes are Raman active.

E10.26(a) See the comment in the solution to Exercise 10.25(a). A_1, B_1 and B_2 are infrared active; all modes are Raman active.

E10.27(a) The ratio of coefficients *A/B* is

$$\frac{A}{B} = \frac{8\pi h \nu^3}{c^3} \text{ [Eqn. 10.55, \textit{Further Information} 10.1]}$$

The frequency is

$$\nu = \frac{c}{\lambda} \quad \text{so} \quad \frac{A}{B} = \frac{8\pi h}{\lambda^3}$$

(a) $$\frac{A}{B} = \frac{8\pi(6.626\times10^{-34}\,\text{J s})}{(70.8\times10^{-12}\,\text{m})^3} = \boxed{0.0469\,\text{J m}^{-3}\,\text{s}}$$

(b) $$\frac{A}{B} = \frac{8\pi h}{\lambda^3} = \frac{8\pi(6.626\times10^{-34}\,\text{J s})}{(500\times10^{-9}\,\text{m})^3} = \boxed{1.33\times10^{-13}\text{J m}^{-3}\text{s}}$$

(c) $$\frac{A}{B} = \frac{8\pi h}{\lambda^3} = 8\pi h\tilde{\nu}^3 = 8\pi\{6.626\times10^{-34}\,\text{J s}\times[3000\,\text{cm}^{-1}\times(10^2\,\text{m}^{-1}/1\,\text{cm}^{-1})]^3\}$$

$$= \boxed{4.50\times10^{-16}\,\text{J m}^{-3}\,\text{s}}$$

COMMENT. Comparison of these ratios shows that the relative importance of spontaneous transitions decreases as the frequency decreases. The quotient $\frac{A}{B}$ has units. A unitless quotient is $\frac{A}{B\rho}$ with ρ given by eqn. 10.51.

Question. What are the ratios $\frac{A}{B\rho}$ for the radiation of (a) through (c) and what additional conclusions can you draw from these results?

Solutions to problems

Solutions to numerical problems

P10.1 Rotational line separations are $2\tilde{B}$ in wavenumber units, $2\tilde{B}c$ in frequency units, and $(2\tilde{B})^{-1}$ in wavelength units. Hence the transitions are separated by $\boxed{596\text{GHZ}}$, $\boxed{19.9\text{ cm}^{-1}}$, and $\boxed{0.503\text{ mm}}$.

Ammonia is a symmetric rotor (Section 10.1), and we know that

$$\tilde{B} = \frac{\hbar}{4\pi c I_\perp}\,[10.11]$$

and from Table 10.1,

$$I_\perp = m_A R^2(1-\cos\theta) + \left(\frac{m_A m_B}{m}\right) R^2(1+2\cos\theta)$$

$m_A = 1.6735\times10^{-27}\,\text{kg}, m_B = 2.3252\times10^{-26}\,\text{kg}$, and $m = 2.8273\times10^{-26}\,\text{kg}$ with $R = 101.4$ pm and $\theta = 106°47'$, which gives

$$I_\perp = (1.6735\times10^{-27}\,\text{kg})\times(101.4\times10^{-12}\,\text{m})^2\times(1-\cos106°47')$$

$$+\left(\frac{(1.6735\times10^{-27})\times(2.3252\times10^{-26}\,\text{kg}^2)}{2.8273\times10^{-26}\,\text{kg}}\right)$$

$$\times(101.4\times10^{-12}\,\text{m})^2\times(1+2\cos\ 106°47')$$

$$= 2.815\overline{8}\times10^{-47}\,\text{kg m}^2$$

Therefore,

$$\tilde{B} = \frac{1.05457\times10^{-34}\text{ Js}}{(4\pi)\times(2.9979\times10^{8}\text{ m s}^{-1})\times(2.815\bar{8}\times10^{-47}\text{ kg m}^2)} = 994.1\text{ m}^{-1} = \boxed{9.941\text{ cm}^{-1}}$$

which is in accord with the data.

P10.3 Rotation about any axis perpendicular to the C_6 axis may be represented in its essentials by rotation of the pseudolinear molecule in Figure 10.3 about the x-axis in the figure.

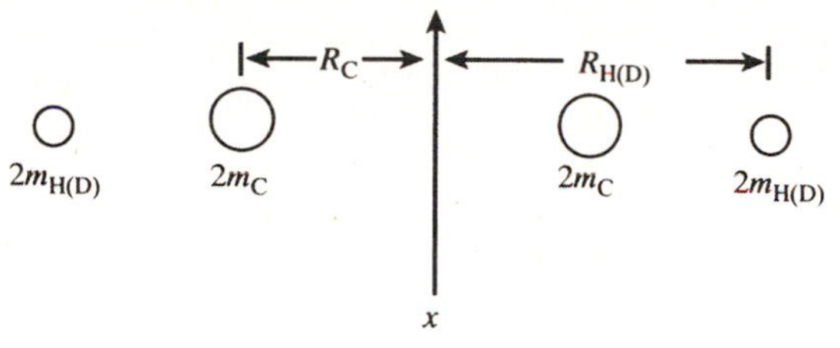

Figure 10.3

The data allow for a determination of R_C and $R_{H(D)}$, which may be decomposed into R_{CC} and $R_{CH(D)}$.

$$I_H = 4m_H R_H^2 + 4m_C R_C^2 = 147.59\times10^{-47}\text{kg m}^2$$

$$I_D = 4m_D R_D^2 + 4m_C R_C^2 = 178.45\times10^{-47}\text{kg m}^2$$

Subtracting I_H from I_D (assume that $R_H = R_D$) yields

$$4(m_D - m_H)R_H^2 = 30.86\times10^{-47}\text{ kg m}^2$$

$$4(2.01410\text{ u} - 1.0078\text{ u})\times(1.66054\times10^{-27}\text{ kg u}^{-1})\times(R_H^2) = 30.86\times10^{-47}\text{ kg m}^2$$

$$R_H^2 = 4.616\bar{9}\times10^{-20}\text{m}^2 \qquad R_H = 2.149\times10^{-10}\text{m}$$

$$R_C^2 = R_C^2 = \frac{(147.59\times10^{-47}\text{ kg m}^2)-(4m_H R_H^2)}{4m_C}$$

$$= \frac{(147.59\times10^{-47}\text{ kg m}^2)-(4)\times(1.0078\text{ u})\times(1.66054\times10^{-27}\text{ kg u}^{-1})\times(4.616\bar{9}\times10^{-20}\text{ m}^2)}{(4)\times(12.011\text{ u})\times(1.66054\times10^{-27}\text{ kg u}^{-1})}$$

$$= 1.4626\times10^{-20}\text{ m}^2$$

$$R_C = 1.209\times10^{-10}\text{ m}$$

Figure 10.4 shows the relation between R_H, R_C, R_{CC}, and R_{CH}.

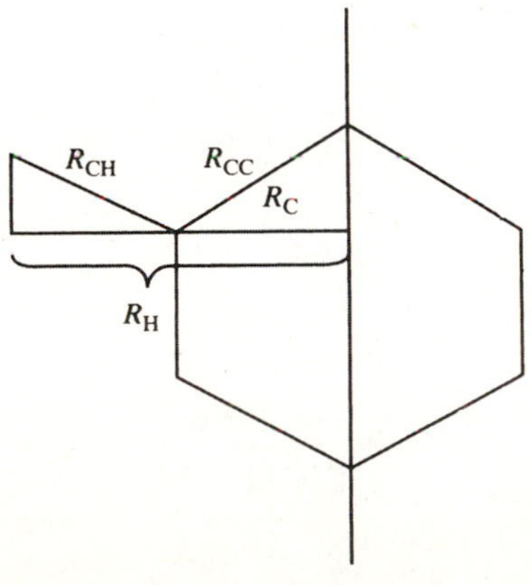

Figure 10.4

$$R_{CC} = \frac{R_C}{\cos 30°} = \frac{1.209\times10^{-10}\,\text{m}}{0.8660} = 1.396\times10^{-10}\,\text{m} = \boxed{139.6\ \text{pm}}$$

$$R_{CH} = \frac{R_H - R_C}{\cos 30°} = \frac{0.940\times10^{-10}}{0.8660} = 1.08\bar{5}\times10^{-10} = \boxed{108.\bar{5}\ \text{pm}}$$

$$R_{CD} = R_{CH}$$

COMMENT. These values are very close to the interatomic distances quoted by Herzberg in *Electronic Spectra and Electronic Structure of Polyatomic Molecules,* p. 666, which are 139.7 and 108.4 pm respectively.

P10.5 $$\tilde{\nu} = 2\tilde{B}(J+1)\ [10.17a] = 2\tilde{B}$$

Hence, $\tilde{B}(^1\text{HCl}) = 10.4392\,\text{cm}^{-1}$, $\tilde{B}(^2\text{HCl}) = 5.3920\,\text{cm}^{-1}$

$$\tilde{B} = \frac{\hbar}{4\pi c I}[10.5] \qquad I = m_{\text{eff}}R^2\ [\text{Table 10.1}]$$

$$R^2 = \frac{\hbar}{4\pi c m_{\text{eff}}\tilde{B}} \qquad \frac{\hbar}{4\pi c} = 2.79927\times10^{-44}\ \text{kg m}$$

$$m_{\text{eff}}(\text{HCl}) = \left(\frac{(1.007825\,\text{u})\times(34.96885\,\text{u})}{(1.007825\,\text{u})+(34.96885\,\text{u})}\right)\times(1.66054\times10^{-27}\ \text{kg u}^{-1})$$
$$= 1.62665\times10^{-27}\ \text{kg}$$

$$m_{\text{eff}}(\text{DCl}) = \left(\frac{(2.0140\,\text{u})\times(34.96885\,\text{u})}{(2.0140\,\text{u})+(34.96885\,\text{u})}\right)\times(1.66054\times10^{-27}\ \text{kg u}^{-1})$$
$$= 3.1622\times10^{-27}\ \text{kg}$$

$$R^2(\text{HCl}) = \frac{2.79927\times10^{-44}\ \text{kg m}}{(1.62665\times10^{-27}\ \text{kg})\times(1.04392\times10^{3}\ \text{m}^{-1})} = 1.64848\times10^{-20}\ \text{m}^2$$

$$R(\text{HCl}) = 1.28393\times10^{-10}\ \text{m} = \boxed{128.393\,\text{pm}}$$

$$R^2(^2\text{HCl}) = \frac{2.79927\times10^{-44}\ \text{kg m}}{(3.1622\times10^{-27}\ \text{kg})\times(5.3920\times10^{2}\ \text{m}^{-1})} = 1.6417\times10^{-20}\ \text{m}^2$$

$$R(^2\text{HCl}) = 1.2813\times10^{-10}\ \text{m} = \boxed{128.13\,\text{pm}}$$

COMMENT. Since the effects of centrifugal distortion have not been taken into account, the number of significant figures in the calculated values of *R* above should be no greater than 4, despite the fact that the data are precise to 6 figures.

P10.7 From the equation for a linear rotor in Table 10.1, it is possible to show that $I\times m = m_a m_c(R+R')^2 + m_a m_b R^2 + m_b m_c R'^2$.

Thus, $$I(^{16}\text{O}^{12}\text{C}^{32}\text{S}) = \left(\frac{m(^{16}\text{O})m(^{32}\text{S})}{m(^{16}\text{O}^{12}\text{C}^{32}\text{S}}\right)\times(R+R')^2 + \left(\frac{m(^{12}\text{C})\{m(^{16}\text{O})R^2 + m(^{32}\text{S})R'^2\}}{m(^{16}\text{O}^{12}\text{C}^{32}\text{S})}\right)$$

$$I(^{16}\text{O}^{12}\text{C}^{34}\text{S}) = \left(\frac{m(^{16}\text{O})m(^{34}\text{S})}{m(^{16}\text{O}^{12}\text{C}^{34}\text{S}}\right)\times(R+R')^2 + \left(\frac{m(^{12}\text{C})\{m(^{16}\text{O})R^2 + m(^{34}\text{S})R'^2\}}{m(^{16}\text{O}^{12}\text{C}^{34}\text{S})}\right)$$

$m(^{16}\text{O}) = 15.9949\ \text{u}$, $m(^{12}\text{C}) = 12.0000\,\text{u}$, $m(^{32}\text{S}) = 31.9721\text{u}$, and $m(^{34}\text{S}) = 33.9679\,\text{u}$.

Hence,

$$I(^{16}\mathrm{O}^{12}\mathrm{C}^{32}\mathrm{S})/\mathrm{u} = (8.5279)\times(R+R')^2 + (0.20011)\times(15.9949R^2 + 31.9721R'^2)$$

$$I(^{16}\mathrm{O}^{12}\mathrm{C}^{34}\mathrm{S})/\mathrm{u} = (8.7684)\times(R+R')^2 + (0.19366)\times(15.9949R^2 + 33.9679R'^2)$$

The spectral data provide the experimental values of the moments of inertia based on the relation $\nu = 2c\tilde{B}(J+1)$ [10.17a] with $\tilde{B} = \dfrac{\hbar}{4\pi cI}$ [10.5]. These values are set equal to the above equations, which are then solved for R and R'. The mean values of I obtained from the data are

$$I(^{16}\mathrm{O}^{12}\mathrm{C}^{32}\mathrm{S}) = 1.37998\times10^{-45}\ \mathrm{kg\ m^2}$$

$$I(^{16}\mathrm{O}^{12}\mathrm{C}^{34}\mathrm{S}) = 1.41460\times10^{-45}\ \mathrm{kg\ m^2}$$

Therefore, after conversion of the atomic mass units to kilograms, the equations we must solve are

$$1.37998\times10^{-45}\ \mathrm{m^2} = (1.4161\times10^{-26})\times(R+R')^2 + (5.3150\times10^{-27}R^2) + (1.0624\times10^{-26}R'^2)$$

$$1.41460\times10^{-45}\ \mathrm{m^2} = (1.4560\times10^{-26})\times(R+R')^2 + (5.1437\times10^{-27}R^2) + (1.0923\times10^{-26}R'^2)$$

These two equations may be solved for R and R'. They are tedious but straightforward to solve by hand. Exercise 10.7(b) illustrates the details of the solution. Readily available mathematical software can be used to quickly give the result. The outcome is $R = \boxed{116.28\ \mathrm{pm}}$ and $R' = \boxed{155.97\ \mathrm{pm}}$. These values may be checked by direct substitution into the equations.

COMMENT. The starting point of this problem is the actual experimental data on spectral line positions. Exercise 10.7(b) is similar to this problem; it starts, however, from given values of the rotational constants B, which were themselves obtained from the spectral line positions. So the results for R and R' are expected to be and are essentially identical.

Question. What are the rotational constants calculated from the data on the positions of the absorption lines?

P10.9 Examination of the wavenumbers of the transitions allows identifications to be made as shown in the table below. The J value in the table is the J value of the rotational states in the $v = 0$ vibrational level. The notation (R) or (P) after the J values refers to the R and P branches of the vibration-rotation spectrum.

$\tilde{\nu}$	2998.05	2981.05	2963.35	2944.99	2925.92	2906.25	2865.14	2843.63	2821.59	2799.00
J	5(R)	4(R)	3(R)	2(R)	1(R)	0(R)	1(P)	2(P)	3(P)	4(P)

The values of $\tilde{B}_0$ and $\tilde{B}_1$ are determined from the equations for the combination differences, eqns. 10.44(a) and (b).

$\Delta_0 = \tilde{\nu}_R(J-1) - \tilde{\nu}_P(J+1) = 4\tilde{B}_0(J+\tfrac{1}{2})$, and

$\Delta_1 = \tilde{\nu}_R(J) - \tilde{\nu}_P(J) = 4\tilde{B}_1(J+\tfrac{1}{2})$

Substituting values for $\tilde{\nu}_R$ and $\tilde{\nu}_P$ from the above table for J = 1, 2, 3, and 4, we obtain the following values for Δ_0 and Δ_1.

J	1	2	3	4
$J + ½$	3/2	5/2	7/2	9/2
Δ_0/cm^{-1}	62.62	104.33	145.99	
Δ_1/cm^{-1}	60.78	101.36	141.76	182.05
$\tilde{B}_0/\text{cm}^{-1}$	10.437	10.433	10.428	
$\tilde{B}_1/\text{cm}^{-1}$	10.130	10.136	10.126	10.113

An average of these values gives $\boxed{\tilde{B}_0 = 10.433\ \text{cm}^{-1}}$ and $\boxed{\tilde{B}_1 = 10.126\ \text{cm}^{-1}}$. Values for $\tilde{B}_0$ and $\tilde{B}_1$ can also be obtained by plotting the combination differences, Δ_0 and Δ_1, against $J + ½$; the slopes give $4\tilde{B}_0$ and $4\tilde{B}_1$, respectively.

P10.11 Plot frequency against J as in Figure 10.5.

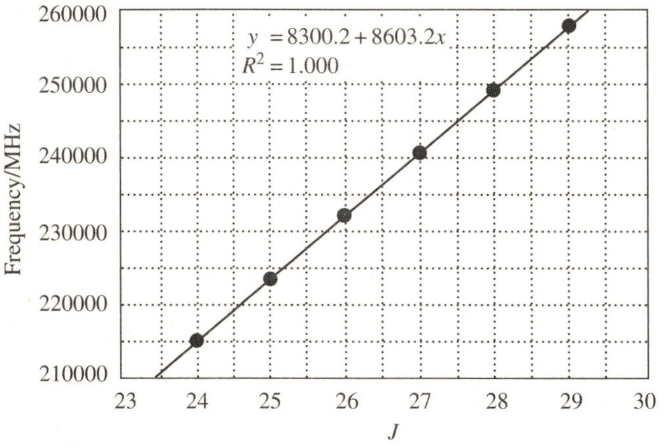

Figure 10.5

The rotational constant is related to the wavenumbers of observed transitions by

$$\tilde{\nu} = 2\tilde{B}(J+1) = \frac{\nu}{c} \quad \text{so} \quad \nu = 2\tilde{B}c(J+1)$$

A plot of ν versus J, then, has a slope of $2\tilde{B}c$. From Figure 10.5, the slope is 8603 MHZ, so

$$\tilde{B} = \frac{8603\times10^{6}\,\text{s}^{-1}}{2(2.988\times10^{8}\,\text{m s}^{-1})} = \boxed{14.35\ \text{m}^{-1}}$$

The most highly populated energy level is roughly

$$J_{\max} = \left(\frac{kT}{2hc\tilde{B}}\right)^{1/2} - \frac{1}{2}$$

so $$J_{\max} = \left(\frac{(1.381\times10^{-23}\,\text{J K}^{-1})\times(298\ \text{K})}{(6.626\times10^{-34}\,\text{J s})\times(8603\times10^{6}\text{s}^{-1})}\right)^{1/2} - \frac{1}{2} = \boxed{26} \text{ at 298 K}$$

and $$J_{\max} = \left(\frac{(1.381\times10^{-23}\,\text{J K}^{-1})\times(100\ \text{K})}{(6.626\times10^{-34}\,\text{J s})\times(8603\times10^{6}\text{s}^{-1})}\right)^{1/2} - \frac{1}{2} = \boxed{15} \text{ at 100 K}$$

P10.13 For IF, the rotational constant $\tilde{B}_e = 0.27971\ \text{cm}^{-1}$ and $a = 0.187\ \text{m}^{-1} = 0.00187\ \text{cm}^{-1}$. Values for $\tilde{B}_0$ and $\tilde{B}_1$ are calculated from $\tilde{B}_v = \tilde{B}_e - a(v + \tfrac{1}{2})$.

$$\tilde{B}_0 = \tilde{B}_e - \tfrac{1}{2}a = 0.27971\ \text{cm}^{-1} - \tfrac{1}{2}(0.00187\ \text{cm}^{-1}) = \boxed{0.27877\overline{5}\ \text{cm}^{-1}}$$

$$\tilde{B}_1 = \tilde{B}_e - \tfrac{3}{2}a = 0.27971\ \text{cm}^{-1} - \tfrac{3}{2}(0.00187\ \text{cm}^{-1}) = \boxed{0.27690\overline{5}\ \text{cm}^{-1}}$$

The wavenumbers of the $J' \to 3$ transitions of the P and R branches of the spectrum are given by eqn. 10.43.

$$\tilde{\nu}_P(J) = \tilde{\nu} - (\tilde{B}_1 + \tilde{B}_0)J + (\tilde{B}_1 - \tilde{B}_0)J^2 \quad \text{and} \quad \tilde{\nu}_R(J) = \tilde{\nu} + (\tilde{B}_1 + \tilde{B}_0)(J+1) + (\tilde{B}_1 - \tilde{B}_0)(J+1)^2$$

When anharmonicities are present $\tilde{\nu}$ in the formulas above is replaced by

$$\Delta\tilde{G}(v) = \tilde{\nu} - 2(v+1)x_e\tilde{\nu} \ [10.38]$$

For $v = 0$,

$$\Delta\tilde{G}(v) = \tilde{\nu} - 2x_e\tilde{\nu} = 610.258\ \text{cm}^{-1} - 2\times 3.141\ \text{cm}^{-1} = 603.976\ \text{cm}^{-1}$$

For the P branch $J' = J = 4$, and for the R branch $J' = J = 2$. Substituting all these values into eqn. 10.43, we obtain

$$\tilde{\nu}_P(J) = \Delta\tilde{G}(v) - (\tilde{B}_1 + \tilde{B}_0)J + (\tilde{B}_1 - \tilde{B}_0)J^2 = \boxed{601.723\ \text{cm}^{-1}}$$

$$\tilde{\nu}_R(J) = \Delta\tilde{G}(v) + (\tilde{B}_1 + \tilde{B}_0)(J+1) + (\tilde{B}_1 - \tilde{B}_0)(J+1)^2 = \boxed{605.626\ \text{cm}^{-1}}$$

The dissociation energy of the IF molecule may be obtained from $\tilde{D}_e = \dfrac{\tilde{\nu}^2}{4x_e\tilde{\nu}}$ [10.36] and the relation $\tilde{D}_0 = \tilde{D}_e - \frac{1}{2}\tilde{\nu} + \frac{1}{4}x_e\tilde{\nu}$ [10.36] if a Morse potential energy is assumed.

Substituting the values given for $\tilde{\nu}$ and $\tilde{\nu}x_e$, we obtain

$$\tilde{D}_e = 29641\ \text{cm}^{-1} \ \text{and} \ \boxed{\tilde{D}_0 = 29337\ \text{cm}^{-1}}$$

P10.15 See Figure 10.6 for a plot of $\Delta\tilde{G}(v) = \tilde{\nu} - 2(v+1)x_e\tilde{\nu}$ [10.38] against $v + 1$.

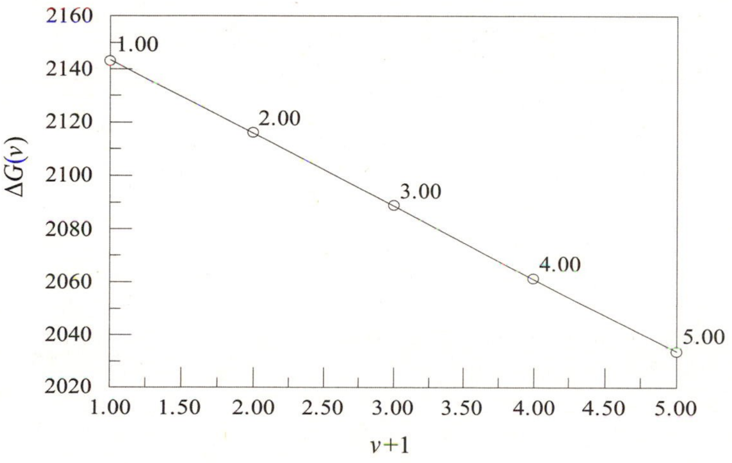

Figure 10.6

The intercept gives $\boxed{\tilde{\nu} = 2170.8\ \text{cm}^{-1}}$ and the slope gives $2x_e\tilde{\nu} = 27.4\ \text{cm}^{-1}$; thus, $\boxed{x_e\tilde{\nu} = 13.7\ \text{cm}^{-1}}$.

P10.17 $\tilde{D}_0 = \tilde{D}_e - \tilde{\nu}'$ with $\tilde{\nu}' = \frac{1}{2}\tilde{\nu}' - \frac{1}{4}x_e\tilde{\nu}'$ [Section 10.9]

(a) ^{1}HCl: $\tilde{\nu}' = \{(1494.9) - (\frac{1}{4})\times(52.05)\},\text{cm}^{-1} = 1481.8\,\text{cm}^{-1}$, or $0.184\,\text{eV}$

Hence, $\tilde{D}_0 = 5.33 - 0.18 = \boxed{5.15\text{ eV}}$.

(b) ^{2}HCl $\dfrac{2m_{\text{eff}}\omega x_e}{\text{h}} = a^2[10.36]$, so $\tilde{\nu}x_e \propto \dfrac{1}{m_{\text{eff}}}$ as a is a constant. We also have $\tilde{D}_e = \dfrac{\tilde{\nu}^2}{4x_e\tilde{\nu}}$; so $\tilde{\nu}^2 \propto \dfrac{1}{m_{\text{eff}}}$, implying $\tilde{\nu} \propto \dfrac{1}{m_{\text{eff}}^{1/2}}$. Reduced masses were calculated in Exercises 10.16(a) and (b), and we can write

$$\tilde{\nu}(^2\text{HCl}) = \left(\frac{m_{\text{eff}}(^1\text{HCl})}{m_{\text{eff}}(^2\text{HCl})}\right)^{1/2} \times \tilde{\nu}(^1\text{HCl}) = (0.7172)\times(2989.7\,\text{cm}^{-1}) = 2144.2\,\text{cm}^{-1}$$

$$x_e\tilde{\nu}(^2\text{HCl}) = \left(\frac{m_{\text{eff}}(^1\text{HCl})}{m_{\text{eff}}(^2\text{HCl})}\right) \times x_e\tilde{\nu}(^1\text{HCl}) = (0.5144)\times(52.05\,\text{cm}^{-1}) = 26.77\,\text{cm}^{-1}$$

$$\tilde{\nu}'(^2\text{HCl}) = (\tfrac{1}{2})\times(2144.2) - (\tfrac{1}{4})\times(26.77\,\text{cm}^{-1}) = 1065.4\,\text{cm}^{-1},\quad 0.132\,\text{eV}$$

Hence, $\tilde{D}_0(^2\text{HCl}) = (5.33 - 0.132)\,\text{eV} = \boxed{5.20\text{ eV}}$.

P10.19 (a) In the harmonic approximation,

$$\tilde{D}_e = \tilde{D}_0 + \tfrac{1}{2}\tilde{\nu} \quad \text{so} \quad \tilde{\nu} = 2(\tilde{D}_e - \tilde{D}_0)$$

$$\tilde{\nu} = \frac{2(1.51\times10^{-23}\,\text{J} - 2\times10^{-26}\,\text{J})}{(6.626\times10^{-34}\,\text{J s})\times(2.998\times10^{8}\,\text{m s}^{-1})} = \boxed{152\,\text{m}^{-1}}$$

The force constant is related to the vibrational frequency by

$$\omega = \left(\frac{k}{m_{\text{eff}}}\right)^{1/2} = 2\pi\nu = 2\pi c\tilde{\nu} \quad \text{so} \quad k = (2\pi c\tilde{\nu})^2 m_{\text{eff}}$$

The effective mass is

$$m_{\text{eff}} = \tfrac{1}{2}m = \tfrac{1}{2}(4.003\text{ u})\times(1.66\times10^{-27}\,\text{kg u}^{-1}) = 3.32\times10^{-27}\,\text{kg}$$

$$k = \left[2\pi(2.998\times10^{8}\text{ ms}^{-1})\times(152\,\text{m}^{-1})\right]^2 \times (3.32\times10^{-27}\,\text{kg})$$

$$= \boxed{2.72\times10^{-4}\,\text{kg s}^{-2}}$$

The moment of inertia is

$$I = m_{\text{eff}}R_e^2 = (3.32\times10^{-27}\,\text{kg})\times(297\times10^{-12}\,\text{m})^2 = \boxed{2.93\times10^{-46}\,\text{kg m}^2}$$

The rotational constant is

$$\tilde{B} = \frac{\hbar}{4\pi c I} = \frac{1.0546\times10^{-34}\ \mathrm{J\ s}}{4\pi(2.998\times10^{8}\mathrm{ms^{-1}})\times(2.93\times10^{-46}\mathrm{kg\ m^2})} = \boxed{95.5\ \mathrm{m^{-1}}}$$

(b) In the Morse potential,

$$x_e = \frac{\tilde{\nu}}{4\tilde{D}_e} \quad \text{and} \quad \tilde{D}_e = \tilde{D}_0 + \frac{1}{2}\left(1 - \frac{1}{2}x_e\right)\tilde{\nu} = \tilde{D}_0 + \frac{1}{2}\left(1 - \frac{\tilde{\nu}}{8\tilde{D}_e}\right)\tilde{\nu}$$

This rearranges to a quadratic equation in $\tilde{\nu}$:

$$\frac{\tilde{\nu}^2}{16\tilde{D}_e} - \frac{1}{2}\tilde{\nu} + \tilde{D}_e - \tilde{D}_0 = 0 \quad \text{so} \quad \tilde{\nu} = \frac{\frac{1}{2} - \sqrt{\left(\frac{1}{2}\right)^2 - \frac{4(\tilde{D}_e - \tilde{D}_0)}{16\tilde{D}_e}}}{2(16\tilde{D}_e)^{-1}}$$

$$\tilde{\nu} = 4\tilde{D}_e\left(1 - \sqrt{\frac{\tilde{D}_0}{\tilde{D}_e}}\right)$$

$$= \frac{4(1.51\times10^{-23}\ \mathrm{J})}{(6.626\times10^{-34}\mathrm{J\ s})\times(2.998\times10^{8}\ \mathrm{m\ s^{-1}})}\left(1 - \sqrt{\frac{2\times10^{-26}\ \mathrm{J}}{1.51\times10^{-23}\ \mathrm{J}}}\right)$$

$$= \boxed{293\ \mathrm{m^{-1}}}$$

and $$x_e = \frac{(293\ \mathrm{m^{-1}})\times(6.626\times10^{-34}\mathrm{J\ s})\times(2.998\times10^{8}\mathrm{m\ s^{-1}})}{4(1.51\times10^{-23}\mathrm{J})} = \boxed{0.96}$$

P10.21 (a) Vibrational wavenumbers ($\tilde{\nu}$/cm^{-1}) computed by PC Spartan Pro™ at several levels of theory are tabulated below, along with experimental values.

	A_1	A_1	B_2
Semi-empirical PM3	412	801	896
SCF 6-316G**	592	1359	1569
Density functional	502	1152	1359
Experimental	525	1151	1336

The vibrational modes are shown graphically in Figure 10.7.

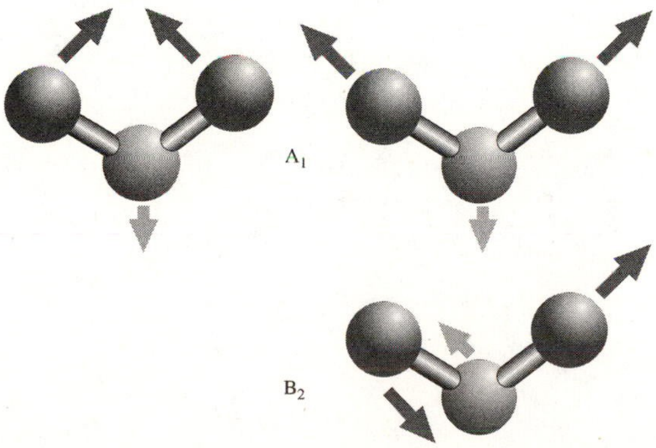

Figure 10.7

(b) The wavenumbers computed by density functional theory agree quite well with experiment. Agreement of the semi-empirical and SCF values with experiment is not as good. In this molecule, experimental wavenumbers can be correlated rather easily to computed vibrational modes even where the experimental and computed wavenumbers disagree substantially. Often, as in this case, computational methods that do a poor job of computing absolute transition wavenumbers still put transitions in proper order by wavenumber. That is, the modeling software systematically overestimates (as in this SCF computation) or underestimates (as in this semi-empirical computation) the wavenumbers, thus keeping them in the correct order. Group theory is another aid in the assignment of transitions: it can classify modes as forbidden, allowed only in particular polarizations, and so on. Also, visual examination of the modes of motion can help to classify many modes as predominantly bond-stretching, bond-bending, or with an internal rotation; these different modes of vibration can be correlated to quite different ranges of wavenumbers (stretches highest, especially stretches involving hydrogen atoms, and internal rotations lowest).

P10.23 Summarize the six observed vibrations according to their wavenumbers ($\tilde{\nu}/\text{cm}^{-1}$).

IR	870	1370	2869	3417
Raman	877	1408	1435	3407

(a) If H_2O_2 were linear, it would have $3N-5=\boxed{7}$ vibrational modes.

(b) Follow the flow chart in Fig. 7.7. Structure **6** is not linear, there is only one C_n axis (a C_2), and there is a σ_h; the point group is $\boxed{C_{2h}}$. Structure **7** is not linear, and there is only one C_n axis (a C_2), no σ_h, but two σ_v; the point group is $\boxed{C_{2v}}$. Structure **8** is not linear, and there is only one C_n axis (a C_2), no σ_h, and no σ_v; the point group is $\boxed{C_2}$.

(c) The exclusion rule applies to structure **6** because it has a center of inversion: no vibrational modes can be both IR and Raman active. So structure **6** is inconsistent with the observations. The vibrational modes of structure **7** span $3A_1+A_2+2B_2$. (The full basis of 12 cartesian coordinates spans $4A_1+2A_2+2B_1+4B_2$; remove translations and rotations.) The C_{2v} character table says that five of these modes are IR active $(3A_1+2B_2)$ and all are Raman active. All the modes of structure **8** are both IR and Raman active. (A look at the character table shows that both symmetry species are IR and Raman active, so determining the symmetry species of the normal modes does not help here.) Both structures **7** and **8** have more active modes than were observed. This is consistent with the observations. After all, group theory can tell us only whether the transition moment *must* be zero by symmetry; it does not tell us whether the transition moment is sufficiently strong to be observed under experimental conditions.

Solutions to theoretical problems

P10.25

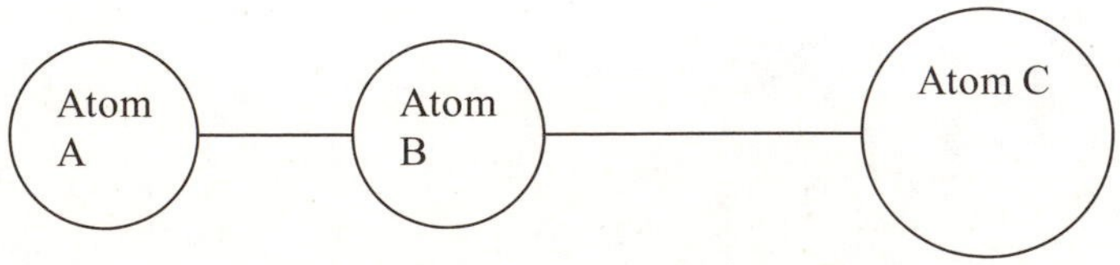

Figure 10.8

Let us assume that atom C is the most massive. Then the center of mass, CM, will be located at a distance, D, from atom B. In the notation of Table 10.1, we must have the relation

$m_A(R+D)+m_B D=m_C(R'-D)$, which may be rearranged into

$D(m_A+m_B+m_C)=m_C R'-m_A R$

Solving for D, we obtain $D=\dfrac{m_C R'-m_A R}{m}$, where $m=m_A+m_B+m_C$. Expanding,

$$I=\sum_i m_i r_i^2 \,[10.1]$$

$$I=m_A R^2+m_A D^2+2RDm_A+m_B D^2+m_C R'^2+m_C D^2-2R'Dm_C$$

$$=m_A R^2+m_C R'^2+D^2(m_A+m_B+m_C)+2D(m_A R-m_C R')$$

After substituting the above formula for D and using $m=m_A+m_B+m_C$, we obtain

$$I=m_A R^2+m_C R'^2+m\left(\frac{m_C R'-m_A R}{m}\right)^2+2\left(\frac{m_C R'-m_A R}{m}\right)\times(m_A R-m_C R')$$

$$=m_A R^2+m_C R'^2+\frac{1}{m}(m_C R'-m_A R)^2-\frac{2}{m}(m_C R'-m_A R)^2$$

$$=m_A R^2+m_C R'^2-\frac{1}{m}(m_C R'-m_A R)^2$$

$$=\boxed{m_A R^2+m_C R'^2-\frac{1}{m}(m_A R-m_C R')^2}$$

QED.

P10.27 If we apply the selection rules $\Delta J=\pm1$ and $\Delta K=0$ to the formula for the rotational terms given in the problem, we obtain for the frequencies of the allowed transitions the expression

$$\nu_{J+1,K\leftarrow J,K}=F(J+1,K)-F(J,K)=2B(J+1)-4D_J(J+1)^3-2D_{JK}(J+1)K^2$$

In terms of wavenumbers, the following expression is similar:

$$\tilde{\nu}_{J+1,K\leftarrow J,K}=\tilde{F}(J+1,K)-\tilde{F}(J,K)=2\tilde{B}(J+1)-4\tilde{D}_J(J+1)^3-2\tilde{D}_{JK}(J+1)K^2$$

To work with the latter expression one must convert the data given in frequency units to wavenumbers. Here we solve the problem in frequency units using the former expression. We note that A and D_K drop out of the expression for the transition frequencies; hence these constants cannot be determined from the data given. Examination of the data suggests that the identification of the transitions as shown in the table below can be made.

Transition	1	2	3	4	5
Transition frequency, ν/GHz	51.0718	102.1426	102.1408	153.2103	153.2076
Transition quantum numbers	$K=0$ $J=0\rightarrow1$	$K=0$ $J=1\rightarrow2$	$K=1$ $J=1\rightarrow2$	$K=0$ $J=2\rightarrow3$	$K=1$ $J=2\rightarrow3$
Transition frequency expression	$2B-4D_J$	$4B-32D_J$	$4B-32D_J-4D_{JK}$	$6B-108D_J$	$6B-108D_J-6D_{JK}$

Examination of these expressions reveals that the difference in transition frequencies between transitions 3 and 2 and between transitions 5 and 4 yields the value of D_{JK} directly.

$$\boxed{D_{JK} = 4.5 \times 10^2\ \text{kHz}}$$

B and D_J can be found from simultaneous solution of the equations for transitions 1 and 2 and also from transitions 2 and 4. The average value of D_J obtained in this way is $\boxed{D_J = 56\ \text{kHz}}$. The value of B obtained from transition 1 is then $\boxed{B = 25.5360\ \text{GHz}}$. If desired, these results in frequency units, Hz, can be converted to units of wavenumber, cm^{-1}, by division by c, the velocity of light, expressed in units of cm s^{-1}.

P10.29 $N \propto g\text{e}^{-E/kT}$ [Boltzmann distribution, Chapter 13]

$$N_J \propto g_J\text{e}^{-E_J/kT} \propto (2J+1)\text{e}^{-hc\tilde{B}J(J+1)/kT} \quad [g_J = 2J+1 \text{ for a diatomic rotor}]$$

The maximum population occurs when

$$\frac{\text{d}}{\text{d}J}N_J \propto \left\{2-(2J+1)^2\times\left(\frac{hc\tilde{B}}{kT}\right)\right\}\text{e}^{-hc\tilde{B}J(J+1)/kT} = 0$$

and, since the exponential can never be zero at a finite temperature, then

$$(2J+1)^2\times\left(\frac{hc\tilde{B}}{kT}\right) = 2$$

or when $\boxed{J_{\text{max}} = \left(\frac{kT}{2hc\tilde{B}}\right)^{1/2} - \frac{1}{2}}$

For ICl, with $\frac{kT}{hc} = 207.22\ \text{cm}^{-1}$ (inside front cover)

$$J_{\text{max}} = \left(\frac{207.22\ \text{cm}^{-1}}{0.2284\ \text{cm}^{-1}}\right)^{1/2} - \frac{1}{2} = \boxed{30}$$

For a spherical rotor, $N_J \propto (2J+1)^2\text{e}^{-hc\tilde{B}J(J+1)/kT}$ $[g_J = (2J+1)^2]$, and the greatest population occurs when

$$\frac{\text{d}N_J}{\text{d}J} \propto \left(8J+4-\frac{hc\tilde{B}(2J+1)^3}{kT}\right)\text{e}^{-hc\tilde{B}J(J+1)/kT} = 0$$

which occurs when

$$4(2J+1) = \frac{hc\tilde{B}(2J+1)^3}{kT}$$

or at $\boxed{J_{\text{max}} = \left(\frac{kT}{hc\tilde{B}}\right)^{1/2} - \frac{1}{2}}$

For CH_4, $J_{\text{max}} = \left(\frac{207.22\ \text{cm}^{-1}}{5.24\ \text{cm}^{-1}}\right)^{1/2} - \frac{1}{2} = \boxed{6}$

P10.31 The energy levels of a Morse oscillator, expressed as wavenumbers, are given by

$$\tilde{G}(v) = \left(\tilde{v}+\tfrac{1}{2}\right)\tilde{v} - \left(v+\tfrac{1}{2}\right)^2 x_e\tilde{v} = \left(v+\tfrac{1}{2}\right)\tilde{v} - \left(v+\tfrac{1}{2}\right)^2 \tilde{v}^2/4\tilde{D}_e$$

States are bound only if the energy is less than the well depth, $\tilde{D}_e$, also expressed as a wavenumber:

$$\tilde{G}(v) < \tilde{D}_e \quad \text{or} \quad \left(v+\tfrac{1}{2}\right)\tilde{v} - \left(v+\tfrac{1}{2}\right)^2\tilde{v}^2/4\tilde{D}_e < \tilde{D}_e$$

Solve for the maximum value of v by making the inequality into an equality.

$$\left(v+\tfrac{1}{2}\right)^2\tilde{v}/4\tilde{D}_e - \left(v+\tfrac{1}{2}\right)\tilde{v} + \tilde{D}_e = 0$$

Multiplying through by $4\tilde{D}_e$ results in an expression that can be factored by inspection into

$$\left[\left(v+\tfrac{1}{2}\right)\tilde{v} - 2\tilde{D}_e\right]^2 = 0 \quad \text{so} \quad v+\tfrac{1}{2} = 2\tilde{D}_e/\tilde{v} \quad \text{and} \quad v = \boxed{2\tilde{D}_e/\tilde{v} - \tfrac{1}{2}}.$$

Of course, v is an integer, so its maximum value is really the greatest integer less than this quantity.

P10.33 We work with eqn. 10.45, which gives the transition energies for the S and O branches of the vibrational Raman spectra. Transitions having $\tilde{v}_S(J-2)$ and $\tilde{v}_O(J+2)$ have a common upper state; hence, the corresponding combination difference, Δ_0, is a function of B_0 only. Likewise, transitions $\tilde{v}_S(J)$ and $\tilde{v}_O(J)$ have a common lower state and the combination difference, Δ_1, is a function of B_1 only. Using eqn. 10.45 we obtain

$$\tilde{v}_S(J-2) = \tilde{v}_i - \tilde{v} - 4\tilde{B}_0(J-2) - 6\tilde{B}_0$$

$$\tilde{v}_O(J+2) = \tilde{v}_i - \tilde{v} + 4\tilde{B}_0(J+2) - 2\tilde{B}_0$$

Taking the difference between $\tilde{v}_O(J+2)$ and $\tilde{v}_S(J-2)$, we obtain for the combination difference $\boxed{\Delta_0 = \tilde{v}_O(J+2) - \tilde{v}_S(J-2) = 8\tilde{B}_0(J+\tfrac{1}{2})}$. In a similar manner we can obtain $\boxed{\Delta_1 = \tilde{v}_O(J) - \tilde{v}_S(J) = 8\tilde{B}_1(J+\tfrac{1}{2})}$.

Solutions to applications

P10.35 No solution. This is a report left for the student.

P10.37 The question of whether to use CN or CH within the interstellar cloud of constellation Ophiuchus for the determination of the temperature of the cosmic background radiation depends on which one has a rotational spectrum that best spans blackbody radiation of 2.726 K. Given $\tilde{B}_0(\text{CH}) = 14.19\ \text{cm}^{-1}$, the rotational constant that is needed for the comparative analysis may be calculated from the 226.9 GHz spectral line of the Orion Nebula. Assuming that the line is for the $^{12}C^{14}N$ isotopic species and $J+1 \leftarrow J = 1$, which gives a reasonable estimate of the CN bond length (117.4 pm), the CN rotational constant is calculated as follows.

$$\tilde{B}_0 = B/c = \frac{v}{2c(J+1)} = \frac{v}{4c} \qquad (1)$$

$$= 1.892\ \text{cm}^{-1} \qquad (2)$$

Blackbody radiation at 2.726 K can be plotted against radiation wavenumber with suitable transformation of the Planck distribution law.

$$\rho(\tilde{v}) = \frac{8\pi hc\tilde{v}^3}{e^{hc\tilde{v}/kT} - 1}$$

Spectral absorption lines of $^{12}C^{14}N$ and $^{12}C^{1}H$ are calculated with eqn. 10.17(a).

$$\tilde{\nu}(J+1 \leftarrow J) = 2\tilde{B}(J+1) \qquad J = 0, 1, 2, 3.......$$

The cosmic background radiation and molecular absorption lines are shown in Figure 10.9. It is evident that only CN spans the background radiation.

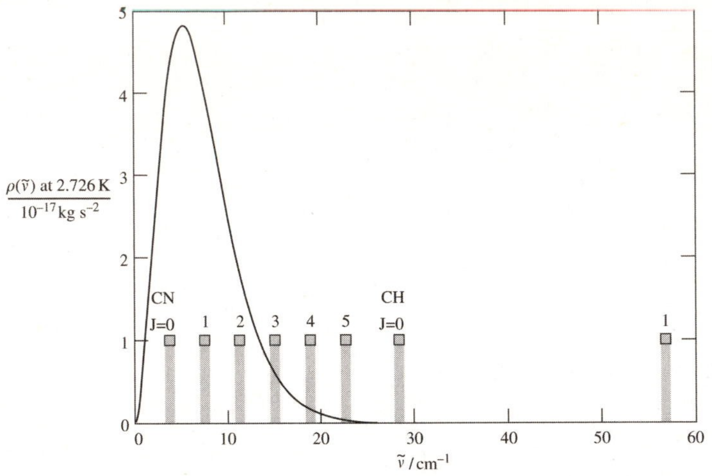

Figure 10.9

P10.39 (a) The H_3^+ molecule is held together by a two-electron, three-center bond, and hence its structure is expected to be an equilateral triangle. Looking at Figure 10.10 and using the law of cosines:

$$R^2 = 2R_C^2 - 2R_c^2\cos(180° - 2\theta)$$
$$= 2R_C^2(1 - \cos(120°)) = 3R_C^2$$

Therefore

$$R_C = R/\sqrt{3}$$
$$I_C = 3mR_C^2 = 3m(R/\sqrt{3})^2 = mR^2$$
$$I_B = 3mR_B = 2m(R/2)^2 = mR^2/2$$

Therefore

$$I_C = 2I_B$$

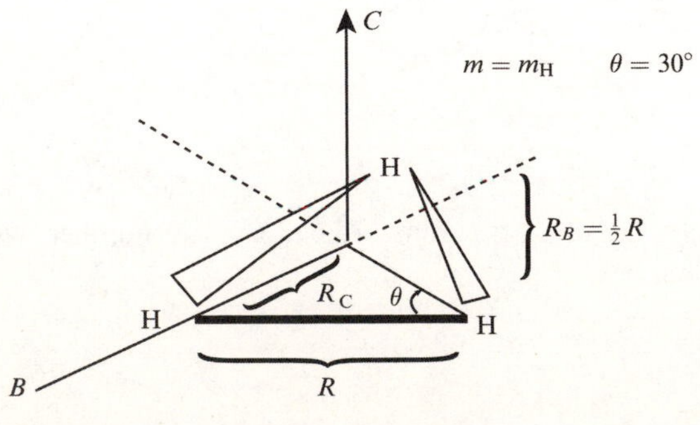

Figure 10.10

(b) $$\widetilde{B}=\frac{\hbar}{4\pi c I_B}=\frac{2\hbar}{4\pi c m R^2}=\frac{h}{2\pi c m R^2}\ [10.11]$$

$$R=\left(\frac{\hbar}{2\pi c m\widetilde{B}}\right)^{1/2}=\left(\frac{\hbar N_A}{2\pi c M_H\widetilde{B}}\right)^{1/2}$$

$$\left(\frac{(1.0546\times10^{-34}\,\mathrm{J\,s})\times(6.0221\times10^{23}\,\mathrm{mol^{-1}})\times\left(\frac{10^{-2}\,\mathrm{m}}{\mathrm{cm}}\right)}{2\pi(2.998\times10^{8}\,\mathrm{m\,s^{-1}})\times(0.001008\,\mathrm{kg\,mol^{-1}})\times(43.55\,\mathrm{cm^{-1}})}\right)^{1/2}$$

$$=8.764\times10^{-11}\,\mathrm{m}=\boxed{87.64\,\mathrm{pm}}$$

Alternatively the rotational constant $\widetilde{C}$ can be used to calculate R.

$$\widetilde{C}=\frac{\hbar}{4\pi c I_C}=\frac{\hbar}{4\pi c m R^2}\ [10.11]$$

$$R=\left(\frac{\hbar}{4\pi c m\widetilde{C}}\right)^{1/2}=\left(\frac{\hbar N_A}{4\pi c M_H\widetilde{C}}\right)^{1/2}$$

$$=\left(\frac{(1.0546\times10^{-34}\,\mathrm{J\,s})\times(6.0221\times10^{23}\,\mathrm{mol^{-1}})\times\left(\frac{10^{-2}\,\mathrm{m}}{\mathrm{cm}}\right)}{4\pi(2.998\times10^{8}\,\mathrm{m\,s^{-1}})\times(0.001008\,\mathrm{kg\,mol^{-1}})\times(20.71\,\mathrm{cm^{-1}})}\right)$$

$$=8.986\times10^{-11}\,\mathrm{m}=\boxed{89.86\,\mathrm{pm}}$$

The values of R calculated with either the rotational constant $\widetilde{C}$ or the rotational constant $\widetilde{B}$ differ slightly. We approximate the bond length as the average of these two.

$$\langle R\rangle\approx\frac{(87.64+89.86)\,\mathrm{pm}}{2}=\boxed{88.7\,\mathrm{pm}}$$

(c) $$\widetilde{B}=\frac{\hbar}{2\pi c m R^2}=\frac{(1.0546\times10^{-34}\,\mathrm{J\,s})\times(6.0221\times10^{23}\,\mathrm{mol^{-1}})\times\left(\frac{10^{-2}\,\mathrm{m}}{\mathrm{cm}}\right)}{2\pi(2.998\times10^{8}\,\mathrm{m\,s^{-1}})\times(0.001008\,\mathrm{kg\,mol^{-1}})\times(87.32\times10^{-12}\,\mathrm{m})^2}$$

$$=\boxed{43.87\,\mathrm{cm^{-1}}}$$

$$\widetilde{C}=\tfrac{1}{2}\widetilde{B}=\boxed{21.93\,\mathrm{cm^{-1}}}$$

(d) $$\frac{1}{m_{\mathrm{eff}}}=\frac{3}{m}\quad\text{or}\quad m_{\mathrm{eff}}=\tfrac{1}{3}m$$

Since $m_D=2m_H, m_{\mathrm{eff,D}}=2m_H/3$,

$$\tilde{\nu}_2(\mathrm{D_3^+})=\left(\frac{m_{\mathrm{eff}}(\mathrm{H_3})}{m_{\mathrm{eff}}(\mathrm{D_3})}\right)^{1/2}\tilde{\nu}_2(\mathrm{H_3})\ [10.31]$$

$$=\left(\frac{m_H/3}{2m_H/3}\right)^{1/2}\tilde{\nu}_2(\mathrm{H_3})=\frac{\tilde{\nu}_2(\mathrm{H_2})}{2^{1/2}}$$

$$=\frac{2521.6\,\mathrm{cm^{-1}}}{2^{1/2}}=\boxed{1783.0\,\mathrm{cm^{-1}}}$$

Since $\tilde{B}$ and $\tilde{C} \propto \dfrac{1}{m}$, where m = mass of H or D,

$$\tilde{B}(D_3^+) = \tilde{B}(H_3^+) \times \frac{M_H}{M_D} = 43.55\,\text{cm}^{-1} \times \left(\frac{1.008}{2.014}\right) = \boxed{21.80\,\text{cm}^{-1}}$$

$$\tilde{C}(D_3^+) = \tilde{C}(H_3^+) \times \frac{M_H}{M_D} = 20.71\,\text{cm}^{-1} \times \left(\frac{1.008}{2.014}\right) = \boxed{10.37\,\text{cm}^{-1}}$$

11 Electronic spectroscopy

Answers to discussion questions

D11.1 The ground electronic configuration of dioxygen, $1\sigma_g^{\,2}1\sigma_u^{\,2}2\sigma_g^{\,2}1\pi_u^{\,4}1\pi_g^{\,2}$, is discussed in Section 5.4(d), and the determination of the term symbol $^3\Sigma_g^-$ is described in Section 11.3(a). The term symbol Σ represents a total orbital angular momentum about the internuclear axis of zero. This happens because for every π orbital electron with $\lambda = +1$, there is a π orbital electron with $\lambda = -1$. For example, the two $1\pi_g$ electrons are, according to Hund's rules, in separate degenerate orbitals for which one orbital has $\lambda = +1$ and the other has $\lambda = -1$. Except for the two $1\pi_g$ electrons, electrons have paired α and β spins, which results in zero contribution to the total spin angular momentum. According to Hund's rules, the two $1\pi_g$ electrons (i.e., $1\pi_g^{\,1}1\pi_g^{\,1}$) have parallel spins in the ground state. They provide a total spin angular momentum of $S = ½ + ½ = 1$ and a spin multiplicity of $2S + 1 = 3$, which appears as the left superscript 3. The term symbol indicates a gerade total symmetry because electrons are paired in the ungerade molecular orbitals and u × u = g, and the resultant symmetry of electrons in different molecular orbitals must therefore be given by g × g = g. The π orbitals change sign upon reflection in the plane that contains the internuclear axis. Consequently, the term symbol has the superscript – to indicate that the molecular wavefunction for O_2 changes sign upon reflection in the plane containing the nuclei.

D11.3 A band head is the convergence of the frequencies of electronic transitions with increasing rotational quantum number J. They result from the rotational structure superimposed on the vibrational structure of the electronic energy levels of the diatomic molecule. (See Figs. 11.10 and 11.13 in the textbook.) To understand how a band head arises, one must examine the equations describing the transition frequencies (eqn. 11.9). As seen from the analysis in Section 11.3(e), convergence can arise only when terms in both $(B' - B)$ and $(B' + B)$ occur in the equation. Because only a term in $(B' - B)$ occurs for the Q branch, no band head can arise for that branch.

D11.5 (a) The transition intensity is proportional to the square of the transition dipole moment. We initially suspect that the transition dipole moment should increase as the length L of the alternating carbon-to-carbon double/single/double/single . . . bond sequence of the polyene is increased, and consequently, the transition intensity should also increase as L increases. To test this hypothesis, consider that the polyene has N pi electrons that fill the first $n = N/2$ quantum states of the particle in a one-dimensional box of length $L = Nd = 2nd$ where d is the average carbon-to-carbon bond length. (The length choice $L = Nd$ adds half a bond length at each end to the distance between the two end-carbon nuclei.) The transition dipole moment of this model is

$$\mu_x = \int_0^L \psi_{n_f} x \psi_{n_i}\,dx$$

We quickly find a selection rule for transitions with the substitution $x = f(x) + L/2$ where $f(x) = x - L/2$ is a function of ungerade symmetry w/r/t inversion through the center of symmetry at $x = L/2$, and we note that the wavefunctions have alternating gerade and ungerade symmetry as n increases. Then

$$\mu_x = \int_0^L \psi_{n_f}\{f(x)+L/2\}\psi_{n_i}\,dx = \int_0^L \psi_{n_f} f(x)\psi_{n_i}\,dx + (L/2)\int_0^L \psi_{n_f}\psi_{n_i}\,dx$$

The last integral vanishes because the wavefunctions of different energy levels are orthogonal, which leaves

$$\mu_x = \int_0^L \psi_{n_f} f(x) \psi_{n_i} \, dx$$

Because the integrand factor $f(x)$ has ungerade symmetry, the product $\psi_{n_f}\psi_{n_i}$ of an allowed transition must also have ungerade symmetry so that the total symmetry of the integrand has u × u = g symmetry and the integral can be non-zero. Thus, the lowest energy transition is $n + 1 \leftarrow n$ where $n = n_i$. It is now convenient to return to the original transition dipole moment integral and substitute the wavefunctions for the final and initial states:

$$\mu_x = \frac{2}{L}\int_0^L \sin\left(\frac{(n+1)\pi x}{L}\right) x \sin\left(\frac{n\pi x}{L}\right) dx \; [2.6b]$$

We need not evaluate the integral exactly because the development of a method to improve the intensity of a dye requires only a knowledge of the approximate relationship between u_x and either n or L (proportional properties for a polyene), so we recognize that $n + 1 \sim n$ for a large polyene and we make the estimate

$$\mu_x \propto \frac{1}{L}\int_0^L x \sin^2\left(\frac{n\pi x}{L}\right) dx \quad \text{(The integral is found in standard mathematical tables.)}$$

$$\propto L$$

The model confirms the hypothesis that as the transition dipole moment is increased, so the transition intensity is increased by increasing the length of the polyene.

(b) Since $E_n = \frac{n^2 h^2}{8m_e L^2}$ [2.6a], $\Delta E = \frac{(2n+1)h^2}{8m_e L^2}$ $[\Delta n = +1] \sim \frac{nh^2}{4m_e L^2}$ for large n.

But $L = 2nd$ is the length of the chain where d is the average carbon–carbon interatomic distance. Hence

$$\Delta E \propto \frac{1}{L}$$

Therefore, the transition moves toward the red as L is increased. When white light is used to illuminate the dyed object, the color absorbed is the **complementary color** to the reflected, observed color. Newton's color wheel, shown in Figure 11.1, usefully displays complementary and observed colors. For example, draw a line from complementary violet through the circle centre to find the observed color of green-yellow. As the polyene length is increased, the complementary color progresses from violet to indigo to blue to green and so on, while the observed color progresses from green-yellow to yellow-orange to orange-red to red-violet to violet-indigo. We say that the apparent color of the dye shifts toward blue.

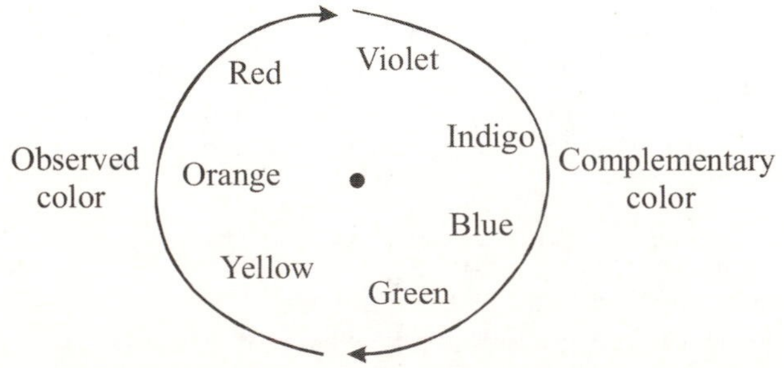

Figure 11.1

D11.7 Several characteristics of fluorescence are consistent with the accepted mechanism: (1) Fluorescence ceases as soon as the source of illumination is removed; (2) the time scale of fluorescence, ~10^{-9} s, is typical of a process in which the rate-determining step is a spontaneous radiative transition between states of the same multiplicity—slower than a stimulated transition but faster than phosphorescence; (3) fluorescence occurs at longer wavelength (lower frequency) than the inducing radiation; (4) its vibrational structure is characteristic of that of a transition from the ground vibrational level of the excited electronic state to the vibrational levels of the ground electronic state; (5) the fluorescence spectrum is observed to shift and in some cases be quenched by interactions with the solvent.

D11.9 See the table below for a summary of the characteristics of laser radiation that result in its many advantages for chemical and biochemical investigations. Two important applications of lasers in chemistry have been to Raman spectroscopy and to the development of time-resolved spectroscopy.

Characteristics of Laser Radiation and Their Chemical Applications

Characteristic	Advantage	Application
High power	Multiphoton processes	Nonlinear spectroscopy
		Saturation spectroscopy
	Low detector noise	Improved sensitivity
	High scattering intensity	Raman spectroscopy
Monochromatic	High resolution	Spectroscopy
	State selection	Isotope separation
		Photochemically precise
		State-to-state reaction dynamics
Collimated beam	Long path lengths	Sensitivity
	Forward scattering observable	Nonlinear Raman spectroscopy
Coherent	Interference between separate beams	Coherent anti-Stokes Raman spectroscopy (CARS) (Section 10.14c)
Pulsed	Precise timing of excitation	Fast reactions
		Relaxation
		Energy transfer

Modern Raman spectroscopy (Section 10.4) utilizes an intense excitation beam of a laser to increase the intensity of scattered radiation and thereby to increase measurement sensitivity. The monochromaticity of laser radiation is also a great advantage, for it makes possible the observation of scattered light that differs by only fractions of reciprocal centimetres from the incident radiation. Such high resolution is particularly useful for observing the rotational structure of Raman lines because rotational transitions are of the order of a few reciprocal centimeters. Monochromaticity also allows observations to be made very close to absorption frequencies, giving rise to the techniques of Fourier-transform Raman spectroscopy and resonance Raman spectroscopy (Section 10.14b).

Time-resolved laser spectroscopy can be used to study the dynamics of chemical reactions. Laser pulses are used to obtain the absorption, emission, and Raman spectrum of reactants, intermediates, products, and even transition states of reactions. When we want to study the rates at which energy is transferred from one mode to another in a molecule, we need femotosecond and picosecond pulses. These time scales are available from mode-locked lasers, and their development has opened up the possibility of examining the details of chemical reactions at a level that would have been unimaginable before.

Solutions to exercises

E11.1(a) The reduction in intensity obeys the Beer–Lambert law introduced in Section 11.2.

$$\log\frac{I}{I_0} = -\log\frac{I_0}{I} = -\varepsilon[\mathrm{J}]l \quad [11.3 \text{ and } 11.4]$$
$$= (-723\ \mathrm{dm^3\ mol^{-1}\ cm^{-1}})\times(4.25\times10^{-3}\ \mathrm{mol\ dm^{-3}})\times(0.250\ \mathrm{cm})$$
$$= -0.768$$

Hence, $\frac{I}{I_0} = 10^{-0.768} = 0.171$, and the reduction in intensity is $\boxed{82.9 \text{ percent}}$.

E11.2(a) $$\log\frac{I}{I_0} = -\log\frac{I_0}{I} = -\varepsilon[\mathrm{J}]l \quad [11.3, 11.4]$$

Hence, $$\varepsilon = -\frac{1}{[\mathrm{J}]l}\log\frac{I}{I_0} = -\frac{\log(0.181)}{(1.39\times10^{-4}\,\mathrm{mol\ dm^{-3}})\times(1.00\ \mathrm{cm})} = \boxed{5.34\times10^{3}\ \mathrm{dm^3\ mol^{-1}\ cm^{-1}}}.$$

E11.3(a) $$\log T = -A = -\varepsilon[\mathrm{J}]l \quad [11.1, 11.3, 11.4]$$

$$[\mathrm{J}] = -\frac{1}{\varepsilon l}\log T = \frac{-\log(1-0.385)}{(386\ \mathrm{dm^3\ mol^{-1}\ cm^{-1}})\times(0.500\ \mathrm{cm})}$$

$$= \boxed{1.09\ \mathrm{mmol\ dm^{-3}}}$$

E11.4(a) $$\mathcal{A} = \int_{\text{band}}\varepsilon(\tilde{\nu})\,\mathrm{d}\tilde{\nu} \quad [11.5] = \int_{\tilde{\nu}_\mathrm{i}}^{\tilde{\nu}_\mathrm{f}}\varepsilon(\tilde{\nu})\,\mathrm{d}\tilde{\nu}$$

Since $\tilde{\nu} = \lambda^{-1}$ and $\tilde{\nu}/\mathrm{cm^{-1}} = 10^7/(\lambda/\mathrm{nm})$,

$$\tilde{\nu}_\mathrm{i}/\mathrm{cm^{-1}} = 10^7/(300) = 3.3\times10^4$$

$$\tilde{\nu}_\mathrm{peak}/\mathrm{cm^{-1}} = 10^7/(270) = 3.7\times10^4$$

$$\tilde{\nu}_\mathrm{f}/\mathrm{cm^{-1}} = 10^7/(220) = 4.5\times10^4$$

The positions of the wavenumber end points and peak (max) of the band are schematically presented in Figure 11.2. It is apparent that, because of the relative position of the peak, the molar absorption coefficient is not symmetrically distributed around the peak wavenumber. The distribution is skewed toward higher wavenumbers. However, a reasonable estimate of the area under the curve may be approximated by adding the areas of triangle 1 and triangle 2 shown as dashed lines in the figure.

$$\mathcal{A} = \int_{\tilde{\nu}_\mathrm{i}}^{\tilde{\nu}_\mathrm{f}}\varepsilon(\tilde{\nu})\,\mathrm{d}\tilde{\nu} = \text{area 1} + \text{area 2}$$

$$= \frac{1}{2}(\tilde{\nu}_\mathrm{peak} - \tilde{\nu}_\mathrm{i})\varepsilon_\mathrm{peak} + \frac{1}{2}(\tilde{\nu}_\mathrm{f} - \tilde{\nu}_\mathrm{peak})\varepsilon_\mathrm{peak} = \frac{1}{2}(\tilde{\nu}_\mathrm{f} - \tilde{\nu}_\mathrm{i})\varepsilon_\mathrm{peak}$$

$$= \frac{1}{2}\times(1.2\times10^4\ \mathrm{cm^{-1}})\times2.21\times10^4\ \mathrm{dm^3\ mol^{-1}\ cm^{-1}}$$

$$= \boxed{1.3\times10^{8}\ \mathrm{dm^3\ mol^{-1}\ cm^{-2}}}$$

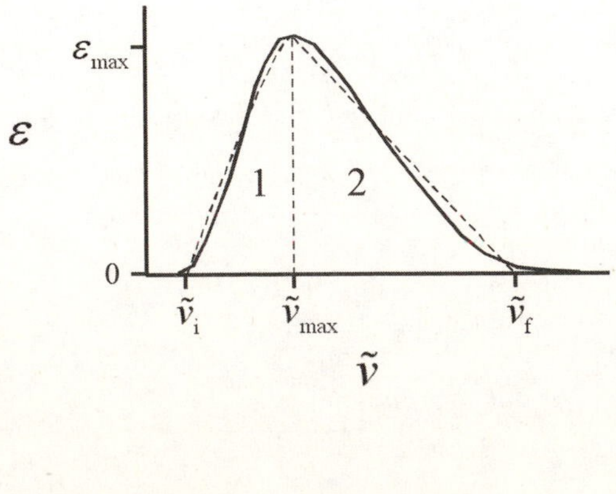

Figure 11.2

E11.5(a) $\varepsilon = -\frac{1}{[\mathrm{J}]l}\log\frac{I}{I_0}$ [11.3,11.4] with $l = 0.20$ cm

We use this formula to draw up the following table.

$[\mathrm{Br_2}]$ / mol dm^{-3}	0.0010	0.0050	0.0100	0.0500	
I/I_0	0.814	0.356	0.127	3.0×10^{-5}	
ε / (dm^3 mol^{-1} cm^{-1})	447	449	448	452	Mean: $44\bar{9}$

Hence, the molar absorption coefficient is $\varepsilon = \boxed{450\ \mathrm{dm^3\ mol^{-1}\ cm^{-1}}}$.

E11.6(a) $\varepsilon = -\frac{1}{[\mathrm{J}]l}\log\frac{I}{I_0}$ [11.3, 11.4] $= \frac{-1}{(0.010\ \mathrm{mol\ dm^{-3}})\times(0.20\ \mathrm{cm})}\log(0.48) = \boxed{15\bar{9}\ \mathrm{dm^3\ mol^{-1}\ cm^{-1}}}$

$$T = \frac{I}{I_0} = 10^{-[\mathrm{J}]\varepsilon l}\ [11.1, 11.2]$$

$$= 10^{(-0.010\ \mathrm{mol\ dm^{-3}})\times(15\bar{9}\ \mathrm{dm^3\ mol^{-1}cm^{-1}})\times(0.40\ \mathrm{cm})} = 10^{-0.63\bar{6}} = 0.23,\text{ or } \boxed{23\text{ percent}}$$

E11.7(a) The Beer–Lambert law [11.3, 11.4] is

$$\log\frac{I}{I_0} = -\varepsilon[\mathrm{J}]l \quad \text{so} \quad l = -\frac{1}{\varepsilon[\mathrm{J}]}\log\frac{I}{I_0}$$

For water, $[\mathrm{H_2O}] \approx \frac{1.00\ \mathrm{kg/dm^3}}{18.02\ \mathrm{g\ mol^{-1}}} = 55.5\ \mathrm{mol\ dm^{-3}}$

and $\varepsilon[\mathrm{J}] = (55.5\ \mathrm{mol\ dm^{-3}})\times(6.2\times10^{-5}\ \mathrm{dm^3\ mol^{-1}\ cm^{-1}}) = 3.4\times10^{-3}\ \mathrm{cm^{-1}}$, so $\frac{1}{\varepsilon[\mathrm{J}]} = 2.9$ m.

Hence, $l/\mathrm{m} = -2.9\times\log\frac{I}{I_0}$.

(a) $\frac{I}{I_0} = 0.50,\ l = -2.9\ \mathrm{m}\times\log(0.50) = \boxed{0.87\ \mathrm{m}}$

(b) $\frac{I}{I_0} = 0.1,\ l = -2.9\ \mathrm{m}\times\log(0.10) = \boxed{2.9\ \mathrm{m}}$

E11.8(a) The left superscript of the dihydrogen excited state $^3\Pi_u$ is the value of $2S + 1 = 3$, so $S = 1$, which means that the two electrons of H_2 are parallel ($S = s_1 + s_2 = ½ + ½ = 1$) in this excited state. The symbol Π indicates that the total orbital angular momentum around the molecular axis is $|\Lambda| = 1$. Since Λ is the sum of the individual electron orbital angular momentum quantum numbers around the molecular axis, we see that one of the unpaired electrons must be in a σ orbital ($\lambda = 0$) and the other electron must be in a π orbital ($\lambda = 1$). The excited state has ungerade overall parity. Since a σ bonding orbital has gerade symmetry while a π bonding orbital has ungerade symmetry and g × u = u, we deduce that one possible electron configuration is $\boxed{1\sigma_g^1 1\pi_u^1}$ (see Fig. 5.29 of text).

E11.9(a) The $1\sigma_g^2 1\sigma_u^2 1\pi_u^3 1\pi_g^1$ valence configuration has two unpaired electrons so $S = s_1 + s_2 = ½ + ½ = 1$ and the spin multiplicity is given by $2S + 1 = 2(1) + 1 = \boxed{3}$. Because u × u = g and g × g = g, the net parity of two electrons paired in an orbital is always gerade. Consequently, the overall parity is found by multiplying the parity of unpaired electrons. For this configuration, u × g = $\boxed{\mathrm{u}}$.

E11.10(a) The electronic spectrum selection rules concerned with changes in angular momentum are (Section 11.3b)

$$\Delta\Lambda = 0,\ \pm1 \quad \Delta S = 0 \quad \Delta\Sigma = 0 \quad \Delta\Omega = 0,\ \pm1 \quad \text{where } \Omega = \Lambda + \Sigma$$

Λ gives the total orbital angular momentum about the internuclear axis, and Σ gives the total spin angular momentum about the internuclear axis. The ± superscript selection rule for reflection in the plane along the internuclear axis is $+\leftrightarrow+$ or $-\leftrightarrow-$ (i.e., $+\leftrightarrow-$ is forbidden). The **Laporte selection rule** states that for a centrosymmetric molecule (a molecule with a center of inversion), the only allowed transitions are transitions that are accompanied by a change of parity: $u\leftrightarrow g$.

(a) The changes in the transition $^2\Pi \leftrightarrow {}^2\Pi$ are $\Delta\Lambda = 0$, $\Delta S = 0$, $\Delta\Sigma = 0$, and $\Delta\Omega = 0$, so the transition is $\boxed{\text{allowed}}$.
(b) The changes in the transition $^1\Sigma \leftrightarrow {}^1\Sigma$ are $\Delta\Lambda = 0$, $\Delta S = 0$, $\Delta\Sigma = 0$, and $\Delta\Omega = 0$, so the transition is $\boxed{\text{allowed}}$.
(c) The changes in the transition $\Sigma \leftrightarrow \Delta$ are $\Delta\Lambda = 2$, so the transition is $\boxed{\text{forbidden}}$.
(d) The transition $\Sigma^+ \leftrightarrow \Sigma^-$ is $\boxed{\text{forbidden}}$ because $+\leftrightarrow-$.
(e) The transition $\Sigma^+ \leftrightarrow \Sigma^+$ is $\boxed{\text{allowed}}$ because $\Delta\Lambda = 0$ and $+\leftrightarrow+$.

E11.11(a) We begin by evaluating the normalization constants N_0 and N_v.

$$N_0^2 = \frac{1}{\int_{-\infty}^{\infty} e^{-2ax^2}\,dx} = \left(\frac{2a}{\pi}\right)^{1/2} \text{(standard integral)};\quad N_0 = \left(\frac{2a}{\pi}\right)^{1/4}$$

Likewise, $$N_v^2 = \frac{1}{\int_{-\infty}^{\infty} e^{-2b(x-x_0)^2}\,dx} = \left(\frac{2b}{\pi}\right)^{1/2};\quad N_v = \left(\frac{2b}{\pi}\right)^{1/4}$$

Furthermore, we can easily check that

$$ax^2 + b(x-x_0)^2 = z^2 + \frac{ab}{a+b}x_0^2 \quad \text{where} \quad z = (a+b)^{1/2}x - \frac{b}{(a+b)^{1/2}}x_0 \quad \text{and} \quad dx = \frac{1}{(a+b)^{1/2}}dz$$

Then the vibration overlap integral between the vibrational wavefunction in the upper and lower electronic states is

$$S(v,0) = \langle v|0\rangle = N_0N_v\int_{-\infty}^{\infty} e^{-ax^2}e^{-b(x-x_0)^2}\,dx = N_0N_v\int_{-\infty}^{\infty} e^{-\left\{ax^2+b(x-x_0)^2\right\}}dx$$

$$= \frac{N_0N_v}{(a+b)^{1/2}}\int_{-\infty}^{\infty} e^{-\left\{z^2+\frac{ab}{a+b}x_0^2\right\}}dz = \frac{N_0N_v}{(a+b)^{1/2}}e^{-\frac{ab}{a+b}x_0^2}\int_{-\infty}^{\infty} e^{-z^2}dz = N_0N_v\left(\frac{\pi}{a+b}\right)^{1/2}e^{-\frac{ab}{a+b}x_0^2}$$

$$= \left(\frac{2a}{\pi}\right)^{1/4}\left(\frac{2b}{\pi}\right)^{1/4}\left(\frac{\pi}{a+b}\right)^{1/2}e^{-\frac{ab}{a+b}x_0^2} = (4ab)^{1/4}\left(\frac{1}{a+b}\right)^{1/2}e^{-\frac{ab}{a+b}x_0^2}$$

For the case $b = a/2$, this simplifies to

$$S(v,0) = \frac{2}{\left(3\sqrt{2}\right)^{1/2}}e^{-ax_0^2/3}$$

The Franck–Condon factor is

$$|S(v,0)|^2 = \boxed{\frac{2\sqrt{2}}{3}e^{-2ax_0^2/3}}$$

E11.12(a)

$$\psi_0 = \left(\frac{2}{L}\right)^{1/2}\sin\left(\frac{\pi x}{L}\right) \quad \text{for } 0 \le x \le L \text{ and 0 elsewhere}$$

$$\psi_v = \left(\frac{2}{L}\right)^{1/2}\sin\left\{\frac{\pi}{L}\left(x-\frac{L}{4}\right)\right\} \quad \text{for } \frac{L}{4} \le x \le \frac{5L}{4} \text{ and 0 elsewhere}$$

$$S(v,0) = \langle v|0\rangle = \frac{2}{L}\int_{L/4}^{L}\sin\left(\frac{\pi x}{L}\right)\sin\left\{\frac{\pi}{L}\left(x-\frac{L}{4}\right)\right\}dx$$

The above integral is recognized as the standard integral (see math handbook):

$$\int \sin(ax)\sin(ax+b)\,\mathrm{d}x = \frac{x}{2}\cos(b) - \frac{\sin(2ax+b)}{4a}$$

with the transformations $a = \pi/L$ and $b = -\pi/4$. Thus,

$$S(v,0) = \frac{2}{L}\left[\frac{x}{2}\cos\left(-\frac{\pi}{4}\right) - \frac{\sin(2\pi x/L - \pi/4)}{4\pi/L}\right]_{x=L/4}^{x=L} = \frac{2}{L}\left[\frac{x}{2}\cos\left(\frac{\pi}{4}\right) - \frac{\sin(2\pi x/L - \pi/4)}{4\pi/L}\right]_{x=L/4}^{x=L}$$

$$= \left[x\cos\left(\frac{\pi}{4}\right) - \frac{\sin(2\pi x - \pi/4)}{2\pi}\right]_{x=1/4}^{x=1}$$

$$= \cos\left(\frac{\pi}{4}\right) - \frac{\sin(2\pi - \pi/4)}{2\pi} - \left[\frac{1}{4}\cos\left(\frac{\pi}{4}\right) - \frac{\sin(\pi/2 - \pi/4)}{2\pi}\right]$$

$$= \frac{3}{4}\cos\left(\frac{\pi}{4}\right) - \frac{\sin(7\pi/4)}{2\pi} + \frac{\sin(\pi/4)}{2\pi} = \frac{3}{4}\cos\left(\frac{\pi}{4}\right) + \frac{\sin(\pi/4)}{2\pi} + \frac{\sin(\pi/4)}{2\pi}$$

$$= \frac{1}{4}\cos\left(\frac{\pi}{4}\right)\left\{3 + \frac{4}{\pi}\right\} = \frac{\sqrt{2}}{8}\left\{3 + \frac{4}{\pi}\right\}$$

The Franck–Condon factor is

$$|S(v,0)|^2 = \boxed{\frac{1}{32}\left(3 + \frac{4}{\pi}\right)^2}$$

E11.13(a) P branch $(\Delta J = -1)$: $\tilde{\nu}_P(J) = \tilde{\nu} - (\tilde{B}' + \tilde{B})J + (\tilde{B}' - \tilde{B})J^2$ [11.9a]

When the bond is shorter in the excited state than in the ground state, $\tilde{B}' > \tilde{B}$ and $\tilde{B}' - \tilde{B}$ are positive. In this case, the lines of the P branch appear at successively decreasing energies as J increases, begin to converge, go through a head at J_{head}, begin to increase with increasing J, and become greater than $\tilde{\nu}$ when $J > (\tilde{B}' + \tilde{B})/(\tilde{B}' - \tilde{B})$ (see Section 11.3e; the quadratic shape of the $\tilde{\nu}_P$ against J curve is called the Fortrat parabola). This means that $\tilde{\nu}_P(J)$ is a minimum when $J = J_{\text{head}}$. It is reasonable to deduce that J_{head} is the closest integer to $\boxed{\tfrac{1}{2}(\tilde{B}' + \tilde{B})/(\tilde{B}' - \tilde{B})}$ because it takes twice as many J values to reach the minimum line of the P branch and to return to $\tilde{\nu}$. We can also find J_{head} by finding the minimum of the Fortrat parabola: $\mathrm{d}\tilde{\nu}_P/\mathrm{d}J = 0$ when $J = J_{\text{head}}$.

$$\frac{\mathrm{d}\tilde{\nu}_P}{\mathrm{d}J} = \frac{\mathrm{d}}{\mathrm{d}J}\left\{\tilde{\nu} - (\tilde{B}' + \tilde{B})J + (\tilde{B}' - \tilde{B})J^2\right\} = -(\tilde{B}' + \tilde{B}) + 2(\tilde{B}' - \tilde{B})J$$

$$-(\tilde{B}' + \tilde{B}) + 2(\tilde{B}' - \tilde{B})J_{\text{head}} = 0$$

$$J_{\text{head}} = \frac{(\tilde{B}' + \tilde{B})}{2(\tilde{B}' - \tilde{B})}$$

E11.14(a) Since $\tilde{B}' < \tilde{B}$ and $\tilde{B}' - \tilde{B}$ is negative, the R branch shows a head at the closest integer to the value of $\boxed{\tfrac{1}{2}(\tilde{B}' + \tilde{B})/|(\tilde{B}' - \tilde{B})| - 1}$ (see E11.13b).

$$\frac{(\tilde{B}' + \tilde{B})}{2|(\tilde{B}' - \tilde{B})|} - 1 = \frac{(0.3101 + 0.3540)}{2|0.3101 - 0.3540|} - 1 = 6.6$$

$$J_{\text{head}} = \boxed{7}$$

E11.15(a) When the R branch has a head, J_{head} is the closest integer to $\tfrac{1}{2}(\tilde{B}' + \tilde{B})/|(\tilde{B}' - \tilde{B})| - 1$ (see E11.13b). Thus, if we are given only that $J_{\text{head}} = 1$ and $\tilde{B} = 60.80\ \text{cm}^{-1}$, we know only that

$$0.5 < \tfrac{1}{2}(\tilde{B}' + \tilde{B})/(\tilde{B} - \tilde{B}') - 1 < 1.5$$

because the fractional value of a $\frac{1}{2}(\tilde{B}'+\tilde{B})/(\tilde{B}'-\tilde{B})-1$ calculation must be rounded-off to give the integer value J_{head}. Algebraic manipulation of the inequality yields

$$\frac{\{1+2(0.5)\}\tilde{B}}{\{3+2(0.5)\}} < \tilde{B}' < \frac{\{1+2(1.5)\}\tilde{B}}{\{3+2(1.5)\}}$$

$$\frac{\tilde{B}}{2} < \tilde{B}' < \frac{2\tilde{B}}{3}$$

$$\boxed{30.4\ \text{cm}^{-1} < \tilde{B}' < 40.5\ \text{cm}^{-1}}$$

When $\tilde{B}' < \tilde{B}$, the bond length in the electronically excited state is $\boxed{\text{greater}}$ than that in the ground state.

Here's an alternative solution that gives the same answer with insight into the band head concept: At the head of an R band, $\tilde{\nu}_{J_{head}} > \tilde{\nu}_{J_{head}-1}$ where $\tilde{\nu}_{J_{head}-1}$ is the transition $J_{head} \leftarrow J = J_{head}-1$. Substitution of eqn. 11.9(c) into this inequality yields the relation $\tilde{B}' > J_{head}\tilde{B}/(J_{head}+1)$. Similarly, $\tilde{\nu}_{J_{head}} > \tilde{\nu}_{J_{head}+1}$ where $\tilde{\nu}_{J_{head}+1}$ is the transition $J_{head}+2 \leftarrow J = J_{head}+1$. Substitution of eqn. 11.9(c) into this inequality yields the relation $\tilde{B}' < (J_{head}+1)\tilde{B}/(J_{head}+2)$. Consequently,

$$J_{head}\tilde{B}/(J_{head}+1) < \tilde{B}' < (J_{head}+1)\tilde{B}/(J_{head}+2)$$

E11.16(a) The transition wavenumber is $\tilde{\nu} = \frac{1}{\lambda} = \frac{1}{700\ \text{nm}} = 14\times10^{3}\ \text{cm}^{-1}$.

Water molecules are weak ligand field splitters, so we expect the d^5 electrons of Fe^{3+} to have the $t_{2g}^3e_g^2$ high-spin ground-state configuration in the octahedral $[Fe(H_2O)_6]^{3+}$ complex. The *d*-orbital electron spins are expected to be parallel with $S = 5/2$ and $2S + 1 = 6$ by Hund's maximum multiplicity rule. We also expect that $P > \Delta_O$ where P is the energy of repulsion for pairing two electrons in an orbital. A *d-d* transition to the $t_{2g}^4e_g^1$ octahedral excited state is expected to be both spin and parity forbidden and therefore to have a very small molar absorption coefficient. This transition releases the energy Δ_O and requires the energy P needed to pair two electrons within a t_{2g} orbital. Thus, $\tilde{\nu} = P - \Delta_O$ and $\boxed{\Delta_O = P - \tilde{\nu}}$. Using the typical value $P \sim 28 \times 10^3\ \text{cm}^{-1}$ yields the estimate $\Delta_O \sim \boxed{14 \times 10^3\ \text{cm}^{-1}}$. See F. A. Cotton and G. Wilkinson, *Advanced Inorganic Chemistry*, 4th ed. (New York: Wiley-Interscience, 1980), p. 646, for electron-pairing energies.

E11.17(a) The normalized wavefunctions are

$$\psi_i = \left(\frac{1}{a}\right)^{1/2} \quad \text{for } 0 \le x \le a \text{ and 0 elsewhere}$$

$$\psi_f = \left(\frac{1}{b-\frac{1}{2}a}\right)^{1/2} \quad \text{for } \tfrac{1}{2}a \le x \le b \text{ and 0 elsewhere}$$

$$\int \psi_f x \psi_i\, dx = \left(\frac{1}{a}\right)^{1/2}\left(\frac{1}{b-\frac{1}{2}a}\right)^{1/2}\int_{\frac{1}{2}a}^{a} x\, dx = \left(\frac{1}{a}\right)^{1/2}\left(\frac{1}{b-\frac{1}{2}a}\right)^{1/2}\frac{x^2}{2}\Bigg|_{x=\frac{1}{2}a}^{x=a}$$

$$= \left(\frac{1}{a}\right)^{1/2}\left(\frac{1}{b-\frac{1}{2}a}\right)^{1/2}\left(\frac{3a^2}{8}\right) = \boxed{\frac{3}{8}\left(\frac{a^3}{b-\frac{1}{2}a}\right)^{1/2}}$$

E11.18(a) The normalized wavefunctions are

$$\psi_i = \left(\frac{1}{a\sqrt{\pi}}\right)^{1/2} e^{-x^2/2a^2} \quad \text{for } -\infty \le x \le \infty \text{ and width } a$$

$$\psi_f = \left(\frac{1}{a\sqrt{\pi}}\right)^{1/2} e^{-(x-a/2)^2/2a^2} \quad \text{for } -\infty \le x \le \infty \text{ and width } a$$

$$\int \psi_f x \psi_i \, dx = \left(\frac{1}{a\sqrt{\pi}}\right)\int_{-\infty}^{\infty} x\,e^{-x^2/2a^2}\,e^{-(x-a/2)^2/2a^2}\,dx = \left(\frac{1}{a\sqrt{\pi}}\right)\int_{-\infty}^{\infty} x\,e^{-\{x^2+(x-a/2)^2\}/2a^2}\,dx$$

Since $x^2 + (x-a/2)^2 = \left(2^{1/2}x - \frac{a}{2^{3/2}}\right)^2 + \frac{a^2}{8}$, let $z = 2^{1/2}x - \frac{a}{2^{3/2}}$ and $x = 2^{-1/2}z + \frac{a}{4}$.

Then $dx = 2^{-1/2}dz$, and substitution gives

$$\int \psi_f x \psi_i \, dx = \left(\frac{e^{-1/16}}{a\sqrt{2\pi}}\right)\int_{-\infty}^{\infty}\left(2^{-1/2}z + \frac{a}{4}\right)e^{-z^2/2a^2}\,dz$$

$$= \left(\frac{e^{-1/16}}{a\sqrt{2\pi}}\right)\left\{2^{-1/2}\int_{-\infty}^{\infty} z e^{-z^2/2a^2}\,dz + \frac{a}{4}\int_{-\infty}^{\infty} e^{-z^2/2a^2}\,dz\right\}$$

The factors within the first integral have ungerade and gerade symmetry. Because u × g = u, the integrand has ungerade symmetry and the first integral is necessarily zero (the integral of an ungerade function over a symmetric interval equals zero).

$$\int \psi_f x \psi_i \, dx = \left(\frac{e^{-1/16}}{a\sqrt{2\pi}}\right)\left\{\frac{a}{4}\int_{-\infty}^{\infty} e^{-z^2/2a^2}\,dz\right\} = \left(\frac{e^{-1/16}}{4\sqrt{2\pi}}\right)\left(2a^2\pi\right)^{1/2}$$

$$= \boxed{\tfrac{1}{4}e^{-1/16}a}$$

E11.19(a) The weak absorption at 30,000 cm^{-1} (330 nm) is typical of a carbonyl chromophore of an enol. The assignment is $\pi^*\leftarrow n$ where a non-bonding electron comes from one of the two lone pairs of the oxygen valence. The two lone pairs of oxygen are in sp^2 hybrid orbitals, which define the *xy* plane that contains the σ bond of the carbonyl. The π^* molecular orbital is perpendicular to this plane. There is little overlap between the *n* and π^* orbitals, thereby producing a low value for the dipole transition integral and a low molar absorption coefficient.

The strong absorption at 46,950 cm^{-1} (213 nm) has the $\pi^*\leftarrow\pi$ assignment. The conjugation of the π bonds of the ethenic chromophore and the carbonyl chromophore causes this transition to be shifted to lower energies w/r/t both the $\pi^*\leftarrow\pi$ transition of ethene (165 nm) and the $\pi^*\leftarrow\pi$ transition of propanone (190 nm). This shift can be understood in terms of the simple Hückel theory of π molecular orbitals using the butadiene π energy model shown in text Figure 5.36 and Figure 11.3 below. The figure demonstrates a broad principle (see Section 9.9a): the difference between neighboring energy levels becomes smaller as the number of adjacent, overlapping orbitals becomes larger.

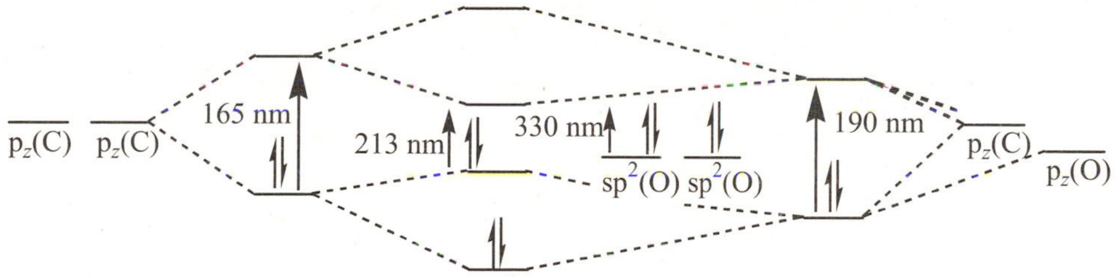

Figure 11.3

E11.20(a) The propanone weak absorption at $\boxed{\text{280 nm has the } \pi^* \leftarrow n \text{ assignment}}$ where a non-bonding electron comes from one of the two lone pairs of the oxygen valence. See E11.19(a) for the explanation of the weakness of this absorption band. The strong absorption at $\boxed{\text{189 nm has the } \pi^*\leftarrow\pi \text{ assignment}}$. Both bands are associated with the carbonyl chromophore, and they can be identified in Figure 11.3.

E11.21(a) Because the molecular response is proportional to the square of the electric field, a field of angular frequency ω elicits the response $\cos^2 \omega t = \frac{1}{2}\{1+\cos(2\omega t)\}$ (standard formula for a power of a trigonometric function; see math handbook) and we conclude that the effect is equivalent to an incident field of 2ω.

E11.22(a) (a) Vibrational energy spacings of the **lower** state are determined by the spacing of the peaks of fluorescence spectrum A of benzophenone. $\boxed{\nu \approx 1800\,\text{cm}^{-1}}$

(b) The peaks of A give no information about the spacing of the upper vibrational levels (without a detailed analysis of the intensities of the lines).

E11.23(a) When the steeply repulsive section of the O_2 potential energy curve for the excited state lies slightly toward the short side of the equilibrium bond length of the ground state and the minimum of the excited state lies to the longer side (as shown in text Fig. 11.31), a great many excited vibrational states overlap with the lowest energy vibration of the ground state. The **Franck–Condon factor** is appreciable for many vertical transitions (see text Fig. 11.10) and the absorption band is broad. Furthermore, predissociation to the unbound $^5\Pi_u$ state shortens the lifetime of excited vibrational states. This causes the high resolution lines of the corresponding vibrational-rotational transitions to be broad through the Heisenberg uncertainty principle $\Delta E \Delta t \geq \hbar / 2$.

E11.24(a) Only an integral number of half-wavelengths fit into the cavity. These are the **resonant modes.**

$\lambda = 2L/n$ [11.11] where n is an integer and L is the length of the cavity.

The resonant frequencies are given by $\nu = c/\lambda = nc/2L$. The lowest energy resonant modes (n = 1) in a 30-cm cavity are $\boxed{\lambda = 60\text{ cm } (\nu = 500\text{ MHz})}$.

E11.25(a) Referring to Example 11.3 of the text, we have

$$P_{\text{peak}} = E_{\text{pulse}}/t_{\text{pulse}} \quad \text{and} \quad P_{\text{average}} = E_{\text{total}}/t = E_{\text{pulse}} \times \nu_{\text{repetition}}$$

where $\nu_{\text{repetition}}$ is the pulse repetition rate.

$$t_{\text{pulse}} = E_{\text{pulse}} / P_{\text{peak}} = \frac{0.10\text{ mJ}}{5.0\text{ MW}} = \boxed{20\text{ ps}}$$

$$\nu_{\text{repetition}} = P_{\text{average}} / E_{\text{pulse}} = \frac{7.0\text{ kW}}{0.10\text{ mJ}} = \boxed{70\text{ MHz}}$$

E11.26(a) This Mathcad worksheet simulates the output of a mode-locked laser. The radiation intensity is shown in text *Justification* 11.5 to be proportional to the function $f(t, N)$ of the worksheet. The plots demonstrate that the superposition of a great many modes creates very narrow spikes separated by $t = 2L/c$.

$\mathbf{L} := 30\cdot\text{cm}$ $\quad \mathbf{c} := 299792458\text{m}\cdot\text{s}^{-1}$ $\quad \text{ns} := 10^{-9}\cdot\text{s}$

$t_{max} := 5\cdot 2\cdot L\cdot c^{-1}$ $\quad t_{max} = 1.001\times 10^{-8}\text{s}$

$i_{max} := 1000$ $\quad i := 1 .. i_{max}$ $\quad t_i := \dfrac{i\cdot t_{max}}{i_{max}}$

$$f(t, N) := \left(\frac{\sin(N\cdot\pi c\cdot t\cdot 2^{-1}\cdot L^{-1})}{\sin(\pi c\cdot t\cdot 2^{-1}\cdot L^{-1})}\right)^2$$

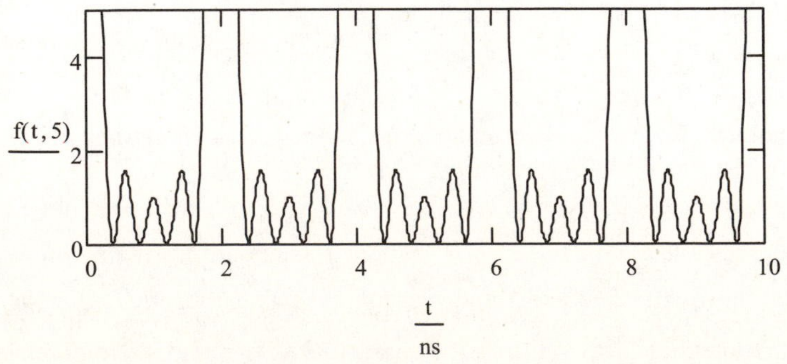

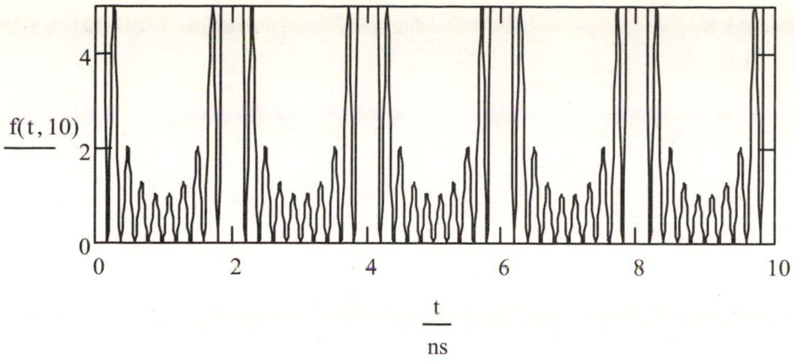

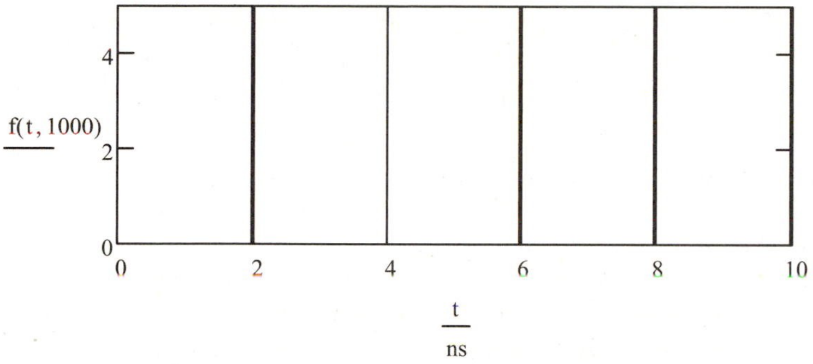

E11.27(a) The Nd-YAG laser transition at 1064 nm is very efficient and capable of substantial power output. Frequency doubling of the infrared band yields green light at 532 nm, and frequency quadrupling gives 266 nm ultraviolet radiation. The latter radiation can initiate the fast chemical reaction.

Solutions to problems

Solutions to numerical problems

P11.1 Solutions that have identical transmittance must have identical values of the absorbance [11.3] and identical values of $\varepsilon[\mathrm{J}]l$ [11.4]. Consequently,

$$[\mathrm{J}]_{\text{cell 2}} = [\mathrm{J}]_{\text{cell 1}}\, l_{\text{cell 1}} / l_{\text{cell 2}}$$
$$= 25\ \mu\text{g dm}^{-3} \times (1.55\ \text{cm})/(1.18\ \text{cm}) = \boxed{33\ \mu\text{g dm}^{-3}}$$

P11.3 For a photon to induce a spectroscopic transition, the transition moment $\langle\mu\rangle$ must be non-zero, a requirement that leads to electronic spectrum selection rules concerned with changes in angular momentum. The rules for a homonuclear diatomic are (Section 11.3b)

$$\Delta\Lambda = 0,\ \pm 1 \quad \Delta S = 0 \quad \Delta\Sigma = 0 \quad \Delta\Omega = 0,\ \pm 1 \quad \text{where } \Omega = \Lambda + \Sigma$$

Λ gives the total orbital angular momentum about the internuclear axis and Σ gives the total spin angular momentum about the internuclear axis. The ± superscript selection rule for reflection in the plane along the internuclear axis is $+\leftrightarrow+$ or $-\leftrightarrow-$ ($+\leftrightarrow-$ is forbidden). The **Laporte selection rule** states that for a centrosymmetric molecule (those with a center of inversion), the only allowed transitions are transitions that are accompanied by a change of parity: $\mathrm{u}\leftrightarrow\mathrm{g}$.

The electric-dipole transition $\boxed{{}^2\Sigma_g^+ \leftarrow {}^2\Sigma_u^+ \text{ is allowed}}$ because none of the above rules negates the possibility of this event. You may also wish to reach this conclusion by direct examination of the dipole transition moment integral, $\int\psi_f^*\mu\psi_i\,\mathrm{d}\tau$, where the dipole moment operator has components proportional to the Cartesian coordinates. The integral vanishes unless the integrand or at least some part of the integrand belongs to the totally symmetric representation (A_{1g}; see Chapter 7). To find the symmetry species of the integrand, we multiply the characters of its factors. Homonuclear diatomic molecules and ions belong to the $D_{\infty h}$ point group and the $D_{\infty h}$ character table tell us the symmetry species of each integrand factor.

$$\psi_f : A_{1g}(\Sigma_g^+)$$
$$\mu_z : A_{1u}(\Sigma_u^+)$$
$$\psi_i : A_{1u}(\Sigma_u^+)$$

Symmetry product: $A_{1g} \times A_{1u} \times A_{1u} = A_{1g} \times A_{1g} = A_{1g}$

Since the integrand spans A_{1g}, the transition ${}^2\Sigma_g^+ \leftarrow {}^2\Sigma_u^+$ is allowed.

An electric-dipole transition from a ${}^2\Sigma_u^+$ ground state to a Π_u excited state is forbidden by the Laporte selection rule.

Finally, we check the possibility of a transition from a ${}^2\Sigma_u^+$ ground state to a ${}^2\Pi_g$ excited state by finding whether or not the integrand of the transition integral spans the totally symmetric representation. The symmetry product for the μ_x component is $E_{1g} \times A_{1u} \times A_{1u} = E_{1g} \times A_{1g} = E_{1g}$. Since the integrand does not span A_{1g}, the transition is forbidden for z-polarized light. The symmetry product for both the μ_x and μ_y components is $E_{1g} \times E_{1u} \times A_{1u} = E_{1g} \times E_{1g}$. Since the species product $E_{1g} \times E_{1g}$ has an angular dependence and therefore does not contain the totally symmetric representation, the transition is forbidden for x- and y-polarized light. You may also wish to show this by application of the orthogonality theorem to find the coefficient of A_{1g} in the integrand.

$D_{\infty h}$	E	$\infty C_2'$	$2C_\phi$	i	$\infty\sigma_v$	$2S_\phi$
$\Sigma_u^+(A_{1u})$	1	−1	1	−1	1	−1
μ_x or $\mu_y(E_{1u})$	2	0	2 cos ϕ	−2	0	2 cos ϕ
$\Pi_g(E_{1g})$	2	0	2 cos ϕ	2	0	−2 cos ϕ
Integrand	4	0	4 cos^2 ϕ	4	0	4 cos^2 ϕ

The orthogonality theorem gives the coefficient of A_{1g} in the integrand as

$$c_{A_{1g}} = (1/h)\Sigma_c\, g(C)\chi(C) = [4+0+2(4\cos^2\phi)+4+0+2(4\cos^2\phi)]/h$$

Since the group order h, which equals infinity, does not cancel with a numerator factor, $c_{A_{1g}} = 0$.

P11.5 The ionization is $HBr \rightarrow HBr^+ + e^-$ with the accompanying electronic energy change given by the equation $I_i + \Delta E_{v' \leftarrow v=0} = h\nu - \frac{1}{2}m_e v^2$. This modified form of eqn. 5.19(b) accounts for the possibility of an excitation change in the vibrational energy in going from the ground electronic vibrational state $v = 0$, in which a majority of molecules start, to the ionized electronic vibrational state $v' = 0, 1, 2 ...$ The vibrational transition $v' = 0 \leftarrow v = 0$ is called an **adiabatic transition**. Figure 11.4 shows the potential energy relationships between the ground electronic state and two possible ionized electronic states.

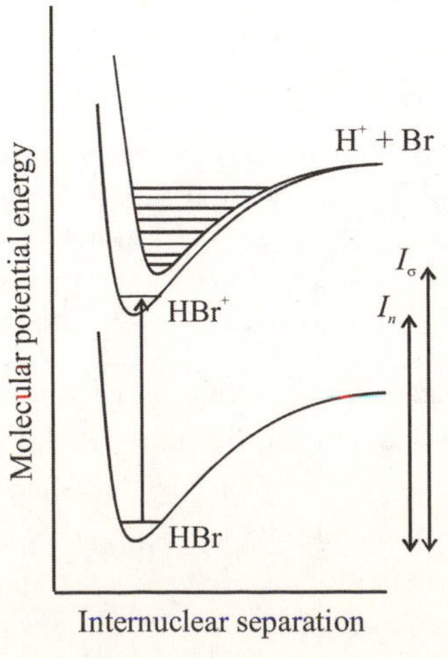

Figure 11.4

(a) The photoelectron spectrum band between 15.2 eV and 16.2 eV is the ejection of a bonding σ electron. Loss of this electron reduces the bond order from 1 to ½, reduces the magnitude of the bond force constant, and lengthens the equilibrium bond length of the ionized molecule. The electronic transition is labeled as I_σ in Figure 11.4. The longer bond length of the ionized state cause the Franck–Condon factor for the adiabatic transition ($v' = 0 \leftarrow v = 0$) to be small. This is the lowest energy transition of the band at about 15.3 eV. The increasing spectral intensity for the $v' = 1 \leftarrow v = 0$ and $v' = 2 \leftarrow v = 0$ transitions indicates that these vertical transitions have successively larger Franck–Condon factors (see text Fig. 11.10). The separation of lines (~0.162 eV) corresponds to an ionized vibrational wavenumber of about 1300 cm^{-1}, which is considerably lower than the 2648.98 cm^{-1} of the neutral ground state. The presence of one unpaired electron in the bonding σ orbital means that the ionized molecule is in a $^2\Sigma^+$ state.

(b) The lines between 11.6 eV and 12.3 eV involve transitions of a non-bonding electron of the chlorine p valence subshell to two very closely spaced electronic states of the ionized molecule. The ionization energy of these states is labeled as I_n in Figure 11.4. The unpaired electron of the ionized state makes it a doublet with spin-orbit coupling producing $j = |l + s|,...,|l - s| = |1 + ½|, |1 - ½| = {}^3/_2, {}^1/_2$. Consequently, the term symbols of these states are $^2\Pi_{1/2}$ and $^2\Pi_{3/2}$, and Hund's rule predicts that $^2\Pi_{3/2}$ is lowest in energy because the subshell is more than half-filled. Excitation of a non-bonding electron does not affect the molecular bond, nor does it affect the bonding force constant or the equilibrium bond length. Only the vertical, adiabatic transition ($v' = 0 \leftarrow v = 0$) has an appreciable Franck–Condon factor. The transition $v' = 1 \leftarrow v = 0$ of the $^2\Pi_{3/2}$ transition has a very small vertical, vibrational overlap integral; it cannot be seen in the spectrum because it lies below the $^2\Pi_{1/2}$ adiabatic transition at 12.0 ev. The transition $v' = 1 \leftarrow v = 0$ of the $^2\Pi_{1/2}$ transition has a very small vertical, vibrational overlap integral located at 12.3 eV. The 0.3 eV line separation corresponds to an ionized vibrational wavenumber of about 2400 cm^{-1}. This is consistent with the vibrational wavenumber of the ground state (2648.98 cm^{-1}) and confirms the expectation that excitation of a non-bonding electron does not affect the σ bond.

P11.7 The spectrum gives the peak and half-height points:

$$\varepsilon_{\text{peak}} = 250 \text{ dm}^3 \text{ mol}^{-1} \text{ cm}^{-1},\ \lambda_{\text{peak}} = 284 \text{ nm } \left(\tilde{\nu} = 35200 \text{ cm}^{-1}\right)$$

$$\varepsilon_{1/2} = 125 \text{ dm}^3 \text{ mol}^{-1} \text{ cm}^{-1},\ \lambda_{1/2} = 305 \text{ nm } \left(\tilde{\nu} = 32800 \text{ cm}^{-1}\right)$$
$$\lambda_{1/2} = 265 \text{ nm } \left(\tilde{\nu} = 37700 \text{ cm}^{-1}\right)$$

We estimate that the wavenumber band has a normal Gaussian shape.

$\varepsilon = \varepsilon_{\text{max}} \text{e}^{-\left(\tilde{\nu}-\tilde{\nu}_{\text{peak}}\right)^2/a^2}$ where a is a constant related to the half-width $\Delta\tilde{\nu}_{1/2} = (37700-32800) \text{ cm}^{-1} = 4900 \text{ cm}^{-1}$

$$A = \int_{\text{band}} \varepsilon(\tilde{\nu})\, \text{d}\tilde{\nu} \quad [11.5] = \varepsilon_{\text{max}} \int_{-\infty}^{\infty} \text{e}^{-\left(\tilde{\nu}-\tilde{\nu}_{\text{peak}}\right)^2/a^2}\, \text{d}\tilde{\nu}$$
$$= \varepsilon_{\text{max}} a\sqrt{\pi} \quad \text{(standard integral)}$$

The relationship between the half-width and a is found by evaluation of the line shape at $\varepsilon(\tilde{\nu}_{1/2}) = \varepsilon_{\text{max}}/2$.

$$\varepsilon_{\text{max}}/2 = \varepsilon_{\text{max}} \text{e}^{-\left(\tilde{\nu}_{1/2}-\tilde{\nu}_{\text{peak}}\right)^2/a^2}$$
$$\ln(1/2) = -\left(\tilde{\nu}_{1/2} - \tilde{\nu}_{\text{peak}}\right)^2/a^2$$
$$a^2 = \frac{\left(\tilde{\nu}_{1/2} - \tilde{\nu}_{\text{peak}}\right)^2}{\ln(2)} = \frac{\left(\Delta\tilde{\nu}_{1/2}/2\right)^2}{\ln(2)}$$
$$a = \frac{\Delta\tilde{\nu}_{1/2}}{2\sqrt{\ln 2}}$$

Thus,

$$A = \boxed{\tfrac{1}{2}\Delta\tilde{\nu}_{1/2}\varepsilon_{\max}\sqrt{\pi/\ln(2)}} = 1.0645\,\Delta\tilde{\nu}_{1/2}\varepsilon_{\max}$$

$$A = \tfrac{1}{2}\left(4900\ \text{cm}^{-1}\right)\times\left(250\ \text{dm}^3\ \text{mol}^{-1}\ \text{cm}^{-1}\right)\sqrt{\pi/\ln(2)} = \boxed{1.3\bar{0}\times10^6\ \text{dm}^3\ \text{mol}^{-1}\ \text{cm}^{-2}}$$

Since the dipole moment components transform as $A_1(z)$, $B_1(x)$, and $B_2(y)$, excitations from A_1 to A_1, B_1, and B_2 terms are allowed.

P11.9 The anthracene vapour fluorescence spectrum gives the vibrational splitting of the lower state. The wavelengths stated correspond to the wavenumbers 22 730, 24 390, 25 640, 27 030 cm^{-1}, indicating spacings of 1660, 1250, and 1390 cm^{-1}. The absorption spectrum spacing gives the separation of the vibrational levels of the upper state. The wavenumbers of the absorption peaks are 27 800, 29 000, 30 300, and 32 800 cm^{-1}. The vibrational spacings are therefore 1200, 1300, and 2500 cm^{-1}. The data are compatible with the deactivation of the excited-state vibrational modes before spontaneous emission returns the molecule to the ground electronic state. This produces a fluorescence band of lower energy than the energy of the absorption band. Furthermore, while the absorption band has a vibrational progression that depends on vibrational modes of the excited state, the fluorescence band has a vibrational progression that depends on vibrational modes of the ground state. The absorption and fluorescence spectra are not mirror images.

P11.11 (a) The molar concentration corresponding to 1 molecule per cubic μm is

$$\frac{n}{V} = \frac{1}{6.022\times10^{23}\ \text{mol}^{-1}}\times\frac{\left(10^6\ \mu\text{m m}^{-1}\right)^3}{\left(1.0\ \mu\text{m}^3\right)\left(10\ \text{dm m}^{-1}\right)^3} = \boxed{1.7\times10^{-9}\ \text{mol dm}^{-3}}$$

i.e., nanomolar concentrations.

(b) An impurity of a compound of molar mass 100 g mol^{-1} present at 1.0×10^{-7} kg per 1.00 kg water can be expected to be present at a level of N molecules per cubic μm where N is

$$N = \frac{1.0\times10^{-7}\ \text{kg impurity}}{1.00\ \text{kg water}}\times\frac{6.022\times10^{23}\ \text{mol}^{-1}}{100\times10^{-3}\ \text{kg impurity mol}^{-1}}\times\left(1.0\times10^3\ \text{kg water m}^{-3}\right)\times\left(10^{-6}\ \text{m}\right)^3$$

$$N = \boxed{6.0\times10^2}.$$

Pure as it seems, the solvent is much too contaminated for single-molecule spectroscopy.

P11.13 The light-scattering equation

$$\frac{I_0}{I_\theta} = a + b\times\left(\frac{I_0}{I_\theta}\sin^2(\theta/2)\right) \quad \text{where} \quad a = \left(Kc_\text{M}M\right)^{-1} \quad \text{and} \quad b = \frac{16\pi^2R^2}{5\lambda^2}$$

has a linear form when I_0/I_θ is plotted against $\frac{I_0}{I_\theta}\sin^2(\theta/2)$. The intercept a and slope b are determined with a linear regression fit of this plot. Knowing a and b, the above equations can be used to calculate M and R. We begin by drawing up the requisite data table, preparing a plot (shown in Figure 11.5), and calculating the linear regression fit.

$\theta/°$	15.0	45.0	70.0	85.0	90.0
$100\times I_0/I_\theta$	4.20	4.37	4.63	4.83	4.90
$100\times I_0/I_\theta\times\sin^2(\theta/2)$	0.0716	0.640	1.52	2.20	2.45

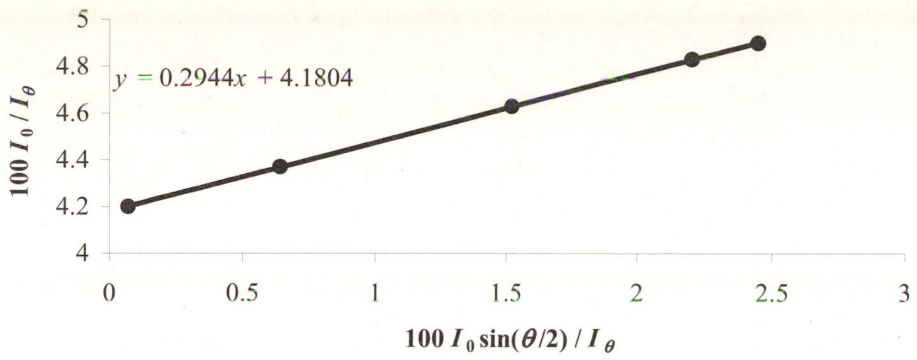

Figure 11.5

Thus, $a = 0.0418$ and $b = 0.294$.

$$M = (Kc_{\mathrm{M}}a)^{-1} = \{(2.40\times10^{-2}\ \mathrm{mol\ m^3\ kg^{-2}})\times(2.0\ \mathrm{kg\ m^{-3}})\times(0.0418)\}^{-1}$$

$$= \boxed{498\ \mathrm{kg\ mol^{-1}}}$$

$$R = (5b)^{1/2}\lambda/(4\pi) = (5\times0.294)^{1/2}\times(532\ \mathrm{nm})/(4\pi)$$

$$= \boxed{51.3\ \mathrm{nm}}$$

Solutions to theoretical problems

P11.15 The derivation is similar to that of the Beer–Lambert law in *Justification* 11.1, but let dx be the infinitesimal thickness of the layer containing the absorbing species J. Then

$$\mathrm{d}I = -\kappa[\mathrm{J}]I\mathrm{d}x$$

$$\frac{\mathrm{d}I}{I} = -\kappa[\mathrm{J}]\mathrm{d}x$$

$$\int_{I_0}^{I}\frac{\mathrm{d}I}{I} = -\kappa\int_0^l[\mathrm{J}]\mathrm{d}x = -\kappa[\mathrm{J}]_0\int_0^l \mathrm{e}^{-x/\lambda}\mathrm{d}x$$

$$\ln(I)\Big|_{I=I_0}^{I=I} = \kappa\lambda[\mathrm{J}]_0\,\mathrm{e}^{-x/\lambda}\Big|_{x=0}^{x=l}$$

$$-\ln\frac{I_0}{I} = -\kappa\lambda[\mathrm{J}]_0(1-\mathrm{e}^{-l/\lambda})$$

$$(\ln 10)\log\frac{I_0}{I} = \kappa\lambda[\mathrm{J}]_0(1-\mathrm{e}^{-l/\lambda})$$

Let $A = \log\frac{I_0}{I}$ [11.3] and $\varepsilon' = \kappa\lambda/(\ln 10)$.

Then

$$\boxed{A = \varepsilon'[\mathrm{J}]_0(1-\mathrm{e}^{-l/\lambda})}$$

In the case for which $l/\lambda >> 1$, $\mathrm{e}^{-l/\lambda} = 0$ and $A = \varepsilon'[\mathrm{J}]_0$.

In the case for which $l/\lambda << 1$, $\mathrm{e}^{-l/\lambda} \sim 1 + (-l/\lambda) + (-l/\lambda)^2/2$ (Taylor series truncated after 2nd order).

Thus,

$$A = \varepsilon'[\mathrm{J}]\left(1-\mathrm{e}^{-l/\lambda}\right)$$
$$A = \varepsilon'[\mathrm{J}]\left(1-1+l/\lambda-(l/\lambda)^2/2\right)$$
$$A = \frac{\varepsilon' l[\mathrm{J}]}{\lambda}(1-l/2\lambda)$$
$$\boxed{A = \varepsilon l[J]\times(1-l/2\lambda)} \text{ when } l/\lambda \ll 1$$

P11.17 Suppose that non-absorbing species B (pyridine in this problem) is progressively added to a solution of light-absorbing species A (I_2 in this problem). Furthermore, suppose that A and B form a complex of light-absorbing species AB. Suppose that all absorbance measurements are made at the equilibrium $A + B \rightleftharpoons AB$. By the conservation of mass, the sum of the equilibrium concentrations of the absorbing species, $[A]_e$ + $[AB]_e$, is a constant for all additions of B. That is, the sum of concentrations of all species that contain A must equal the concentration of A present before any B is added, $[A]_0$. The absorbance at any wavelength is

$$\begin{aligned}A_\lambda &= \varepsilon_{A,\lambda} l[\mathrm{A}]_e + \varepsilon_{AB,\lambda} l[\mathrm{AB}]_e \\ &= \varepsilon_{A,\lambda} l([\mathrm{A}]_0 - [\mathrm{AB}]_e) + \varepsilon_{AB,\lambda} l[\mathrm{AB}]_e \\ &= \varepsilon_{A,\lambda} l[\mathrm{A}]_0 + (\varepsilon_{AB,\lambda} - \varepsilon_{A,\lambda}) l[\mathrm{AB}]_e\end{aligned}$$

The first term to the right in the above equation is a constant for all additions of B; the second term is generally not constant, and consequently, A_λ generally varies with the addition of B. However, at any wavelength for which $\varepsilon_{AB,\lambda} = \varepsilon_{A,\lambda}$, the second term vanishes, making A_λ a constant at all additions of B. The wavelength at which this happens is called the **isosbestic point**. It is characterized by the equilibrium between absorbing species and $\varepsilon_{AB,\lambda} = \varepsilon_{A,\lambda}$.

P11.19 We need to establish whether the transition dipole moments

$$\mu_{fi} = \int \psi_f^* \mu \psi_i \, \mathrm{d}\tau \ [10.56]$$

connecting the states 1 and 2 and the states 1 and 3 are zero or non-zero. The particle-in-a-box wavefunctions are $\psi_n = (2/L)^{1/2}\sin(n\pi x/L)$ [2.6b].

Thus $$\mu_{2,1} \propto \int \sin\left(\frac{2\pi x}{L}\right) x \sin\left(\frac{\pi x}{L}\right)\mathrm{d}x \propto \int x\left[\cos\left(\frac{\pi x}{L}\right) - \cos\left(\frac{3\pi x}{L}\right)\right]\mathrm{d}x$$

and $$\mu_{3,1} \propto \int \sin\left(\frac{3\pi x}{L}\right) x \sin\left(\frac{\pi x}{L}\right)\mathrm{d}x \propto \int x\left[\cos\left(\frac{2\pi x}{L}\right) - \cos\left(\frac{4\pi x}{L}\right)\right]\mathrm{d}x$$

having used $\sin\alpha\sin\beta = \frac{1}{2}\cos(\alpha-\beta) - \frac{1}{2}\cos(\alpha+\beta)$. Both of these integrals can be evaluated using the standard form

$$\int x(\cos ax)\,\mathrm{d}x = \frac{1}{a^2}\cos ax + \frac{x}{a}\sin ax$$

$$\int_0^L x\cos\left(\frac{\pi x}{L}\right)\mathrm{d}x = \frac{1}{(\pi/L)^2}\cos\left(\frac{\pi x}{L}\right)\bigg|_0^L + \frac{x}{(\pi/L)}\sin\left(\frac{\pi x}{L}\right)\bigg|_0^L = -2\left(\frac{L}{\pi}\right)^2 \neq 0$$

$$\int_0^L x\cos\left(\frac{3\pi x}{L}\right)\mathrm{d}x = \frac{1}{(3\pi/L)^2}\cos\left(\frac{3\pi x}{L}\right)\bigg|_0^L + \frac{x}{(3\pi/L)}\sin\left(\frac{3\pi x}{L}\right)\bigg|_0^L = -2\left(\frac{L}{3\pi}\right)^2 \neq 0$$

Thus $\mu_{2,1} \neq 0$.

In a similar manner, $\mu_{3,1} = 0$.

COMMENT. A general formula for μ_{fi} applicable to all possible particle-in-a-box transitions may be derived. The result is $(n=f, m=i)$

$$\mu_{nm} = -\frac{eL}{\pi^2}\left[\frac{\cos(n-m)\pi - 1}{(n-m)^2} - \frac{\cos(n+m)\pi - 1}{(n+m)^2}\right]$$

For m and n both even or both odd numbers, $\mu_{nm} = 0$; if one is even and the other odd, $\mu_{nm} \neq 0$. See also Problem 11.22.

Question. Can you establish the general relation for μ_{nm} above?

P11.21 (a) Ethene (ethylene) belongs to D_{2h}. In this group the x, y, and z components of the dipole moment transform as B_{3u}, B_{2u}, and B_{1u}, respectively. The π orbital is B_{1u} (like z, the axis perpendicular to the plane) and π^* is B_{3g}. Since $B_{3g} \times B_{1u} = B_{2u}$ and $B_{2u} \times B_{2u} = A_{1g}$, the transition is $\boxed{\text{allowed}}$ (and is y-polarized).

(b) Regard the CO group with its attached groups as locally C_{2v}. The dipole moment has components that transform as $A_1(z)$, $B_1(x)$, and $B_2(y)$, with the z-axis along the C=O direction and x perpendicular to the R_2CO plane. The n orbital is p_y (in the R_2CO plane) and hence transforms as B_2. The π^* orbital is p_x (perpendicular to the R_2CO plane) and hence transforms as B_1. Since $\Gamma_f \times \Gamma_i = B_1 \times B_2 = A_2$, but no component of the dipole moment transforms as A_2, the transition is $\boxed{\text{forbidden}}$.

P11.23 (a) The Beer–Lambert Law is

$$A = \log\frac{I_0}{I} = \varepsilon[\mathrm{J}]l$$

The absorbed intensity is

$$I_{\text{abs}} = I_0 - I \quad \text{so} \quad I = I_0 - I_{\text{abs}}$$

Substitute this expression into the Beer–Lambert law and solve for I_{abs}:

$$\log\frac{I_0}{I_0 - I_{\text{abs}}} = \varepsilon[\mathrm{J}]l \quad \text{so} \quad I_0 - I_{\text{abs}} = I_0 \times 10^{-\varepsilon[\mathrm{J}]l}$$

and $$I_{\text{abs}} = \boxed{I_0 \times \left(1 - 10^{-\varepsilon[\mathrm{J}]l}\right)}$$

(b) The problem states that $I_f(\tilde{\nu}_f)$ is proportional to ϕ_f and to $I_{\text{abs}}(\tilde{\nu})$, so

$$I_f(\tilde{\nu}_f) \propto \phi_f\, I_0(\tilde{\nu}) \times \left(1 - 10^{-\varepsilon[\mathrm{J}]l}\right)$$

If the exponent is small, we can expand $1 - 10^{-\varepsilon[\mathrm{J}]l}$ in a power series:

$$10^{-\varepsilon[\mathrm{J}]l} = (\mathrm{e}^{\ln 10})^{-\varepsilon[\mathrm{J}]l} \approx 1 - \varepsilon[\mathrm{J}]l \ln 10 + \cdots$$

and $$I_f(\tilde{\nu}_f) \propto \boxed{\phi_f\, I_0(\tilde{\nu})\, \varepsilon[\mathrm{J}]l}$$

Solutions to applications

P11.25 Fraction transmitted to the retina is

$$(1 - 0.30) \times (1 - 0.25) \times (1 - 0.09) \times 0.57 = 0.272$$

The number of photons focused on the retina in 0.1 s is

$$0.272 \times 40\,\text{mm}^2 \times 0.1\,\text{s} \times 4 \times 10^3\,\text{mm}^{-2}\,\text{s}^{-1} = \boxed{4.4 \times 10^3}$$

—more than what one might have guessed.

P11.27 The integrated absorption coefficient is

$$A = \int_{\text{band}} \varepsilon(\tilde{\nu})\,\mathrm{d}\tilde{\nu} \ [11.5]$$

If we can express ε as an analytical function of $\tilde{\nu}$, we can carry out the integration analytically. Following the hint in the problem, we seek to fit ε to an exponential function, which means that a plot of $\ln\varepsilon$ versus $\tilde{\nu}$ ought to be a straight line (Figure 11.6). So if $\ln\varepsilon = m\tilde{\nu} + b$, then

$$\varepsilon = \exp(m\tilde{\nu})\exp(b) \quad \text{and} \quad A = \left(\mathrm{e}^{b}/m\right)\left\{\exp(m\tilde{\nu}_{\mathrm{f}}) - \exp(m\tilde{\nu}_{\mathrm{i}})\right\}$$

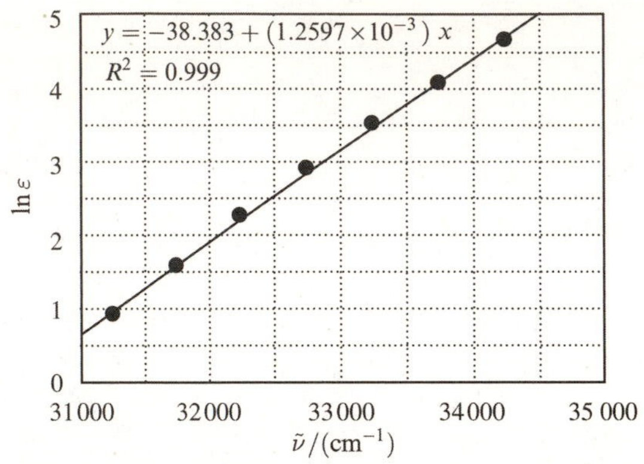

Figure 11.6

We draw up the following table, find the best-fit line, and make the plot of Figure 11.6. The linear regression fit yields the values of m and b for the computation of the integrated absorption coefficient.

λ/nm	$\varepsilon/(\mathrm{dm^3\ mol^{-1}\ cm^{-1}})$	$\tilde{\nu}/\mathrm{cm^{-1}}$	$\ln\varepsilon/(\mathrm{dm^3\ mol^{-1}\ cm^{-1}})$
292.0	1512	34248	4.69
296.3	865	33748	4.13
300.8	477	33248	3.54
305.4	257	32748	2.92
310.1	135.9	32248	2.28
315.0	69.5	31746	1.61
320.0	34.5	31250	0.912

So $$A = \frac{\mathrm{e}^{-38.383}}{1.26\times10^{-3}\ \mathrm{cm}}\left[\exp\left(\frac{1.26\times10^{-3}\ \mathrm{cm}}{290\times10^{-7}\ \mathrm{cm}}\right) - \exp\left(\frac{1.26\times10^{-3}\ \mathrm{cm}}{320\times10^{-7}\ \mathrm{cm}}\right)\right]\mathrm{dm^3\ mol^{-1}\ cm^{-1}}$$

$$= \boxed{1.24\times10^{5}\ \mathrm{dm^3\ mol^{-1}\ cm^{-2}}}$$

P11.29 In Figure 11.7

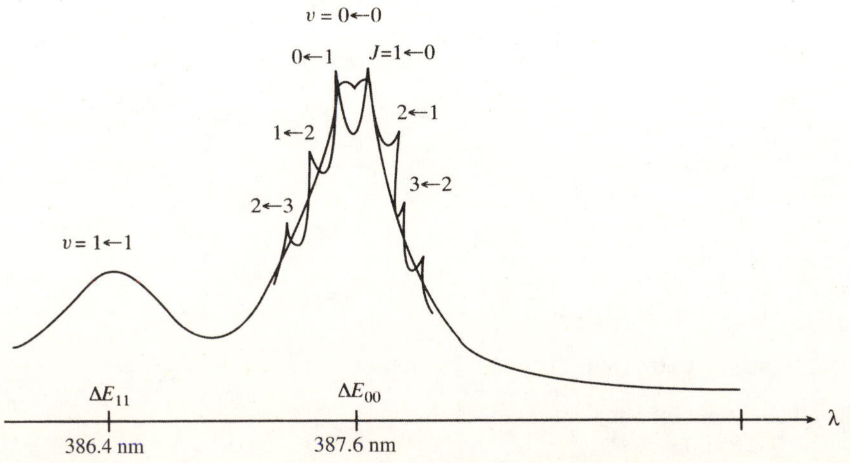

Figure 11.7

$$\Delta E_{11} = \frac{hc}{\lambda_{11}} = \frac{hc}{386.4\,\text{nm}} = 5.1409\times10^{-19}\ \text{J} = 3.2087\,\text{eV}$$

and

$$\Delta E_{00} = \frac{hc}{\lambda_{00}} = \frac{hc}{387.6\,\text{nm}} = 5.1250\times10^{-19}\ \text{J} = 3.1987\,\text{eV}$$

Energy of excited singlet, S_1: $E_1(\upsilon, J) = V_1 + (\upsilon + \tfrac{1}{2})\tilde{\nu}_1 hc + J(J+1)\tilde{B}_1 hc$

Energy of ground singlet, S_0: $E_0(\upsilon,\ J) = V_0 + (\upsilon + \tfrac{1}{2})\tilde{\nu}_0 hc + J(J+1)\tilde{B}_0 hc$

The midpoint of the 0–0 band corresponds to the forbidden Q branch ($\Delta J = 0$) with $J = 0$ and $\upsilon = 0 \leftarrow 0$.

$$\Delta E_{00} = E_1(0,0) - E_0(0,0) = (V_1 - V_0) + \tfrac{1}{2}(\tilde{\nu}_1 - \tilde{\nu}_0)hc \qquad (1)$$

The midpoint of the 1–1 band corresponds to the forbidden Q branch ($\Delta J = 0$) with $J = 0$ and $\upsilon = 1 \leftarrow 1$.

$$\Delta E_{11} = E_1(1,0) - E_0(1,0) = (V_1 - V_0) + \tfrac{3}{2}(\tilde{\nu}_1 - \tilde{\nu}_0)hc \qquad (2)$$

Multiplying eqn. 1 by 3 and subtracting eqn. 2 gives

$$3\Delta E_{00} - \Delta E_{11} = 2(V_1 - V_0)$$

$$\begin{aligned}V_1 - V_0 &= \frac{1}{2}(3\Delta E_{00} - \Delta E_{11})\\ &= \frac{1}{2}\{3(5.1250) - (5.1409)\}\times10^{-19}\ \text{J}\\ &= 5.1171\times10^{-19}\ \text{J} = \boxed{3.1938\ \text{eV}}\end{aligned} \qquad (3)$$

This is the potential energy difference between S_0 and S_1.

Equations (1) and (3) may be solved for $\tilde{\nu}_1 - \tilde{\nu}_0$.

$$\begin{aligned}\tilde{\nu}_1 - \tilde{\nu}_0 &= 2\{\Delta E_{00} - (V_1 - V_0)\}\\ &= 2\{5.1250 - 5.1171\}\times10^{-19}\ \text{J}/hc\\ &= 1.5800\times10^{-21}\ \text{J} = 0.0098615\ \text{eV} = \boxed{79.538\ \text{cm}^{-1}}\end{aligned}$$

The $\tilde{\nu}_1$ value can be determined by analyzing the band head data for which $J + 1 \leftarrow J$.

$$\begin{aligned}\Delta E_{10}(J) &= E_1(0,J) - E_0(1,J+1)\\ &= V_1 - V_0 + \frac{1}{2}(\tilde{\nu}_1 - 3\tilde{\nu}_0)hc + J(J+1)\tilde{B}_1 hc - (J+1)\times(J+2)\tilde{B}_0 hc\end{aligned}$$

$$\Delta E_{00}(J) = V_1 - V_0 + \frac{1}{2}(\tilde{\nu}_1 - \tilde{\nu}_0)hc + J(J+1)\tilde{B}_1 hc - (J+1)\times(J+2)\tilde{B}_0 hc$$

Therefore,

$$\Delta E_{00}(J) - \Delta E_{10}(J) = \tilde{\nu}_0 hc$$

$$\Delta E_{00}(J_{\text{head}}) = \frac{hc}{388.3\,\text{nm}} = 5.1158\times10^{-19}\,\text{J}$$

$$\Delta E_{10}(J_{\text{head}}) = \frac{hc}{421.6\text{nm}} = 4.7117\times10^{-19}\,\text{J}$$

$$\begin{aligned}\tilde{\nu}_0 &= \frac{\Delta E_{00}(J) - \Delta E_{10}(J)}{hc}\\ &= \frac{(5.1158 - 4.7117)\times10^{-19}\,\text{J}}{hc}\\ &= \frac{4.0410\times10^{-20}\,\text{J}}{hc} = 0.25222\ \text{eV} = \boxed{2034.3\ \text{cm}^{-1}}\end{aligned}$$

$$\tilde{\nu}_1 = \tilde{\nu}_0 + 79.538\,\text{cm}^{-1}$$

$$= (2034.3 + 79.538)\,\text{cm}^{-1} = \boxed{2113.8\ \text{cm}^{-1} = \frac{4.1990\times10^{-20}\,\text{J}}{hc}}$$

$$\frac{I_{1-1}}{I_{0-0}} \approx \frac{\mathrm{e}^{-E_1(1,0)/kT_{\text{eff}}}}{\mathrm{e}^{-E_1(0,0)/kT_{\text{eff}}}} = \mathrm{e}^{-(E_1(1,0)-E_1(0,0))/kT_{\text{eff}}}$$

$$\approx \mathrm{e}^{-hc\tilde{\nu}_1/kT_{\text{eff}}}$$

$$\ln\left(\frac{I_{1-1}}{I_{0-0}}\right) = -\frac{hc\tilde{\nu}_1}{kT_{\text{eff}}}$$

$$T_{\text{eff}} = \frac{hc\tilde{\nu}_1}{k\ln\left(\dfrac{I_{0-0}}{I_{1-1}}\right)} = \frac{4.1990\times10^{-20}\,\text{J}}{(1.38066\times10^{-23}\,\text{J K}^{-1})\ln(10)} = \boxed{1321\ \text{K}}$$

The relative population of the $v = 0$ and $v = 1$ vibrational states is the inverse of the relative intensities of the transitions from those states; hence $\dfrac{1}{0.1} = \boxed{10}$.

It would seem that with such a high effective temperature, more than eight of the rotational levels of the S_1 state should have a significant population. But the spectra of molecules in comets are never as clearly resolved as those obtained in the laboratory, and that is most probably the reason why additional rotational structure does not appear in these spectra.

12 Magnetic resonance

Answers to discussion questions

D12.1 Refer to eqn. 12.13(b), which describes the condition for resonance. The resonance field and resonance frequency are directly proportional to each other. High magnetic fields imply high frequencies and vice versa. The intensity of an NMR transition depends on a number of factors, two of which are seen in eqn. 12.14(a): one is the strength of the magnetic field itself and the other is the population difference between the energy levels of the α and β spins, which is also proportional to the magnetic field as shown by eqn. 12.14(b). Hence the intensity of the signal is proportional to $\mathcal{B}_0^2$. Consequently an increase in external field results in an even larger increase in signal intensity. See Section 12.4 and *Justification* 12.1 for a more detailed discussion. Chemical shifts as expressed in eqn. 12.15 are also seen to be proportional to B_0. Finally, as discussed in Section 12.6, the use of high magnetic fields simplifies the appearance of the complex spectra of macromolecules and allows them to be interpreted more readily. Although spin-spin splittings in NMR are not field dependent, strongly coupled (second-order) spectra are simplified at high fields because the chemical shifts increase and individual groups of nuclei become identifiable again.

D12.3 Before the application of a pulse the magnetization vector, $\boldsymbol{M}$, points along the direction of the static external magnetic field $\mathcal{B}_0$. There are more α spins than β spins. When we apply a rotating magnetic field $\mathcal{B}_1$ at right angles to the static field, the magnetization vector as seen in the rotating frame begins to precess about the $\mathcal{B}_1$ field with angular frequency $\omega_1 = \gamma\mathcal{B}_1$. The angle through which M rotates is $\theta = \gamma\mathcal{B}_1 t$, where t is the time for which the $\mathcal{B}_1$ pulse is applied. When $t = \pi/2\gamma\mathcal{B}_1$, $\theta = \pi/2 = 90°$, and $\boldsymbol{M}$ has rotated into the xy plane. Now there are equal numbers of α and β spins. A 180° pulse applied for a time $\pi/\gamma\mathcal{B}_1$ rotates $\boldsymbol{M}$ antiparallel to the static field. Now there are more β spins than α spins. A population inversion has occurred.

D12.5 For example, at room temperature, the tumbling rate of benzene, the small molecule, in a mobile solvent may be close to the Larmor frequency, and hence its spin-lattice relaxation time will be short. As the temperature increases, the tumbling rate may increase well beyond the Larmor frequency, resulting in an increased spin-lattice relaxation time.

For the large molecule (like a polymer) at room temperature, the tumbling rate may be well below the Larmor frequency, but with increasing temperature it will approach the Larmor frequency due to the increased thermal motion of the molecule combined with the decreased viscosity of the solvent. Therefore, the spin-lattice relaxation time may decrease.

D12.7 The basic COSY experiment uses the simplest of all two-dimensional pulse sequences: a single 90° pulse to excite the spins at the end of the preparation period and a mixing period containing just a second 90° pulse.

The key to the COSY technique is the effect of the second 90° pulse, which can be illustrated by consideration of the four energy levels of an AX system (as shown in Figs. 12.11 and 12.12 of the text). At thermal equilibrium, the population of the αAαX level is the greatest, and that of βAβX level is the smallest; the other two levels have the same energy and an intermediate population. After the first 90° pulse, the spins are no longer at thermal equilibrium. If a second 90° pulse is applied at a time t_1 that is short compared to the spin-lattice relaxation time T_1, the extra input of energy causes further changes in the populations of the four states. The changes in populations will depend on how far the individual magnetizations have precessed during the evolution period.

For simplicity, let us consider a COSY experiment in which the second 90° pulse is split into two selective pulses, one applied to X and one to A. Depending on the evolution time t_1, the 90° pulse that excites X may leave the population differences across each of the two X transitions unchanged, inverted, or somewhere in between. Consider the extreme case in which one population difference is inverted and the other unchanged. The 90° pulse that excites A will now generate an FID in which one of the two A transitions has increased in intensity and the other has decreased. The overall effect is that precession of the X spins during the evolution period determines the amplitudes of the signals from the A spins obtained during the detection period. As the evolution time t_1 is increased, the intensities of the signals from A spins oscillate at rates determined by the frequencies of the two X transitions.

This transfer of information between spins is at the heart of two-dimensional NMR spectroscopy and leads to the correlation of different signals in a spectrum. In this case, information transfer tells us that there is a scalar coupling between A and X. If we conduct a series of experiments in which t_1 is incremented, Fourier transformation of the FIDs on t_2 yields a set of spectra $I(\nu_1, \nu_2)$ in which the A signal amplitudes oscillate as a function of t_1. A second Fourier transformation, this time on t_1, converts these oscillations into a two-dimensional spectrum $I(\nu_1, \nu_2)$. The signals are spread out in ν_1 according to their precession frequencies during the detection period. Thus, if we apply the COSY pulse sequence to our AX spin system, the result is a two-dimensional spectrum that contains four groups of signals centred on the two chemical shifts in ν_1 and ν_2. Each group will show fine structure, consisting of a block of four signals separated by J_{AX}. The diagonal peaks are signals centerd on (δ_A δ_A) and (δ_X δ_X) and lie along the diagonal $\nu_1 = \nu_2$. They arise from signals that did not change chemical shift between t_1 and t_2. The cross peaks (or *off-diagonal peaks*) are signals centred on (δ_A δ_X) and (δ_X δ_A) and owe their existence to the coupling between A and X. Consequently, cross peaks in COSY spectra allow us to map the couplings between spins and to trace out the bonding network in complex molecules. Fig. 12.47 of the textbook shows a simple example of a proton COSY spectrum of 1-nitropropane.

D12.9 The hyperfine parameter a due to a nucleus in an aromatic radical, which is easily measured from the splittings of the lines in the EPR spectrum of the radical, as illustrated in Fig. 12.51 of the textbook, for the benzene anion radical, can be related to the spin density ρ of the unpaired electron on the nuclei in the aromatic radical. For the hyperfine splitting due to protons in aromatic systems, the relationship required is the McConnell equation, eqn. 12.44. The process of obtaining ρ from the McConnell equation is illustrated in *A Brief Illustration* following eqn. 12.44. For nuclei other than protons in aromatic radicals similar, although more complicated, equations arise; but in all cases the spin densities can be related to the coefficients of the basis functions used to describe the molecular orbital of the unpaired electron.

Solutions to exercises

E12.1(a) Work with eqn. 12.11. $\gamma\hbar = g_I\mu_N$ $\hbar$ has units J s and μ_N has units J T^{-1}; therefore γ has units

$$\frac{\text{J T}^{-1}}{\text{J s}} = \boxed{\text{s}^{-1}\ \text{T}^{-1}}.$$

E12.2(a) The magnitude of the angular momentum is given by $\{I(I+1)\}^{1/2}\hbar$. For a proton $I = 1/2$; hence

$$\text{magnitude} = \frac{\sqrt{3}}{2}\hbar = \boxed{9.133\times10^{-35}\ \text{J s}}.$$ The components along the z-axis are

$$\pm\frac{1}{2}\hbar = \boxed{\pm5.273\times10^{-35}\ \text{J s}}.$$ The angles that the projections of the angular momentum make with

the z-axis are $\theta = \pm\cos^{-1}\dfrac{\hbar/2}{\sqrt{3}\hbar/2} = \boxed{\pm0.9553\ \text{rad} = \pm54.74°}$.

E12.3(a) The resonance frequency is equal to the Larmor frequency of the proton and is given by

$$\nu = \nu_L = \frac{\gamma\mathcal{B}_0}{2\pi}\,[12.9] \text{ with } \gamma = \frac{g_I\mu_N}{\hbar}\,[12.11]$$

Hence $\nu = \dfrac{g_I\mu_N\mathcal{B}_0}{h} = \dfrac{(5.5857)\times(5.0508\times10^{-27}\,\text{J T}^{-1})\times(13.5\text{ T})}{6.626\times10^{-34}\,\text{J s}} = \boxed{574\text{ MHz}}$

E12.4(a) $E_{m_I} = -\gamma\hbar\mathcal{B}_0 m_I\ [12.10b] = -g_I\mu_N\mathcal{B}_0 m_I\ [12.11, \gamma\hbar = g_I\mu_N]$

$$m_I = \frac{3}{2}, \frac{1}{2}, -\frac{1}{2}, -\frac{3}{2}$$

$$E_{m_I} = (-0.4289)\times(5.051\times10^{-27}\text{ J T}^{-1})\times(6.800\text{ T}\times m_I) = \boxed{-1.473\times10^{-26}\text{ J}\times m_I}$$

E12.5(a) The energy level separation is

$$\Delta E = h\nu \quad \text{where } \nu = \frac{\gamma\text{B}_0}{2\pi}\ [12.9]$$

So

$$\nu = \frac{(6.73\times10^{7}\,\text{T}^{-1}\text{s}^{-1})\times15.4\,\text{T}}{2\pi} = 1.65\times10^{8}\text{ s}^{-1}$$
$$= 1.65\times10^{8}\text{ Hz} = \boxed{165\,\text{MHz}}$$

E12.6(a) (a) By a calculation similar to that in Exercise 12.3(a), we can show that a 600-MHz NMR spectrometer operates in a magnetic field of 14.1 T. Thus

$$\Delta E = \gamma\hbar\mathcal{B}_0 = h\nu_L = h\nu \text{ at resonance}$$
$$= (6.626\times10^{-34}\,\text{Js})\times(6.00\times10^{8}\text{ s}^{-1}) = \boxed{3.98\times10^{-25}\text{ J}}$$

(b) A 600-MHz NMR spectrometer means that 600 MHz is the resonance frequency for protons for which the magnetic field is 14.1 T. In high-field NMRs, it is the field, not the frequency, that is fixed, so for the deuteron

$$\nu = \frac{g_I\mu_N\mathcal{B}_0}{h}\ [\text{Exercise 12.3a}]$$
$$= \frac{(0.8575)\times(5.051\times10^{-27}\,\text{JT}^{-1})\times(14.1\text{T})}{6.626\times10^{-34}\,\text{Js}} = 9.21\overline{6}\times10^{7}\text{ Hz} = 92.1\overline{6}\,\text{MHz}$$

$$\Delta E = h\nu = (6.626\times10^{-34}\text{ J s})\times(9.21\overline{6}\times10^{7}\text{ s}^{-1}) = \boxed{6.11\times10^{-26}\text{ J}}$$

Thus the separation in energy is larger for the proton $\boxed{\text{(a)}}$.

E12.7(a) $\Delta E = h\nu = \gamma\hbar\mathcal{B} = g_I\mu_N\mathcal{B}$ [solution to exercise 12.3a]

Hence, $\mathcal{B} = \dfrac{h\nu}{g_I\mu_N} = \dfrac{(6.626\times10^{-34}\text{ J Hz}^{-1})\times(90.0\times10^{6}\text{ Hz})}{(1.793)\times(5.051\times10^{-27}\text{ J T}^{-1})} = \boxed{6.59\text{ T}}$

E12.8(a) In all cases, the selection rule $\Delta m_I = \pm1$ is applied; hence

$$\mathcal{B}_0 = \frac{h\nu}{g_I\mu_N} = \frac{6.626\times10^{-34}\,\text{J Hz}^{-1}}{5.0508\times10^{-27}\,\text{J T}^{-1}}\times\frac{\nu}{g_I}$$
$$= (1.3119\times10^{-7})\times\frac{(\nu/\text{Hz})}{g_I}\text{T} = (0.13119)\times\frac{(\nu/\text{MHz})}{g_I}\text{T}$$

Refer to the following table.

$\mathcal{B}_0$ / T	(a)^{1}H	(b)^{2}H	(c)^{13}C
g_I	5.5857	0.85745	1.4046
(i) 800 MHz	18.8	123	74.9
(ii) 500 MHz	11.7	76.6	46.8

COMMENT. Magnetic fields above 30 T have not yet been obtained for use in NMR spectrometers. So most of these calculated values have no current significance. As discussed in the solution to Exercise 12.6 (both a and b), it is the field, not the frequency, that is fixed in high-field NMR spectrometers. Thus an NMR spectrometer that is called a 500-MHz spectrometer refers to the resonance frequency for protons and has a magnetic field fixed at 11.7 T.

Question. What are the resonance frequencies of these nuclei in 250-MHz and 500-MHz spectrometers? Refer to Exercise 12.6.

E12.9(a) The ground state has

$$m_I + \frac{1}{2} = \alpha \text{ spin}, \quad m_I = -\frac{1}{2} = \beta \text{ spin}$$

Hence, with

$$\delta N = N_\alpha - N_\beta$$

$$\frac{\delta N}{N} = \frac{N_\alpha - N_\beta}{N_\alpha + N_\beta} = \frac{N_\alpha - N_\alpha e^{-\Delta E/kT}}{N_\alpha + N_\alpha e^{-\Delta E/kT}} \quad [\textit{Justification } 12.1]$$

$$= \frac{1-e^{-\Delta E/kT}}{1+e^{-\Delta E/kT}} \approx \frac{1-(1-\Delta E/kT)}{1+1} \approx \frac{\Delta E}{2kT} = \frac{g_I\mu_N\mathcal{B}_0}{2kT} \quad [\text{for } \Delta E = kT]$$

That is, $\dfrac{\delta N}{N} \approx \dfrac{g_I\mu_N\mathcal{B}_0}{2kT} = \dfrac{(5.5857)\times(5.0508\times10^{-27}\,\text{J T}^{-1})\times(\mathcal{B}_0)}{(2)\times(1.38066\times10^{-23}\,\text{J T}^{-1})\times(298\text{ K})} \approx 3.43\times10^{-6}\,\text{B}_0/\text{T}$

(a) $\mathcal{B}_0 = 0.3$ T, $\delta N/N = \boxed{1\times10^{-6}}$

(b) $\mathcal{B}_0 = 1.5$ T, $\delta N/N = \boxed{5.1\times10^{-6}}$

(c) $\mathcal{B}_0 = 10$ T, $\delta N/N = \boxed{3.4\times10^{-5}}$

E12.10(a)

$$\delta N \approx \frac{Ng_I\mu_N\mathcal{B}_0}{2kT} [\text{Exercise 12.9a}] = \frac{Nh\nu}{2kT}$$

Thus, $\delta N \propto \nu$

$$\frac{\delta N(800\text{MHz})}{\delta N(60\text{MHz})} = \frac{800\text{MHz}}{60\text{MHz}} = \boxed{13}$$

This ratio is not dependent on the nuclide as long as the approximation $\Delta E = kT$ holds (Exercise 12.9a).

E12.11(a) (a) $\delta = \dfrac{\nu - \nu^\circ}{\nu^\circ}\times10^6$ [12.18]

Since both ν and ν° depend on the magnetic field in the same manner, namely,

$$\nu = \frac{g_I\mu_N\mathcal{B}}{h} \quad \text{and} \quad \nu^\circ = \frac{g_I\mu_N\mathcal{B}_0}{h} \quad [\text{Exercise 12.3(a)}]$$

δ is $\boxed{\text{independent}}$ of both $\mathcal{B}$ and ν.

(b) Rearranging [12.18], we see $\nu - \nu^\circ = \nu^\circ \delta \times 10^{-6}$ and we see that the relative chemical shift is

$$\frac{\nu - \nu^\circ(800\ \text{MHz})}{\nu - \nu^\circ(60\ \text{MHz})} = \frac{(800\ \text{MHz})}{(60\ \text{MHz})} = \boxed{13}$$

COMMENT. This direct proportionality between $\nu - \nu^\circ$ and ν° is one of the major reasons for operating an NMR spectrometer at the highest frequencies possible.

E12.12(a) $\mathcal{B}_{\text{loc}} = (1-\sigma)\mathcal{B}_0$ [12.16]

$$|\Delta\mathcal{B}_{\text{loc}}| = |(\Delta\sigma)|\mathcal{B}_0 \approx |[\delta(\text{CH}_3) - \delta(\text{CHO})]|\mathcal{B}_0 \quad \left[|\Delta\sigma| \approx \left|\frac{\nu-\nu^\circ}{\nu^\circ}\right|\right]$$

$$= |(2.20 - 9.80)| \times 10^{-6}\mathcal{B}_0 = 7.60 \times 10^{-6}\mathcal{B}_0$$

(a) $\mathcal{B}_0 = 1.5\ \text{T}$, $|\Delta\mathcal{B}_{\text{loc}}| = 7.60 \times 10^{-6} \times 1.5\ \text{T} = \boxed{11\,\mu\text{T}}$

(b) $\mathcal{B}_0 = 15\ \text{T}$, $|\Delta\mathcal{B}_{\text{loc}}| = \boxed{110\,\mu\text{T}}$

E12.13(a) $\nu - \nu^\circ = \nu^\circ\delta \times 10^{-6}$ [12.18]

$$|\Delta\nu| \equiv (\nu - \nu^\circ)(\text{CHO}) - (\nu - \nu^\circ)(\text{CH}_3)$$
$$= \nu^\circ[\delta(\text{CHO}) - \delta(\text{CH}_3)] \times 10^{-6}$$
$$= (9.80 - 2.20) \times 10^{-6}\nu^\circ = 7.60 \times 10^{-6}\nu^\circ$$

(a) $\nu^\circ = 250\ \text{MHz}$, $|\Delta\nu| = 7.60 \times 10^{-6} \times 250\ \text{MHz} = 1.90\ \text{kHz}$

The spectrum is shown in Figure 12.1 with the value of $|\Delta\nu|$ as calculated above.

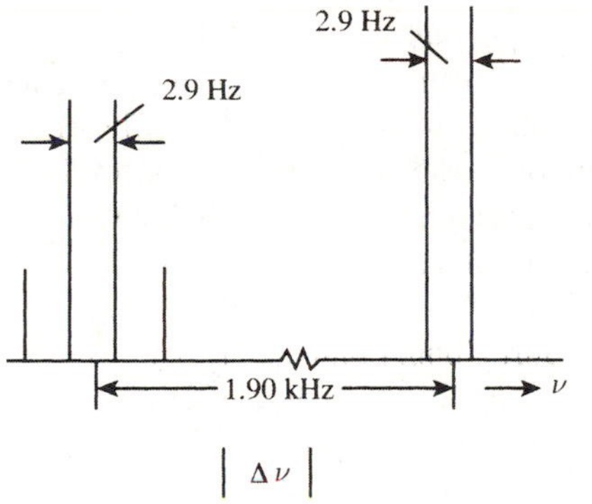

Figure 12.1

(b) $\nu^\circ = 800\ \text{MHz}$, $|\Delta\nu| = 6.08\ \text{MHz}$

When the frequency is changed to 800 MHz, the $|\Delta\nu|$ changes to 6.08 kHz. The fine structure (the splitting within groups) remains the same as spin–spin splitting is unaffected by the strength of the applied field. However, the intensity of the lines increases by a factor of 3.2 because $\delta N/N \propto \nu$ (Exercise 12.10a).

The observed splitting pattern is that of an AX_3 (or A_3X) species, the spectrum of which is described in Section 12.6.

E12.14(a) $$\tau \approx \frac{\sqrt{2}}{\pi\Delta\nu} \text{[12.29, with } \delta\nu \text{ written as } \Delta\nu\text{]}$$

$$\Delta\nu = \nu^{\circ}(\delta' - \delta)\times 10^{-6} \text{ [Exercise 12.13a]}$$

Then $$\tau \approx \frac{\sqrt{2}}{\pi\nu_0(\delta' - \delta)\times 10^{-6}}$$

$$\approx \frac{\sqrt{2}}{(\pi)\times(550\times 10^{6}\,\text{Hz})\times(4.8-2.7)\times 10^{-6}} \approx 3.9\times 10^{-4}\,\text{s}$$

Therefore, the signals merge when the lifetime of each isomer is less than about $\boxed{0.39\ \text{ms}}$, corresponding to a conversion rate of about $\boxed{2.6\times 10^{3}\,\text{s}^{-1}}$.

E12.15(a) The four equivalent ^{19}F nuclei ($I = \frac{1}{2}$) give a single line. However, the ^{10}B nucleus ($I = 3$, 19.6 percent abundant) splits this line into $2\times 3+1=7$ lines, and the ^{11}B nucleus ($I = \frac{3}{2}$, 80.4 percent abundant) splits it into $2\times\frac{3}{2}+1=4$ lines. The splitting arising from the ^{11}B nucleus will be larger than that arising from the ^{10}B nucleus (since its magnetic moment is larger by a factor of 1.5; Table 12.2). Moreover, the total intensity of the four lines due to the ^{11}B nuclei will be greater (by a factor of $80.4/19.6 \approx 4$) than the total intensity of the seven lines due to the ^{10}B nuclei. The individual line intensities will be in the ratio $\frac{7}{4}\times 4 = 7$ ($\frac{4}{7}$ the number of lines and about four times as abundant). The spectrum is sketched in Figure 12.2.

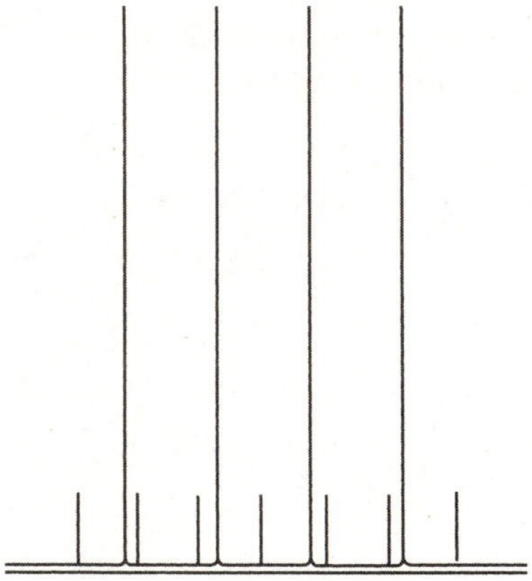

Figure 12.2

E12.16(a) $$\nu = \frac{g_I\mu_N\mathcal{B}}{h} \text{ [solution to Exercise 12.3a]}$$

Hence, $$\frac{\nu(^{19}\mathbf{F})}{\nu(^{1}\mathbf{H})} = \frac{g(^{19}\mathbf{F})}{g(^{1}\mathbf{H})}$$

or $$\nu(^{19}\text{F}) = \frac{5.2567}{5.5857}\times 800\,\text{MHz} = \boxed{753\ \text{MHz}}$$

The proton resonance consists of 2 lines $(2\times\frac{1}{2}+1)$ and the ^{19}F resonance of 3 lines $[2\times(2\times\frac{1}{2})+1]$. The intensities are in the ratio 1:2:1 (Pascal's triangle for two equivalent spin $\frac{1}{2}$ nuclei; Section 12.6). The lines are spaced $\frac{5.5857}{5.2567} = 1.06$ times greater in the fluorine region than in the proton region. The spectrum is sketched in Figure 12.3.

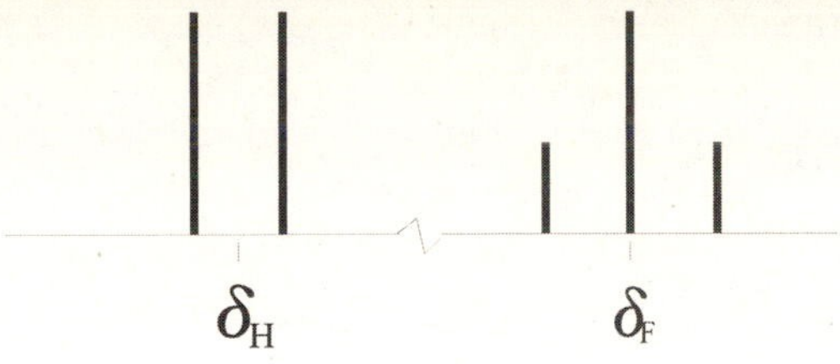

Figure 12.3

E12.17(a) The A, M, and X resonances lie in distinctively different groups. The A resonance is split into a 1:2:1 triplet by the M nuclei, and each line of that triplet is split into a 1:4:6:4:1 quintet by the X nuclei (with $J_{AM} > J_{AX}$). The M resonance is split into a 1:3:3:1 quartet by the A nuclei and each line is split into a quintet by the X nuclei (with $J_{AM} > J_{MX}$). The X resonance is split into a quartet by the A nuclei and then each line is split into a triplet by the M nuclei (with $J_{AX} > J_{MX}$). The spectrum is sketched in Figure 12.4.

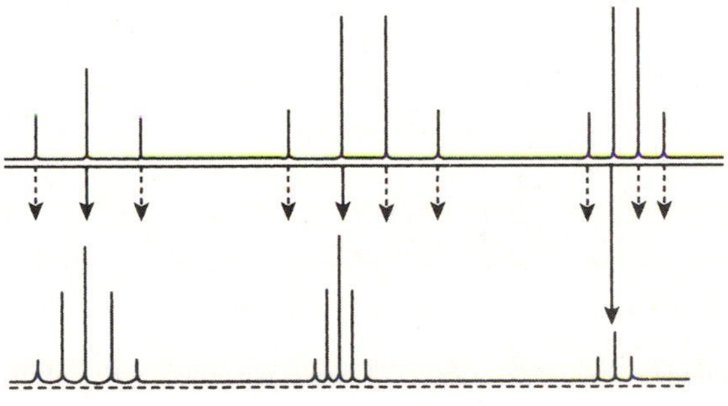

Figure 12.4

E12.18(a) (a) If there is rapid rotation about the axis, the H nuclei are both chemically and magnetically equivalent.

(b) Since $J_{cis} \neq J_{trans}$, the H nuclei are chemically but not magnetically equivalent.

E12.19(a) Analogous to precession of the magnetization vector in the laboratory frame due to the presence of $\mathcal{B}_0$, that is,

$$\nu_L = \frac{\gamma \mathcal{B}_0}{2\pi} \quad [12.9]$$

there is a precession in the rotating frame due to the presence of $\mathcal{B}_1$, namely,

$$\nu_L = \frac{\gamma \mathcal{B}_1}{2\pi} \quad \text{or} \quad \omega_1 = \gamma \mathcal{B}_1 \quad [\omega = 2\pi\nu]$$

Since ω is an angular frequency, the angle through which the magnetization vector rotates is

$$\theta = \gamma \mathcal{B}_1 t = \frac{g_I \mu_N}{\hbar} \mathcal{B}_1 t$$

and
$$\mathcal{B}_1 = \frac{\theta \hbar}{g_I \mu_N t} = \frac{(\frac{\pi}{2}) \times (1.055 \times 10^{-34}\ \mathrm{J\,s})}{(5.586) \times (5.051 \times 10^{-27}\ \mathrm{J\,T^{-1}}) \times (1.0 \times 10^{-5}\ \mathrm{s})} = \boxed{5.9 \times 10^{-4}\ \mathrm{T}}$$

A 180° pulse requires $2 \times 10\ \mu\mathrm{s} = \boxed{20\ \mu\mathrm{s}}$.

E12.20(a) (a)
$$\mathcal{B}_0 = \frac{h\nu}{g_I \mu_N} = \frac{(6.626 \times 10^{-34}\ \mathrm{J\,Hz^{-1}}) \times (9 \times 10^9\ \mathrm{Hz})}{(5.5857) \times (5.051 \times 10^{-27}\ \mathrm{J\,T^{-1}})} = \boxed{2 \times 10^2\ \mathrm{T}}$$

(b)
$$\mathcal{B}_0 = \frac{h\nu}{g_e \mu_B} = \frac{(6.626 \times 10^{-34}\ \mathrm{J\,Hz^{-1}}) \times (300 \times 10^6\ \mathrm{Hz})}{(2.0023) \times (9.274 \times 10^{-24}\ \mathrm{J\,T^{-1}})} = \boxed{10\ \mathrm{mT}}$$

COMMENT. Because of the sizes of these magnetic fields, neither experiment seems feasible.

Question. What frequencies are required to observe electron resonance in the magnetic field of a 300-MHz NMR magnet and nuclear resonance in the field of a 9-GHz (g = 2.00) ESR magnet? Are these experiments feasible?

E12.21(a)

$$g = \frac{h\nu}{\mu_B \mathcal{B}_0} \quad [12.41]$$

We shall often need the value

$$\frac{h}{\mu_B} = \frac{6.62608\times10^{-34}\,\mathrm{J\,Hz^{-1}}}{9.27402\times10^{-24}\,\mathrm{J\,T^{-1}}} = 7.14478\times10^{-11}\,\mathrm{T\,Hz^{-1}}$$

Then, in this case,

$$g = \frac{(7.14478\times10^{-11}\,\mathrm{T\,Hz^{-1}})\times(9.2231\times10^{9}\,\mathrm{Hz})}{329.12\times10^{-3}\,\mathrm{T}} = \boxed{2.0022}$$

E12.22(a)

$$a = \mathcal{B}(\text{line }3) - \mathcal{B}(\text{line }2) = \mathcal{B}(\text{line }2) - \mathcal{B}(\text{line }1)$$

$$\left.\begin{aligned} \mathcal{B}_3 - \mathcal{B}_2 &= (334.8 - 332.5)\,\mathrm{mT} = 2.3\,\mathrm{mT} \\ \mathcal{B}_2 - \mathcal{B}_1 &= (332.5 - 330.2)\,\mathrm{mT} = 2.3\,\mathrm{mT} \end{aligned}\right\} a = \boxed{2.3\ \mathrm{mT}}$$

Use the center line to calculate g.

$$g = \frac{h\nu}{\mu_B \mathcal{B}_0} = (7.14478\times10^{-11}\,\mathrm{T\,Hz^{-1}})\times\frac{9.319\times10^{9}\,\mathrm{Hz}}{332.5\times10^{-3}\,\mathrm{T}} = \boxed{2.002\overline{5}}$$

E12.23(a) The center of the spectrum will occur at $332.5\,\mathrm{mT}$. Proton 1 splits the line into two components with separation $2.0\,\mathrm{mT}$ and hence at $332.5 \pm 1.0\,\mathrm{mT}$. Proton 2 splits these two hyperfine lines into two, each with separation $2.6\,\mathrm{mT}$, and hence the lines occur at $332.5 \pm 1.0 \pm 1.3\,\mathrm{mT}$. The spectrum therefore consists of four lines of $\boxed{\text{equal intensity}}$ at the fields $\boxed{330.2\,\mathrm{mT},\ 332.2\,\mathrm{mT},\ 332.8\,\mathrm{mT},\ \text{and}\ 334.8\,\mathrm{mT}}$.

E12.24(a) We construct Figure 12.5(a) for CH_3 and Figure 12.5(b) for CD_3. The predicted intensity distribution is determined by counting the number of overlapping lines of equal intensity from which the hyperfine line is constructed.

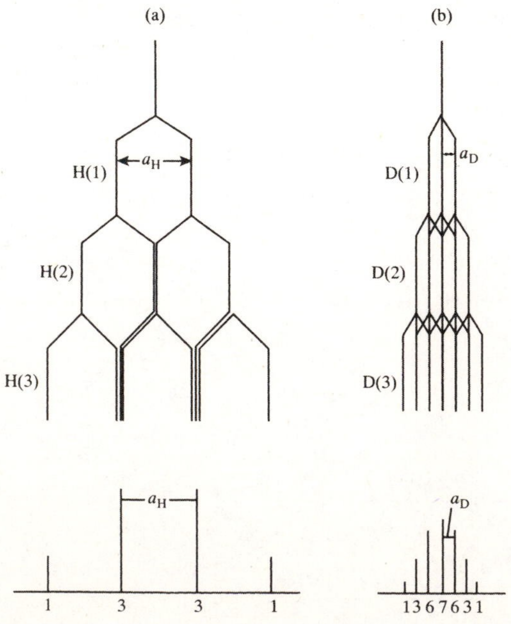

Figure 12.5

E12.25(a) $$\mathcal{B}_0 = \frac{h\nu}{g\mu_B} = \frac{7.14478\times10^{-11}}{2.0025}\text{T Hz}^{-1}\times\nu\ [\text{Exercise 12.21}\mathbf{a}] = 35.68\,\text{mT}\times(\nu/\text{GHz})$$

(a) $\nu = 9.313\,\text{GHz},\quad \mathcal{B}_0 = \boxed{332.3\text{ mT}}$

(b) $\nu = 33.80\,\text{GHz},\quad \mathcal{B}_0 = 1206\,\text{mT} = \boxed{1.206\,\text{T}}$

E12.26(a) Since the number of hyperfine lines arising from a nucleus of spin I is $2I+1$, we solve $2I+1=4$ and find that $\boxed{I=\frac{3}{2}}$.

COMMENT. Four lines of equal intensity could also arise from two inequivalent nuclei with $I=\frac{1}{2}$.

E12.27(a) The X nucleus produces six lines of equal intensity. The pair of H nuclei in XH_2 split each of these lines into a $1:2:1$ triplet (Figure 12.6a). The pair of D nuclei ($I=1$) in XD_2 split each line into a $1:2:3:2:1$ quintet (Fig. 12.6b). The total number of hyperfine lines observed is then $6\times3=18$ in XH_2 and $6\times5=30$ in XD_2.

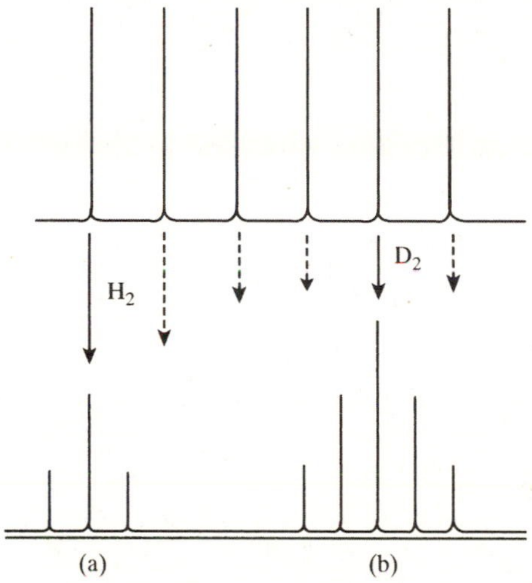

Figure 12.6

Solutions to problems

Solutions to numerical problems

P12.1 (a) $\mu = g_I\mu_N\,|I|\quad [\mu_N = 5.05079\times10^{-27}\ \text{JT}^{-1}]$

Using the formulas

$$\text{Sensitivity ratio}(\nu) = \frac{R_\nu(\text{nuclide})}{R_\nu(^1\text{H})} = \frac{2}{3}(I+1)\left[\frac{\mu(\text{nuclide})}{\mu(^1\text{H})}\right]$$

$$\text{Sensitivity ratio}(\mathcal{B}) = \frac{R_\text{B}(\text{nuclide})}{R_\text{B}(^1\text{H})} = \frac{1}{6}\left(\frac{I+1}{I^2}\right)\left[\frac{\mu(\text{nuclide})}{\mu(^1\text{H})}\right]^3$$

we construct the following table.

Nuclide	Spin I	μ/μ_N	Sensitivity Ratio (ν)	Sensitivity Ratio ($\mathcal{B}$)
^{2}H	1	0.85745	0.409	0.00965
^{13}C	$\frac{1}{2}$	0.7023	0.251	0.01590
^{14}N	1	0.40356	0.193	0.00101
^{19}F	$\frac{1}{2}$	2.62835	0.941	0.83350
^{31}P	$\frac{1}{2}$	1.1317	0.405	0.06654
^{1}H	$\frac{1}{2}$	2.79285		

(b) $\mu = \gamma \text{h} I = g_I \mu_N \,|\, I \,|$

Hence $\gamma = \dfrac{\mu}{\hbar I}$

At constant frequency,

$$R_\nu \propto (I+1)\mu\omega_0^2 \quad \text{or} \quad R_\nu \propto (I+1)\mu \quad [\omega_0 \text{ is constant between the nuclei.}]$$

Thus

$$\text{Sensitivity ratio}(\nu) = \frac{R_\nu(\text{nuclide})}{R_\nu(^1\text{H})}$$

$$= \frac{2}{3}(I+1)\left[\frac{\mu(\text{nuclide})}{\mu(^1\text{H})}\right] = \frac{2}{3}(I+1)\left[\frac{\mu(\text{nuclide})/\mu_N}{\mu(^1\text{H})/\mu_N}\right]$$

as above. Substituting $\omega_0 = \gamma\mathcal{B}_0$ and $\gamma = \dfrac{\mu}{\text{h}I}$, $\omega_0 = \dfrac{\mu\mathcal{B}_0}{\text{h}I}$, so

$$R_B \propto \frac{(I+1)\mu^3\mathcal{B}_0^2}{I^2}$$

$$\text{Sensitivity ratio}(\mathcal{B}) = \frac{R_B(\text{nuclide})}{R_B(^1\text{H})} = \frac{1}{6}\left(\frac{I+1}{I^2}\right)\left[\frac{\mu(\text{nuclide})}{\mu(^1\text{H})}\right]^3$$

$$= \frac{1}{6}\left(\frac{I+1}{I^2}\right)\left[\frac{\mu(\text{nuclide})/\mu_N}{\mu(^1\text{H})/\mu_N}\right]^3$$

as in part (a).

P12.3 The envelopes of maxima and minima of the curve are determined by T_2 through eqn. 12.30, but the time interval between the maxima of this decaying curve corresponds to the reciprocal of the frequency difference $\Delta\nu$ between the pulse frequency ν_0 and the Larmor frequency ν_L, that is, $\Delta\nu = |\nu_0 - \nu_L|$.

$$\Delta\nu = \frac{1}{0.12\,\text{s}} = 8.3\,\text{s}^{-1} = 8.3\,\text{Hz}$$

Therefore the Larmor frequency is $\boxed{400\times10^6\ \text{Hz}\pm8\,\text{Hz}}$.

According to eqns. 12.30 and 12.34, the intensity of the maxima in the FID curve decays exponentially as e^{-t/T_2}. Therefore T_2 corresponds to the time at which the intensity has been reduced to 1/e of the original value. In the text figure, this corresponds to a time slightly before the fourth maximum has occurred, or about $\boxed{0.29\,\text{s}}$.

P12.5 (a) The *Lorentzian function* in terms of angular frequencies is

$$I_L(\omega) = \frac{S_0T_2}{1+T_2^2(\omega-\omega_0)^2}$$

The maximum in this function occurs when $\omega=\omega_0$. Hence $I_{L,max} = S_0T_2$ and

$$I_L(\Delta\omega_{1/2}) = \frac{I_{L,max}}{2} = \frac{S_0T_2}{2} = \frac{S_0T_2}{1+T_2^2(\omega_{1/2}-\omega_0)^2} = \frac{S_0T_2}{1+T_2^2(\Delta\omega)^2}$$

where $\Delta\omega = \frac{1}{2}\Delta\omega_{½}$; hence $2 = 1+T_2^2(\Delta\omega)^2$ and $\Delta\omega = 1/T_2$. Therefore $\boxed{\Delta\omega_{½} = \frac{2}{T_2}}$.

(b) The *Gaussian function* in terms of angular frequencies is

$$I_G(\omega) = S_0T_2 e^{-T_2^2(\omega-\omega_0)^2}$$

The maximum in this function occurs when $\omega=\omega_0$. Hence $I_{G,max} = S_0T_2$ and

$$I_G(\Delta\omega_{1/2}) = \frac{I_{G,max}}{2} = \frac{S_0T_2}{2} = S_0T_2 e^{-T_2^2(\omega_{½}-\omega_0)^2} = S_0T_2 e^{-T_2^2(\Delta\omega)^2}$$

where $\Delta\omega = \frac{1}{2}\Delta\omega_{½}$; thus $\ln 2 = T_2^2(\Delta\omega)^2$ and $\Delta\omega = (\ln 2)^{½}/T_2$.

Therefore $\boxed{\Delta\omega_{½} = \frac{2(\ln 2)^{½}}{T_2}}$.

(c) If we choose the same values of S_0, T_2, and ω_0 for both functions, we can rewrite them as

$$I_L(\omega) = L(x) \propto \frac{1}{1+x^2} \quad \text{and} \quad I_G(\omega) = G(x) \propto e^{-x^2} \quad \text{where } x = T_2(\omega-\omega_0)$$

These functions are plotted against x in the following Mathcad worksheet. Note that the Lorentzian function is slightly sharper in the center, although this is difficult to discern with the scale of x used in Figure 12.7, and decreases much more slowly in the wings beyond the half amplitude points. Note that the functions plotted in the figure are not normalized but are matched at their peak amplitude in order to more clearly display the differences in their shapes. If the curves had been normalized, the areas under the two curves would be equal, but the peak height in the Lorentzian would be lower than the Gaussian peak height.

$$x := -5, -4.95 .. 5$$

$$L(x) := \frac{1}{1+x^2} \qquad G(x) := e^{-x^2}$$

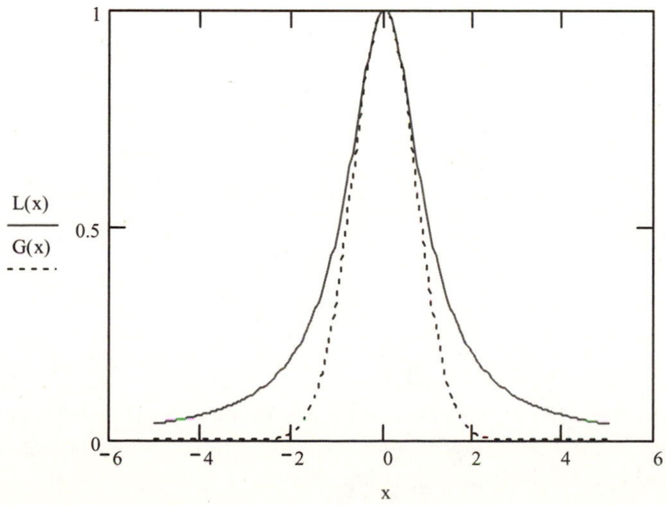

Figure 12.7

P12.7 (a) The Karplus equation [12.27] for $^3J_{HH}$ is a linear equation in $\cos\phi$ and $\cos 2\phi$. The experimentally determined equation for $^3J_{SnSn}$ is a linear equation in $^3J_{HH}$. In general, if $F(f)$ is linear in f, and if $f(x)$ is linear in x, then $F(x)$ is linear. So we expect $^3J_{SnSn}$ to be linear in $\cos\phi$ and $\cos 2\phi$. This is demonstrated in (b).

(b) $$^3J_{SnSn}/\text{Hz} = 78.86(^3J_{HH}/\text{Hz}) + 27.84$$

Inserting the Karplus equation for $^3J_{HH}$ we obtain

$$^3J_{SnSn}/\text{Hz} = 78.86\{A + B\cos\phi + C\cos 2\phi\} + 27.84$$

Using $A = 7$, $B = -1$, and $C = 5$, we obtain

$$^3J_{SnSn}/\text{Hz} = \boxed{580 - 79\cos\phi + 395\cos 2\phi}$$

The plot of $^3J_{SnSn}$ is shown in Figure 12.8.

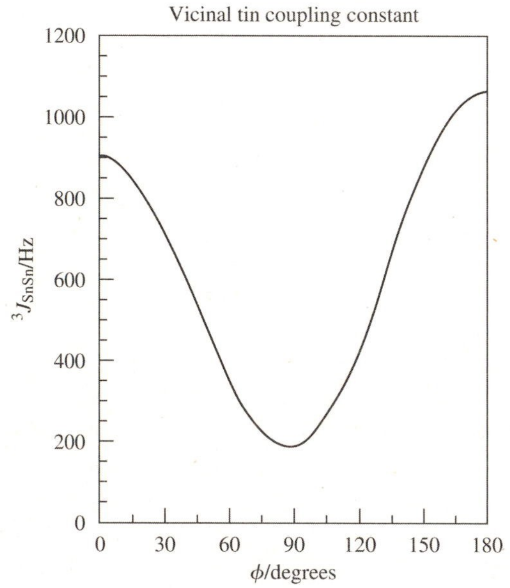

Figure 12.8

(c) A staggered configuration (Figure 12.9) with the SnMe3 groups *trans* to each other is the preferred configuration. The $SnMe_3$ repulsions are then at a minimum.

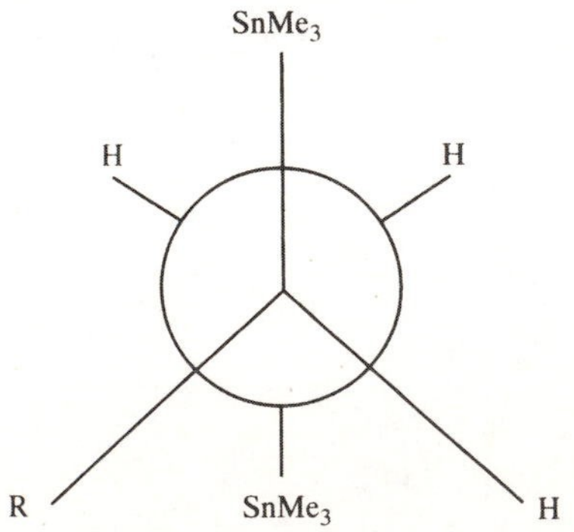

Figure 12.9

P12.9

$$\nu = \frac{g_e\mu_B\mathcal{B}}{h} = \frac{(2.00)\times(9.274\times10^{-24}\ \text{J T}^{-1})\times(1.0\times10^{3}\ \text{T})}{6.626\times10^{-34}\ \text{J s}} = \boxed{2.8\times10^{13}\ \text{Hz}}$$

This frequency is in the infrared region of the electromagnetic spectrum and hence is comparable to the frequencies and energies of molecular vibrations; it is much greater than those of molecular rotations and far less than those of molecular electronic motion.

P12.11 Refer to Figure 12.6 in the solution to Exercise 12.24(a). The width of the CH_3 spectrum is $3a_H = \boxed{6.9\text{ mT}}$. The width of the CD_3 spectrum is $6a_D$. It seems reasonable to assume, since the hyperfine interaction is an interaction of the magnetic moments of the nuclei with the magnetic moment of the electron, that the strength of the interactions is proportional to the nuclear moments.

$$\mu = g_I\mu_N I \quad \text{or} \quad \mu_z = g_I\mu_N m_I \ [12.11, 12.10b]$$

and thus nuclear magnetic moments are proportional to the nuclear g-values; hence

$$a_D \approx \frac{0.85745}{5.5857} \times a_H = 0.1535a_H = 0.35\,\text{mT}$$

Therefore, the overall width is $6a_D = \boxed{2.1\,\text{mT}}$.

P12.13 Write $P(\text{N}2s) = \dfrac{5.7\text{mT}}{55.2\text{mT}} = \boxed{0.10}$ (10 percent of its time)

$$P(\text{N}2p_z) = \frac{1.3\,\text{mT}}{3.4\,\text{mT}} = \boxed{0.38} \ \text{(38 percent of its time)}$$

The total probability is

(a) $P(\text{N}) = 0.10 + 0.38 = \boxed{0.48}$ (48 percent of its time).

(b) $P(\text{O}) = 1 - P(\text{N}) = \boxed{0.52}$ (52 percent of its time).

The hybridization ratio is

$$\frac{P(\text{N}2p)}{P(\text{N}2s)} = \frac{0.38}{0.10} = \boxed{3.8}$$

The unpaired electron therefore occupies an orbital that resembles an sp^3 hybrid on N, in accord with the radical's nonlinear shape.

From the discussion in Section 5.2 we can write

$$a^2 = \frac{1+\cos\phi}{1-\cos\phi}$$

$$b^2 = 1 - a^2 = \frac{-2\cos\phi}{1-\cos\phi}$$

$$\lambda = \frac{b'^2}{a'^2} = \frac{-1\cos\phi}{1+\cos\phi}, \text{ implying that } \cos\phi = \frac{\lambda}{2+\ell}$$

Then, since $\lambda = 3.8$, $\cos\phi = -0.66$, so $\phi = \boxed{131°}$.

P12.15 When spin label molecules approach to within 800 pm, orbital overlap of the unpaired electrons and dipolar interactions between magnetic moments cause an exchange coupling interaction between the spins (Figure 12.10). The electron exchange process occurs at a rate that increases as concentration increases. Thus the process has a lifetime that is too long at low concentrations to affect the "pure" ESR signal. As the concentration increases, the linewidths increase until the triplet coalesces into a broad singlet. Further increase of the concentration decreases the exchange lifetime and therefore the linewidth of the singlet.

When spin labels within biological membranes are highly mobile, they may approach closely and the exchange interaction may provide the ESR spectra with information that mimics the moderate and high-concentration signals above.

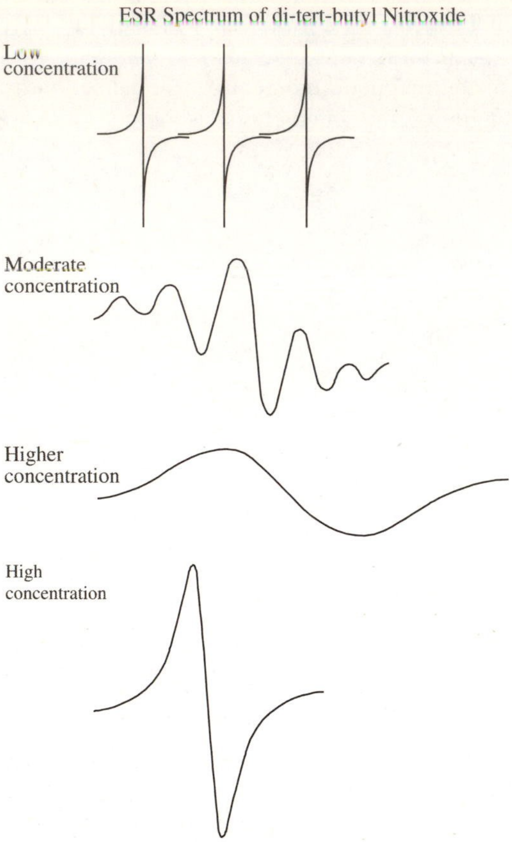

Figure 12.10

Solutions to theoretical problems

P12.17 The first table below displays experimental ^{13}C chemical shifts and computed* atomic charges on the carbon atom para to a number of substituents in substituted benzenes. Two sets of charges are shown, one derived by fitting the electrostatic potential and the other by Mulliken population anlysis.

Substituent	CH_3	H	CF_3	CN	NO_2
Δ	128.4	128.5	128.9	129.1	129.4
Electrostatic charge/*e*	−0.1273	−0.0757	−0.0227	−0.0152	−0.0541
Mulliken charge/*e*	−0.1089	−0.1021	−0.0665	−0.0805	−0.0392

*Semi-empirical, PM3 level, PC Spartan Pro™

In Problem 6.7 we have recalculated net charges at a higher level than the semi-empirical PM3 level displayed above. The following table obtained from the solution to Problem 6.7 displays both the experimental and calculated* (HF-SCF/6-311G*) ^{13}C chemical shifts and computed* atomic charges on the carbon atom para to a number of substituents in substituted benzenes. Three sets of charges are shown, one derived by fitting the electrostatic potential, another by Mulliken population analysis, and the other by the method of "natural" charges.

Substituent	CH_3	H	CF_3	CN	NO_2
δ_{exp}	128.4	128.5	128.9	129.1	129.4
$\delta_{calc}{}^*$	134.4	133.5	132.1	138.8	141.8
Electrostatic charge*/*e*	−0.240	−0.135	−0.138	−0.102	−0.116
Mulliken charge*/*e*	−0.231	−0.217	−0..205	−0.199	−0.182
Natural charge*/*e*	−0.199	−0.183	−0.158	−0.148	−0.135

Spartan '06™; HF-SCF/6-311G

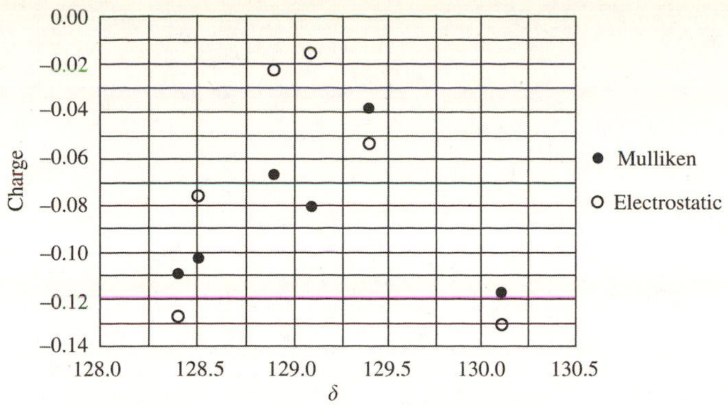

Figure 12.11

(a) Neither set of charges correlates well with the chemical shifts; however some correlation is apparent, particularly for the Mulliken charges.

(b) The diamagnetic local contribution to shielding is roughly proportional to the electron density on the atom. The extent to which the para-carbon atom is affected by electron-donating or electron-withdrawing groups on the other side of the benzene ring is reflected in the net charge on the atom. If the diamagnetic local contribution dominated, then the more positive the atom, the greater the deshielding and the greater the chemical shift δ would be. That no good correlation is observed leads to several possible hypotheses: for example, the diamagnetic local contribution is not the dominant contribution in these molecules (or not in all of these molecules), or the computation is not sufficiently accurate to provide meaningful atomic charges. See the solution to Problem 6.7 for additional discussion.

P12.19 Eqn. 12.28 may be written

$$\mathcal{B} = k(1-3\cos^2\theta)$$

where k is a constant independent of angle. Thus

$$\mathcal{B} \propto \int_0^{\pi}(1-3\cos^2\theta)\sin\theta\,d\theta\int_0^{2\pi}d\phi$$

$$\propto \int_1^{-1}(1-3x^2)\,dx\times 2\pi \;[x=\cos\theta, dx=-\sin\theta\,d\theta]$$

$$\propto (x-x^3)\Big|_1^{-1}=0$$

P12.21 The shape of spectral line $I(\omega)$ is related to the free induction decay signal $G(t)$ by

$$I(\omega) = a\,\mathrm{Re}\int_0^{\infty} G(t)e^{i\omega t}\,dt$$

where a is a constant and Re means take the real part of what follows. Calculate the lineshape corresponding to an oscillating, decaying function.

$$G(t) = \cos\omega_0 t\,e^{-t/\tau}$$

$$I(\omega) = a\,\mathrm{Re}\int_0^{\infty} G(t)e^{i\omega t}\,dt$$

$$= a\,\mathrm{Re}\int_0^{\infty}\cos\omega_0 t e^{-t/\tau+i\omega t}\,dt$$

$$= \frac{1}{2}a\,\mathrm{Re}\int_0^{\infty}(e^{-i\omega_0 t}+e^{i\omega_0 t})e^{-t/\tau+i\omega t}\,dt$$

$$= \frac{1}{2}a\,\mathrm{Re}\int_0^{\infty}\{e^{i(\omega_0+\omega+i/\tau)t}+e^{-i(\omega_0-\omega-i/\tau)t}\}\,dt$$

$$= -\frac{1}{2}a\,\mathrm{Re}\left[\frac{1}{i(\omega_0+\omega+i/\tau)}-\frac{1}{i(\omega_0-\omega-i/\tau)}\right]$$

when ω and ω_0 are similar to magnetic resonance frequencies (or higher), only the second term in brackets is significant $\left(\text{because} \dfrac{1}{(\omega_0+\omega)} = 1 \text{ but} \dfrac{1}{(\omega_0-\omega)} \text{may be large if } \omega \approx \omega_0\right)$. Therefore

$$I(\omega) \approx \frac{1}{2} a \,\mathrm{Re} \frac{1}{\mathrm{i}(\omega_0-\omega)+1/\tau}$$

$$= \frac{1}{2} a \,\mathrm{Re} \frac{-\mathrm{i}(\omega_0-\omega)+1/\tau}{(\omega_0-\omega)^2+1/\tau^2} = \frac{1}{2} a \frac{1/\tau}{(\omega_0-\omega)^2+1/\tau^2}$$

$$= \boxed{\left(\frac{1}{2}\right)\frac{a\tau}{1+(\omega_0-\omega)^2\tau^2}}$$

which is a Lorentzian line centered on ω_0 of amplitude $\frac{1}{2}a\tau$ and width $\dfrac{2}{\tau}$ at half-height.

P12.23 $\quad {}^3J_{\mathrm{HH}} = A + B\cos\phi + C\cos 2\phi \quad [12.27]$

$$\frac{\mathrm{d}}{\mathrm{d}\phi}\left({}^3J_{\mathrm{HH}}\right) = -B\sin\phi - 2C\sin 2\phi = 0$$

This equation has a number of solutions:

$$\phi = 0,\ \phi = n\pi,\ \phi = \pi - \arccos\left(\frac{B}{4C}\right) = \arccos\left(\frac{B}{4C}\right)$$

The first two are trivial solutions.

If $\quad \phi = \arccos\left(\dfrac{B}{4C}\right)$ then, $\sin\phi = \sqrt{1-\dfrac{B}{16C^2}}$

$$\sin 2\phi = 2\sin\phi\cos\phi = 2\sqrt{1-\frac{B^2}{16C^2}}\left(\frac{B}{4C}\right)$$

$$B\sin\phi + 2C\sin 2\phi = B\sqrt{1-\frac{B^2}{16C^2}} + 4C\sqrt{1-\frac{B^2}{16C^2}}\left(\frac{B}{4C}\right) = 0$$

So $\dfrac{B}{4C} = \cos\phi$ clearly satisfies the condition for an extremum.

The second derivative is

$$\frac{\mathrm{d}^2}{\mathrm{d}\phi^2}\left({}^3J_{\mathrm{HH}}\right) = -B\cos\phi - 4C\cos 2\phi = -B\cos\phi - 4C(2\cos^2\phi - 1)$$

$$= -B\left(\frac{B}{4C}\right) - 4C\left(2\frac{B^2}{16C^2} - 1\right) = -\frac{B^2}{4C} - \frac{2B^2}{4C} + 4C$$

This quantity is positive if

$$16C^2 > 3B^2$$

This is certainly true for typical values of B and C, namely, $B = -1$ Hz and C = 5 Hz. Therefore the condition for a minimum is as stated, namely, $\boxed{\cos\phi = B/4C}$.

Solutions to applications

P12.25 Methionine-105 is in the vicinity of both typtophan-28 and tyrosine-23, but the latter two residues are not in the vicinity of each other. The methionine residue may lay between them as represented in Figure 12.12.

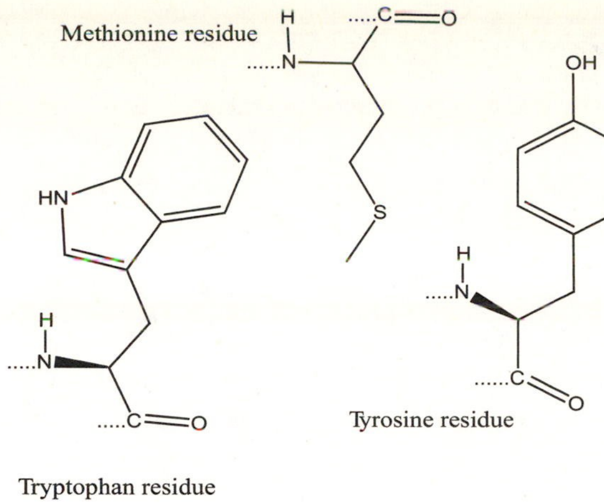

Figure 12.12

P12.27 See Figure 12.13. Only the H(N) and H(C_α) protons and the H(C_α) and H(C_β) protons are expected to show coupling. This results in a simple COSY spectrum with only two off-diagonals, one at (8.25 ppm, 4.35 ppm) and the other at (4.35 ppm, 1.39 ppm).

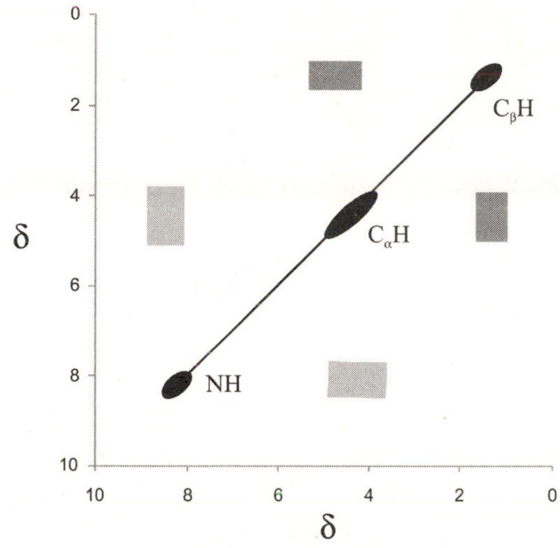

Figure 12.13

P12.29 Assume that the radius of the disk is 1 unit. The volume of each slice is proportional to length of slice × δ_x.

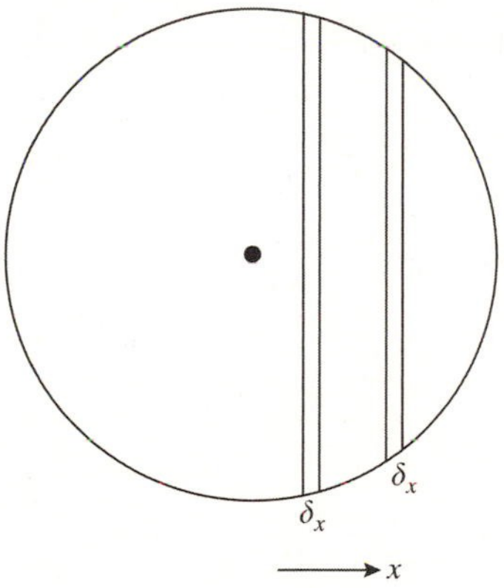

Figure 12.14

length of slice at $x = 2 \sin \theta$

$x = \cos \theta$

$\theta = \arccos x$

ranges from −1 to +1

length of slice at $x = 2 \sin(\arccos x)$

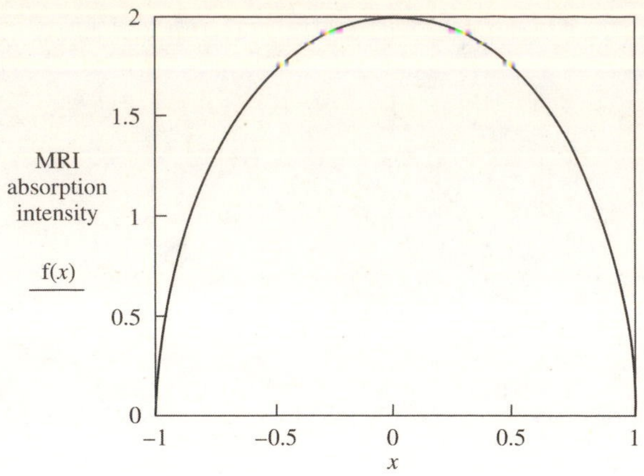

Figure 12.15

Plot $f(x) = 2\sin(\arccos x)$ against x between the limits −1 and +1. The plot is shown in Figure 12.15. The volume at each value of x is proportional to $f(x)$ and the intensity of the MRI signal is proportional to the volume, so the figure above represents the absorption intensity for the MRI image of the disk.

PART 4 Molecular thermodynamics

13 The Boltzmann distribution

Answers to discussion questions

D13.1 The molecular partition is roughly equal to the number of physically distinct states thermally accessible to a molecule at a given temperature. At low temperatures, very little energy is available, so only the lowest energy states of a molecule are accessible; therefore, as the temperature approaches absolute zero, the partition function approaches the degeneracy of the molecule's ground state. The higher the temperature, the greater the Boltzmann weighting factor $e^{-\beta\varepsilon}$ and the more accessible a state of energy ε becomes. Thus the number of accessible states increases with temperature. (The partition function is only "roughly equal" to the number of accessible states because at any non-zero temperature, each state is accessible with a finite probability proportional to $e^{-\beta\varepsilon}$. Thus, to state whether or not a given state is "accessible" at a given temperature is somewhat arbitrary, so an exactly counted number of accessible states is also arbitrary. Because the partition function is the sum of all of these fractional probabilities, it is a good estimate of this number.) See Section 13.2.

D13.3 The **population** of a state is the number of molecules of a sample that are in that state, on average. The population of a state is the number of molecules in the sample times the probability of the state. The **configuration** of a system is a list of populations in order of the energy of the corresponding states. For example, $\{N-3, 2, 1, 0, ...\}$ is a possible configuration of a system of N molecules in which all but three of the molecules are in the ground state, two molecules are in the next lowest state, one is in the next state, and so on. The **weight** of a given configuration is the number of ways the configuration can be made up. In our example, the single molecule in the second excited state could be any of the system's N molecules, so there are N ways to arrange that state alone. The weight of the configuration $\{N_0, N_1, N_2,...\}$ is

$$\mathcal{W} = \frac{N!}{N_0!N_1!N_2!\ \cdots} \quad (13.1)$$

When N is large (as it is for any macroscopic sample), the most probable configuration is so much more probable than all other possible configurations that it is the system's dominant configuration. See Section 13.1.

D13.5 Because this chapter focuses on the application of statistics to the distribution of physical states in systems that contain a large number of atoms or molecules, we begin with a statistical answer: the thermodynamic temperature is the one quantity that determines the most probable populations of those states in systems at thermal equilibrium (Section 13.1b). As a consequence, the temperature provides a necessary condition for thermal equilibrium; a system is at thermal equilibrium only if all of its sub-systems have the same temperature. Note that this is not a circular definition of temperature, for thermal equilibrium is not defined by uniformity of temperature: systems whose sub-systems can exchange energy tend toward thermal equilibrium. In this context, sub-systems can be different materials placed in contact (such as a block of copper in a beaker of water) or they can be more abstract (such as rotational and vibrational modes of motion).

Finally, the equipartition theorem allows us to connect the temperature of statistical thermodynamics to the empirical concept of temperature developed long beforehand. Temperature is a measure of the intensity of thermal energy directly proportional to the mean energy for each quadratic contribution to the energy (provided that the temperature is sufficiently high; see Discussion Question 13.7)

D13.7 The equipartition theorem says that the average value of each quadratic contribution to the energy is the same and equal to $(1/2)kT = 1/(2\beta)$. This result agrees with the mean energy computed from a partition function, $\mathrm{d}q/\mathrm{d}\beta$, at sufficiently high temperatures. The partition function yields the general result for mean energy. Mathematically, evaluating $\mathrm{d}q/\mathrm{d}\beta$ involves summing over states. In mathematical terms, the equipartition theorem is valid when the temperature is high enough for the discrete sum to be well approximated by a continuous integral. In physical terms, this corresponds to temperatures sufficiently high for the difference between energy levels to be small compared to kT.

D13.9 Identical particles can be regarded as distinguishable when they are localized as in a crystal lattice where we can assign a set of coordinates to each particle. Strictly speaking, it is the lattice site that carries the set of coordinates, but as long as the particle is fixed to the site, it too can be considered distinguishable.

Solutions to exercises

E13.1(a) The weight is given by

$$\mathcal{W} = \frac{N!}{N_0!N_1!N_2!\cdots} = \frac{16!}{0!1!2!3!8!0!0!0!0!2!}$$

This can be simplified by removing the common factor of 8! from the numerator and denominator and noting that 0! = 1! = 1:

$$\mathcal{W} = \frac{16\times15\times14\times13\times12\times11\times10\times9}{2\times(3\times2)\times2} = \boxed{21621600}$$

E13.2(a) (a) $8! = 8\times7\times6\times5\times4\times3\times2\times1 = \boxed{40320}$ exactly.

(b) According to Stirling's approximation,

$$\ln x! \approx x\ln x - x$$

so $\ln 8! \approx 8\ln 8 - 8 = 8.636$ and $8! \approx \mathrm{e}^{8.636} = \boxed{5.63\times10^3}$

(c) According to Stirling's better approximation,

$$x! \approx (2\pi)^{1/2}x^{x+1/2}\mathrm{e}^{-x} \quad \text{so} \quad 8! \approx (2\pi)^{1/2}8^{8.5}\mathrm{e}^{-8} = \boxed{3.99\times10^4}$$

E13.3(a) For two non-degenerate levels,

$$\frac{N_2}{N_1} = \frac{\mathrm{e}^{-\beta\varepsilon_2}}{\mathrm{e}^{-\beta\varepsilon_1}} = \mathrm{e}^{-\beta(\varepsilon_2-\varepsilon_1)} = \mathrm{e}^{-\beta\Delta\varepsilon} = \mathrm{e}^{-\Delta\varepsilon/kT} \qquad \left[13.7 \text{ with } \beta = \frac{1}{kT}\right]$$

Hence, as $T\to\infty$, $\frac{N_2}{N_1} = \mathrm{e}^{-0} = \boxed{1}$. That is, the two levels would become equally populated.

E13.4(a) For two non-degenerate levels,

$$\frac{N_2}{N_1} = \frac{\mathrm{e}^{-\beta\varepsilon_2}}{\mathrm{e}^{-\beta\varepsilon_1}} = \mathrm{e}^{-\beta(\varepsilon_2-\varepsilon_1)} = \mathrm{e}^{-\beta\Delta\varepsilon} = \mathrm{e}^{-\Delta\varepsilon/kT} \qquad \left[13.7 \text{ with } \beta = \frac{1}{kT}\right],$$

so $\ln\frac{N_2}{N_1} = -\frac{\Delta\varepsilon}{kT}$ and $T = -\frac{\Delta\varepsilon}{k\ln\frac{N_2}{N_1}}$

Thus $T = -\dfrac{6.626\times10^{-34}\ \text{J s}\times2.998\times10^{10}\ \text{cm s}^{-1}\times400\ \text{cm}^{-1}}{1.381\times10^{-23}\ \text{J K}^{-1}\times\ln(1/3)} = \boxed{524\ \text{K}}$

E13.5(a) See Example 13.1. The ratio of populations of a particular state at the $J = 5$ level to the population of the non-degenerate $J = 0$ level is

$$\frac{N_{5,M_J}}{N_0} = \frac{e^{-\beta\varepsilon_5}}{e^{-\beta\varepsilon_0}} = e^{-\beta(\varepsilon_5-\varepsilon_0)} = e^{-(\varepsilon_5-\varepsilon_0)/kT} \quad \left[13.7 \text{ with } \beta = \frac{1}{kT}\right]$$

Because all the states of a degenerate level are equally likely, the ratio of populations of a particular *level* is

$$\frac{N_5}{N_0} = \frac{g_5 e^{-\beta\varepsilon_5}}{g_0 e^{-\beta\varepsilon_0}} = \frac{g_5}{g_0} e^{-(\varepsilon_5-\varepsilon_0)/kT}$$

The degeneracy of linear rotor energy levels is

$$g_J = (2J + 1)$$

and its energy levels are

$$\varepsilon_J = hc\tilde{B}J(J+1) \quad [\text{Section 10.2(c)}]$$

Thus, using $kT/hc = 207.224\ \text{cm}^{-1}$ at 298.15 K,

$$\frac{N_5}{N_0} = \frac{g_5}{g_0} e^{-5(5+1)hc\tilde{B}/kT} = \frac{(2\times5+1)}{(2\times0+1)} e^{-5(5+1)\times2.71\ \text{cm}^{-1}/207.224\ \text{cm}^{-1}} = \boxed{7.43}$$

E13.6(a) For two non-degenerate levels,

$$\frac{N_2}{N_1} = \frac{e^{-\beta\varepsilon_2}}{e^{-\beta\varepsilon_1}} = e^{-\beta(\varepsilon_2-\varepsilon_1)} = e^{-\beta\Delta\varepsilon} = e^{-\Delta\varepsilon/kT} \quad \left[13.7 \text{ with } \beta = \frac{1}{kT}\right]$$

so, assuming that other states (if any) are negligibly populated,

$$\ln\frac{N_2}{N_1} = -\frac{\Delta\varepsilon}{kT} \quad \text{and} \quad T = -\frac{\Delta\varepsilon}{k\ln\dfrac{N_2}{N_1}}$$

Thus $T = -\dfrac{6.626\times10^{-34}\ \text{J s}\times2.998\times10^{10}\ \text{cm s}^{-1}\times540\ \text{cm}^{-1}}{1.381\times10^{-23}\ \text{J K}^{-1}\times\ln(10/90)} = \boxed{35\bar{4}\ \text{K}}$

E13.7(a) (a) The thermal wavelength is

$$\Lambda = \frac{h}{(2^1\ mkT)^{1/2}} \quad [13.15\text{b}]$$

We need the molecular mass, not the molar mass:

$$m = \frac{150\times10^{-3}\ \text{kg mol}^{-1}}{6.022\times10^{23}\ \text{mol}^{-1}} = 2.49\times10^{-25}\ \text{kg}$$

$$\text{So } \Lambda = \frac{6.626\times10^{-34}\ \text{J s}}{(2^{1}\times2.49\times10^{-25}\ \text{kg}\times1.381\times10^{-23}\ \text{J K}^{-1}\times T)^{1/2}} = \frac{1.43\times10^{-10}\ \text{m}}{(T/\text{K})^{1/2}}$$

(i) $T = 300$ K: $\Lambda = \dfrac{1.43\times10^{-10}\ \text{m}}{(300)^{1/2}} = \boxed{8.23\times10^{-12}\ \text{m}} = \boxed{8.23\ \text{pm}}$

(ii) $T = 3000$ K: $\Lambda = \dfrac{1.43\times10^{-10}\ \text{m}}{(3000)^{1/2}} = \boxed{2.60\times10^{-12}\ \text{m}} = \boxed{2.60\ \text{pm}}$

(b) The translational partition function is

$$q^{\text{T}} = \frac{V}{\Lambda^3}\ [13.18\text{b}]$$

(i) $T = 300$ K: $q^{\text{T}} = \dfrac{1.00\ \text{cm}^3}{(8.23\times10^{-12}\ \text{m})^3} = \dfrac{(1.00\times10^{-2}\ \text{m})^3}{(8.23\times10^{-12}\ \text{m})^3} = \boxed{1.79\times10^{27}}$

(ii) $T = 3000$ K: $q^{\text{T}} = \dfrac{(1.00\times10^{-2}\ \text{m})^3}{(2.60\times10^{-12}\ \text{m})^3} = \boxed{5.67\times10^{28}}$

E13.8(a) $q^{\text{T}} = \dfrac{V}{\Lambda^3}$ [13.18b], implying that $\dfrac{q}{q'} = \left(\dfrac{\Lambda'}{\Lambda}\right)^3$.

However, $\Lambda = \dfrac{h}{(2^{1}\ mkT)^{1/2}} \propto \dfrac{1}{m^{1/2}}$ so $\dfrac{q}{q'} = \left(\dfrac{m}{m'}\right)^{3/2}$.

Therefore, $\dfrac{q_{\text{H}_2}}{q_{\text{He}}} = \left(\dfrac{2\times1.008}{4.003}\right)^{3/2} = \boxed{0.3574}$.

E13.9(a) The high-temperature expression for the rotational partition function of a linear molecule is

$$q^{\text{R}} = \frac{kT}{\sigma hc\tilde{B}}[13.21\text{b}],\quad \tilde{B} = \frac{\hbar}{4\pi cI}\ [10.5],\quad I = \mu R^2\ [\text{Table 10.1}]$$

Hence $q = \dfrac{8\pi^2 kTI}{\sigma h^2} = \dfrac{8\pi^2 kT\mu R^2}{\sigma h^2}$

For O_2, $\mu = \frac{1}{2}m(\text{O}) = \frac{1}{2}\times16.00\ \text{u} = 8.00\ \text{u}$, and $\sigma = 2$; therefore,

$$q = \frac{(8\pi^2)\times(1.381\times10^{-23}\ \text{J K}^{-1})\times(300\ \text{K})\times(8.00\times1.6605\times10^{-27}\ \text{kg})\times(1.2075\times10^{-10}\ \text{m})^2}{(2)\times(6.626\times10^{-34}\ \text{J s})^2}$$

$= \boxed{72.2}$

E13.10(a) The high-temperature expression for the rotational partition function of a non-linear molecule is [Table 13.1]

$$q^{\text{R}} = \frac{1.027}{\sigma}\frac{(T/\text{K})^{3/2}}{(\tilde{A}\tilde{B}\tilde{C}/\text{cm}^{-3})^{1/2}} = \frac{1.027\times(T/\text{K})^{3/2}}{(3.1752\times0.3951\times0.3505)^{1/2}} = 1.549\times(T/\text{K})^{3/2}$$

(a) At 25°C, $q^{\mathrm{R}} = 1.549\times(298)^{3/2} = \boxed{7.97\times10^{3}}$.

(b) At 100°C, $q^{\mathrm{R}} = 1.549\times(373)^{3/2} = \boxed{1.12\times10^{4}}$.

E13.11(a) The rotational partition function of a non-symmetrical linear molecule is

$$q^{\mathrm{R}} = \sum_J (2J+1)\mathrm{e}^{-hc\tilde{B}J(J+1)/kT} \qquad \left[\text{Example 13.2 with } \beta = \frac{1}{kT}\right]$$

Use $$\frac{hc\tilde{B}}{k} = \frac{6.626\times10^{-34}\ \mathrm{J\,s}\times2.998\times10^{10}\ \mathrm{cm\,s^{-1}}\times1.931\ \mathrm{cm^{-1}}}{1.381\times10^{-23}\ \mathrm{J\,K^{-1}}} = 2.778\ \mathrm{K}$$

so $$q^{\mathrm{R}} = \sum_J (2J+1)\mathrm{e}^{-2.778\ \mathrm{K}\times J(J+1)/T}$$

Use a spreadsheet or other mathematical software to evaluate the terms of the sum and to sum the terms until they converge. The high-temperature expression is

$$q^{\mathrm{R}} = \frac{kT}{hc\tilde{B}} = \frac{T}{2.778\ \mathrm{K}}$$

The explicit and high-temperature expressions are compared in Figure 13.1. The high-temperature expression reaches 95% of the explicit sum at $\boxed{18\ \mathrm{K}}$.

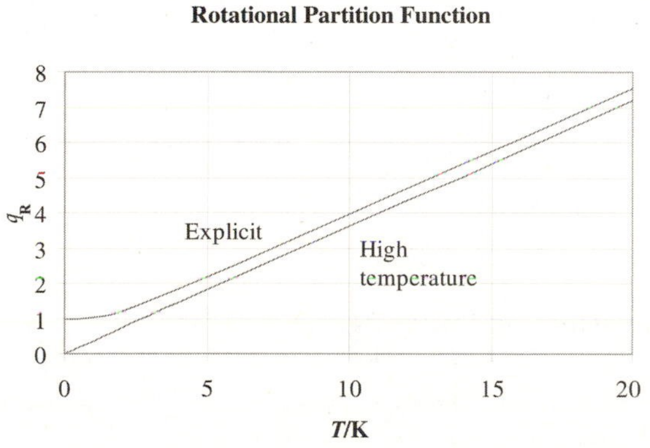

Figure 13.1

E13.12(a) The rotational partition function of a spherical rotor molecule, ignoring nuclear statistics, is

$$q^{\mathrm{R}} = \sum_J g_J \mathrm{e}^{-\varepsilon_J^{\mathrm{R}}/kT}[13.9\mathrm{b}] = \sum_J (2J+1)^2 \mathrm{e}^{-hc\tilde{B}J(J+1)/kT}[10.6]$$

Use $$\frac{hc\tilde{B}}{k} = \frac{6.626\times10^{-34}\ \mathrm{J\,s}\times2.998\times10^{10}\ \mathrm{cm\,s^{-1}}\times5.241\ \mathrm{cm^{-1}}}{1.381\times10^{-23}\ \mathrm{J\,K^{-1}}} = 7.539\ \mathrm{K}$$

so $$q^{\mathrm{R}} = \sum_J (2J+1)^2 \mathrm{e}^{-7.539\ \mathrm{K}\times J(J+1)/T}$$

Use a spreadsheet or other mathematical software to evaluate the terms of the sum and to sum the terms until they converge. The high-temperature expression is eqn. 13.22, neglecting σ and with $\tilde{A} = \tilde{B} = \tilde{C}$:

$$q^{\mathrm{R}} = \pi^{1/2}\left(\frac{kT}{hc\tilde{B}}\right)^{3/2} = \pi^{1/2}\left(\frac{T}{7.539\ \mathrm{K}}\right)^{3/2}$$

The explicit and high-temperature expressions are compared in Figure 13.2. The high-temperature expression reaches 95% of the explicit sum at 37 K.

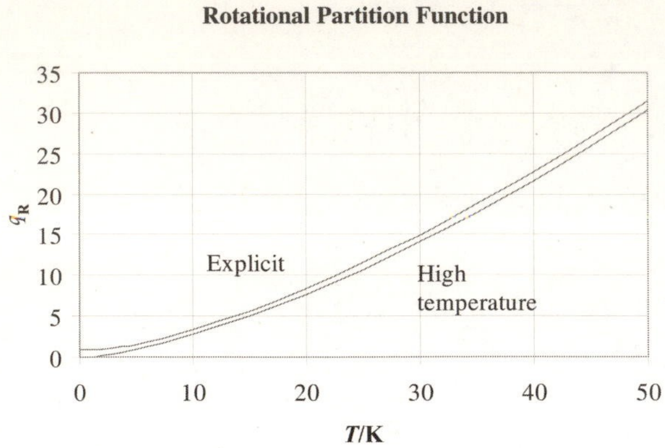

Figure 13.2

E13.13(a) The rotational partition function of a symmetric rotor molecule, ignoring nuclear statistics, is

$$q^{R} = \sum_{J,K} g_{J,K} e^{-\varepsilon^{R}_{J,K}/kT} [13.9b] = \sum_{J=0} (2J+1) e^{-hc\tilde{B}J(J+1)/kT} \left(1 + 2\sum_{K=1}^{J} e^{-hc(\tilde{A}-\tilde{B})K^2/kT}\right) [10.10]$$

Use $\dfrac{hc\tilde{B}}{k} = \dfrac{6.626\times10^{-34}\text{ J s}\times2.998\times10^{10}\text{ cm s}^{-1}\times0.443\text{ cm}^{-1}}{1.381\times10^{-23}\text{ J K}^{-1}} = 0.637\text{ K}$ and

$$\frac{hc(\tilde{A}-\tilde{B})}{k} = \frac{6.626\times10^{-34}\text{ J s}\times2.998\times10^{10}\text{ cm s}^{-1}\times(5.097-0.443)\text{ cm}^{-1}}{1.381\times10^{-23}\text{ J K}^{-1}} = 6.694\text{ K}$$

so $$q^{R} = \sum_{J=0} (2J+1) e^{-0.637\text{ K}\times J(J+1)/T} \left(1 + 2\sum_{K=1}^{J} e^{-6.694\text{ K}\times K^2/T}\right)$$

Write a brief computer program or use other mathematical software to evaluate the terms of the sum and to sum the terms until they converge. Nested sums are straightforward to program in languages such as BASIC or FORTRAN, whereas spreadsheets are more unwieldy. Compare the results of the direct sum with the high-temperature expression, eqn. 13.20(b), with $\tilde{B} = \tilde{C}$:

$$q^{R} = \left(\frac{\pi}{\tilde{A}}\right)^{1/2} \left(\frac{kT}{hc}\right)^{3/2} \frac{1}{\tilde{B}}$$

The explicit and high-temperature expressions are compared in Figure 13.3. The high-temperature expression reaches 95% of the explicit sum at 4.5 K.

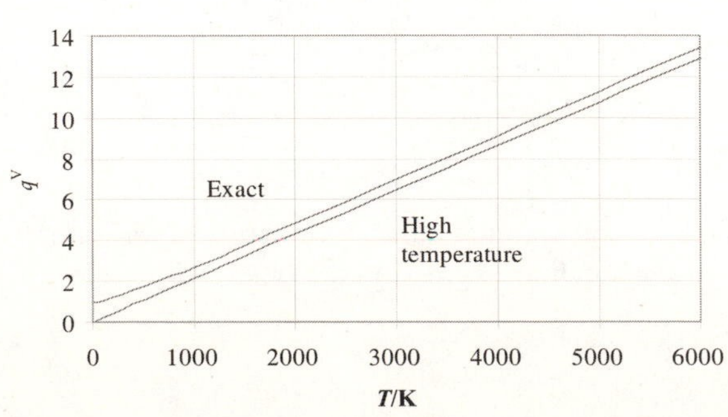

Figure 13.3

E13.14(a) The symmetry number is the order of the rotational subgroup of the group to which a molecule belongs (except for linear molecules, for which $\sigma = 2$ if the molecule has inversion symmetry and 1 otherwise).

(a) CO: Full group $C_{\infty v}$; subgroup C_1; hence $\sigma = \boxed{1}$

(b) O_2: Full group $D_{\infty h}$; subgroup C_2; $\sigma = \boxed{2}$

(c) H_2S: Full group C_{2v}; subgroup C_2; $\sigma = \boxed{2}$

(d) SiH_4: Full group T_d; subgroup T; $\sigma = \boxed{12}$

(e) $CHCl_3$: Full group C_{3v}; subgroup C_3; $\sigma = \boxed{3}$

E13.15(a) Ethene has four indistinguishable atoms that can be interchanged by rotations, so $\sigma = 4$. The rotational partition function of a non-linear molecule is [Table 13.1]

$$q^{\mathrm{R}} = \frac{1.0270}{\sigma}\frac{(T/\mathrm{K})^{3/2}}{(\tilde{A}\tilde{B}\tilde{C}/\mathrm{cm}^{-3})^{1/2}} = \frac{1.0270\times(298.15)^{3/2}}{4\times(4.828\times1.0012\times0.8282)^{1/2}} = \boxed{660.6}$$

E13.16(a) The partition function for a mode of molecular vibration is

$$q^{\mathrm{V}} = \sum_{v} \mathrm{e}^{-vhc\tilde{\nu}/kT} = \frac{1}{1-\mathrm{e}^{-hc\tilde{\nu}/kT}} \text{ [13.24 with } \beta = 1/kT]$$

Use $\quad \dfrac{hc\tilde{\nu}}{k} = \dfrac{6.626\times10^{-34}\ \mathrm{J\,s}\times2.998\times10^{10}\ \mathrm{cm\,s^{-1}}\times323.2\ \mathrm{cm^{-1}}}{1.381\times10^{-23}\ \mathrm{J\,K^{-1}}} = 464.9\ \mathrm{K}$

So $\quad q^{\mathrm{V}} = \sum_{v} \mathrm{e}^{-vhc\tilde{\nu}/kT} = \dfrac{1}{1-\mathrm{e}^{-464.9\ \mathrm{K}/T}}$

The high-temperature expression is

$$q^{\mathrm{V}} = \frac{kT}{hc\tilde{\nu}} = \frac{T}{464.9\ \mathrm{K}} \text{ [13.25]}$$

The explicit and high-temperature expressions are compared in Figure 13.4. The high-temperature expression reaches 95% of the explicit sum at $\boxed{4500\ \mathrm{K}}$.

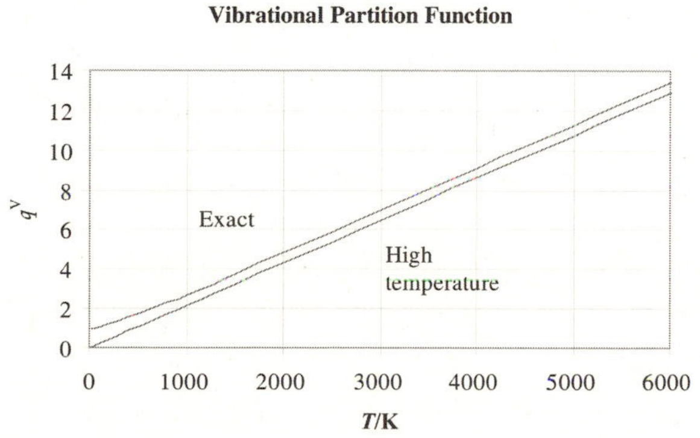

Figure 13.4

E13.17(a) The partition function for a mode of molecular vibration is

$$q^{\mathrm{V}} = \frac{1}{1-\mathrm{e}^{-hc\tilde{\nu}/kT}} \text{ [13.24 with } \beta = 1/kT]$$

and the overall vibrational partition function is the product of the partition functions of the individual modes. (See Example 13.6.) We draw up the following table

Mode	1	2	3	4
$\tilde{\nu}$ / cm^{-1}	658	397	397	1535
$hc\tilde{\nu}/kT$	1.893	1.142	1.142	4.416
$q^{\mathrm{V}}_{\mathrm{mode}}$	1.177	1.469	1.469	1.012

The overall vibrational partition function is

$$q^{\mathrm{V}}_{\mathrm{mode}} = 1.177 \times 1.469 \times 1.469 \times 1.012 = \boxed{2.571}$$

E13.18(a) The partition function for a mode of molecular vibration is

$$q^{\mathrm{V}} = \frac{1}{1-\mathrm{e}^{-hc\tilde{\nu}/kT}} \quad [13.24 \text{ with } \beta = 1/kT]$$

and the overall vibrational partition function is the product of the partition functions of the individual modes. (See Example 13.6.) We draw up the following table, including the degeneracy of each level.

Mode	1	2	3	4
$\tilde{\nu}$ / cm^{-1}	459	217	776	314
g_{mode}	1	2	3	3
$hc\tilde{\nu}/kT$	1.320	0.624	2.232	0.903
$q^{\mathrm{V}}_{\mathrm{mode}}$	1.364	2.15	1.120	1.681

The overall vibrational partition function is

$$q^{\mathrm{V}}_{\mathrm{mode}} = 1.364 \times 2.15^2 \times 1.120^3 \times 1.681^3 = \boxed{42.3}$$

E13.19(a)

$$q = \sum_{\text{levels}} g_j \mathrm{e}^{-\beta\varepsilon_j}[13.9\mathrm{b}] = \sum_{\text{levels}} g_j \mathrm{e}^{-hc\tilde{\nu}_j/kT} = 4 + \mathrm{e}^{-hc\tilde{\nu}_1/kT} + 2\mathrm{e}^{-hc\tilde{\nu}_2/kT}$$

where

$$\frac{hc\tilde{\nu}_j}{kT} = \frac{6.626\times10^{-34}\ \mathrm{J\,s}\times2.998\times10^{10}\ \mathrm{cm\,s^{-1}}\times\tilde{\nu}_j}{1.381\times10^{-23}\ \mathrm{J\,K^{-1}}\times1900\ \mathrm{K}} = 7.571\times10^{-4}\times(\tilde{\nu}_j/\mathrm{cm^{-1}})$$

Therefore

$$q = 4 + \mathrm{e}^{-7.571\times10^{-4}\times2500} + 2\mathrm{e}^{-7.571\times10^{-4}\times3500} = 4 + 0.151 + 2\times0.0707 = \boxed{4.292}$$

The individual terms in the last expression are the relative populations of the levels, namely 2×0.0707 to 0.151 to 4 (second excited level to first to ground) or $\boxed{0.0353 \text{ to } 0.0377 \text{ to } 1}$.

E13.20(a) The mean energy is

$$\langle\varepsilon\rangle = \frac{1}{q}\sum_i \varepsilon_i \mathrm{e}^{-\beta\varepsilon_i}[13.33] = \frac{\sum_i \varepsilon_i \mathrm{e}^{-\beta\varepsilon_i}}{\sum_i \mathrm{e}^{-\beta\varepsilon_i}} = \frac{\varepsilon\mathrm{e}^{-\beta\varepsilon}}{1+\mathrm{e}^{-\beta\varepsilon}}$$

where the last expression specializes to two non-degenerate levels. Substitute

$$\varepsilon = hc\tilde{\nu} = 6.626\times10^{-34}\ \text{J s}\times2.998\times10^{10}\ \text{cm s}^{-1}\times500\ \text{cm}^{-1} = 9.93\times10^{-21}\ \text{J}$$

and $$\beta\varepsilon = \frac{\varepsilon}{kT} = \frac{9.93\times10^{-21}\ \text{J}}{1.381\times10^{-23}\ \text{J K}^{-1}\times298\ \text{K}} = 2.41$$

so $$\langle\varepsilon\rangle = \frac{9.93\times10^{-21}\ \text{J}\times\text{e}^{-2.41}}{1+\text{e}^{-2.41}} = \boxed{8.16\times10^{-22}\ \text{J}}$$

E13.21(a) The mean energy is

$$\langle\varepsilon\rangle = \frac{1}{q}\sum_{\text{states}}\varepsilon_i\text{e}^{-\beta\varepsilon_i}[13.33] = \frac{1}{q}\sum_{\text{levels}}g_i\varepsilon_i\text{e}^{-\beta\varepsilon_i} = \frac{1}{q}\sum_J(2J+1)\varepsilon_J\text{e}^{-\varepsilon_J/kT}$$

$$\varepsilon_J = hc\tilde{B}J(J+1) = 6.626\times10^{-34}\ \text{J s}\times2.998\times10^{10}\ \text{cm s}^{-1}\times1.931\ \text{cm}^{-1}\times J(J+1)$$

$$= J(J+1)\times3.836\times10^{-23}\ \text{J}$$

and $$\frac{\varepsilon_J}{k} = \frac{J(J+1)\times3.836\times10^{-23}\ \text{J}}{1.381\times10^{-23}\ \text{J K}^{-1}} = J(J+1)\times2.778\ \text{K}$$

so $$\langle\varepsilon\rangle = \frac{1}{q}\sum_J J(J+1)(2J+1)\times3.836\times10^{-23}\ \text{J}\times\text{e}^{-J(J+1)\times2.778\ \text{K}/T}$$

Use a spreadsheet or other mathematical software to evaluate the terms of the sum and to sum the terms until they converge. For the partition function, see Exercise 13.11(a).

The equipartition value is simply kT (i.e., $kT/2$ for each rotational degree of freedom). The explicit and equipartition expressions are compared in Figure 13.5. The explicit sum reaches 95% of the equipartition value at about $\boxed{18.5\ \text{K}}$.

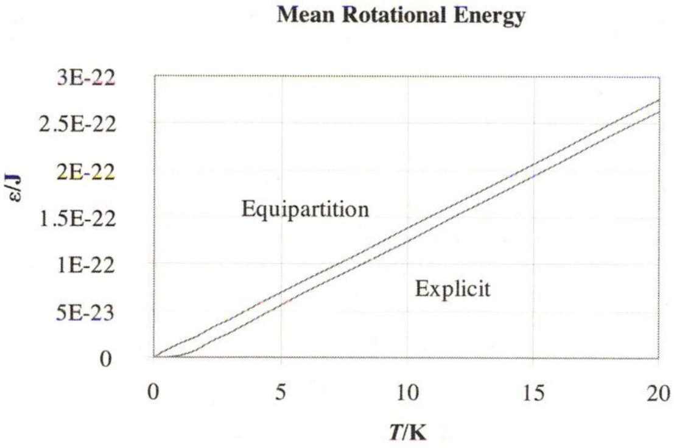

Figure 13.5

E13.22(a) The mean energy is

$$\langle\varepsilon\rangle = \frac{1}{q}\sum_{\text{states}}\varepsilon_i\text{e}^{-\beta\varepsilon_i}[13.33] = \frac{1}{q}\sum_{\text{levels}}g_i\varepsilon_i\text{e}^{-\beta\varepsilon_i} = \frac{1}{\sigma q}\sum_J(2J+1)^2\varepsilon_J\text{e}^{-\varepsilon_J/kT}\ [10.7]$$

Note that the sum over levels is restricted by nuclear statistics; to avoid multiple counting, we sum over all J without restriction and divide the result by the symmetry number σ.

$$\varepsilon_J = hc\tilde{B}J(J+1) = 6.626\times10^{-34}\ \text{J s}\times2.998\times10^{10}\ \text{cm s}^{-1}\times5.241\ \text{cm}^{-1}\times J(J+1)$$

$$= J(J+1)\times1.041\times10^{-22}\ \text{J}$$

and $$\frac{\varepsilon_J}{k} = \frac{J(J+1)\times1.041\times10^{-22}\ \text{J}}{1.381\times10^{-23}\ \text{J K}^{-1}} = J(J+1)\times7.539\ \text{K}$$

so $$\langle\varepsilon\rangle = \frac{1}{\sigma q}\sum_J J(J+1)(2J+1)^2\times1.041\times10^{-22}\ \text{J}\times \text{e}^{-J(J+1)\times7.539\ \text{K}/T}$$

Use a spreadsheet or other mathematical software to evaluate the terms of the sum and to sum the terms until they converge. For σq, see Exercise 13.12(a). The quantity evaluated explicitly in that exercise is σq, because we computed the partition function without taking the symmetry number into account; in effect, the sum evaluated here **and** the sum evaluated in the earlier exercise contain factors of σ, which cancel.

The equipartition value is simply $3kT/2$ (i.e., $kT/2$ for each rotational degree of freedom). The explicit and equipartition expressions are compared in Figure 13.6. The explicit sum reaches 95% of the equipartition value at about $\boxed{25\ \text{K}}$.

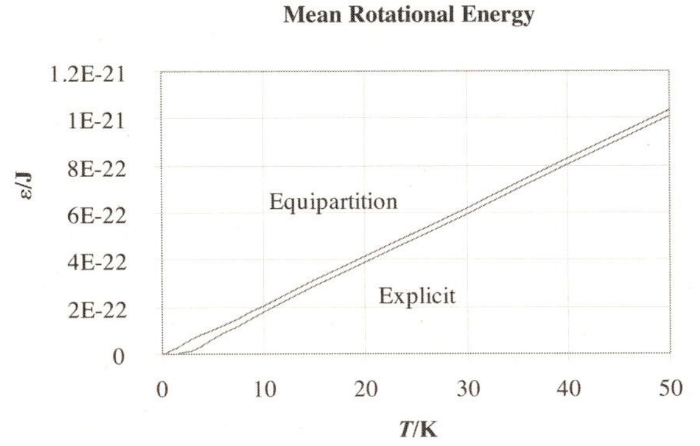

Figure 13.6

E13.23(a) The mean energy is

$$\langle\varepsilon\rangle = \frac{1}{q}\sum_{\text{states}} \varepsilon_i \text{e}^{-\beta\varepsilon_i}[13.33] = \frac{1}{q}\sum_{\text{levels}} g_i\varepsilon_i \text{e}^{-\beta\varepsilon_i}$$

$$= \frac{1}{\sigma q}\sum_{J=0}(2J+1)\text{e}^{-hc\tilde{B}J(J+1)/kT}\left(\sum_{K=-J}^{J}\varepsilon_{J,K}\text{e}^{-hc(\tilde{A}-\tilde{B})K^2/kT}\right)[10.10]$$

Note that the sum over levels is restricted by nuclear statistics; to avoid multiple counting, we sum over all J without restriction and divide the result by the symmetry number σ. (See Exercise 13.22a.)

$$\varepsilon_{J,K} = hc\{\tilde{B}J(J+1)+(\tilde{A}-\tilde{B})K^2\}$$

Use $$hc\tilde{B} = 6.626\times10^{-34}\ \text{J s}\times2.998\times10^{10}\ \text{cm s}^{-1}\times0.443\ \text{cm}^{-1} = 8.80\times10^{-24}\ \text{J}$$

$$\frac{hc\tilde{B}}{k} = \frac{8.80\times10^{-24}\ \text{J}}{1.381\times10^{-23}\ \text{J K}^{-1}} = 0.637\ \text{K}$$

$$hc(\tilde{A}-\tilde{B}) = 6.626\times10^{-34}\ \text{J s}\times2.998\times10^{10}\ \text{cm s}^{-1}\times(5.097-0.443)\ \text{cm}^{-1}$$

$$= 9.245\times10^{-23}\ \text{J}$$

and $$\frac{hc(\tilde{A}-\tilde{B})}{k}=\frac{9.245\times10^{-23}\ \text{J}}{1.381\times10^{-23}\ \text{J K}^{-1}}=6.694\ \text{K}$$

so $$\langle\varepsilon\rangle=\frac{1}{\sigma q}\sum_{J=0}(2J+1)\text{e}^{-0.637\ \text{K}\times J(J+1)/T}$$

$$\times\sum_{K=-J}^{J}\{J(J+1)\times8.80\times10^{-24}\ \text{J}+K^2\times9.245\times10^{-23}\ \text{J}\}\text{e}^{-6.694\ \text{K}\times K^2/T}$$

Write a brief computer program or use other mathematical software to evaluate the terms of the sum and to sum the terms until they converge. Nested sums are straightforward to program in languages such as BASIC or FORTRAN, whereas spreadsheets are more unwieldy.

Compare the results of the direct sum with the equipartition value, namely $3kT/2$. The explicit and equipartition expressions are compared in Figure 13.7. The explicit sum reaches 95% of the equipartition value at about 4.5 K.

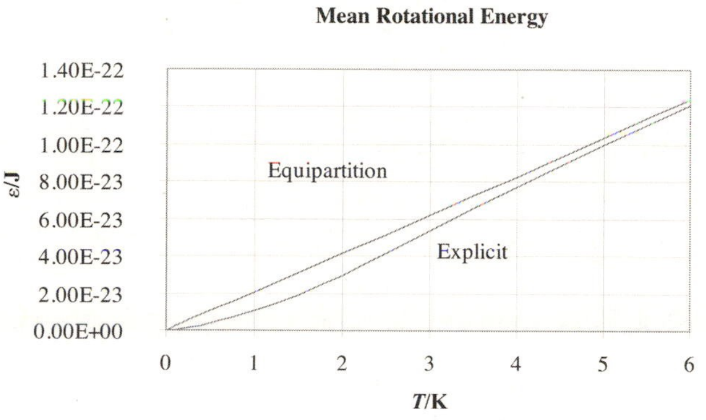

Figure 13.7

E13.24(a) The mean vibrational energy is

$$\langle\varepsilon^{\text{V}}\rangle=\frac{hc\tilde{\nu}}{\text{e}^{hc\tilde{\nu}/kT}-1}\quad[13.39\text{ with }\beta=1/kT]$$

Use $$hc\tilde{\nu}=6.626\times10^{-34}\ \text{J s}\times2.998\times10^{10}\ \text{cm s}^{-1}\times323.2\ \text{cm}^{-1}=6.420\times10^{-21}\ \text{J}$$

and $$\frac{hc\tilde{\nu}}{k}=\frac{6.420\times10^{-21}\ \text{J}}{1.381\times10^{-23}\ \text{J K}^{-1}}=464.9\ \text{K}$$

so $$\langle\varepsilon^{\text{V}}\rangle=\frac{6.420\times10^{-21}\ \text{J}}{\text{e}^{464.9\ \text{K}/T}-1}$$

The equipartition value is simply kT for a single vibrational mode. The explicit and equipartition values are compared in Figure 13.8. The explicit expression reaches 95% of the equipartition value at 4600 K.

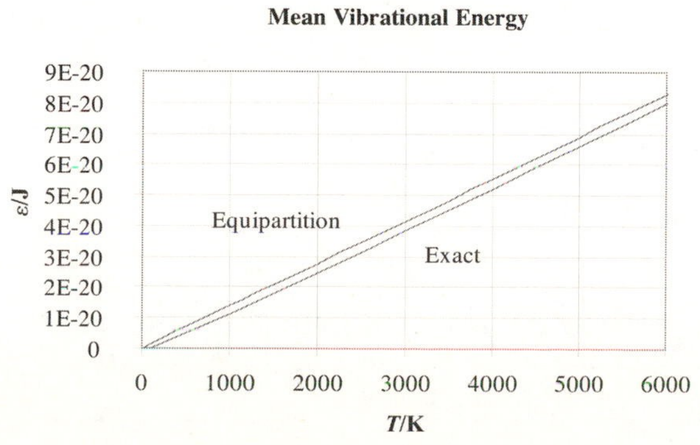

Figure 13.8

E13.25(a) The mean vibrational energy per mode is

$$\langle \varepsilon^{\mathrm{V}} \rangle = \frac{hc\tilde{\nu}}{\mathrm{e}^{hc\tilde{\nu}/kT}-1} \quad [13.39 \text{ with } \beta = 1/kT]$$

We draw up the following table.

Mode	1	2	3	4
$\tilde{\nu}$ / cm^{-1}	658	397	397	1535
$hc\tilde{\nu}$ / (10^{-21} J)	13.07	7.89	7.89	30.49
$(hc\tilde{\nu}/k)$ / K	946	571	571	2208

So $$\langle \varepsilon^{\mathrm{V}} \rangle = \frac{1.307\times10^{-20}\ \mathrm{J}}{\mathrm{e}^{946\ \mathrm{K}/T}-1} + 2\times\frac{7.89\times10^{-21}\ \mathrm{J}}{\mathrm{e}^{571\ \mathrm{K}/T}-1} + \frac{3.049\times10^{-20}\ \mathrm{J}}{\mathrm{e}^{2208\ \mathrm{K}/T}-1}$$

The equipartition value is simply $4kT$, that is, kT per vibrational mode. The explicit and equipartition values are compared in Figure 13.9. The explicit expression reaches 95% of the equipartition value at $\boxed{10500\ \mathrm{K}}$.

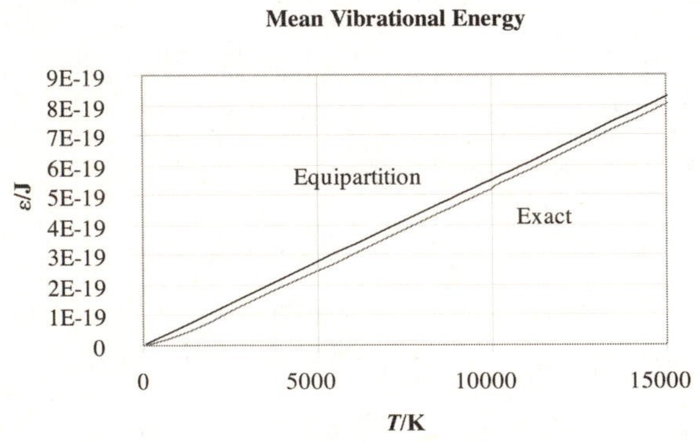

Figure 13.9

E13.26(a) The mean vibrational energy per mode is

$$\langle \varepsilon^{\mathrm{V}} \rangle = \frac{hc\tilde{\nu}}{\mathrm{e}^{hc\tilde{\nu}/kT}-1} \quad [13.39 \text{ with } \beta = 1/kT]$$

We draw up the following table.

Mode	1	2	3	4
$\tilde{\nu}$ / cm^{-1}	459	217	776	314
Degeneracy	1	2	3	3
$hc\tilde{\nu}$ / (10^{-21} J)	9.12	4.31	15.42	6.24
$(hc\tilde{\nu}/k)$ / K	660	312	1116	452

So $$\langle \varepsilon^{\mathrm{V}} \rangle = \frac{9.12\times10^{-21}\ \mathrm{J}}{\mathrm{e}^{660\ \mathrm{K}/T}-1} + 2\times\frac{4.31\times10^{-21}\ \mathrm{J}}{\mathrm{e}^{312\ \mathrm{K}/T}-1} + 3\times\frac{1.542\times10^{-20}\ \mathrm{J}}{\mathrm{e}^{1116\ \mathrm{K}/T}-1} + 3\times\frac{6.24\times10^{-21}\ \mathrm{J}}{\mathrm{e}^{452\ \mathrm{K}/T}-1}$$

The equipartition value is simply $9kT$, that is, kT per vibrational mode. The explicit and equipartition values are compared in Figure 13.10. The explicit expression reaches 95% of the equipartition value at $\boxed{6500\ \text{K}}$.

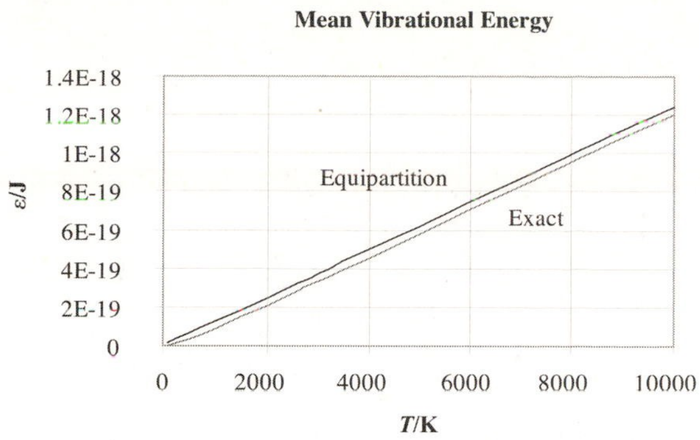

Figure 13.10

E13.27(a)

$$\langle\varepsilon\rangle = -\frac{1}{q}\frac{\mathrm{d}q}{\mathrm{d}\beta}[13.34] = -\frac{1}{q}\frac{\mathrm{d}}{\mathrm{d}\beta}\left(4+\mathrm{e}^{-\beta\varepsilon_1}+2\mathrm{e}^{-\beta\varepsilon_2}\right) = -\frac{1}{q}\left(-\varepsilon_1\mathrm{e}^{-\beta\varepsilon_1}-2\varepsilon_2\mathrm{e}^{-\beta\varepsilon_2}\right)$$

$$= \frac{hc}{q}\left(\tilde{\nu}_1\mathrm{e}^{-\beta hc\tilde{\nu}_1}+2\tilde{\nu}_2\mathrm{e}^{-\beta hc\tilde{\nu}_2}\right) = \frac{hc}{q}\left(\tilde{\nu}_1\mathrm{e}^{-hc\tilde{\nu}_1/kT}+2\tilde{\nu}_2\mathrm{e}^{-hc\tilde{\nu}_2/kT}\right)$$

Use $\dfrac{hc\tilde{\nu}_j}{kT} = 7.571\times10^{-4}\times(\tilde{\nu}_j/\text{cm}^{-1})$ and $q = 4.292$ [Exercise 13.19a]

Thus
$$\langle\varepsilon\rangle = \frac{6.626\times10^{-34}\ \text{J s}\times2.998\times10^{10}\ \text{cm s}^{-1}}{4.292}$$
$$\times\left(2500\ \text{cm}^{-1}\times\mathrm{e}^{-7.571\times10^{-4}\times2500}+2\times3500\ \text{cm}^{-1}\times\mathrm{e}^{-7.571\times10^{-4}\times3500}\right)$$
$$= \boxed{4.033\times10^{-21}\ \text{J}}$$

E13.28(a) The partition function is

$$q = \sum_i \mathrm{e}^{-\beta\varepsilon_i} = \boxed{1+\mathrm{e}^{-2\mu_\mathrm{B}\beta\mathcal{B}}} \quad \text{[energies measured from lower state]}$$

The mean energy is [13.34]

$$\langle\varepsilon\rangle = -\frac{1}{q}\frac{\mathrm{d}q}{\mathrm{d}\beta} = \boxed{\frac{2\mu_\mathrm{B}\mathcal{B}\mathrm{e}^{-2\mu_\mathrm{B}\beta\mathcal{B}}}{1+\mathrm{e}^{-2\mu_\mathrm{B}\beta\mathcal{B}}}}$$

Alternatively, if we measure energy with respect to zero magnetic field (rather than to ground-state energy), then [13.35a]

$$\langle\varepsilon\rangle = \varepsilon_\mathrm{gs} - \frac{1}{q}\frac{\mathrm{d}q}{\mathrm{d}\beta} = \boxed{-\mu_\mathrm{B}\mathcal{B}+\frac{2\mu_\mathrm{B}\mathcal{B}\mathrm{e}^{-2\mu_\mathrm{B}\beta\mathcal{B}}}{1+\mathrm{e}^{-2\mu_\mathrm{B}\beta\mathcal{B}}}}$$

We write $x = 2\mu_\mathrm{B}\beta\mathcal{B}$, so the partition function becomes

$$q = 1+\mathrm{e}^{-x}$$

and the mean energy, scaled by the energy separation, becomes

$$\frac{\langle\varepsilon\rangle}{2\mu_\mathrm{B}\mathcal{B}} = -\frac{1}{2}+\frac{\mathrm{e}^{-x}}{1+\mathrm{e}^{-x}} = -\frac{1}{2}+\frac{1}{\mathrm{e}^{x}+1}$$

The functions are plotted in Figs. 13.11(a) and (b). The effect of increasing the magnetic field is to concentrate population into the lower level.

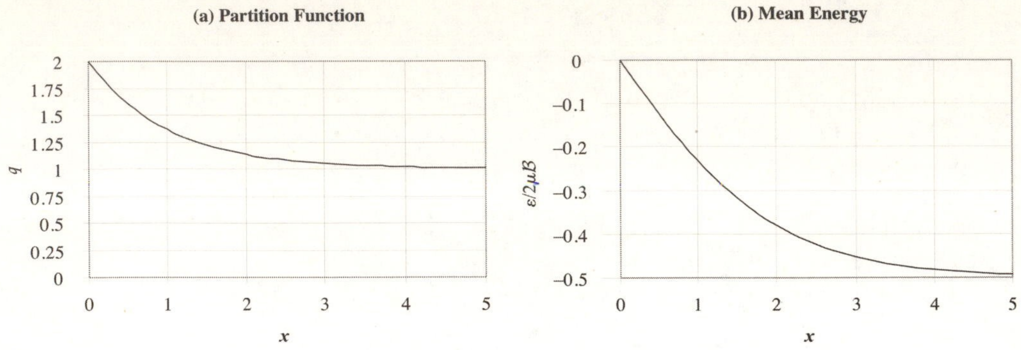

Figure 13.11

The relative populations are

$$\frac{N_+}{N_-} = e^{-\beta\Delta\varepsilon}[13.7] = e^{-x}$$

$$x = 2\mu_B \beta \mathcal{B} = \frac{2\mu_B \mathcal{B}}{kT} = \frac{2\times 9.274\times 10^{-24}\ \text{J T}^{-1}\times 1.0\ \text{T}}{(1.381\times 10^{-23}\ \text{J K}^{-1})T} = \frac{1.343}{T/\text{K}}$$

(a) $T = 4K,\quad \dfrac{N_+}{N_-} = e^{-1.343/4} = \boxed{0.71}$

(b) $T = 298\ \text{K},\quad \dfrac{N_+}{N_-} = e^{-1.343/298} = \boxed{0.996}$

E13.29(a) Inclusion of a factor of $1/N!$ is necessary when considering indistinguishable particles. Because of their translational freedom, gases are collections of indistinguishable particles. Solids are collections of particles that are distinguishable by their positions. The factor must be included in calculations on (a) $\boxed{\text{He gas}}$, (b) $\boxed{\text{CO gas}}$, and (d) $\boxed{H_2O \text{ vapor}}$, but not in (c) solid CO.

Solutions to problems

Solutions to numerical problems

P13.1 (a) $\mathcal{W} = \dfrac{N!}{N_0!N_1!N_2!\cdots}[13.1] = \dfrac{5!}{0!5!0!0!0!} = \boxed{1}$

(b) We draw up the following table.

0	ε	2ε	3ε	4ε	5ε	$\mathcal{W} = \dfrac{N!}{n_1!n_2!\ldots}$
4	0	0	0	0	1	5
3	1	0	0	1	0	20
3	0	1	1	0	0	20
2	2	0	1	0	0	30
2	1	2	0	0	0	30
1	3	1	0	0	0	20
0	5	0	0	0	0	1

The most probable configurations are $\boxed{\{2, 2, 0, 1, 0, 0\}}$ and $\boxed{\{2, 1, 2, 0, 0, 0\}}$.

P13.3 Listing all possible configurations for a 20-particle system would be very time-consuming and tedious indeed; however, listing representative configurations for a given total energy is manageable if done systematically. By "representative," list only one configuration that has a given weight.

For example, consider systems that have a total energy of 10ε, where ε is the separation between adjacent energy levels. There are many distinct configurations that have 17 particles in the ground state and one particle in each of three different states, including configurations in which the singly occupied levels are (ε, 2ε, 7ε), (2ε, 3ε, 5ε), and (ε, 4ε, 5ε). Any one of these configurations is "representative" of all of them, however, because they all have the same weight, namely,

$$\mathcal{W} = \frac{N!}{N_0!N_1!N_2!\cdots}\;[13.1] = \frac{20!}{17!1!1!1!} = 20\times19\times18 = 6840$$

So it is sufficient to enumerate just one of the configurations whose occupancy numbers are 17, 1, 1, and 1 (to make sure that one such configuration exists consistent with the desired total energy). A systematic way of keeping track of representative configurations is to lower maximum occupancy numbers. The next set of occupancy numbers to look at would be 16, 4; then 16, 3, 1; 16, 2, 2; 16, 2, 1, 1; and 16, 1, 1, 1, 1. We would eliminate 16, 4, because no such occupancy numbers yield a total energy of 10ε.

Set up a spreadsheet to generate one of each kind of representative configuration, while keeping the total number of particles constant at 20 and total energy constant (at 10ε for the moment). Again, the most systematic way to do this is to give the highest occupancy numbers to the lowest energy available states. (This rule also generates the most "exponential" configurations.) For example, out of several configurations corresponding to occupancy numbers of 16, 3, and 1, the one used is $N_0 = 16$, $N_1 = 3$, and $N_7 = 1$.

Next examine systems of total energy 10ε, 15ε, and 20ε. The most probable configurations and corresponding weights are

10ε: $\{N_0 = 12, N_1 = 6, N_2 = 2, N_3 = 0, \ldots\}$ $\qquad \mathcal{W} = \dfrac{20!}{12!6!2!} = 3527160$

15ε: $\{N_0 = 10, N_1 = 6, N_2 = 3, N_3 = 1, N_4 = 0, \ldots\}$ $\qquad \mathcal{W} = \dfrac{20!}{10!6!3!1!} = 155195040$

20ε: $\{N_0 = 10, N_1 = 4, N_2 = 3, N_3 = 2, N_4 = 1, N_5 = 0, \ldots\}$ $\qquad \mathcal{W} = \dfrac{20!}{10!4!3!2!1!} = 2327925600$

and $\{N_0 = 9, N_1 = 6, N_2 = 2, N_3 = 2, N_4 = 1, N_5 = 0, \ldots\}$ $\qquad \mathcal{W} = \dfrac{20!}{9!6!2!2!1!} = 2327925600$

As the total energy increases, the most probable configuration has more occupied levels and occupancy of higher energy levels.

The Boltzmann distribution would predict the following relative probabilities for equally spaced energy levels above the ground state:

$$\frac{p_j}{p_0} = e^{-\beta j\varepsilon} \quad \text{so} \quad \ln\frac{p_j}{p_0} = -\beta j\varepsilon$$

Thus, a plot of the natural log of relative probability vs. the ordinal number of the energy level (j) should be a straight line whose slope is $-\beta\varepsilon$. Using the non-zero occupancy numbers from the configuration, the plots are indeed roughly linear. Furthermore, the value of β decreases with increasing total energy, corresponding to an increase in temperature.

P13.5 According to the "integral" approximation

$$q^{\mathrm{T}} = \frac{X}{\Lambda} = \frac{(2\pi mkT)^{1/2}X}{h} \quad [13.15\mathrm{b}]$$

and hence for an H atom in a one-dimensional 100-nm box, when $q^{\mathrm{T}} = 10$,

$$T = \left(\frac{1}{2\pi mk}\right)\times\left(\frac{q^{\mathrm{T}}h}{X}\right)^2$$
$$= \left(\frac{1}{2\pi\times 1.008\times 1.6605\times 10^{-27}\,\mathrm{kg}\times 1.381\times 10^{-23}\,\mathrm{J\,K^{-1}}}\right)\times\left(\frac{10\times 6.626\times 10^{-34}\,\mathrm{J\,s}}{100\times 10^{-9}\,\mathrm{m}}\right)^2$$
$$= \boxed{0.030\,\mathrm{K}}$$

The exact partition function in one dimension is

$$q^{\mathrm{T}} = \sum_{n=1}^{\infty} \mathrm{e}^{-(n^2-1)h^2\beta/8mL^2} = \mathrm{e}^{h^2\beta/8mL^2}\sum_{n=1}^{\infty}(\mathrm{e}^{-h^2\beta/8mL^2})^{n^2}$$

For the H atom,

$$\frac{h^2\beta}{8mL^2} = \frac{(6.626\times 10^{-34}\,\mathrm{J\,s})^2}{8\times 1.008\times 1.6605\times 10^{-27}\,\mathrm{kg}\times 1.381\times 10^{-23}\,\mathrm{J\,K^{-1}}\times 0.030\,\mathrm{K}\times(100\times 10^{-9}\,\mathrm{m})^2}$$
$$= 7.9\times 10^{-3}$$

and $\quad \mathrm{e}^{h^2\beta/8mL^2} = \mathrm{e}^{7.9\times 10^{-3}} = 1.008\,, \quad \mathrm{e}^{-h^2\beta/8mL^2} = \mathrm{e}^{-7.9\times 10^{-3}} = 0.992$

Then $\quad q^{\mathrm{T}} = 1.008\times\sum_{n=1}^{\infty}(0.992)^{n^2} = 1.008\times(0.992+0.969+0.932+\cdots) = \boxed{9.57}$

COMMENT. Even under these conditions, the integral approximation is less than 5% from the explicit sum.

P13.7 If the electronic states were in thermal equilibrium with the translational states, then the temperature would be the same for both. The ratio of electronic states at 300 K would be

$$\frac{N_1}{N_0} = \frac{g_1\mathrm{e}^{-\varepsilon_1/kT}}{g_0\mathrm{e}^{-\varepsilon_0/kT}} = \frac{4}{2}\times\mathrm{e}^{-\Delta\varepsilon/kT} = 2\mathrm{e}^{-hc\tilde{\nu}/kT} = 2\mathrm{e}^{-\{(1.4388\times 450)/300\}} = 0.23$$

The observed ratio is $\frac{0.30}{0.70} = 0.43$. Hence the populations are $\boxed{\text{not at equilibrium}}$.

P13.9 (a) First, evaluate the partition function

$$q = \sum_j g_j\mathrm{e}^{-\beta\varepsilon_j}\,[13.9\mathrm{b}] = \sum_j g_j\mathrm{e}^{-hc\beta\tilde{\nu}_j}$$

At 3287°C = 3560 K, $hc\beta = \frac{1.43877\,\mathrm{cm\,K}}{3560\,\mathrm{K}} = 4.041\times 10^{-4}\,\mathrm{cm}$

$$q = 5 + 7\mathrm{e}^{-\{(4.041\times 10^{-4}\,\mathrm{cm})\times(170\,\mathrm{cm^{-1}})\}} + 9\mathrm{e}^{-\{(4.041\times 10^{-4}\,\mathrm{cm})\times(387\,\mathrm{cm^{-1}})\}} + 3\mathrm{e}^{-\{(4.041\times 10^{-4}\,\mathrm{cm})\times(6557\,\mathrm{cm^{-1}})\}}$$
$$= 5 + 7\times(0.934) + 9\times(0.855) + 3\times(0.0707) = 19.444$$

The fractions of molecules in the various energy levels are

$$p_j = \frac{g_j\mathrm{e}^{-\beta\varepsilon_j}}{q}\,[13.8] = \frac{g_j\mathrm{e}^{-hc\beta\tilde{\nu}_j}}{q}$$

$$p(^3\mathrm{F}_2) = \frac{5}{19.444} = \boxed{0.257} \qquad p(^3\mathrm{F}_3) = \frac{(7)\times(0.934)}{19.444} = \boxed{0.336}$$

$$p(^3\mathrm{F}_4) = \frac{(9)\times(0.855)}{19.444} = \boxed{0.396} \qquad p(^4\mathrm{F}_1) = \frac{(3)\times(0.0707)}{19.444} = \boxed{0.011}$$

COMMENT. $\sum_j p_j = 1$. Note that the most highly populated level is not the lowest level.

P13.11 The partition function is

$$q = \sum_j g_j e^{-\beta\varepsilon_j}[13.9b] = \sum_j g_j e^{-hc\beta\tilde{\nu}_j} = \sum_j g_j e^{-hc\tilde{\nu}_j/kT}$$

g is the degeneracy, given by

$$g = 2J + 1$$

where J is the level of the term (displayed as the subscript in the term symbol).

At 298 K, $\dfrac{hc\tilde{\nu}_j}{kT} = \dfrac{1.43877\,\text{cm K}\times\tilde{\nu}_j}{298\,\text{K}} = 4.83\times10^{-3}\times\tilde{\nu}_j/\text{cm}^{-1}$

so $$q = 1 + 3e^{-4.83\times10^{-3}\times557.1} + 5e^{-4.83\times10^{-3}\times1410.0} + 5e^{-4.83\times10^{-3}\times7125.3} + e^{-4.83\times10^{-3}\times16367.3} = \boxed{1.209}$$

At 1000 K, $\dfrac{hc\tilde{\nu}_j}{kT} = \dfrac{1.43877\,\text{cm K}\times\tilde{\nu}_j}{1000\,\text{K}} = 1.439\times10^{-3}\times\tilde{\nu}_j/\text{cm}^{-1}$

so $$q = 1 + 3e^{-1.439\times10^{-3}\times557.1} + 5e^{-1.439\times10^{-3}\times1410.0} + 5e^{-1.439\times10^{-3}\times7125.3} + e^{-1.439\times10^{-3}\times16367.3} = \boxed{3.004}$$

P13.13 The partition function is

$$q = \sum_v e^{-\beta\varepsilon_v}[13.9b] = \sum_v e^{-hc\tilde{\nu}_v/kT}$$

(a) At 100 K, $\dfrac{hc\tilde{\nu}_j}{kT} = \dfrac{1.43877\,\text{cm K}\times\tilde{\nu}_j}{100\,\text{K}} = 1.44\times10^{-2}\times\tilde{\nu}_j/\text{cm}^{-1}$

so $$q = 1 + e^{-0.0144\times213.30} + e^{-0.0144\times425.39} + e^{-0.0144\times636.27} + e^{-0.0144\times845.93} = \boxed{1.049}$$

(b) At 298 K, $\dfrac{hc\tilde{\nu}_j}{kT} = \dfrac{1.43877\,\text{cm K}\times\tilde{\nu}_j}{298\,\text{K}} = 4.83\times10^{-3}\times\tilde{\nu}_j/\text{cm}^{-1}$

so $$q = 1 + e^{-4.83\times10^{-3}\times213.30} + e^{-4.83\times10^{-3}\times425.39} + e^{-4.83\times10^{-3}\times636.27} + e^{-4.83\times10^{-3}\times845.93} = \boxed{1.548}$$

The fractions of molecules at the various levels are

$$p_v = \frac{e^{-\beta\varepsilon_v}}{q}[13.8] = \frac{e^{-hc\tilde{\nu}_v/kT}}{q}$$

So $$p_0 = \frac{1}{q} = \text{(a)}\ \boxed{0.953},\quad \text{(b)}\ \boxed{0.645}$$

$$p_1 = \frac{e^{-hc\tilde{\nu}_1/kT}}{q} = \text{(a)}\ \boxed{0.044},\quad \text{(b)}\ \boxed{0.230}$$

$$p_2 = \frac{e^{-hc\tilde{\nu}_2/kT}}{q} = \text{(a)}\ \boxed{0.002},\quad \text{(b)}\ \boxed{0.083}$$

COMMENT. Eqn 13.24 gives a closed-form expression for the vibrational partition function, based on an infinite ladder of harmonic oscillator states. The explicit sum will deviate slightly from this number because an actual molecule has a finite number of slightly anharmonic vibrational states below the bond dissociation energy.

Solutions to theoretical problems

P13.15 (a) The probability of finding a molecule in state j is

$$p_j = \frac{N_j}{N} = \frac{e^{-\beta\varepsilon_j}}{q} \quad [13.8]$$

In the systems under consideration, ε is both the mean energy and the energy difference between adjacent levels, so

$$p_j = \frac{e^{-j\beta\varepsilon}}{q}$$

which implies that

$$-j\beta\varepsilon = \ln N_j - \ln N + \ln q \quad \text{and} \quad \ln N_j = \ln N - \ln q - j\beta\varepsilon = \boxed{\ln\frac{N}{q} - \frac{j\varepsilon}{kT}}$$

Thus, a plot of $\ln N_j$ against j should be a straight line with slope $-\varepsilon/kT$.

We draw up the following table using the information in Problem 13.2.

j	0	1	2	3
N_j	4	2	2	1
$\ln N_j$	1.39	0.69	0.69	0

[most probable configuration]

These are points plotted in Figure 13.12 (full line). The slope is –0.416, and since $\frac{\varepsilon}{hc} = 50\,\text{cm}^{-1}$, the slope corresponds to a temperature.

$$T = \frac{(50\,\text{cm}^{-1}) \times (2.998 \times 10^{10}\,\text{cm s}^{-1}) \times (6.626 \times 10^{-34}\,\text{J s})}{(0.416) \times (1.381 \times 10^{-23}\,\text{J K}^{-1})} = \boxed{163\ \text{K}}$$

(A better estimate, 104 K, represented by the dashed line in Figure 13.12, is found in Problem 13.16.)

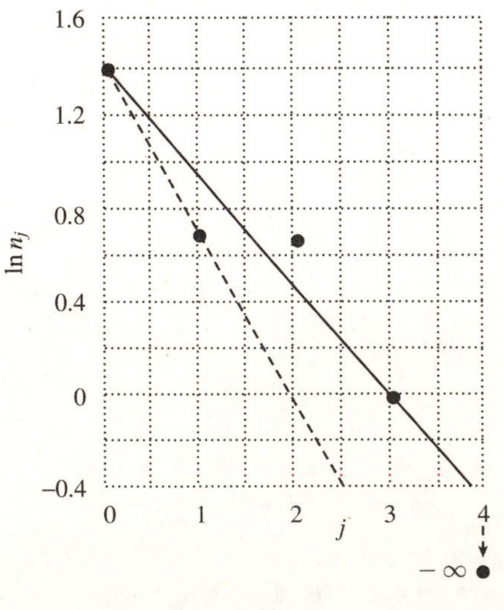

Figure 13.12

(b) Choose one of the weight 2520 configurations and one of the weight 504 configurations, and draw up the following table.

	J	0	1	2	3	4
$\mathcal{W} = 2520$	N_j	4	3	1	0	1
	$\ln N_j$	1.39	1.10	0	$-\infty$	0
$\mathcal{W} = 504$	N_j	6	0	1	1	1
	$\ln N_j$	1.79	$-\infty$	0	0	0

Inspection confirms that these data give very crooked lines

P13.17 Mean values of any variable can be found by weighting possible values of the variable by the probability of that value. Thus, the mean of the square of energy is

$$\langle \varepsilon^2 \rangle = \sum_j p_j \varepsilon_j^{\ 2} = \sum_j \frac{\mathrm{e}^{-\beta\varepsilon_j}}{q} \varepsilon_j^{\ 2}$$

Note that

$$-\frac{\mathrm{d}}{\mathrm{d}\beta}\mathrm{e}^{-\beta\varepsilon_j} = \varepsilon_j \mathrm{e}^{-\beta\varepsilon_j} \quad \text{so} \quad \frac{\mathrm{d}^2}{\mathrm{d}\beta^2}\mathrm{e}^{-\beta\varepsilon_j} = \varepsilon_j^{\ 2}\mathrm{e}^{-\beta\varepsilon_j}$$

so $$\langle \varepsilon^2 \rangle = \frac{1}{q}\sum_j \frac{\mathrm{d}^2}{\mathrm{d}\beta^2}\mathrm{e}^{-\beta\varepsilon_j} = \frac{1}{q}\frac{\mathrm{d}^2}{\mathrm{d}\beta^2}\sum_j \mathrm{e}^{-\beta\varepsilon_j} = \frac{1}{q}\frac{\mathrm{d}^2 q}{\mathrm{d}\beta^2}$$

Thus $$\langle \varepsilon^2 \rangle^{1/2} = \boxed{\left(\frac{1}{q}\frac{\mathrm{d}^2 q}{\mathrm{d}\beta^2}\right)^{1/2}} \quad \text{and} \quad \Delta\varepsilon = (\langle \varepsilon^2 \rangle - \langle \varepsilon \rangle^2)^{1/2} = \boxed{\left\{\frac{1}{q}\frac{\mathrm{d}^2 q}{\mathrm{d}\beta^2} - \left(\frac{1}{q}\frac{\mathrm{d}q}{\mathrm{d}\beta}\right)^2\right\}^{1/2}}$$

For a harmonic oscillator, we have a closed-form expression for the partition function,

$$q = \frac{1}{1-\mathrm{e}^{-\beta hc\tilde{\nu}}} \quad [13.24]$$

so $$\frac{\mathrm{d}q}{\mathrm{d}\beta} = \frac{-hc\tilde{\nu}\mathrm{e}^{-\beta hc\tilde{\nu}}}{(1-\mathrm{e}^{-\beta hc\tilde{\nu}})^2} \quad \text{and} \quad \frac{\mathrm{d}^2 q}{\mathrm{d}\beta^2} = \frac{(hc\tilde{\nu})^2\mathrm{e}^{-\beta hc\tilde{\nu}}(1+\mathrm{e}^{-\beta hc\tilde{\nu}})}{(1-\mathrm{e}^{-\beta hc\tilde{\nu}})^3}$$

$$\Delta\varepsilon = \left\{\frac{(hc\tilde{\nu})^2\mathrm{e}^{-\beta hc\tilde{\nu}}(1+\mathrm{e}^{-\beta hc\tilde{\nu}})}{(1-\mathrm{e}^{-\beta hc\tilde{\nu}})^2} - \left(\frac{-hc\tilde{\nu}\mathrm{e}^{-\beta hc\tilde{\nu}}}{1-\mathrm{e}^{-\beta hc\tilde{\nu}}}\right)^2\right\}^{1/2}$$

After some algebra, the expression becomes

$$\Delta\varepsilon = \boxed{\dfrac{hc\tilde{\nu}\mathrm{e}^{-\beta hc\tilde{\nu}/2}}{1-\mathrm{e}^{-\beta hc\tilde{\nu}}}} = \boxed{\dfrac{hc\tilde{\nu}}{2\sinh(\beta hc\tilde{\nu}/2)}}$$

P13.19 Basic thermodynamic relationships will be introduced in Chapters 14 and 15. Among the key quantities is the internal energy, U, where

$$U = U(0) + \langle E \rangle \ [14.1\text{b}]$$

Here, $U(0)$ is the value of the energy at $T = 0$. The relationship between the internal energy and other macroscopic thermodynamic functions is

$$\mathrm{d}U = T\mathrm{d}S - p\mathrm{d}V \ [15.54]$$

where S is the entropy, a main topic of Chapter 15. Already we can see a connection between U and the canonical partition function, Q, on the one hand (via eqn. 13.46), and to macroscopic system variables like pressure and temperature, on the other. However, the thermodynamic function most directly related to Q is the Helmholtz energy, A, where

$$A = U - TS \ [15.37]$$

The relationship of A to system variables, therefore, is

$$\mathrm{d}A = \mathrm{d}U - T\mathrm{d}S - S\mathrm{d}T = T\mathrm{d}S - p\mathrm{d}V - T\mathrm{d}S - S\mathrm{d}T = -p\mathrm{d}V - S\mathrm{d}T$$

so $\left(\dfrac{\partial A}{\partial V}\right)_{T,N} = -p$

The simple and direct relationship between A and Q is

$$A = A(0) - kT\ln Q \ [15.38\text{a}]$$

Thus $p = kT\left(\dfrac{\partial \ln Q}{\partial V}\right)_{T,N}$

$$= kT\left(\frac{\partial \ln(q^N/N!)}{\partial V}\right)_{T,N}$$

$$= kT\left(\frac{\partial[N\ln q - \ln N!]}{\partial V}\right)_{T,N} = NkT\left(\frac{\partial \ln q}{\partial V}\right)_{T,N}$$

The only contribution to the molecular partition function that depends on V is translation:

$$q = q^{\mathrm{T}}q^{\mathrm{R}}q^{\mathrm{V}}q^{\mathrm{E}} \ [13.14] = \frac{V}{\Lambda^3}q^{\mathrm{R}}q^{\mathrm{V}}q^{\mathrm{E}} \ [13.18\text{b}]$$

so $$p = NkT\left(\frac{\partial \ln(Vq^{\mathrm{R}}q^{\mathrm{V}}q^{\mathrm{E}}/\Lambda^3)}{\partial V}\right)_{T,N}$$

$$= NkT\left(\frac{\partial[\ln V + \ln(q^{\mathrm{R}}q^{\mathrm{V}}q^{\mathrm{E}}/\Lambda^3)]}{\partial V}\right)_{T,N} = NkT\left(\frac{\partial \ln V}{\partial V}\right)_{T,N}$$

$$= \frac{NkT}{V} \quad \text{or} \quad \boxed{pV = NkT = nRT}$$

Solutions to applications

P13.21 If the atmosphere were at equilibrium, then the Boltzmann distribution would apply, so the relative populations per unit volume would be

$$\frac{N_i}{N_j} = \frac{\mathrm{e}^{-\beta\varepsilon_i}}{\mathrm{e}^{-\beta\varepsilon_j}} = \mathrm{e}^{-(\varepsilon_j-\varepsilon_i)/kT} \qquad \left[13.7 \text{ with } \beta = \frac{1}{kT}\right]$$

What distinguishes the states and energies of molecules in a planet's gravitational field is the distance r from the center of the planet. The energy is gravitational, measured from the ground-state energy.

$$\varepsilon(r) = V(r) - V(r_0) = -GMm\left(\frac{1}{r} - \frac{1}{r_0}\right)$$

(Note that the ground-state energy is literally the energy at the ground—or more precisely, at the lowest point of the atmosphere, r_0.)

So $$\frac{N(r)}{N(r_0)} = \mathrm{e}^{-\{V(r)-V(r_0)\}/kT}$$

Far from the planet, we have

$$\lim_{r\to\infty} \varepsilon(r) = -GMm\left(0 - \frac{1}{r_0}\right) = \frac{GMm}{r_0}$$

and $$\lim_{r\to\infty} \frac{N(r)}{N(r_0)} = \mathrm{e}^{-GMm/r_0kT}$$

Hence, if the atmosphere were at equilibrium, the farther one ventured from the planet, the concentration of molecules would tend toward a non-zero fraction of the concentration at the surface. This is obviously not the current distribution for planetary atmospheres where the corresponding limit is zero. Consequently, we may conclude that no planet's atmosphere, including Earth's, is at equilibrium.

P13.23 Each protein-binding site can be represented as a distinct box into which a ligand, L, may bind. All possible configurations are shown in the following table and the configuration count of i indistinguishable ligands being placed in n distinguishable sites is seen to be given by the combinatorial

$$C(n,i) = \frac{n!}{(n-1)!\ i!}$$

L			
	L		
		L	
			L
L	L		
L		L	
L			L
	L	L	
	L		L
		L	L
L	L	L	
L	L		L
L		L	L
	L	L	L
L	L	L	L

PL_0: $C(4,0)=\dfrac{4!}{(4-0)!0!}=1$ configuration

PL_1: $C(4,1)=\dfrac{4!}{(4-1)!1!}=4$ configurations

PL_2: $C(4,2)=\dfrac{4!}{(4-2)!2!}=6$ configurations

PL_3: $C(4,3)=\dfrac{4!}{(4-3)!3!}=4$ configurations

PL_4: $C(4,4)=\dfrac{4!}{(4-4)!4!}=1$ configuration

P13.25 (a) According to Impact *I*13.1, $p_i = q_i/q$ and

$$\frac{q}{q_0}=1+\frac{\sigma s\{s^{n+1}-(n+1)s^n+1\}}{(s-1)^2}\quad[13.30]$$

So p_0 is just the reciprocal of this quantity. For $I>0$,

$$p_i=\frac{(n-i+1)\sigma s^i}{q/q_0}$$

(Recall that $i = 0$ corresponds to the completely helical form, a form with only one possible configuration and the reference form, so there are no factors of σ, s, or n involved in counting its configurations.) The probabilities for the three different values of stability parameter s are plotted in Figure 13.13.

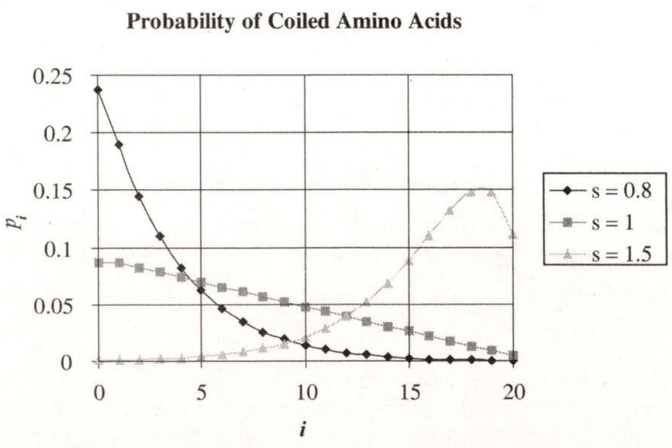

Figure 13.13

The fraction of molecules with i coiled residues depends dramatically on the value of s. When $s < 1$, low values of i are observed, but large i values are observed when $s > 1$. Thus, when $s < 1$, the polypeptide is largely helical, and when $s > 1$, it is more of a random coil. The same sort of distribution can be seen in Fig. 13.15 of the main text, but the smaller value of σ there makes the completely helical form even more likely than it is here when $s < 1$.

(b) The mean value of the number of coiled residues is

$$\langle i \rangle = \sum_{i=1}^{n} i p_i$$

This number is plotted against the stability parameter in Figure 13.14. (From left to right, the curves are for $\sigma = 0.05$, 0.01, 0.001, and 0.0001, respectively. Plot symbols are shown at the s values specified in part (a), namely 0.8, 1, and 1.5.) The plot shows that for $s < 0.5$ the model polypeptide is helical and little changed as s is varied. At the other extreme, for $s > 1.5$ the polypeptide is largely a random coil that changes only slightly with variance of s. The mean number of coiled residues changes rapidly in the middle range of $0.5 > s > 1.5$, giving an overall sigmoidal dependence on s.

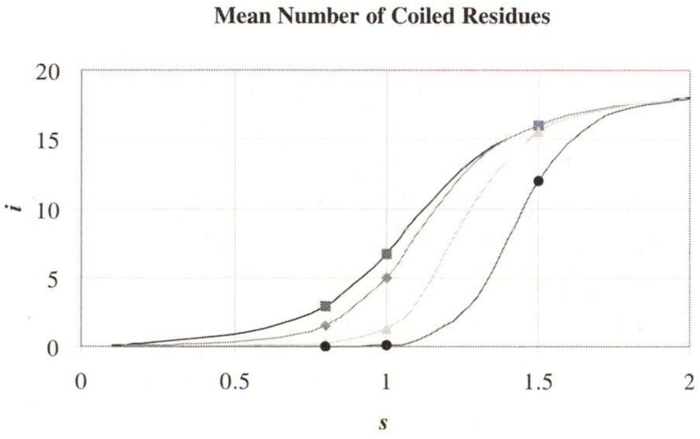

Figure 13.14

14 The first law of thermodynamics

Answers to discussion questions

D14.1 In physical chemistry, the universe is considered to be divided into two parts: the system and its surroundings. In thermodynamics, the system is the object of interest; it is separated from its surroundings, the rest of the universe, by a boundary. The characteristics of the boundary determine whether the system is open, closed, or isolated. An open system has a boundary that permits the passage of both matter and energy. A closed system has a boundary that allows the passage of energy but not of matter. Closed systems can be either adiabatic or diathermic. The former do not allow the transfer of energy as a result of a temperature difference, but the latter do. An isolated system is one with a boundary that allows the transfer of neither matter nor energy between the system and the surroundings.

In thermodynamics, the state of a system is characterized by state functions, thermodynamic properties that are independent of the previous history of the system. In quantum mechanics the state of a system is characterized by a time-independent wave function.

D14.3 A state function is a thermodynamic property the value of which is independent of the history of the system. Examples of state functions are the properties of pressure, temperature, internal energy, enthalpy and, as well as the properties of entropy, Gibbs energy, and Helmholtz energy to be discussed fully in Chapter 15. The differentials of state functions are exact differentials. Hence, we can use the mathematical properties of exact differentials to draw far-reaching conclusions about the relations between physical properties and to establish connections that were unexpected but turn out to be very significant. One practical importance of these results is that we can obtain the value of a property we require from the combination of measurements of other properties without actually having to measure the required property itself, the measurement of which might be very difficult.

D14.5 When a system is subjected to constant pressure conditions, and only expansion work can occur, the energy supplied as heat is the change in enthalpy of the system. Thus enthalpy changes in the system can be determined by measuring the amount of heat supplied. The calorimeters used for measuring these heats and hence the enthalpy changes of the systems being studied are constant pressure calorimeters. A simple example is a thermally insulated vessel (a coffee cup calorimeter) open to the atmosphere: the heat released in the reaction is determined by measuring the change in temperature of the contents. For a combustion reaction, a constant-pressure flame calorimeter may be used, as illustrated in textbook Fig. 14.19 where a certain amount of substance burns in a supply of oxygen and the rise in temperature is monitored. The most sophisticated way of measuring enthalpy changes is to use a differential scanning calorimeter, described in detail in *Impact* 14.1.

D14.7 One can use the general expression for π_T proved in Section 15.7 (eqn. 15.58) to derive its specific form for a van der Waals gas as given in Exercise 14.29(a), that is, $\pi_T = a/V_m^2$. (The derivation is carried out in Example 15.7.) For an isothermal expansion in a van der Waals gas, $dU_m = (a/V_m)^2 dV_m$. Hence $\Delta U_m = -a(1/V_{m,2} - 1/V_{m,1})$. This derivation is given in the solution to Exercise 14.29(a). This formula corresponds to what one would expect for a real gas. As the molecules get closer and closer, the molar volume gets smaller and smaller, and the energy of attraction gets larger and larger.

D14.9 Equations of state can be thought of as expressions for the pressure of a gas in terms of the state functions, n, V, and T. They are obtained from the expression for the pressure in terms of the canonical partition function given in eqn. 15.61 in Section 15.9(b).

$$p = kT\left(\frac{\partial \ln Q}{\partial V}\right)_T$$

Partition functions for perfect and imperfect gases are different. That for the perfect gas is given by $Q = q^N / N!$ with $q = V / \Lambda^3$. There is no one form for imperfect gases. One example is shown in *Self-test* 15.5. Another example, which can be shown to lead to the van der Waals equation of state, is

$$Q = \frac{1}{N!}\left(\frac{2\pi mkT}{h^2}\right)^{3N/2} (V - Nb)e^{aN^2/kTV}$$

For the case of the perfect gas there are no molecular features in the partition function, but for imperfect gases there are repulsive and attractive features in the partition function that are related to the structure of the molecules.

Solutions to exercises

E14.1(a) $C_{V,\mathrm{m}} = \frac{1}{2}(3 + \nu_\mathrm{R}^* + 2\nu_\mathrm{V}^*)R$ [14.27]

with a mode active if $T > \theta_\mathrm{M}$

(a) $\nu_\mathrm{R}^* = 2, \ \nu_\mathrm{V}^* \approx 1$

Hence $C_{V,\mathrm{m}} = \frac{1}{2}(3+2+2\times1)R = \boxed{\frac{7}{2}R}$ [experimental = 3.4R] and the energy is

$$\frac{7}{2}RT = \frac{7}{2}\times 8.314\ \mathrm{J\ K^{-1}\ mol^{-1}} \times 298.15\ \mathrm{K} = 8671\ \mathrm{J\ mol^{-1}} = \boxed{8.671\ \mathrm{kJ\ mol^{-1}}}$$

Note that I_2 has quite a low vibrational wavenumber, so

$$\theta_\mathrm{V} = \frac{hc\tilde{\nu}}{k} = \frac{(6.626\times10^{-34}\ \mathrm{J\ s})(2.998\times10^{10}\ \mathrm{cm\ s^{-1}})(214\ \mathrm{cm^{-1}})}{1.381\times10^{-23}\ \mathrm{J\ K^{-1}}} = 308\ \mathrm{K}$$

The temperature specified for this exercise is less than this but only slightly. Looking at Fig. 14.16 in the text suggests that the mode is closer to active than inactive when the temperature is even close to the vibrational temperature. Thus applying the equipartition theorem may not be adequate in this case.

(b) $\nu_\mathrm{R}^* = 3, \ \nu_\mathrm{V}^* \approx 0$

Hence $C_{V,\mathrm{m}} = \frac{1}{2}(3+3+0)R = \boxed{3R}$ [experimental = 3.2R] and the energy is

$$E = 3RT = 3\times 8.314\ \mathrm{J\ K^{-1}\ mol^{-1}} \times 298.15\ \mathrm{K} = 7436\ \mathrm{J\ mol^{-1}} = \boxed{7.436\ \mathrm{kJ\ mol^{-1}}}$$

(c) $\nu_\mathrm{R}^* = 3, \ \nu_\mathrm{V}^* \approx 4$

Hence $C_{V,\mathrm{m}} = \frac{1}{2}(3+3+2\times4)R = \boxed{7R}$ [experimental = 8.8R] and the energy is

$$E = 7RT = 8\times 8.314\ \mathrm{J\ K^{-1}\ mol^{-1}} \times 298.15\ \mathrm{K} = \boxed{17.35\ \mathrm{kJ\ mol^{-1}}}$$

Note that data from G. H. Herzberg, *Molecular Spectra and Molecular Structure II. Infrared and Raman Spectra of Polyatomic Molecules* (Princeton, NJ: van Nostrand, 1945, pp. 364–365) give low vibrational wavenumbers for four modes of benzene; they are included above. There are 26 more vibrational modes we have neglected, taking them to be inactive. Slight activity from these modes accounts for the difference of about 1.8R in the heat capacity between the experimental value and our estimate.

E14.2(a) (a) Pressure, (b) temperature, and (d) enthalpy are state functions.

E14.3(a) The physical definition of work is $dw = -F\,dz$ [14.5].

In a gravitational field the force is the weight of the object, which is $F = mg$.

If g is constant over the distance the mass moves, dw may be integrated to give the total work

$$w = -\int_{z_i}^{z_f} F\,dz = -\int_{z_i}^{z_f} mg\,dz = -mg(z_f - z_i) = -mgh \quad \text{where} \quad h = (z_f - z_i)$$

On earth: $w = -(60\ \text{kg})\times(9.81\ \text{m s}^{-2})\times(6.0\ \text{m}) = -3.5\times10^3\ \text{J} = \boxed{3.5\times10^3\ \text{J needed}}$

On the moon: $w = -(60\ \text{kg})\times(1.60\ \text{m s}^{-2})\times(6.0\ \text{m}) = -5.7\times10^2\ \text{J} = \boxed{5.7\times10^2\ \text{J needed}}$

E14.4(a) This is an expansion against a constant external pressure; hence $w = -p_{ex}\Delta V$ [14.8].

$$p_{ex} = (1.0\,\text{atm})\times(1.013\times10^5\ \text{Pa atm}^{-1}) = 1.0\bar{1}\times10^5\ \text{Pa}$$

The change in volume is the cross-sectional area times the linear displacement,

$$\Delta V = (50\,\text{cm}^2)\times(15\,\text{cm})\times\left(\frac{1\,\text{m}}{100\,\text{cm}}\right)^3 = 7.5\times10^{-4}\ \text{m}^3$$

so $w = -(1.01\times10^5\ \text{Pa})\times(7.5\times10^{-4}\ \text{m}^3) = \boxed{-75\,\text{J}}$ as $1\ \text{Pa m}^3 = 1\ \text{J}$

E14.5(a) For all cases $\Delta U = 0$, since the internal energy of a perfect gas depends only on temperature. From the definition of enthalpy, $H = U + pV$, so $\Delta H = \Delta U + \Delta(pV) = \Delta U + \Delta(nRT)$ (perfect gas). Hence, $\Delta H = 0$ as well, at constant temperature for all processes in a perfect gas.

(a) $\boxed{\Delta U = \Delta H = 0}$

$$w = -nRT\ln\left(\frac{V_f}{V_i}\right) \quad [14.12]$$

$$= -(1.00\,\text{mol})\times(8.314\,\text{J K}^{-1}\,\text{mol}^{-1})\times(293\,\text{K})\times\ln\left(\frac{30.0\,\text{dm}^3}{10.0\,\text{dm}^3}\right)$$

$$= -2.68\times10^3\ \text{J} = \boxed{-2.68\,\text{kJ}}$$

$$q = \Delta U - w\ [\text{first law}] = 0 + 2.68\,\text{kJ} = \boxed{+2.68\,\text{kJ}}$$

(b) $\boxed{\Delta U = \Delta H = 0}$

$$w = -p_{ex}\Delta V \quad [14.8] \qquad \Delta V = (30.0 - 10.0)\,\text{dm}^3 = 20.0\,\text{dm}^3$$

p_{ex} can be computed from the perfect gas law

$$pV = nRT$$

so $$p_{ex} = p_f = \frac{nRT}{V_f} = \frac{(1.00\,\text{mol})\times(0.08206\,\text{dm}^3\,\text{atm K}^{-1}\,\text{mol}^{-1})\times(293\,\text{K})}{30.0\,\text{dm}^3} = 0.801\,\text{atm}$$

$$w = -(0.801\,\text{atm})\times\left(\frac{1.013\times10^5\ \text{Pa}}{1\,\text{atm}}\right)\times(20.0\,\text{dm}^3)\times\left(\frac{1\,\text{m}^3}{10^3\ \text{dm}^3}\right)$$

$$= -1.62\times10^3\ \text{Pa m}^3 = -1.62\times10^3\ \text{J} = \boxed{-1.62\,\text{kJ}}$$

$$q = \Delta U - w = 0 + 1.62\,\text{kJ} = \boxed{+1.62\,\text{kJ}}$$

(c) $\boxed{\Delta U = \Delta H = 0}$

Free expansion is expansion against no force, so $\boxed{w=0}$ and $q = \Delta U - w = 0 - 0 = \boxed{0}$.

COMMENT. An isothermal free expansion of a perfect gas is also adiabatic.

E14.6(a)
$$C_p = \frac{q_p}{\Delta T}\,[14.33c] = \frac{229\,\text{J}}{2.55\,\text{K}} = 89.8\,\text{J}\,\text{K}^{-1}$$

so $C_{p,\text{m}} = (89.8\ \text{J K}^{-1})/(3.0\ \text{mol}) = \boxed{30\ \text{J K}^{-1}\ \text{mol}^{-1}}$

For a perfect gas, $C_{p,\text{m}} - C_{V,\text{m}} = R$ [14.35]

$$C_{V,\text{m}} = C_{p,\text{m}} - R = (30 - 8.3)\,\text{J}\,\text{K}^{-1}\,\text{mol}^{-1} = \boxed{22\,\text{J}\,\text{K}^{-1}\,\text{mol}^{-1}}$$

E14.7(a) For a perfect gas at constant volume,

$$\frac{p}{T} = \frac{nR}{V} = \text{constant}, \quad \text{hence} \quad \frac{p_1}{T_1} = \frac{p_2}{T_2}$$

$$p_2 = \left(\frac{T_2}{T_1}\right) \times p_1 = \left(\frac{400\,\text{K}}{300\,\text{K}}\right) \times (1.00\,\text{atm}) = \boxed{1.33\,\text{atm}}$$

$$\Delta U = nC_{V,\text{m}}\Delta T\ [2.16b] = (n)\times\left(\frac{3}{2}R\right)\times(400\,\text{K} - 300\,\text{K})$$

$$= (1.00\,\text{mol})\times\left(\frac{3}{2}\right)\times(8.314\,\text{J}\,K^{-1}\,\text{mol}^{-1})\times(100\,\text{K})$$

$$= 1.25\times10^3\ \text{J} = \boxed{+1.25\,\text{kJ}}$$

$\boxed{w=0}$ [constant volume] $\qquad q = \Delta U - w\,[\text{first law}] = 1.25\,\text{kJ} - 0 = \boxed{+1.25\,\text{kJ}}$

E14.8(a) (a) $q = \Delta H$, since pressure is constant.

$$\Delta H = \int_{T_i}^{T_f} \text{d}H, \quad \text{d}H = nC_{p,\text{m}}\text{d}T$$

$$\text{d}(H/\text{J}) = \{20.17 + 0.3665(T/\text{K})\}\text{d}(T/\text{K})$$

$$\Delta(H/\text{J}) = \int_{T_i}^{T_f}(H/\text{J}) = \int_{298}^{473}\{20.17 + 0.3665(T/\text{K})\}\text{d}(T/\text{K})$$

$$= (20.17)\times(373-298) + \left(\frac{0.3665}{2}\right)\times\left(\frac{T}{\text{K}}\right)^2\Bigg|_{298}^{373}$$

$$= (1.51\bar{3}\times10^3) + (9.22\bar{2}\times10^3) = 1.07\times10^4$$

$q = \Delta H = \boxed{1.07\times10^4\,\text{J}} = \boxed{+10.7\ \text{kJ}}$ for 1.00 mol

$w = -p_{\text{ex}}\Delta V$ [14.8] where $p_{\text{ex}} = p$

$$w = -p\Delta V = -\Delta(pV)\ [\text{constant pressure}] = -\Delta(nRT)\ [\text{perfect gas}] = -nR\Delta T$$
$$= (-1.00\,\text{mol})\times(8.314\,\text{J}\,\text{K}^{-1}\,\text{mol}^{-1})\times(373\,\text{K} - 298\,\text{K}) = \boxed{-0.624\times10^3\ \text{J}} = \boxed{-0.624\ \text{kJ}}$$

$$\Delta U = q + w = (10.7\,\text{kJ}) - (0.624\text{kJ}) = \boxed{+10.1\,\text{kJ}}$$

(b) The energy and enthalpy of a perfect gas depend on temperature alone; hence it does not matter whether the temperature change is brought about at constant volume or constant pressure; ΔH and ΔU are the same.

$$\Delta H = \boxed{+10.7\ \text{kJ}},\quad \Delta U = \boxed{+10.1\ \text{kJ}}$$

Under constant volume, $w = \boxed{0}$.

$$q = \Delta U - w = \boxed{+10.1\ \text{kJ}}$$

E14.9(a)

$$q_p = C_p\Delta T\ [14.33\text{c}] = nC_{p,\text{m}}\Delta T = (3.0\ \text{mol})\times(29.4\ \text{J K}^{-1}\ \text{mol}^{-1})\times(25\ \text{K}) = \boxed{+2.2\ \text{kJ}}$$

$$\Delta H = q_p\ [14.29\text{b}] = \boxed{+2.2\ \text{kJ}}$$

$$\Delta U = \Delta H - \Delta(pV)\ [\text{From } H \equiv U + pV] = \Delta H - \Delta(nRT)\ [\text{perfect gas}] = \Delta H - nR\Delta T$$
$$= (2.2\ \text{kJ}) - (3.0\ \text{mol})\times(8.314\ \text{J K}^{-1}\ \text{mol}^{-1})\times(25\ \text{K}) = (2.2\ \text{kJ}) - (0.62\ \text{kJ}) = \boxed{+1.6\ \text{kJ}}$$

E14.10(a) The partition function is

$$q = \sum_j g_j \text{e}^{-\beta\varepsilon_j}$$

with degeneracies $g_j = 2J + 1$

so $$q = 4 + 2\text{e}^{-\beta\varepsilon}\quad [g(^2\text{P}_{3/2}) = 4,\ g(^2\text{P}_{1/2}) = 2]$$

$$U - U(0) = -\frac{N}{q}\frac{\text{d}q}{\text{d}\beta} = \frac{N\varepsilon \text{e}^{-\beta\varepsilon}}{2 + \text{e}^{-\beta\varepsilon}}$$

$$C_V = \left(\frac{\partial U}{\partial T}\right)_V = -k\beta^2\left(\frac{\partial U}{\partial \beta}\right)_V = \frac{2R(\varepsilon\beta)^2 \text{e}^{-\beta\varepsilon}}{(2 + \text{e}^{-\beta\varepsilon})^2}[N = N_\text{A}]$$

(a) Therefore, since at 500 K $\beta\varepsilon = 2.53\overline{5}$,

$$C_{V,\text{m}}/R = \frac{(2)\times(2.53\overline{5})^2\times(\text{e}^{-2.53\overline{5}})}{(2 + \text{e}^{-2.53\overline{5}})^2} = \boxed{0.236}$$

(b) At 900 K, when $\beta\varepsilon = 1.408$,

$$C_{V,\text{m}}/R = \frac{(2)\times(1.408)^2\times(\text{e}^{-1.408})}{(2 + \text{e}^{-1.408})^2} = \boxed{0.193}$$

COMMENT. $C_{V,m}$ is smaller at 900 K than at 500 K, for then the temperature is higher than the peak in the "two-level" heat capacity curve.

E14.11(a) Assuming that all rotational modes are active, we can draw up the following table for $C_{V,\text{m}}, C_{p,\text{m}}$ and γ with and without active vibrational modes.

	$C_{V,\text{m}}$	$C_{p,\text{m}}$	γ	Exptl	
$NH_3(\nu_V^* = 0)$	$3R$	$4R$	1.33	1.31	$\boxed{\text{closer}}$
$NH_3(\nu_V^* = 6)$	$9R$	$10R$	1.11		
$CH_4(\nu_V^* = 0)$	$3R$	$4R$	1.33	1.31	$\boxed{\text{closer}}$
$CH_4(\nu_V^* = 9)$	$12R$	$13R$	1.08		

The experimental values are obtained from Tables 14.5 and 14.6 assuming $C_{p,\mathrm{m}} = C_{V,\mathrm{m}} + R$. It is clear from the comparison in the above table that the vibrational modes are not active. This is confirmed by the experimental vibrational wavenumbers (see Herzberg, *Molecular Spectra and Molecular Structure II. Infrared and Raman Spectra of Polyatomic Molecules*, Princeton, NJ: Van Nostrand, 1945) all of which are much greater than kT at 298 K.

E14.12(a) We need to calculate $\Delta\varepsilon^2 = \langle\varepsilon^2\rangle - \langle\varepsilon\rangle^2$[14.25]. We assume that argon may be treated as a perfect gas and follow the results of Example 14.1. The root mean square deviation of the molecular energy is then

$$\sqrt{\Delta\varepsilon^2} = \sqrt{\tfrac{1}{2}k^2T^2} = \frac{kT}{\sqrt{2}} = \frac{1.381\times10^{-23}\ \mathrm{J\ K^{-1}}\times 298\ \mathrm{K}}{\sqrt{2}} = \boxed{2.91\times10^{-21}\ \mathrm{J}}$$

E14.13(a) For reversible adiabatic expansion,

$$T_\mathrm{f} = T_\mathrm{i}\left(\frac{V_\mathrm{i}}{V_\mathrm{f}}\right)^{1/c} \quad [14.37\mathrm{a}]$$

where $$c = \frac{C_{V,\mathrm{m}}}{R} = \frac{C_{p,\mathrm{m}} - R}{R} = \frac{(20.786 - 8.3145)\ \mathrm{J\ K^{-1}\ mol^{-1}}}{8.3145\ \mathrm{J\ K^{-1}\ mol^{-1}}} = 1.500$$

so the final temperature is

$$T_\mathrm{f} = (273.15\,\mathrm{K})\times\left(\frac{1.0\,\mathrm{dm^3}}{3.0\,\mathrm{dm^3}}\right)^{1/1.500} = \boxed{13\bar{1}\ \mathrm{K}}$$

E14.14(a) In an adiabatic process, the initial and final pressures are related by (eqn. 14.40)

$$p_\mathrm{f}V_\mathrm{f}^{\gamma} = p_\mathrm{i}V_\mathrm{i}^{\gamma}$$

where $$\gamma = \frac{C_p}{C_V} = \frac{C_V + nR}{C_V} = \frac{20.8\ \mathrm{J\ K^{-1}} + (1.0\ \mathrm{mol})(8.31\ \mathrm{J\ K^{-1}\ mol^{-1}})}{20.8\ \mathrm{J\ K^{-1}}} = 1.40$$

Find V_i from the perfect gas law.

$$V_\mathrm{i} = \frac{nRT_\mathrm{i}}{p_\mathrm{i}} = \frac{(1.0\ \mathrm{mol})\times(8.31\ \mathrm{J\ K^{-1}\ mol^{-1}})\times(300\ \mathrm{K})}{4.25\ \mathrm{atm}}\times\frac{1\ \mathrm{atm}}{1.013\times10^5\ \mathrm{Pa}}$$

$$V_\mathrm{i} = 5.7\bar{9}\times10^{-3}\ \mathrm{m^3}$$

so $$V_\mathrm{f} = V_\mathrm{i}\left(\frac{p_\mathrm{i}}{p_\mathrm{f}}\right)^{1/\gamma} = (5.7\bar{9}\times10^{-3}\ \mathrm{m^3})\times\left(\frac{4.25\ \mathrm{atm}}{2.50\ \mathrm{atm}}\right)^{1/1.40} = \boxed{0.0084\bar{6}\ \mathrm{m^3}}$$

Find the final temperature from the perfect gas law.

$$T_\mathrm{f} = \frac{p_\mathrm{f}V_\mathrm{f}}{nR} = \frac{(2.50\ \mathrm{atm})\times(0.0084\bar{6}\ \mathrm{m^3})}{(1.0\ \mathrm{mol})\times(8.31\ \mathrm{J\ K^{-1}\ mol^{-1}})}\times\frac{1.013\times10^5\ \mathrm{Pa}}{1\ \mathrm{atm}}$$

$$T_\mathrm{f} = \boxed{25\bar{7}\ \mathrm{K}}$$

Adiabatic work is (eqn. 14.36)

$$w = C_V\Delta T = (20.8\ \mathrm{J\ K^{-1}})\times(25\bar{7} - 300)\ \mathrm{K} = \boxed{-0.89\times10^3\ \mathrm{J}}$$

E14.15(a) Reversible adiabatic work is

$$w = C_V\Delta T\ [14.36] = n(C_{p,\mathrm{m}} - R)\times(T_\mathrm{f} - T_\mathrm{i})$$

where the temperatures are related by

$$T_f = T_i\left(\frac{V_i}{V_f}\right)^{1/c} \quad [14.37a] \quad \text{where} \quad c = \frac{C_{V,m}}{R} = \frac{C_{p,m} - R}{R} = 3.463$$

So $$T_f = [(27.0 + 273.15)\,\text{K}] \times \left(\frac{500 \times 10^{-3}\,\text{dm}^3}{3.00\,\text{dm}^3}\right)^{1/3.463} = 179\,\text{K}$$

and $$w = \left(\frac{2.45\,\text{g}}{44.0\,\text{g mol}^{-1}}\right) \times [(37.11 - 8.3145)\,\text{J K}^{-1}\text{mol}^{-1}] \times (179 - 300)\,\text{K} = \boxed{-194\,\text{J}}$$

E14.16(a) For reversible adiabatic expansion,

$$p_f V_f^{\gamma} = p_i V_i^{\gamma} \; [14.40], \quad \text{so } p_f = p_i\left(\frac{V_i}{V_f}\right)^{\gamma} = (67.4\,\text{kPa}) \times \left(\frac{0.50\,\text{dm}^3}{2.0\,\text{dm}^3}\right)^{1.4} = \boxed{9.7\,\text{kPa}}$$

E14.17(a) At constant pressure,

$$q = \Delta H = n\Delta_{\text{vap}}H^{\ominus} = (0.75\,\text{mol}) \times (30.0\,\text{kJ mol}^{-1}) = \boxed{22.5\,\text{kJ}}$$

and $$w = -p\Delta V \approx -pV_{\text{vapor}} = -nRT = -(0.75\,\text{mol}) \times (8.3145\,\text{J K}^{-1}\,\text{mol}^{-1}) \times (250\,\text{K})$$

$$w = -1.6 \times 10^3\,\text{J} = \boxed{-1.6\,\text{kJ}}$$

$$\Delta U = w + q = -1.6 + 22.5\,\text{kJ} = \boxed{20.9\,\text{kJ}}$$

COMMENT. Because the vapor is treated as a perfect gas, the specific value of the external pressure provided in the statement of the exercise does not affect the numerical value of the answer.

E14.18(a) The reaction is

$$C_6H_5C_2H_5(l) + \frac{21}{2}O_2(g) \rightarrow 8\,CO_2(g) + 5\,H_2O(l)$$

$$\Delta_c H^{\ominus} = 8\Delta_f H^{\ominus}(CO_2, g) + 5\Delta_f H^{\ominus}(H_2O, l) - \Delta_f H^{\ominus}(C_6H_5C_2H_5, l)$$

$$= [(8) \times (-393.51) + (5) \times (-285.83) - (-12.5)]\,\text{kJ mol}^{-1}$$

$$= \boxed{-4564.7\,\text{kJ mol}^{-1}}$$

E14.19(a) First $\Delta_f H[(CH_2)_3, g]$ is calculated, and then that result is used to calculate $\Delta_r H$ for the isomerization.

$$(CH_2)_3(g) + \frac{9}{2}O_2(g) \rightarrow 3\,CO_2(g) + 3\,H_2O(l) \quad \Delta_c H = -2091\,\text{kJ mol}^{-1}$$

$$\Delta_f H[(CH_2)_3, g] = -\Delta_c H + 3\Delta_f H(CO_2, g) + 3\Delta_f H(H_2O, g)$$

$$= [+2091 + (3) \times (-393.51) + (3) \times (-285.83)]\,\text{kJ mol}^{-1}$$

$$= \boxed{+53\,\text{kJ mol}^{-1}}$$

$$(CH_2)_3(g) \rightarrow C_3H_6(g) \quad \Delta_r H = ?$$

$$\Delta_r H = \Delta_f H(C_3H_6, g) - \Delta_f H[(CH_2)_3, g]$$

$$= (20.42 - 53)\,\text{kJ mol}^{-1} = \boxed{-33\,\text{kJ mol}^{-1}}$$

E14.20(a) Because $\Delta_f H^{\ominus}(H^+, \text{aq}) = 0$, the whole of $\Delta_f H^{\ominus}$ (HCl, aq) is ascribed to $\Delta_f H^{\ominus}(\text{Cl}^-, \text{aq})$. Therefore, $\Delta_f H^{\ominus}(\text{Cl}^-,\text{aq}) = \boxed{-167 \text{ kJ/mol}^{-1}}$.

E14.21(a) For naphthalene, the reaction is $C_{10}H_8(s) + 12O_2(g) \rightarrow 10CO_2(g) + 4H_2O(l)$.

A bomb calorimeter gives $q_V = n\Delta_c U^{\ominus}$ rather than $q_p = n\Delta_c H^{\ominus}$; thus we need

$$\Delta_c U^{\ominus} = \Delta_c H^{\ominus} - \Delta n_g RT \text{ [14.31]}, \quad \Delta n_g = -2\,\text{mol}$$

$$\Delta_c H^{\ominus} = -5157\,\text{kJ mol}^{-1} \text{ [Table 14.5]}$$

$$\Delta_c U^{\ominus} = (-5157\,\text{kJ mol}^{-1}) - (-2)\times(8.3\times10^{-3}\,\text{kJ K}^{-1}\,\text{mol}^{-1})\times(298\,\text{K})$$

$$= \boxed{-5152\,\text{kJ mol}^{-1}}$$

$$|q| = |q_V| = |n\Delta_c U^{\ominus}| = \left(\frac{120\times10^{-3}\,\text{g}}{128.18\,\text{g mol}^{-1}}\right)\times(5152\,\text{kJ mol}^{-1}) = 4.82\bar{3}\,\text{kJ}$$

$$C = \frac{|q|}{\Delta T} = \frac{4.82\bar{3}\,\text{kJ}}{3.05\,\text{K}} = \boxed{1.58\,\text{kJ K}^{-1}}$$

When phenol is used the reaction is

$$C_6H_5OH(s) + \frac{15}{2}O_2(g) \rightarrow 6CO_2(g) + 3H_2O(l)$$

$$\Delta_c H^{\ominus} = -3054\,\text{kJ mol}^{-1} \text{ [Table 14.5]}$$

$$\Delta_c U^{\ominus} = \Delta_c H^{\ominus} - \Delta n_g RT, \; \Delta n_g = -\frac{3}{2}$$

$$= (-3054\,\text{kJ mol}^{-1}) + \left(\frac{3}{2}\right)\times(8.314\times10^{-3}\,\text{kJ K}^{-1}\,\text{mol}^{-1})\times(298\,\text{K})$$

$$= -3050\,\text{kJ mol}^{-1}$$

$$|q| = \frac{150\times10^{-3}\,\text{g}}{94.12\,\text{g mol}^{-1}}\times(3050\,\text{kJ mol}^{-1}) = 4.86\,\text{kJ}$$

$$\Delta T = \frac{|q|}{C} = \frac{4.86\,\text{kJ}}{1.58\,\text{kJ K}^{-1}} = \boxed{+3.08\,\text{K}}$$

COMMENT. In this case $\Delta_c U^{\ominus}$ and $\Delta_c H^{\ominus}$ differed by about 0.1 per cent. Thus, to within 3 significant figures, it would not have mattered if we had used $\Delta_c H^{\ominus}$ instead of $\Delta_c U^{\ominus}$, but for very precise work it would.

E14.22(a) (a) reaction(3) = (–2) × reaction(1) + reaction(2) and $\Delta n_g = -1$

The enthalpies of reactions are combined in the same manner as the equations (Hess's law).

$$\Delta_r H^{\ominus}(3) = (-2)\times\Delta_r H^{\ominus}(1) + \Delta_r H^{\ominus}(2)$$

$$= [(-2)\times(-184.62) + (-483.64)]\,\text{kJ mol}^{-1}$$

$$= \boxed{-114.40\,\text{kJ mol}^{-1}}$$

$$\Delta_r U^{\ominus} = \Delta_r H^{\ominus} - \Delta n_g RT \;\; [14.31] = (-114.40\,\text{kJ mol}^{-1}) - (-1)\times(2.48\,\text{kJ mol}^{-1})$$

$$= \boxed{-111.92\,\text{kJ mol}^{-1}}$$

(b) $\Delta_f H^\ominus$ refers to the formation of one mole of the compound, hence

$$\Delta_f H^\ominus(\mathrm{J}) = \frac{\Delta_r H^\ominus(\mathrm{J})}{\nu_\mathrm{J}}$$

$$\Delta_f H^\ominus(\mathrm{HCl,g}) = \frac{-184.62}{2}\ \mathrm{kJ\ mol^{-1}} = \boxed{-92.31\ \mathrm{kJ\ mol^{-1}}}$$

$$\Delta_f H^\ominus(\mathrm{H_2O,g}) = \frac{-483.64}{2}\ \mathrm{kJ\ mol^{-1}} = \boxed{-241.82\ \mathrm{kJ\ mol^{-1}}}$$

E14.23(a) $\Delta_r H^\ominus = \Delta_r U^\ominus + \Delta n_g RT$ [14.31]; $\Delta n_g = +2$

$$= (-1373\ \mathrm{kJ\ mol^{-1}}) + 2\times(2.48\ \mathrm{kJ\ mol^{-1}}) = \boxed{-1368\ \mathrm{kJ\ mol^{-1}}}$$

COMMENT. As a number of these exercises have shown, the use of $\Delta_r H$ as an approximation for $\Delta_r U$ is often valid.

E14.24(a) (a) $\Delta_r H^\ominus = \sum_{\text{Products}} \nu\Delta_f H^\ominus - \sum_{\text{Reactants}} \nu\Delta_f H^\ominus$ [14.45]

$$\Delta_r H^\ominus(298\ \mathrm{K}) = [(-110.53)-(-241.82)]\ \mathrm{kJ\ mol^{-1}} = \boxed{+131.29\ \mathrm{kJ\ mol^{-1}}}$$

$$\Delta_r U^\ominus(298\ \mathrm{K}) = \Delta_r H^\ominus(298\ \mathrm{K}) - \Delta n_g RT\ [14.31]$$

$$= (131.29\ \mathrm{kJ\ mol^{-1}}) - (1)\times(2.48\ \mathrm{kJ\ mol^{-1}}) = \boxed{+128.81\ \mathrm{kJ\ mol^{-1}}}$$

(b) $\Delta_r H^\ominus(478K) = \Delta_r H^\ominus(298K) + (T_2 - T_1)\Delta_r C_p^\ominus$ [Example 14.6]

$$\Delta_r C_p^\ominus = C_{p,m}^\ominus(\mathrm{CO,g}) + C_{p,m}^\ominus(\mathrm{H_2,g}) - C_{p,m}^\ominus(\mathrm{C,gr}) - C_{p,m}^\ominus(\mathrm{H_2O,g})$$

$$= (29.14 + 28.82 - 8.53 - 33.58)\times10^{-3}\ \mathrm{kJ\ K^{-1}\ mol^{-1}}$$

$$= 15.85\times10^{-3}\ \mathrm{kJ\ K^{-1}\ mol^{-1}}$$

$$\Delta_r H^\ominus(478\ \mathrm{K}) = (131.29\ \mathrm{kJ\ mol^{-1}}) + (15.85\times10^{-3}\ \mathrm{kJ\ K^{-1}\ mol^{-1}})\times(180\ \mathrm{K})$$

$$= (131.29 + 2.85)\ \mathrm{kJ\ mol^{-1}} = \boxed{+134.14\ \mathrm{kJ\ mol^{-1}}}$$

$$\Delta_r U^\ominus(478\ \mathrm{K}) = \Delta_r H^\ominus(478\ \mathrm{K}) - (1)\times(8.31\times10^{-3}\ \mathrm{kJ\ K^{-1}\ mol^{-1}})\times(478\ \mathrm{K})$$

$$= (134.14 - 3.97)\ \mathrm{kJ\ mol^{-1}} = \boxed{+130.17\ \mathrm{kJ\ mol^{-1}}}$$

COMMENT. The differences in both $\Delta_r H^\ominus$ and $\Delta_r U^\ominus$ between the two temperatures are small and justify the use of the approximation that $\Delta_r C_p^\ominus$ is a constant.

E14.25(a) For the reaction $CH_4(g) + 2O_2(g) \rightarrow CO_2(g) + 2H_2O(g)$,

$$\Delta_r H^\ominus = \Delta_f H^\ominus(\mathrm{CO_2,g}) + 2\times\Delta_f H^\ominus(\mathrm{H_2O,g}) - \Delta_f H^\ominus(\mathrm{CH_4,g})$$

To calculate the enthalpy of reaction at 500 K, we first calculate its value at 298 K using data in Tables 14.5 and 14.6.

$$\Delta_r H^\ominus(298\ \mathrm{K}) = -393.51\ \mathrm{kJ\ mol^{-1}} + 2\times(-241.82\ \mathrm{kJ\ mol^{-1}}) - (-74.81\ \mathrm{kJ\ mol^{-1}}) = -802.34\ \mathrm{kJ\ mol^{-1}}$$

Then using data on the heat capacities of all the reacting substances, we can calculate the change in enthalpy, ΔH, of each substance as the temperature increases from 298 K to 500 K. The enthalpy of reaction at 500 K could be obtained by adding all these enthalpy changes to the enthalpy of reaction at 298 K:

$$\Delta_r H^\ominus(500\text{ K}) = \Delta_r H^\ominus(298\text{ K}) + \Delta H(CO_2, g) + 2\times\Delta H(H_2O, g) - \Delta H(CH_4, g) - 2\times\Delta H(O_2, g)$$

However, we need not calculate each individual ΔH value. It is more efficient to proceed as follows using eqn. 14.47:

$$\Delta_r H^\ominus(500\text{ K}) = \Delta_r H^\ominus(298\text{ K}) + \int_{298\text{K}}^{500\text{K}} \Delta_r C_p^\ominus \mathrm{d}T \quad [14.47]$$

We will express the temperature dependence of the heat capacities in the form of the equation given in Problem 14.20 as data for the heat capacities of the substances involved in this reaction are available only in that form. They are not available for all the substances in the form of the equation of Table 14.2.

If $C_{p,\mathrm{m}}^\ominus = \alpha + \beta T + \gamma T^2$ then $\Delta_r C_p^\ominus = \Delta\alpha + \Delta\beta T + \Delta\gamma T^2$ with $\Delta\alpha = \sum_J \nu_J \alpha_J$, etc.

Using the data given in Problem 14.20, we calculate

$$\Delta\alpha = (26.86 + 2\times30.36 - 14.16 - 2\times25.72)\text{ J K}^{-1}\text{ mol}^{-1} = 21.98\text{ J K}^{-1}\text{ mol}^{-1}$$

$$\Delta\beta = (6.97 + 2\times9.61 - 75.50 - 2\times12.98)\text{ mJ K}^{-2}\text{ mol}^{-1} = -75.27\text{ mJ K}^{-2}\text{ mol}^{-1}$$

$$\Delta\gamma = [-0.820 + 2\times1.184 - (-17.99) - 2\times(-3.862)]\ \mu\text{J K}^{-3}\text{ mol}^{-1} = 27.262\ \mu\text{J K}^{-3}\text{ mol}^{-1}$$

Integrating eqn. 14.47 between T_1(298 K) and T_2(500 K), we obtain

$$\Delta_r H^\ominus(500\text{ K}) = \Delta_r H^\ominus(298\text{ K}) + \Delta\alpha(500\text{ K} - 298\text{ K}) + \frac{1}{2}\Delta\beta[(500\text{ K})^2 - (298\text{ K})^2]$$

$$+\frac{1}{3}\Delta\gamma[(500\text{ K})^3 - (298\text{ K})^3] = \boxed{-803.07\text{ kJ mol}^{-1}}$$

E14.26(a) Since enthalpy is a state function, $\Delta_r H$ for the process (see Figure 14.1)

$$Mg^{2+}(g) + 2Cl(g) + 2e^- \rightarrow MgCl_2(aq)$$

is independent of path; therefore the change in enthalpy for the path on the left is equal to the change in enthalpy for the path on the right. All numerical values are in kJ mol^{-1}.

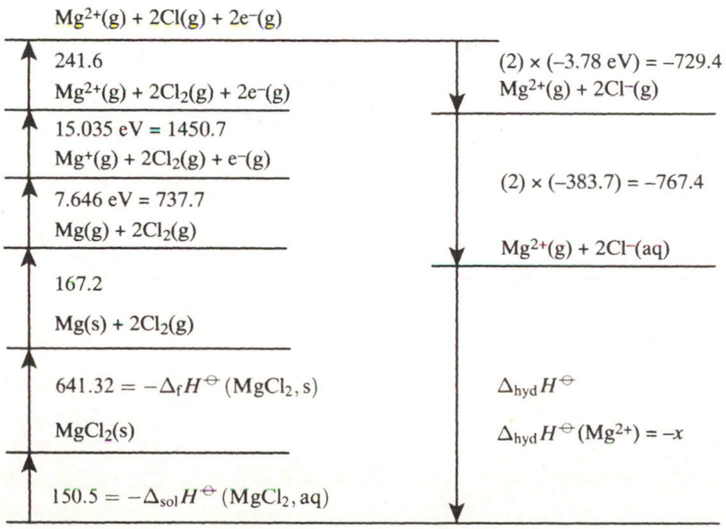

Figure 14.1

The cycle is the distance traversed upward along the left plus the distance traversed downward on the right. The sum of these distances is zero. Note that $E_{ea} = -\Delta_{eg}H$. Therefore, following the cycle up the left and down the right and using kJ units,

$$-(-150.5)-(-641.32)+(167.2)+(241.6)+(737.7+1450.7)$$

$$+2\times(-364.\bar{7})+2\times(-383.7)+\Delta_{hyd}H(Mg^{2+})=0$$

which yields $\Delta_{hyd}H(Mg^{2+}) = \boxed{-1892\,kJ\,mol^{-1}}$.

E14.27(a) The Joule-Thomson coefficient μ is the ratio of temperature change to pressure change under conditions of isenthalpic expansion. So

$$\mu = \left(\frac{\partial T}{\partial p}\right)_H [14.65] = \lim_{\Delta p\to 0}\left(\frac{\Delta T}{\Delta p}\right)_H \approx \frac{\Delta T}{\Delta p} \quad [\mu \text{ constant over temperature range}]$$

$$\mu = \frac{-22\text{ K}}{-31\text{atm}} = \boxed{0.71\text{ K atm}^{-1}}$$

E14.28(a) See Example 15.6 and Exercise E14.29(a). The internal pressure of a van der Waals gas is $\pi_T = a/V_m^2$. The molar volume can be estimated from the perfect gas equation. We will assume that the temperature of 298 K given in the statement of the exercise was a typographical error and instead do the calculation for 398 K. Water is not a vapor at 1.00 bar and 298 K.

$$V_m = \frac{RT}{p} = \frac{0.08206\text{ dm}^3\text{ atm K}^{-1}\text{ mol}^{-1}\times 398\text{ K}}{1.00\text{ bar}\times\left(\frac{1.000\text{ atm}}{1.013\text{ bar}}\right)} = 33.09\text{ dm}^3\text{ mol}^{-1}$$

$$\pi_T = \frac{a}{V_m^2} = \frac{5.464\text{ atm dm}^6\text{ mol}^{-2}}{(33.09\text{ dm}^3\text{ mol}^{-1})^2} = 4.99\times10^{-3}\text{ atm} = \boxed{5.06\text{ mbar}}$$

E14.29(a) The internal energy is a function of temperature and volume, $U_m = U_m(T, V_m)$, so

$$dU_m = \left(\frac{\partial U_m}{\partial T}\right)_{V_m} dT + \left(\frac{\partial U_m}{\partial V_m}\right)_T dV_m \qquad [\pi_T = (\partial U_m/\partial V_m)_T]$$

For an isothermal expansion, $dT = 0$; hence,

$$dU_m = \left(\frac{\partial U_m}{\partial V_m}\right)_T dV = \pi_T\, dV_m = \frac{a}{V_m^2} dV_m$$

$$\Delta U_m = \int_{V_{m,1}}^{V_{m,2}} dU_m = \int_{V_{m,2}}^{V_{m,1}} \frac{a}{V_m^2} dV_m = a\int_{1.00\text{ dm}^3\text{ mol}^{-1}}^{20.0\text{ dm}^3\text{ mol}^{-1}} \frac{dV_m}{V_m^2} = -\frac{a}{V_m}\Bigg|_{1.00\text{ dm}^3\text{ mol}^{-1}}^{20.0\text{ dm}^3\text{ mol}^{-1}}$$

$$= -\frac{a}{20.0\text{ dm}^3\text{ mol}^{-1}} + \frac{a}{1.00\text{ dm}^3\text{ mol}^{-1}} = \frac{19.0a}{20.0\text{ dm}^3\text{ mol}^{-1}} = 0.950\,a\text{ mol dm}^{-3}$$

From Table 8.7, $a = 1.352\text{ dm}^6\text{ atm mol}^{-1}$.

$$\Delta U_m = (0.950\text{ mol dm}^{-3})\times(1.352\text{ dm}^6\text{ atm mol}^{-2})$$

$$= (1.28\bar{4}\text{ dm}^3\text{ atm mol}^{-1})\times\left(\frac{1\text{ m}}{10\text{ dm}}\right)^3\times\left(\frac{1.013\times10^5\text{ Pa}}{\text{atm}}\right) = \boxed{+130.\bar{1}\text{ J mol}^{-1}}$$

$$w = -\int p\,\mathrm{d}V_\mathrm{m} \quad \text{where} \quad p = \frac{RT}{V_\mathrm{m}-b} - \frac{a}{V_\mathrm{m}^2} \text{ for a van der Waals gas. Hence,}$$

$$w = -\int\left(\frac{RT}{V_\mathrm{m}-b}\right)\mathrm{d}V_\mathrm{m} + \int\frac{a}{V_\mathrm{m}^2}\mathrm{d}V_\mathrm{m} = -q + \Delta U_\mathrm{m}$$

Therefore,

$$q = \int_{1.00\,\mathrm{dm^3\,mol^{-1}}}^{20.0\,\mathrm{dm^3\,mol^{-1}}}\left(\frac{RT}{V_\mathrm{m}-b}\right)\mathrm{d}V_\mathrm{m} = RT\ln(V_\mathrm{m}-b)\Big|_{1.00\,\mathrm{dm^3\,mol^{-1}}}^{20.0\,\mathrm{dm^3\,mol^{-1}}}$$

$$= (8.314\,\mathrm{J\,K^{-1}\,mol^{-1}})\times(298\,\mathrm{K})\times\ln\left(\frac{20.0-3.9\times10^{-2}}{1.00-3.9\times10^{-2}}\right) = \boxed{+7.52\times10^3\,\mathrm{J\,mol^{-1}}}$$

and $\quad w = -q + \Delta U_\mathrm{m} = -(7.52\times10^3\,\mathrm{J\,mol^{-1}}) + (130.\bar{1}\,\mathrm{J\,mol^{-1}}) = \boxed{-7.39\times10^3\,\mathrm{J\,mol^{-1}}}$

E14.30(a) $\quad \alpha = \left(\frac{1}{V}\right)\left(\frac{\partial V}{\partial T}\right)_p$ [14.58]; $\quad \alpha_{320} = \left(\frac{1}{V_{320}}\right)\left(\frac{\partial V}{\partial T}\right)_{p,320}$

$$\left(\frac{\partial V}{\partial T}\right)_p = V_{300}(3.9\times10^{-4}/\mathrm{K} + 2.96\times10^{-6}T/\mathrm{K}^2)$$

$$\left(\frac{\partial V}{\partial T}\right)_{p,320} = V_{300}(3.9\times10^{-4}/\mathrm{K} + 2.96\times10^{-6}\times320/\mathrm{K}) = 1.34\times10^{-3}\,\mathrm{K}^{-1}V_{300}$$

$$V_{320} = V_{300}\{(0.75) + (3.9\times10^{-4})\times(320) + (1.48\times10^{-6})\times(320)^2\} = (V_{300})\times(1.02\bar{6})$$

so $\quad \alpha_{320} = \left(\frac{1}{V_{320}}\right)\left(\frac{\partial V}{\partial T}\right)_{p,320} = \left(\frac{1}{1.02\bar{6}V_{300}}\right)\times(1.34\times10^{-3}\,\mathrm{K}^{-1}\,V_{300})$

$$\alpha_{320} = \frac{1.34\times10^{-3}\,\mathrm{K}^{-1}}{1.02\bar{6}} = \boxed{1.31\times10^{-3}\,\mathrm{K}^{-1}}$$

COMMENT. Knowledge of the density at 300 K is not required to solve this exercise, but it would be required to obtain numerical values of the volumes at the two temperatures.

E14.31(a) The isothermal compressibility is

$$\kappa_T = -\left(\frac{1}{V}\right)\left(\frac{\partial V}{\partial p}\right)_T \text{ [14.59] so } \left(\frac{\partial V}{\partial p}\right)_T = -\kappa_T V$$

At constant temperature,

$$\mathrm{d}V = \left(\frac{\partial V}{\partial p}\right)_T \mathrm{d}p, \quad \text{so} \quad \mathrm{d}V = -\kappa_T V\,\mathrm{d}p \quad \text{or} \quad \frac{\mathrm{d}V}{V} = -\kappa_T\,\mathrm{d}p$$

Substituting $V = \frac{m}{\rho}$ yields $\mathrm{d}V = -\frac{m}{\rho^2}\mathrm{d}\rho$; $\frac{\mathrm{d}V}{V} = -\frac{\mathrm{d}\rho}{\rho} = -\kappa_T\mathrm{d}p$.

Therefore, $\frac{\delta\rho}{\rho} \approx \kappa_T\delta p$.

For $\frac{\delta\rho}{\rho} = 0.10\times10^{-2} = 1.0\times10^{-3}$, $\delta p \approx \frac{1.0\times10^{-3}}{\kappa_T} = \frac{1.0\times10^{-3}}{4.96\times10^{-5}\,\mathrm{atm}^{-1}} = \boxed{2.0\times10^3\,\mathrm{atm}}$.

E14.32(a) The isothermal Joule-Thomson coefficient is

$$\mu_T = \left(\frac{\partial H_m}{\partial p}\right)_T = -\mu C_{p,m}\ [14.69] = (-0.25\,\text{K atm}^{-1})\times(29\,\text{J K}^{-1}\,\text{mol}^{-1}) = \boxed{-7.2\,\text{J atm}^{-1}\,\text{mol}^{-1}}$$

$$dH = n\left(\frac{\partial H_m}{\partial p}\right)_T dp = -n\mu C_{p,m}\,dp$$

$$\Delta H = \int_{p_1}^{p_2}(-n\mu C_{p,m})\,dp = -n\mu C_{p,m}(p_2 - p_1)\ [\mu \text{ and } C_p \text{ are constant}]$$

$$\Delta H = -(10.0\,\text{mol})\times(+7.2\,\text{J atm}^{-1}\,\text{mol}^{-1})\times(-85\,\text{atm}) = +6.1\,\text{kJ}$$

so $q(\text{supplied}) = +\Delta H = \boxed{+6.1\,\text{kJ}}$.

Solutions to problems

Assume that all gases are perfect unless stated otherwise. Unless otherwise stated, thermochemical data are for 298 K.

Solutions to numerical problems

P14.1 The coefficient of thermal expansion is

$$\alpha = \frac{1}{V}\left(\frac{\partial V}{\partial T}\right)_p \approx \frac{\Delta V}{V\Delta T} \quad \text{so} \quad \Delta V \approx \alpha V \Delta T$$

This change in volume is equal to the change in height (sea level rise, Δh) times the area of the ocean (assuming that area remains constant). We will use α of pure water, although the oceans are complex solutions. For a 2°C rise in temperature,

$$\Delta V = (2.1\times10^{-4}\,\text{K}^{-1})\times(1.37\times10^{9}\,\text{km}^3)\times(2.0\,\text{K}) = 5.8\times10^{5}\,\text{km}^3$$

so $$\Delta h = \frac{\Delta V}{A} = 1.6\times10^{-3}\,\text{km} = \boxed{1.6\,\text{m}}$$

Since the rise in sea level is directly proportional to the rise in temperature, $\Delta T = 1°\text{C}$ would lead to $\Delta h = \boxed{0.80\,\text{m}}$ and $\Delta T = 3.5°\text{C}$ would lead to $\Delta h = \boxed{2.8\,\text{m}}$.

COMMENT. More detailed models of climate change predict somewhat smaller rises but the same order of magnitude.

P14.3 To explore which of the two proposed equations best fits the data, we have used PSI-PLOT®. The parameters obtained with the fitting process to eqn. 14.34, along with their standard deviations, are given in the following table.

Parameters	a	$b/10^{-3}\,\text{K}^{-1}$	$c/10^{5}\,\text{K}^{2}$
Values	28.796	27.89	−1.490
Std dev of parameter	0.820	0.91	0.6480

The correlation coefficient is 0.99947. The parameters and their standard deviations obtained with the fitting process to the suggested alternate equation are as follows.

Parameters	α	$\beta/10^{-3}\,\text{K}^{-1}$	$\gamma/10^{-6}\,\text{K}^{-2}$
Values	24.636	38.18	−6.495
Std dev of parameter	0.437	1.45	1.106

The correlation coefficient is 0.99986. It appears that the alternate form for the heat capacity equation fits the data slightly better, but there is very little difference.

P14.5

$$q^{\mathrm{E}} = \sum_j g_j \mathrm{e}^{-\beta\varepsilon_j} = 2 + 2\mathrm{e}^{-\beta\varepsilon},\ \varepsilon = \Delta\varepsilon = 121.1\,\mathrm{cm}^{-1}$$

$$U_{\mathrm{m}} - U_{\mathrm{m}}(0) = -\frac{N_{\mathrm{A}}}{q^{\mathrm{E}}}\left(\frac{\partial q^{\mathrm{E}}}{\partial\beta}\right)_V = \frac{2N_{\mathrm{A}}\varepsilon \mathrm{e}^{-\beta\varepsilon}}{q^{\mathrm{E}}} = 1$$

$$C_{V,\mathrm{m}} = -k\beta^2\left(\frac{\partial U_{\mathrm{m}}}{\partial\beta}\right)_V \quad [14.26]$$

Let $x = \beta\varepsilon$; then $\mathrm{d}\beta = \frac{1}{\varepsilon}\mathrm{d}x$.

$$C_{V,\mathrm{m}} = -k\left(\frac{x}{\varepsilon}\right)^2 \varepsilon\frac{\partial}{\partial x}\left(\frac{N_{\mathrm{A}}\varepsilon \mathrm{e}^{-x}}{1+\mathrm{e}^{-x}}\right) = -N_{\mathrm{A}}kx^2 \times \frac{\partial}{\partial x}\left(\frac{\mathrm{e}^{-x}}{1+\mathrm{e}^{-x}}\right) = R\left(\frac{x^2\mathrm{e}^{-x}}{(1+\mathrm{e}^{-x})^2}\right)$$

Therefore,

$$C_{V,\mathrm{m}}/R = \frac{x^2\mathrm{e}^{-x}}{(1+\mathrm{e}^{-x})^2},\quad x = \beta\varepsilon$$

We then draw up the following table.

T / K	100	298	600
(kT/hc) / cm^{-1}	69.5	207	417
x	1.741	0.585	0.290
$C_{V,\mathrm{m}}/R$	0.385	0.0786	0.0206
$C_{V,\mathrm{m}}$ / (J K^{-1} mol^{-1})	3.20	0.654	0.171

COMMENT. Note that the double degeneracies do not affect the results because the two factors of 2 in q cancel when U is formed. In the range of temperature specified, the electronic contribution to the heat capacity decreases with increasing temperature.

P14.7 The energy expression for a particle on a ring is

$$E = \frac{\eta^2 m_l^2}{2I} \quad [\text{Section 3.3}]$$

Therefore

$$q = \sum_{m=-\infty}^{\infty} \mathrm{e}^{-m_l^2\eta^2/2IkT} = \sum_{m=-\infty}^{\infty} \mathrm{e}^{-\beta m_l^2\eta^2/2I}$$

The summation may be approximated by an integration.

$$q \approx \frac{1}{\sigma}\int_{-\infty}^{\infty} \mathrm{e}^{-m_l^2\eta^2/2IkT}\,\mathrm{d}m_l = \frac{1}{\sigma}\left(\frac{2IkT}{\eta^2}\right)^{1/2}\int_{-\infty}^{\infty}\mathrm{e}^{-x^2}\mathrm{d}x \approx \frac{1}{\sigma}\left(\frac{2\pi IkT}{\eta^2}\right)^{1/2} = \frac{1}{\sigma}\left(\frac{2\pi I}{\eta^2\beta}\right)^{1/2}$$

$$U - U(0) = -N\frac{\partial \ln q}{\partial\beta} = -N\frac{\partial}{\partial\beta}\ln\frac{1}{\sigma}\left(\frac{2\pi I}{\eta^2\beta}\right)^{1/2} = \frac{N}{2\beta} = \frac{1}{2}NkT = \frac{1}{2}RT\ (N = N_{\mathrm{A}})$$

$$C_{V,\mathrm{m}} = \left(\frac{\partial U_{\mathrm{m}}}{\partial T}\right)_V = \frac{1}{2}R = \boxed{4.2\ \mathrm{J\,K^{-1}\,mol^{-1}}}$$

P14.9 The vibrational temperature is defined by $k\theta_V = hc\tilde{\nu}$, so a vibration with θ_V less than 1000 K has a wavenumber less than

$$\tilde{\nu} = \frac{k\theta_V}{hc} = \frac{(1.381\times10^{-23}\ \mathrm{J\,K^{-1}})\times(1000\ \mathrm{K})}{(6.626\times10^{-34}\ \mathrm{J\,s})\times(2.998\times10^{10}\ \mathrm{cm\,s^{-1}})} = 695.2\ \mathrm{cm^{-1}}$$

There are seven such wavenumbers listed among those for C_{60}: two T_{1u}, a T_{2u}, a G_u, and three H_u. The number of *modes* involved, ν_V^*, must take into account the degeneracy of these vibrational energies.

$$\nu_V^* = 2(3)+1(3)+1(4)+3(5) = \boxed{28}$$

The molar heat capacity of a molecule is roughly

$$C_{V,\mathrm{m}} = \frac{1}{2}(3+\nu_R^*+2\nu_V^*)R\,[14.27] = \frac{1}{2}(3+3+2\times28)R = 31R = 31(8.3145\ \mathrm{J\,mol^{-1}\,K^{-1}})$$

$$= \boxed{258\ \mathrm{J\,mol^{-1}\,K^{-1}}}$$

P14.11 $$w = -\int_{V_1}^{V_2} p\,\mathrm{d}V \quad \text{with} \quad p = \frac{nRT}{V-nb} - \frac{n^2a}{V^2}\ \text{[van der Waals equation]}$$

Therefore,

$$w = -nRT\int_{V_1}^{V_2}\frac{\mathrm{d}V}{V-nb} + n^2a\int_{V_1}^{V_2}\frac{\mathrm{d}V}{V^2} = \boxed{-nRT\ln\left(\frac{V_2-nb}{V_1-nb}\right) - n^2a\left(\frac{1}{V_2}-\frac{1}{V_1}\right)}$$

This expression can be interpreted more readily if we assume that $V >> nb$, which is certainly valid at all but the highest pressures. Then using the first term of the Taylor series expansion,

$$\ln(1-x) = -x - \frac{x^2}{2}\cdots \quad \text{for } |x| << 1$$

$$\ln(V-nb) = \ln V + \ln\left(1-\frac{nb}{V}\right) \approx \ln V - \frac{nb}{V}$$

and after substitution,

$$w \approx -nRT\ln\left(\frac{V_2}{V_1}\right) + n^2bRT\left(\frac{1}{V_2}-\frac{1}{V_1}\right) - n^2a\left(\frac{1}{V_2}-\frac{1}{V_1}\right)$$

$$\approx -nRT\ln\left(\frac{V_2}{V_1}\right) - n^2(a-bRT)\left(\frac{1}{V_2}-\frac{1}{V_1}\right)$$

$$\approx +w_0 - n^2(a-bRT)\left(\frac{1}{V_2}-\frac{1}{V_1}\right) = \text{perfect gas value + van der Waals correction}$$

w_0, the perfect gas value, is negative in expansion and positive in compression. Considering the correction term, in expansion $V_2 > V_1$, so $\left(\frac{1}{V_2}-\frac{1}{V_1}\right) < 0$. If attractive forces predominate, $a > bRT$, and the work done by the van der Waals gas is less in magnitude (less negative) than the perfect gas, the gas cannot easily expand. If repulsive forces predominate, $bRT > a$, and the work done by the van der Waals gas is greater in magnitude than the perfect gas, the gas easily expands. In the numerical calculations, consider a doubling of the initial volume.

(a) $$w_0 = -nRT\ln\left(\frac{V_f}{V_i}\right) = (-1.0\,\text{mol}^{-1})\times(8.314\,\text{J K}^{-1}\,\text{mol}^{-1})\times(298\,\text{K})\times\ln\left(\frac{2.0\,\text{dm}^3}{1.0\,\text{dm}^3}\right)$$

$$w_0 = -1.7\bar{2}\times10^3\,\text{J} = \boxed{-1.7\,\text{kJ}}$$

(b) $$w = w_0 - (1.0\,\text{mol})^2\times[0-(5.11\times10^{-2}\,\text{dm}^3\,\text{mol}^{-1})\times(8.314\,\text{J K}^{-1}\text{mol}^{-1})\times(298\,\text{K})]$$
$$\times\left(\frac{1}{2.0\,\text{dm}^3}-\frac{1}{1.0\,\text{dm}^3}\right) = (-1.7\bar{2}\times10^3\,\text{J})-(63\,\text{J}) = -1.7\bar{8}\times10^3\,\text{J} = \boxed{-1.8\,\text{kJ}}$$

(c) $$w = w_0 - (1.0\,\text{mol})^2\times(4.2\,\text{dm}^6\,\text{atm mol}^{-2})\times\left(\frac{1}{2.0\,\text{dm}^3}-\frac{1}{1.0\,\text{dm}^3}\right)$$

$$w = w_0 + 2.1\,\text{dm}^3\,\text{atm}$$
$$= (-1.7\bar{2}\times10^3\,\text{J}) + (2.1\,\text{dm}^3\,\text{atm})\times\left(\frac{1\,\text{m}}{10\,\text{dm}}\right)^3\times\left(\frac{1.01\times10^5\,\text{Pa}}{1\,\text{atm}}\right)$$
$$= (-1.7\bar{2}\times10^3\,\text{J}) + (0.21\times10^3\,\text{J}) = \boxed{-1.5\,\text{kJ}}$$

Schematically, the indicator diagrams for cases (a), (b), and (c) would appear as in Figure 14.2. For case (b) the pressure is always greater than the perfect gas pressure, and for case (c) the pressure is always less. Therefore,

$$\int_{V_1}^{V_2} p\,\mathrm{d}V(\text{c}) < \int_{V_1}^{V_2} p\,\mathrm{d}V(\text{a}) < \int_{V_1}^{V_2} p\,\mathrm{d}V(\text{b})$$

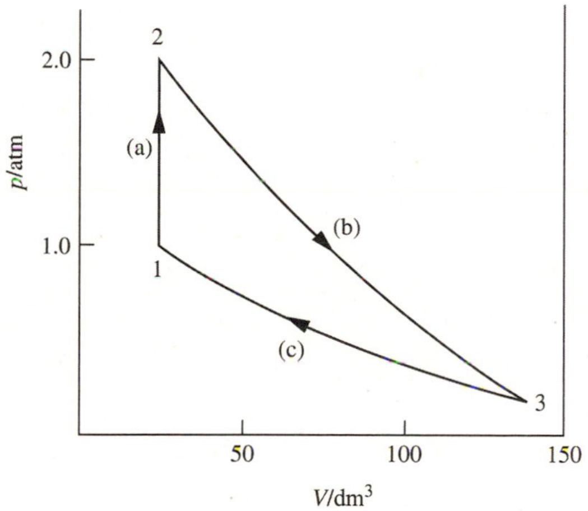

Figure 14.2

P14.13 $Cr(C_6H_6)_2(s) \rightarrow Cr(s) + 2C_6H_6(g) \quad \Delta n_g = +2\,\text{mol}$

$$\Delta_r H^\ominus = \Delta_r U^\ominus + 2RT, \text{ from [14.31]}$$
$$= (8.0\,\text{kJ mol}^{-1}) + (2)\times(8.314\,\text{J K}^{-1}\,\text{mol}^{-1})\times(583\,\text{K}) = \boxed{+17.7\,\text{kJ mol}^{-1}}$$

In terms of enthalpies of formation,

$$\Delta_r H^\ominus = (2)\times\Delta_f H^\ominus(\text{benzene}, 583\,\text{K}) - \Delta_f H^\ominus(\text{metallocene}, 583\,\text{K})$$

or $$\Delta_r H^\ominus(\text{metallocene}, 583\,\text{K}) = 2\Delta_f H^\ominus(\text{benzene}, 583\,\text{K}) - 17.7\,\text{kJ mol}^{-1}$$

The enthalpy of formation of benzene gas at 583 K is related to its value at 298 K by

$$\Delta_f H^\ominus(\text{benzene}, 583\,\text{K}) = \Delta_f H^\ominus(\text{benzene}, 298\,\text{K})$$
$$+(T_b - 298\,\text{K})C_{p,m}(\text{l}) + \Delta_{\text{vap}}H^\ominus + (583\,\text{K} - T_b)C_{p,m}(\text{g})$$
$$-6\times(583\,\text{K} - 298\,\text{K})C_{p,m}(\text{gr}) - 3\times(583\,\text{K} - 298\,\text{K})C_{p,m}(\text{H}_2,\text{g})$$

where T_b is the boiling temperature of benzene (353 K). We shall assume that the heat capacities of graphite and hydrogen are approximately constant in the range of interest and use their values from Table 14.6.

$$\begin{aligned}\Delta_f H^\ominus(\text{benzene}, 583\,\text{K}) &= (49.0\,\text{kJ mol}^{-1}) + (353-298)\,\text{K}\times(136.1\,\text{J K}^{-1}\,\text{mol}^{-1})\\ &\quad + (30.8\,\text{kJ mol}^{-1}) + (583-353)\,\text{K}\times(81.67\,\text{J K}^{-1}\,\text{mol}^{-1})\\ &\quad - (6)\times(583-298)\,\text{K}\times(8.53\,\text{J K}^{-1}\,\text{mol}^{-1})\\ &\quad - (3)\times(583-298)\,\text{K}\times(28.82\,\text{J K}^{-1}\,\text{mol}^{-1})\\ &= \{(49.0)+(7.49)+(18.78)+(30.8)-(14.59)-(24.64)\}\,\text{kJ mol}^{-1}\\ &= +66.8\,\text{kJ mol}^{-1}\end{aligned}$$

Therefore, $\Delta_f H^\ominus(\text{metallocene}, 583\,\text{K}) = (2\times 66.8 - 17.7)\,\text{kJ mol}^{-1} = \boxed{+116.0\,\text{kJ mol}^{-1}}$.

P14.15 (a) and (b). The table displays computed enthalpies of formation (semi-empirical, PM3 level, PC Spartan Pro), enthalpies of combustion based on them (and on experimental enthalpies of formation of $H_2O(l)$ and $CO_2(g)$, –285.83 and –393.51 kJ mol^{-1}, respectively), experimental enthalpies of combustion (Table 14.5), and the relative error in enthalpy of combustion.

Compound	$\Delta_f H^\ominus$ / kJ mol^{-1}	$\Delta_c H^\ominus$ / kJ mol^{-1}(calc.)	$\Delta_c H^\ominus$ / kJ mol^{-1}(expt.)	% error
$CH_4(g)$	–54.45	–910.72	–890	2.33
$C_2H_6(g)$	–75.88	–1568.63	–1560	0.55
$C_3H_8(g)$	–98.84	–2225.01	–2220	0.23
$C_4H_{10}(g)$	–121.60	–2881.59	–2878	0.12
$C_5H_{12}(g)$	–142.11	–3540.42	–3537	0.10

The combustion reactions can be expressed as

$$C_nH_{2n+2}(g) + \left(\frac{3n+1}{2}\right)O_2(g) \rightarrow nCO_2(g) + (n+1)H_2O(l)$$

The enthalpy of combustion, in terms of enthalpies of reaction, is

$$\Delta_c H^\ominus = n\Delta_f H^\ominus(CO_2) + (n+1)\Delta_f H^\ominus(H_2O) - \Delta_f H^\ominus(C_nH_{2n+2})$$

where we have left out $\Delta_f H^\ominus(O_2) = 0$. The % error is defined as

$$\%\ \text{error} = \frac{\Delta_c H^\ominus(\text{calc}) - \Delta_c H^\ominus(\text{expt.})}{\Delta_c H^\ominus(\text{expt.})}\times 100\%$$

The agreement is quite good.

(c) If the enthalpy of combustion is related to the molar mass by

$$\Delta_c H^\ominus = k[M/(\text{g mol}^{-1})]^n$$

then one can take the natural log of both sides to obtain

$$\ln|\Delta_c H^\ominus| = \ln|k| + n\ln M/(\text{g mol}^{-1})$$

Thus, if one plots ln $|\Delta_c H^{\ominus}|$ against ln $[M/(\text{g mol}^{-1})]$, one ought to obtain a straight line with slope n and y-intercept ln $|k|$. Draw up the following table.

Compound	$M/(\text{g mol}^{-1})$	$\Delta_c H$ / kJ mol^{-1}	ln $M/(\text{g mol}^{-1})$	ln $\left|\Delta_c H^{\ominus}/\text{kJ mol}^{-1}\right|$
CH_4(g)	16.04	–910.72	2.775	6.814
C_2H_6(g)	30.07	–1568.63	3.404	7.358
C_3H_8(g)	44.10	–2225.01	3.786	7.708
C_4H_{10}(g)	58.12	–2881.59	4.063	7.966
C_5H_{12}(g)	72.15	–3540.42	4.279	8.172

The plot is shown in Figure 14.3.

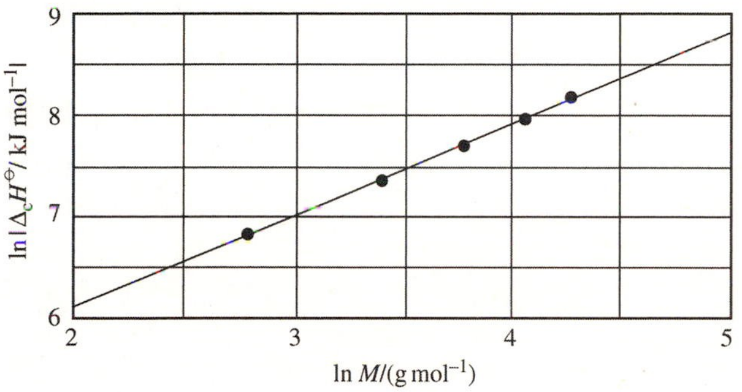

Figure 14.3

The linear least-squares fit equation is

$$\ln|\Delta_c H^{\ominus}/\text{kJ mol}^{-1}| = 4.30 + 0.903 \ln M/(\text{g mol}^{-1}) \quad R^2 = 1.00$$

These compounds support the proposed relationships, with

$$n = \boxed{0.903} \quad \text{and} \quad k = -e^{4.30}\ \text{kJ mol}^{-1} = \boxed{-73.7\ \text{kJ mol}^{-1}}$$

The agreement of these theoretical values of k and n with the experimental values obtained in P14.14 is rather good.

P14.17 We must relate the formation of $DyCl_3$

$$\text{Dy(s)} + 1.5\text{Cl}_2\text{(g)} \rightarrow \text{DyCl}_3\text{(s)}$$

to the three reactions for which for which we have information. This reaction can be seen as a sequence of reaction (2), three times reaction (3), and the reverse of reaction (1), so

$$\Delta_f H^{\ominus}(\text{DyCl}_3, \text{s}) = \Delta_r H^{\ominus}(2) + 3\Delta_r H^{\ominus}(3) - \Delta_r H^{\ominus}(1)$$

$$\Delta_f H^{\ominus}(\text{DyCl}_3, \text{s}) = [-699.43 + 3(-158.31) - (-180.06)]\ \text{kJ mol}^{-1}$$

$$= \boxed{-994.30\ \text{kJ mol}^{-1}}$$

P14.19 (a) The probability distribution of rotational energy levels is the Boltzmann factor of each level, weighted by the degeneracy, over the partition function

$$p_J^{\text{R}}(T) = \frac{g(J)e^{-\varepsilon_J/kT}}{q^{\text{R}}} = \frac{(2J+1)e^{-hc\tilde{B}J(J+1)/kT}}{\sum_{J=0}(2J+1)e^{-hc\tilde{B}J(J+1)/kT}} \quad [13.19]$$

It is conveniently plotted against J at several temperatures using mathematical software. This distribution at 100 K is shown in Figure 14.4 as both a bar plot and a line plot.

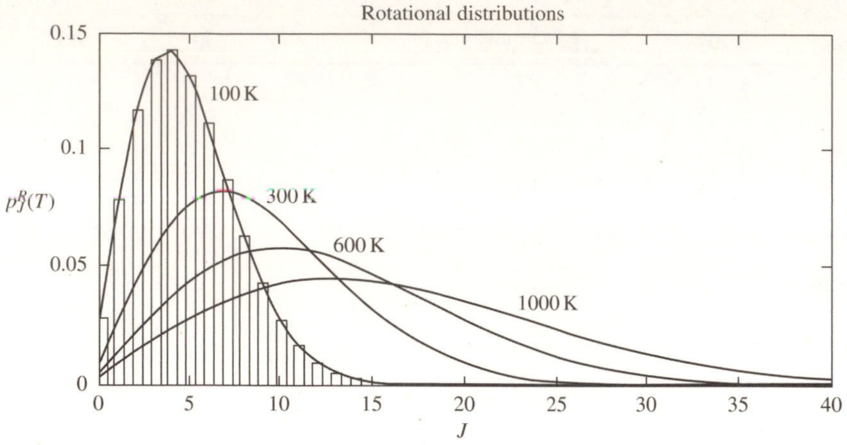

Figure 14.4

The plots show that higher rotational states become more heavily populated at higher temperature. Even at 100 K, the most populated state has 4 quanta of rotational energy; it is elevated to 13 quanta at 1000 K.

Values of the vibrational state probability distribution,

$$p_v^V(T) = \frac{e^{-\varepsilon_J/kT}}{q^V} = e^{-vhc\tilde{v}/kT}(1 - e^{-hc\tilde{v}/kT}) \quad [13.24]$$

are conveniently tabulated against v at several temperatures. Computations may be discontinued when values drop below some small number like 10^{-7}.

		$p_v^V(T)$		
v	100 K	300 K	600 K	1000 K
0	1	1	0.095	0.956
1	2.77×10^{-14}	3.02×10^{-5}	5.47×10^{-3}	0.042
2		9.15×10^{-10}	3.01×10^{-5}	1.86×10^{-3}
3			1.65×10^{-7}	8.19×10^{-5}
4				3.61×10^{-6}
5				1.59×10^{-7}

Only the state $v = 0$ is appreciably populated below 1000 K and even at 1000 K only 4% of the molecules have 1 quanta of vibrational energy.

(b) The classical (equipartition) rotational partition function is

$$q_{\text{classical}}^R(T) = \frac{kT}{hcB} = \frac{T}{\theta_R} \quad [13.21b]$$

where θ_R is the rotational temperature. We would expect the partition function to be well approximated by this expression for temperatures much greater than the rotational temperature.

$$\theta_R = \frac{hcB}{k} = \frac{(6.626\times10^{-34}\text{ J s})\times(2.998\times10^{10}\text{ cm s}^{-1})\times(1.931\text{ cm}^{-1})}{1.381\times10^{-23}\text{J K}^{-1}}$$

$$\theta_R = 2.779\,\text{K}$$

In fact $\theta_R \ll T$ for all temperatures of interest in this problem (100 K or more). Agreement between the classical expression and the explicit sum is indeed good, as Figure 14.5 confirms. The figure displays the percentage deviation $(q_{\text{classical}}^R - q^R)100/q^R$. The maximum deviation is about –0.9% at 100 K and the magnitude decreases with increasing temperature.

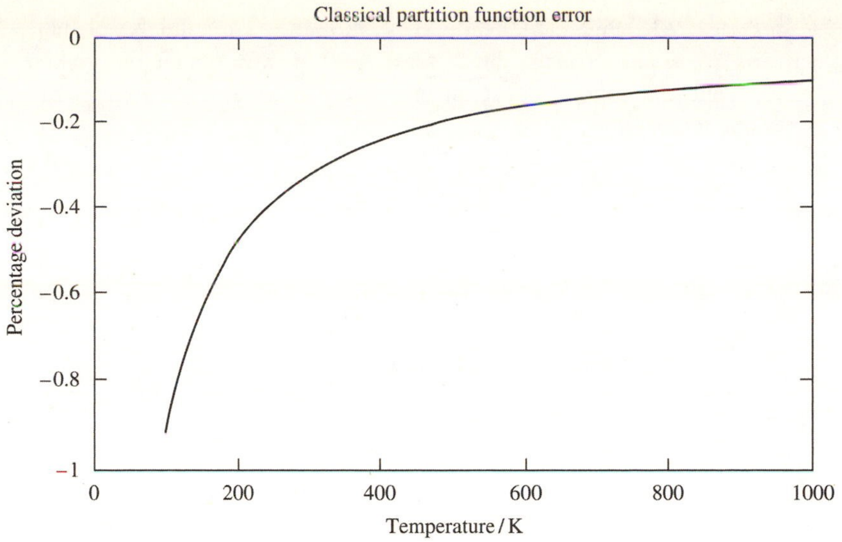

Figure 14.5

(c) The translational, rotational, and vibrational contributions to the total energy are specified by eqns. 13.36(b), 13.37(b), and 13.39, respectively. As molar quantities, they are

$$U^{\mathrm{T}} = \frac{3}{2}RT, \quad U^{\mathrm{R}} = RT, \quad U^{\mathrm{V}} = \frac{N_{\mathrm{A}}hc\tilde{\nu}}{\mathrm{e}^{hc\tilde{\nu}/kT} - 1}$$

The contributions to the difference in energy from its 100 K value are $\Delta U^{\mathrm{T}}(T) = U^{\mathrm{T}}(T) - U^{\mathrm{T}}(100\mathrm{K})$, and so on. Figure 14.6 shows the individual contributions to $\Delta U(T)$. Translational motion contributes 50% more than the rotational motion because it has 3 quadratic degrees of freedom compared to 2 quadratic degrees of freedom for rotation. Very little change occurs in the vibrational energy because very high temperatures are required to populate $v = 1, 2, \ldots$ states (see part a).

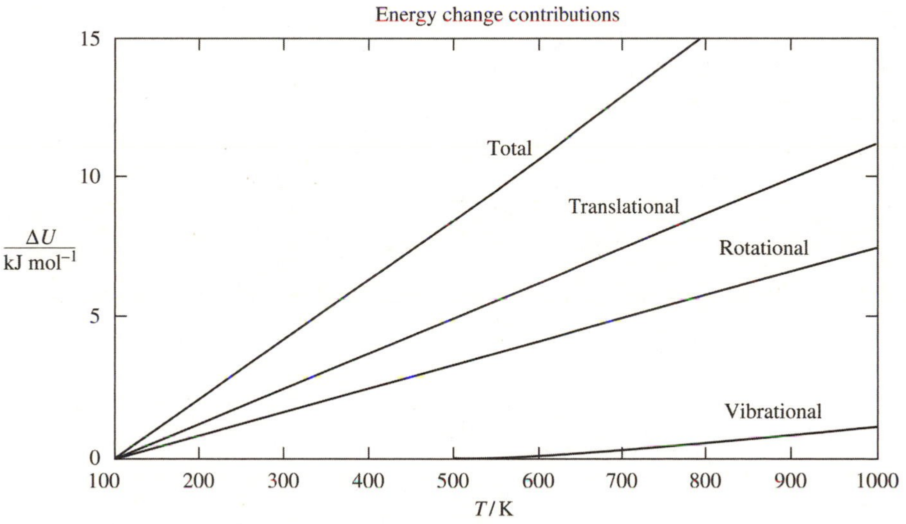

Figure 14.6

$$C_{V,\mathrm{m}}(T) = \left(\frac{\partial U(T)}{\partial T}\right)_V = \left(\frac{\partial}{\partial T}\right)_V (U^{\mathrm{T}} + U^{\mathrm{R}} + U^{\mathrm{V}})$$

$$= \frac{3}{2}R + R + \frac{\mathrm{d}U^{\mathrm{V}}}{\mathrm{d}T} = \frac{5}{2}R + \frac{\mathrm{d}U^{\mathrm{V}}}{\mathrm{d}T}$$

The derivative $\mathrm{d}U^{\mathrm{V}}/\mathrm{d}T$ may be evaluated numerically with numerical software (we advise exploration of the technique) or it may be evaluated analytically using the equation for C_{V} in Example 14.2:

$$C_{V,\mathrm{m}}^{\mathrm{V}} = \frac{\mathrm{d}U^{\mathrm{V}}}{\mathrm{d}T} = R\left\{\frac{\theta_{\mathrm{V}}}{T}\left(\frac{\mathrm{e}^{-\theta_{\mathrm{V}}/2T}}{1 - \mathrm{e}^{-\theta_{\mathrm{V}}/T}}\right)\right\}^2$$

where $\theta_V = hc\tilde{\nu}/k = 3122$ K. Figure 14.7 shows the ratio of the vibrational contribution to the sum of translational and rotational contributions. Below 300 K, vibrational motion makes a small, perhaps negligible, contribution to the heat capacity. The contribution is about 10% at 600 K and grows with increasing temperature.

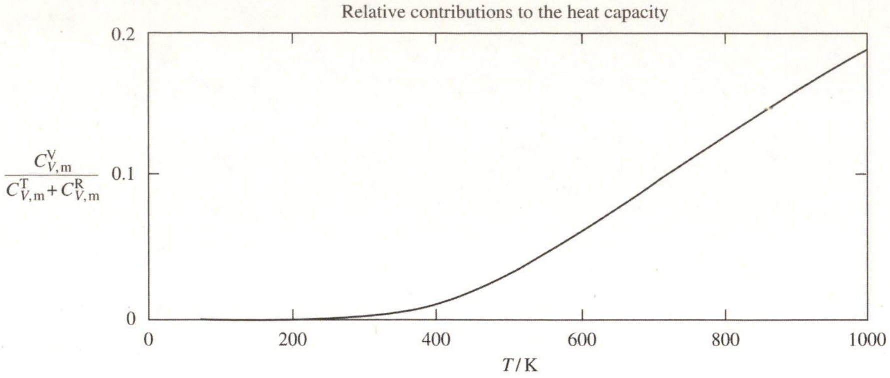

Figure 14.7

P14.21 To calculate the enthalpy of the protein's unfolding, we need to determine the area under the plot of $C_{p,ex}$ against T, from the baseline value of $C_{p,ex}$ at T_1, the start of the process, to the baseline value of $C_{p,ex}$ at T_2, the end of the process. We are provided with an illustration that shows the plot, but no numerical values are provided. Approximate numerical values can be extracted from the plot, and then the value of the integral $\Delta H = \int_{T_1}^{T_2} C_{p,ex}\,dT$ can be obtained by numerical evaluation of the area under the curve. The first two columns in the table below show the data estimated from the curve; the last column gives the approximate area under the curve from the beginning of the process to the end. The final value, $\boxed{1889 \text{ kJ mol}^{-1}}$, is the enthalpy of unfolding of the protein. The four significant figures shown are not really justified because of the imprecise estimation process involved.

θ / °C	$C_{p,ex}$/kJ K^{-1} mol^{-1}	ΔH/kJ mol^{-1}
30	20	0
40	23	215
50	26	460
54	28	567
56	33	626
57	40	663
58	46	706
59	52	755
60	58	810
61	63	870
62	70	937
63	80	1011
64	89	1096
64.5	90	1141
65	85	1185
66	80	1267
67	68	1342
68	60	1405
69	52	1461
70	47	1511
72	41	1598
74	37	1676
80	36	1889

Solutions to theoretical problems

P14.23 We work with eqn. 14.26 in the form $C_V = -\frac{N}{kT^2}\left(\frac{\partial\langle\varepsilon\rangle}{\partial\beta}\right)_V$ with $\langle\varepsilon\rangle$ given by $\langle\varepsilon\rangle = \frac{\sum_i \varepsilon_i e^{-\beta\varepsilon_i}}{\sum_i e^{-\beta\varepsilon_i}}$.

Differentiating this expression with respect to β, we obtain

$$\left(\frac{\partial\langle\varepsilon\rangle}{\partial\beta}\right)_V = -\frac{\sum_i \varepsilon_i^2 e^{-\beta\varepsilon_i}}{\sum_i e^{-\beta\varepsilon_i}} + \frac{\left(\sum_i \varepsilon_i e^{-\beta\varepsilon_i}\right)^2}{\sum_i e^{-2\beta\varepsilon_i}},$$ which can be rewritten as

$$-\frac{\sum_i \varepsilon_i^2 e^{-\beta\varepsilon_i}}{q} + \left(\frac{\sum_i \varepsilon_i e^{-\beta\varepsilon_i}}{q}\right)^2 \quad \text{or} \quad -\langle\varepsilon^2\rangle + \langle\varepsilon\rangle^2 = -\Delta\varepsilon^2$$

After substituting into the expression above for C_V, we obtain

$$C_V = -\frac{N}{kT^2}\left(\frac{\partial\langle\varepsilon\rangle}{\partial\beta}\right)_V = -\frac{N}{kT^2}(-\Delta\varepsilon^2) = \boxed{\frac{N\Delta\varepsilon^2}{kT^2}}$$, which was to be proved.

P14.25 We begin the derivation with the full expression for the molecular rotational partition function.

$$q^{\mathrm{R}} = \frac{1}{\sigma}\sum_J (2J+1)e^{-\frac{hc\tilde{B}}{kT}J(J+1)} [13.19] = \frac{1}{\sigma}\sum_J (2J+1)e^{-\frac{\theta_{\mathrm{R}}}{T}J(J+1)} \quad [\theta_{\mathrm{R}} = hc\tilde{B}/k]$$

We first note that no single analytical expression for this summation can be obtained that is valid at all temperatures. The summation process must be continued until additional terms change the sum to less than some pre-defined limit of accuracy, say 0.2%. However, various approximate analytical expressions that are valid within certain temperature ranges can be derived. Taken together, these expressions can give values that are valid within small error limits over the entire temperature range.

One of these approximations applies the Euler–Maclaurin summation formula to the expression for q^{R}. We obtain an expression that is accurate to about 0.1% when $\frac{\theta_{\mathrm{R}}}{T} \equiv \frac{1}{x} \le 0.7$.

$$q^{\mathrm{R}} = \frac{T}{\sigma\theta_{\mathrm{R}}}\left\{1 + \frac{1}{3}\frac{\theta_{\mathrm{R}}}{T} + \frac{1}{15}\left(\frac{\theta_{\mathrm{R}}}{T}\right)^2 + \frac{4}{315}\left(\frac{\theta_{\mathrm{R}}}{T}\right)^3 + \cdots\right\}$$

$$q^{\mathrm{R}} = \frac{x}{\sigma}\left\{1 + \frac{1}{3x} + \frac{1}{15x^2} + \frac{4}{315x^3} + \cdots\right\}$$

However, when $\frac{1}{x} \equiv \frac{\theta_{\mathrm{R}}}{T} \ge 0.7$, we must use the full summation expression for q^{R}, but the first five terms are usually sufficient for better than 0.1% accuracy. Fewer terms are needed as $1/x$ increases.

$$q^{\mathrm{R}} = \frac{1}{\sigma}(1 + 3e^{-2/x} + 5e^{-6/x} + 7e^{-12/x} + 9e^{-20/x} + \cdots)$$

The energy is calculated from $U_{\mathrm{m}} - U_{\mathrm{m}}(0) = kT^2\left[\frac{\partial \ln Q}{\partial T}\right]_V$,

and the heat capacity is calculated from $C_{V,\mathrm{m}} = 2kT\left[\frac{\partial \ln Q}{\partial T}\right]_V + kT^2\left[\frac{\partial^2 \ln Q}{\partial T^2}\right]_V$.

The canonical partition function for a linear rotor is $Q^{R} = (q^{R})^{N}$. Substituting this expression for Q^{R} into the expression for $C_{V,m}$ and performing the differentiations term by term, we obtain the final expressions for $C_{V,m}$: (1) using q^{R} from the Euler–Maclaurin expansion,

$$\boxed{C_{V,m} = R\left\{1+\frac{1}{45}\left(\frac{\theta_R}{T}\right)^2+\frac{16}{945}\left(\frac{\theta_R}{T}\right)^3+\cdots\right\}}, \text{ or}$$

$$C_{V,m} = R\left\{1+\frac{1}{45x^2}+\frac{16}{945x^3}+\ldots\right\} \quad (1)$$

(0.3% accuracy when $\frac{1}{x} \le 0.65$, or $\frac{T}{\theta_R} \ge 1.54$), and (2) using q^{R} from the full summation

$$\boxed{C_{V,m} = R\left\{\left(\frac{1}{xq^{R}}\right)^2 12e^{-2/x}(1+15e^{-4/x}+20e^{-6/x}+84e^{-10/x}+175e^{-12/x}+105e^{-16/x}+\cdots)\right\}}$$

$$\boxed{C_{V,m} = R\left\{\left(\frac{1}{xq^{R}}\right)^2 12e^{-2/x}(1+15e^{-4/x}+20e^{-6/x}+84e^{-10/x}+175e^{-12/x}+105e^{-16/x}+\cdots)\right\}} \; x \equiv \frac{T}{\theta_R} \quad (2)$$

(0.2% accuracy when $\frac{1}{x} \ge 0.65$ or $\frac{T}{\theta_R} \le 1.54$).

Examination of the data in Table 10.2 reveals that only in the case of H_2 and possibly the hydrogen halides is this latter expression, eqn. (2), needed. The case of H_2 is treated separately in Problem 14.33.

An alternative but equivalent approach would be to use the expression derived in Problem 14.32. Again a decision would have to be made as to when to terminate the summation in order to obtain the desired level of accuracy.

To complete these calculations and to plot the heat capacity, a spreadsheet program such as Excel or a computer algebra system such as Mathcad is very useful. See Figure 14.8 for a plot of $C_m(x) \equiv \frac{C_{V,m}}{R}$ [eqn. (2) above] against $x \equiv \frac{T}{\theta_R}$.

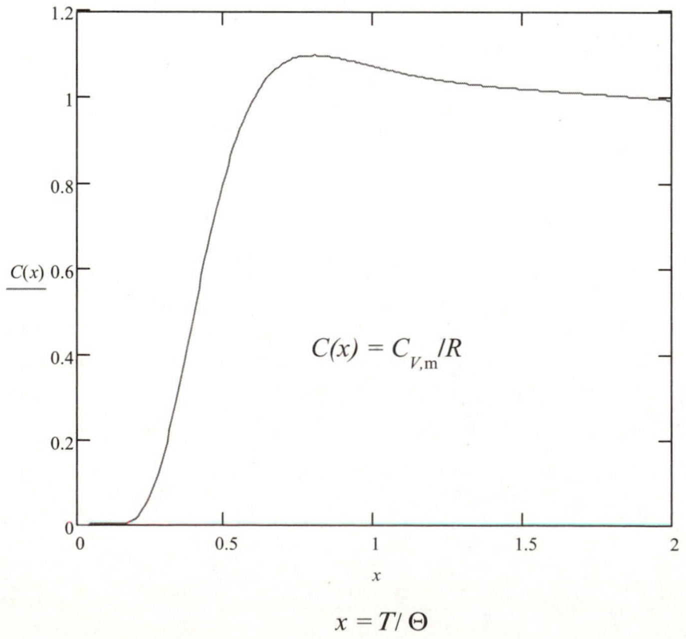

Figure 14.8

P14.27 The partition function for one dimension of vibration of a harmonic oscillator is

$q^{\mathrm{V}} = \dfrac{1}{1-\mathrm{e}^{-\beta hc\tilde{\nu}}}$ [13.24] $= \dfrac{1}{1-\mathrm{e}^{-\theta_{\mathrm{E}}/T}}$. The energy is calculated from $U_{\mathrm{m}} - U_{\mathrm{m}}(0) = kT^2\left[\dfrac{\partial \ln Q}{\partial T}\right]_V$ with $Q^{\mathrm{V}} = (q^{\mathrm{V}})^N$. Performing the differentiation, we obtain $U_{\mathrm{m}} - U_{\mathrm{m}}(0) = Nk\theta_{\mathrm{E}}\left(\dfrac{1}{\mathrm{e}^{\theta_{\mathrm{E}}/T}-1}\right)$. The heat capacity is calculated from $C_{V,\mathrm{m}} = \left(\dfrac{\partial U_{\mathrm{m}}}{\partial T}\right)_V$. Performing this differentiation we obtain $C_{V,\mathrm{m}} = Nk\left\{\left(\dfrac{\theta_{\mathrm{E}}}{T}\right)^2 \dfrac{\mathrm{e}^{\theta_{\mathrm{E}}/T}}{(\mathrm{e}^{\theta_{\mathrm{E}}/T}-1)^2}\right\}$. For three independent dimensions of oscillation as in one mole of a crystalline solid, we multiply by 3 and set $Nk = R$, finally obtaining the Einstein formula $\boxed{C_{V,\mathrm{m}} = 3R\left\{\left(\dfrac{\theta_{\mathrm{E}}}{T}\right)^2 \dfrac{\mathrm{e}^{\theta_{\mathrm{E}}/T}}{(\mathrm{e}^{\theta_{\mathrm{E}}/T}-1)^2}\right\}}$. The heat capacity $C_{V,\mathrm{m}}/3R$ is plotted against $x = T/\theta_{\mathrm{E}}$ in Figure 14.9.

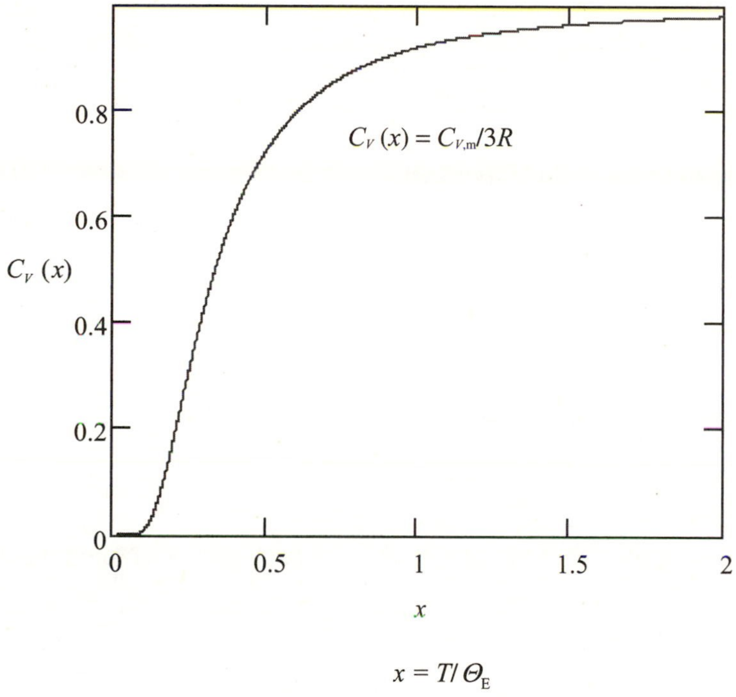

Figure 14.9

P14.29 (a) θ_{V} and θ_{R} are the constant factors in the numerators of the negative exponents in the sums that are the partition functions for vibration and rotation. They have the dimensions of temperature, which occurs in the denominator of the exponents. So high temperature means that $T \gg \theta_{\mathrm{V}}$ or θ_{R}, and only then does the exponential become substantial. Thus θ_{V} and θ_{R} are measures of the temperature at which higher vibrational and rotational states, respectively, become significantly populated.

$$\theta_{\mathrm{R}} = \frac{hc\tilde{B}}{k} = \frac{(2.998\times10^{10}\,\mathrm{cm\,s^{-1}})\times(6.626\times10^{-34}\,\mathrm{J\,s})\times(60.864\,\mathrm{cm^{-1}})}{(1.381\times10^{-23}\,\mathrm{J\,K^{-1}})} = \boxed{87.55\,\mathrm{K}}$$

and $$\theta_{\mathrm{V}} = \frac{hc\tilde{\nu}}{k} = \frac{(6.626\times10^{-34}\,\mathrm{J\,s})\times(4400.39\,\mathrm{cm^{-1}})\times(2.998\times10^{10}\,\mathrm{cm\,s^{-1}})}{(1.381\times10^{-23}\,\mathrm{J\,K^{-1}})} = \boxed{6330\,\mathrm{K}}$$

(b) and (c) These parts of the solution were performed with Mathcad and are reproduced on the following pages.

Objective: To calculate the equilibrium constant K(T) and $C_p(T)$ for dihydrogen at high temperature for a system made with n mol H_2 at 1 bar.

$$H_2(g) \rightleftharpoons 2H(g)$$

At equilibrium, the degree of dissociation, α, and the equilibrium amounts of H_2 and atomic hydrogen are related by the expressions

$$n_{H_2} = (1-\alpha)n \quad \text{and} \quad n_H = 2\alpha n$$

The equilibrium mole fractions are

$$x_{H_2} = (1-\alpha)n/\{(1-\alpha)n + 2\alpha n\} = (1-\alpha)/(1+\alpha)$$

$$x_H = 2\alpha n/\{(1-\alpha)n + 2\alpha n\} = 2\alpha/(1+\alpha)$$

The partial pressures are

$$p_{H_2} = (1-\alpha)p/(1+\alpha) \quad \text{and} \quad p_H = 2\alpha p/(1+\alpha)$$

The equilibrium constant is

$$K(T) = \frac{(p_H/p^{\ominus})^2}{(p_{H_2}/p^{\ominus})} = 4\alpha^2 \frac{(p/p^{\ominus})}{(1-\alpha^2)} = \frac{4\alpha^2}{(1-\alpha^2)} \quad \text{where } p = p^{\ominus} = 1 \text{ bar}$$

The above equation is easily solved for α,

$$\boxed{\alpha = (K/(K+4))^{1/2}}$$

The heat capacity at constant volume for the equilibrium mixture is

$$C_V(\text{mixture}) = n_H C_{V,m}(H) + n_{H_2} C_{V,m}(H_2)$$

The heat capacity at constant volume per mole of dihydrogen used to prepare the equilibrium mixture is

$$C_V = C_V(\text{mixture})/n = \{n_H C_{V,m}(H) + n_{H_2} C_{V,m}(H_2)\}/n$$

$$= \boxed{2\alpha C_{V,m}(H) + (1-\alpha) C_{V,m}(H_2)}$$

The formula for the heat capacity at constant pressure per mole of dihydrogen used to prepare the equilibrium mixture (C_p) can be deduced from the molar relationship.

$$C_{p,m} = C_{V,m} + R$$

$$C_p = \{n_H C_{p,m}(H) + n_{H_2} C_{p,m}(H_2)\}/n$$

$$= \frac{n_H}{n}\{C_{V,m}(H) + R\} + \frac{n_{H_2}}{n}\{C_{V,m}(H_2) + R\}$$

$$= \frac{n_H C_{V,m}(H) + n_{H_2} C_{V,m}(H_2)}{n} + R\left(\frac{n_H + n_{H_2}}{n}\right)$$

$$= C_V + R(1+\alpha)$$

Calculations

J = joule	s = second	kJ = 1000 J
mol = mole	g = gram	bar = 1×10^5 Pa
$h = 6.62608 \times 10^{-34}$ J s	$c = 2.9979 \times 10^8$ m s^{-1}	$k = 1.38066 \times 10^{-23}$ J K^{-1}
$R = 8.31451$ J K^{-1} mol^{-1}	$N_A = 60.2214 \times 10^{23}$ mol^{-1}	$p^{\ominus} = 1$ bar

Molecular properties of H_2

$$v = 4400.39\ \text{cm}^{-1} \qquad \tilde{B} = 60.864\ \text{cm}^{-1} \qquad D = 432.1\ \text{kJ mol}^{-1}$$

$$m_H = \frac{1\ \text{g mol}^{-1}}{N_A} \qquad m_{H_2} = 2m_H$$

$$\theta_V = \frac{hc\tilde{v}}{k} \qquad \theta_R = \frac{hc\tilde{B}}{k}$$

Computation of $K(T)$ and $\alpha(T)$

$$N = 200 \quad i = 0, \ldots, N \quad T_i = 500K + \frac{i \times 5500\ \text{K}}{N}$$

$$\Lambda_{Hi} = \frac{h}{(2\pi m_H kT_i)^{1/2}} \qquad \Lambda_{H_2 i} = \frac{h}{(2\pi m_{H_2} kT_i)^{1/2}}$$

$$q_{Vi} = \frac{1}{1 - e^{-(\theta_V/T_i)}} \qquad q_{Ri} = \frac{T_i}{2\theta_R}$$

$$\boxed{K_{eqi} = \frac{kT_i\left(\Lambda_{H_2 i}\right)^3 e^{-(D/RT_i)}}{p^{\ominus} q_{Vi} q_{Ri}\left(\Lambda_{Hi}\right)^6}} \qquad \alpha_i = \left(\frac{K_{eqi}}{K_{eqi} + 4}\right)^{1/2}$$

See Figure 14.10.

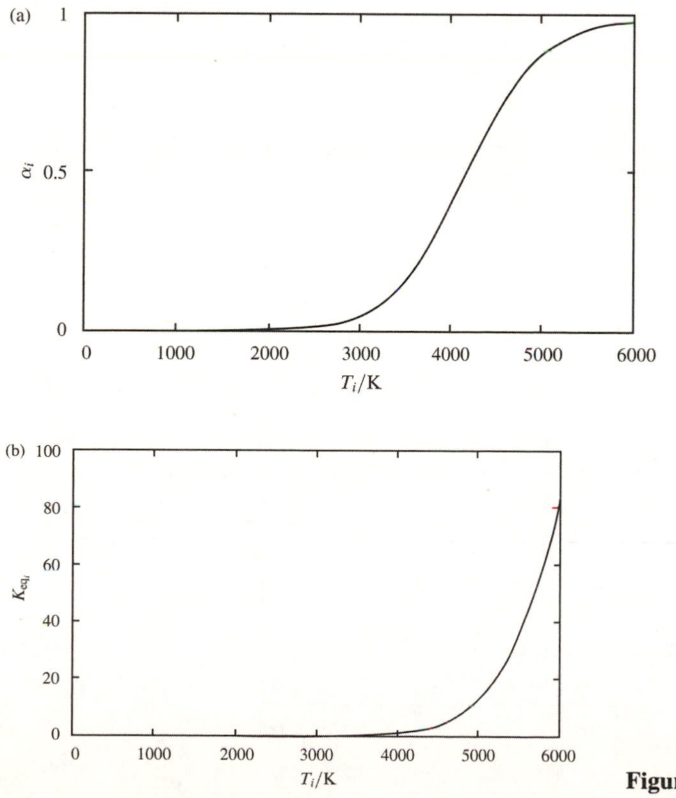

Figure 14.10

Heat capacity at constant volume per mole of dihydrogen used to prepare the equilibrium mixture is (see Figure 14.11).

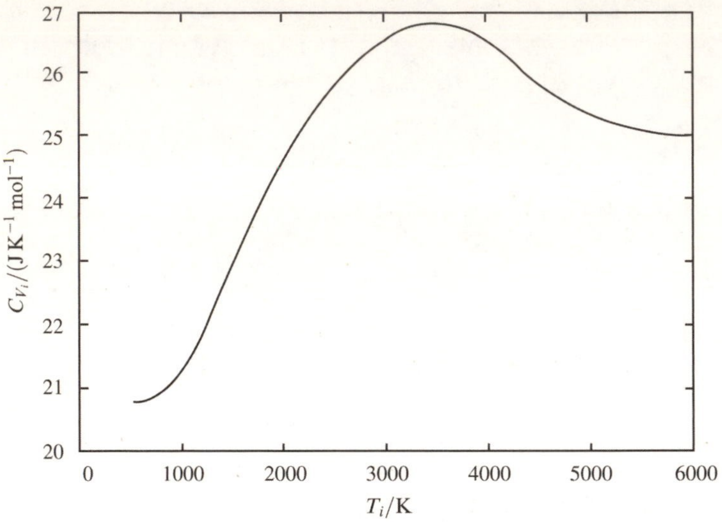

Figure 14.11

$$C_V(\mathrm{H}) = \boxed{1.5R}$$

$$C_V(\mathrm{H}_{2_i}) = \boxed{2.5R + \left[\frac{\theta_\mathrm{V}}{T_i} \times \frac{\mathrm{e}^{-(\theta_\mathrm{V}/2T_i)}}{1-\mathrm{e}^{\theta_\mathrm{V}/T_i}}\right]^2 R} \quad C_{V_i} = 2\alpha_i C_V(\mathrm{H}) + (1-\alpha_i)C_V(\mathrm{H}_{2_i})$$

The heat capacity at constant pressure per mole of dihydrogen used to prepare the equilibrium mixture is (see Figure 14.12).

$$C_{p_i} = C_{V_i} + R(1+\alpha_i)$$

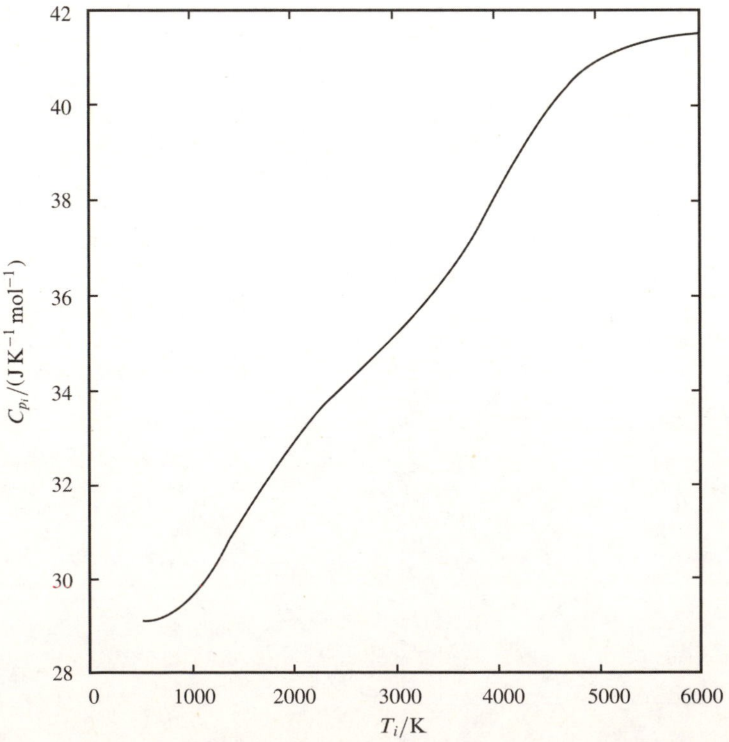

Figure 14.12

P14.31 We are given $q=\sum_j \mathrm{e}^{-\beta\varepsilon_j}$, $\dot{q}=\sum_j \beta\varepsilon_j \mathrm{e}^{-\beta\varepsilon_j}$, $\ddot{q}=\sum_j (\beta\varepsilon_j)^2 \mathrm{e}^{-\beta\varepsilon_j}$. But note that the degeneracy factor, $g=(2J+1)$, should be included in these expressions. The energy level expression for linear and spherical rotors is $\varepsilon_J = hc\tilde{B}J(J+1)$. Inserting this expression and the degeneracy factor into the expression for q above, we obtain for the rotational partition function including the degeneracy factor

$$q_{\mathrm{R}} = \frac{1}{\sigma}\sum_J (2J+1)\mathrm{e}^{-\frac{hc\tilde{B}}{kT}J(J+1)} \text{ [13.19], which at 298 K becomes } q_{\mathrm{R}} = \frac{1}{\sigma}\sum_J (2J+1)\mathrm{e}^{-\frac{\tilde{B}}{207.22\ \mathrm{cm}^{-1}}J(J+1)}$$

The symmetry number for the molecule must be obtained from the point group of the molecule; it is equal to the number of distinct proper rotations of the molecule plus the identity operation. For HCl that is 1, but for CH_4 it is 12. For $\dot{q}^{\mathrm{R}}$ and $\ddot{q}^{\mathrm{R}}$ we obtain at 298 K

$$\dot{q}^{\mathrm{R}} = \frac{1}{\sigma}\sum_J \left(\frac{\tilde{B}}{207.22\ \mathrm{cm}^{-1}}J(J+1)\right)(2J+1)\mathrm{e}^{-\frac{\tilde{B}}{207.22\ \mathrm{cm}^{-1}}J(J+1)}$$

$$\ddot{q}^{\mathrm{R}} = \frac{1}{\sigma}\sum_J \left(\frac{\tilde{B}}{207.22\ \mathrm{cm}^{-1}}J(J+1)\right)^2(2J+1)\mathrm{e}^{-\frac{\tilde{B}}{207.22\ \mathrm{cm}^{-1}}J(J+1)}$$

(a) For HCl, $\tilde{B}=10.593\ \mathrm{cm}^{-1}$. When this value is inserted into the expressions above, the results of these summations are

$$\boxed{q^{\mathrm{R}}=19.899,\ \dot{q}^{\mathrm{R}}=19.558,\ \text{and}\ \ \ddot{q}^{\mathrm{R}}=576.536}$$

(b) For CCl_4, $\tilde{B}=5.797\ \mathrm{cm}^{-1}$. When this value is inserted into the expressions above along with $\sigma=12$, the results of the summations are

$$\boxed{q^{\mathrm{R}}=3.007,\ \dot{q}^{\mathrm{R}}=2.979,\ \text{and}\ \ \ddot{q}^{\mathrm{R}}=118.5}$$

P14.33 The partition functions for the ortho- and para- forms of H_2 are

$$q^{\mathrm{R}}_{\mathrm{ortho}} = 3\sum_{J=1,3,5\ldots}(2J+1)\mathrm{e}^{-\frac{\theta_{\mathrm{R}}}{T}J(J+1)} \quad [\theta_{\mathrm{R}} = hc\tilde{B}/k = 87.6\ \mathrm{K}]$$

$$q^{\mathrm{R}}_{\mathrm{para}} = \sum_{J=0,2,4\ldots}(2J+1)\mathrm{e}^{-\frac{\theta_{\mathrm{R}}}{T}J(J+1)}$$

To conform to the notation of Problem 14.32, we can rewrite these expressions as

$$q^{\mathrm{R}}_{\mathrm{ortho}} = 3\sum_{J=1,3,5\ldots} g(J)\mathrm{e}^{-\beta\varepsilon(J)}$$

$$q^{\mathrm{R}}_{\mathrm{para}} = \sum_{J=0,2,4\ldots} g(J)\mathrm{e}^{-\beta\varepsilon(J)} \qquad \text{with } g(J)=2J+1 \quad\text{and}\quad \varepsilon(J)=hc\tilde{B}J(J+1)$$

To simplify the summation process and in order to use the formulas derived in the solution to P14.32, we define a new quantum number K by $J=2K+1$ for the ortho- case and by $J=2K$ for the para- case. Then

$$g_{\mathrm{o}}(K)=4K+3 \quad\text{and}\quad \varepsilon_{\mathrm{o}}(K)=hc\tilde{B}[(2K+1)(2K+2)] \text{ for ortho-hydrogen and}$$

$$g_{\mathrm{p}}(K)=4K+1 \quad\text{and}\quad \varepsilon_{\mathrm{p}}(K)=hc\tilde{B}[2K(2K+1)] \text{ for para-hydrogen.}$$

With these modifications the partition functions can now be rewritten as

$$q^{\mathrm{R}}_{\mathrm{ortho}} = 3\sum_{K=0,1,2\ldots} g_{\mathrm{o}}(K)\mathrm{e}^{-\beta\varepsilon_{\mathrm{o}}(K)}$$

$$q^{\mathrm{R}}_{\mathrm{para}} = \sum_{K=0,1,2\ldots} g_{\mathrm{p}}(K)\mathrm{e}^{-\beta\varepsilon_{\mathrm{p}}(K)}$$

Now the final expression for $C_{V,\mathrm{m}}/R$ derived in P14.32, with K in place of J, $g(K)$ in place of $g(J)$, and $\varepsilon(K)$ in place of $\varepsilon(J)$, can be used to calculate the heat capacities. With these substitutions the expression becomes

$$\frac{C_{V,\mathrm{m}}}{R}=\frac{1}{q}\sum_K g(K)\beta^2\varepsilon^2(K)\mathrm{e}^{-\beta\varepsilon(K)}-\frac{1}{q^2}\left(\sum_K g(K)\beta\varepsilon(K)\mathrm{e}^{-\beta\varepsilon(K)}\right)^2$$

This expression needs to be evaluated separately for both the para- and ortho- forms of hydrogen. It is most easily evaluated with a spreadsheet program such as Excel or a CAS system such as Mathcad.

For the ortho- case, $C_\mathrm{o}\equiv\dfrac{C_{V,\mathrm{m}}}{R}=\dfrac{1}{q_\mathrm{o}}\sum_K g_\mathrm{o}b_\mathrm{o}^2\mathrm{e}^{-b_\mathrm{o}}-\dfrac{1}{q_\mathrm{o}^2}\left(\sum_K g_\mathrm{o}b_\mathrm{o}\mathrm{e}^{-b_\mathrm{o}}\right)^2$ [eqn. 1], in which

$$b_\mathrm{o}=\beta\varepsilon_\mathrm{o}=\frac{\theta_\mathrm{R}}{T}(2K+1)(2K+2)=\frac{1}{x}(2K+1)(2K+2)\quad x\equiv\frac{T}{\theta_\mathrm{R}}$$

For the para- case, $C_\mathrm{p}\equiv\dfrac{C_{V,\mathrm{m}}}{R}=\dfrac{1}{q_\mathrm{p}}\sum_K g_\mathrm{p}b_\mathrm{p}^2\mathrm{e}^{-b_\mathrm{p}}-\dfrac{1}{q_\mathrm{p}^2}\left(\sum_K g_\mathrm{p}b_\mathrm{p}\mathrm{e}^{-b_\mathrm{p}}\right)^2$ [eqn. 2], in which

$$b_\mathrm{p}=\beta\varepsilon_\mathrm{p}=\frac{\theta_\mathrm{R}}{T}[2K(2K+1)]=\frac{1}{x}[2K(2K+1)]\quad x\equiv T/\theta_\mathrm{R}$$

In the Mathcad worksheet below the notation is as follows: $x\equiv T/\theta_\mathrm{R}$; for ortho-hydrogen, q_o is its partition function, $C_{\mathrm{o}1}$ and $C_{\mathrm{o}2}$ are the first and second terms on the right in eqn (1), and $C_\mathrm{o}=C_{V,\mathrm{m}}/R$ s its heat capacity in units of R. Similarly, for para-hydrogen, q_p is its partition function, $C_{\mathrm{p}1}$ and $C_{\mathrm{p}2}$ are the first and second terms on the right in eqn. (2), and $C_\mathrm{p}=C_{V,\mathrm{m}}/R$ is its heat capacity in units of R.

$$\mathrm{i}:=0..24\qquad \mathrm{x_{start}}:=0.2\qquad \mathrm{x_{end}}:=5.0$$

$$\mathrm{x_i}:=\mathrm{x_{start}}+(\mathrm{x_{end}}-\mathrm{x_{start}})\cdot\frac{\mathrm{i}}{24}\qquad \mathrm{x_i}=\mathrm{T}/\Theta$$

$$\mathrm{q_{o_i}}:=\sum_{\mathrm{K}=0}^{6}\left[(4\cdot\mathrm{K}+3)\cdot\mathrm{e}^{-\frac{1}{\mathrm{x_i}}\cdot(2\cdot\mathrm{K}+1)\cdot(2\cdot\mathrm{K}+2)}\right]$$

$$\mathrm{C_{o1_i}}:=\left(\frac{1}{\mathrm{q_{o_i}}}\right)\cdot\sum_{\mathrm{K}=0}^{6}\left[(4\cdot\mathrm{K}+3)\left[\left[\frac{1}{\mathrm{x_i}}(2\cdot\mathrm{K}+1)(2\cdot\mathrm{K}+2)\right]^2\cdot\mathrm{e}^{-\frac{1}{\mathrm{x_i}}(2\cdot\mathrm{K}+1)(2\cdot\mathrm{K}+2)}\right]\right]$$

$$\mathrm{C_{o2_i}}:=-\left(\frac{1}{\mathrm{q_{o_i}}}\right)^2\left[\sum_{\mathrm{K}=0}^{6}\left[(4\cdot\mathrm{K}+3)\left[\frac{1}{\mathrm{x_i}}(2\cdot\mathrm{K}+1)(2\cdot\mathrm{K}+2)\cdot\mathrm{e}^{-\frac{1}{\mathrm{x_i}}(2\cdot\mathrm{K}+1)(2\cdot\mathrm{K}+2)}\right]\right]\right]^2$$

$$\mathrm{C_{o_i}}:=\mathrm{C_{o1_i}}+\mathrm{C_{o2_i}}$$

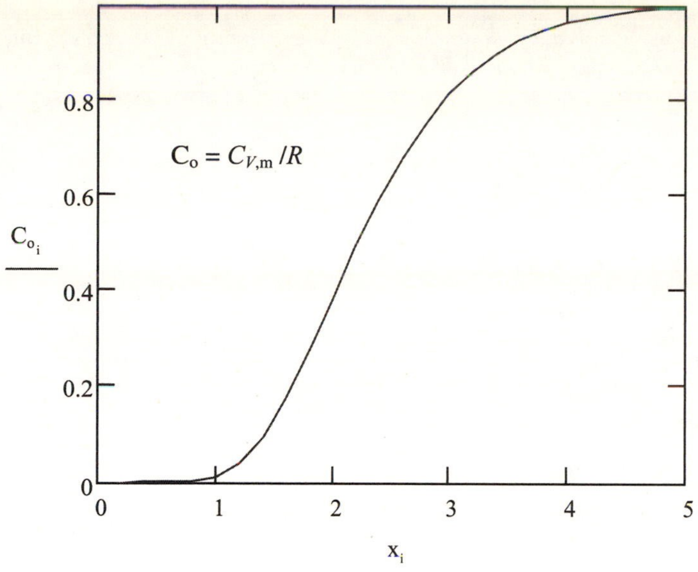

$x = T/\Theta$

$$q_{p_i} := \sum_{K=0}^{6} \left[(4 \cdot K+1) \cdot e^{-\frac{1}{x_i} \cdot (2 \cdot K+1) \cdot (2 \cdot K)} \right]$$

$$C_{p1_i} := \left(\frac{1}{q_{p_i}} \right) \cdot \sum_{K=0}^{6} \left[(4 \cdot K+1) \left[\left[\frac{1}{x_i} (2 \cdot K+1)(2 \cdot K) \right]^2 \cdot e^{-\frac{1}{x_i}(2 \cdot K+1)(2 \cdot K)} \right] \right]$$

$$C_{p2_i} := -\left(\frac{1}{q_{p_i}} \right)^2 \left[\sum_{K=0}^{6} \left[(4 \cdot K+1) \left[\frac{1}{x_i} (2 \cdot K+1)(2 \cdot K) \cdot e^{-\frac{1}{x_i}(2 \cdot K+1)(2 \cdot K)} \right] \right] \right]^2$$

$$C_{p_i} := C_{p1_i} + C_{p2_i}$$

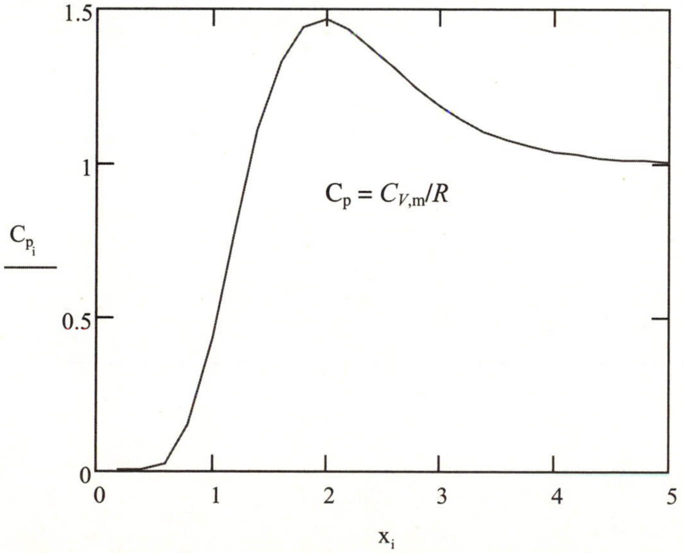

$x = T/\Theta$

P14.35 The differential equation to be solved is $\frac{\mathrm{d}}{\mathrm{d}L}\left(\frac{\mathrm{e}^{-\beta\varepsilon_i}}{q(\beta)}\right)=0$. We differentiate this quotient with respect to L and obtain

$$\frac{\mathrm{d}}{\mathrm{d}L}\left(\frac{\mathrm{e}^{-\beta\varepsilon_i}}{q(\beta)}\right)=\frac{1}{q}\frac{\mathrm{d}(\mathrm{e}^{-\beta\varepsilon_i})}{\mathrm{d}L}-\frac{\mathrm{e}^{-\beta\varepsilon_i}}{q^2}\frac{\mathrm{d}q}{\mathrm{d}L}$$

In an adiabatic process, β changes with L. Thus

$$\frac{\mathrm{d}}{\mathrm{d}L}\mathrm{e}^{-\beta\varepsilon_i}=\frac{\partial\mathrm{e}^{-\beta\varepsilon_i}}{\partial\beta}\frac{\mathrm{d}\beta}{\mathrm{d}L}+\frac{\partial\mathrm{e}^{-\beta\varepsilon_i}}{\partial\varepsilon_i}\frac{\mathrm{d}\varepsilon_i}{\mathrm{d}L}$$

Performing the differentiations and using $\varepsilon_i=\frac{\gamma_i}{L^2}$, we obtain

$$\frac{\mathrm{d}}{\mathrm{d}L}\mathrm{e}^{-\beta\varepsilon_i}=-\varepsilon_i\mathrm{e}^{-\beta\varepsilon_i}\frac{\mathrm{d}\beta}{\mathrm{d}L}+\frac{2\beta\varepsilon_i}{L}\mathrm{e}^{-\beta\varepsilon_i}$$

Relative to the lowest level, $q=\left(\frac{2\pi m}{h^2\beta}\right)^{\frac{1}{2}}L$ for a particle in a one-dimensional box.

Differentiating q with respect to L while recognizing again that β changes with L, we obtain (omitting a couple of algebraic steps)

$$\frac{\mathrm{d}q}{\mathrm{d}L}=\frac{q}{L}-\frac{q}{2\beta}\frac{\mathrm{d}\beta}{\mathrm{d}L}$$

Putting this all together, we have

$$\frac{\mathrm{d}}{\mathrm{d}L}\left(\frac{\mathrm{e}^{-\beta\varepsilon_i}}{q(\beta)}\right)=\frac{1}{q}\left\{-\varepsilon_i\mathrm{e}^{-\beta\varepsilon_i}\frac{\mathrm{d}\beta}{\mathrm{d}L}+\frac{2\beta\varepsilon_i}{L}\mathrm{e}^{-\beta\varepsilon_i}\right\}-\frac{\mathrm{e}^{-\beta\varepsilon_i}}{q^2}\left\{\frac{q}{L}-\frac{q}{2\beta}\frac{\mathrm{d}\beta}{\mathrm{d}L}\right\}$$

By inspection we can see that this equals zero when $\frac{\mathrm{d}\beta}{\mathrm{d}L}=\frac{2\beta}{L}$.

Thus $\frac{\mathrm{d}\beta}{\beta}=2\frac{\mathrm{d}L}{L}$ and $\ln\beta=2\ln L+\text{constant}$ or $\ln\beta=\ln L^2+$ constant, or $\ln\frac{\beta}{L^2}=$ constant

and hence $\frac{\beta}{L^2}=$ constant. Since $\beta=\frac{1}{kT}$ and recognizing that k is a constant, we finally obtain

$\frac{1}{TL^2}=$ constant or $TL^2=$ constant or $LT^{\frac{1}{2}}=$ constant.

Q. E. D.

P14.37 $$C_V=\left(\frac{\partial U}{\partial T}\right)_V$$

$$\boxed{\left(\frac{\partial C_V}{\partial V}\right)_T=\left(\frac{\partial}{\partial V}\left(\frac{\partial U}{\partial T}\right)_V\right)_T=\left(\frac{\partial}{\partial T}\left(\frac{\partial U}{\partial V}\right)_T\right)_V}$$ [derivatives may be taken in any order]

$\left(\frac{\partial U}{\partial V}\right)_T=0$ for a perfect gas [Section 14.10]

Hence, $\boxed{\left(\frac{\partial C_V}{\partial V}\right)_T=0}$

Likewise, $C_p = \left(\frac{\partial H}{\partial T}\right)_p$, so $\boxed{\left(\frac{\partial C_p}{\partial p}\right)_T = \left(\frac{\partial}{\partial p}\left(\frac{\partial H}{\partial T}\right)_p\right)_T = \left(\frac{\partial}{\partial T}\left(\frac{\partial H}{\partial p}\right)_T\right)_p}$

$$\left(\frac{\partial H}{\partial p}\right)_T = 0 \text{ for a perfect gas}$$

Hence, $\left(\frac{\partial C_p}{\partial p}\right)_T = 0$

P14.39 Using the Euler's chain relation and the reciprocal identity [*Mathematical Background* 8],

$$\left(\frac{\partial p}{\partial T}\right)_V = -\left(\frac{\partial p}{\partial V}\right)_T\left(\frac{\partial V}{\partial T}\right)_p$$

Substituting into the given expression for $C_p - C_V$,

$$C_p - C_V = -T\left(\frac{\partial p}{\partial V}\right)_T\left(\frac{\partial V}{\partial T}\right)_p^2$$

Using the reciprocal identity again,

$$C_p - C_V = -\frac{T\left(\frac{\partial V}{\partial T}\right)_p^2}{\left(\frac{\partial V}{\partial p}\right)_T}$$

For a perfect gas, $pV = nRT$, so

$$\left(\frac{\partial V}{\partial T}\right)_p^2 = \left(\frac{nR}{p}\right)^2 \quad \text{and} \quad \left(\frac{\partial V}{\partial p}\right)_T = -\frac{nRT}{p^2}$$

so $$C_p - C_V = \frac{-T\left(\frac{nR}{p}\right)^2}{-\frac{nRT}{p^2}} = \boxed{nR}$$

P14.41 (a) $V = V(p,T)$; hence, $\mathrm{d}V = \boxed{\left(\frac{\partial V}{\partial p}\right)_T \mathrm{d}p + \left(\frac{\partial V}{\partial T}\right)_p \mathrm{d}T}$.

Likewise $p = p(V,T)$, so $\mathrm{d}p = \boxed{\left(\frac{\partial p}{\partial V}\right)_T \mathrm{d}V + \left(\frac{\partial p}{\partial T}\right)_V \mathrm{d}T}$.

(b) We use $\alpha = \left(\frac{1}{V}\right)\left(\frac{\partial V}{\partial T}\right)_p$ [14.58] and $\kappa_T = -\left(\frac{1}{V}\right)\left(\frac{\partial V}{\partial p}\right)_T$ [14.50] and obtain

$$\mathrm{d}\ln V = \frac{1}{V}\mathrm{d}V = \left(\frac{1}{V}\right)\left(\frac{\partial V}{\partial p}\right)_T \mathrm{d}p + \left(\frac{1}{V}\right)\left(\frac{\partial V}{\partial T}\right)_p \mathrm{d}T = \boxed{-\kappa_T\,\mathrm{d}p + \alpha\,\mathrm{d}T}$$

Likewise $$\mathrm{d}\ln p = \frac{\mathrm{d}p}{p} = \frac{1}{p}\left(\frac{\partial p}{\partial V}\right)_T \mathrm{d}V + \frac{1}{p}\left(\frac{\partial p}{\partial T}\right)_V \mathrm{d}T$$

We express $\left(\frac{\partial p}{\partial V}\right)_T$ in terms of κ_T:

$$\kappa_T = -\frac{1}{V}\left(\frac{\partial V}{\partial p}\right)_T = -\left[V\left(\frac{\partial p}{\partial V}\right)_T\right]^{-1} \quad \text{so} \quad \left(\frac{\partial p}{\partial V}\right)_T = -\frac{1}{\kappa_T V}$$

We express $\left(\frac{\partial p}{\partial T}\right)_V$ in terms of κ_T and α:

$$\left(\frac{\partial p}{\partial T}\right)_V\left(\frac{\partial T}{\partial V}\right)_P\left(\frac{\partial V}{\partial p}\right)_T = -1 \quad \text{so} \quad \left(\frac{\partial p}{\partial T}\right)_V = -\frac{(\partial V/\partial T)_p}{(\partial V/\partial p)_T} = \frac{\alpha}{\kappa_T}$$

so $$\mathrm{d}\ln p = -\frac{\mathrm{d}V}{p\kappa_T V} + \frac{\alpha\,\mathrm{d}T}{p\kappa_T} = \boxed{\frac{1}{p\kappa_T}\left(\alpha\,\mathrm{d}T - \frac{\mathrm{d}V}{V}\right)}$$

P14.43 $$\mu \equiv \left(\frac{\partial T}{\partial p}\right)_H \quad [14.65]$$

Use of Euler's chain relation [*Mathematical Background* 8] yields

$$\mu = -\frac{\left(\frac{\partial H_\mathrm{m}}{\partial p}\right)_T}{C_{p,\mathrm{m}}} \quad [14.67]$$

$$\left(\frac{\partial H_\mathrm{m}}{\partial p}\right)_T = \left(\frac{\partial U_\mathrm{m}}{\partial p}\right)_T + \left[\frac{\partial (pV_\mathrm{m})}{\partial p}\right]_T = \left(\frac{\partial U_\mathrm{m}}{\partial V_\mathrm{m}}\right)_T\left(\frac{\partial V_\mathrm{m}}{\partial p}\right)_T + \left[\frac{\partial (pV_\mathrm{m})}{\partial p}\right]_T$$

Use the virial expansion of the van der Waals equation in terms of p. Now let us evaluate some of these derivatives.

$$\left(\frac{\partial U_\mathrm{m}}{\partial V_\mathrm{m}}\right)_T = \left(\frac{\partial U}{\partial V}\right)_T = \pi_T = \frac{a}{V_\mathrm{m}^2} \quad \text{[Exercise 14.29]}$$

$$pV_\mathrm{m} = RT\left[1 + \frac{1}{RT}\left(b - \frac{a}{RT}\right)p + \cdots\right]$$

$$\left[\frac{\partial (pV_\mathrm{m})}{\partial p}\right]_T \approx b - \frac{a}{RT}, \quad \left(\frac{\partial V_\mathrm{m}}{\partial p}\right)_T \approx -\frac{RT}{p^2}$$

Substituting $\left(\frac{\partial H}{\partial p}\right)_T \approx \left(\frac{a}{V_m^2}\right)\times\left(-\frac{RT}{p^2}\right) + \left(b - \frac{a}{RT}\right) \approx \frac{-aRT}{(pV_\mathrm{m})^2} + \left(b - \frac{a}{RT}\right)$

Since $\left(\frac{\partial H}{\partial p}\right)_T$ is in a sense a correction term, that is, it approaches zero for a perfect gas, little error will be introduced by the approximation, $(pV_\mathrm{m})^2 = (RT)^2$.

Thus $\left(\frac{\partial H}{\partial p}\right)_T \approx \left(-\frac{a}{RT}\right) + \left(b - \frac{a}{RT}\right) = \left(b - \frac{2a}{RT}\right)$ and $\mu = \dfrac{\left(\frac{2a}{RT} - b\right)}{C_{p,\mathrm{m}}}$

P14.45 $$\alpha=\frac{1}{V}\left(\frac{\partial V}{\partial T}\right)_p=\frac{1}{V\left(\frac{\partial T}{\partial V}\right)_p}$$ [*Mathematical Background* 8]

$$\alpha=\frac{1}{V}\times\frac{1}{\left(\frac{T}{V-nb}\right)-\left(\frac{2na}{RV^3}\right)\times(V-nb)}$$ [Problem 14.44]

$$=\boxed{\frac{(RV^2)\times(V-nb)}{(RTV^3)-(2na)\times(V-nb)^2}}$$

$$\kappa_T=-\frac{1}{V}\left(\frac{\partial V}{\partial p}\right)_T=\frac{-1}{V\left(\frac{\partial p}{\partial V}\right)_T}$$ [reciprocal identity]

$$\kappa_T=-\frac{1}{V}\times\frac{1}{\left(\frac{-nRT}{(V-nb)^2}\right)+\left(\frac{2n^2a}{V^3}\right)}$$ [Problem 14.44]

$$=\boxed{\frac{V^2(V-nb)^2}{nRTV^3-2n^2a(V-nb)^2}}$$

Then $\frac{\kappa_T}{\alpha}=\frac{V-nb}{nR}$, implying that $\kappa_T R=\alpha(V_m-b)$.

Alternatively, from the definitions of α and κ_T above,

$$\frac{\kappa_T}{\alpha}=\frac{-\left(\frac{\partial V}{\partial p}\right)_T}{\left(\frac{\partial V}{\partial T}\right)_p}=\frac{-1}{\left(\frac{\partial p}{\partial V}\right)_T\left(\frac{\partial V}{\partial T}\right)_p}$$ [reciprocal identity]

$$=\left(\frac{\partial T}{\partial p}\right)_V$$ [Euler chain relation]

$$=\frac{V-nb}{nR}$$ [Problem 14.44],

$$\kappa_T R=\frac{\alpha(V-nb)}{n}$$

Hence, $\kappa_T R=\alpha(V_m-b)$.

P14.47 (a) $$\mu=-\frac{1}{C_p}\left(\frac{\partial H}{\partial p}\right)_T=\frac{1}{C_p}\left\{T\left(\frac{\partial V_m}{\partial T}\right)_p-V_m\right\}$$ [14.67 and Problem 14.43]

$$V_m=\frac{RT}{p}+aT^2 \quad \text{so} \quad \left(\frac{\partial V_m}{\partial T}\right)_p=\frac{R}{p}+2aT$$

$$\mu=\frac{1}{C_p}\left\{\frac{RT}{p}+2aT^2-\frac{RT}{p}-aT^2\right\}=\boxed{\frac{aT^2}{C_p}}$$

(b) $$C_V=C_p-\alpha TV_m\left(\frac{\partial p}{\partial T}\right)_V=C_p-T\left(\frac{\partial V_m}{\partial T}\right)_p\left(\frac{\partial p}{\partial T}\right)_V$$

But $$p=\frac{RT}{V_m-aT^2}$$

$$\left(\frac{\partial p}{\partial T}\right)_V = \frac{R}{V_m - aT^2} - \frac{RT(-2aT)}{(V_m - aT^2)^2}$$

$$= \frac{R}{(RT/p)} + \frac{2aRT^2}{(RT/p)^2} = \frac{p}{T} + \frac{2ap^2}{R}$$

Therefore,

$$C_V = C_p - T\left(\frac{R}{p} + 2aT\right) \times \left(\frac{p}{T} + \frac{2ap^2}{R}\right)$$

$$= C_p - \frac{RT}{p}\left(1 + \frac{2apT}{R}\right) \times \left(1 + \frac{2apT}{R}\right) \times \left(\frac{p}{T}\right)$$

$$\boxed{C_V = C_p - R\left(1 + \frac{2apT}{R}\right)^2}$$

Solutions to applications

P14.49 (a) One major limitation of Hooke's law is that it applies to displacements from a single equilibrium value of the end-to-end distance. In fact, if a DNA molecule or any other macromolecular chain that is susceptible to strong non-bonding intramolecular interactions is disturbed sufficiently from one equilibrium configuration, it is likely to settle into a different equilibrium configuration, a so-called "local minimum" in potential energy. Hooke's law is a good approximation for systems that have a single equilibrium configuration corresponding to a single minimum in potential energy. Another limitation is the assumption that it is just as easy (or as difficult) to move the ends away from each other in any direction. In fact, the intramolecular interactions would be quite different depending on whether one were displacing an end along the chain or outward from the chain. (See Figure 14.13.)

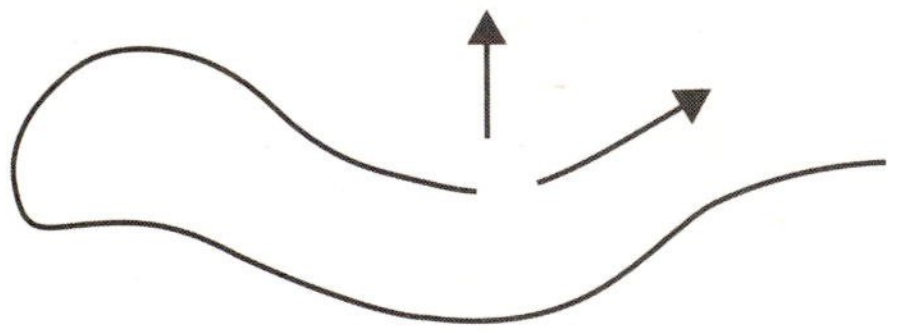

Figure 14.13

Work is $dw = -F dx = +k_F x\, dx$. This integrates to

$$w = \int_0^{x_f} k_F x dx = \frac{1}{2}k_F x^2\Big|_0^{x_f} = \boxed{\frac{1}{2}k_F x_f^{\,2}}$$

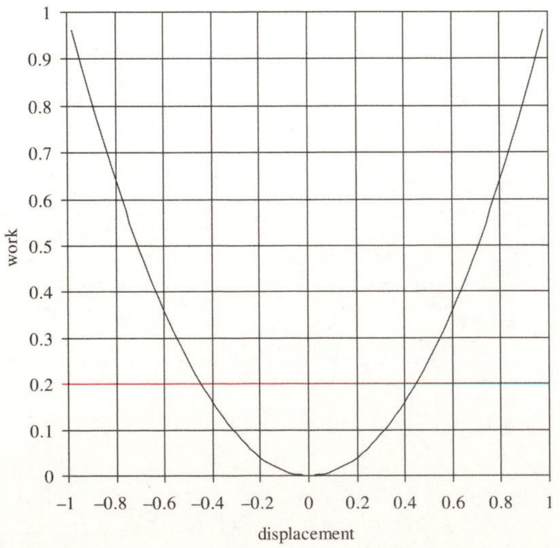

Figure 14.14

(b) One obvious limitation is that the model treats only displacements along the chain, not displacements that take an end away from the chain. (See Figure 14.13.)

(i) The displacement is twice the persistence length, so

$x = 2l$, $n = 2$, $\nu = n/N = 2/200 = 1/100$

and $$|F| = \frac{kT}{2l}\ln\left(\frac{1+\nu}{1-\nu}\right) = \frac{(1.381\times10^{-23}\ \mathrm{J\,K^{-1}})(298\ \mathrm{K})}{2\times45\times10^{-9}\ \mathrm{m}}\ln\left(\frac{1.01}{0.99}\right) = \boxed{9.1\times10^{-16}\ \mathrm{N}}$$

(ii) Figure 14.15 displays a plot of force versus displacement for Hooke's law and for the one-dimensional freely jointed chain. For small displacements, the plots very nearly coincide. However, for large displacements, the magnitude of the force in the one-dimensional model grows much faster. In fact, in the one-dimensional model, the magnitude of the force approaches infinity for a finite displacement, namely a displacement the size of the chain itself ($|\nu| = 1$). (For Hooke's law, the force approaches infinity only for infinitely large displacements.)

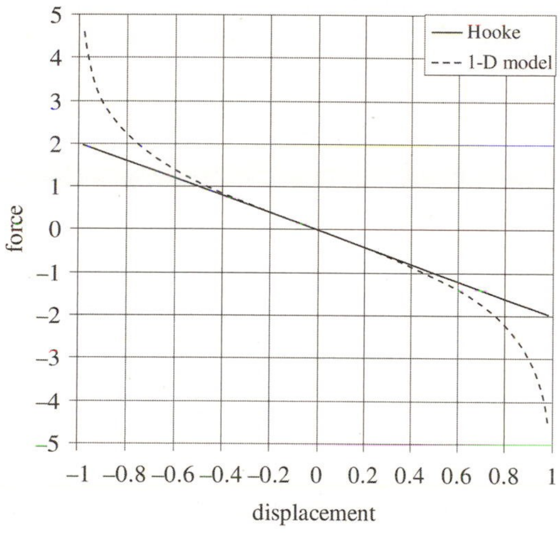

Figure 14.15

(iii) Work is $$\mathrm{d}w = -F\,\mathrm{d}x = \frac{kT}{2l}\ln\left(\frac{1+\nu}{1-\nu}\right)\mathrm{d}x = \frac{kNT}{2}\ln\left(\frac{1+\nu}{1-\nu}\right)\mathrm{d}\nu$$

This integrates to

$$w = \int_0^{\nu_f}\frac{kNT}{2}\ln\left(\frac{1+\nu}{1-\nu}\right)\mathrm{d}\nu = \frac{kNT}{2}\int_0^{\nu_f}[\ln(1+\nu)-\ln(1-\nu)]\mathrm{d}\nu$$

$$= \frac{kNT}{2}[(1+\nu)\ln(1+\nu)-\nu+(1-\nu)\ln(1-\nu)+\nu]\Big|_0^{\nu_f}$$

$$= \boxed{\frac{kNT}{2}[(1+\nu_f)\ln(1+\nu_f)+(1-\nu_f)\ln(1-\nu_f)]}$$

(iv) The expression for work is well behaved for displacements less than the length of the chain; however, for $\nu_f = \pm1$, we must be a bit more careful, for the expression above is indeterminate at these points. In particular, for expansion to the full length of the chain

$$w = \lim_{\nu\to 1}\frac{kNT}{2}[(1+\nu)\ln(1+\nu)+(1-\nu)\ln(1-\nu)]$$

$$= \frac{kNT}{2}\left[(1+1)\ln(1+1)+\lim_{\nu\to 1}(1-\nu)\ln(1-\nu)\right] = \frac{kNT}{2}\left[2\ln 2+\lim_{\nu\to 1}\frac{\ln(1-\nu)}{(1-\nu)^{-1}}\right]$$

where we have written the indeterminate term in the form of a ratio in order to apply l'Hôpital's rule. Focusing on the problematic limit and taking the required derivatives of numerator and denominator yields

$$\lim_{\nu\to 1}\frac{\ln(1-\nu)}{(1-\nu)^{-1}} = \lim_{\nu\to 1}\frac{-(1-\nu)^{-1}}{(1-\nu)^{-2}} = \lim_{\nu\to 1}[-(1-\nu)] = 0$$

Therefore $w = \frac{kNT}{2}(2\ln 2) = \boxed{kNT\ln 2}$

(c) For $\nu \ll 1$, the natural log can be expanded: $\ln(1+\nu) \approx \nu$ and $\ln(1-\nu) \approx -\nu$. Therefore,

$$|F| = \frac{kT}{2l}\ln\left(\frac{1+\nu}{1-\nu}\right) = \frac{kT}{2l}[\ln(1+\nu)-\ln(1-\nu)]$$

$$\approx \frac{kT}{2l}[\nu-(-\nu)] = \frac{\nu kT}{l} = \frac{nkT}{Nl} = \frac{xkT}{Nl^2}$$

(d) Figure 14.15 already suggested what the derivation in part (c) confirms: that the one-dimensional chain model and Hooke's law have the same behavior for small displacements.

Part (c) allows us to identify $\frac{kT}{Nl^2}$ as the Hooke's law force constant.

P14.51 (a) $q = n\Delta_c H^\ominus = \frac{1.5\,\text{g}}{342.3\,\text{g mol}^{-1}} \times (-5645\,\text{kJ mol}^{-1}) = \boxed{-25\,\text{kJ}}$

(b) Effective work available is $\approx 25\,\text{kJ}\times 0.25 = 6.2\,\text{kJ}$.

Because $w = mgh$ and $m \approx 65$ kg,

$$h \approx \frac{6.2\times 10^3\,\text{J}}{65\,\text{kg}\times 9.81\,\text{m s}^{-2}} = \boxed{9.7\,\text{m}}$$

(c) The energy released as heat is

$$q = -\Delta_r H = -n\Delta_c H^\ominus = -\left(\frac{2.5\,\text{g}}{180\,\text{g mol}^{-1}}\right)\times(-2808\,\text{kJ mol}^{-1}) = \boxed{39\,\text{kJ}}$$

(d) If one-quarter of this energy were available as work, a 65-kg person could climb to a height h given by

$$\frac{1}{4}q = w = mgh \quad \text{so} \quad h = \frac{q}{4\,mg} = \frac{39\times 10^3\,\text{J}}{4(65\,\text{kg})\times(9.8\,\text{m s}^{-2})} = \boxed{15\,\text{m}}$$

P14.53 (a) The Joule–Thomson coefficient is related to the given data by

$$\mu = -(1/C_p)(\partial H/\partial p)_T = -(-3.29\times10^3 \text{ J mol}^{-1} \text{ MPa}^{-1})/(110.0 \text{ J K}^{-1} \text{ mol}^{-1})$$

$$= \boxed{29.9 \text{ K MPa}^{-1}}$$

(b) The Joule–Thomson coefficient is defined as

$$\mu = (\partial T/\partial p)_H \approx (\Delta T/\Delta p)_H$$

Assuming that the expansion is a Joule–Thomson constant-enthalpy process, we have

$$\Delta T = \mu\Delta p = (29.9 \text{ K MPa}^{-1})\times[(0.5-1.5)\times10^{-1} \text{ MPa}] = \boxed{-2.99 \text{ K}}$$

15 The second law of thermodynamics

Answers to discussion questions

D15.1 We must remember that the second law of thermodynamics states only that the total entropy of both the system (here, the molecules organizing themselves into cells) and the surroundings (here, the medium) must increase in a naturally occurring process. It does not state that entropy must increase in a portion of the universe that interacts with its surroundings. In this case, the cells grow by using chemical energy from their surroundings (the medium), and in the process the increase in the entropy of the medium outweighs the decrease in entropy of the system. Hence, the second law is not violated.

D15.3 Everyday experience indicates that the direction of spontaneous change in an isolated system is accompanied by the dispersal of the total energy of the system. For example, for a gas expanding freely and spontaneously into a vacuum, the process is accompanied by a dispersal of energy and matter. It is easy to calculate the increase in the thermodynamic entropy that accompanies this process. For a perfect gas this entropy change is given by the formula $\Delta S = nR\ln\frac{V_\mathrm{f}}{V_\mathrm{i}}$ [eqn. 15.26], which is clearly positive if V_f is greater than V_i. The molecular interpretation of this thermodynamic result is based on the identification of entropy with molecular disorder. An increase in disorder results from the chaotic dispersal of matter and energy, and the only changes that can take place within an isolated system (the universe) are those in which this kind of dispersal occurs. This interpretation of entropy in terms of dispersal and disorder allows for a direct connection of the thermodynamic entropy to the statistical entropy through the Boltzmann formula $S = k\ln W$, where W is the number of microstates, the number of ways in which the molecules of the system can be arranged while keeping the total energy constant. The concept of the number of microstates makes quantitative the more ill-defined qualitative concepts of "disorder" and "the dispersal of matter and energy" used above to give a physical feel for the concept of entropy. A more "disorderly" distribution of energy and matter corresponds to a greater number of microstates associated with the same total energy.

D15.5 It was demonstrated in *Justification* 13.5 that for distinguishable particles the relationship between the canonical, Q, and the molecular, q, partition functions is $Q = q^N$. But if all the molecules are identical and free to move through space, we cannot distinguish between them and the relation $Q = q^N$ is not valid. The correct relationship is instead $Q = \frac{q^N}{N!}$. This can be demonstrated as follows. Suppose that molecule 1 is in some state a, 2 in b, and 3 in c. Then one member of the ensemble of these three molecules has energy $E = \varepsilon_a + \varepsilon_b + \varepsilon_c$. If the molecules are indistinguishable, this member of the ensemble is indistinguishable from one in which molecule 1 is in state b, 2 in c, and 3 in a, or some other permutation. Altogether in this case there are 3! = 6 indistinguishable such permutations, and in general for N molecules, there are N! such indistinguishable permutations. Hence the number of products of molecular partition functions that must be included in the canonical partition function is reduced by the factor $1/N$!

In general, the entropy is given by $S = \frac{U - U(0)}{T} + k \ln Q$. For distinguishable particles using $Q = q^N$ the entropy becomes $S = \frac{U - U(0)}{T} + Nk \ln q$. For indistinguishable particles, using $Q = \frac{q^N}{N!}$, the entropy becomes

$$S = \frac{U - U(0)}{T} + Nk \ln q - k \ln N!$$

D15.7 See *Justification* 13.2 for the derivation from the Boltzmann distribution of the formula for the partition function of a perfect gas, which is given by $q = \frac{V}{\Lambda^3}$, with $\Lambda = \frac{h}{(2\pi mkT)^{1/2}}$. The general expression for the entropy is $S = \frac{U - U(0)}{T} + k \ln Q$ [15.4]. For indistinguishable, non-interacting particles, $Q = \frac{q^N}{N!}$. After insertion of this expression for Q, we obtain for the entropy [see *Justification* 15.1] $S = \frac{U - U(0)}{T} + Nk \ln \frac{q\mathrm{e}}{N}$ [15.2b]. From this expression, we derive the Sackur-Tetrode equation, $S_\mathrm{m} = R \ln\left(\frac{V_\mathrm{m}\mathrm{e}^{5/2}}{N_\mathrm{A}\Lambda^3}\right)$ [15.5a], for the entropy of a perfect gas. See *Justification* 15.2 for the derivation. Because the molar volume appears in the numerator, the molar entropy increases in proportion to the natural log of the molar volume. In terms of the Boltzmann distribution, this is natural: large containers have more closely spaced energy levels than do small containers, so more states are thermally accessible. Because temperature appears in the numerator (the denominator of Λ), the molar entropy increases with temperature. The reason for this behavior from the point of view of the Boltzmann distribution is that more energy levels become accessible as temperature increases.

D15.9 Because solutions of cations cannot be prepared in the absence of anions and vice versa, in order to assign numerical values to the entropies of ions in solution, we arbitrarily assign the value of zero to the standard entropy of H^+ ions in water at all temperatures, i.e., $S^\ominus(\mathrm{H}^+, \mathrm{aq}) = 0$. With this choice, the entropies of ions in water are values relative to the hydrogen ion in water; hence they may be either positive or negative. Ion entropies vary as expected to the degree to which the ions order the water molecules around them in solution. Small, highly charged ions induce local structure in the surrounding water, and the disorder of the solution is decreased more than for the case of large, singly charged ions.

D15.11 All the thermodynamic properties of a system that we have encountered, U, H, S, A, and G, can be used as the criteria for the spontaneity of a process under specific conditions. The criteria are derived directly from the fundamental relation of thermodynamics, which is a combination of the first and second laws, namely,

$$-\mathrm{d}U - p_\mathrm{ext}\mathrm{d}V + \mathrm{d}w_\text{non-pV} + T\mathrm{d}S \geq 0$$

The inequality sign gives the criteria for the spontaneity of a process; the equality gives the criteria for equilibrium.

The specific conditions we are interested in and the criteria that follow from inserting these conditions into the fundamental relation are the following:

(1) Constant U and V, no work at all

$$\mathrm{d}S_{U,V} \geq 0$$

(2) Constant S and V, no work at all

$$\mathrm{d}U_{S,V} \leq 0$$

(3) Constant S and p, no work at all

$$dH_{S,p} \le 0$$

(4) Constant T

$$dA_T \le dw$$

(5) Constant T and V, only non-pV work

$$dA_{T,V} \le dw_{\text{non-pV}}$$

(6) Constant T and V, no work at all

$$dA_{T,V} \le 0$$

(7) Constant T and p, $p = p_{\text{ext}}$

$$dG_{T,p} \le dw_{\text{non-pV}}$$

(8) Constant T and p, no non-pV work

$$dG_{T,p} \le 0$$

Solutions to exercises

Assume that all gases are perfect and that data refer to 298.15 K unless otherwise stated.

E15.1(a) All spontaneous processes are irreversible processes, which implies through eqn. 15.19, the Clausius inequality, that $\Delta S_{\text{tot}} = \Delta S_{\text{sys}} + \Delta S_{\text{surr}} > 0$ for all spontaneous processes. In this case, $\Delta S_{\text{tot}} = 0$. Therefore, the process is $\boxed{\text{not spontaneous.}}$

E15.2(a) $\boxed{I_2(g)}$ will have the higher standard molar entropy at 298 K primarily because ΔS_{fus} and ΔS_{vap} are greater for I_2. At 298 K, I_2 is a solid in its standard state.

E15.3(a)
$$S_m^{\ominus} = R\ln\left(\frac{e^{5/2}kT}{p^{\ominus}\Lambda^3}\right) \quad [15.5\text{b with } p = p^{\ominus}]$$

(a)
$$\Lambda = \frac{h}{(2\pi mkT)^{1/2}} = \frac{6.626\times10^{-34}\ \text{J s}}{[(2\pi)\times(4.003)\times(1.6605\times10^{-27}\ \text{kg})\times(1.381\times10^{-23}\ \text{J K}^{-1}T)]^{1/2}}$$
$$= \frac{8.726\times10^{-10}\ \text{m}}{(T/\text{K})^{1/2}}$$

$$S_m^{\ominus} = R\ln\left(\frac{(e^{5/2})\times(1.381\times10^{-23}\ \text{J K}^{-1}T)}{(1.013\times10^5\ \text{Pa})\times(8.726\times10^{-10}\ \text{m})^3}\right)\times\left(\frac{T}{\text{K}}\right)^{3/2} = R\ln(2.499\times(T/\text{K})^{5/2})$$

$$T = 298.15\ \text{K},\quad S_m^{\ominus} = (8.314\ \text{J K}^{-1}\ \text{mol}^{-1})\times\ln(2.499\times(298)^{5/2})$$
$$\boxed{= 126\ \text{J K}^{-1}\ \text{mol}^{-1}}$$

(b)
$$\Lambda = \frac{h}{(2\pi mkT)^{1/2}} = \frac{6.626\times10^{-34}\ \text{J s}}{[(2\pi)\times(131.29)\times(1.6605\times10^{-27}\ \text{kg})\times(1.381\times10^{-23}\ \text{J K}^{-1}T)]^{1/2}}$$
$$= \frac{1.524\times10^{-10}\ \text{m}}{(T/\text{K})^{1/2}}$$

$$S_m^{\ominus} = R\ln\left(\frac{(e^{5/2})\times(1.381\times10^{-23}\ \text{J K}^{-1}T)}{(1.013\times10^5\ \text{Pa})\times(1.524\times10^{-10}\ \text{m})^3}\right)\times\left(\frac{T}{\text{K}}\right)^{3/2} = R\ln(469.1\times(T/\text{K})^{5/2})$$

$$T = 298.15\text{ K},\quad S_m^{\ominus} = (8.314\text{ J K}^{-1}\text{ mol}^{-1})\times\ln(469.1\times(298)^{5/2})$$
$$\boxed{= 169\text{ J K}^{-1}\text{ mol}^{-1}}$$

E15.4(a) From the solution to exercise E15.3(a) we have for helium

$$S_m^{\ominus} = (8.314\text{ J K}^{-1}\text{ mol}^{-1})\times\ln(2.499\times(T)^{5/2}) = 169\text{ J K}^{-1}\text{ mol}^{-1}.$$

We solve for T: $\boxed{T = 2.35\times10^3\text{ K}}$

E15.5(a) The rotational partition function of a non-linear molecule is

$$q^{\mathrm{R}} = \frac{1.0270}{\sigma}\,\frac{(T/\mathrm{K})^{3/2}}{(ABC/\mathrm{cm}^{-3})^{1/2}}\ \text{[Table 13.1]}$$
$$= \frac{1.0270\times298^{3/2}}{(2)\times(27.878\times14.509\times9.287)^{1/2}}\ [\sigma = 2\text{ from Table 13.3}] = \boxed{43.1}$$

The high-temperature approximation is valid if $T > \theta_{\mathrm{R}}$, where $\theta_{\mathrm{R}} = \dfrac{hc(ABC)^{1/3}}{k}$.

$$\theta_{\mathrm{R}} = \frac{hc(ABC)^{1/3}}{k}$$
$$= \frac{(6.626\times10^{-34}\text{ J s})\times(2.998\times10^{10}\text{ cm s}^{-1})\times[(27.878)\times(14.509)\times(9.287)\text{ cm}^{-3}]^{1/3}}{1.38\times10^{-23}\text{ J K}^{-1}}$$
$$= \boxed{22.36\text{ K}}$$

Thus the high-temperature approximation is valid.

$q^{\mathrm{R}} = 43.1$ [Exercise 15.5(a)]

All the rotational modes of water are fully active at 25°C [Exercise 15.5a]; therefore,

$$U_m^{\mathrm{R}} - U_m^{\mathrm{R}}(0) = E^{\mathrm{R}} = \frac{3}{2}RT$$

$$S_m^{\mathrm{R}} = \frac{E^{\mathrm{R}}}{T} + R\ln q^{\mathrm{R}}$$
$$= \frac{3}{2}R + R\ln 43.1 = \boxed{43.76\text{ J K}^{-1}\text{ mol}^{-1}}$$

COMMENT. Division of q^{R} by $N_{\mathrm{A}}!$ is not required for the internal contributions; internal motions may be thought of as localized (distinguishable). It is the overall canonical partition function, which is a product of internal and external contributions, that is divided by $N_{\mathrm{A}}!$

E15.6(a) We assume that the upper nine of the $(2\times\frac{9}{2}+1) = 10$ spin-orbit states of the ion lie at an energy much greater than kT at 1 K; hence, since the spin degeneracy of Co^{2+} is 4 (the ion is a spin quartet), $q = 4$. The contribution to the entropy is

$$R\ln q = (8.314\text{ J K}^{-1}\text{mol}^{-1})\times(\ln 4) = \boxed{11.5\text{ J K}^{-1}\text{ mol}^{-1}}$$

E15.7(a)

$$\Delta S = nR\ln\left(\frac{V_{\mathrm{f}}}{V_{\mathrm{i}}}\right)\ \text{[Example 15.1]}$$
$$= \left(\frac{15\text{ g}}{44\text{ g/mol}}\right)\times8.314\text{ J K}^{-1}\text{ mol}^{-1}\times\ln\left(\frac{3.0}{1.0}\right) = \boxed{3.1\text{ J K}^{-1}}$$

E15.8(a) The molar entropy of a collection of oscillators is given by

$$S_m = \frac{U_m - U_m(0)}{T} + k\ln Q \ [15.4] = \frac{N_A\langle\varepsilon\rangle}{T} + R\ln q$$

where $\langle\varepsilon\rangle = \frac{hc\tilde{\nu}}{e^{\beta hc\tilde{\nu}} - 1} = k\frac{\theta}{e^{\theta/T} - 1}$ [13.39], $q = \frac{1}{1 - e^{-\beta hc\tilde{\nu}}} = \frac{1}{1 - e^{-\theta/T}}$ [13.24]

and θ_V is the vibrational temperature $hc\tilde{\nu}/k$. Thus

$$S_m = \frac{R(\theta/T)}{e^{\theta/T} - 1} - R\ln(1 - e^{-\theta/T})$$

The vibrational entropy of formic acid is the sum of contributions of this form from each of its nine normal modes. The table below shows results from a spreadsheet programmed to compute S_m/R at a given temperature for the normal-mode wavenumbers of formic acid.

		$T = 298$ K		$T = 500$ K	
$\tilde{\nu}$ / cm^{-1}	θ_V/K	T/θ_V	S_m/R	T/θ_V	S_m/R
638	918.0	0.324635	1.624247	0.544689	4.340205
1033	1486.3	0.200501	0.341767	0.33641	1.769583
625	899.3	0.331387	1.707377	0.556019	4.471129
1105	1589.9	0.187436	0.254961	0.31449	1.500608
1229	1768.3	0.168525	0.153044	0.28276	1.127318
1387	1995.6	0.149327	0.079118	0.250549	0.779719
1770	2546.7	0.117015	0.015427	0.196334	0.312684
2943	4234.4	0.070376	8.53E-05	0.118081	0.016529
3570	5136.5	0.058016	4.95E-06	0.097342	0.003239
			4.176032		14.32101

(a) At 298 K, $S_m = 4.176R = \boxed{34.72\ \text{J mol}^{-1}\ \text{K}^{-1}}$.

(b) At 500 K, $S_m = 14.32R = \boxed{119.06\ \text{J mol}^{-1}\ \text{K}^{-1}}$.

COMMENT. These calculated values are the vibrational contributions to the standard molar entropy of formic acid. The total molar entropy would also include translational and rotational contributions, but without knowledge of the rotational constants, the total molar entropy cannot be calculated.

E15.9(a) (a) For H_2 we need to evaluate the expression for the rotational partition function term by term since $\tilde{B}$ is a significant fraction of kT at 298 K; hence the high-temperature approximation is not valid at that temperature.

$$q_R = \frac{1}{\sigma}\sum_J (2J+1)e^{-\frac{hc\tilde{B}}{kT}J(J+1)} \ [13.19], \ \sigma = 2 \text{ for } H_2$$

At 298.15 K, $kT/hc = 207.22$ cm^{-1} and q_R can be rewritten

$$q_R = \frac{1}{2}\sum_J (2J+1)e^{-\frac{\tilde{B}}{207.22\ \text{cm}^{-1}}J(J+1)} = \frac{1}{2}\sum_J (2J+1)e^{-0.29372J(J+1)}$$

q_R as a function of J converges rapidly to the value 1.8794.

We can now use $S_m = \frac{U_m - U_m(0)}{T} + k\ln Q$ [15.4] $= \frac{U_m - U_m(0)}{T} + R\ln q_R$

$$U_m - U_m(0) = N_A\langle\varepsilon^R\rangle = -\frac{1}{q^R}\left(\frac{\partial q^R}{\partial\beta}\right)_V$$

$$U_m - U_m(0) = \frac{1}{2q^R}N_A hc\tilde{B}\sum_{J=1}(2J+1)[J(J+1)]e^{-\beta hc\tilde{B}J(J+1)}$$

$$U_m - U_m(0) = \frac{N_A}{2\times1.8794}\times1.2090\times10^{-21}\ \mathrm{J}\times\sum_{J=1}(2J+1)[J(J+1)]e^{-0.29372J(J+1)}$$

$$U_m - U_m(0) = \frac{N_A}{2\times1.8794}\times1.2090\times10^{-21}\ \mathrm{J}\times11.5162 = 2.2307\ \mathrm{kJ/mol^{-1}}$$

$$S_m = \frac{U_m - U_m(0)}{T} + R\ln q_R = \frac{2230.7\ \mathrm{J/mol^{-1}}}{298.15\ \mathrm{K}} + 8.314\ \mathrm{J\,K^{-1}\,mol^{-1}}\times\ln1.8794$$

$$= \boxed{12.73\ \mathrm{J\,K^{-1}\,mol^{-1}}}$$

(b) For Cl_2, the high-temperature approximation is valid at 298 K and we can write

$$q_R = \frac{kT}{\sigma hc\tilde{B}} = \frac{207.22\ \mathrm{cm^{-1}}}{2\tilde{B}} = \frac{207.22\ \mathrm{cm^{-1}}}{2\times0.2441\ \mathrm{cm^{-1}}} = 424.5$$

We can now use $S_m = \frac{U_m - U_m(0)}{T} + k\ln Q$ [15.4] $= \frac{U_m - U_m(0)}{T} + R\ln q_R$ with $U_m - U_m(0) = 2RT$.

Therefore,

$$S_m = \frac{2RT}{T} + R\ln q_R = 2\times8.314\ \mathrm{J\,K^{-1}\,mol^{-1}} + 8.314\ \mathrm{J\,K^{-1}\,mol^{-1}}\times\ln424.5 = \boxed{66.94\ \mathrm{J\,K^{-1}\,mol^{-1}}}$$

COMMENT. In the solution to part (a) above we have used the symmetry number approach in treating the case of H_2. This effectively assumes that hydrogen is a 50/50 mixture of the nuclear ortho- and para- forms. If the calculation had been done separately for the para- and ortho- forms, the results would have been

$$12.61\ \mathrm{J\,K^{-1}\,mol^{-1}}\ \text{and}\ 12.86\ \mathrm{J\,K^{-1}\,mol^{-1}}$$

respectively. In actuality, equilibrium H_2 is not a 50/50 mixture of the ortho- and para- forms. Instead it is approximately a 3/1 mixture. See the solutions to Problems 14.26, 14.29, and 14.33 for further analysis of the H_2 case.

E15.10(a) Efficiency, η, is $\frac{\text{work performed}}{\text{heat absorbed}} = \frac{|w|}{q_h}$ [15.15] $= \frac{3.00\ \mathrm{kJ}}{10.00\ \mathrm{kJ}} = 0.300$. For an ideal heat engine we have $\eta_{rev} = 1 - \frac{T_c}{T_h}$ [15.17] $= 0.300 = 1 - \frac{T_c}{273.16\ \mathrm{K}}$. Solving for T_c, we obtain $\boxed{T_c = 191.2\ \mathrm{K}}$ as the temperature of the organic liquid.

E15.11(a) Assume that the block is so large that its temperature does not change significantly as a result of the heat transfer. Then

$$\Delta S = \int_i^f \frac{dq_{rev}}{T}\ [15.10] = \frac{1}{T}\int_i^f dq_{rev}\ [\text{constant } T] = \frac{q_{rev}}{T}$$

(a) $\Delta S = \frac{100\times10^3\ \mathrm{J}}{273.15\,\mathrm{K}} = \boxed{366\ \mathrm{J\,K^{-1}}}$ (b) $\Delta S = \frac{100\times10^3\ \mathrm{J}}{323.15\,\mathrm{K}} = \boxed{309\,\mathrm{J\,K^{-1}}}$

E15.12(a) Trouton's rule in the form $\Delta_{vap}H^{\ominus} = T_b\times85\ \mathrm{J\,K^{-1}\,mol^{-1}}$ can be used to obtain approximate enthalpies of vaporization. For benzene,

$$\Delta_{vap}H^{\ominus} = (273.2+80.1)\mathrm{K}\times85\ \mathrm{J\,K^{-1}\,mol^{-1}} = \boxed{30.0\ \mathrm{kJ/mol^{-1}}}$$

E15.13(a) $$S_m(T_f) = S_m(T_i) + \int_{T_i}^{T_f} \frac{C_{V,m}}{T}\,dT\,[15.30, \text{ with } C_{V,m} \text{ in place of } C_p]$$

If we assume that neon is a perfect gas, then $C_{V,m}$ can be taken to be constant and given by $C_{V,m} = C_{p,m} - R$.

$$C_{p,m} = 20.786\,\text{J K}^{-1}\,\text{mol}^{-1}\,[\text{Table 14.6}]$$
$$= (20.786 - 8.314)\ \text{J K}^{-1}\ \text{mol}^{-1}$$
$$= 12.472\ \text{J K}^{-1}\ \text{mol}^{-1}$$

Integrating, we obtain

$$S_m(500\,\text{K}) = S_m(298\,\text{K}) + C_{V,m}\ln\frac{T_f}{T_i}$$
$$= (146.22\,\text{J K}^{-1}\,\text{mol}^{-1}) + (12.472\,\text{J K}^{-1}\,\text{mol}^{-1})\ln\left(\frac{500\,\text{K}}{298\,\text{K}}\right)$$
$$= (146.22 + 6.45)\,\text{J K}^{-1}\,\text{mol}^{-1} = \boxed{152.67\,\text{J K}^{-1}\,\text{mol}^{-1}}$$

E15.14(a) Since entropy is a state function, ΔS may be calculated from the most convenient path, which in this case corresponds to constant-pressure heating followed by constant-temperature compression.

$$\Delta S = nC_{p,m}\ln\left(\frac{T_f}{T_i}\right)\ [15.30, \text{at } p_i] + nR\ln\left(\frac{V_f}{V_i}\right)\ [15.24, \text{at } T_f]$$

Since pressure and volume are inversely related (Boyle's law), $\frac{V_f}{V_i} = \frac{p_i}{p_f}$. Hence,

$$\Delta S = nC_{p,m}\ln\left(\frac{T_f}{T_i}\right) - nR\ln\left(\frac{p_f}{p_i}\right) = (3.00\,\text{mol})\times\frac{5}{2}\times(8.314\,\text{J K}^{-1}\,\text{mol}^{-1})\times\ln\left(\frac{398\,\text{K}}{298\,\text{K}}\right)$$
$$-(3.00\,\text{mol})\times(8.314\,\text{J K}^{-1}\,\text{mol}^{-1})\times\ln\left(\frac{5.00\,\text{atm}}{1.00\,\text{atm}}\right)$$
$$= (18.0\bar{4} - 40.1\bar{4})\,\text{J K}^{-1} = -22.1\,\text{J K}^{-1}$$

Though ΔS (system) is negative, the process can still occur spontaneously if ΔS (total) is positive.

E15.15(a) For an adiabatic reversible process, $q = q_{rev} = \boxed{0}$.

$$\Delta S = \int_i^f \frac{dq_{rev}}{T} = \boxed{0}$$

E15.16(a) Since the container is isolated, the heat flow is zero and therefore $\boxed{\Delta H = 0}$; since the masses of the blocks are equal, the final temperature must be their mean temperature, 25°C. Specific heat capacities are heat capacities per gram and are related to the molar heat capacities by

$$C_s = \frac{C_m}{M}\quad [C_{p,m} \approx C_{V,m} = C_m]$$

So $$nC_m = mC_s\ [nM = m]$$

$$\Delta S = mC_s\ln\left(\frac{T_f}{T_i}\right)[15.30]$$

$$\Delta S_1 = (1.00\times10^3\,\text{g})\times(0.385\,\text{J K}^{-1}\,\text{g}^{-1})\times\ln\left(\frac{298\,\text{K}}{323\,\text{K}}\right) = -31.0\,\text{J K}^{-1}$$

$$\Delta S_2 = (1.00\times10^3\,\text{g})\times(0.385\,\text{J K}^{-1}\,\text{g}^{-1})\times\ln\left(\frac{298\,\text{K}}{273\,\text{K}}\right) = 33.7\,\text{J K}^{-1}$$

$$\Delta S_{tot} = \Delta S_1 + \Delta S_2 = \boxed{+2.7\ \mathrm{J\,K^{-1}}}$$

COMMENT. The positive value of ΔS_{tot} corresponds to a spontaneous process.

E15.17(a) (a) $\Delta S(\mathrm{gas}) = nR\ln\frac{V_f}{V_i}[15.24] = \left(\frac{14\,\mathrm{g}}{28.02\,\mathrm{g\,mol^{-1}}}\right)\times(8.314\,\mathrm{J\,K^{-1}\,mol^{-1}})\times(\ln 2)$

$$= \boxed{+2.9\,\mathrm{J\,K^{-1}}}$$

$\Delta S(\mathrm{surroundings}) = \boxed{-2.9\,\mathrm{J\,K^{-1}}}$ [overall zero entropy production]

$\Delta S(\mathrm{total}) = \boxed{0}$ [reversible process]

(b) $\Delta S(\mathrm{gas}) = \boxed{+2.9\ \mathrm{J\,K^{-1}}}$ [S is a state function]

$\Delta S(\mathrm{surroundings}) = \boxed{0}$ [surroundings do not change]

$\Delta S(\mathrm{total}) = \boxed{+2.9\,\mathrm{J\,K^{-1}}}$

(c) $\Delta S(\mathrm{gas}) = \boxed{0}$ [$q_{rev} = 0$]

$\Delta S(\mathrm{surroundings}) = \boxed{0}$ [no heat is transferred to surroundings]

$\Delta S(\mathrm{total}) = \boxed{0}$

E15.18(a) (a) $\Delta_{vap}S = \frac{\Delta_{vap}H}{T_b} = \frac{29.4\times10^3\,\mathrm{J\,mol^{-1}}}{334.88\,\mathrm{K}} = \boxed{+87.8\,\mathrm{J\,K^{-1}\,mol^{-1}}}$

(b) If the vaporization occurs reversibly, $\Delta S_{tot} = 0$, so $\Delta S_{surr} = \boxed{-87.8\,\mathrm{J\,K^{-1}\,mol^{-1}}}$.

E15.19(a)

$$\Delta S = nC_p(\mathrm{H_2O,\ s})\ln\frac{T_f}{T_i} + n\frac{\Delta_{fus}H}{T_{fus}} + nC_p(\mathrm{H_2O,\ l})\ln\frac{T_f}{T_i} + n\frac{\Delta_{vap}H}{T_{vap}} + nC_p(\mathrm{H_2O,\ g})\ln\frac{T_f}{T_i}$$

$$n = \frac{10.0\ \mathrm{g}}{18.02\ \mathrm{g\ mol^{-1}}} = 0.555\ \mathrm{mol}$$

$$\Delta S = 0.555\ \mathrm{mol}\times38.02\ \mathrm{J\ K^{-1}\ mol^{-1}}\times\ln\frac{273.15}{263.15} + 0.555\ \mathrm{mol}\times\frac{6.008\ \mathrm{kJ/mol^{-1}}}{273.15\ \mathrm{K}}$$

$$+0.555\ \mathrm{mol}\times75.291\ \mathrm{J\ K^{-1}\ mol^{-1}}\times\ln\frac{373.15}{273.15}$$

$$+0.555\ \mathrm{mol}\times\frac{44.016\ \mathrm{kJ/mol^{-1}}}{373.15\ \mathrm{K}} + 0.555\ \mathrm{mol}\times33.58\ \mathrm{J\ K^{-1}\ mol^{-1}}\times\ln\frac{388.15}{373.15}$$

$$\boxed{\Delta S = 92.2\ \mathrm{J\ K^{-1}}}$$

COMMENT. This calculation is based on the assumption that the heat capacities remain constant over the range of temperatures involved and that the enthalpy of vaporization at 298.15 K given in Table 14.3 can be applied to the vaporization at 373.15 K. Neither assumption is strictly valid. Therefore, the calculated value is only approximate.

E15.20(a) In each case, $S_m = R \ln s$ where s is the number of orientations of about equal energy that the molecule can adopt. Therefore,

(a) $S_m = R \ln 3 = 8.3145 \text{ J K}^{-1} \text{ mol}^{-1} \times \ln 3 = \boxed{9.13 \text{ J K}^{-1} \text{ mol}^{-1}}$

(b) $S_m = R \ln 5 = 8.3145 \text{ J K}^{-1} \text{ mol}^{-1} \times \ln 5 = \boxed{13.4 \text{ J K}^{-1} \text{ mol}^{-1}}$

(c) $S_m = R \ln 6 = 8.3145 \text{ J K}^{-1} \text{ mol}^{-1} \times \ln 6 = \boxed{14.9 \text{ J K}^{-1} \text{ mol}^{-1}}$

E15.21(a) In each case $\Delta_r S^\ominus = \sum_{\text{Products}} \nu S_m^\ominus - \sum_{\text{Products}} \nu S_m^\ominus$ [15.32]

with S_m values obtained from Tables 14.5 and 14.6.

(a)
$$\Delta_r S = 2S_m^\ominus(\text{CH}_3\text{COOH, l}) - 2S_m^\ominus(\text{CH}_3\text{CHO, g}) - S_m^\ominus(\text{O}_2\text{, g})$$
$$= [(2\times159.8) - (2\times250.3) - 205.14]\,\text{J K}^{-1}\text{ mol}^{-1} = \boxed{-386.1\,\text{J K}^{-1}\text{ mol}^{-1}}$$

(b)
$$\Delta_r S^\ominus = 2S_m^\ominus(\text{AgBr, s}) + S_m^\ominus(\text{Cl}_2\text{, g}) - 2S_m^\ominus(\text{AgCl, s}) - S_m^\ominus(\text{Br}_2\text{, l})$$
$$= [(2\times107.1) + (223.07) - (2\times96.2) - (152.23)]\,\text{J K}^{-1}\text{ mol}^{-1}$$
$$= \boxed{+92.6\,\text{J K}^{-1}\text{ mol}^{-1}}$$

(c)
$$\Delta_r S^\ominus = S_m^\ominus(\text{HgCl}_2\text{, s}) - S_m^\ominus(\text{Hg, l}) - S_m^\ominus(\text{Cl}_2\text{, g})$$
$$= [146.0 - 76.02 - 223.07]\,\text{J K}^{-1}\text{ mol}^{-1} = \boxed{-153.1\,\text{J K}^{-1}\text{ mol}^{-1}}$$

E15.22(a) In each case we use

$$\Delta_r G^\ominus = \Delta_r H^\ominus - T\Delta_r S^\ominus \text{ [Example 15.6]}$$

along with

$$\Delta_r H^\ominus = \sum_{\text{Products}} \nu\Delta_f H^\ominus - \sum_{\text{Reactants}} \nu\Delta_f H^\ominus \text{[14.43b]}$$

(a)
$$\Delta_r H^\ominus = 2\Delta_f H^\ominus(\text{CH}_3\text{COOH, l}) - 2\Delta_f H^\ominus(\text{CH}_3\text{CHO, g})$$
$$= [2\times(-484.5) - 2\times(-166.19)]\,\text{kJ mol}^{-1} = -636.6\bar{2}\,\text{kJ mol}^{-1}$$

$$\Delta_r G^\ominus = -636.6\bar{2}\,\text{kJ mol}^{-1} - (298.15\,\text{K})\times(-386.1\,\text{J K}^{-1}\text{ mol}^{-1}) = \boxed{-521.5\,\text{kJ mol}^{-1}}$$

(b)
$$\Delta_r H^\ominus = 2\Delta_f H^\ominus(\text{AgBr, s}) - 2\Delta_f H^\ominus(\text{AgCl, s})$$
$$= [2\times(-100.37) - 2\times(-127.07)]\,\text{kJ mol}^{-1} = +53.40\,\text{kJ mol}^{-1}$$

$$\Delta_r G^\ominus = +53.40\,\text{kJ mol}^{-1} - (298.15\,\text{K})\times(+92.6)\,\text{J K}^{-1}\text{ mol}^{-1} = \boxed{+25.8\,\text{kJ mol}^{-1}}$$

(c)
$$\Delta_r H^\ominus = \Delta_f H^\ominus(\text{HgCl}_2\text{, s}) = -224.3\,\text{kJ mol}^{-1}$$

$$\Delta_r G^\ominus = -224.3\,\text{kJ mol}^{-1} - (298.15\,\text{K})\times(-153.1\,\text{J K}^{-1}\text{ mol}^{-1}) = \boxed{-178.7\,\text{kJ mol}^{-1}}$$

E15.23(a) In each case $\Delta_r G^\ominus = \sum_{\text{Products}} \nu\Delta_f G^\ominus - \sum_{\text{Reactants}} \nu\Delta_f G^\ominus$[15.51]

with $\Delta_f G^\ominus(\text{J})$ values from Tables 14.5 and 14.6.

(a)
$$\Delta_r G^\ominus = 2\Delta_f G^\ominus(\text{CH}_3\text{COOH, l}) - 2\Delta_f G^\ominus(\text{CH}_3\text{CHO, g})$$
$$= [2\times(-389.9) - 2\times(-128.86)]\,\text{kJ mol}^{-1} = \boxed{-522.1\,\text{kJ mol}^{-1}}$$

(b) $\Delta_r G^\ominus = 2\Delta_f G^\ominus(\text{AgBr, s}) - 2\Delta_f G^\ominus(\text{AgCl, s}) = [2\times(-96.90) - 2\times(-109.79)]\,\text{kJ mol}^{-1}$

$$= \boxed{+25.78\,\text{kJ mol}^{-1}}$$

(c) $\Delta_r G^\ominus = \Delta_f G^\ominus(\text{HgCl}_2\text{, s}) = \boxed{-178.6\,\text{kJ mol}^{-1}}$

COMMENT. In each case, these values of $\Delta_r G^\ominus$ agree closely with the calculated values in Exercise 15.22(a).

E15.24(a) $\Delta_r G^\ominus = \Delta_r H^\ominus - T\Delta_r S^\ominus$ [Example 15.6] $\Delta_r H^\ominus = \sum_{\text{Products}} \nu\Delta_f H^\ominus - \sum_{\text{Reactants}} \nu\Delta_f H^\ominus$ [14.43b]

$$\Delta_r S^\ominus = \sum_{\text{Products}} \nu S_m^\ominus - \sum_{\text{Reactants}} \nu S_m^\ominus \text{[15.32]}$$

$$\Delta_r H^\ominus = 2\Delta_f H^\ominus(\text{H}_2\text{O, l}) - 4\Delta_f H^\ominus(\text{HI, g}) = \{2\times(-285.83) - 4\times(+26.48)\}\,\text{kJ mol}^{-1}$$
$$= -677.58\,\text{kJ mol}^{-1}$$

$$\Delta_r S = 2S_m^\ominus(\text{I}_2\text{, s}) + 2S_m^\ominus(\text{H}_2\text{O, l}) - 4S_m^\ominus(\text{HI, g}) - S_m^\ominus(\text{O}_2\text{, g})$$
$$= [(2\times116.135) + (2\times69.91) - (4\times206.59) - (205.14)]\,\text{J K}^{-1}\,\text{mol}^{-1}$$
$$= -659.41\,\text{J K}^{-1}\,\text{mol}^{-1} = -0.65941\,\text{kJ K}^{-1}\,\text{mol}^{-1}$$

$$\Delta_r G^\ominus = -677.58\,\text{kJ mol}^{-1} - (298.15\,\text{K})\times(-0.65941\,\text{kJ K}^{-1}\,\text{mol}^{-1}) = \boxed{-480.98\,\text{kJ mol}^{-1}}$$

Question. Repeat the calculation based on $\Delta_f G$ data of Table 14.6. What difference, if any, is there from the value above?

E15.25(a) The formation reaction for ethyl acetate is

$$4\text{C(s)} + 4\text{H}_2\text{(g)} + \text{O}_2\text{(g)} \rightarrow \text{CH}_3\text{COOC}_2\text{H}_5\text{(l)}$$

$$\Delta_f G^\ominus = \Delta_f H^\ominus - T\Delta_f S^\ominus \text{ [Example 15.6]}$$

$\Delta_f H^\ominus$ is to be obtained from $\Delta_c H^\ominus$ for ethyl acetate and data from Tables 14.5 and 14.6.

Thus

$$\text{CH}_3\text{COOC}_2\text{H}_5\text{(l)} + 5\text{O}_2\text{(g)} \rightarrow 4\text{CO}_2\text{(g)} + 4\text{H}_2\text{O(l)}$$

$$\Delta_c H^\ominus = 4\Delta_f H^\ominus(\text{CO}_2\text{, g}) + 4\Delta_f H^\ominus(\text{H}_2\text{O, l}) - \Delta_f H^\ominus(\text{CH}_3\text{COOC}_2\text{H}_5\text{(l)})$$

$$\Delta_f H^\ominus(\text{CH}_3\text{COOC}_2\text{H}_5\text{(l)}) = 4\Delta_f H^\ominus(\text{CO}_2\text{, g}) + 4\Delta_f H^\ominus(\text{H}_2\text{O, l}) - \Delta_c H^\ominus$$
$$= [4\times(-393.51) + 4\times(-285.83) - (-2231)]\,\text{kJ mol}^{-1}$$
$$= -486.\bar{4}\,\text{kJ mol}^{-1}$$

$$\Delta_r S^\ominus = \sum_{\text{Products}} \nu S_m^\ominus - \sum_{\text{Reactants}} \text{[15.32]}$$

$$\Delta_f S = S_m^\ominus(\text{CH}_3\text{COOC}_2\text{H}_5\text{(l)}) - 4S_m^\ominus(\text{C,s}) - 4S_m^\ominus(\text{H}_2\text{,g}) - S_m^\ominus(\text{O}_2\text{,g})$$

$$= [259.4 - (4\times5.740) - (4\times130.68) - (205.14)]\,\text{J K}^{-1}\,\text{mol}^{-1}$$

$$= -491.4\,\text{J K}^{-1}\,\text{mol}^{-1}$$

Hence $\Delta_r G^\ominus = -486.\bar{4}\,\text{kJ mol}^{-1} - (298.15\,\text{K})\times(-491.\bar{4}\,\text{J K}^{-1}\,\text{mol}^{-1}) = \boxed{-340\text{ kJ mol}^{-1}}$

E15.26(a) $$CH_4(g)+2O_2(g)\rightarrow CO_2(g)+2H_2O(l)$$

$$\Delta_r G^{\ominus} = \sum_{\text{Products}} \nu\Delta_f G^{\ominus} - \sum_{\text{Rectants}} \nu\Delta_f G^{\ominus}[15.51]$$

$$\Delta_r G^{\ominus} = \Delta_f G^{\ominus}(CO_2, g) + 2\Delta_f G^{\ominus}(H_2O, l) - \Delta_f G^{\ominus}(CH_4, g)$$

$$= \{-394.36 + (2\times -237.13) - (-50.72)\}\,\text{kJ mol}^{-1} = -817.90\,\text{kJ mol}^{-1}$$

Therefore, the maximum non-expansion work is $\boxed{817.90\,\text{kJ mol}^{-1}}$ since $|w_{add}| = |\Delta G|$.

E15.27(a) $$\Delta G = nRT\ln\left(\frac{p_f}{p_i}\right)[15.66] = nRT\ln\left(\frac{V_i}{V_f}\right)\ \text{[Boyle's law]}$$

$$\Delta G = (2.5\times 10^{-3}\,\text{mol})\times(8.314\ \text{J K}^{-1}\,\text{mol}^{-1})\times(300\,\text{K})\times\ln\left(\frac{42}{600}\right) = \boxed{-17\,\text{J}}$$

E15.28(a) $$\left(\frac{\partial G}{\partial T}\right)_p = -S\,[15.60];\quad \text{hence}\quad \left(\frac{\partial G_f}{\partial T}\right)_p = -S_f,\quad \text{and}\quad \left(\frac{\partial G_i}{\partial T}\right)_p = -S_i$$

$$\Delta S = S_f - S_i = -\left(\frac{\partial G_f}{\partial T}\right)_p + \left(\frac{\partial G_i}{\partial T}\right)_p = -\left(\frac{\partial(G_f - G_i)}{\partial T}\right)_p$$

$$= -\left(\frac{\partial \Delta G}{\partial T}\right)_p = -\frac{\partial}{\partial T}\left(-85.40\,\text{J} + 36.5\,\text{J}\times\frac{T}{\text{K}}\right)$$

$$= \boxed{-36.5\,\text{J K}^{-1}}$$

E15.29(a) $dG = -S\,dT + V\,dp$ [15.59]; at constant T, $dG = V\,dp$; therefore

$$\Delta G = \int_{p_i}^{p_f} V\,dp$$

The change in volume of a condensed phase under isothermal compression is given by the isothermal compressibility (eqn. 14.59).

$$\kappa_T = -\frac{1}{V}\left(\frac{\partial V}{\partial p}\right)_T = 76.8\times 10^{-6}\,\text{atm}^{-1}\quad \text{[Table 14.7]}$$

This small isothermal compressibility (typical of condensed phases) tells us that we can expect a small change in volume from even a large increase in pressure. So we can make the following approximations to obtain a simple expression for the volume as a function of the pressure:

$$\kappa_T \approx -\frac{1}{V}\left(\frac{V - V_i}{p - p_i}\right) \approx -\frac{1}{V_i}\left(\frac{V - V_i}{p}\right)\quad \text{so}\quad V = V_i(1 - \kappa_T p)$$

where V_i is the volume at 1 atm, namely, the sample mass over the density, m/ρ.

$$\Delta G = \int_{1\,\text{atm}}^{3000\,\text{atm}} \frac{m}{\rho}(1 - \kappa_T p)\,dp$$

$$= \frac{m}{\rho}\left(\int_{1\,\text{atm}}^{3000\,\text{atm}} dp - \kappa_T\int_{1\,\text{atm}}^{3000\,\text{atm}} p\,dp\right)$$

$$= \frac{m}{\rho}\left(p\Big|_{1\,\text{atm}}^{3000\,\text{atm}} - \kappa_T\frac{p^2}{2}\Big|_{1\,\text{atm}}^{3000\,\text{atm}}\right)$$

$$= \frac{35 \text{ g}}{0.789 \text{ g cm}^{-3}}(2999 \text{ atm} - (76.8\times10^{-6} \text{ atm}^{-1})\times(9.00\times10^{6} \text{ atm}^{2})/2)$$

$$= 44.\bar{4}\,\text{cm}^3 \times \left(\frac{1\,\text{m}}{100\text{ cm}}\right)^3 \times 2653\,\text{atm}\times(1.013\times10^5\text{ Pa atm}^{-1})$$

$$= 1.2\times10^4\text{ J} = \boxed{12\,\text{kJ}}$$

E15.30(a) $\Delta G = nV_\text{m}\Delta p\,[15.65] = V\Delta p$

$$\Delta G = (1.0\text{ dm}^3)\times\left(\frac{1\text{m}^3}{10^3\text{ dm}^3}\right)\times(99\text{ atm})\times(1.013\times10^5\text{ Pa}) = 10\text{ kPa m}^3 = \boxed{+10\,\text{kJ}}$$

E15.31(a) $\Delta G_\text{m} = RT\ln\frac{p_\text{f}}{p_\text{i}}\,[15.66] = (8.314\text{ J K}^{-1}\text{ mol}^{-1})\times(298\,\text{K})\times\ln\left(\frac{100.0}{1.0}\right) = \boxed{+11\text{ kJ mol}^{-1}}$

E15.32(a) In each case the contribution to G is given by

$G - G(0) = -nRT\ln q$ [15.40; also see *Comment* to the solution to Exercise 15.5a]

Therefore, we first evaluate q^R and q^V

$$q^\text{R} = \frac{0.6950}{\sigma}\frac{T/K}{\tilde{B}/\text{cm}^{-1}}\quad[\text{Table13.1}, \sigma = 2]$$

$$= \frac{0.6950\times(298)}{(2)\times(0.3902)} = 265$$

$$q^\text{V} = \left(\frac{1}{1-\text{e}^{-a}}\right)\times\left(\frac{1}{1-\text{e}^{-b}}\right)^2\times\left(\frac{1}{1-\text{e}^{-c}}\right)\quad[\text{Table 13.1}]$$

with

$$a = \frac{(1.4388)\times(1388.2)}{298} = 6.70\bar{2}$$

$$b = \frac{(1.4388)\times(667.4)}{298} = 3.22\bar{2}$$

$$c = \frac{(1.4388)\times(2349.2)}{298} = 11.3\bar{4}$$

Hence

$$q^\text{V} = \frac{1}{1-\text{e}^{-6.702}}\times\left(\frac{1}{1-\text{e}^{-3.222}}\right)^2\times\frac{1}{1-\text{e}^{-11.34}} = 1.08\bar{6}$$

Therefore, the rotational contribution to the molar Gibbs energy is

$$-RT\ln q^\text{R} = -8.314\text{ J K}^{-1}\text{ mol}^{-1}\times298\text{ K}\times\ln 265$$

$$= \boxed{-13.8\text{ kJ mol}^{-1}}$$

and the vibrational contribution is

$$-RT\ln q^\text{V} = -8.314\text{ J K}^{-1}\text{ mol}^{-1}\times298\text{ K}\times\ln 1.08\bar{6} = \boxed{-0.20\text{ kJ mol}^{-1}}$$

E15.33(a) The partition function is

$$q = \sum_j g_j\,\text{e}^{-\beta\varepsilon_j}$$

with degeneracies $g_j = 2J + 1$

so $\quad q = 4 + 2e^{-\beta\varepsilon}[g(^2P_{3/2}) = 4,\ g(^2P_{1/2}) = 2]$

(a) At 500 K

$$\beta\varepsilon = \frac{(1.4388\,\text{cm K})\times(881\,\text{cm}^{-1})}{500\,\text{K}} = 2.53\bar{5}$$

Therefore, the contribution to G_m is

$$G_m - G_m(0) = -RT\ \ln q\ \text{[15.40; also see the \textit{Comment} to Exercise 15.5(a)]}$$

$$-RT\ \ln q = -(8.314\ \text{J K}^{-1}\,\text{mol}^{-1})\times(500\,\text{K})\times\ln(4+2\times e^{-2.535})$$
$$= -(8.314\ \text{J K}^{-1}\,\text{mol}^{-1})\times(500\,\text{K})\times(\ln 4 + 0.158) = \boxed{-6.42\,\text{kJ mol}^{-1}}$$

(b) At 900 K,

$$\beta\varepsilon = \frac{(1.4388\,\text{cm K})\times(881\,\text{cm}^{-1})}{900\,\text{K}} = 1.40\bar{8}$$

Therefore, the contribution to G_m is

$$G_m - G_m(0) = -RT\ \ln q$$

$$-RT\ \ln q = -(8.314\ \text{J K}^{-1}\,\text{mol}^{-1})\times(900\,\text{K})\times\ln(4+2\times e^{-1.408})$$
$$= -(8.314\ \text{J K}^{-1}\,\text{mol}^{-1})\times(900\,\text{K})\times(\ln 4 + 0.489) = \boxed{-14.0\,\text{kJ mol}^{-1}}$$

Solutions to problems

Solutions to numerical problems

P15.1 The absorption lines are the values of differences in adjacent rotational terms. Using eqns. 10.6, 10.7, and 10.8,

$$F(J+1) - F(J) = \frac{E(J+1) - E(J)}{hc} = 2\tilde{B}(J+1)$$

for $J = 0, 1, \ldots$. Therefore, we can find the rotational constant and reconstruct the energy levels from the data of Problem 13.12. To make use of all of the data, one would plot the wavenumbers, which represent $F(J+1) - F(J)$, versus J; the slope of that linear plot is $2\tilde{B}$. However, in this case, plotting the data is not necessary because inspection of the data shows that the lines in the spectrum are equally spaced with a separation of 21.19 cm^{-1}, so that is the slope:

$$\text{slope} = 21.19\ \text{cm}^{-1} = 2\tilde{B} \quad \text{and hence} \quad \tilde{B} = 10.59\bar{5}\ \text{cm}^{-1}$$

The partition function is

$$q = \sum_{J=0}^{\infty}(2J+1)e^{-\beta E(J)} \quad \text{where} \quad E(J) = hc\tilde{B}J(J+1)\ [10.6]$$

and the factor $(2J + 1)$ is the degeneracy of the energy levels. Making these substitutions into the partition function q, we obtain

$$q^R = \sum_J (2J+1)e^{-\frac{hc\tilde{B}}{kT}J(J+1)}[13.19] = \sum_J (2J+1)e^{-\frac{\theta_R}{T}J(J+1)} \quad [\theta_R = hc\tilde{B}/k]$$

For HCl, $\theta_R = hc\tilde{B}/k = 15.244$ K. Defining $x \equiv \dfrac{T}{\theta_R}$, q^R can be rewritten

$$q^R = \sum_J (2J+1)e^{-J(J+1)/x}$$

At temperatures above about 30 K, the high-temperature approximation for q^{R} would be adequate to calculate the molar entropy, but at lower temperatures the summation needs to be performed.

The molar entropy is calculated from $S_m = \frac{U_m - U_m(0)}{T} + k \ln Q$ [15.4] $= \frac{U_m - U_m(0)}{T} + R \ln q^{R}$ and the molar energy from

$$U_m - U_m(0) = N_A \langle \varepsilon^{R} \rangle = -\frac{1}{q^{R}}\left(\frac{\partial q^{R}}{\partial \beta}\right)_V$$

$$U_m - U_m(0) = \frac{1}{q^{R}} N_A hc\tilde{B} \sum_{J=1} (2J+1)[J(J+1)]e^{-J(J+1)/x} \text{ or since } hc\tilde{B} = k\theta_R$$

$$U_m - U_m(0) = \frac{1}{q^{R}} R\theta_R \sum_{J=1} (2J+1)[J(J+1)]e^{-J(J+1)/x}$$

Substituting into the expression for the entropy we obtain

$$S_m = \frac{1}{q^{R}} R \frac{1}{x} \sum_{J=1} (2J+1)[J(J+1)]e^{-J(J+1)/x} + R \ln q^{R}$$

S_m and q^{R} are best evaluated with a spreadsheet program such as Excel or a CAS system such as Mathcad. Here we have used Mathcad. See the Mathcad worksheet that follows. S_m/R is plotted as a function of $x \equiv \frac{T}{\theta_R}$ in Figure 15.1.

$$i := 0 \ldots 47$$

$$x_{start} := 0.2 \qquad x_{end} := 9.6$$

$$x_1 := x_{start} + (x_{end} - x_{start}) \cdot \frac{i}{47} \qquad x_i = T/\Theta$$

$$q_i := \sum_{J=0}^{6} \left[(2 \cdot J + 1) \cdot e^{-J \cdot \frac{J+1}{x_i}} \right] \qquad q_i = q^{R}$$

$$U_i := \left(\frac{1}{q_i}\right) \left[\frac{1}{x_i} \sum_{j=0}^{6} \left[(2 \cdot J + 1) \cdot J \cdot (J+1) \cdot e^{-\left(J \cdot \frac{J+1}{x_i}\right)} \right] \right] \qquad U_i = U_m / RT$$

$$\mathbf{S}_i := 8.314\ (U_i + \ln(q_1)) \qquad S_i = S_m\ R/R$$

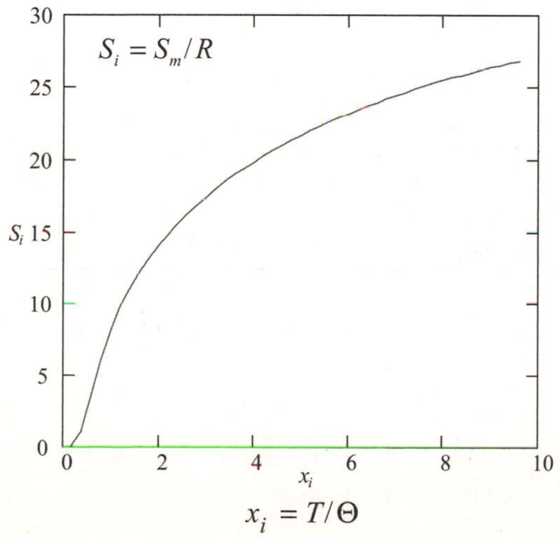

Figure 15.1

P15.3

$$\Delta S_{\rm m} = \int_{T_1}^{T_2} \frac{C_{p,\rm m}\,\mathrm{d}T}{T}[15.29] = \int_{T_1}^{T_2}\left(\frac{a+bT}{T}\right)\mathrm{d}T = a\ln\left(\frac{T_2}{T_1}\right) + b(T_2 - T_1)$$

$$a = 91.47\,\mathrm{J\,K^{-1}\,mol^{-1}}, \qquad b = 7.5\times10^{-2}\,\mathrm{J\,K^{-2}\,mol^{-1}}$$

$$\Delta S_{\rm m} = (91.47\,\mathrm{J\,K^{-1}\,mol^{-1}})\times\ln\left(\frac{300\ \mathrm{K}}{273\ \mathrm{K}}\right) + (0.075\,\mathrm{J\,K^{-2}\,mol^{-1}})\times(27\,\mathrm{K})$$

$$= \boxed{10.\overline{7}\,\mathrm{J\,K^{-1}\,mol^{-1}}}$$

P15.5 First, determine the final state in each section. In section B, the volume was halved at constant temperature, so the pressure was doubled: $p_{\rm B,f} = 2p_{\rm B,i}$. The piston ensures that the pressures are equal in both chambers, so $p_{\rm A,f} = 2p_{\rm B,i} = 2p_{\rm A,i}$. From the perfect gas law,

$$\frac{T_{\rm A,f}}{T_{\rm A,i}} = \frac{p_{\rm A,f}V_{\rm A,f}}{p_{\rm A,i}V_{\rm A,i}} = \frac{(2p_{\rm A,i})\times(3.00\,\mathrm{dm^3})}{(p_{\rm A,i})\times(2.00\,\mathrm{dm^3})} = 3.00 \qquad \text{so} \qquad T_{\rm A,f} = 900\ \mathrm{K}$$

(a)

$$\Delta S_{\rm A} = nC_{V,\rm m}\ln\left(\frac{T_{\rm A,f}}{T_{\rm A,i}}\right)[15.30] + nR\ln\left(\frac{V_{\rm A,f}}{V_{\rm A,i}}\right)[15.24]$$

$$\Delta S_{\rm A} = (2.0\,\mathrm{mol})\times(20\,\mathrm{J\,K^{-1}\,mol^{-1}})\times\ln 3.00 + (2.00\,\mathrm{mol})\times(8.314\,\mathrm{J\,K^{-1}\,mol^{-1}})\times\ln\left(\frac{3.00\,\mathrm{dm^3}}{2.00\,\mathrm{dm^3}}\right)$$

$$= \boxed{50.7\ \mathrm{J\,K^{-1}}}$$

$$\Delta S_{\rm B} = nR\ln\left(\frac{V_{\rm B,f}}{V_{\rm B,i}}\right) = (2.00\,\mathrm{mol})\times(8.314\,\mathrm{J\,K^{-1}\,mol^{-1}})\times\ln\left(\frac{1.00\,\mathrm{dm^3}}{2.00\,\mathrm{dm^3}}\right)$$

$$= \boxed{-11.5\,\mathrm{J\,K^{-1}}}$$

(b) The Helmholtz free energy is defined as $A = U - TS$ [3.29]. Because section B is isothermal, $\Delta U = 0$ and $\Delta(TS) = T\Delta S$, so

$$\Delta A_{\rm B} = -T_{\rm B}\Delta S_{\rm B} = -(300\ \mathrm{K})(-11.5\ \mathrm{J\,K^{-1}}) = 3.46\times10^3\ \mathrm{J} = \boxed{+3.46\ \mathrm{kJ}}$$

In section A, we cannot compute $\Delta(TS)$, so we cannot compute ΔU. ΔA $\boxed{\text{is indeterminate}}$ in both magnitude and sign. We know that in a perfect gas, U depends only on temperature; moreover, $U(T)$ is an increasing function of T, for $\frac{\partial U}{\partial T} = C$ (heat capacity), which is positive; since $\Delta T > 0$, $\Delta U > 0$ as well. But $\Delta(TS) > 0$ too, since both the temperature and the entropy increase.

(c) Likewise, under constant-temperature conditions,

$$\Delta G = \Delta H - T\Delta S$$

In section B, $\Delta H_{\rm B} = 0$ (constant temperature, perfect gas), so

$$\Delta G_{\rm B} = -T_{\rm B}\Delta S_{\rm B} = -(300\,\mathrm{K})\times(-11.5\,\mathrm{J\,K^{-1}}) = \boxed{3.46\times10^3\ \mathrm{J}}$$

$\Delta G_{\rm A}$ is $\boxed{\text{indeterminate}}$ in both magnitude and sign.

(d) $\Delta S(\text{total system}) = \Delta S_{\rm A} + \Delta S_{\rm B} = (50.7 - 11.5)\,\mathrm{J\,K^{-1}} = \boxed{+39.2\ \mathrm{J\,K^{-1}}}$

If the process has been carried out reversibly, as assumed in the statement of the problem, we can say

$$\Delta S(\text{system}) + \Delta S(\text{surroundings}) = 0$$

Hence, $\Delta S(\text{surroundings}) = \boxed{-39.2\,\mathrm{J\,K^{-1}}}$

Question. Can you design this process such that heat is added to section A reversibly?

P15.7

	q	w	$\Delta U = \Delta H$	ΔS	ΔS_{sur}	ΔS_{tot}
Path (a)	2.74 kJ	–2.74 kJ	0	9.13 J K^{-1}	–9.13 J K^{-1}	0
Path (b)	1.66 kJ	–1.66 kJ	0	9.13 J K^{-1}	–5.53 J K^{-1}	3.60 J K^{-1}

Path (a)

$$w = -nRT\ln\left(\frac{V_{\text{f}}}{V_{\text{i}}}\right)[14.12] = -nRT\ln\left(\frac{p_{\text{i}}}{p_{\text{f}}}\right)\text{ [Boyle's law]}$$

$$= -(1.00\,\text{mol})\times(8.314\,\text{J K}^{-1}\,\text{mol}^{-1})\times(300\,\text{K})\times\ln\left(\frac{3.00\,\text{atm}}{1.00\,\text{atm}}\right) = -2.74\times10^{3}\text{ J}$$

$$= \boxed{-2.74\,\text{kJ}}$$

$$\Delta H = \Delta U = \boxed{0}\text{ [isothermal process in perfect gas]}$$

$$q = \Delta U - w = 0-(-2.74\,\text{kJ}) = \boxed{+2.74\,\text{kJ}}$$

$$\Delta S = \frac{q_{\text{rev}}}{T}\text{[isothermal]} = \frac{2.74\times10^{3}\text{ J}}{300\,\text{K}} = \boxed{+9.13\text{ J K}^{-1}}$$

$$\Delta S_{\text{tot}} = \boxed{0}\text{ [reversible process]}$$

$$\Delta S_{\text{tot}} = \Delta S_{\text{sur}} = \Delta S_{\text{tot}} - \Delta S = 0 - 9.13\,\text{J K}^{-1} = \boxed{-9.13\text{ J K}^{-1}}$$

Path (b)

$$w = -p_{\text{ex}}(V_{\text{f}} - V_{\text{i}}) = -p_{\text{ex}}\left(\frac{nRT}{p_{\text{f}}} - \frac{nRT}{p_{\text{i}}}\right) = -nRT\left(\frac{p_{\text{ex}}}{p_{\text{f}}} - \frac{p_{\text{ex}}}{p_{\text{i}}}\right)$$

$$= -(1.00\text{ mol})\times(8.314\text{ J K}^{-1})\times(300\text{ K})\times\left(\frac{1.00\text{ atm}}{1.00\text{ atm}} - \frac{1.00\text{ atm}}{3.00\text{ atm}}\right)$$

$$= -1.66\times10^{3}\text{ J} = \boxed{-1.66\text{ kJ}}$$

$$\Delta H = \Delta U = \boxed{0}\text{ [isothermal process in perfect gas]}$$

$$q = \Delta U - w = 0-(-1.66\,\text{kJ}) = \boxed{+1.66\,\text{kJ}}$$

$$\Delta S = \frac{q_{\text{rev}}}{T}\text{[isothermal]} = \frac{2.74\times10^{3}\text{ J}}{300\,\text{K}} = \boxed{+9.13\text{ J K}^{-1}}$$

(*Note*: One can arrive at this by using q from Path (a) as the reversible path, or one can simply use ΔS from Path (a), realizing that entropy is a state function.)

$$\Delta S_{\text{sur}} = \frac{q_{\text{sur}}}{T_{\text{sur}}} = \frac{-q}{T_{\text{sur}}} = \frac{-1.66\times10^{3}\text{ J}}{300\,\text{K}} = \boxed{-5.53\,\text{J K}^{-1}}$$

$$\Delta S_{\text{tot}} = \Delta S + \Delta S_{\text{sur}} = (9.13-5.53)\,\text{J K}^{-1} = \boxed{+3.60\,\text{J K}^{-1}}$$

P15.9 ΔS depends on only the initial and final states, so we can use $\Delta S = nC_{p,\mathrm{m}} \ln \frac{T_\mathrm{f}}{T_\mathrm{i}}$ [15.30].

Since $q = nC_{p,\mathrm{m}}(T_\mathrm{f} - T_\mathrm{i}),\ T_\mathrm{f} = T_\mathrm{i} + \frac{q}{nC_{p,\mathrm{m}}} = T_\mathrm{i} + \frac{I^2Rt}{nC_{p,\mathrm{m}}}$ $[q = ItV = I^2Rt]$

That is, $\Delta S = nC_{p,\mathrm{m}} \ln\left(1 + \frac{I^2Rt}{nC_{p,\mathrm{m}}T_\mathrm{i}}\right)$

Since $n = \frac{500\ \mathrm{g}}{63.5\ \mathrm{g\ mol^{-1}}} = 7.87\ \mathrm{mol}$

$$\Delta S = (7.87\ \mathrm{mol})\times(24.4\ \mathrm{J\ K^{-1}\ mol^{-1}})\times\ln\left(1 + \frac{(1.00\ \mathrm{A})^2\times(1000\ \Omega)\times(15.0\ \mathrm{s})}{(7.87)\times(24.4\ \mathrm{J\ K^{-1}})\times(293\ \mathrm{K})}\right)$$

$$= (192\ \mathrm{J\ K^{-1}})\times(\ln 1.27) = \boxed{+45.4\ \mathrm{J\ K^{-1}}}$$

$[1\ \mathrm{J} = 1\ \mathrm{AVs} = 1\ \mathrm{A^2\Omega\ s}]$

For the second experiment, no change in state occurs for the copper; hence, $\Delta S(\text{copper}) = 0$. However, for the water, considered as a large heat sink,

$$\Delta S(\text{water}) = \frac{q}{T} = \frac{I^2Rt}{T} = \frac{(1.00\ \mathrm{A})^2\times(1000\ \Omega)\times(15.0\ \mathrm{s})}{293\ \mathrm{K}} = \boxed{+51.2\ \mathrm{J\ K^{-1}}}$$

P15.11 Consider the temperature as a function of pressure and enthalpy:

$$T = T(p,H) \quad \text{so} \quad \mathrm{d}T = \left(\frac{\partial T}{\partial p}\right)_H \mathrm{d}p + \left(\frac{\partial T}{\partial H}\right)_p \mathrm{d}H$$

The Joule–Thomson expansion is a constant enthalpy process (Section 14.11). Hence,

$$\mathrm{d}T = \left(\frac{\partial T}{\partial p}\right)_H \mathrm{d}p = \mu\,\mathrm{d}p$$

$$\Delta T = \int_{p_i}^{p_f} \mu\,\mathrm{d}p = \mu\Delta p \quad [\mu \text{ is constant}]$$

$$= (0.21\,\mathrm{K\,atm^{-1}})\times(1.00\,\mathrm{atm} - 100\,\mathrm{atm}) = \boxed{-21\ \mathrm{K}}$$

$$T_\mathrm{f} = T_\mathrm{i} + \Delta T = (373 - 21)\,\mathrm{K} = 352\,\mathrm{K}\ [\text{Mean } T = 363\,\mathrm{K}]$$

Consider the entropy as a function of temperature and pressure: $S = S(T, p)$.

Therefore, $\mathrm{d}S = \left(\frac{\partial S}{\partial T}\right)_p \mathrm{d}T + \left(\frac{\partial S}{\partial p}\right)_T \mathrm{d}p$

$$\left(\frac{\partial S}{\partial T}\right)_p = \frac{C_p}{T} \qquad \left(\frac{\partial S}{\partial p}\right)_T = -\left(\frac{\partial V}{\partial T}\right)_p \qquad \text{[Table 15.5]}$$

For $V_\mathrm{m} = \frac{RT}{p}(1 + Bp)$

$$\left(\frac{\partial V_\mathrm{m}}{\partial T}\right)_p = \frac{R}{p}(1 + Bp)$$

Then $\mathrm{d}S_{\mathrm{m}} = \dfrac{C_{p,\mathrm{m}}}{T}\mathrm{d}T - \dfrac{R}{p}(1+Bp)\,\mathrm{d}p$

or $\mathrm{d}S_{\mathrm{m}} = \dfrac{C_{p,\mathrm{m}}}{T}\mathrm{d}T - \dfrac{R}{p}\mathrm{d}p - RB\,\mathrm{d}p$

On integration,

$$\Delta S_{\mathrm{m}} = \int_1^2 \mathrm{d}S_{\mathrm{m}} = C_{p,\mathrm{m}}\ln\left(\frac{T_2}{T_1}\right) - R\ln\left(\frac{p_2}{p_1}\right) - RB(p_2-p_1)$$

$$= \frac{5}{2}R\ln\left(\frac{352}{373}\right) - R\ln\left(\frac{1}{100}\right) - R\left(-\frac{0.525\ \mathrm{atm}^{-1}}{363}\right)\times(-99\ \mathrm{atm})$$

$$= \boxed{+35.9\ \mathrm{J\,K^{-1}\,mol^{-1}}}$$

P15.13 $$S_{\mathrm{m}}(T) = S_{\mathrm{m}}(0) + \int_0^T \frac{C_{p,\mathrm{m}}\,\mathrm{d}T}{T} \quad [15.29]$$

From the data, draw up the following table.

T / K	10	15	20	25	30	50
$\frac{C_{p,\mathrm{m}}}{T}$ / (J K^{-2} mol^{-1})	0.28	0.47	0.540	0.564	0.550	0.428
T / K	70	100	150	200	250	298
$\frac{C_{p,\mathrm{m}}}{T}$ / (J K^{-2} mol^{-1})	0.333	0.245	0.169	0.129	0.105	0.089

Plot $C_{p,\mathrm{m}}/T$ against T (Figure 15.2). This has been done on two scales. The region 0 to 10 K has been constructed using $C_{p,\mathrm{m}} = aT^3$, fitted to the point at $T = 10$ K, at which $C_{p,\mathrm{m}} = 2.8$ J K^{-1} mol^{-1}, so $a = 2.8\times10^{-3}\ \mathrm{J\,K^{-4}\,mol^{-1}}$. The area can be determined (primitively) by counting squares. Area A = 38.28 J K^{-1} mol^{-1}. Area B up to 0°C = 25.60 J K^{-1} mol^{-1}; area B up to 25°C = 27.80 J K^{-1} mol^{-1}.

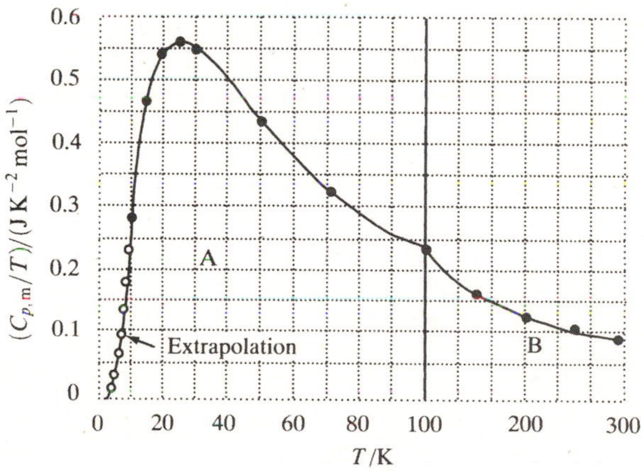

Figure 15.2

Hence,

(a) $S_{\mathrm{m}}(273\ \mathrm{K}) = S_{\mathrm{m}}(0) + \boxed{63.88\ \mathrm{J\,K^{-1}\,mol^{-1}}}$

(b) $S_{\mathrm{m}}(298\ \mathrm{K}) = S_{\mathrm{m}}(0) + \boxed{66.08\ \mathrm{J\,K^{-1}\,mol^{-1}}}$

P15.16 $S_m(T) = S_m(0) + \int_0^T \frac{C_{p,m}\,dT}{T}$ [15.29]

Perform a graphical integration by plotting $C_{p,m}/T$ against T and determining the area under the curve. Draw up the following table. (The last two columns come from determining areas under the curves described below.)

T/K	$\frac{C_{p,m}}{J\,K^{-1}\,mol^{-1}}$	$\frac{C_{p,m}/T}{J\,K^{-2}\,mol^{-1}}$	$\frac{S_m^{\ominus} - S_m^{\ominus}(0)}{J\,K^{-1}\,mol^{-1}}$	$\frac{H_m^{\ominus} - H_m^{\ominus}(0)}{kJ\,mol^{-1}}$
0.00	0.00	0.00	0.00	0.00
10.00	2.09	0.21	0.80	0.01
20.00	14.43	0.72	5.61	0.09
30.00	36.44	1.21	15.60	0.34
40.00	62.55	1.56	29.83	0.85
50.00	87.03	1.74	46.56	1.61
60.00	111.00	1.85	64.62	2.62
70.00	131.40	1.88	83.29	3.84
80.00	149.40	1.87	102.07	5.26
90.00	165.30	1.84	120.60	6.84
100.00	179.60	1.80	138.72	8.57
110.00	192.80	1.75	156.42	10.44
150.00	237.60	1.58	222.91	19.09
160.00	247.30	1.55	238.54	21.52
170.00	256.50	1.51	253.79	24.05
180.00	265.10	1.47	268.68	26.66
190.00	273.00	1.44	283.21	29.35
200.00	280.30	1.40	297.38	32.13

Plot $C_{p,m}$ against T (Figure 15.3a). Extrapolate to $T = 0$ using $C_{p,m} = aT^3$ fitted to the point at $T =$ 10 K, which gives $a = 2.09$ mJ K^{-2} mol^{-1}. Determine the area under the graph up to each T and plot A_m against T (Figure 15.3b).

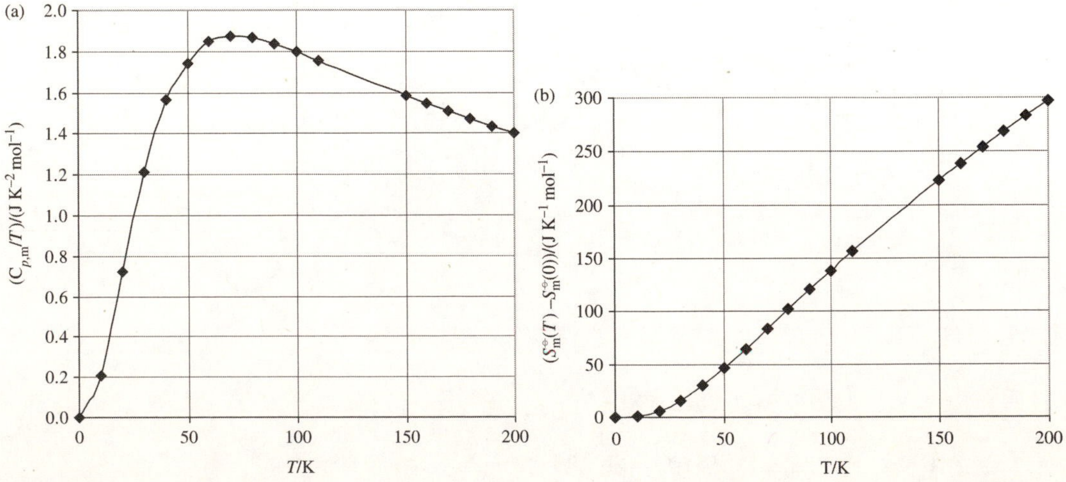

Figure 15.3

The molar enthalpy is determined in a similar manner from a plot of $C_{p,\mathrm{m}}$ against T by determining the area under the curve (Figure 15.4).

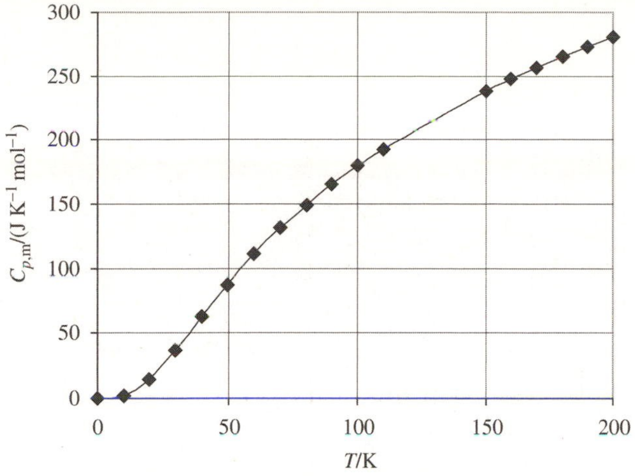

Figure 15.4

$$H_\mathrm{m}^\ominus(200\,\mathrm{K}) - H_\mathrm{m}^\ominus(0) = \int_0^{200\,\mathrm{K}} C_{p,\mathrm{m}}\,\mathrm{d}T = \boxed{32.1\ \mathrm{kJ\,mol^{-1}}}$$

P15.17 The entropy at 200 K is calculated from

$$S_\mathrm{m}^\ominus(200\ \mathrm{K}) = S_\mathrm{m}^\ominus(100\ \mathrm{K}) + \int_{100\ \mathrm{K}}^{200\ \mathrm{K}} \frac{C_{p,\mathrm{m}}\,\mathrm{d}T}{T}$$

The integrand may be evaluated at each data point; the transformed data appear below. The numerical integration can be carried out by a standard procedure such as the trapezoid rule (taking the integral within any interval as the mean value of the integrand times the length of the interval). Programs for performing this integration are readily available for personal computers. Many graphing calculators will also perform this numerical integration.

T / K	100	120	140	150	160	180	200
$C_{p,\mathrm{m}}$ / (J K^{-1} mol^{-1})	23.00	23.74	24.25	24.44	24.61	24.89	25.11
$\frac{C_{p,\mathrm{m}}}{T}$ / (J K^{-2} mol^{-1})	0.230	0.1978	0.1732	0.1629	0.1538	0.1383	0.1256

Integration by the trapezoid rule yields

$$S_\mathrm{m}^\ominus(200\ \mathrm{K}) = (29.79 + 16.81)\ \mathrm{J\,K^{-1}\ mol^{-1}} = \boxed{46.60\ \mathrm{J\,K^{-1}\ mol^{-1}}}$$

Taking $C_{p,\mathrm{m}}$ constant yields

$$S_\mathrm{m}^\ominus(200\ \mathrm{K}) = S_\mathrm{m}^\ominus(100\ \mathrm{K}) + C_{p,\mathrm{m}}\ \ln\,(200\ \mathrm{K}/100\ \mathrm{K})$$

$$= [29.79 + 24.44\ \ln(200\ /100\ \mathrm{K})]\ \mathrm{J\,K^{-1}\ mol^{-1}} = \boxed{46.60\ \mathrm{J\,K^{-1}\ mol^{-1}}}$$

The difference is slight.

P15.19 The Gibbs–Helmholtz equation [15.62b] may be recast into an analogous equation involving ΔG and ΔH since

$$\left(\frac{\partial \Delta G}{\partial T}\right)_p = \left(\frac{\partial G_\mathrm{f}}{\partial T}\right)_p - \left(\frac{\partial G_\mathrm{i}}{\partial T}\right)_p$$

and $\Delta H = H_f - H_i$

Thus, $\left(\frac{\partial}{\partial T}\frac{\Delta_r G^\ominus}{T}\right)_p = -\frac{\Delta_r H^\ominus}{T^2}$

$$\mathrm{d}\left(\frac{\Delta_r G^\ominus}{T}\right) = \left(\frac{\partial}{\partial T}\frac{\Delta_r G^\ominus}{T}\right)_p \mathrm{d}T \text{ [constant pressure]} = -\frac{\Delta_r H^\ominus}{T^2}\mathrm{d}T$$

$$\Delta\left(\frac{\Delta_r G^\ominus}{T}\right) = -\int_{T_c}^{T}\frac{\Delta_r H^\ominus\,\mathrm{d}T}{T^2}$$

$$\approx -\Delta_r H^\ominus \int_{T_c}^{T}\frac{\mathrm{d}T}{T^2} = \Delta_r H^\ominus\left(\frac{1}{T}-\frac{1}{T_c}\right) \quad [\Delta_r H^\ominus \text{ assumed constant}]$$

Therefore,

$$\frac{\Delta_r G^\ominus(T)}{T} - \frac{\Delta_r G^\ominus(T_c)}{T_c} \approx \Delta_r H^\ominus\left(\frac{1}{T}-\frac{1}{T_c}\right)$$

so $$\Delta_r G^\ominus(T) = \frac{T}{T_c}\Delta_r G^\ominus(T_c) + \left(1-\frac{T}{T_c}\right)\Delta_r H^\ominus(T_c)$$

$$= \tau\Delta_r G^\ominus(T_c) + (1-\tau)\Delta_r H^\ominus(T_c) \quad \text{where} \quad \tau = \frac{T}{T_c}$$

For the reaction $N_2(g) + 3H_2(g) \rightarrow 2NH_3(g)$, $\Delta_r G^\ominus = 2\Delta_f G^\ominus(NH_3, g)$.

(a) At 500 K, $\tau = \frac{500}{298} = 1.67\overline{8}$

so $$\Delta_r G^\ominus(500\,\mathrm{K}) = \{(1.67\overline{8})\times 2\times(-16.45) + (1-1.67\overline{8})\times 2\times(-46.11)\}\,\mathrm{kJ\,mol^{-1}}$$

$$\boxed{= -7\,\mathrm{kJ\,mol^{-1}}}$$

(b) At 1000 K, $\tau = \frac{1000}{298} = 3.35\overline{6}$

so $$\Delta_r G^\ominus(1000\,\mathrm{K}) = \{(3.35\overline{6})\times 2\times(-16.45) + (1-3.35\overline{6})\times 2\times(-46.11)\}\,\mathrm{kJ\,mol^{-1}}$$

$$\boxed{= +107\,\mathrm{kJ\,mol^{-1}}}$$

P15.21 The molar entropy is given by

$$S_m = \frac{U_m - U_m(0)}{T} + R\left(\ln\frac{q_m}{N_A} - 1\right) \quad \text{where} \frac{U_m - U_m(0)}{T} = -N_A\left(\frac{\partial \ln q}{\partial \beta}\right)_V$$

where $\frac{U_m - U_m(0)}{T} = -N_A\left(\frac{\partial \ln q}{\partial \beta}\right)_V$ and $\frac{q_m}{N_A} = \frac{q_m^{\,T}}{N_A}q^R q^V q^E$

The energy term $U_m - U_m(0)$ works out to be

$$U_m - U_m(0) = N_A[\langle\varepsilon^T\rangle + \langle\varepsilon^R\rangle + \langle\varepsilon^V\rangle + \langle\varepsilon^E\rangle]$$

Translation:

$$\frac{q_{\rm m}^{\rm T\ominus}}{N_{\rm A}} = 2.561\times10^{-2}(T/{\rm K})^{5/2}\times(M/{\rm g\,mol^{-1}})^{3/2} \text{ [Table 13.1]}$$

$$= 2.561\times10^{-2}\times(298)^{5/2}\times(38.00)^{3/2} = 9.20\times10^{6}$$

and $\langle\varepsilon^{\rm T}\rangle = \frac{3}{2}kT$

Rotation of a linear molecule:

$$q^{\rm R} = \frac{0.6950}{\sigma}\times\frac{T/{\rm K}}{\tilde{B}/{\rm cm^{-1}}} \text{ [Table 13.1]}$$

The rotational constant is

$$\tilde{B} = \frac{\hbar}{4\pi cI} = \frac{\hbar}{4\pi c\mu R^2}$$

$$= \frac{(1.0546\times10^{-34}\,{\rm J\,s})\times(6.022\times10^{23}\,{\rm mol^{-1}})}{4\pi(2.998\times10^{10}\,{\rm cm\,s^{-1}})\times(\frac{1}{2}\times19.00\times10^{-3}\,{\rm kg\,mol^{-1}})\times(190.0\times10^{-12}\,{\rm m})^2}$$

$$= 0.4915\,{\rm cm^{-1}}$$

so $q^{\rm R} = \frac{0.6950}{2}\times\frac{298}{0.4915} = 210.\bar{7}$

Also $\langle\varepsilon^{\rm R}\rangle = kT$

Vibration:

$$q^{\rm V} = \frac{1}{1-{\rm e}^{-hc\tilde{\nu}/kT}} = \frac{1}{1-\exp\left(\dfrac{-1.4388(\tilde{\nu}/{\rm cm^{-1}})}{T/{\rm K}}\right)} = \frac{1}{1-\exp\left(\dfrac{-1.4388(450.0)}{298}\right)}$$

$$= 1.129$$

$$\langle\varepsilon^{\rm V}\rangle = \frac{hc\tilde{\nu}}{{\rm e}^{hc\tilde{\nu}/kT}-1} = \frac{(6.626\times10^{-34}\,{\rm J\,s})\times(2.998\times10^{10}\,{\rm cm\,s^{-1}})\times(450.0\,{\rm cm^{-1}})}{\exp\left(\dfrac{1.4388(450.0)}{298}\right)-1}$$

$$= 1.149\times10^{-21}\,{\rm J}$$

The Boltzmann factor for the lowest-lying excited electronic state is

$$\exp\left(\frac{-(1.609\,{\rm eV})\times(1.602\times10^{-19}\,{\rm J\,eV^{-1}})}{(1.381\times10^{-23}\,{\rm J\,K^{-1}})\times(298\,{\rm K})}\right) = 6\times10^{-28}$$

so we can take $q^{\rm E}$ to equal the degeneracy of the ground state, namely 2, and $\langle\varepsilon^{\rm E}\rangle$ to be zero. Putting it all together yields

$$\frac{U_{\rm m}-U_{\rm m}(0)}{T} = \frac{N_{\rm A}}{T}\left(\frac{3}{2}kT+kT+1.149\times10^{-21}\,{\rm J}\right) = \frac{5}{2}R+\frac{N_{\rm A}(1.149\times10^{-21}\,{\rm J})}{T}$$

$$= (2.5)\times(8.3145\,{\rm J\,mol^{-1}\,K^{-1}}) + \frac{(6.022\times10^{23}\,{\rm mol^{-1}})\times(1.149\times10^{-21}\,{\rm J})}{298\,{\rm K}}$$

$$= 23.11\,{\rm J\,mol^{-1}\,K^{-1}}$$

$$R\left(\ln\frac{q_{\rm m}}{N_{\rm A}}-1\right) = (8.3145\,{\rm J\,mol^{-1}\,K^{-1}})\times\{\ln[(9.20\times10^{6})\times(210.7)\times(1.129)\times(2)]-1\}$$

$$= 176.3\,{\rm J\,mol^{-1}\,K^{-1}} \quad\text{and}\quad S_{\rm m}^{\circ} = \boxed{199.4\,{\rm J\,mol^{-1}\,K^{-1}}}$$

P15.23

$$w_{\text{add, max}} = \Delta_r G \ [15.49b]$$

$$\Delta_r G^{\ominus}(37°\text{C}) = \tau \Delta_r G^{\ominus}(T_c) + (1-\tau)\Delta_r H^{\ominus}(T_c) \quad \left[\text{Problem 15.19, } \tau = \frac{T}{T_c}\right]$$

$$= \left(\frac{310\,\text{K}}{298.15\,\text{K}}\right)\times(-6333\,\text{kJ mol}^{-1}) + \left(1-\frac{310\,\text{K}}{298.15\,\text{K}}\right)\times(-5797\,\text{kJ mol}^{-1})$$

$$= -6354\,\text{kJ mol}^{-1}$$

The difference is

$$\Delta_r G^{\ominus}(37°\text{C}) - \Delta_r G^{\ominus}(T_c) = \{-6354-(-6333)\}\,\text{kJ mol}^{-1} = \boxed{-21\,\text{kJ mol}^{-1}}$$

Therefore an additional 21 kJ mol^{-1} of non-expansion work may be done at the higher temperature.

COMMENT. As shown by Problem 15.18, increasing the temperature does not necessarily increase the maximum non-expansion work. The relative magnitude of $\Delta_r G^{\ominus}$ and $\Delta_r H^{\ominus}$ is the determining factor.

P15.25 In effect, we are asked to compute the maximum work extractable from a gallon of octane, assuming that the internal combustion engine is a reversible heat engine operating between the specified temperatures, and to equate that quantity of energy with gravitational potential energy of a 1000-kg mass. The efficiency is

$$\eta = \frac{|w|}{|q_h|}[15.15] = \frac{|w|}{|\Delta H|} = \eta_{\text{rev}} = 1-\frac{T_c}{T_h} \ [15.17] \quad \text{so} \quad |w| = |\Delta H|\left(1-\frac{T_c}{T_h}\right)$$

$$|\Delta H| = 5512\times10^3\ \text{J mol}^{-1}\times 1.00\ \text{gal}\times\frac{3.00\times10^3\ \text{g}}{1\ \text{gal}}\times\frac{1\ \text{mol}}{114.23\ \text{g}} = 1.4\bar{4}\times10^8\ \text{J}$$

so $$|w| = 1.44\bar{8}\times10^8\ \text{J}\times\left(1-\frac{1073\ \text{K}}{2273\ \text{K}}\right) = 7.64\bar{2}\times10^7\ \text{J}$$

If this work is converted completely to potential energy, it could lift a 1000-kg object to a height h given by $|w| = mgh$, so

$$h = \frac{|w|}{mg} = \frac{7.64\bar{2}\times10^7\ \text{J}}{(1000\ \text{kg})(9.81\ \text{m s}^{-2})} = 7.79\times10^3\ \text{m} = \boxed{7.79\ \text{km}}$$

Solutions to theoretical problems

P15.27 The partition function for a two-level system with energy separation ε is

$$q = 1 + e^{-\varepsilon/kT} = 1 + e^{-\beta\varepsilon}$$

The entropy is calculated from

$$S = \frac{E}{T} + Nk\ln q = k\beta E + Nk\ln q$$

$$E = -\frac{N}{q}\frac{\mathrm{d}q}{\mathrm{d}\beta} = \frac{N\varepsilon}{1+\mathrm{e}^{\beta\varepsilon}}$$

$$S = Nk\left[\frac{\beta\varepsilon}{1+\mathrm{e}^{\beta\varepsilon}} + \ln(1+\mathrm{e}^{-\beta\varepsilon})\right]$$

We now plot $\frac{S}{Nk} = \frac{S}{R}$ (for $N = N_A$) against $x \equiv \beta\varepsilon = \frac{\varepsilon}{kT}$ and against $y \equiv \frac{1}{x} = \frac{kT}{\varepsilon}$. These plots are shown in the following Mathcad worksheets.

$x := -10, -9.95 \ldots 10 \qquad y := -10, -9.95 \ldots 10$

$$S(x) := \left(\frac{x}{1+\mathrm{e}^{x}}\right) + \ln(1+\mathrm{e}^{-x})$$

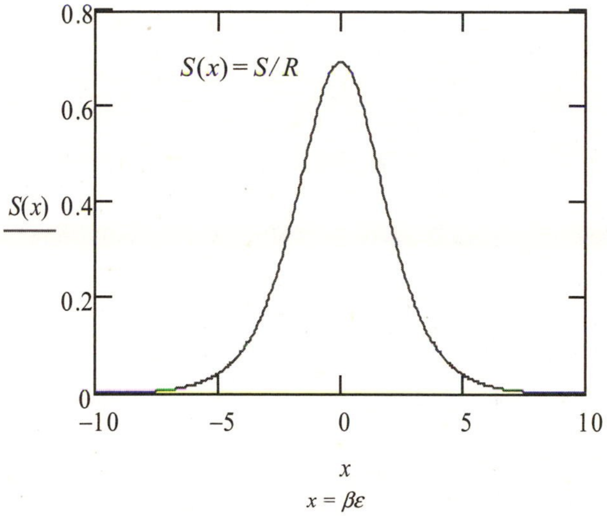

Figure 15.5(a)

$$S(y) := \left(\frac{1}{y}\right)\left(\frac{1}{1+\mathrm{e}^{\frac{1}{y}}}\right) + \ln\left(1+\mathrm{e}^{\frac{-1}{y}}\right)$$

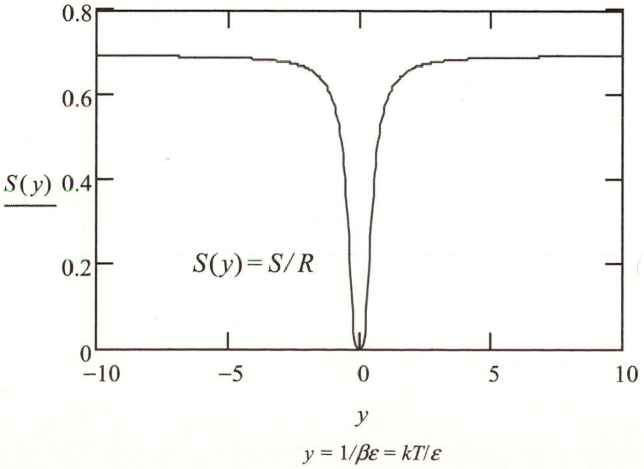

Figure 15.5(b)

Physically, the increase in entropy for $T > 0$ corresponds to the increasing accessibility of the upper state, and the decrease for $T < 0$ corresponds to the shift towards population of the upper state alone as more energy is packed into the system.

P15.29 Let us write Newton's law of cooling as follows:

$$\frac{dT}{dt} = -A(T - T_s)$$

where A is a constant characteristic of the system and T_S is the temperature of the surroundings. The negative sign appears because we assume that $T > T_S$. Separating variables $\frac{dT}{T - T_S} = -A\,dt$, and integrating, we obtain $\ln(T - T_S) = -At + K$, where K is a constant of integration.

Let T_i be the initial temperature of the system when $t = 0$. Then

$$K = \ln(T_i - T_S)$$

Introducing this expression for K gives

$$\ln\left(\frac{T - T_S}{T_i - T_S}\right) = -At \text{ or } T = T_S + (T_i - T_S)e^{-At}$$

$$\frac{dS}{dt} = \frac{d}{dt}\left(C \ln \frac{T}{T_i}\right) = \frac{d}{dt}(C \ln T)$$

From the above expression for T, we obtain $\ln T = \ln T_S - At \ln(T_i - T_S)$. Substituting $\ln t$, we obtain $\boxed{\frac{dS}{dt} = -CA \ln(T_i - T_S)}$, where now T_i can be interpreted as any temperature T during the course of the cooling process.

P15.31 The Otto cycle is represented in Figure 15.6. Assume 1 mole of air.

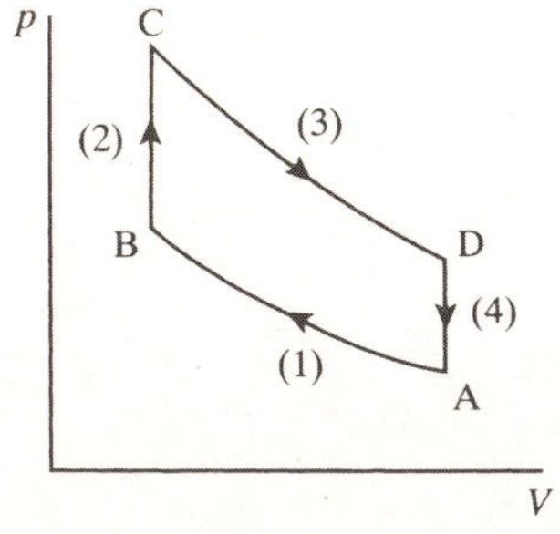

Figure 15.6

$$\eta = \frac{|w|_{\text{cycle}}}{|q_2|} \quad [15.15]$$

$$w_{\text{cycle}} = w_1 + w_3 = \Delta U_1 + \Delta U_3 \; [q_1 = q_3 = 0] = C_V(T_B - T_A) + C_V(T_D - T_C) \quad [14.36]$$

$$q_2 = \Delta U_2 = C_V(T_C - T_B)$$

$$\eta = \frac{|T_B - T_A + T_D - T_C|}{|T_C - T_B|} = 1 - \left(\frac{T_D - T_A}{T_C - T_B}\right)$$

We know that

$$\frac{T_A}{T_B} = \left(\frac{V_B}{V_A}\right)^{1/c} \quad \text{and} \quad \frac{T_D}{T_C} = \left(\frac{V_C}{V_D}\right)^{1/c} \quad [14.37a]$$

Since $V_B = V_C$ and $V_A = V_D$, $\dfrac{T_A}{T_B} = \dfrac{T_D}{T_C}$ or $T_D = \dfrac{T_A T_C}{T_B}$

then $$\eta = 1 - \frac{\dfrac{T_A T_C}{T_B} - T_A}{T_C - T_B} = 1 - \frac{T_A}{T_B} \quad \text{or} \quad \boxed{\eta = 1 - \left(\frac{V_B}{V_A}\right)^{1/c}}$$

Given that $C_{p,m} = {}^7/_2R$, we have $C_{V,m} = {}^5/_2R$ [14.35] and $c = \dfrac{2}{5}$.

For $\dfrac{V_A}{V_B} = 10$, $\eta = 1 - \left(\dfrac{1}{10}\right)^{2/5} = \boxed{0.47}$

$$\Delta S_1 = \Delta S_3 = \Delta S_{\text{sur},1} = \Delta S_{\text{sur},3} = \boxed{0} \quad \text{[adiabatic reversible steps]}$$

$$\Delta S_2 = C_{V,m} \ln\left(\frac{T_C}{T_B}\right)$$

At constant volume, $\left(\dfrac{T_C}{T_B}\right) = \left(\dfrac{p_C}{p_B}\right) = 5.0.$

$$\Delta S_2 = \left(\frac{5}{2}\right) \times (8.314\,\text{J K}^{-1}\,\text{mol}^{-1}) \times (\ln 5.0) = \boxed{+33\,\text{J K}^{-1}}$$

$$\Delta S_{\text{sur},2} = -\Delta S_2 = \boxed{-33\,\text{J K}^{-1}}$$

$$\Delta S_4 = -\Delta S_2 \left[\frac{T_C}{T_D} = \frac{T_B}{T_A}\right] = \boxed{-33\,\text{J K}^{-1}}$$

$$\Delta S_{\text{sur},4} = -\Delta S_4 = \boxed{+33\,\text{J K}^{-1}}$$

P15.33 Einstein solid: $C_{V,m}(T) = 3Rf_E(T)$ $\quad f_E(T) = \left(\dfrac{\theta_E}{T}\right)^2 \left(\dfrac{e^{\theta/2T}}{e^{\theta/T} - 1}\right)^2$

$$S_m = \int_0^T (C_{V,m}/T)\mathrm{d}T = 3R\int_0^T [f_E(T)/T]\mathrm{d}T$$

$$= 3R\int_0^T \left(\frac{\theta_E}{T}\right)^2 \left(\frac{1}{T}\right) \left\{\frac{e^{\theta/T}}{(1 - e^{\theta/T})^2}\right\} \mathrm{d}T$$

Set $x = \theta_E / T$, then $dT = -T^2 dx / \theta_E$; hence $\frac{S_m}{3R} = \int_{\theta_E/T}^{\infty}\left[\frac{xe^x}{(1-e^x)^2}\right]dx$ .This integral can be evaluated analytically. The result is $\frac{S_m}{3R} = \left[\frac{x}{e^x-1} - \ln(1-e^{-x})\right]$ with $x \equiv \frac{\theta_E}{T}$.

A more instructive alternative derivation starting from the partition function follows. The partition function or one dimension of vibration of an harmonic oscillator is

$$q^V = \frac{1}{1-e^{-\beta hc\tilde{\nu}}} \; [13.24] = \frac{1}{1-e^{-\theta_E/T}}$$

The energy is calculated from

$$U_m - U_m(0) = kT^2\left[\frac{\partial \ln Q}{\partial T}\right]_V \quad \text{with } Q^V = (q^V)^N$$

Performing the differentiation we obtain

$$U_m - U_m(0) = R\theta_E\left(\frac{1}{e^{\theta_E/T}-1}\right)$$

We have used $N = N_A$ and $N_A k = R$. The molar entropy is calculated from

$$S_m = \frac{U_m - U_m(0)}{T} + k\ln Q \; [15.4] = \frac{U_m - U_m(0)}{T} + R\ln q^V$$

Substituting for $U_m - U_m(0)$ and q^V and multiplying by 3 for the 3 dimensions of vibration in a crystal, we obtain $\frac{S_m}{3R} = \left[\frac{x}{e^x-1} - \ln(1-e^{-x})\right]$ with $x \equiv \frac{\theta_E}{T}$. Note that this is the same expression obtained above. We now plot this result against $y \equiv \frac{1}{x} = \frac{T}{\Theta_E}$.

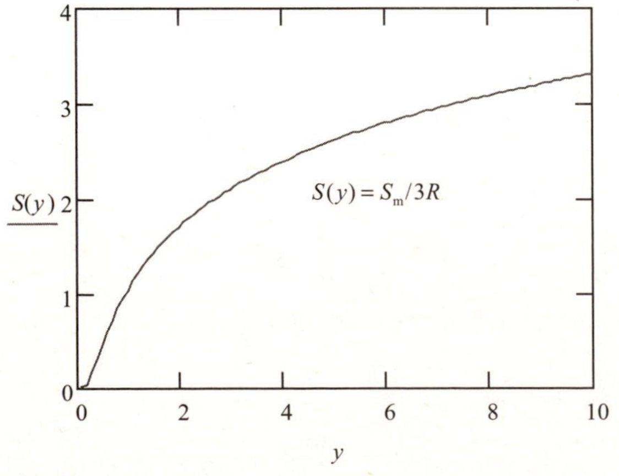

$y = T/\Theta_E$

Figure 15.7(a)

Debye solid: $C_{V,\text{m}}(T) = 3Rf_{\text{D}}(T) \quad f_{\text{D}}(T) = 3\left(\frac{T}{\theta_{\text{D}}}\right)^3 \int_0^{\theta_{\text{D}}/T} \frac{x^4 e^x}{(e^x - 1)^2} dx$

$$S_{\text{m}} = \int_0^T (C_{V,\text{m}}/T)dT = 3R\int_0^T [f_{\text{D}}(T)/T]dT$$

The function $f_{\text{D}}(T)$ can be integrated only numerically, not analytically, the integration yielding a table of numbers, one for each value of $\frac{\theta_{\text{D}}}{T}$. The entropy requires a second integration, which would be the area under a plot of $C_{V,\text{m}}$ against T. This double integration can certainly be accomplished rather easily with available software such as Mathcad, and we do this below. However we could again proceed as we did in the alternative derivation of the entropy equation for the Einstein solid, starting from the partition function for the Debye solid. This derivation is complex, but in essence it is conceptually the same as the calculation above for the case of the Einstein solid. The derivation finally yields an equation for the entropy that involves only a single numerical integration.

$$\frac{S}{3R} = \frac{3}{u^3}\int_0^u \left[\frac{x}{e^x - 1} - \ln(1 - e^{-x})\right] x^2 dx \text{ with } u \equiv \frac{\theta_{\text{D}}}{T}$$

In Figure 15.7b, we plot this function against $y = \frac{1}{u} = \frac{T}{\Theta_{\text{D}}}$.

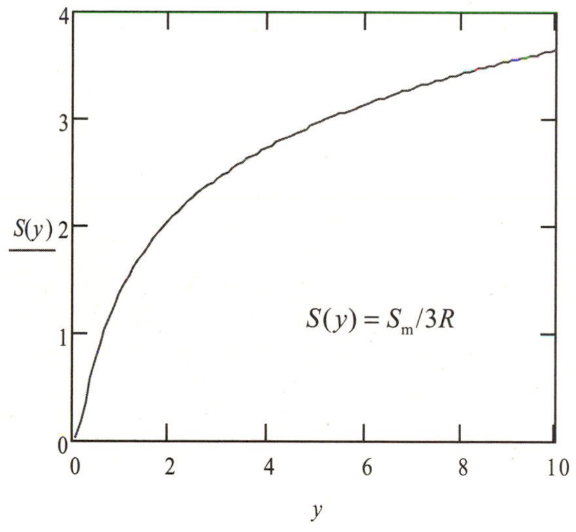

$y = T/\Theta_{\text{D}}$

Figure 15.7(b)

The entropy can be calculated by double integration of the equation

$$\frac{S_{\text{m}}}{3R} = \int_0^y 3y^2 \left[\int_0^{\frac{1}{y}} \frac{x^4 e^x}{(e^x - 1)^2} dx\right] dy$$

The plot obtained in Figure 15.7(c) is identical to the one in Figure 15.7(b). Also note that the Debye plots are not much different from the Einstein plot. It is a little steeper at low values of y.

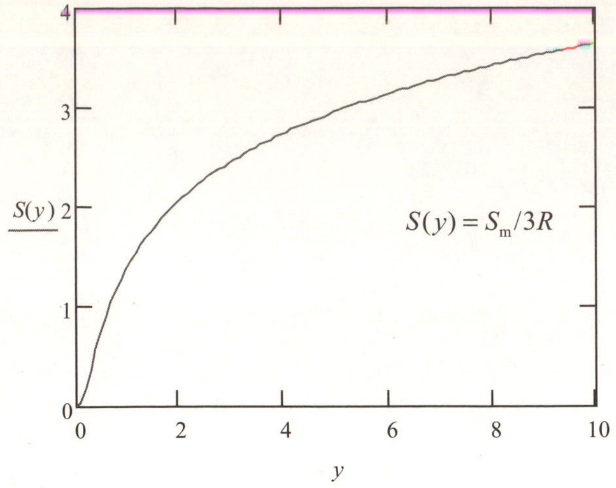

$y = T/\Theta_D$

Figure 15.7(c)

P15.35 $H \equiv U + pV$

$$dH = dU + p\,dV + V\,dp = T\,dS - p\,dV\ [3.43] + p\,dV + V\,dp = T\,dS + V\,dp$$

Since H is a state function, dH is exact, and it follows that

$$\left(\frac{\partial H}{\partial S}\right)_p = T \quad \text{and} \quad \boxed{\left(\frac{\partial V}{\partial S}\right)_p = \left(\frac{\partial T}{\partial p}\right)_S}$$

Similarly, $A \equiv U - TS$

$$dA = dU - T\,dS - S\,dT = T\,dS - p\,dV\ [3.43] - T\,dS - S\,dT = -p\,dV - S\,dT$$

Since dA is exact,

$$\boxed{\left(\frac{\partial S}{\partial V}\right)_T = \left(\frac{\partial p}{\partial T}\right)_V}$$

P15.37 If $S = S(T, p)$

then $dS = \left(\frac{\partial S}{\partial T}\right)_p dT + \left(\frac{\partial S}{\partial p}\right)_T dp$

$$T\,dS = T\left(\frac{\partial S}{\partial T}\right)_p dT + T\left(\frac{\partial S}{\partial p}\right)_T dp$$

Use $\left(\frac{\partial S}{\partial T}\right)_p = \left(\frac{\partial S}{\partial H}\right)_p\left(\frac{\partial H}{\partial T}\right)_p = \frac{1}{T}\times C_p \quad \left[\left(\frac{\partial H}{\partial S}\right)_p = T, \text{Problem 15.35}\right]$

$$\left(\frac{\partial S}{\partial p}\right)_T = -\left(\frac{\partial V}{\partial T}\right)_p \text{ [Maxwell relation]}$$

Hence, $T\,dS = C_p\,dT - T\left(\frac{\partial V}{\partial T}\right)_p dp = \boxed{C_p dT - \alpha TV\,dp}$

For reversible, isothermal compression, $TdS = dq_{rev}$ and $dT = 0$; hence

$$dq_{rev} = -\alpha TV\, dp$$

$$q_{rev} = \int_{p_i}^{p_f} -\alpha TV\, dp = \boxed{-\alpha TV\Delta p}\ [\alpha \text{ and } V \text{ assumed constant}]$$

For mercury,

$$q_{rev} = (-1.82\times10^{-4}\,\mathrm{K^{-1}})\times(273\,\mathrm{K})\times(1.00\times10^{-4}\,\mathrm{m^{-3}})\times(1.0\times10^{8}\,\mathrm{Pa}) = \boxed{-0.50\,\mathrm{kJ}}$$

P15.39 (a) $$U - U(0) = -\frac{N}{q}\frac{\partial q}{\partial \beta} = \frac{N}{q}\sum \varepsilon_j e^{-\beta\varepsilon_j} = \frac{NkT}{q}\dot{q} = \boxed{nRT\left(\frac{\dot{q}}{q}\right)}$$

$$S = \frac{U - U(0)}{T} + nR\ln(q+1) = \boxed{nR\left(\frac{\dot{q}}{q} + \ln(eq)\right)}$$

Note that division of q by N in the expression for the entropy is not required for internal degrees of freedom; division of q by N is required only for the total molecular partition function, which includes translational degrees of freedom.

(b) At 5000 K, $\dfrac{kT}{hc} = 3475\,\mathrm{cm^{-1}}$. We form the sums

$$q = \sum_j e^{-\beta\varepsilon_j} = 1 + e^{-21850/3475} + 3e^{-21870/3475} + \cdots = 1.0167$$

$$\dot{q} = \sum_j \frac{\varepsilon_j}{kT} e^{-\beta\varepsilon_j} = \frac{hc}{kT}\sum_j \tilde{\nu}_j e^{-\beta\varepsilon_j}$$

$$= \left(\frac{1}{3475}\right)\times\{0 + 21850\,e^{-21850/3475} + 3\times21870\,e^{-21870/3475} + \Lambda\} = 0.1057$$

$$\ddot{q} = \sum_j \left(\frac{\varepsilon_j}{kT}\right)^2 e^{-\beta\varepsilon_j} = \left(\frac{hc}{kT}\right)^2 \sum_j \tilde{\nu}_j^{\,2} e^{-\beta\varepsilon_j}$$

$$= \left(\frac{1}{3475}\right)^2 \times\{0 + 21850^2\,e^{-21850/3475} + 3\times21870^2\,e^{-21870/3475} + \Lambda\} = 0.6719$$

The electronic contribution to the standard molar entropy is

$$S_m = R\left(\frac{\dot{q}}{q} + \ln(eq)\right) = 8.314\ \mathrm{J\,K^{-1}\,mol^{-1}}\times\left(\frac{0.1057}{1.0167} + \ln(2.71828\times1.0167)\right)$$

$$= \boxed{9.316\ \mathrm{J\,K^{-1}\,mol^{-1}}}$$

P15.41 Begin with the partition function of an oscillator (Table 13.3).

$$q = \frac{1}{1-e^{-x}}, \quad x = \frac{\theta_V}{T} = hc\tilde{\nu}\beta = \hbar\omega\beta$$

The molar internal energy, molar entropy, and molar Helmholtz energy are obtained from the partition function as follows:

$$U - U(0) = -\frac{N}{q}\left(\frac{\partial q}{\partial \beta}\right)_V = -N(1-e^{-x})\frac{d}{d\beta}(1-e^{-x})^{-1} = \frac{N\eta\omega e^{-x}}{1-e^{-x}} = \boxed{\frac{N\eta\omega}{e^{x}-1}}$$

$$S = \frac{U - U(0)}{T} + nR\ln q = \frac{Nkxe^{-x}}{1-e^{-x}} - Nk\ln(1-e^{-x})$$

$$= \boxed{Nk\left(\frac{x}{e^x - 1} - \ln(1-e^{-x})\right)}$$

$$A - A(0) = G - G(0) = -nRT\ln q = \boxed{NkT\ \ln(1-e^{-x})}$$

The functions are plotted in Figure 15.8.

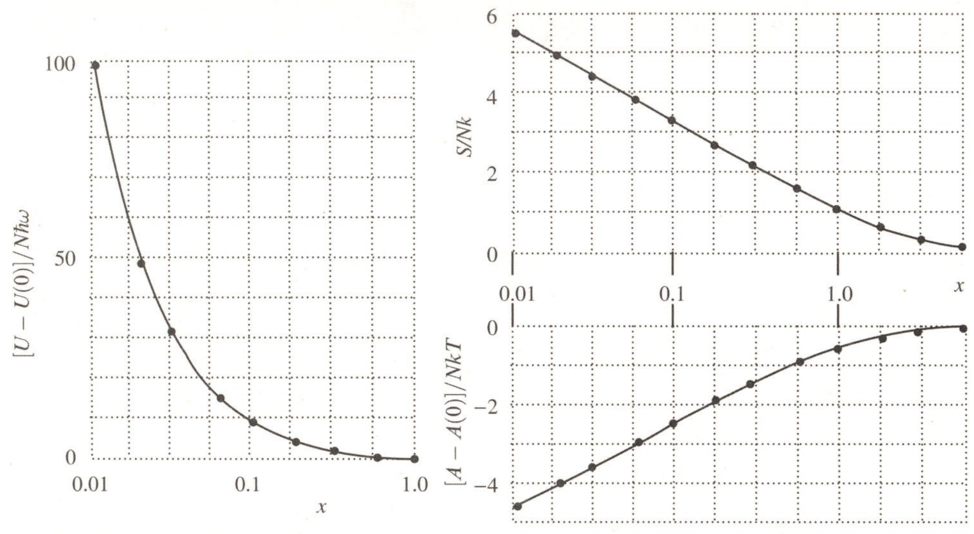

Figure 15.8

P15.43 We evaluate β by comparing calculated and experimental values for thermodynamic properties. The calculated values are obtained from the theoretical formulas for these properties, all of which are expressed in terms of the parameter β. So there can be as many ways of identifying β as there are thermodynamic properties. One way is through the energy, as shown in Section 13.1. Another is through the pressure, as demonstrated in Section 15.9. Another yet is through the entropy, Section 15.2, and this approach to the identification may be the most fundamental.

When plotted against T in model systems such as a two-level system, q and U show sharp discontinuities on passing through $T = 0$. Only S is continuous. However, when plotted against β, all three are continuous on passing through $\beta = 0$. See the solution to Problem 14.48. Hence β seems a more natural variable than T. Another way of looking at this is to consider τ, defined as $\tau = 1/\beta = kT$, to be the fundamental temperature, and then the "fundamental" constant k appears merely as a scale factor.

Solutions to applications

P15.45 $S = k\ln W$ [also see Exercises 15.20(a) and (b)]

so $S = k\ln 4^N = Nk\ln 4$

$$= (5\times10^8)\times(1.38\times10^{-23}\ \text{J K}^{-1})\times\ln 4 = \boxed{9.57\times10^{-15}\ \text{J K}^{-1}}$$

Question. Is this a large residual entropy? The answer depends on what comparison is made. Multiply the answer by Avogadro's number to obtain the molar residual entropy, $5.76\times10^9\ \text{J K}^{-1}\ \text{mol}^{-1}$, surely a large number—but then DNA is a macromolecule. The residual entropy per mole of base pairs may be a more reasonable quantity to compare to molar residual entropies of small molecules. To obtain that answer, divide the molecule's entropy by the number of base pairs before multiplying by N_A. The result is 11.5 J K^{-1} mol^{-1}, a quantity more in line with examples discussed in parts (a) and (b) of Exercise 15.20.

P15.47 (a) At constant temperature,

$$\Delta_r G = \Delta_r H - T\Delta_r S \quad \text{so} \quad \Delta_r S = \frac{\Delta_r H - \Delta_r G}{T}$$

and
$$\Delta_r S = \frac{[-20-(-31)]\text{ kJ mol}^{-1}}{310\text{ K}} = +0.035\text{ kJ K}^{-1}\text{ mol}^{-1} = \boxed{+35\text{ J K}^{-1}\text{ mol}^{-1}}$$

(b) The power density P is

$$P = \frac{|\Delta_r G|\, n}{V}$$

where n is the number of moles of ATP hydrolyzed per second

$$n = \frac{N}{N_A} = \frac{10^6\text{ s}^{-1}}{6.02\times10^{23}\text{ mol}^{-1}} = 1.6\bar{6}\times10^{-18}\text{ mol s}^{-1}$$

and V is the volume of the cell

$$V = \frac{4}{3}\pi r^3 = \frac{4}{3}\pi(10\times10^{-6}\text{m})^3 = 4.1\bar{9}\times10^{-15}\text{ m}^3$$

Thus
$$P = \frac{|\Delta_r G|\, n}{V} = \frac{(31\times10^3\text{ J mol}^{-1})\times(1.6\bar{6}\times10^{-18}\text{ mol s}^{-1})}{4.1\bar{9}\times10^{-15}\text{ m}^3} = \boxed{12\text{ W m}^{-3}}$$

This is orders of magnitude less than the power density of a computer battery, which is about

$$P_{\text{battery}} = \frac{15\text{ W}}{100\text{ cm}^3}\times\left(\frac{100\text{ cm}}{1\text{ m}}\right)^3 = \boxed{1.5\times10^4\text{ W m}^{-3}}$$

(c) Simply make a ratio of the magnitudes of the free energies.

$$\frac{14.2\text{ kJ (mol glutamine)}^{-1}}{31\text{ kJ (mol ATP)}^{-1}} = \boxed{0.46\,\frac{\text{mol ATP}}{\text{mol glutamine}}}$$

P15.49 The Gibbs–Helmholtz equation is

$$\frac{\partial}{\partial T}\left(\frac{\Delta G}{T}\right) = -\frac{\Delta H}{T^2}$$

so for a small temperature change,

$$\Delta\left(\frac{\Delta_r G^\ominus}{T}\right) = \frac{\Delta_r H^\ominus}{T^2}\Delta T \quad \text{and} \quad \frac{\Delta_r G_2^\ominus}{T_2} = \frac{\Delta_r G_1^\ominus}{T_1} - \frac{\Delta_r H^\ominus}{T^2}\Delta T$$

so
$$\int d\frac{\Delta_r G^\ominus}{T} = -\int\frac{\Delta_r H^\ominus dT}{T^2} \quad \text{and} \quad \frac{\Delta_r G_{190}^\ominus}{T_{190}} = \frac{\Delta_r G_{220}^\ominus}{T_{220}} + \Delta_r H^\ominus\left(\frac{1}{T_{190}} - \frac{1}{T_{220}}\right)$$

$$\Delta_r G_{190}^\ominus = \Delta_r G_{220}^\ominus\frac{T_{190}}{T_{220}} + \Delta_r H^\ominus\left(1 - \frac{T_{190}}{T_{220}}\right)$$

For the monohydrate,

$$\Delta_r G^{\ominus}_{190} = (46.2\,\text{kJ mol}^{-1}) \times \left(\frac{190\,\text{K}}{220\,\text{K}}\right) + (127\,\text{kJ mol}^{-1}) \times \left(1 - \frac{190\,\text{K}}{220\,\text{K}}\right)$$

$$\Delta_r G^{\ominus}_{190} = \boxed{57.2\,\text{kJ mol}^{-1}}$$

For the dehydrate,

$$\Delta_r G^{\ominus}_{190} = (69.4\,\text{kJ mol}^{-1}) \times \left(\frac{190\,\text{K}}{220\,\text{K}}\right) + (188\,\text{kJ mol}^{-1}) \times \left(1 - \frac{190\,\text{K}}{220\,\text{K}}\right)$$

$$\Delta_r G^{\ominus}_{190} = \boxed{85.6\,\text{kJ mol}^{-1}}$$

For the trihydrate,

$$\Delta_r G^{\ominus}_{190} = (93.2\,\text{kJ mol}^{-1}) \times \left(\frac{190\,\text{K}}{220\,\text{K}}\right) + (237\,\text{kJ mol}^{-1}) \times \left(1 - \frac{190\,\text{K}}{220\,\text{K}}\right)$$

$$\Delta_r G^{\ominus}_{190} = \boxed{112.8\,\text{kJ mol}^{-1}}$$

16 Physical equilibria

Answers to discussion questions

D16.1

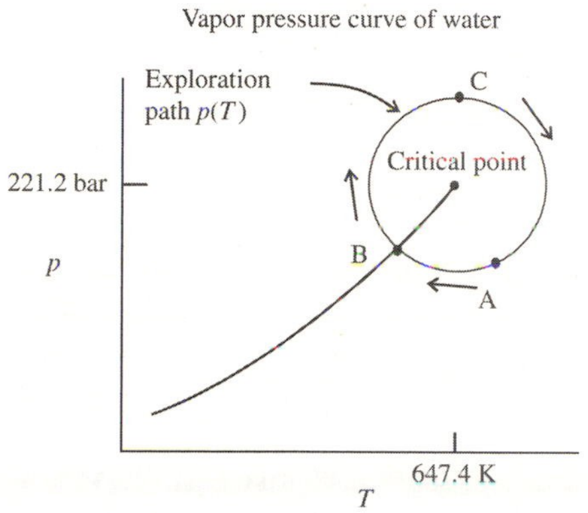

Figure 16.1

Refer to Figure 16.1 above and Fig. 16.6 in the text. Starting at point A and continuing clockwise on path $p(T)$ toward point B, we see only a gaseous phase within the container with water at pressures and temperatures $p(T)$. On reaching point B on the vapor pressure curve, liquid appears on the bottom of the container and a phase boundary or meniscus is evident between the liquid and the less dense gas above it. The liquid and gaseous phases are at equilibrium at this point. Proceeding clockwise away from the vapor pressure curve, the meniscus disappears, and the system becomes wholly liquid. Continuing along $p(T)$ to point C, at the critical temperature no abrupt changes are observed in the isotropic fluid. Before point C is reached, it is possible to return to the vapor pressure curve and a liquid–gas equilibrium by reducing the pressure isothermally. Continuing clockwise from point C along path $p(T)$ back to point A, no phase boundary is observed even though we now consider the water to have returned to the gaseous state. Additionally, if the pressure is isothermally reduced at any point after point C, it is impossible to return to a liquid–gas equilibrium.

When the path $p(T)$ is chosen to be very close to the critical point, the water appears opaque. At near-critical conditions, densities and refractive indices of the liquid and gas phases are nearly identical. Furthermore, molecular fluctuations cause spatial variations of densities and refractive indices on a scale large enough to strongly scatter visible light. This is called critical opalescence.

D16.3 The fundamental equation of chemical thermodynamics (eqn. 16.5) is a general expression that describes how the Gibbs energy changes with conditions (i.e., with pressure, temperature, and composition). Although eqn. 16.5 is general with respect to the behavior of the Gibbs function, the Gibbs function is only an indicator of spontaneous change under conditions of constant temperature and pressure. Under those conditions, the behavior of the Gibbs function is given by eqn. 16.6. At constant temperature and pressure, a system's Gibbs function decreases when the system undergoes spontaneous change, and it remains constant when the system is at equilibrium.

D16.5 The Clapeyron equation (eqn. 16.12) is exact and applies rigorously to all first-order phase transitions. It shows how pressure and temperature vary with respect to each other (temperature or pressure) along the phase boundary line, and in that sense, it defines the phase boundary line.

The Clausius–Clapeyron equation (eqn. 16.14) serves the same purpose, but it is not exact; its derivation involves approximations, in particular the assumptions that the perfect gas law holds and that the volume of condensed phases can be neglected in comparison to the volume of the gaseous phase. It applies only to phase transitions between the gaseous state and condensed phases.

Eqn. 16.13 is an exact corollary of the Clapeyron equation for a first-order phase transition, for it substitutes $\Delta_{trs}H/T$ for $\Delta_{trs}S$, quantities that are equal when two phases are at equilibrium. Enthalpy of transition and temperature are quantities easily measured in the laboratory.

D16.7 Mathematically, the reason is clear. The dependence of chemical potential on pressure, expressed in eqn. 16.10, follows from the chemical potential's status as a molar Gibbs energy and from the dependence of the Gibbs energy on conditions as expressed in the fundamental equation of chemical thermodynamics (eqn. 16.5).

The question implies that the dependence of chemical potential on pressure is not intuitive for an incompressible phase. Perhaps the dependence is more comprehensible for a compressible phase, because increasing pressure does work on a compressible phase. In that case, one can imagine a compressible phase (a vapor, for example) in equilibrium with an incompressible phase (say, a liquid) in a closed container. As the pressure increases, the chemical potential of the vapor increases, yet the chemical potentials of the two phases in equilibrium remain equal; therefore, the chemical potential of the liquid must also increase with pressure.

D16.9 A regular solution has excess entropy of zero but an excess enthalpy that is non-zero and dependent on composition, perhaps in the manner of eqn. 16.32. We can think of a regular solution as one in which the different molecules of the solution are distributed randomly, as in an ideal solution, but have different energies of interaction with each other. The parameter ξ is a measure of the difference between A-B interactions and A-A interactions. If $\xi < 0$, A-B interactions are energetically more favorable than A-A interactions; in that case, mixing is exothermic and even more spontaneous in all proportions than for an ideal solution. If, on the other hand, $\xi > 0$, A-B interactions are energetically less favorable than A-A interactions. In this case, mixing is endothermic. If ξ is not too large, mixing can still be spontaneous in all proportions. If ξ is larger, mixing is only spontaneous when one component dominates the composition. (See eqn. 16.33.)

D16.11 Phase: a state of matter that is uniform throughout, not only in chemical composition but also in physical state.
Constituent: any chemical species present in the system.
Component: a chemically independent constituent of the system. It is best understood in relation to the phrase "number of components," which is the minimum number of independent species necessary to define the composition of all the phases present in the system.
Degree of freedom (or variance): the number of intensive variables that can be changed without disturbing the number of phases in equilibrium.

Solutions to exercises

E16.1(a) See Figure 16.2.

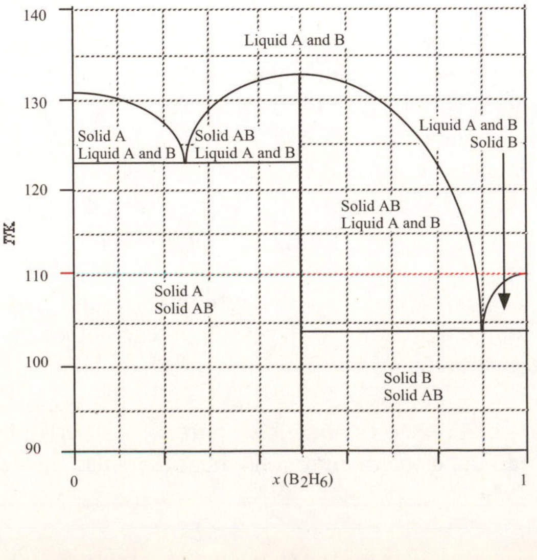

Figure 16.2

E16.2(a) Refer to Fig. 16.50 of the text. At the lowest temperature shown on the phase diagram, there are two liquid phases, a water-rich phase ($x_B = 0.07$) and a methylpropanol-rich phase ($x_B = 0.88$); the latter phase is about 10 times as abundant as the former (lever rule). On heating, the compositions of the two phases change, the water-rich phase increasing significantly in methylpropanol and the methylpropanol-rich phase more gradually increasing in water. (Note how the composition of the left side of the diagram changes more with temperature than does the composition of the right.) The relative proportions of the phases continue to be given by the lever rule. Just before the isopleth intersects the phase boundary, the methylpropanol-rich phase ($x_B = 0.8$) is in equilibrium with a vanishingly small water-rich phase ($x_B = 0.28$). Then the phases merge, and the single-phase region is encountered with $x_B = 0.8$.

E16.3(a) At equilibrium, the chemical potential of a substance is the same throughout a sample, regardless of how many phases are present [Section 16.3(b)]. Thus

$$\boxed{\mu_W(s) = \mu_W(l)} \quad \text{and} \quad \boxed{\mu_E(s) = \mu_E(l)}$$

E16.4(a) The Gibbs-Duhem equation relates changes in the chemical potential of a mixture's components to the composition:

$$d\mu_E = -\frac{n_W}{n_E} d\mu_W \quad [16.8]$$

For a small macroscopic change,

$$\delta\mu_E = -\frac{n_W}{n_E}\delta\mu_W = -\frac{0.60}{0.40}\times 0.25\ \text{J mol}^{-1} = \boxed{-0.38\ \text{J mol}^{-1}}$$

E16.5(a) The temperature dependence of the chemical potential is given by

$$\left(\frac{\partial \mu}{\partial T}\right)_p = -S_m \quad [16.9]$$

For a small macroscopic change,

$$\delta\mu = -S_m\delta T = (-69.91\ \text{J mol}^{-1}\ \text{K}^{-1}) \times (5\ \text{K}) = \boxed{-3\times 10^2\ \text{J mol}^{-1}}$$

E16.6(a) The pressure dependence of the chemical potential is given by

$$\left(\frac{\partial \mu}{\partial p}\right)_T = V_m \quad [16.10]$$

So
$$\Delta\mu = \int_{1.0\ \text{bar}}^{100\ \text{kbar}} \left(\frac{\partial \mu}{\partial p}\right)_T dp = \int_{1.0\ \text{bar}}^{100\ \text{kbar}} V_m dp = V_m \Delta p$$

$$= \frac{18.0\ \text{g}}{\text{mol}} \times \frac{\text{cm}^3}{0.997\ \text{g}} \times \left(\frac{1\ \text{m}}{100\ \text{cm}}\right)^3 \times 99\ \text{kbar} \times \frac{10^3 \times 10^5\ \text{Pa}}{1\ \text{kbar}} = \boxed{1.79\times 10^5\ \text{J mol}^{-1}}.$$

E16.7(a) The Clausius-Clapeyron equation gives the variation of vapor pressure with temperature if the vapor is a perfect gas.

$$\frac{d\ln p}{dT} = \frac{\Delta_{vap}H}{RT^2} \quad [16.14] \quad \text{so} \quad \int d\ln p = \int \frac{\Delta_{vap}H}{RT^2} dT$$

Assume that $\Delta_{vap}H$ is independent of temperature.

$$\ln\frac{p_2}{p_1} = \frac{\Delta_{vap}H}{R}\left(\frac{1}{T_1} - \frac{1}{T_2}\right)$$

$$\frac{1}{T_2}=\frac{1}{T_1}-\frac{R}{\Delta_{\text{vap}}H}\ln\frac{p_2}{p_1}=\frac{1}{297.25\text{ K}}+\frac{8.3145\text{ J K}^{-1}\text{ mol}^{-1}}{28.7\times10^3\text{ J mol}^{-1}}\times\ln\left(\frac{80.0}{53.3}\right)$$
$$=3.24\overline{7}\times10^{-3}\text{ K}^{-1},$$

$$T_2=\frac{1}{3.24\overline{7}\times10^{-3}\text{ K}^{-1}}=308.\overline{0}\text{ K}=\boxed{39°\text{C}}$$

E16.8(a) The Clapeyron equation relates the pressure and temperature of phase boundaries to state functions:

$$\frac{dp}{dT}=\frac{\Delta_{\text{trs}}S}{\Delta_{\text{trs}}V}\quad[16.12]$$

so $$\Delta_{\text{fus}}S=\Delta_{\text{fus}}V\times\left(\frac{dp}{dT}\right)\approx\Delta_{\text{fus}}V\times\frac{\Delta p}{\Delta T}$$

where the approximation holds if $\Delta_{\text{fus}}S$ and $\Delta_{\text{fus}}V$ are independent of temperature.

$$\Delta_{\text{fus}}S=[(163.3-161.0)\times(10^{-2}\text{ m})^3\text{ mol}^{-1}]\times\left(\frac{(100-1)\text{ atm}\times(1.013\times10^5\text{ Pa atm}^{-1})}{(351.26-350.75)\text{ K}}\right)$$

$$=\boxed{+45\text{ J K}^{-1}\text{ mol}^{-1}}.$$

$$\Delta_{\text{fus}}H=T_f\Delta_{\text{fus}}S=(350.75\text{ K})\times(45\text{ J K}^{-1}\text{ mol}^{-1})$$
$$=\boxed{+1.59\times10^4\text{ J mol}^{-1}}=\boxed{+15.9\text{ kJ mol}^{-1}}.$$

E16.9(a) The Clausius-Clapeyron equation gives the variation of vapor pressure with temperature if the vapor is a perfect gas.

$$\frac{d\ln p}{dT}=\frac{\Delta_{\text{vap}}H}{RT^2}\quad[16.14]$$

so $$\int d\ln p=\int\frac{\Delta_{\text{vap}}H}{RT^2}dT\qquad\text{and}\qquad\ln p=\text{constant}-\frac{\Delta_{\text{vap}}H}{RT}$$

Therefore, $\Delta_{\text{vap}}H=(1501.8\text{ K})\times R=(1501.8\text{ K})\times(8.3145\text{ J mol}^{-1}\text{ K}^{-1})$

$$\Delta_{\text{vap}}H=\boxed{+12487\text{ J mol}^{-1}}=\boxed{+12.487\text{ kJ mol}^{-1}}$$

E16.10(a) Use eqn. 16.13(a) and assume that $\Delta_{\text{fus}}H$ and $\Delta_{\text{fus}}V$ are independent of temperature:

$$\frac{dp}{dT}=\frac{\Delta_{\text{fus}}H}{T_f\Delta_{\text{fus}}V}\approx\frac{\Delta p}{\Delta T}$$

Thus $$\Delta T\approx\frac{T_f\Delta_{\text{fus}}V}{\Delta_{\text{fus}}H}\times\Delta p$$

Because the change in molar volume upon fusion (melting) is

$$\Delta_{\text{fus}}V=(78.11\text{ g mol}^{-1})\times\left(\frac{1}{0.879\text{ g cm}^{-3}}-\frac{1}{0.891\text{ g cm}^{-3}}\right)\times\left(\frac{\text{m}}{100\text{ cm}}\right)^3$$
$$=1.2\overline{0}\times10^{-6}\text{ m}^3\text{ mol}^{-1}$$

the change in melting temperature is

$$\Delta T\approx\frac{(278.6\overline{5}\text{ K})\times(1.2\overline{0}\times10^{-6}\text{ m}^3\text{ mol}^{-1})}{10.59\times10^3\text{ J mol}^{-1}}\times(10.0\times10^3-1)\text{ bar}\times\frac{10^5\text{ Pa}}{\text{bar}}$$
$$=31.\overline{5}\text{ K}.$$

Therefore, at 10.0 kbar, $T_f=(5.5+31.\overline{5})°\text{C}=\boxed{37°\text{C}}$

E16.11(a) Assume perfect gas behavior.

$$pV = nRT = \frac{mRT}{M} \quad \text{so} \quad m = \frac{pVM}{RT}$$

where $V = 4.0\text{ m} \times 4.0\text{ m} \times 3.0\text{ m} = 48\text{ m}^3$

(a) $$m = \frac{(3.2\times10^3\text{ Pa})\times(48\text{ m}^3)\times(18.02\text{ g mol}^{-1})}{(8.3145\text{ J K}^{-1}\text{ mol}^{-1})\times(298.15\text{K})} = \boxed{1.1\times10^3\text{ g}} = \boxed{1.1\text{ kg}}$$

(b) $$m = \frac{(13.1\times10^3\text{ Pa})\times(48\text{ m}^3)\times(78.11\text{ g mol}^{-1})}{(8.3145\text{ J K}^{-1}\text{ mol}^{-1})\times(298.15\text{K})} = \boxed{2.0\times10^4\text{ g}} = \boxed{20.\text{ kg}}$$

E16.12(a) (a) According to Trouton's rule (Section 15.3b),

$$\Delta_{\text{vap}}H = (85\text{ J K}^{-1}\text{ mol}^{-1})\times T_{\text{b}} = (85\text{ J K}^{-1}\text{ mol}^{-1})\times(76.8+273.15)\text{ K}$$

$$= \boxed{2.9\overline{7}\times10^4\text{ J mol}^{-1}} = \boxed{29.\overline{7}\text{ kJ mol}^{-1}}$$

(b) The Clausius-Clapeyron equation [16.14] integrates to

$$\ln\frac{p_2}{p_1} = \frac{\Delta_{\text{vap}}H}{R}\left(\frac{1}{T_1}-\frac{1}{T_2}\right) \quad \text{[Exercise 16.7(a)]}$$

Let T_1 be the normal boiling temperature, 350.0 K; then $p_1 = 1$ bar. Thus at 25°C,

$$\ln\frac{p_2}{\text{bar}} = \left(\frac{2.9\overline{7}\times10^3\text{ J mol}^{-1}}{8.3145\text{ J K}^{-1}\text{ mol}^{-1}}\right)\times\left(\frac{1}{350.0\text{ K}}-\frac{1}{298\text{ K}}\right) = -1.77\overline{6}$$

so $\quad p_2 = \boxed{0.169\text{ bar}}$

At $\quad 70°\text{C}, \ln\frac{p_2}{\text{bar}} = \left(\frac{2.9\overline{7}\times10^3\text{ J mol}^{-1}}{8.3145\text{ J K}^{-1}\text{ mol}^{-1}}\right)\times\left(\frac{1}{350.0\text{ K}}-\frac{1}{343\text{ K}}\right) = -0.20\overline{3}$

so $\quad p_2 = \boxed{0.82\text{ bar}}$

E16.13(a) $$\Delta T \approx \frac{T_{\text{f}}\Delta_{\text{fus}}V}{\Delta_{\text{fus}}H}\times\Delta p \quad \text{[Exercise 16.10(a)]},$$

where $$\Delta_{\text{fus}}V = (18.02\text{ g mol}^{-1})\times\left(\frac{1}{1.00\text{ g cm}^{-3}}-\frac{1}{0.92\text{ g cm}^{-3}}\right)\times\left(\frac{\text{m}}{100\text{ cm}}\right)^3$$

$$= -1.5\overline{7}\times10^{-6}\text{ m}^3\text{ mol}^{-1}$$

so $$\Delta T \approx \frac{(273.15\text{ K})\times(-1.5\overline{7}\times10^{-6}\text{ m}^3\text{ mol}^{-1})}{6008\text{ J mol}^{-1}}\times(50-1)\text{ bar}\times\frac{10^5\text{ Pa}}{\text{bar}} = -0.35\text{ K}$$

Therefore, at 50 bar, $T_{\text{f}} = \boxed{-0.35°\text{C}} = (273.15 - 0.35)\text{ K} = \boxed{272.80\text{ K}}$.

E16.14(a) $\Delta_{\text{mix}}G = nRT(x_{\text{A}}\ln x_{\text{A}} + x_{\text{B}}\ln x_{\text{B}})$ [16.18].

The mole fractions are equal, so $x_{\text{A}} = x_{\text{B}} = 0.50$. Because the gases are perfect, $pV = nRT$. Therefore,

$$\Delta_{\text{mix}}G = (pV)\times(0.50\ln0.50 + 0.50\ln0.50) = pV\ln0.50$$

$$= (1.5\text{ bar})\times\left(\frac{10^5\text{ Pa}}{\text{bar}}\right)\times10.0\times(10^{-1}\text{ m})^3\times(\ln0.50) = -1.0\overline{4}\times10^3\text{ J} = \boxed{-1.0\overline{4}\text{ kJ}}$$

$$\Delta_{\text{mix}}S = -nR(x_A \ln x_A + x_B \ln x_B)\ [16.20] = \frac{-\Delta_{\text{mix}}G}{T} = -\frac{-1.04\times10^3\ \text{J}}{298\ \text{K}}$$

$$= \boxed{+3.5\ \text{J K}^{-1}}$$

E16.15(a) $\Delta_{\text{mix}}G = nRT(x_A \ln x_A + x_B \ln x_B + x_C \ln x_C)$ [16.18 with three components]

We need to determine mole fractions from the given mass percentages. Assume a 100-g sample:

$$n_N = \frac{76\ \text{g}}{28.02\ \text{g mol}^{-1}} = 2.71\ \text{mol} \qquad n_O = \frac{23\ \text{g}}{32.00\ \text{g mol}^{-1}} = 0.72\ \text{mol}$$

$$n_{Ar} = \frac{1\ \text{g}}{39.95\ \text{g mol}^{-1}} = 0.03\ \text{mol} \qquad \text{and} \qquad n_{\text{total}} = 3.46\ \text{mol}$$

Thus $x_N = n_N / n_{\text{total}} = 2.71 / 3.46 = 0.78$; similarly $x_o = 0.21$ and $x_{Ar} = 0.007$, and for 1 mole of air at 298 K,

$$\Delta_{\text{mix}}G = (8.3145\ \text{J mol}^{-1}\ \text{K}^{-1}) \times (298\ \text{K})$$

$$\times (0.78 \ln 0.78 + 0.21 \ln 0.21 + 0.007 \ln 0.007)$$

$$\Delta_{\text{mix}}G = \boxed{-1.37\times10^3\ \text{J mol}^{-1}} = \boxed{-1.37\ \text{kJ mol}^{-1}}$$

$$\Delta_{\text{mix}}S = \frac{-\Delta_{\text{mix}}G}{T}\ [\text{Exercise 16.14(a)}] = -\frac{-1.37\times10^3\ \text{J mol}^{-1}}{298\ \text{K}} = \boxed{+4.6\ \text{J K}^{-1}\ \text{mol}^{-1}}$$

For a perfect gas, $\boxed{\Delta_{\text{mix}}H = 0}$ [no intermolecular interactions before or after mixing].

E16.16(a) Hexane and heptane form nearly ideal solutions, therefore eqn. 16.20 applies.

$$\Delta_{\text{mix}}S = -nR(x_A \ln x_A + x_B \ln x_B)$$

(a) To maximize $\Delta_{\text{mix}}S$, differentiate the equation with respect to x_A and look for the value of x_A which makes the derivative vanish. Since $x_B = 1 - x_A$, we need to differentiate

$$\Delta_{\text{mix}}S = -nR\{x_A \ln x_A + (1 - x_A) \ln (1 - x_A)\}$$

This yields

$$\frac{d\Delta_{\text{mix}}S}{dx_A} = -nR\left\{\ln x_A + \frac{x_A}{x_A} - \ln(1-x_A) - \frac{(1-x_A)}{(1-x_A)}\right\} = -nR \ln\frac{x_A}{1-x_A}$$

which is zero when $x_A = \boxed{\frac{1}{2}} = x_B$. Hence, the maximum entropy of mixing occurs for the preparation of a mixture that contains equal mole fractions of the two components.

(b) An equimolar mixture still maximizes the entropy of mixing; we only have to express the concentrations by mass fraction rather than mole fraction:

$$n_A = n_B \qquad \text{so} \qquad \frac{m_A}{M_A} = \frac{m_B}{M_B} = \frac{m_{\text{total}} - m_A}{M_B}$$

Solve for m_A in terms of m_{total}:

$$m_A = m_{\text{total}}\left(\frac{M_A}{M_A + M_B}\right)$$

But m_A / m_{total} is just the mass fraction of A (hexane, say). So the mass fractions are

$$\frac{m_{\text{hex}}}{m_{\text{total}}} = \frac{86.17\ \text{g mol}^{-1}}{(86.17 + 100.20)\ \text{g mol}^{-1}} = \boxed{0.4624} \qquad \text{and} \qquad \frac{m_{\text{hep}}}{m_{\text{total}}} = 1 - 0.4624 = \boxed{0.5376}$$

E16.17(a) Check whether $p_{\text{HCl}} / x_{\text{HCl}}$ is constant; if so, that is the Henry's law constant [16.26a].

x_{HCl}	0.005	0.012	0.019
$(p_{\text{HCl}} / \text{kPa}) / x_{\text{HCl}}$	6.4×10^3	6.4×10^3	6.4×10^3

Hence, $K_B = \boxed{6.4 \times 10^3 \text{ kPa}}$.

E16.18(a) The vapor pressures of components A and B may be expressed in terms of their compositions in the liquid (mole fractions x_A and x_B) or in the vapor (mole fractions y_A and y_B). The pressures are the same whatever the expression; hence the expressions can be set equal to each other and solved for the composition.

$$p_A = y_A p = 0.350p = x_A p_A^* = x_A \times (76.7 \text{ kPa})$$

$$p_B = y_B p = (1 - y_A)p = 0.650p = x_B p_B^* = (1 - x_A) \times (52.0 \text{ kPa})$$

Therefore, $\dfrac{y_A p}{y_B p} = \dfrac{x_A p_A^*}{x_B p_B^*}$

Hence $\dfrac{0.350}{0.650} = \dfrac{76.7 x_A}{52.0(1 - x_A)}$

which solves to $x_A = \boxed{0.268}$ and $x_B = 1 - x_A = \boxed{0.732}$.

Also, since $0.350p = x_A p_A^*$,

$$p = \frac{x_A p_A^*}{0.350} = \frac{(0.268) \times (76.7 \text{ kPa})}{0.350} = \boxed{58.6 \text{ kPa}}$$

E16.19(a) In Exercise 16.17(a), the Henry's law constant, K_B, was determined for concentrations expressed in mole fractions. To use that value, $K_B = 6.4 \times 10^3$ kPa, we must express the molality given here as a mole fraction. For convenience, assume a sample with 1 kg solvent ($GeCl_4$). Thus

$$n(\text{GeCl}_4) = \frac{1000 \text{ g}}{214.44 \text{ g mol}^{-1}} = 4.663 \text{ mol} \qquad \text{and} \qquad n(\text{HCl}) = 0.15 \text{ mol}$$

Therefore, $x_{\text{HCl}} = \dfrac{0.15 \text{ mol}}{0.15 \text{ mol} + 4.663 \text{ mol}} = 0.031$

$$p_{\text{HCl}} = (0.031) \times (6.4 \times 10^3 \text{ kPa}) = \boxed{2.0 \times 10^2 \text{ kPa}}$$

E16.20(a) We assume that the solvent, benzene (A), is ideal and obeys Raoult's law; then

$$p_A = x_A p_A^* \quad \text{and} \quad x_A = \frac{n_A}{n_A + n_B}$$

Hence $p_A = \dfrac{n_A p_A^*}{n_A + n_B}$; which solves to

$$n_B = \frac{n_A (p_A^* - p_A)}{p_A}$$

Then, since $n_B = \dfrac{m_B}{M_B}$, where m_B is the mass of B (solute) present,

$$M_B = \frac{m_B p_A}{n_A (p_A^* - p_A)} = \frac{m_B M_A p_A}{m_A (p_A^* - p_A)}$$

From the data,

$$M_B = \frac{(19.0 \text{ g}) \times (78.11 \text{ g mol}^{-1}) \times (51.5 \text{ kPa})}{(500 \text{ g}) \times (53.3 - 51.5) \text{ kPa}} = \boxed{85 \text{ g mol}^{-1}}$$

E16.21(a) With concentrations expressed in molality, Henry's law is $p_B = b_B K_B$ [16.26b].

Solving for b, the solubility, we have $b_B = \dfrac{p_B}{K_B}$.

(a) $$b = \frac{0.10 \text{ atm}}{3.01\times10^3 \text{ kPa kg mol}^{-1}} \times \frac{101 \text{ kPa}}{1 \text{ atm}} = \boxed{3.4\times10^{-3} \text{ mol kg}^{-1}}$$

(b) $$b = \frac{1.00 \text{ atm}}{3.01\times10^3 \text{ kPa kg mol}^{-1}} \times \frac{101 \text{ kPa}}{1 \text{ atm}} = \boxed{3.4\times10^{-2} \text{ mol kg}^{-1}}$$

E16.22(a) The best value of the molar mass is obtained from values of the data extrapolated to zero concentration, since it is under this condition that the van't Hoff equation applies.

$$\Pi = [\text{B}]RT = \frac{n_B RT}{V} \text{ [16.28]} \quad \text{so} \quad \Pi = \frac{m_B RT}{M_B V} = \frac{cRT}{M_B} \quad \text{where} \quad c = \frac{m}{V}$$

At the same time, the osmotic pressure is the hydrostatic pressure:

$$\Pi = \rho g h, \quad \text{so} \quad h = \left(\frac{RT}{\rho g M_B}\right)c$$

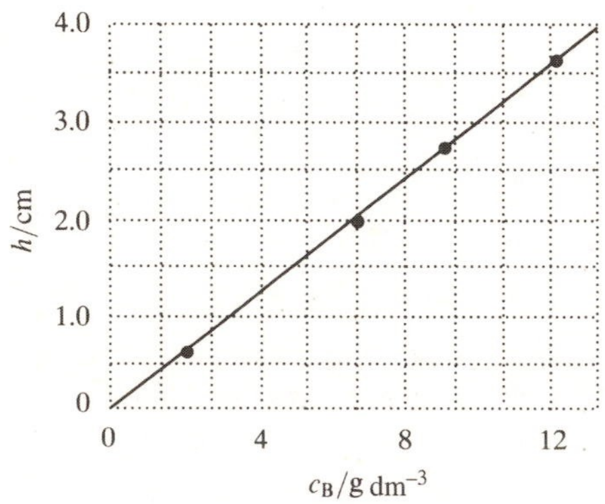

Figure 16.3

Thus, a plot of h against c should be a straight line with slope $\dfrac{RT}{\rho g M_B}$. Figure 16.3 shows the plot.

The slope of the best-fit line is

$$\frac{RT}{\rho g M_B} = \frac{0.2854 \text{ cm}}{\text{g dm}^{-3}} = \frac{0.2854\times(10^{-2} \text{ m})}{(10^{-3} \text{ kg})\times(10^{-1} \text{ m})^{-3}} = 2.854\times10^{-3} \text{ m}^4 \text{ kg}^{-1}$$

Therefore,

$$M_B = \frac{RT}{(\rho g)\times\text{slope}}$$
$$= \frac{(8.3145 \text{ J K}^{-1} \text{ mol}^{-1})\times(298.15 \text{K})}{(1.004\times10^3 \text{ kg m}^{-3})\times(9.807 \text{ m s}^{-2})\times(2.854\times10^{-3} \text{ m}^4 \text{ kg}^{-1})} = \boxed{88.2 \text{ kg mol}^{-1}}$$

E16.23(a) (a) The number-average molar mass is

$$\overline{M}_n = \frac{N_1 M_1 + N_2 M_2}{N_1 + N_2} = \frac{\left(\dfrac{m_1}{M_1}\right)M_1 + \left(\dfrac{m_2}{M_2}\right)M_2}{\left(\dfrac{m_1}{M_1}\right) + \left(\dfrac{m_2}{M_2}\right)} = \frac{m_1 + m_2}{\left(\dfrac{m_1}{M_1}\right) + \left(\dfrac{m_2}{M_2}\right)}$$

Assume 100 g (a convenient sample size for mass percent information).

$$\overline{M}_n = \frac{100 \text{ g}}{\left(\dfrac{30 \text{ g}}{30 \text{ kg mol}^{-1}}\right) + \left(\dfrac{70 \text{ g}}{15 \text{ kg mol}^{-1}}\right)} = \boxed{18 \text{ kg mol}^{-1}}$$

(b) The weight-average molar mass is

$$\overline{M}_w = \frac{m_1 M_1 + m_2 M_2}{m_1 + m_2} = \frac{(30)\times(30) + (70)\times(15)}{100}\ \text{kg mol}^{-1} = \boxed{20\ \text{kg mol}^{-1}}$$

(c) The heterogeneity index is $\dfrac{\overline{M}_w}{\overline{M}_n} = \dfrac{20}{18} = \boxed{1.1}$.

E16.24(a) The excess enthalpy in a regular solution is

$$H^E = n\xi RTx_A x_B = n\xi RTx_A(1 - x_A)$$

The maximum value occurs at the composition where $x_A(1 - x_A)$ is at a maximum. That maximum occurs at $x_A = x_B = {}^1/_2$. Now we can find ξ:

$$\xi = \frac{H^E_{max}/n}{RTx_A x_B} = \frac{700\ \text{J mol}^{-1}}{(8.3145\ \text{J K}^{-1}\ \text{mol}^{-1})(298\ \text{K})(0.500)^2} = \boxed{1.13}$$

Phase separation occurs if $\xi > 2$, so $\boxed{\text{no phase separation}}$ occurs at this temperature.

E16.25(a) For A (Raoult's law basis; concentration in mole fraction),

$$a_A = \frac{p_A}{p_A^*}[16.35] = \frac{250\ \text{Torr}}{300\ \text{Torr}} = \boxed{0.833} \qquad \gamma_A = \frac{a_A}{x_A} = \frac{0.833}{0.9} = \boxed{0.9\overline{3}}$$

For B (Henry's law basis; concentration in mole fraction),

$$a_B = \frac{p_B}{K_B}[16.42] = \frac{25\ \text{Torr}}{200\ \text{Torr}} = \boxed{0.125} \qquad \gamma_B = \frac{a_B}{x_B} = \frac{0.125}{0.1} = \boxed{1.\overline{25}}$$

For B (Henry's law basis; concentration in molality), we use an equation analogous to 16.42 but with a modified Henry's law constant K'_B:

$$a_B = \frac{p_B}{K'_B b^\ominus} \quad \text{with } p_B = b_B K'_B \text{ analogous to } p_B = x_B K_B$$

K'_B and K_B are related by equating the two expressions for p_B:

$$p_B = b_B K'_B = x_B K_B \quad \text{so} \quad K'_B = \frac{x_B K_B}{b_B} = \frac{0.1\times 200\ \text{Torr}}{2.22\ \text{mol kg}^{-1}} = 9.\overline{0}\ \text{Torr kg mol}^{-1}$$

Then $$a_B = \frac{25\ \text{Torr}}{9.\overline{0}\ \text{Torr}} = \boxed{2.\overline{8}} \qquad \gamma_B = \frac{a_B}{b_B/b^\ominus} = \frac{2.8}{2.22} = \boxed{1.25}$$

COMMENT. The two methods for the "solute" B give different values for the activities. This is reasonable since the reference states are different and therefore the chemical potentials in the reference states are also different.

Question. What are the activity and activity coefficient of B in the ideal solution (Raoult's law) basis?

E16.26(a) Activities and activity coefficients on the ideal-solution (Raoult's law) basis are defined by

$$a_A = \frac{p_A}{p_A^*} \quad \text{and} \quad \gamma_A = \frac{a_A}{x_A}$$

So we need partial pressures. Dalton's law relates the vapor-phase mole fractions to partial pressures.

$$y_A = \frac{p_A}{p_A + p_M} = \frac{p_A}{101.3\ \text{kPa}} = 0.516$$

So $p_A = 0.516 \times 101.3\ \text{kPa} = 52.3\ \text{kPa}$ and $p_M = (101.3 - 52.3)\ \text{kPa} = 49.0\ \text{kPa}$

$$a_A = \frac{p_A}{p_A^*} = \frac{52.3\,\text{kPa}}{105\,\text{kPa}} = \boxed{0.498} \qquad a_M = \frac{p_M}{p_M^*} = \frac{49.0\,\text{kPa}}{73.5\,\text{kPa}} = \boxed{0.667}$$

$$\gamma_A = \frac{a_A}{x_A} = \frac{0.498}{0.400} = \boxed{1.24} \qquad \gamma_M = \frac{a_M}{x_M} = \frac{0.667}{0.600} = \boxed{1.11}.$$

E16.27(a) The excess enthalpy in a regular solution is

$$H^E = n\xi RTx_Ax_B = n\xi RTx_A(1 - x_A)$$

The maximum excess enthalpy occurs at $x_A = x_B = {}^1/_2$ [Exercise 16.24(a)].

$$\xi = \frac{H^E_{\max}/n}{RTx_Ax_B} = \frac{800\ \text{J mol}^{-1}}{(8.3145\ \text{J K}^{-1}\ \text{mol}^{-1})(293\ \text{K})(0.500)^2} = 1.31$$

The Margules equations [16.45] give the natural logarithms of activity coefficients:

$$\ln\gamma_A = \xi x_B^2 \qquad \text{and} \qquad \ln\gamma_B = \xi x_A^2$$

Because $x_A = x_B$, $\ln\gamma_A = \ln\gamma_B = 1.31 \times (0.500)^2 = 0.328$ and $\gamma_A = \gamma_B = 1.39$.

Finally, $a_A = x_A\gamma_A = 0.500 \times 1.39 = \boxed{0.694} = a_B$.

E16.28(a) $$I = \frac{1}{2}\sum_i (b_i/b^\ominus)z_i^2 \quad [16.57]$$

For a salt of formula M_pX_q, $b_M = pb$ and $b_X = qb$,

so $$I = \frac{1}{2}\left(pz_+^2 + qz_-^2\right)\left(\frac{b}{b^\ominus}\right)$$

$$I(\text{KCl}) = \frac{1}{2}(1\times1+1\times1)\left(\frac{b_{\text{KCl}}}{b^\ominus}\right) = \frac{b_{\text{KCl}}}{b^\ominus}$$

$$I(\text{CuSO}_4) = \frac{1}{2}(1\times2^2+1\times2^2)\left(\frac{b_{\text{CuSO}_4}}{b^\ominus}\right) = 4\left(\frac{b_{\text{CuSO}_4}}{b^\ominus}\right)$$

$$I = I(\text{KCl}) + I(\text{CuSO}_4) = \frac{b_{\text{KCl}}}{b^\ominus} + 4\left(\frac{b_{\text{CuSO}_4}}{b^\ominus}\right) = 0.15 + 4\times0.35 = \boxed{1.55}$$

E16.29(a) The ionic strength of the original solution is $I(\text{KNO}_3) = 0.250$ [$b/b^\ominus$ for a singly-charged salt]. So the added salt must contribute an additional ionic strength of 0.200.

(a) $$I(\text{Ca(NO}_3)_2) = \frac{1}{2}(2^2+2\times1^2)\frac{b}{b^\ominus} = 3\frac{b}{b^\ominus} = 0.200$$

so $$b = \frac{0.200}{3}b^\ominus = 0.0667\,\text{mol kg}^{-1}$$

and $$m(\text{Ca(NO}_3)_2) = 0.0667\ \text{mol kg}^{-1} \times 0.800\ \text{kg} \times 261.32\ \text{g mol}^{-1} = \boxed{8.75\ \text{g}}$$

(b) An additional 0.200 mol NaCl must be added per kilogram of solvent, so

$$m(\text{NaCl}) = 0.200\ \text{mol kg}^{-1} \times 0.800\ \text{kg} \times 58.44\ \text{g mol}^{-1} = \boxed{9.35\ \text{g}}$$

E16.30(a) The solution is dilute, so use the Debye–Hückel limiting law.

$$\log\gamma_\pm = -|z_+z_-|AI^{1/2} \quad [16.56]$$

$$I = \frac{1}{2}\sum_i z_i^2\left(\frac{b_i}{b^\ominus}\right) = \frac{1}{2}\{(2^2\times0.0050) + (1\times0.0050\times2) + (1\times0.0040) + (1\times0.0040)\}$$

$$= \boxed{0.0190}$$

$CaCl_2$: $\log \gamma_\pm = -2 \times 1 \times 0.509 \times (0.0190)^{1/2} = -0.140$ so $\gamma_\pm = \boxed{0.72}$

NaF: $\log \gamma_\pm = -1 \times 1 \times 0.509 \times (0.0190)^{1/2} = -0.070$ so $\gamma_\pm = \boxed{0.85}$

E16.31(a) The extended Debye–Hückel law (eqn. 16.59), with the parameter C set equal to zero, is

$$\log \gamma_\pm = -\frac{A|z_+z_-|I^{1/2}}{1+BI^{1/2}}$$

Solve for B.

$$B = -\left(\frac{1}{I^{1/2}} + \frac{A|z_+z_-|}{\log \gamma_\pm}\right) = -\left(\frac{1}{(b/b^\ominus)^{1/2}} + \frac{0.509}{\log \gamma_\pm}\right)$$

Draw up the following table.

b / (mol kg^{-1})	5.0×10^{-3}	10.0×10^{-3}	20.0×10^{-3}
$\gamma_\pm$	0.930	0.907	0.879
B	2.01	2.01	2.02

The values of B are constant, illustrating that the extended law fits these activity coefficients with $B = \boxed{2.01}$.

Solutions to problems

Solutions to numerical problems

P16.1 The data are plotted in Figure 16.4. From the graph, the vapor in equilibrium with a liquid of composition (a) $x_M = 0.25$ is determined from the tie line labeled a in the figure extending from $x_M = 0.25$ to $\boxed{y_M = 0.36}$, (b) $x_O = 0.25$ is determined from the tie line labeled b in the figure extending from $x_M = 0.75$ to $\boxed{y_M = 0.81}$.

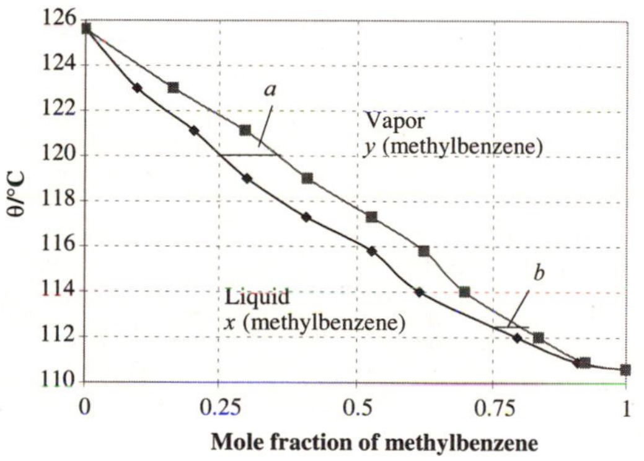

Figure 16.4

P16.3 The phase diagram is shown in Figure 16.5. Point symbols are plotted at the given data points. The lines are schematic at best.

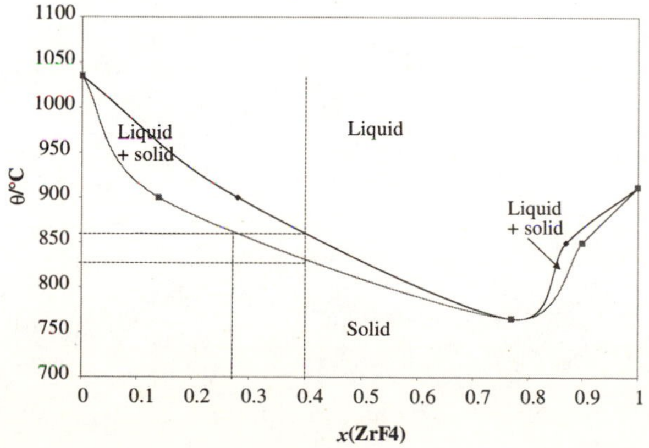

Figure 16.5

At 860°C, a solid solution with $x(ZrF_4) = 0.27$ appears. The solid solution continues to form, and its ZrF_4 content increases until it reaches $x(ZrF_4) = 0.40$ at 830°C. At that temperature and below, the entire sample is solid.

P16.5 The phase diagram is drawn in Figure 16.6.

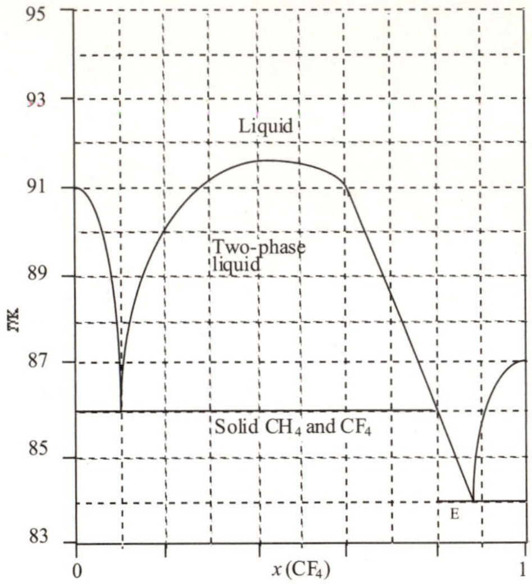

Figure 16.6

P16.7 (a) The Clapeyron equation [16.12] is

$$\frac{dp}{dT} = \frac{\Delta_{vap}S}{\Delta_{vap}V} = \frac{\Delta_{vap}H}{T_b\Delta_{vap}V}$$

$$= \frac{22.0\times10^3\ \text{J mol}^{-1}}{(380\ \text{K})\times(13.5\times10^{-3} - 1.20\times10^{-4})\ \text{m}^3\ \text{mol}^{-1}} = \boxed{+4.33\times10^3\ \text{Pa K}^{-1}}$$

(b) The Clausius-Clapeyron equation is

$$\frac{d\ln p}{dT} = \frac{\Delta_{vap}H}{RT^2} \quad [16.14],$$

so $$\frac{dp}{dT} = \frac{p\Delta_{vap}H}{RT^2} \times = \frac{(10^5\ \text{Pa})\times(22.0\times10^3\ \text{J mol}^{-1})}{(8.3145\ \text{J K}^{-1}\ \text{mol}^{-1})\times(380\ \text{K})^2} = +1.83\times10^3\ \text{Pa K}^{-1}$$

The percentage error is $\boxed{0.7 \text{ percent}}$.

P16.9 The Clausius-Clapeyron equation,

$$\frac{d\ln p}{dT} = \frac{\Delta_{vap}H}{RT^2} \quad [16.14]$$

yields upon indefinite integration

$$\ln p = \text{constant} - \frac{\Delta_{vap}H}{RT}$$

Therefore, plot $\ln p$ against $1/T$ and identify $-\Delta_{vap}H/R$ as the slope of the best-fit line. Construct the following table.

T / K	250	270	280	290	300	310	330
1000 K / T	4.00	3.70	3.57	3.45	3.33	3.23	3.03
p / kPa	0.036	0.514	1.662	4.918	13.43	34.1	182.9
ln (p / kPa)	−3.32	−0.666	0.508	1.593	2.597	3.529	5.209

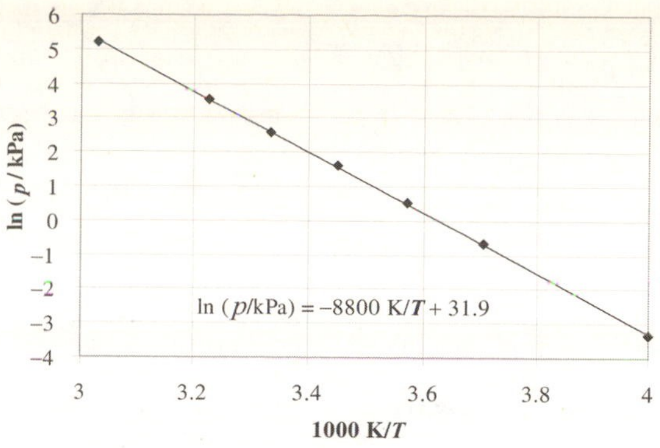

Figure 16.7

(a) The points are plotted in Figure 16.7. The slope is –8800 K, so

$$\Delta_{vap}H = -8.3145\ \text{J mol}^{-1}\ \text{K}^{-1} \times (-8800\ \text{K}) = \boxed{+7.32 \times 10^4\ \text{J mol}^{-1}} = \boxed{+73.2\ \text{kJ mol}^{-1}}$$

(b) The normal boiling point occurs at $p = 1$ bar = 100 kPa. Solving the equation of the best-fit line for T yields

$$T = \frac{8800\ \text{K}}{31.9 - \ln(p/\text{kPa})} = \frac{8800\ \text{K}}{31.9 - \ln(100)} = \boxed{322\ \text{K}}$$

P16.11 Refer to Example 16.3, using eqn. 16.29 with [J] = c/M and $\Pi = \rho gh$:

$$\frac{\Pi}{c} = \frac{RT}{M}\left(1 + \frac{B}{M}c + \cdots\right) = \frac{RT}{M} + \frac{RTB}{M^2}c + \cdots$$

where c is the mass concentration of the polymer. Therefore, plot Π / c against c. The y-intercept is RT/M and the slope is RTB/M^2. The transformed data to plot are given in the following table.

c / (mg cm^{-3})	1.33	2.10	4.52	7.18	9.87
Π / Pa	30	51	132	246	390
(Π / c) / (Pa mg^{-1}	22.6	24.3	29.2	34.3	39.5

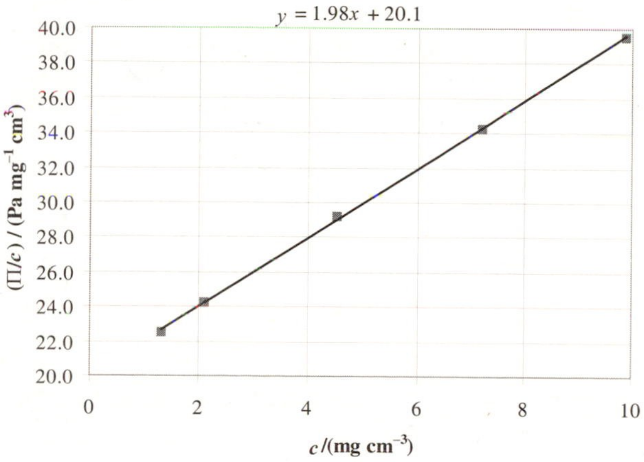

Figure 16.8

The plot is shown in Figure 16.8. The y-intercept is 20.1 Pa mg^{-1} cm^3. The slope is 1.98 Pa mg^{-2} cm^6. Therefore,

$$M = \frac{RT}{20.1\ \text{Pa mg}^{-1}\ \text{cm}^3} = \frac{8.3145\ \text{J K}^{-1}\ \text{mol}^{-1} \times 303\ \text{K}}{20.1\ \text{Pa mg}^{-1}\ \text{cm}^3} \times \frac{10^{-3}\ \text{g}}{1\ \text{mg}} \times \left(\frac{1\ \text{cm}}{10^{-2}\ \text{m}}\right)^3$$

$$= \boxed{1.25 \times 10^5\ \text{g mol}^{-1}}$$

$$B = \frac{M^2}{RT} \times 1.98 \text{ Pa mg}^{-2} \text{ cm}^6 = \frac{M}{\left(\frac{RT}{M}\right)} \times 1.98 \text{ Pa mg}^{-2} \text{ cm}^6$$

$$= \frac{1.25\times10^5 \text{ g mol}^{-1} \times 1.98 \text{ Pa mg}^{-2} \text{ cm}^6}{20.1 \text{ Pa mg}^{-1} \text{ cm}^3}$$

$$= 1.23\times10^4 \text{ g mol}^{-1} \text{ mg}^{-1} \text{ cm}^3 \times \frac{1 \text{ mg}}{10^{-3} \text{ g}} \times \left(\frac{10^{-1} \text{ dm}}{1 \text{ cm}}\right)^3 = \boxed{1.23\times10^4 \text{ dm}^3 \text{ mol}^{-1}}$$

P16.13 The partial pressure of a solvent is related to its activity by

$$p_A = a_A p_A^* \text{ [16.35]} = \gamma_A x_A p_A^* \text{ [16.37]}$$

so $$\gamma_A = \frac{p_A}{x_A p_A^*} = \frac{y_A p}{x_A p_A^*}$$

(*Note:* Here x and y are mole *fractions*. In the table in the statement of the problem, x and y are mole *percentages*.)

Sample calculation at 80 K:

$$\gamma(O_2, 80\text{ K}) = \frac{0.11\times100\text{ kPa}}{0.34\times225\text{ Torr}} \times \frac{760\text{ Torr}}{101.325\text{ kPa}} = 1.079$$

Summary:

T / K	77.3	78	80	82	84	86	88	90.2
$\gamma(O_2)$	—	0.877	1.079	1.039	0.995	0.993	0.990	0.987

To within the experimental uncertainties, the solution appears to be ideal ($\gamma \approx 1$). The low value at 78 K may be caused by non-ideality; however, the larger relative uncertainty in $y(O_2)$ is probably the origin of the low value. A temperature–composition diagram is shown in Figure 16.9(a). The near ideality of this solution is, however, best shown in the pressure–composition diagram of Figure 16.9(b). There the liquid line is essentially straight, as predicted for an ideal solution.

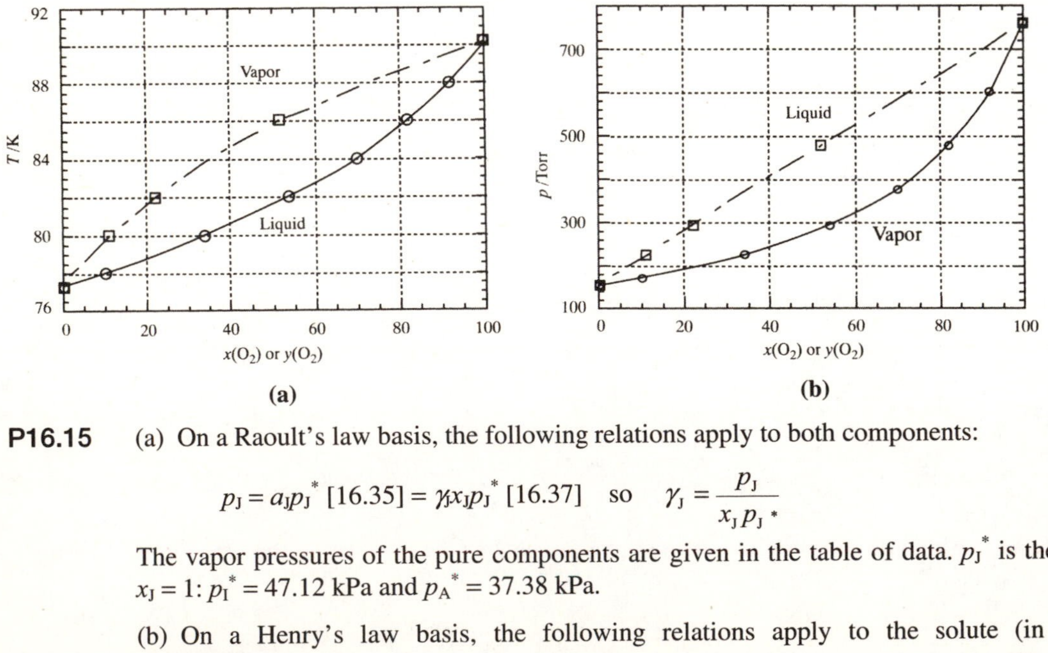

Figure 16.9

P16.15 (a) On a Raoult's law basis, the following relations apply to both components:

$$p_J = a_J p_J^* \text{ [16.35]} = \gamma_J x_J p_J^* \text{ [16.37]} \quad \text{so} \quad \gamma_J = \frac{p_J}{x_J p_J^*}$$

The vapor pressures of the pure components are given in the table of data. p_J^* is the value of p_J at $x_J = 1$: $p_I^* = 47.12$ kPa and $p_A^* = 37.38$ kPa.

(b) On a Henry's law basis, the following relations apply to the solute (in this problem, iodoethane):

$$a_B = \frac{p_B}{K_B} \text{ [16.42]} \quad \text{so} \quad \gamma_B = \frac{p_B}{x_B K_B}$$

(The solvent activity remains based on Raoult's law, so γ_A would remain as in part a.) Henry's law constants are determined by plotting the data and extrapolating the low concentration data (i.e., the best straight line in the neighborhood of $x_B = 0$) all the way across the plot to $x_B = 1$. The data are plotted in Figure 16.10, which also displays p_I^*, p_A^*, K_I, and K_A . $K_I = 64.4$ kPa.

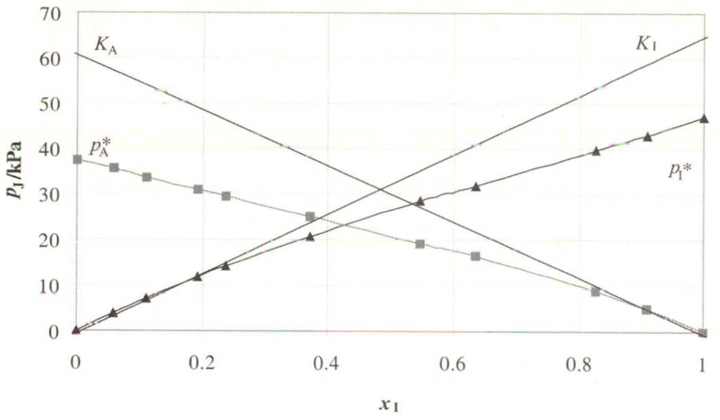

Figure 16.10

x_I	p_I/kPa	p_A/kPa	γ_I(Raoult)	γ_A(Raoult)	γ_I(Henry)
0	0	37.38†	—	1	1
0.0579	3.73	35.48	1.367	1.008	1.000
0.1095	7.03	33.64	1.362	1.011	0.997
0.1918	11.7	30.85	1.295	1.021	0.947
0.2353	14.05	29.44	1.267	1.030	0.927
0.3718	20.72	25.04	1.183	1.066	0.865
0.5478	28.44	19.23	1.102	1.138	0.806
0.6349	31.88	16.39	1.066	1.201	0.779
0.8253	39.58	8.88	1.018	1.360	0.744
0.9093	43.00	5.09	1.004	1.501	0.734
1.0000	47.12‡	0	1	—	0.731
†Value of p_A*	‡Value of p_I*				

Question. In this problem both I and A were treated as solvents, but only I was treated as a solute. Extend the table by including a column for γ_A(Henry).

P16.17 (a) The volume of an ideal mixture is

$$V_{\text{ideal}} = n_1 V_{\text{m},1} + n_2 V_{\text{m},2}$$

so the volume of a real mixture is

$$V = V_{\text{ideal}} + V^{\text{E}}$$

We have an expression for excess molar volume in terms of mole fractions. To compute partial molar volumes, we need an expression for the actual excess volume as a function of moles.

$$V^{\text{E}} = (n_1 + n_2)V_{\text{m}}^{\text{E}} = \frac{n_1 n_2}{n_1 + n_2}\left(a_0 + \frac{a_1(n_1 - n_2)}{n_1 + n_2}\right)$$

so $$V = n_1 V_{\text{m},1} + n_2 V_{\text{m},2} + \frac{n_1 n_2}{n_1 + n_2}\left(a_0 + \frac{a_1(n_1 - n_2)}{n_1 + n_2}\right)$$

The partial molar volume of propionic acid is

$$V_1 = \left(\frac{\partial V}{\partial n_1}\right)_{p,T,n_2} = V_{\text{m},1} + \frac{a_0 n_2^2}{(n_1 + n_2)^2} + \frac{a_1(3n_1 - n_2)n_2^2}{(n_1 + n_2)^3}$$

$$\boxed{V_1 = V_{\text{m},1} + a_0 x_2^2 + a_1(3x_1 - x_2)x_2^2}$$

The partial molar volume of oxane is

$$\boxed{V_2 = V_{m,2} + a_0 x_1^2 + a_1(x_1 - 3x_2)x_1^2}$$

(b) We need the molar volumes of the pure liquids

$$V_{m,1} = \frac{M_1}{\rho_1} = \frac{74.08\ \text{g mol}^{-1}}{0.97174\ \text{g cm}^{-3}} = 76.23\ \text{cm}^3\ \text{mol}^{-1}$$

and $$V_{m,2} = \frac{86.13\ \text{g mol}^{-1}}{0.86398\ \text{g cm}^{-3}} = 99.69\ \text{cm}^3\ \text{mol}^{-1}$$

In an equimolar mixture, the partial molar volume of propionic acid is

$$V_1 = 76.23 + (-2.4697)\times(0.5)^2 + (0.0608)\times\{3(0.5) - 0.5\}\times(0.5)^2\ \text{cm}^3\ \text{mol}^{-1}$$
$$= \boxed{75.63\ \text{cm}^3\ \text{mol}^{-1}}.$$

and the partial molar volume of oxane is

$$V_2 = 99.69 + (-2.4697)\times(0.5)^2 + (0.0608)\times\{0.5 - 3(0.5)\}\times(0.5)^2\ \text{cm}^3\ \text{mol}^{-1}$$
$$= \boxed{99.06\ \text{cm}^3\ \text{mol}^{-1}}$$

P16.19 According to the Debye–Hückel limiting law,

$$\log \gamma_\pm = -0.509\,|z_+z_-|\,I^{1/2}\ [16.56] = -0.509\left(\frac{b}{b^\ominus}\right)^{1/2}$$

We draw up the following table:

b / (mmol kg^{-1})	1.0	2.0	5.0	10.0	20.0
$I^{1/2}$	0.032	0.045	0.071	0.100	0.141
$\gamma_\pm$(calc)	0.964	0.949	0.920	0.889	0.847
$\gamma_\pm$(exp)	0.9649	0.9519	0.9275	0.9024	0.8712
log $\gamma_\pm$(calc)	−0.0161	−0.0228	−0.0360	−0.0509	−0.0720
log $\gamma_\pm$(calc)	−0.0155	−0.0214	−0.0327	−0.0446	−0.0599

The points are plotted against $I^{1/2}$ in Figure 16.11. Note that the limiting slopes of the calculated and experimental curves coincide. We use the extended Debye–Hückel law (eqn. 16.59) with $C = 0$. A sufficiently good value of B in the extended law may be obtained by assuming that the constant A in the extended law is the same as A in the limiting law. Using the data at 20.0 mmol kg^{-1} we can solve for B.

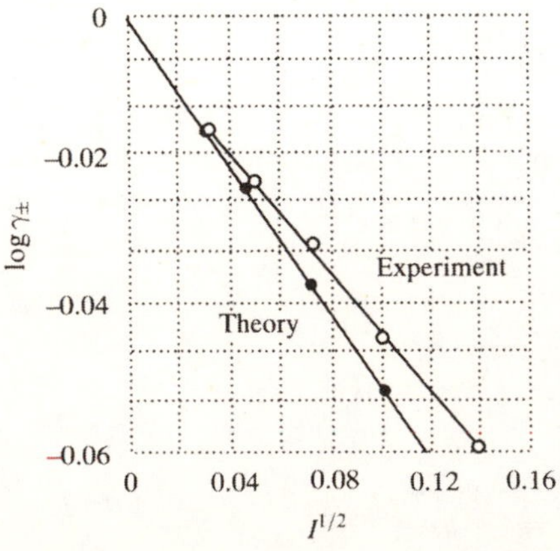

Figure 16.11

$$B = -\frac{A}{\log\gamma_\pm} - \frac{1}{I^{1/2}} = -\frac{0.509}{(-0.0599)} - \frac{1}{0.141} = 1.40\overline{5}$$

Thus,

$$\log \gamma_{\pm} = -\frac{0.509 I^{1/2}}{1 + 1.40\bar{5} I^{1/2}}$$

To determine whether or not the fit is improved, we use the data at 10.0 mmol kg^{-1}

$$\log \gamma_{\pm} = \frac{-(0.509)\times(0.100)}{(1)+(1.405)\times(0.100)} = -0.0446$$

which fits the data almost exactly. The fits to the other data points will also be almost exact.

Solutions to theoretical problems

P16.21 Consider two phases α and β, in mechanical contact with each other, constituting an isolated system. Making the system isolated allows us to confine ourselves to interactions *between* the phases and to neglect any surroundings external to those phases. The general condition of equilibrium in an isolated system is $dS = 0$. Hence,

(a) $\quad dS = dS_{\alpha} + dS_{\beta} = 0$

Energy may be transferred between phases and the volumes of the phases may also change; however, $U_{\alpha} + U_{\beta} = \text{constant}$ and $V_{\alpha} + V_{\beta} = \text{constant}$. Hence,

(b) $\quad dU_{\beta} = -dU_{\alpha}$

(c) $\quad dV_{\beta} = -dV_{\alpha}$

For each phase, eqn. 15.54 applies:

$$dU_{\alpha} = T_{\alpha} dS_{\alpha} - p_{\alpha} dV_{\alpha}$$

so $$dS_{\alpha} = \frac{1}{T_{\alpha}} dU_{\alpha} + \frac{p_{\alpha}}{T_{\alpha}} dV_{\alpha}$$

Using this result for both phases as well as conditions (b) and (c), condition (a) becomes

$$dS = \left(\frac{1}{T_{\alpha}} - \frac{1}{T_{\beta}}\right) dU_{\alpha} + \left(\frac{p_{\alpha}}{T_{\alpha}} - \frac{p_{\beta}}{T_{\beta}}\right) dV_{\alpha} = 0$$

The only way in which this expression may, in general, equal zero is for

$$\frac{1}{T_{\alpha}} - \frac{1}{T_{\beta}} = 0 \quad \text{and} \quad \frac{p_{\alpha}}{T_{\alpha}} - \frac{p_{\beta}}{T_{\beta}} = 0$$

Therefore, $\boxed{T_{\alpha} = T_{\beta} \text{ and } p_{\alpha} = p_{\beta}}$.

P16.23 The barometric formula gives the variation of pressure with height:

$$p(h) = p_0 e^{-Mgh/RT} \text{ [Problem 13.20]}$$

In Exercise 16.7(a), we obtained from the Clausius-Clapeyron equation

$$\ln \frac{p}{p_1} = \frac{\Delta_{\text{vap}} H}{R}\left(\frac{1}{T_1} - \frac{1}{T}\right)$$

so, letting T_1 be the normal boiling point T_b, so $p_1 = 1$ bar, we have an expression for the variation of vapor pressure with temperature:

$$p(T) = p_1 e^{-\chi}, \text{ where } \chi = \frac{\Delta_{vap}H}{R}\left(\frac{1}{T} - \frac{1}{T_b}\right)$$

In this expression, let T_h be the boiling temperature at altitude h. Boiling occurs when the vapor pressure is equal to the ambient pressure, that is, when $p(h) = p(T_h)$. Setting the two expressions equal to each other, with $p_0 = p_1 = 1$ bar, implies that

$$e^{-Mgh/RT} = \exp\left\{-\frac{\Delta_{vap}H}{R} \times \left(\frac{1}{T_h} - \frac{1}{T_b}\right)\right\}$$

It follows that $\dfrac{1}{T_h} = \dfrac{1}{T_b} + \dfrac{Mgh}{T\Delta_{vap}H}$ where T is the ambient temperature and M the molar mass of the air. For water at 3000 m, using $M = 29$ g mol^{-1},

$$\frac{1}{T_h} = \frac{1}{373\text{ K}} + \frac{29\times10^{-3}\text{ kg mol}^{-1} \times 9.81\text{ m s}^{-2} \times 3.000\times10^{3}\text{ m}}{293\text{ K} \times 40.7\times10^{3}\text{ J mol}^{-1}}$$

$$= \frac{1}{373\text{K}} + \frac{1}{1.40\times10^{4}\text{K}} = 2.75\times10^{-3}\text{ K}^{-1}$$

Hence, $T_h = \boxed{363\text{ K } (90°\text{C})}$.

P16.25 The question asks us to express derivatives of the chemical potential in statistical thermodynamic terms, so we start with the chemical potential itself, which is the partial molar Gibbs function.

$$G = G(0) - nRT\ln\frac{q}{N}\ [15.40]$$

so $$\mu = \left(\frac{\partial G}{\partial n}\right)_{p,T} = \frac{\partial G(0)}{\partial n} - RT\ln\frac{q}{N} - nRT\frac{\partial}{\partial n}\left(\ln\frac{q}{N}\right)$$

How does q/N depend on n (at constant p and T)? $N = nN_A$. q depends on molecular parameters and temperature and (in the translational partition function) volume. Recall that

$$q^{T} = \frac{V}{\Lambda^3}\ [13.18b]$$

so $$\frac{q}{N} = \frac{q^{T}q^{R}q^{V}q^{E}}{N} = \frac{Vq^{R}q^{V}q^{E}}{N\Lambda^3} = \frac{kTq^{R}q^{V}q^{E}}{p\Lambda^3}$$

where we used the perfect gas law in the form $pV = NkT$ in the last step. Thus, for a perfect gas, the partial derivative of $\ln q/N$ with respect to n at constant p and T vanishes, and

$$\mu = \left(\frac{\partial G}{\partial n}\right)_{p,T} = \frac{\partial G(0)}{\partial n} - RT\ln\frac{q}{N}$$

(a) We are to show that eqn. 16.9,

$$\left(\frac{\partial \mu}{\partial T}\right)_{p,n} = -S_m$$

is consistent with eqns. 15.40 and 15.2,

$$S = \frac{U - U(0)}{T} + Nk \ln \frac{qe}{N} = \frac{U - U(0)}{T} + nR + nR \ln \frac{q}{N}$$

We differentiate our statistical-thermodynamic expression of the chemical potential:

$$\left(\frac{\partial \mu}{\partial T}\right)_{p,n} = \frac{\partial}{\partial T}\left(\frac{\partial G(0)}{\partial n} - RT \ln \frac{q}{N}\right)_{p,n} = -R\left(\ln \frac{q}{N}\right) - RT \frac{\partial}{\partial T}\left(\ln \frac{q}{N}\right)_{p,n}$$

$$= -R \ln \frac{q}{N} - RT\left(\frac{\partial \ln q}{\partial T}\right)_{p,n}$$

We must show that this equals S/n. To bridge the gap between our expressions, we need a statistical-thermodynamic expression for internal energy. For a collection of independent particles,

$$U - U(0) = N\langle\varepsilon\rangle = -N\left(\frac{\partial \ln q}{\partial \beta}\right)_V [13.35\text{b}] = -N\left(\frac{\partial \ln q}{\partial T}\right)_V \left(\frac{\partial T}{\partial \beta}\right)_V$$

$$= -N\left(\frac{\partial \ln q}{\partial T}\right)_V \left(\frac{\partial \beta}{\partial T}\right)^{-1} = NkT^2\left(\frac{\partial \ln q}{\partial T}\right)_V$$

Note that what is held constant here is volume, not pressure as in the chemical potential derivative, so we need the following property of partial derivatives based on eqn. MB 8.3:

$$\left(\frac{\partial \ln q}{\partial T}\right)_V = \left(\frac{\partial \ln q}{\partial T}\right)_p + \left(\frac{\partial \ln q}{\partial p}\right)_T \left(\frac{\partial p}{\partial T}\right)_V$$

$$= \left(\frac{\partial \ln q}{\partial T}\right)_p + \left\{\frac{\partial}{\partial p}\left(\ln \frac{nRTq^{\text{R}}q^{\text{V}}q^{\text{E}}}{p\Lambda^3}\right)\right\}_T \left\{\frac{\partial}{\partial T}\left(\frac{nRT}{V}\right)\right\}_V$$

$$= \left(\frac{\partial \ln q}{\partial T}\right)_p - \frac{1}{p} \times \frac{nR}{V} = \left(\frac{\partial \ln q}{\partial T}\right)_p - \frac{1}{T}$$

Thus $$\frac{U - U(0)}{T} = nRT\left(\frac{\partial \ln q}{\partial T}\right)_p - nR$$

which allows us to identify our statistical-thermodynamic expression for $\partial\mu/\partial T$:

$$\left(\frac{\partial \mu}{\partial T}\right)_{p,n} = -R \ln \frac{q}{N} - \frac{U - U(0)}{nT} + R = -\frac{S}{n} = \boxed{-S_m}$$

(b) We are to show that eqn. 16.10

$$\left(\frac{\partial \mu}{\partial p}\right)_{T,n} = V_m$$

is consistent with statistical thermodynamic results based on eqn. 15.40. So we differentiate

$$\left(\frac{\partial \mu}{\partial p}\right)_{T,n} = \frac{\partial}{\partial p}\left(\frac{\partial G(0)}{\partial n} - RT\ln\frac{q}{N}\right)_{T,n} = -RT\left(\frac{\partial}{\partial p}(\ln q - \ln N)\right)_{T,n}$$

$$= -RT\left(\frac{\partial \ln q}{\partial p}\right)_{T,n} = -RT\times\left(-\frac{1}{p}\right) = \frac{V}{n} = \boxed{V_{\mathrm{m}}}$$

P16.27 $$G^{\mathrm{E}} = RTx(1-x)\{0.4857 - 0.1077(2x-1) + 0.0191(2x-1)^2\}$$

A regular solution is defined by $S^{\mathrm{E}} = 0$. Thus, a solution is regular if $G^{\mathrm{E}} = H^{\mathrm{E}}$. Because an ideal solution has zero enthalpy of mixing, $H^{\mathrm{E}} = \Delta_{\mathrm{mix}}H$; therefore, H^{E} is readily measured experimentally by calorimetry. Thus, the test of whether the solution is regular amounts to measuring $\Delta_{\mathrm{mix}}H$ and determining how closely it matches the expression for G^{E}.

To test whether the given expression is consistent with the regular solution model given by eqn. 16.32, we must compare the above expression for G^{E} with the model expression for H^{E} :

$$H^{\mathrm{E}} = \xi RTx(x-1)$$

Begin the comparison by plotting G^{E}/RT against x, shown as the solid curve in Figure 16.12, which resembles the curves shown in Figs. 16.38 and 16.39 of the text. The resemblance suggests that the given G^{E} is consistent with the model regular solution. To make the comparison more salient, plot the model H^{E}/RT against x on the same graph, selecting ξ so that the two curves have the same maximum. The maximum in the model function is $\xi/4$, and it occurs at $x = 0.5$. So let $\xi = 4G^{\mathrm{E}}(\max) = 4\times 0.123 = 0.497$.

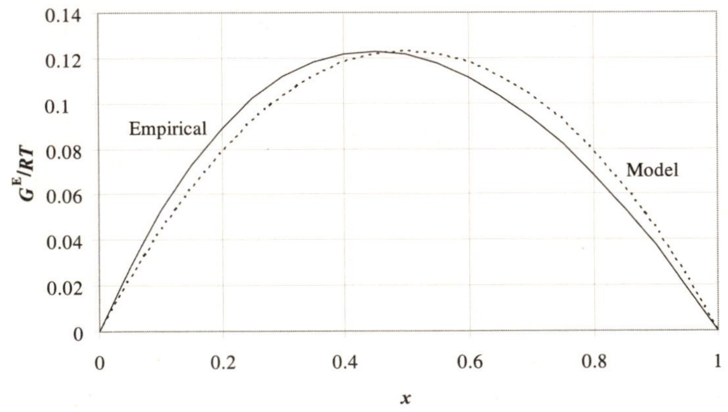

Figure 16.12

$\boxed{G^{\mathrm{E}} \text{ is reasonably consistent with the model regular solution}}$ over the entire range of composition.

P16.29 The analogous integral is

$$V_{\mathrm{P}}(x_{\mathrm{T}}) - V_{\mathrm{P}}(0) = -\int_{V_{\mathrm{T}}(0)}^{V_{\mathrm{T}}(x_{\mathrm{T}})} \frac{x_{\mathrm{T}}}{1-x_{\mathrm{T}}}\mathrm{d}V_{\mathrm{T}}$$

We should now plot $x_{\mathrm{T}}/(1-x_{\mathrm{T}})$ against V_{T} and estimate the integral. We must integrate up to $x_{\mathrm{T}} = 0.5$, which corresponds to $x_{\mathrm{T}}/(1-x_{\mathrm{T}}) = 1$. So draw up the following table including all data points with $x_{\mathrm{T}} <$ 0.5 and one point beyond so as to define the plot of the integrand over the interval of integration.

x_{T}	0	0.194	0.385	0.559
V_{T} ($\mathrm{cm^3\,mol^{-1}}$)	73.99	75.29	76.50	77.55
$x_{\mathrm{A}}/(1-x_{\mathrm{A}})$	0	0.241	0.626	1.266

The points are plotted in Figure 16.13. To estimate the upper limit, we draw a smooth curve through the data points. The value of V_T where $x_T = 0.5$ is 77.1 cm^3 mol^{-1}.

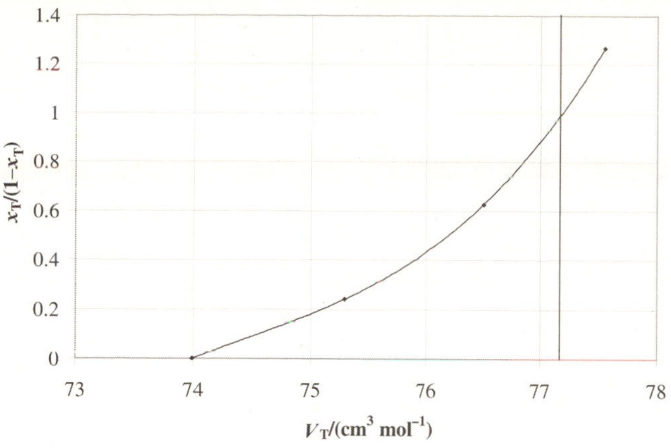

Figure 16.13

The integral is the area under the curve. Applying the trapezoid rule yields 1.39 cm^3 mol^{-1} for the integral, so

$$V_P(x_T = 0.5) = V_P(x_T = 0) - 1.39\ \text{cm}^3\ \text{mol}^{-1}$$

We need $V_P(x_T = 0)$, that is, the partial molar volume of propanone in a "mixture" containing no trichloromethane. $V_P(x_T = 0)$ is the partial molar volume of pure propanone, i.e., the molar volume of propanone.

$$V_m(\text{propanone}) = \frac{M}{\rho} = \frac{58.08\ \text{g mol}^{-1}}{0.7857\ \text{g cm}^{-3}} = 73.92\ \text{cm}^3\ \text{mol}^{-1}$$

$$V_P(x_T = 0.5) = (73.92 - 1.39)\ \text{cm}^3\ \text{mol}^{-1} = \boxed{72.53\ \text{cm}^3\ \text{mol}^{-1}}$$

P16.31 Retrace the argument in *Justification* 16.3. Exactly the same process applies with a_A in place of x_A. At equilibrium,

$$\mu_A^*(p) = \mu_A(a_A, p + \Pi)$$

which implies that, with $\mu = \mu^* + RT \ln a$ for a real solution,

$$\int_p^{p+\Pi} V_m\, dp = -RT \ln a_A$$

For an incompressible solution, the integral evaluates to ΠV_m, so $\Pi V_m = -RT \ln a_A$.

Thus, the osmotic coefficient ϕ, defined in Problem 16.30 as $\phi = -(x_A/x_B) \ln a_A$,

$$\Pi V_m = \frac{x_B}{x_A}\phi RT = \frac{n_B}{n_A}\phi RT$$

For a dilute solution, $n_A V_m \approx V$, so $\Pi V \approx n_B \phi RT$,

and therefore, with $[B] = \frac{n_B}{V}$, $\boxed{\Pi = \phi[B]RT}$.

P16.33 At the boiling point of the solution (T_b) the chemical potential of the liquid solvent is equal to that of the solvent vapor:

$$\mu_A(l, T_b) = \mu_A(g, T_b)$$

Substituting the ideal-solution expressions for these chemical potentials yields

$$\mu_A^{\ominus}(l, T_b) + RT_b \ln x_A = \mu_A^{\ominus}(g, T_b) + RT_b \ln(p_A / p^{\ominus})$$

The last term vanishes because $p_A = p°$ at the boiling point of the solution. We are interested in comparing T_b to T_b^*, the boiling point of the pure solvent, so we apply eqn. 16.9 to both pure-substance chemical potentials:

$$\left(\frac{\partial \mu}{\partial T}\right)_p = -S_m \quad \text{so} \quad \mu(T_b) \approx \mu(T_b^*) - (T_b - T_b^*)S_m \quad \text{if } T_b - T_b^* \text{ is small}$$

Thus we can express chemical potential at the boiling point of the solution (T_b), in terms of differences from the chemical potential at the boiling point of the pure substance (T_b^*):

$$\mu_A^\ominus(\mathrm{l},T_b^*) - (T_b - T_b^*)S_m(\mathrm{l}) + RT_b \ln x_A = \mu_A^\ominus(\mathrm{g},T_b^*) - (T_b - T_b^*)S_m(\mathrm{g})$$

Now $\mu_A^\ominus(\mathrm{l},T_b^*) = \mu_A^\ominus(\mathrm{g},T_b^*)$ [pure liquid and vapor are in equilibrium at T_b^*], so eliminating them from the equation and gathering the entropy terms on the left, we have

$$(T_b - T_b^*)\{S_m(\mathrm{g}) - S_m(\mathrm{l})\} \approx -RT_b \ln x_A = -RT_b \ln(1 - x_B) \approx RT_b x_B \; [x_B \text{ small}]$$

The differences on the left side of the equation are simply ΔT_b and $\Delta_{vap}S$, so we have

$$\Delta T_b \approx \frac{RT_b}{\Delta_{vap}S} x_B \approx \frac{RT_b^*}{\Delta_{vap}S} x_B [T_b^* \approx T_b] \approx \boxed{\frac{R(T_b^*)^2}{\Delta_{vap}H} x_B} [\Delta_{vap}H = T_b^* \Delta_{vap}S]$$

Solutions to applications

P16.35 In this case it is convenient to rewrite the Henry's law expression as

$$\text{mass of } N_2 = p_{N_2} \times \text{mass of } H_2O \times K_{N_2}$$

At 4.0 atm air, $p_{N_2} = 0.78 \times 4.0 \text{ atm} = 3.1 \text{ atm}$

so $\quad \text{mass of } N_2 = 3.1 \text{ atm} \times 100 \text{ g } H_2O \times 0.18\ \mu\text{g } N_2/(\text{g } H_2O \text{ atm}) = \boxed{56\ \mu\text{g } N_2}$

At 1.0 atm air, $p_{N_2} = 0.78$ atm and mass of $N_2 = \boxed{14\ \mu\text{g } N_2}$.

In fatty tissue the increase in N_2 concentration from 1 atm to 4 atm is

$$4 \times (56 - 14)\ \mu\text{g } N_2 = \boxed{1.7 \times 10^2\ \mu\text{g } N_2}$$

P16.37 (i) Below a denaturant concentration of 0.1, only the native and unfolded forms are stable; the "molten globule" form is not stable.

(ii) At a denaturant concentration of 0.15, only the native form is stable below a temperature of about 0.65. At temperature 0.65, the native and molten-globule forms are at equilibrium. Heating above 0.65 causes all native forms to become molten-globules. At temperature 0.85, equilibrium between molten-globule and unfolded protein is observed, and above this temperature only the unfolded form is stable.

P16.39 (a) The phase boundary is plotted in Figure 16.14.

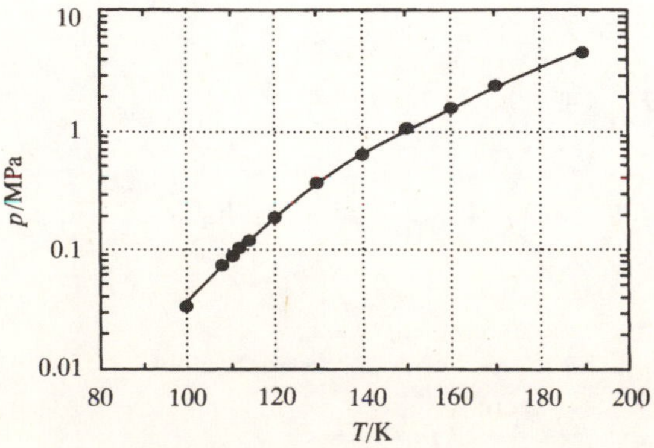

Figure 16.14

(b) The normal boiling point is the temperature at which the liquid is in equilibrium with the standard pressure of 1 bar (0.1 MPa). Interpolation of the plotted points gives $\boxed{T_b = 112\text{ K}}$.
(c) The slope of the liquid–vapor coexistence curve is given by

$$\frac{dp}{dT} = \frac{\Delta_{vap}H}{T\Delta_{vap}V} \text{ [16.13b]}, \qquad \text{so} \qquad \Delta_{vap}H = (T\Delta_{vap}V)\frac{dp}{dT}$$

The slope can be obtained graphically or by fitting the points nearest the boiling point. Then $dp/dT = 8.14\times10^{-3}\text{ MPa K}^{-1}$, so

$$\Delta_{vap}H = (112\text{ K})\times\left(\frac{(8.89-0.0380)\text{ dm}^3\text{ mol}^{-1}}{1000\text{ dm}^3\text{ m}^{-3}}\right)\times(8.14\text{ kPa K}^{-1}) = \boxed{8.07\text{ kJ mol}^{-1}}$$

P16.41 By the van't Hoff equation [16.28],

$$\Pi = [\text{B}]RT = \frac{cRT}{M}$$

Division by the standard acceleration of gravity, g, gives

$$\frac{\Pi}{g} = \frac{c(R/g)T}{M}$$

(a) This expression can be written in the form

$$\Pi' = \frac{cR'T}{M}$$

which has the same form as the van't Hoff equation, but the new "osmotic pressure," Π', has the dimensions of Π/g:

$$\frac{\text{force area}^{-1}}{\text{length time}^{-2}} = \frac{(\text{mass length time}^{-2})(\text{area}^{-1})}{\text{length time}^{-2}} = \frac{\text{mass}}{\text{area}}$$

This ratio can be specified in units of g cm^{-2}, as specified in the problem. Likewise, the proportionality constant, R', would have the dimensions of R/g:

$$\frac{\text{energy temperature}^{-1}\text{ amount}^{-1}}{\text{length time}^{-2}} = \frac{(\text{mass length}^2\text{ time}^{-2})\text{ temperature}^{-1}\text{ amount}^{-1}}{\text{length time}^{-2}}$$
$$= \text{mass length temperature}^{-1}\text{ amount}^{-1}.$$

This result may be specified in units of $\boxed{\text{g cm K}^{-1}\text{ mol}^{-1}}$.

(b) $$R' = \frac{R}{g} = \frac{8.31447\text{ J K}^{-1}\text{ mol}^{-1}}{9.80665\text{ m s}^{-2}}$$

$$= 0.847840\text{ kg m K}^{-1}\text{ mol}^{-1}\times\left(\frac{10^3\text{ g}}{\text{kg}}\right)\times\left(\frac{\text{cm}}{10^{-2}\text{ m}}\right)$$

$$= \boxed{84784.0\text{ g cm K}^{-1}\text{ mol}^{-1}}$$

We will drop the primes, giving

$$\Pi = \frac{cRT}{M}$$

and use the Π units of g cm^{-2} and the R units of g cm K^{-1} mol^{-1}.

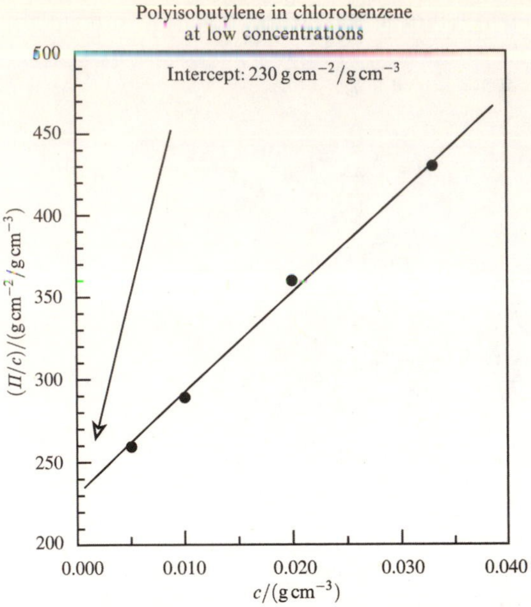

Figure 16.15(a)

We extrapolate the low-concentration portion of the plot of Π/c versus c to $c = 0$ (in effect, employing the virial analogue of the van't Hoff equation, eqn. 16.29, as in Example 16.3). The y-intercept is 230 g cm^{-2} / g cm^{-3} (Figure 16.15a). In the limit of small c, the van't Hoff equation is valid so

$$\frac{RT}{M} = \text{intercept},$$

and $$M = \frac{RT}{\text{intercept}} = \frac{(84784.0\,g\,cm\,K^{-1}\,mol^{-1})\times(298.15\,K)}{(230\,g\,cm^{-2})/(g\,cm^{-3})}$$

$$= \boxed{1.1\times10^{5}\,g\,mol^{-1}}$$

(c) The plot of Π/c versus c for the full concentration range (Figure 16.15b) is highly nonlinear. We can conclude that the solvent is good due to the nonpolar nature of both solvent and solute.

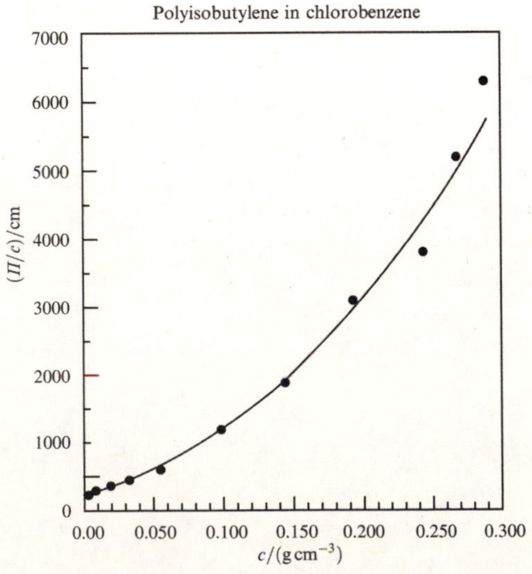

Figure 16.15(b)

(d) The virial analogue to the van't Hoff equation (eqn. 16.29) rearranges to

$$\Pi / c = (RT / M)(1 + B'c + C'c^2 + \cdots)$$

Since RT / M has been determined in part (b) by extrapolation to $c = 0$, it is best to determine the second and third virial coefficients by a linear regression fit.

$$\frac{\{(\Pi / c) / (RT / M)\} - 1}{c} = B' + C'c$$

That is, call the left side of this expression $f(\Pi, c)$ and plot it against c. The best linear fit in the resulting plot (Fig. 16.15c) gives $\boxed{B' = 21.1 \text{ cm}^3 \text{ g}^{-1}}$ and $\boxed{C' = 212 \text{ cm}^6 \text{ g}^{-2}}$. The plot shows a fair amount of scatter.

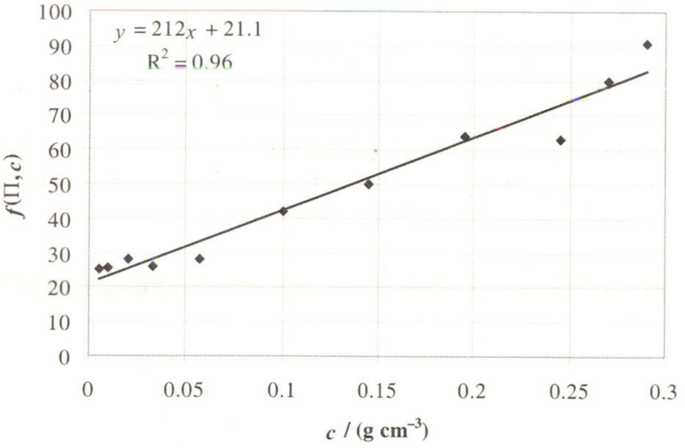

Figure 16.15(c)

(e) Using 1/4 for g and neglecting terms beyond the second power, we may write

$$\frac{\Pi}{c} = \frac{RT}{M}\left(1 + B'c + \frac{B'^2c^2}{4}\right) = \frac{RT}{M}\left(1 + \frac{B'c}{2}\right)^2$$

so $$\left(\frac{\Pi}{c}\right)^{1/2} = \left(\frac{RT}{M}\right)^{1/2}\left(1 + \frac{1}{2}B'c\right)$$

We can solve for B' and then evaluate $g(B')^2 = C'$.

$$B' = \frac{2}{c}\left\{\left(\frac{\Pi / c}{RT / M}\right)^{1/2} - 1\right\}$$

RT / M has been determined above as 230 g cm^{-2} / g cm^{-3}. We analytically solve for B' from one of the data points, say, $\Pi / c = 430$ g cm^{-2} / g cm^{-3} at $c = 0.033$ g cm^{-3}.

$$B' = \frac{2}{0.033 \text{ g cm}^{-3}}\left\{\left(\frac{430 \text{ g cm}^{-2} / \text{g cm}^{-3}}{230 \text{ g cm}^{-2} / \text{g cm}^{-3}}\right)^{1/2} - 1\right\} = 22.\bar{1} \text{ cm}^3 \text{ g}^{-1}$$

and $$C' = g(B')^2 = 0.25 \times (22.\bar{1} \text{ cm}^3 \text{ g}^{-1})^2 = 12\bar{3} \text{ cm}^6 \text{ g}^{-2}$$

Better values of B' and C' can be obtained by plotting $\left(\frac{\Pi / c}{RT / M}\right)^{1/2}$ against c. This plot is shown in Figure 16.15(d). The slope is 14 $cm^3\ g^{-1}$

$$B' = 2 \times \text{slope} = \boxed{28\ cm^3\ g^{-1}} \quad \text{and} \quad C' = g(B')^2 = \boxed{19\bar{6}\ cm^6\ g^{-2}}$$

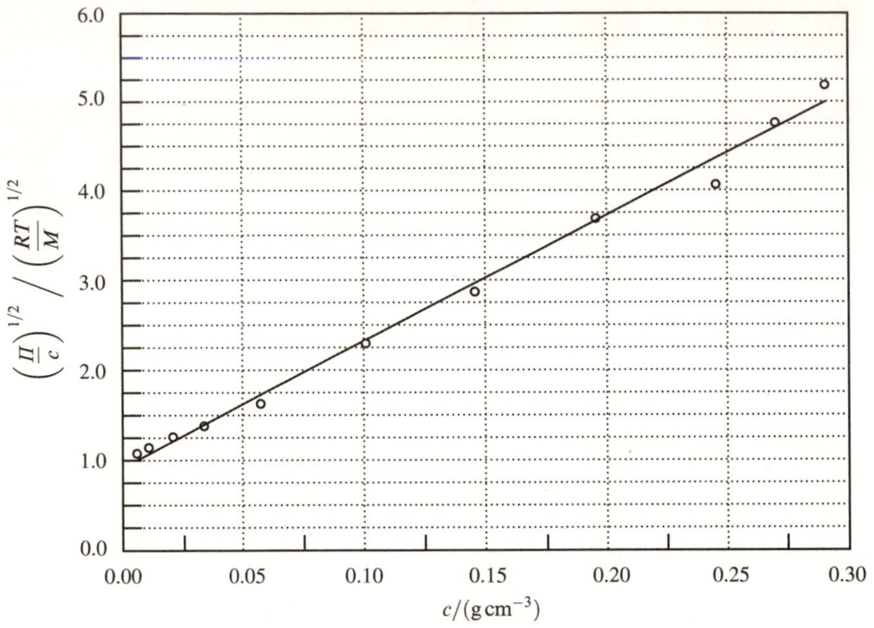

Figure 16.15(d)

The intercept of this plot should theoretically be 1.00, but it is in fact 0.915 with a standard deviation of 0.066. The overall consistency of the values of the parameters confirms that g is roughly 1/4 as assumed.

17 Chemical equilibrium

Answers to discussion questions

D17.1 The position of equilibrium is always determined by the condition that the reaction quotient, Q, must equal the equilibrium constant, K. If the mixing in of an additional amount of reactant or product destroys the equality, then the reacting system will shift so that equality is restored. That implies that some of the added reactant or product must be removed by the reacting system and the amounts of other components will also be affected. These adjustments restore the concentrations to their (new) equilibrium values.

D17.3 See *Justification* 17.2 for a derivation of the general expression (eqn. 17.17b) for the equilibrium constant in terms of the partition functions and difference in molar energy, $\Delta_r E_0$, of the products and reactants in a chemical reaction. The partition functions are functions of temperature, and the ratio of partition functions in eqn. 17.17(b) will therefore vary with temperature. However, the most direct effect of temperature on the equilibrium constant is through the exponential term $e^{-\Delta_r E_0/RT}$. The manner in which both factors affect the magnitudes of the equilibrium constant and its variation with temperature is described in detail for a simple $R \rightleftharpoons P$ gas phase equilibrium in Section 17.3(b) and *Justification* 17.3.

D17.5 (1) Response to change in pressure. The equilibrium constant is independent of pressure, but the individual partial pressures can change as the total pressure changes. This will happen when there is a difference, Δn_g, between the sums of the number of moles of gases on the product and reactant sides of the balanced chemical reaction equation:

$$\Delta n_g = \sum_{J=\text{product gases}} \nu_J - \sum_{J=\text{reactant gases}} |\nu_J|$$

The requirement of an unchanged equilibrium constant implies that the side with the smaller number of moles of gas is favored as pressure increases. To see this, we examine the general reaction equation $0 = \sum_J \nu_J J$ [17.8] is in the special case for which all reactants and products are perfect gases. In this case, the activities of eqn. 17.12 equal the partial pressure of the gaseous species, and therefore

$$a_{J(\text{gas})} = p_J / p^{\ominus} = x_J p / p^{\ominus}$$

where x_J is the mole fraction of gaseous species J. Substitution into eqn. 17.12 and simplification yield a useful equation:

$$K = \left(\prod_J a_J^{\nu_J}\right)_{\text{equilibrium}} = \left(\prod_J x_J^{\nu_J} \left(p/p^{\ominus}\right)^{\nu_J}\right)_{\text{equilibrium}}$$

$$= \left(\prod_J x_J^{\nu_J}\right)_{\text{equilibrium}} \left(\prod_J \left(p/p^{\ominus}\right)^{\nu_J}\right)_{\text{equilibrium}} = \left(\prod_J x_J^{\nu_J}\right)_{\text{equilibrium}} \left(p/p^{\ominus}\right)^{\Delta n_g}$$

$$= K_x \left(p/p^{\ominus}\right)^{\Delta n_g} \quad \text{where} \quad K_x = \left(\prod_J x_J^{\nu_J}\right)_{\text{equilibrium}}$$

K_x is not an equilibrium constant. It is a ratio of product and reactant concentration factors that has a form analogous to the equilibrium constant K. However, whereas K depends on temperature alone, the concentration ratio K_x depends on both temperature and pressure. Solving for K_x provides an equation that directly indicates its pressure dependence.

$$K_x = K\left(p/p^{\ominus}\right)^{-\Delta n_g}$$

This equation indicates that, if $\Delta n_g = 0$ (an equal number of gas moles on both sides of the balanced reaction equation), $K_x = K$ and the concentration ratio has no pressure dependence. An increase in pressure causes no change in K_x and no shift in the concentration equilibrium is observed upon a change in pressure.

However this equation indicates that, if $\Delta n_g < 0$ (fewer moles of gas on the product side of the balanced reaction equation), $K_x = K(p/p^{\ominus})^{|\Delta n_g|}$. Because p is raised to a positive power in this case, an increase in pressure causes K_x to increase. This means that the numerator concentrations (products) must increase while the denominator concentrations (reactants) decrease. The concentrations shift to the product side to reestablish equilibrium when an increase in pressure has stressed the reaction equilibrium. Similarly, if $\Delta n_g > 0$ (fewer moles of gas on the reactant side of the balanced reaction equation), $K_x = K(p/p^{\ominus})^{-|\Delta n_g|}$. Because p is raised to a negative power in this case, the concentrations now shift to the reactant side to reestablish equilibrium when an increase in pressure has stressed the reaction equilibrium.

(2) Response to change in temperature. Equation 17.26(a) shows that K decreases with increasing temperature when the reaction is exothermic; thus the reaction shifts to the left, the opposite occurs in endothermic reactions. See Section 17.5(a) for a more detailed discussion.

Failure of Le Chatelier's principle is very unusual. The expected response to pressure may prove wrong should the gas components be non-perfect, real gases. Should the gases of the reaction side that is favored by Le Chatelier's principle strongly repel while the gas molecules of the other side strongly attract, a compression may shift the reaction in the direction opposite to that expected for perfect gases. This is reflected in the analysis of eqn. 17.12 for the special case in which all products and gases are real gases having activity coefficients that do not equal 1. Coefficients greater than 1 are observed when intermolecular repulsions dominate, and they are less than 1 when attractive forces dominate. In this case, $a_{\text{J(gas)}} = \gamma_J p_J / p^{\ominus} = \gamma_J x_J p / p^{\ominus}$ where γ_J is the activity coefficient of gas J. Substitution and simplification yields

$$K_x = K(p/p^{\ominus})^{-\Delta n_g}/K_\gamma \quad \text{where} \quad K_\gamma = \prod_J \gamma_J^{\nu_J}$$

If $\Delta n_g < 0$ (fewer moles of non-perfect gas on the product side of the balanced reaction equation), a pressure increase will not shift the reaction equilibrium toward the products, which is predicted by Le Chatelier's principle, provided that the pressure increase causes K_γ to increase to the extent that the ratio $(p/p^{\ominus})^{-\Delta n_g}/K_\gamma$ decreases.

D17.7 A galvanic cell uses a spontaneous chemical reaction to generate a potential difference and deliver an electric current to an external device. An electrolytic cell uses an external potential difference to drive a chemical reaction in the cell that is by itself non-spontaneous. In their essential features, these two kinds of cells can be considered opposites of each other, in the sense that an electrolytic cell can be thought of as a galvanic cell operating in the reverse direction. For some electrochemical cells, this is easy to accomplish. We say they are rechargeable. The most common example is the lead–acid battery used in automobiles. For many other cells, however, this kind of reversibility cannot be achieved. A fuel cell, like the galvanic cell, uses a spontaneous chemical reaction to generate a potential difference and deliver an electric current to an external device. Unlike the galvanic cell, the fuel cell must receive reactants from an external storage tank. See the text on engineering, *Impact* 17.2, Fuel cells.

D17.9 The electrochemical series lists metallic elements and hydrogen in the order of their reducing power as measured by their standard potentials in aqueous solution (see Table 17.3). It is used to quickly determine whether one metal can spontaneously displace another from solution at 298 K. Application rule: A low element in the series will displace the cation of a high element in the series. For example, zinc is lower than hydrogen in the series, so zinc will spontaneously react with the hydronium cation, H^+(aq), to form the zinc cation and hydrogen gas: Zn(s) + $2H^+$(aq) $\rightarrow$ Zn^{2+}(aq) + H_2(g).

D17.11 Eqn. 17.41,

$$\Delta_r H^\ominus = -\nu F\left(E^\ominus_{\text{cell}} - T\frac{dE^\ominus_{\text{cell}}}{dT}\right) \quad [17.41]$$

provides a non-calorimetric method for measuring $\Delta_r H^\ominus$. $E^\ominus_{\text{cell}}$ must be measured over a range of temperatures so that the temperature derivative, $dE^\ominus_{\text{cell}}/dT$, may be evaluated by curve fitting the $E^\ominus_{\text{cell}}(T)$ data. Substitution of the $E^\ominus_{\text{cell}}$ and $dE^\ominus_{\text{cell}}/dT$ values into eqn. 17.41 yields the standard reaction enthalpy. The procedure is illustrated in text Example 17.5. In the special case for which $dE^\ominus_{\text{cell}}/dT$ is negligibly small, we may conclude that $\Delta_r S^\ominus \cong 0$ (see eqn. 17.40) and $\Delta_r H^\ominus \cong \Delta_r G^\ominus = -\nu FE^\ominus_{\text{cell}}$.

Solutions to exercises

E17.1(a) Eqn. 17.12 provides equilibrium constants in terms of activities for part (i) with the activities of pure solids and liquids being equal to 1. For part (ii), substitute $a_{\text{solute}} = \gamma_{\text{solute}} b_{\text{solute}}/b^\ominus$ and assume perfect gas behavior with the substitution $a_{\text{gas}} = p_{\text{gas}}/p^\ominus$.

(a) (i) $K = \dfrac{a_{\text{COCl(g)}}a_{\text{Cl(g)}}}{a_{\text{CO(g)}}a_{\text{Cl}_2\text{(g)}}}$ (ii) $K = \dfrac{p_{\text{COCl}}p_{\text{Cl}}}{p_{\text{CO}}p_{\text{Cl}_2}}$

(b) (i) $K = \dfrac{a^2_{\text{SO}_3\text{(g)}}}{a^2_{\text{SO}_2\text{(g)}}a_{\text{O}_2\text{(g)}}}$ (ii) $K = \dfrac{p^2_{\text{SO}_3}p^\ominus}{p^2_{\text{SO}_2}p_{\text{O}_2}}$

(c) (i) $K = \dfrac{a_{\text{FeSO}_4\text{(aq)}}}{a_{\text{PbSO}_4\text{(aq)}}}$ (ii) $K = \dfrac{\gamma_{\text{FeSO}_4}}{\gamma_{\text{PbSO}_4}} \times \dfrac{b_{\text{FeSO}_4}}{b_{\text{PbSO}_4}}$

(d) (i) $K = \dfrac{a^2_{\text{HCl(aq)}}}{a_{\text{H}_2\text{(g)}}}$ (ii) $K = \dfrac{\gamma^2_{\text{HCl(aq)}}b^2_{\text{HCl(aq)}}}{p_{\text{H}_2\text{(g)}}} \times \dfrac{p^\ominus}{(b^\ominus)^2}$.

(e) (i) $K = \dfrac{a_{\text{CuCl}_2\text{(aq)}}}{a^2_{\text{CuCl(aq)}}}$ (ii) $K = \dfrac{\gamma_{\text{CuCl}_2}}{\gamma^2_{\text{CuCl}}} \times \dfrac{b_{\text{CuCl}_2}b^\ominus}{b^2_{\text{CuCl}}}$

E17.2(a) $\nu_{\text{Hg}_2\text{Cl}_2} = -1,\ \nu_{\text{H}_2} = -1,\ \nu_{\text{HCl}} = 2,\ \nu_{\text{Hg}} = 2 \quad (\Delta n_g = -1)$

E17.3(a) Let B = borneol and I = isoborneol; $\text{B} \rightleftharpoons \text{I}$.

$$\Delta_r G = \Delta G^\ominus + RT\ln Q\ [17.9] \quad \text{where} \quad Q = \frac{p_{\text{I}}}{p_{\text{B}}}\ [17.5°]$$

$$p_{\text{B}} = x_{\text{B}}p = \frac{0.15\text{ mol}}{0.15\text{ mol} + 0.30\text{ mol}} \times 600\text{ Torr} = 200\text{ Torr}; \qquad p_{\text{I}} = p - p_{\text{B}} = 400\text{ Torr}$$

$$Q = \frac{400\text{ Torr}}{200\text{ Torr}} = 2.00$$

$$\Delta_r G = (+9.4\text{ kJ mol}^{-1}) + (8.314\text{ J K}^{-1}\text{ mol}^{-1}) \times (503\text{ K}) \times (\ln 2.00) = \boxed{+12.3\text{ kJ mol}^{-1}}$$

This mixture reacts spontaneously to the left (toward the formation of borneol).

E17.4(a) The formation reaction is $2Ag(s) + \frac{1}{2}O_2(g) \rightleftharpoons Ag_2O(s)$.

$$K = \frac{1}{a^{1/2}_{O_2(g)}} = \left(\frac{p^\ominus}{p_{O_2}}\right)^{1/2} \quad (a_{Ag(s)} = a_{Ag_2O(s)} = 1, \text{ and assuming perfect gas behavior, } a_{O_2(g)} = p_{O_2}/p^\ominus.)$$

$$= \left(\frac{10^5\ \text{Pa}}{11.85\ \text{Pa}}\right)^{1/2} = 91.86$$

$$\Delta_r G^\ominus = -RT \ln K \ [17.13]$$
$$= -(8.3145\ \text{J K}^{-1}\ \text{mol}^{-1}) \times (298\ \text{K}) \times (\ln 91.86)$$
$$= \boxed{-11.20\ \text{kJ mol}^{-1}}$$

E17.5(a) The reaction equation is $PbO(s) + CO(g) \rightleftharpoons Pb(s) + CO_2(g)$.

$$\nu_{Pb} = 1 \qquad \nu_{CO_2} = 1 \qquad \nu_{PbO} = -1 \qquad \nu_{CO} = -1$$

(a) $$\Delta_r G^\ominus = \sum_J \nu_J \Delta_f G^\ominus(J) \ [17.10b]$$

$$\Delta_r G^\ominus = \Delta_f G^\ominus(\text{Pb, s}) + \Delta_f G^\ominus(\text{CO}_2\text{, g}) - \Delta_f G^\ominus(\text{PbO, s, red}) - \Delta_f G^\ominus(\text{CO, g})$$
$$= (-394.36\ \text{kJ mol}^{-1}) - (-188.93\ \text{kJ mol}^{-1}) - (-137.17\ \text{kJ mol}^{-1})$$
$$= \boxed{-68.26\ \text{kJ mol}^{-1}}$$

$$\ln K = \frac{-\Delta_r G^\ominus}{RT}\ [17.13] = \frac{+68.26 \times 10^3\ \text{J mol}^{-1}}{(8.3145\ \text{J K}^{-1}\ \text{mol}^{-1}) \times (298\ \text{K})} = 27.55; \quad K = \boxed{9.2 \times 10^{11}}$$

(b) $$\Delta_r H^\ominus = \Delta_f H^\ominus(\text{Pb, s}) + \Delta_f H^\ominus(\text{CO}_2\text{, g}) - \Delta_f H^\ominus(\text{PbO, s, red}) - \Delta_f H^\ominus(\text{CO, g})$$
$$= (-393.51\ \text{kJ mol}^{-1}) - (-218.99\ \text{kJ mol}^{-1}) - (-110.53\ \text{kJ mol}^{-1})$$
$$= \boxed{-63.99\ \text{kJ mol}^{-1}}$$

$$\ln K(400\ \text{K}) = \ln K(298\ \text{K}) - \frac{\Delta_r H^\ominus}{R}\left(\frac{1}{400\ \text{K}} - \frac{1}{298\ \text{K}}\right) [17.28]$$
$$= 27.55 - \left(\frac{-63.99 \times 10^3\ \text{J mol}^{-1}}{8.314\ \text{J K}^{-1}\ \text{mol}^{-1}}\right) \times (-8.55\bar{7} \times 10^{-4}\ \text{K}^{-1}) = 20.9\bar{6}$$

$$K(400\ \text{K}) = \boxed{1.3 \times 10^9}$$

$$\Delta_r G^\ominus(400\ \text{K}) = -RT \ln K(400\ \text{K})\ [17.13] = -(8.3145\ \text{J K}^{-1}\ \text{mol}^{-1}) \times (400\ \text{K}) \times (20.9\bar{6})$$
$$= -6.97 \times 10^4\ \text{J mol}^{-1} = \boxed{-69.7\ \text{kJ mol}^{-1}}$$

E17.6(a) $$CaF_2(s) \rightleftharpoons Ca^{2+}(aq) + 2F^-(aq) \qquad K = 3.9 \times 10^{-11}$$

$$\Delta_r G^\ominus = -RT \ln K$$
$$= -(8.3145\ \text{J K}^{-1}\ \text{mol}^{-1}) \times (298.15\ \text{K}) \times \ln(3.9 \times 10^{-11}) = +59.4\ \text{kJ mol}^{-1}$$
$$= \Delta_f G^\ominus(\text{CaF}_2\text{, aq}) - \Delta_f G^\ominus(\text{CaF}_2\text{, s})$$

$$\Delta_f G^\ominus(\text{CaF}_2\text{, aq}) = \Delta_r G^\ominus + \Delta_f G^\ominus(\text{CaF}_2\text{, s})$$
$$= [59.4 - 1167]\ \text{kJ mol}^{-1} = \boxed{-1108\ \text{kJ mol}^{-1}}$$

E17.7(a) Draw up the following equilibrium table for the reaction equation: $2\,A + B \rightleftharpoons 3\,C + 2\,D$.

	A	B	C	D	Total
Initial amounts/mol	1.00	2.00	0	1.00	4.00
Stated change/mol			+0.90		
Implied change/mol	−0.60	−0.30	+0.90	+0.60	
Equilibrium amounts/mol	0.40	1.70	0.90	1.60	4.60
Mole fractions	0.087	0.370	0.196	0.348	1.001

(a) The mole fractions are given in the table.

(b) $$K_x = \prod_J x_J^{\nu_J}$$

$$K_x = \frac{(0.196)^3 \times (0.348)^2}{(0.087)^2 \times (0.370)} = 0.32\bar{6} = \boxed{0.33}$$

(c) $$p_J = x_J p \qquad p = 1\text{ bar} \qquad p^{\ominus} = 1\text{ bar}$$

Assuming that the gases are perfect, $a_J = p_J / p^{\ominus}$. Hence,

$$K = \frac{(p_C/p^{\ominus})^3 \times (p_D/p^{\ominus})^2}{(p_A/p^{\ominus})^2 \times (p_B/p^{\ominus})}$$

$$= \frac{x_C^3 x_D^2}{x_A^2 x_B} \times \left(\frac{p}{p^{\ominus}}\right)^2 = K_x \quad \text{when } p = 1.00\text{ bar} = \boxed{0.33}$$

(d) $$\Delta_r G^{\ominus} = -RT\ln K = -(8.3145\text{ J K}^{-1}\text{ mol}^{-1}) \times (298\text{ K}) \times (\ln 0.32\bar{6}) = \boxed{+2.8\text{ kJ mol}^{-1}}$$

E17.8(a) $$ATP^{4-}(aq) + H_2O(l) \rightarrow ADP^{3-}(aq) + HPO_4^{2-}(aq) + H_3O^{+}(aq)$$

The biological standard state applies to the state where $a_{H^+} = 10^{-7}$ and all other activities are 1 (i.e., the biological standard state, ⊕, has pH = 7). Hence $Q^{\oplus} = 1\times10^{-7}$.

$$\Delta_r G^{\oplus} = \Delta_r G^{\ominus} + RT\ln Q^{\oplus} \quad [17.9]$$

$$= 10\text{ kJ mol}^{-1} + (8.3145\text{ J K}^{-1}\text{ mol}^{-1}) \times (298\text{ K}) \times \ln(1\times10^{-7})$$

$$= \boxed{-30\text{ kJ mol}^{-1}}$$

E17.9(a) For the chemical reaction $I_2(g) \rightarrow 2\,I(g)$, we use eqn. 17.19a with X = I, $X_2 = I_2$, and $\Delta_r E_0 = D_0(\text{I–I})$ where $D_0(\text{I–I})$ is the dissociation energy of the I–I bond.

$$D_0 = D_e - \frac{1}{2}\tilde{\nu} = 1.5422\text{ eV} \times \frac{8065.5\text{ cm}^{-1}}{1\text{ eV}} - 107.18\text{ cm}^{-1}$$

$$= 1.2331\times10^4\text{ cm}^{-1} = 1.475\times10^5\text{ J mol}^{-1}$$

$$K = \left(\frac{(q^{\ominus}_{I,m})^2}{q^{\ominus}_{I_2,m} N_A}\right) e^{-\Delta E_0/RT} \quad [17.19a]$$

$$q^{\ominus}_{I,m} = q^{T}_{m}(I)q^{E}(I), \qquad q^{E}(I) = 4$$

$$q^{\ominus}_{I_2,m} = q^{T}_{m}(I_2)q^{R}(I_2)q^{V}(I_2)q^{E}(I_2), \quad q^{E}(I_2) = 1$$

$$\frac{q_m^T(I_2)}{N_A} = 2.561\times10^{-2}(T/K)^{5/2}\times(M/g\,mol^{-1})^{3/2} \quad \text{[Table 13.1]}$$

$$= 2.561\times10^{-2}\times1000^{5/2}\times253.8^{3/2} = 3.27\times10^{9}$$

$$\frac{q_m^T(I)}{N_A} = 2.561\times10^{-2}\times1000^{5/2}\times126.9^{3/2} = 1.16\times10^{9}$$

$$q^R(I_2) = \frac{0.6950}{\sigma}\times\frac{T/K}{B/cm^{-1}} = \tfrac{1}{2}\times0.6950\times\frac{1000}{0.0373} = 931\overline{6}$$

$$q^V(I_2) = \frac{1}{1-e^{-a}}, \quad a = 1.4388\frac{\tilde{\nu}/cm^{-1}}{T/K}$$

$$= \frac{1}{1-e^{-1.4388\times214.36/1000}} = 3.77$$

$$K = \frac{(1.16\times10^{9}\times4)^2}{(3.27\times10^{9})\times(9316)\times(3.77)}e^{-1.475\times10^{5}/8.3145/1000} = \boxed{3.70\times10^{-3}}$$

E17.10(a) For the reaction $H_2CO(g) \rightarrow CO(g) + H_2(g)$, $\Delta n_g = 1$.

Assuming that all reactants and products are perfect gases, the activities of eqn. 17.12 equal the partial pressure of the gaseous species, and therefore,

$$a_{J(gas)} = p_J/p^{\ominus} = x_J p/p^{\ominus}$$

where x_J is the mole fraction of gaseous species J. Substitution into eqn. 17.12 and simplification yield a useful equation:

$$K = \left(\prod_J a_J^{\nu_J}\right)_{\text{equilibrium}} = \left(\prod_J x_J^{\nu_J}(p/p^{\ominus})^{\nu_J}\right)_{\text{equilibrium}}$$

$$= \left(\prod_J x_J^{\nu_J}\right)_{\text{equilibrium}}\left(\prod_J (p/p^{\ominus})^{\nu_J}\right)_{\text{equilibrium}} = \left(\prod_J x_J^{\nu_J}\right)_{\text{equilibrium}}(p/p^{\ominus})^{\Delta n_g}$$

$$= K_x(p/p^{\ominus})^{\Delta n_g} \quad \text{where} \quad K_x = \left(\prod_J x_J^{\nu_J}\right)_{\text{equilibrium}}$$

K_x is not an equilibrium constant. It is a ratio of product and reactant concentration factors that has a form analogous to the equilibrium constant K. Therefore,

$$K_x = K\left(\frac{p}{p^{\ominus}}\right)^{-\Delta n_g}$$

Since K depends on temperature alone,

$$\frac{K_x(p_2)}{K_x(p_1)} = \frac{K(p_2/p^{\ominus})^{-\Delta n_g}}{K(p_1/p^{\ominus})^{-\Delta n_g}} = \left(\frac{p_2}{p_1}\right)^{-\Delta n_g}$$

$$\frac{K_x(3.0\text{ bar})}{K_x(1.0\text{ bar})} = \left(\frac{3.0\text{ bar}}{1.0\text{ bar}}\right)^{-1} = 0.33$$

Thus, $\boxed{K_x \text{ is reduced by } 67\%}$ when the pressure is increased from 1.0 bar to 3.0 bar. This is quantification of the Le Chatelier principle that the equilibrium shifts to the left (side that has smaller number of gas moles) to reduce the stress of increased pressure.

E17.11(a) At 1280 K, $\Delta_r G^\ominus = +33\times10^3\,\text{J mol}^{-1}$; thus

$$\ln K_1(1280\,\text{K}) = -\frac{\Delta_r G^\ominus}{RT} = -\frac{33\times10^3\ \text{J mol}^{-1}}{(8.3145\ \text{J K}^{-1}\ \text{mol}^{-1})\times(1280\ \text{K})} = -3.1\overline{0}$$

$$K_1 = \boxed{0.045}$$

$$\ln K_2 = \ln K_1 - \frac{\Delta_r H^\ominus}{R}\left(\frac{1}{T_2}-\frac{1}{T_1}\right)\ [17.28]$$

We look for the temperature T_2 that corresponds to $\ln K_2 = \ln(1) = 0$. This is the crossover temperature. Solving for T_2 from eqn 17.28 with $\ln K_2 = 0$, we obtain

$$\frac{1}{T_2} = \frac{R\ln K_1}{\Delta_r H^\ominus} + \frac{1}{T_1} = \left(\frac{(8.3145\ \text{J K}^{-1}\ \text{mol}^{-1})\times(-3.1\overline{0})}{224\times10^3\ \text{J mol}^{-1}}\right) + \left(\frac{1}{1280\ \text{K}}\right)$$

$$= 6.6\overline{6}\times10^{-4}\ \text{K}^{-1}$$

$$T_2 = \boxed{15\overline{00}\ \text{K}}$$

E17.12(a) Given $\ln K = -1.04 - \frac{1088\ \text{K}}{T} + \frac{1.51\times10^5\ \text{K}^2}{T^2}$,

and since $\frac{\text{d}\ln K}{\text{d}(1/T)} = \frac{-\Delta_r H^\ominus}{R}$ [17.26b],

$$\frac{-\Delta_r H^\ominus}{R} = -1088\ \text{K} + \frac{(2)\times(1.51\times10^5\ \text{K}^2)}{T}$$

Then at 450 K,

$$\Delta_r H^\ominus = \left(1088\ \text{K} - \frac{3.02\times10^5\ \text{K}^2}{450\ \text{K}}\right)\times(8.3145\ \text{J K}^{-1}\ \text{mol}^{-1}) = \boxed{+3.47\ \text{kJ mol}^{-1}}$$

$$\Delta_r G^\ominus = -RT\ln K\ [17.13] = RT\times\left(1.04 + \frac{1088\ \text{K}}{T} - \frac{1.51\times10^5\ \text{K}^2}{T^2}\right)$$

$$= (8.3145\ \text{J K}^{-1}\ \text{mol}^{-1})\times(450\ \text{K})\times\left(1.04 + \frac{1088\ \text{K}}{450\ \text{K}} - \frac{1.51\times10^5\ \text{K}^2}{(450\ \text{K})^2}\right) = +10.15\ \text{kJ mol}^{-1}$$

$$= \Delta_r H^\ominus - T\Delta_r S^\ominus\ [15.50]$$

Therefore, $\Delta_r S^\ominus = \frac{\Delta_r H^\ominus - \Delta_r G^\ominus}{T} = \frac{3.47\ \text{kJ mol}^{-1} - 10.15\ \text{kJ mol}^{-1}}{450\ \text{K}} = \boxed{-14.8\ \text{J K}^{-1}\ \text{mol}^{-1}}$.

E17.13(a)

$$\ln\frac{K_2}{K_1} = -\frac{\Delta_r H^\ominus}{R}\left(\frac{1}{T_2}-\frac{1}{T_1}\right)\ [17.28]$$

Therefore, $\Delta_r H^\ominus = -\frac{R\ln\left(\frac{K_2}{K_1}\right)}{\left(\frac{1}{T_2}-\frac{1}{T_1}\right)}$

$T_2 = 308$ K; hence, with the substitution $K_2/K_1 = \kappa$,

$$\Delta_r H^\ominus = -\frac{(8.3145\ \text{J K}^{-1}\ \text{mol}^{-1})\times(\ln\kappa)}{\left(\frac{1}{308\ \text{K}}-\frac{1}{298\ \text{K}}\right)} = 76.3\ \text{kJ mol}^{-1}\times\ln\kappa$$

Therefore,

(a) $\kappa = 2,\ \Delta_r H^\ominus = (76.3\text{ kJ mol}^{-1}) \times \ln 2 = \boxed{+53\text{ kJ mol}^{-1}}$

(b) $\kappa = \frac{1}{2},\ \Delta_r H^\ominus = (76.3\text{ kJ mol}^{-1}) \times \ln \frac{1}{2} = \boxed{-53\text{ kJ mol}^{-1}}$

E17.14(a) The decomposition reaction is $CaCO_3(s) \rightleftharpoons CaO(s) + CO_2(g)$.

For the purposes of this exercise we assume that the required temperature is the temperature at which $K = 1$, which corresponds to a pressure of 1 bar for the gaseous product. For $K = 1$, $\ln K = 0$ and $\Delta_r G^\ominus = 0$.

$$\Delta_r G^\ominus = \Delta_r H^\ominus - T\Delta_r S^\ominus = 0 \quad \text{when } \Delta_r H^\ominus = T\Delta_r S^\ominus$$

Therefore, the decomposition temperature (when $K = 1$) is

$$T = \frac{\Delta_r H^\ominus}{\Delta_r S^\ominus}$$

$$\Delta_r H^\ominus = \{(-635.09) - (393.51) - (-1206.9)\}\text{ kJ mol}^{-1} = +178.3\text{ kJ mol}^{-1}$$

$$\Delta_r S^\ominus = \{(39.75) + (213.74) - (92.9)\}\text{ J K}^{-1}\text{ mol}^{-1} = +160.6\text{ J K}^{-1}\text{ mol}^{-1}$$

$$T = \frac{178.3\times 10^3\text{ J mol}^{-1}}{160.6\text{ J K}^{-1}\text{ mol}^{-1}} = \boxed{1110\text{ K}}\ (840\ °\text{C})$$

E17.15(a) The cell notation specifies the right and left electrodes. Note that for proper cancelation we must equalize the number of electrons in half-reactions being combined. To calculate the standard cell potential (emf) of the cell, we have used $E^\ominus = E_R^\ominus - E_L^\ominus$, with standard electrode potentials from Table 17.2.

		$E^\ominus$
(a)	R: $2Ag^+(aq) + 2e^- \rightarrow 2Ag(s)$	+0.80 V
	L: $Zn^+(aq) + 2e^- \rightarrow Zn(s)$	–0.76 V
(b)	Overall (R – L): $2Ag^+(aq) + Zn(s) \rightarrow 2Ag(s) + Zn^{2+}(aq)$	+1.56 V
	R: $2H^+(aq) + 2e^- \rightarrow H_2(g)$	0
	L: $Cd^{2+}(aq) + 2e^- \rightarrow Cd(s)$	–0.40 V
(c)	Overall (R – L): $Cd(s) + 2H^+(aq) \rightarrow Cd^{2+}(aq) + H_2(g)$	+0.40 V
	R: $Cr^{3+}(aq) + 3e^- \rightarrow Cr(s)$	–0.74 V
	L: $3[Fe(CN)_6]^{3-}(aq) + 3e^- \rightarrow 3[Fe(CN)_6]^{4-}(aq)$	+0.36 V
(d)	Overall (R – L): $Cr^{3+}(aq) + 3[Fe(CN)_6]^{4-}(aq) \rightarrow Cr(s) + 3[Fe(CN)_6]^{3-}(aq)$	–1.10 V
	R: $Sn^{4+}(aq) + 2e^- \rightarrow Sn^{2+}(aq)$	+0.15 V
	L: $2\ Fe^{3+}(aq) + 2\ e^- \rightarrow 2\ Fe^{2+}(aq)$	+0.77 V
	Overall (R – L): $Sn^{2+}(aq) + 2Fe^{2+}(aq) \rightarrow Sn^{2+}(aq) + 2Fe^{2+}(aq)$	–0.62 V

COMMENT. Cells for which $E^\ominus > 0$ may operate as spontaneous galvanic cells under standard conditions. Cells for which $E^\ominus < 0$ may operate as non-spontaneous electrolytic cells. Recall that $E^\ominus$ informs us of the spontaneity of a cell under standard conditions only. For other conditions we require E.

E17.16(a) The conditions (concentrations, and so on) under which these reactions occur are not given. For the purposes of this exercise we assume standard conditions. The specification of the right and left electrodes is determined by the direction of the reaction as written. As always, in combining half-reactions to form an overall cell reaction, we must write the half-reactions with equal numbers of electrons to ensure proper cancellation. We first identify the half-reactions and then set up the corresponding cell. To calculate the standard cell potential (emf) of the cell, we have used $E^{\ominus} = E_R^{\ominus} - E_L^{\ominus}$, with standard electrode potentials from Table 17.2.

		$E^{\ominus}$
(a)	R: $Pb^{2+}(aq) + 2e^- \rightarrow Pb(s)$	–0.13 V
	L: $Fe^{2+}(aq) + 2e^- \rightarrow Fe(s)$	–0.44 V
	Hence the cell is	
	$Fe(s)\,|\,FeSO_4(aq)\,||\,PbSO_4(aq)\,|\,Pb(s)$	+0.31 V
(b)	R: $Hg_2Cl_2(s) + 2e^- \rightarrow 2Hg(l) + 2Cl^-(aq)$	+0.27 V
	L: $H^+(aq) + e^- \rightarrow \frac{1}{2}H_2(g)$	0
	and the cell is	
	$Pt|H_2(g)|H^+(aq)|Hg_2Cl_2(s)|Hg(l)$	
	or $Pt|H_2(g)|HCl(aq)|Hg_2Cl_2(s)|Hg(l)$	+0.27 V
(c)	R: $O_2(g) + 4H^+(aq) + 4e^- \rightarrow 2H_2O(l)$	+1.23 V
	L: $2H^+(aq) + 2e^- \rightarrow H_2(g)$	0
	and the cell is	
	$Pt|H_2(g)|H^+(aq)\,|\,O_2(g)\,|\,Pt$	+1.23 V

COMMENT. All these cells have $E^{\ominus} > 0$, corresponding to a spontaneous cell reaction under standard conditions. If $E^{\ominus}$ had turned out to be negative, the spontaneous reaction would have been the reverse of the one given, with the right and left electrodes of the cell also reversed.

E17.17(a) $Ag|AgBr(s)|KBr(aq, 0.050\ mol\ kg^{-1})||Cd(NO_3)_2(aq, 0.010\ mol\ kg^{-1})|Cd(s)$

(a) R: $Cd^{2+}(aq) + 2\,e^- \rightarrow Cd(s)$ $\quad E^{\ominus} = -0.40\ V$ (Table 17.2)

L: $AgBr(s) + e^- \rightarrow Ag(s) + Br^-(aq)$ $\quad E^{\ominus} = +0.0713\ V$

R – L: $\boxed{Cd^{2+}(aq) + 2Br^-(aq) + 2\,Ag(s) \rightarrow Cd(s) + 2\,AgBr(s)}$ $\quad E_{cell}^{\ominus} = -0.47\ V$

The cell reaction is not spontaneous toward the right under standard conditions because $E_{cell}^{\ominus} < 0$.

(b) The Nernst equation for the above cell reaction is

$$E_{cell} = E_{cell}^{\ominus} - \frac{RT}{\nu F}\ln Q \ [17.32]$$

where $\nu = 2$ and $Q = \dfrac{1}{a_{Cd^{2+}}a_{Br^-}^2} = \dfrac{1}{\gamma_{Cd^{2+}}\gamma_{Br^-}^2} \times \dfrac{(b^{\ominus})^3}{b_{Cd^{2+}}b_{Br^-}^2} = \dfrac{1}{\gamma_{\pm,R}\gamma_{\pm,L}^2} \times \dfrac{(b^{\ominus})^3}{b_{Cd^{2+}}b_{Br^-}^2}$ [16.53]

$b_{Cd^{2+}} = 0.010\ mol\ kg^{-1}$ for the right-hand electrode and $b_{Br^-} = 0.050\ mol\ kg^{-1}$ for the left-hand electrode.

(c) The ionic strength and mean activity coefficient at the right-hand electrode are

$$I_R = \tfrac{1}{2}\sum_i z_i^2(b_i/b^{\ominus})\ [16.57] = \tfrac{1}{2}\{4(0.010) + 1(.020)\} = 0.030$$

$$\log\gamma_{\pm,R} = -|z_+z_-|AI^{1/2}\ [16.56] = -2\times(0.509)\times(0.030)^{1/2} = -0.176$$

$$\gamma_{\pm,R} = 0.666$$

The ionic strength and mean acitivity coefficient at the left-hand electrode are

$$I_L = \tfrac{1}{2}\sum_i z_i^2(b_i/b^{\ominus})\ [16.57] = \tfrac{1}{2}\{1(0.050) + 1(.050)\} = 0.050$$

$$\log\gamma_{\pm,L} = -|z_+z_-|AI^{1/2}\ [16.56] = -1\times(0.509)\times(0.050)^{1/2} = -0.114$$

$$\gamma_{\pm,L} = 0.769$$

Therefore,

$$Q=\left(\frac{1}{(0.666)\times(0.769)^2}\right)\times\left(\frac{1}{(0.010)\times(0.050)^2}\right)=1.02\times10^5$$

and

$$E_{\text{cell}}=-0.47\text{ V}-\left(\frac{25.693\times10^{-3}\text{ V}}{2}\right)\ln(1.02\times10^5)$$
$$=\boxed{-0.62\text{ V}}$$

E17.18(a) We calculate the standard Gibbs energy with $\Delta_r G^\ominus_{\text{cell}}=-\nu FE^\ominus_{\text{cell}}$ [17.30].

	ν	$E^\ominus_{\text{cell}}$ / V	$\Delta_r G^\ominus_{\text{cell}}$ / kJ mol^{-1}
(a)	2	+1.56	−301
(b)	2	+0.40	−77
(c)	3	−1.10	+318
(d)	2	−0.62	+12$\overline{0}$

E17.19(a) We first recognize the reduction and oxidation processes within the reaction equation and calculate $E^\ominus_{\text{cell}}$ using $E^\ominus=E^\ominus(\text{reduction couple})-E^\ominus(\text{oxidation couple})$ [17.34] with the standard reduction potentials of Table 17.2. Inspection of the balanced reaction equation and comparison with the redox couples also gives the stoichiometric coefficient of the electrons transferred, ν. The equilibrium constant is calculated with

$$K=e^{\nu FE^\ominus/RT}\ [17.33]\quad\text{where}\quad RT/F=25.693\text{ mV at }25\ ^\circ\text{C}$$

(a) $E^\ominus=E^\ominus(Sn^{4+}/Sn^{2+})-E^\ominus(Sn^{2+}/Sn)$ and $\nu=2$

$$E^\ominus=+0.15\text{ V}-(-0.14\text{ V})=+0.29\text{ V}$$

$$K=e^{2\times(0.29)/(0.025693)}=\boxed{6.4\times10^9}$$

(b) $E^\ominus=E^\ominus(Hg^{2+}/Hg)-E^\ominus(Fe^{2+}/Fe)$ and $\nu=2$

$$E^\ominus=+0.86\text{ V}-(-0.44\text{ V})=+1.30\text{ V}$$

$$K=e^{2\times(1.30)/(0.025693)}=\boxed{8.9\times10^{43}}$$

E17.20(a) If half-reaction (c) is the direct sum of half-reactions (a) and (b), the generalization of text Example 17.4 gives

$$E^\ominus(c)=\frac{\nu_a E^\ominus(a)+\nu_b E^\ominus(b)}{\nu_c}$$

Thus,

$$E^\ominus(Ce^{4+}/Ce)=\frac{E^\ominus(Ce^{4+}/Ce^{3+})+3E^\ominus(Ce^{3+}/Ce)}{4}$$
$$=\frac{(+1.61\text{ V})+3(-2.48\text{ V})}{4}$$
$$=\boxed{-1.46\text{ V}}$$

E17.21(a) Zn is lower than mercury in the electrochemical series of Tables 17.2 and 17.3. Consequently, metallic zinc displaces the mercury(II) cation from aqueous solution and elemental mercury cannot spontaneously displace the zinc(II) cation from solution under standard conditions.

An alternative view observes that, since $E^{\ominus}(\mathrm{Zn^{2+}/Zn}) - E^{\ominus}(\mathrm{Hg^{2+}/Hg}) < 0$, the reaction Hg(l) + Zn^{2+}(aq) → Hg^{2+}(aq) + Zn(s) is not spontaneous.

E17.22 (a)

$$\text{R: } \mathrm{Ag^+(aq) + e^- \rightarrow Ag(s)} \qquad +0.80\ \mathrm{V}$$

$$\text{L: } \mathrm{AgI(s) + e^- \rightarrow Ag(s) + I^-(aq)} \qquad -0.15\ \mathrm{V}$$

$$E^{\ominus} = E_R^{\ominus} - E_L^{\ominus} = 0.9509\ \mathrm{V}$$

Overall: $\mathrm{Ag^+(aq) + I^-(aq) \rightarrow AgI(s)}$ $\qquad \nu = 1$

$$\ln K = \frac{\nu F E^{\ominus}}{RT}\ [17.33] = \frac{0.9509\ \mathrm{V}}{25.693\times10^{-3}\ \mathrm{V}} = 37.01\bar{0}$$

$$K = 1.1\bar{8}\times10^{16}$$

(a)
$$K = \frac{a_{\mathrm{AgI(s)}}}{a_{\mathrm{Ag^+(aq)}}\, a_{\mathrm{I^-(aq)}}} = \frac{1}{[\mathrm{Ag^+}][\mathrm{I^-}]} = \frac{1}{[\mathrm{Ag^+}]^2} = 1.1\bar{8}\times10^{16}$$

In the above equation, the activity of the solid equals 1, and since the solution is extremely dilute, the activity coefficients of dissolved ions also equal 1. Solving for the molar ion concentration gives $[\mathrm{Ag^+}] = [\mathrm{I^-}] = 9.2 \times 10^{-9}$ M. AgI has a solubility equal to 9.2×10^{-9} M.

(b) The solubility equilibrium is written as the reverse of the cell reaction. Therefore,

$$K_S = K^{-1} = 1/1.1\bar{8}\times10^{16} = 8.5\times10^{-17}$$

E17.23(a)

$$\mathrm{Hg_2Cl_2(s) + H_2(g) \rightarrow 2\,Hg(l) + 2\,HCl(aq)} \quad \text{and} \quad \nu = 2$$

Within this small 10°C temperature range we can assume that the standard reduction potential is linear in temperature. The two-point method provides the slope.

$$\text{slope: } a = \frac{\mathrm{d}E^{\ominus}_{\mathrm{cell}}}{\mathrm{d}T} = \frac{\Delta E^{\ominus}_{\mathrm{cell}}}{\Delta T} = \frac{(0.2669 - 0.2699)\mathrm{V}}{(303-293)\mathrm{K}} = -3.0\times10^{-4}\ \mathrm{V\,K^{-1}}$$

The linear expression is

$$E^{\ominus}_{\mathrm{cell}}(T) = E_1^{\ominus} + a\times(T - T_1) = 0.2699\ \mathrm{V} + a\times(T - 293\ \mathrm{K})$$
$$= aT + b \quad \text{where} \quad a = -3.0\times10^{-4}\ \mathrm{V\,K^{-1}} \quad \text{and} \quad b = 0.3578\ \mathrm{V}$$

Thus,

$$\Delta_r G^{\ominus}(298\ \mathrm{K}) = -\nu F E^{\ominus}_{\mathrm{cell}}(298\mathrm{K})$$
$$= -2\times(96485\ \mathrm{C\,mol^{-1}})\times\{(-3.0\times10^{-4}\ \mathrm{V\,K^{-1}})\times(298\ \mathrm{K}) + 0.3578\ \mathrm{V}\}$$
$$= \boxed{-52\ \mathrm{kJ\,mol^{-1}}}$$

$$\Delta_r S^{\ominus} = \nu F \frac{\mathrm{d}E^{\ominus}_{\mathrm{cell}}}{\mathrm{d}T}\ [17.40]$$
$$= 2\times(96485\ \mathrm{C\,mol^{-1}})\times(-3.0\times10^{-4}\ \mathrm{V\,K^{-1}})$$
$$= \boxed{-58\ \mathrm{J\,K^{-1}\,mol^{-1}}}$$

$$\Delta_r H^\ominus = \Delta_r G^\ominus + T\Delta_r S^\ominus \ [17.41]$$
$$= (-52\ \text{kJ mol}^{-1}) + (298\ \text{K})\times(-58\ \text{J K}^{-1}\ \text{mol}^{-1})$$
$$= \boxed{-69\ \text{kJ mol}^{-1}}$$

Solutions to problems

Solutions to numerical problems

P17.1 (a) $\Delta_r G^\ominus = -RT\ln K$ [17.13]

$$= -(8.3145\ \text{J K}^{-1}\ \text{mol}^{-1})\times(298\ \text{K})\times(\ln 0.164) = 4.48\times10^3\ \text{J mol}^{-1}$$
$$= \boxed{+4.48\ \text{kJ mol}^{-1}}$$

(b) Draw up the following equilibrium table.

	I_2	Br_2	IBr
Amounts	—	$(1-\alpha)n$	$2\alpha n$
Mole fractions	—	$\dfrac{(1-\alpha)}{(1+\alpha)}$	$\dfrac{2\alpha}{(1+\alpha)}$
Partial pressure $p_J = x_J p$	—	$\dfrac{(1-\alpha)p}{(1+\alpha)}$	$\dfrac{2\alpha p}{(1+\alpha)}$

$$K = \prod_J a_J^{\nu_J}\ [17.12] = \frac{(p_{IBr}/p^\ominus)^2}{p_{Br_2}/p^\ominus}\ \text{[perfect gases]}$$

$$= \frac{\{(2\alpha)^2 p/p^\ominus\}}{(1-\alpha)\times(1+\alpha)} = \frac{(4\alpha^2 p/p^\ominus)}{1-\alpha^2} = 0.164$$

With $p = 0.164\,\text{atm}$, $4\alpha^2 = 1-\alpha^2 \qquad \alpha^2 = 1/5 \qquad \alpha = 0.447$

$$p_{IBr} = \frac{2\alpha}{1+\alpha}\times p = \frac{(2)\times(0.447)}{1+0.447}\times(0.164\ \text{atm}) = \boxed{0.101\ \text{atm}}$$

(c) The equilibrium table needs to be modified as follows.

$$p = p_{I_2} + p_{Br_2} + p_{IBr} \qquad p_{Br_2} = x_{Br_2} p \qquad p_{IBr} = x_{IBr} p \qquad p_{I_2} = x_{I_2} p$$

with $x_{Br_2} = \dfrac{(1-\alpha)n}{(1+\alpha)n + n_{I_2}}$ [n = amount of Br_2 introduced into the container]

and $x_{IBr} = \dfrac{2\alpha n}{(1+\alpha)n + n_{I_2}}$.

K is constructed as above [17.12] but with these modified partial pressures. To complete the calculation, additional data are required, namely, the amount of Br_2 introduced, n, and the equilibrium vapor pressure of $I_2(s)$. n_{I_2} can be calculated from a knowledge of the volume of the container at equilibrium, which is most easily determined by successive approximations since p_{I_2} is small.

What is the partial pressure of IBr(g) if 0.0100 mol of Br_2 (g) is introduced into the container? The partial pressure of I_2 (S) at 25°C is 0.305 Torr.

P17.3 $H_2O(g) \rightarrow H_2(g) + \frac{1}{2} O_2(g)$ $\quad \Delta_r G^{\ominus}(298\text{ K}) = +118.08\text{ kJ mol}^{-1}$ at 2300 K

$$K = e^{-\Delta_r G^{\ominus}/RT} \quad [17.13]$$

$$= e^{-(118080\text{ kJ mol}^{-1})/(8.3145\text{ J K}^{-1}\text{ mol}^{-1})\times(2300\text{ K})}$$

$$= 0.002082$$

Draw up the following equilibrium table.

	$H_2O(g)$	$H_2(g)$	$O_2(g)$	Totals
Amounts at start	n	—	—	n
Amounts at equilibrium	$(1-\alpha)n$	αn	$\frac{1}{2}\alpha n$	$(1+\frac{1}{2}\alpha)n$
Mole fractions at equilibrium	$\dfrac{(1-\alpha)}{(1+\frac{1}{2}\alpha)}$	$\dfrac{\alpha}{(1+\frac{1}{2}\alpha)}$	$\dfrac{\frac{1}{2}\alpha}{(1+\frac{1}{2}\alpha)}$	1

$$K = \prod_J a_J^{\nu_J}\ [17.12] = \frac{(p_{H_2}/p^{\ominus})(p_{O_2}/p^{\ominus})^{1/2}}{(p_{H_2O}/p^{\ominus})} \quad \text{[perfect gases]}$$

$$= \frac{(x_{H_2}p/p^{\ominus})(x_{O_2}p/p^{\ominus})^{1/2}}{(x_{H_2O}p/p^{\ominus})} \quad [p_J = x_J p]$$

$$= \frac{(x_{H_2})(x_{O_2})^{1/2}}{(x_{H_2O})} \quad [p = p^{\ominus} = 1\text{ bar}]$$

$$= \left(\frac{\alpha}{1+\frac{1}{2}\alpha}\right)\times\left(\frac{\frac{1}{2}\alpha}{1+\frac{1}{2}\alpha}\right)^{1/2}\times\left(\frac{1+\frac{1}{2}\alpha}{1-\alpha}\right)$$

$$= (\tfrac{1}{2})^{1/2}\alpha^{3/2}\times\left(\frac{1}{1+\frac{1}{2}\alpha}\right)^{1/2}\times\left(\frac{1}{1-\alpha}\right) \quad \text{(i)}$$

The equilibrium constant is small, and consequently it is reasonable to estimate that $\alpha << 1$. In this case, the factors $1+\frac{1}{2}\alpha$ and $1-\alpha$ in eqn. (i) are approximately equal to 1.

$$K \simeq (\tfrac{1}{2})^{1/2}\alpha^{3/2}$$

$$\alpha \simeq \left\{(\tfrac{1}{2})^{-1/2}K\right\}^{2/3}$$

$$\simeq \left\{(\tfrac{1}{2})^{-1/2}(0.002082)\right\}^{2/3}$$

$$\simeq \boxed{0.02054}$$

Alternatively, the numeric solver of a scientific calculator or software package like Mathcad can be used to acquire the numeric solution to eqn. (i). It is $\alpha = 0.02033$.

P17.5 $H_2O(g) + DCl(g) \rightleftharpoons HDO(g) + HCl(g)$

$$K = \frac{q^{\ominus}(HDO)q^{\ominus}(HCl)}{q^{\ominus}(H_2O)q^{\ominus}(DCl)}e^{-\Delta E_0/RT} \quad [17.14a;\ N_A \text{ factors cancel}]$$

Use partition function expressions from Table 13.1. The ratio of translational partition functions is

$$\frac{q_m^T(HDO)q_m^T(HCl)}{q_m^T(H_2O)q_m^T(DCl)} = \left(\frac{M(HDO)M(HCl)}{M(H_2O)M(DCl)}\right)^{3/2} = \left(\frac{19.02\times 36.46}{18.02\times 37.46}\right)^{3/2} = 1.041$$

The ratio of rotational partition functions is

$$\frac{q^{\mathrm{R}}(\mathrm{HDO})q^{\mathrm{R}}(\mathrm{HCl})}{q^{\mathrm{R}}(\mathrm{H_2O})q^{\mathrm{R}}(\mathrm{DCl})}=\frac{\sigma(\mathrm{H_2O})}{1}\frac{(\tilde{A}(\mathrm{H_2O})\tilde{B}(\mathrm{H_2O})\tilde{C}(\mathrm{H_2O})/\mathrm{cm^{-3}})^{1/2}\,\tilde{B}(\mathrm{DCl})/\mathrm{cm^{-1}}}{(\tilde{A}(\mathrm{HDO})\tilde{B}(\mathrm{HDO})\tilde{C}(\mathrm{HDO})/\mathrm{cm^{-3}})^{1/2}\,\tilde{B}(\mathrm{HCl})/\mathrm{cm^{-1}}}$$

$$=2\times\frac{(27.88\times14.51\times9.29)^{1/2}\times5.449}{(23.38\times9.102\times6.417)^{1/2}\times10.59}=1.707$$

($\sigma=2$ for H_2O; $\sigma=1$ for the other molecules)

The ratio of vibrational partition functions (call it Q) is

$$\frac{q^{\mathrm{V}}(\mathrm{HDO})q^{\mathrm{V}}(\mathrm{HCl})}{q^{\mathrm{V}}(\mathrm{H_2O})q^{\mathrm{V}}(\mathrm{DCl})}=\frac{q(2726.7)q(1402.2)q(3707.5)q(2991)}{q(3656.7)q(1594.8)q(3755.8)q(2145)}=Q$$

where $q(x)=\dfrac{1}{1-\mathrm{e}^{-1.4388x/(T/K)}}$

$$\frac{\Delta E_0}{hc}=\frac{1}{2}\{(2726.7+1402.2+3707.5+2991)-(3656.7+1594.8+3755.8+2145)\}\,\mathrm{cm^{-1}}$$

$$=-162\,\mathrm{cm^{-1}}$$

So the exponent in the energy term is

$$-\Delta E_0/RT=-\frac{\Delta E_0}{kT}=-\frac{hc}{k}\times\frac{\Delta E_0}{hc}\times\frac{1}{T}=-\frac{1.4388\times(-162)}{T/\mathrm{K}}=+\frac{233}{T/\mathrm{K}}$$

Therefore, $K=1.041\times1.707\times Q\times\mathrm{e}^{233/(T/\mathrm{K})}=1.777\,Q\mathrm{e}^{233/(T/\mathrm{K})}$.

We then draw up the following table (using a computer).

T/K	100	200	300	400	500	600	700	800	900	1000
K	18.3	5.70	3.87	3.19	2.85	2.65	2.51	2.41	2.34	2.29

Specifically, (a) $K=\boxed{3.89}$ at 298 K and (b) $K=\boxed{2.41}$ at 800 K.

P17.7 $\mathrm{CH_4(g)}\rightleftharpoons\mathrm{C(s)}+2\,\mathrm{H_2(g)}$

This reaction is the reverse of the formation reaction. Consequently,

$$\Delta_{\mathrm{r}}G^{\ominus}=-\Delta_{\mathrm{f}}G^{\ominus}=-\left(\Delta_{\mathrm{f}}H^{\ominus}-T\Delta_{\mathrm{f}}S^{\ominus}\right)$$

$$=-\left\{-74.85\ \mathrm{kJ\,mol^{-1}}-(298\ \mathrm{K})\times(-80.67\ \mathrm{J\,K^{-1}\,mol^{-1}})\right\}$$

$$=+50.81\ \mathrm{kJ\,mol^{-1}}$$

(a) $K=\mathrm{e}^{-\Delta_{\mathrm{r}}G^{\ominus}/RT}$ [17.13]

$$=\mathrm{e}^{-(5.081\times10^4\ \mathrm{J\,mol^{-1}})/(8.3145\ \mathrm{J\,K^{-1}\,mol^{-1}}\times298\ \mathrm{K})}$$

$$=\boxed{1.24\times10^{-9}}$$

(b) $\Delta_{\mathrm{r}}H^{\ominus}=-\Delta_{\mathrm{f}}H^{\ominus}=74.85\ \mathrm{kJ\,mol^{-1}}$

$$\ln K(50\ ^\circ\mathrm{C})=\ln K(298\ \mathrm{K})-\frac{\Delta_{\mathrm{r}}H^{\ominus}}{R}\left(\frac{1}{323\ \mathrm{K}}-\frac{1}{298\ \mathrm{K}}\right)\ [17.28]$$

$$=-20.508-\left(\frac{7.4850\times10^4\ \mathrm{J\,mol^{-1}}}{8.3145\ \mathrm{J\,K^{-1}\,mol^{-1}}}\right)\times(-2.59\bar{7}\times10^{-4})=-18.17\bar{0}$$

$$K(50\ ^\circ\mathrm{C})=\boxed{1.29\times10^{-8}}$$

P17.9 $CO_2(g) \rightleftharpoons CO(g) + \tfrac{1}{2}\,O_2(g)$

Draw up the following equilibrium table and recognize that, since $\alpha_e \ll 1$, α_e may be neglected when compared to 1 within mole fraction factors.

	CO_2	CO	O_2
Amounts	$(1-\alpha_e)n$	$\alpha_e n$	$\tfrac{1}{2}\alpha_e n$
Mole fractions	$\dfrac{1-\alpha_e}{1+\tfrac{1}{2}\alpha_e}$	$\dfrac{\alpha_e}{1+\tfrac{1}{2}\alpha_e}$	$\dfrac{\tfrac{1}{2}\alpha_e}{1+\tfrac{1}{2}\alpha_e}$
Approximate mole fractions	1	α_e	$\tfrac{1}{2}\alpha_e$

$$K = \prod_J a_J^{\nu_J}\ [17.12] = \frac{(p_{CO}/p^{\ominus})(p_{O_2}/p^{\ominus})^{1/2}}{p_{CO_2}/p^{\ominus}}\ \text{[perfect gases]}$$

$$= \frac{x_{CO}x_{O_2}^{1/2}}{x_{CO_2}}(p/p^{\ominus})^{1/2}\ [p_J = x_J p]$$

$$= \frac{x_{CO}x_{O_2}^{1/2}}{x_{CO_2}}\ [p = p^{\ominus}]$$

$$\approx \alpha_e\left(\tfrac{1}{2}\alpha_e\right)^{1/2}$$

$$\approx \left(\tfrac{1}{2}\alpha_e^{3}\right)^{1/2} \qquad \text{(i)}$$

$$\Delta_r G^{\ominus} = -RT\ln K\ [17.13] \qquad \text{(ii)}$$

The calculated values of K and $\Delta_r G^{\ominus}$ are given in the table below.

T/K	1395	1443	1498
$\alpha_e/10^{-4}$	1.44	2.50	4.71
$K/10^{-6}$	1.22	2.80	7.23
$\Delta_r G^{\ominus}/\ \text{kJ mol}^{-1}$	158	153	147
$\Delta_r S^{\ominus}/\ \text{J K}^{-1}\ \text{mol}^{-1}$	102	102	102

$\Delta_r H^{\ominus}$ can be calculated from any two pairs of K and T.

$$\ln K_2 = \ln K_1 - \frac{\Delta_r H^{\ominus}}{R}\left(\frac{1}{T_2} - \frac{1}{T_1}\right)\ [17.28]$$

Solving for $\Delta_r H^{\ominus}$,

$$\Delta_r H^{\ominus} = \frac{R\ln\left(\dfrac{K_2}{K_1}\right)}{\left(\dfrac{1}{T_1} - \dfrac{1}{T_2}\right)} = \frac{(8.3145\ \text{J K}^{-1}\ \text{mol}^{-1})\times\ln\left(\dfrac{7.23\times10^{-6}}{1.22\times10^{-6}}\right)}{\left(\dfrac{1}{1395\ \text{K}} - \dfrac{1}{1498\ \text{K}}\right)} = \boxed{300.\ \text{kJ mol}^{-1}}$$

$$\Delta_r S^{\ominus} = \frac{\Delta_r H^{\ominus} - \Delta_r G^{\ominus}}{T} \qquad \text{(iii)}$$

K, $\Delta_r G^{\ominus}$, and $\Delta_r S^{\ominus}$ are calculated with eqns. (i), (ii), and (iii).

COMMENT. $\Delta_r S^{\ominus}$ is essentially constant over this temperature range, but it is much different from its value at 25°C. $\Delta_r H^{\ominus}$, however, is only slightly different.

Question. What are the values of $\Delta_r H^{\ominus}$ and $\Delta_r S^{\ominus}$ at 25°C for this reaction?

P17.11 $$\Delta_r G^{\ominus}(H_2CO,g) = \Delta_r G^{\ominus}(H_2CO,l) + \Delta_{vap} G^{\ominus}(H_2CO,l)$$

For $H_2CO(l) \rightleftharpoons H_2CO(g)$, $K(\text{vap}) = \dfrac{p}{p^{\ominus}}$ where $p = 1500$ Torr $= 2.000$ bar and $p^{\ominus} = 1$ bar.

$$\Delta_{vap} G^{\ominus} = -RT \ln K(\text{vap}) = -RT \ln \frac{p}{p^{\ominus}}$$

$$= -(8.3145\ \text{J K}^{-1}\ \text{mol}^{-1}) \times (298\ \text{K}) \times \ln\left(\frac{2.000\ \text{bar}}{1\ \text{bar}}\right) = -1.72\ \text{kJ mol}^{-1}$$

Therefore, for the reaction $CO(g) + H_2(g) \rightleftharpoons H_2CO(g)$,

$$\Delta_r G^{\ominus} = \{(+28.95) + (-1.72)\}\ \text{kJ mol}^{-1} = +27.23\ \text{kJ mol}^{-1}$$

Hence, $K = e^{(-27.23\times10^3\ \text{J mol}^{-1})/(8.3145\ \text{J K}^{-1}\ \text{mol}^{-1})\times(298\ \text{K})} = e^{-10.99} = \boxed{1.69\times10^{-5}}$.

P17.13 Draw up the following table for the reaction $H_2(g) + I_2 \rightleftharpoons 2HI(g)$ $K = 870$.

	H_2	I_2	HI	Total
Initial amounts/mol	0.300	0.400	0.200	0.900
Change/mol	$-x$	$-x$	$+2x$	
Equilibrium amounts/mol	$0.300 - x$	$0.400 - x$	$0.200 + 2x$	0.900
Mole fraction	$\dfrac{0.300 - x}{0.900}$	$\dfrac{0.400 - x}{0.900}$	$\dfrac{0.200 + 2x}{0.900}$	1

$$K = \frac{(p_{HI}/p^{\ominus})^2}{(p_{H_2}/p^{\ominus})(p_{I_2}/p^{\ominus})}\ [\text{perfect gases}]$$

$$= \frac{(x_{HI})^2}{(x_{H_2})(x_{I_2})}\ [p_J = x_J p] = \frac{(0.200 + 2x)^2}{(0.300 - x)(0.400 - x)} = 870\ [\text{given}]$$

Therefore,

$$(0.0400) + (0.800x) + 4x^2 = (870) \times (0.120 - 0.700x + x^2) \quad \text{or} \quad 866x^2 - 609.80x + 104.36 = 0$$

which solves to $x = 0.293$ [$x = 0.411$ is excluded because x cannot exceed 0.300]. The final composition is therefore $\boxed{0.007\ \text{mol}\ H_2}$, $\boxed{0.107\ \text{mol}\ I_2}$, and $\boxed{0.786\ \text{mol HI}}$.

P17.15 If we know $\Delta_r H^{\ominus}$ for the reaction $Cl_2O(g) + H_2O(g) \rightarrow 2\ HOCl(g)$, we can calculate $\Delta_f H^{\ominus}$ (HOCl) from

$$\Delta_r H^{\ominus} = 2\ \Delta_f H^{\ominus}(HOCl,g) - \Delta_f H^{\ominus}(Cl_2O,g) - \Delta_f H^{\ominus}(H_2O,g)$$

We can find $\Delta_r H^{\ominus}$ if we know $\Delta_r G^{\ominus}$ and $\Delta_r S^{\ominus}$ since

$$\Delta_r G^{\ominus} = \Delta_r H - T\Delta_r S$$

And we can find $\Delta_r G^{\ominus}$ from the equilibrium constant.

$$K = \exp(-\Delta_r G^{\ominus}/RT) \quad \text{so} \quad \Delta_r G^{\ominus} = -RT \ln K$$

$$\Delta_r G^{\ominus} = -(8.3145\times10^{-3}\ \text{kJ K}^{-1}\ \text{mol}^{-1}) \times (298\ \text{K}) \ln(8.2\times10^{-2})$$

$$= 6.2\ \text{kJ mol}^{-1}$$

$$\Delta_r H^{\ominus} = \Delta_r G^{\ominus} + T\Delta_r S^{\ominus}$$

$$= 6.2\ \text{kJ mol}^{-1} + (298\ \text{K}) \times (16.38\times10^{-3}\ \text{kJ K}^{-1}\ \text{mol}^{-1}),$$

$$= 11.1\ \text{kJ mol}^{-1}$$

Finally,

$$\Delta_f H^{\ominus}(\text{HOCl,g}) = \tfrac{1}{2}[\Delta_r H^{\ominus} + \Delta_f H^{\ominus}(\text{Cl}_2\text{O,g}) + \Delta_f H^{\ominus}(\text{H}_2\text{O,g})]$$

$$= \tfrac{1}{2}[11.1 + 77.2 + (-241.82)]\ \text{kJ mol}^{-1}$$

$$= \boxed{76.8\ \text{kJ mol}^{-1}}$$

P17.17 $\tfrac{1}{2}\text{N}_2(\text{g}) + \tfrac{3}{2}\ \text{H}_2(\text{g}) \rightarrow \text{NH}_3(\text{g})$ $\quad \sum_{\text{Products}-\text{Reactants,i}} \nu_i = \nu_{\text{NH}_3} + \nu_{\text{N}_2} + \nu_{\text{H}_2} = 1 - \tfrac{1}{2} - \tfrac{3}{2} = -1$

This is the ammonia formation reaction for which we find the standard reaction thermodynamic functions in the text appendix (Table 14.6).

$$\Delta_r H^{\ominus}(298\ \text{K}) = -46.11\ \text{kJ} \quad \text{and} \quad \Delta_r S^{\ominus}(298\ \text{K}) = -99.38\ \text{J K}^{-1}$$

Use appendix information to define functions for the constant-pressure heat capacity of reactants and products (Table 14.2). Define a function $\Delta_r C_p^{\ominus}(T)$ that makes it possible to calculate $\Delta_r C_p$ at 1 bar and any temperature (eqn. 14.48). Define functions that make it possible to calculate the reaction enthalpy and entropy at 1 bar and any temperature (eqns. 14.47 and 15.29).

$$\Delta_r H^{\ominus}(T) = \Delta_r H^{\ominus}(298) + \int_{298.15\text{K}}^{T} \Delta_r C_p^{\ominus}(T)\,\text{d}T$$

$$\Delta_r S^{\ominus}(T) = \Delta_r S^{\ominus}(298) + \int_{298.15\text{K}}^{T} \frac{\Delta_r C_p^{\ominus}(T)}{T}\,\text{d}T$$

(i) For a perfect gas reaction mixture, $\Delta_r H$ is independent of pressure at constant temperature. Consequently, $\Delta_r H(T,p) = \Delta_r H^{\ominus}(T)$. The pressure dependence of the reaction entropy may be evaluated with the expression

$$\Delta_r S(T,p) = \Delta_r S^{\ominus}(T) + \sum_{\text{Products}-\text{Reactants,i}} \nu_i \int_{1\,\text{bar}}^{p} \left(\frac{\partial S_{\text{m,i}}}{\partial p}\right)_T \text{d}p$$

$$= \Delta_r S^{\ominus}(T) - \sum_{\text{Products}-\text{Reactants,i}} \nu_i \int_{1\,\text{bar}}^{p} \left(\frac{\partial V_{\text{m,i}}}{\partial T}\right)_p \text{d}p \quad \text{(Table 15.5)}$$

$$= \Delta_r S^{\ominus}(T) - \sum_{\text{Products}-\text{Reactants,i}} \nu_i \int_{1\,\text{bar}}^{p} \frac{R}{p}\,\text{d}p$$

$$= \Delta_r S^{\ominus}(T) - \left[\sum_{\text{Products}-\text{Reactants,i}} \nu_i\right] R \ln\left(\frac{p}{1\ \text{bar}}\right) = \Delta_r S^{\ominus}(T) - (-1)\ R \ln\left(\frac{p}{1\ \text{bar}}\right)$$

$$= \Delta_r S^{\ominus}(T) + R \ln\left(\frac{p}{1\ \text{bar}}\right)$$

The above two equations make it possible to calculate $\Delta_r G(T,p)$.

$$\Delta_r G(T,p) = \Delta_r H(T,p) - T\Delta_r S(T,p)$$

Once the above functions have been defined on a scientific calculator or with mathematical software on a computer, the root function may be used to evaluate pressure where $\Delta_r G(T,p) = -500\,\text{J}$ at a given temperature.

(a) and (b) perfect gas mixture:

For $T = (450 + 273.15)\,\text{K} = 723.15\ \text{K}$, $\text{root}(\Delta_r G(723.15\ \text{K}, p) + 500\ \text{J}) = \boxed{156.5\ \text{bar}}$.

For $T = (400 + 273.15)\,\text{K} = 673.15\ \text{K}$, $\text{root}(\Delta_r G(673.15\ \text{K}, p) + 500\ \text{J}) = \boxed{81.8\ \text{bar}}$.

(ii) For a van der Waals gas mixture, $\Delta_r H$ does depend upon pressure. The calculation equation is

$$\Delta_r H(T,P) = \Delta_r H^{\ominus}(T) + \sum_{\text{Products-Reactants,i}} \nu_i \int_{1\text{ bar}}^{p} \left(\frac{\partial H_{m,i}}{\partial p}\right)_T \mathrm{d}p$$

[The sum involves the i = 1, 2, 3 gases (NH_3, N_2, or H_2)]

$$= \Delta_r H^{\ominus}(T) + \sum_{\text{Products-Reactants,i}} \nu_i \int_{1\text{ bar}}^{p} \left[V_{m,i} - T\left(\frac{\partial V_{m,i}}{\partial T}\right)_p\right] \mathrm{d}p$$

[The equation of this substitution is proven below.*]

where $(\partial V_{m,i}/\partial T)_p = R(V_{m,i} - b_i)^{-1}(RT(V_{m,i} - b_i)^{-2} - 2a_i V_{m,i}^{-3})^{-1}$ for each gas i (NH_3, N_2, or H_2) and

$$V_{m,i}(T,p) = \text{root}\left(p - \frac{RT}{V_{m,i} - b_i} + \frac{a_i}{V_{m,i}^{\;2}}\right)$$

The functional equation for $\Delta_r S$ calculations is

$$\Delta_r S(T,p) = \Delta_r S^{\ominus}(T) - \sum_{\text{Products-Reactants,i}} \nu_i \int_{1\text{ bar}}^{p} \left(\frac{\partial V_{m,i}}{\partial T}\right)_p \mathrm{d}p$$

where $(\partial V_{m,i}/\partial T)_p$ and $V_{m,i}(T,p)$ are calculated as described above. As usual,

$$\Delta_r G(T,p) = \Delta_r H(T,p) - T\Delta_r S(T,p)$$

(a) and (b) van der Waals gas mixture:

For $T = 723.15$ K, root($\Delta_r G(723.15\text{ K}, p) + 500$ J) = **132.5 bar**.

For $T = 673.15$ K, root($\Delta_r G(673.15\text{ K}, p) + 500$ J) = **73.7 bar**.

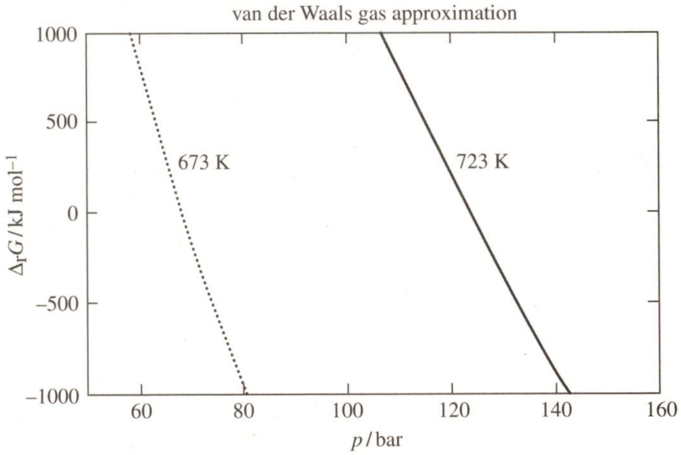

Figure 17.1

(c) $\Delta_r G(T,p)$ isotherms (see Fig. 17.1) **confirm** Le Chatelier's principle. Along an isotherm, $\Delta_r G$ decreases as pressure increases. This corresponds to a shift to the right in the reaction equation and reduces the stress by shifting to the side that has fewer total moles of gas. Additionally, the reaction is exothermic, so Chatelier's principle predicts a shift to the left with an increase in temperature. The isotherms confirm this as an increase in $\Delta_r G$ as temperature is increased at constant pressure.

*Many thermodynamic equations are very useful when deriving desired computation equations. One of them (used above) is $\left(\frac{\partial H}{\partial p}\right)_T = -T\left(\frac{\partial V}{\partial T}\right)_p + V$. To prove this relationship, first use an identity of partial derivatives that involves a change of variable:

$$\left(\frac{\partial H}{\partial p}\right)_T = \left(\frac{\partial H}{\partial S}\right)_p \left(\frac{\partial S}{\partial p}\right)_T + \left(\frac{\partial H}{\partial p}\right)_S$$

We will be able to identify some of these terms if we examine an expression for dH analogous to the fundamental equation of thermodynamics [15.54]. From the definition of enthalpy, we have

$$dH = dU + p\,dV + V\,dp = T\,dS - p\,dV\ [15.54] + p\,dV + V\,dp = T\,dS + V\,dp$$

Compare this expression to the exact differential of H considered as a function of S and p:

$$dH = \left(\frac{\partial H}{\partial S}\right)_p dS + \left(\frac{\partial H}{\partial p}\right)_S dp$$

Thus, $\left(\frac{\partial H}{\partial S}\right)_p = T,\ \left(\frac{\partial H}{\partial p}\right)_S = V$ [dH exact].

Substitution yields $\left(\frac{\partial H}{\partial p}\right)_T = T\left(\frac{\partial S}{\partial p}\right)_T + V = \boxed{-T\left(\frac{\partial V}{\partial T}\right)_p + V}$ [Maxwell relation].

P17.19 The electrode half-reactions and their potentials are

	$E^{\ominus}$
$\mathrm{R:\ Q(aq) + 2H^+(aq) + 2e^- \rightarrow QH_2(aq)}$	0.6994 V
$\mathrm{L:\ Hg_2Cl_2(s) + 2e^- \rightarrow 2Hg(l) + 2Cl^-(aq)}$	0.2676 V
Overall (R – L): $\mathrm{Q(aq) + 2H^+(aq) + 2Hg(l) + 2Cl^-(aq) \rightarrow QH_2(aq) + Hg_2Cl_2(s)}$	0.4318 V

$$Q(\text{reaction quotient}) = \frac{a(\mathrm{QH_2})}{a(\mathrm{Q})a^2(\mathrm{H^+})a^2(\mathrm{Cl^-})}$$

Because quinhydrone is an equimolecular complex of Q and QH_2 and $m(\mathrm{Q}) = m(\mathrm{QH_2})$, and their activity coefficients are assumed to be 1 or to be equal, we have $a(\mathrm{QH_2}) \approx a(\mathrm{Q})$. Thus,

$$Q = \frac{1}{a^2(\mathrm{H^+})a^2(\mathrm{Cl^-})} \quad \text{and} \quad E = E^{\ominus} - \frac{25.7\ \text{mV}}{\nu}\ln Q\ [17.32, 25°\text{C}]$$

$$\ln Q = \frac{\nu(E^{\ominus} - E)}{25.7\ \text{mV}} = \frac{(2)\times(0.4318 - 0.190)\ \text{V}}{25.7\times10^{-3}\ \text{V}} = 18.8\bar{2} \quad \text{and} \quad Q = 1.\bar{4}9\times10^{8}$$

$$a^2(\mathrm{H^+}) = (\gamma_+ b_+)^2; \quad a^2(\mathrm{Cl^-}) = (\gamma_- b_-)^2 \quad [b \equiv b/b^{\ominus}]$$

For HCl(aq), $b_+ = b_- = b$, and if the activity coefficients are assumed equal, $a^2(\mathrm{H^+}) = a^2(\mathrm{Cl^-})$; hence

$$Q = \frac{1}{a^2(\mathrm{H^+})a^2(\mathrm{Cl^-})} = \frac{1}{a^4(\mathrm{H^+})}$$

Thus, $a(\mathrm{H^+}) = \left(\frac{1}{Q}\right)^{1/4} = \left(\frac{1}{1.49\times10^{8}}\right)^{1/4} = 9\times10^{-3}$

$$\text{pH} = -\log a(\mathrm{H^+}) = \boxed{2.0}$$

P17.21 $\mathrm{Pt|H_2(g)|HCl(aq,}b\mathrm{)|Hg_2Cl_2(s)|Hg(l)}$

$$\tfrac{1}{2}\mathrm{Hg_2Cl_2(s)} + \tfrac{1}{2}\mathrm{H_2(g)} \rightarrow \mathrm{Hg(l)} + \mathrm{HCl(aq)} \quad \text{and} \quad \nu = 1$$

$$E = E^{\ominus} - \frac{RT}{F}\ln a(\mathrm{H^+})a(\mathrm{Cl^-})\ [17.32]$$

$$a(\mathrm{H^+}) = \gamma_+ b_+ = \gamma_+ b; \quad a(\mathrm{Cl^-}) = \gamma_- b_- = \gamma_- b \quad [b = b/b^{\ominus} \text{ here and below}]$$

$$a(\mathrm{H^+})a(\mathrm{Cl^-}) = \gamma_+\gamma_- b^2 = \gamma_\pm^2 b^2$$

$$E = E^{\ominus} - \frac{2RT}{F}\ln b - \frac{2RT}{F}\ln \gamma_{\pm} \qquad \text{(i)}$$

Converting from natural logarithms to common logarithms (base 10) in order to introduce the Debye–Hückel expression, we obtain

$$E = E^{\ominus} - \frac{(2.303)\times 2RT}{F}\log b - \frac{(2.303)\times 2RT}{F}\log \gamma_{\pm}$$

$$= E^{\ominus} - (0.1183\text{ V})\log b - (0.1183\text{ V})\log \gamma_{\pm}$$

$$= E^{\ominus} - (0.1183\text{ V})\log b - (0.1183\text{ V})\left[-|z_+z_-|AI^{1/2}\right]$$

$$= E^{\ominus} - (0.1183\text{ V})\log b + (0.1183\text{ V})\times A\times b^{1/2} \quad [I=b]$$

Rearranging,

$$E + (0.1183\text{ V})\log b = E^{\ominus} + \text{constant}\times b^{1/2}$$

Therefore, plot $E + (0.1183\text{ V})\log b$ against $b^{1/2}$, and the intercept at $b = 0$ is $E^{\ominus}/\text{V}$. Draw up the following table.

b/(mmol kg^{-1})	1.6077	3.0769	5.0403	7.6938	10.9474
$(b/b^{\ominus})^{1/2}$	0.04010	0.05547	0.07100	0.08771	0.1046
$E/\text{V} + (0.1183)\log b$	0.27029	0.27109	0.27186	0.27260	0.27337

The points are plotted in Figure 17.2. The intercept is at 0.26840, so $E^{\ominus} = +0.26840\text{ V}$. A least-squares best fit gives $E^{\ominus} = \boxed{+0.26843\text{ V}}$ and a coefficient of determination equal to 0.99895.

For the activity coefficients, we obtain from equation (i)

$$\ln \gamma_{\pm} = \frac{E^{\ominus} - E}{2RT/F} - \ln\frac{b}{b^{\ominus}} = \frac{0.26843 - E/V}{0.05139} - \ln\frac{b}{b^{\ominus}}$$

and we draw up the following table.

b/(mmol kg^{-1})	1.6077	3.0769	5.0403	7.6938	10.9474
E/V	0.60080	0.56825	0.54366	0.52267	0.50532
$\ln \gamma_{\pm}$	–0.03465	–0.05038	–0.06542	–0.07993	–0.09500
$\gamma_{\pm}$	0.9659	0.9509	0.9367	0.9232	0.9094

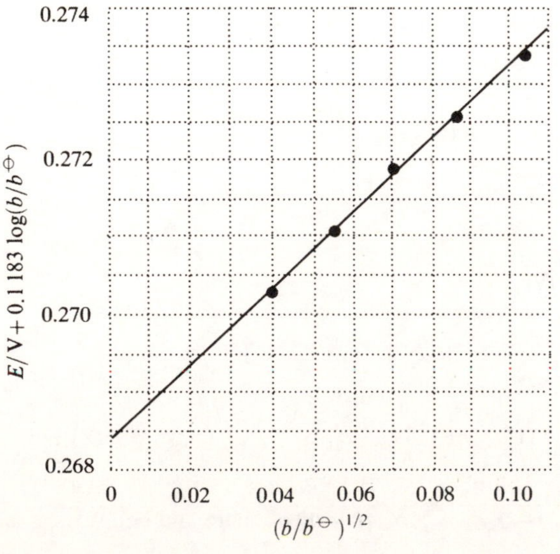

Figure 17.2

P17.23 The cells described in the problem are back-to-back pairs of cells each of the type

$M_xHg(s)|MX(b_2)|AgX(s)|Ag(s)$

R: $AgX(s) + e^- \rightarrow Ag(s) + X^-(b_2)$

L: $M^+(b_2) + e^- \xrightarrow{Hg} M_xHg(s)$ (reduction of M^+ and formation of amalgam)

R − L: $M_xHg(s) + AgX(s) \xrightarrow{Hg} Ag(s) + M^+(b_2) + X^-(b_2)$ and $\nu = 1$

$$Q = \frac{a(M^+)a(X^-)}{a(M_xHg)}$$

$$E = E^{\ominus} - \frac{RT}{F}\ln Q$$

For a pair of such cells back to back,

$$Ag(s)\,|\,AgX(s)\,|\,MX(b_1)|M_xHg(s)\,|\,MX(b_2)\,|\,AgX(s)\,|\,Ag(s)$$

$$E_R = E^{\ominus} - \frac{RT}{F}\ln Q_R \quad \text{and} \quad E_L = E^{\ominus} - \frac{RT}{F}\ln Q_L$$

$$E = E_R - E_L = \frac{-RT}{F}\ln\frac{Q_R}{Q_L} = \frac{RT}{F}\ln\frac{(a(M^+)a(X^-))_L}{(a(M^+)a(X^-))_R}$$

(Note that the unknown quantity $a(M_xHg)$ drops out of the expression for E.)

$$a(M^+)a(X^-) = \left(\frac{\gamma_+ b_+}{b^{\ominus}}\right)\left(\frac{\gamma_- b_-}{b^{\ominus}}\right) = \gamma_\pm^2\left(\frac{b}{b^{\ominus}}\right)^2 \quad (b_+ = b_-)$$

With L = (1) and R = (2), we have

$$E = \frac{2RT}{F}\ln\frac{b_1}{b_2} + \frac{2RT}{F}\ln\frac{\gamma_\pm(1)}{\gamma_\pm(2)}$$

Take $b_2 = 0.09141\ \text{mol kg}^{-1}$ (the reference value), and write $b = b_1/b^{\ominus}$.

$$E = \frac{2RT}{F}\left\{\ln\frac{b}{0.09141} + \ln\frac{\gamma_\pm}{\gamma_\pm(\text{ref})}\right\}$$

For $b = 0.09141$, the extended Debye–Hückel law gives

$$\log\gamma_\pm = -\frac{AI^{1/2}}{1+BI^{1/2}} + CI \quad \text{with} \quad A = 1.461,\ B = 1.70,\ C = 0.20,\ \text{and}\ I = b/b^{\ominus}$$

$$\log\gamma_\pm(\text{ref}) = \frac{(-1.461)\times(0.09141)^{1/2}}{(1)+(1.70)\times(0.09141)^{1/2}} + (0.20)\times(0.09141) = -0.273\bar{5}$$

$$\gamma_\pm(\text{ref}) = 0.532\underline{8}$$

Then $E = (0.05139\ \text{V})\left\{\ln\frac{b}{0.09141} + \ln\frac{\gamma_\pm}{0.5328}\right\}$

$$\ln\gamma_\pm = \frac{E}{0.05139\ \text{V}} - \ln\frac{b}{(0.09141)\times(0.05328)}$$

We then draw up the following table.

$b/(\text{mol/kg}^{-1})$	0.0555	0.09141	0.1652	0.2171	1.040	1.350
E/V	−0.0220	0.0000	0.0263	0.0379	0.1156	0.1336
$\gamma_\pm$	0.572	0.533	0.492	0.469	0.444	0.486

A more precise procedure is described in the original references; for the temperature dependence of $E^{\ominus}(\mathrm{Ag, AgCl, Cl^-})$, see Problem 17.24.

P17.25 Electrochemical cell equation: $\tfrac{1}{2}\mathrm{H_2(g, 1\ bar)} + \mathrm{AgCl(s)} \rightleftharpoons \mathrm{H^+(aq)} + \mathrm{Cl^-(aq)} + \mathrm{Ag(s)}$ with $a(\mathrm{H_2}) = 1\ \mathrm{bar} = p^{\ominus}$ and $a_{\mathrm{Cl^-}} = \gamma_{\mathrm{Cl^-}} b$

Weak acid equilibrium: $\mathrm{BH^+} \rightleftharpoons \mathrm{B} + \mathrm{H^+}$ with $b_{\mathrm{BH^+}} = b_{\mathrm{B}} = b$

and $K_a = a_{\mathrm{B}} a_{\mathrm{H^+}} / a_{\mathrm{BH^+}} = \gamma_{\mathrm{B}}\, b\, a_{\mathrm{H^+}} / \gamma_{\mathrm{BH^+}}\, b = \gamma_{\mathrm{B}}\, a_{\mathrm{H^+}} / \gamma_{\mathrm{BH^+}}$

or $a_{\mathrm{H^+}} = \gamma_{\mathrm{BH^+}} K_a / \gamma_{\mathrm{B}}$

Ionic strength (neglect $b_{\mathrm{H^+}}$ because $b_{\mathrm{H^+}} \ll b$): $I = \tfrac{1}{2}\{z^2_{\mathrm{BH^+}} b_{\mathrm{BH^+}} + z^2_{\mathrm{Cl^-}} b_{\mathrm{Cl^-}}\} = b$

According to the Nernst equation [17.32],

$$E = E^{\ominus} - \frac{RT}{F}\ln\left(\frac{a_{\mathrm{H^+}} a_{\mathrm{Cl^-}}}{p(\mathrm{H_2})/p^{\ominus}}\right) = E^{\ominus} - \frac{RT\ \ln(10)}{F}\log(a_{\mathrm{H^+}} a_{\mathrm{Cl^-}})$$

$$\frac{F}{RT\ln(10)}(E - E^{\ominus}) = -\log(a_{\mathrm{H^+}}\gamma_{\mathrm{Cl^-}} b) = -\log\left(\frac{K_a \gamma_{\mathrm{BH^+}}\gamma_{\mathrm{Cl^-}} b}{\gamma_{\mathrm{B}}}\right) = \mathrm{p}K_a - \log(b) - 2\log(\gamma_\pm)$$

$$\frac{F}{RT\ln(10)}(E - E^{\ominus}) = \mathrm{p}K_a - \log(b) + \frac{2A\sqrt{b}}{1 + B\sqrt{b}} - 2kb \qquad \text{where} \qquad A = 0.5091$$

The expression to the left of the above equality is experimental data and is a function of b. The parameters $\mathrm{p}K_a$, B, and k on the right side are systematically varied with a mathematical regression software until the right side fits the left side in a least-squares sense. The results are

$\boxed{\mathrm{p}K_a = 6.736}$, $\boxed{B = 1.997}$, and $\boxed{k = -0.121}$

The mean activity coefficient is calculated with the equation $\gamma_\pm = 10^{\left(\frac{-AI^{1/2}}{1+BI^{1/2}} + kb\right)}$ for desired values of b and I. Figure 17.3 shows a $\gamma_\pm$ against I plot for $b = 0.04\ \mathrm{mol\ kg^{-1}}$ and $0 \le I \le 0.1$.

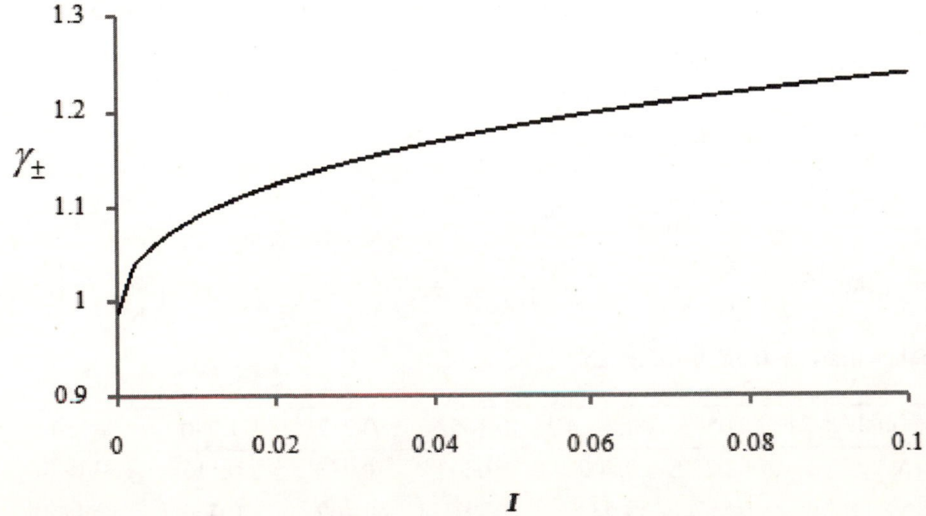

Figure 17.3

P17.27 $$aA(g)+bB(g) \rightleftharpoons cC(g)+dD(g)$$

$$K=\frac{a_{C(g)}^{c}a_{D(g)}^{d}}{a_{A(g)}^{a}a_{B(g)}^{b}}$$

$$=\frac{(p_C/p^{\ominus})^c(p_D/p^{\ominus})^d}{(p_A/p^{\ominus})^a(p_B/p^{\ominus})^b} \quad [\text{perfect gas, } a_J=p_J/p^{\ominus}]$$

$$=\frac{p_C^c p_D^d}{p_A^a p_B^b}(p^{\ominus})^{-\Delta n_g} \quad \text{where} \quad \Delta n_g=c+d-a-b$$

$$=\frac{[C]^c[D]^d}{[A]^a[B]^b}\left(\frac{RT}{p^{\ominus}}\right)^{\Delta n_g} \quad [p_J=n_JRT/V=[J]RT]$$

$$=\frac{([C]/c^{\ominus})^c([D]/c^{\ominus})^d}{([A]/c^{\ominus})^a([B]/c^{\ominus})^b}\left(\frac{c^{\ominus}RT}{p^{\ominus}}\right)^{\Delta n_g} \quad [c^{\ominus}=1\text{ mol dm}^{-3}]$$

$$=K_c(c^{\ominus}RT/p^{\ominus})^{\Delta n_g} \quad \text{where} \quad K_c=\frac{([C]/c^{\ominus})^c([D]/c^{\ominus})^d}{([A]/c^{\ominus})^a([B]/c^{\ominus})^b}$$

P17.29 $$MX(s) \rightleftharpoons M^+(aq)+X^-(aq)$$

$$K_s=a(M^+)a(X^-)=b(M^+)b(X^-)\gamma_{\pm}^2 \quad [b\equiv b/b^{\ominus}]$$

The concentration of the soluble salt, $b(M^+)=S'$, is very low compared to the concentration of the freely soluble salt, [NX] = C. Consequently, $b(X^-)=S'+C=C$ and the ionic strength depends on C alone.

$$I=\tfrac{1}{2}\left(b(N^+)+b(X^-)\right)\ [16.57]=C$$

$$\ln\gamma_{\pm}=2.303\log\gamma_{\pm}=-2.303AI^{1/2}\ [16.56] \quad \text{where} \quad A=0.509$$

$$\gamma_{\pm}=e^{-2.303AI^{1/2}} \quad \text{and} \quad \gamma_{\pm}^2=e^{-4.606AI^{1/2}}$$

Hence, $K_s=S'C\,e^{-4.606AI^{1/2}}$ or $S'=K_s e^{4.606AI^{1/2}}/C$.

Solutions to applications

P17.31 The fractional saturation, s, the fraction of Mb molecules (or Hb molecules) that are oxygenated, is related to the partial pressure of oxygen, p (= p/1 Torr), through the Hill equation,

$$\log\frac{s}{1-s}=v\log p-v\log K_{\text{Hill}} \quad \text{or} \quad p=K_{\text{Hill}}\times\left(\frac{s}{1-s}\right)^{1/v} \quad \text{or} \quad s=\frac{(p/K_{\text{Hill}})^v}{1+(p/K_{\text{Hill}})^v}$$

where v and K_{Hill} are the Hill coefficients.

(a) Let $p_{1/2}$ be the oxygen pressure when $s=\tfrac{1}{2}$. Then

$$p_{1/2}=K_{\text{Hill}}\times\left(\frac{\tfrac{1}{2}}{1-\tfrac{1}{2}}\right)^{1/v}$$
$$=K_{\text{Hill}}$$

Thus, measurement of $p_{½}$ directly yields K_{Hill}. The plots of P17.30 give $p_{½} = 20.0$ Torr for Mb and $p_{½} = 35.0$ Torr for Mb and we conclude that $\boxed{K_{Hill}(\text{Mb}) = 20.0}$ and $\boxed{K_{Hill}(\text{Hb}) = 35.0}$.

The fractional saturation of Mb and Hb at a given p is calculated with the equation

$$s = \frac{(p/K_{\text{Hill}})^{\nu}}{1+(p/K_{\text{Hill}})^{\nu}}$$

The values of K_{Hill} are reported above and the values of ν are given as $\nu(\text{Mb}) = 1$ and $\nu(\text{Mb}) = 2.8$. We draw up a table.

		s	
p / kPa	p / Torr	Mb	Hb
1.0	7.5	0.27	0.01
1.5	11.3	0.36	0.04
2.5	18.8	0.48	0.14
4.0	30.0	0.60	0.39
8.0	60.0	0.75	0.82

(b) Using the K_{Hill} value for Hb reported above and the value $\nu = 4$ in the Hill equation,

$$s = \frac{(p/K_{\text{Hill}})^{\nu}}{1+(p/K_{\text{Hill}})^{\nu}}$$

yields the fractional saturations reported in the following table.

p / kPa	p / Torr	s(Hb)
1.0	7.5	0.002
1.5	11.3	0.01
2.5	18.8	0.08
4.0	30.0	0.35
8.0	60.0	0.89

P17.33 $$\Delta_r G = \Delta_r G^{\ominus} + RT \ln Q \quad [17.9]$$

In Equation 17.9, molar solution concentrations are used with 1 M standard states. The standard state ($\ominus$) pH equals zero in contrast to the biological standard state ($\oplus$) of pH 7. For the ATP hydolysis

$$\text{ATP(aq)} + \text{H}_2\text{O(l)} \rightarrow \text{ADP(aq)} + \text{P}_i^-\text{(aq)} + \text{H}_3\text{O}^+\text{(aq)}$$

we can calculate the standard-state free energy given the biological standard free energy of about −31 kJ mol^{-1} (*Impact on biology* 17.1).

$$\Delta_r G^{\oplus} = \Delta_r G^{\ominus} + RT \ln Q^{\oplus} \quad [17.9]$$

$$\Delta_r G^{\ominus} = \Delta_r G^{\oplus} - RT \ln Q^{\oplus}$$

$$= -31 \text{ kJ mol}^{-1} - (8.3145 \text{ J K}^{-1} \text{ mol}^{-1}) \times (310 \text{ K}) \ln(10^{-7} \text{ M/1 M})$$

$$= +11 \text{ kJ mol}^{-1}$$

This calculation shows that under standard conditions, the hydrolysis of ATP is not spontaneous! It is endergonic.

The calculation of *the* ATP hydrolysis free energy with the cell conditions pH = 7 and [ATP] = [ADP] = $[P_i^-] = 1.0\times10^{-6}$ M is interesting.

$$\Delta_r G = \Delta_r G^\ominus + RT\ln Q = \Delta_r G^\ominus + RT\ln\left(\frac{[\text{ADP}]\times[\text{P}_i^-]\times[\text{H}^+]}{[\text{ATP}]\times(1\ \text{M})^2}\right)$$

$$= +11\ \text{kJ mol}^{-1} + (8.3145\ \text{J K}^{-1}\ \text{mol}^{-1})\times(310\ \text{K})\ln(10^{-6}\times10^{-7})$$

$$= -66\ \text{kJ mol}^{-1}$$

The concentration conditions in biological cells make the hydrolysis of ATP spontaneous and very exergonic. A maximum of 66 kJ of work is available to drive coupled chemical reactions when a mole of ATP is hydrolyzed.

P17.35 Yes, a bacterium can evolve to utilize the ethanol/nitrate pair to exergonically release the free energy needed for ATP synthesis. The ethanol reductant may yield any of the following products:

$$\underset{\text{ethanol}}{CH_3CH_2OH} \rightarrow \underset{\text{ethanal}}{CH_3CHO} \rightarrow \underset{\text{ethanoic acid}}{CH_3COOH} \rightarrow CO_2 + H_2O$$

The nitrate oxidant may receive electrons to yield any of the following products:

$$\underset{\text{nitrate}}{NO_3^-} \rightarrow \underset{\text{nitrite}}{NO_2^-} \rightarrow \underset{\text{dinitrogen}}{N_2} \rightarrow \underset{\text{ammonia}}{NH_3}$$

Oxidation of two ethanol molecules to carbon dioxide and water can transfer 8 electrons to nitrate during the formation of ammonia. The half-reactions and net reaction are

$$2\,[CH_3CH_2OH(l) \rightarrow 2CO_2(g) + H_2O(l) + 4\,H^+(aq) + 4e^-]$$

$$NO_3^-(aq) + 9\,H^+(aq) + 8\,e^- \rightarrow NH_3(aq) + 3\,H_2O(l)$$

$$2\,CH_3CH_2OH(l) + H^+(aq) + NO_3^-(aq) \rightarrow 4\,CO_2(g) + 5\,H_2O(l) + NH_3(aq)$$

$\Delta_r G^\ominus = -2331.29$ kJ for the reaction as written (a Tables 14.5 and 14.6 calculation). Of course, enzymes must evolve that couple this exergonic redox reaction to the production of ATP, which would then be available for carbohydrate, protein, lipid, and nucleic acid synthesis.

P17.37 The half-reactions involved are

$$\text{R:}\quad \text{cyt}_{\text{ox}} + e^- \rightarrow \text{cyt}_{\text{red}} \qquad E^\ominus_{\text{cyt}}$$

$$\text{L:}\quad \text{D}_{\text{ox}} + e^- \rightarrow \text{D}_{\text{red}} \qquad E^\ominus_{\text{D}}$$

The overall cell reaction is

$$\text{R}-\text{L:}\quad \text{cyt}_{\text{ox}} + \text{D}_{\text{red}} \rightleftharpoons \text{cyt}_{\text{red}} + \text{D}_{\text{ox}} \qquad E^\ominus = E^\ominus_{\text{cyt}} - E^\ominus_{\text{D}}$$

(a) The Nernst equation for the cell reaction is

$$E = E^\ominus - \frac{RT}{F}\ln\frac{[\text{cyt}_{\text{red}}][\text{D}_{\text{ox}}]}{[\text{cyt}_{\text{ox}}][\text{D}_{\text{red}}]}.$$

At equilibrium, $E = 0$; therefore,

$$\ln\frac{[\text{cyt}_{\text{red}}]_{\text{eq}}[\text{D}_{\text{ox}}]_{\text{eq}}}{[\text{cyt}_{\text{ox}}]_{\text{eq}}[\text{D}_{\text{red}}]_{\text{eq}}} = \frac{F}{RT}\left(E^\ominus_{\text{cyt}} - E^\ominus_{\text{D}}\right)$$

$$\ln\left(\frac{[\text{D}_{\text{ox}}]_{\text{eq}}}{[\text{D}_{\text{red}}]_{\text{eq}}}\right) = \ln\left(\frac{[\text{cyt}]_{\text{ox}}}{[\text{cyt}]_{\text{red}}}\right) + \frac{F}{RT}(E^\ominus_{\text{cyt}} - E^\ominus_{\text{D}})$$

Thus, a plot of $\ln\left(\frac{[D_{ox}]_{eq}}{[D_{red}]_{eq}}\right)$ against $\ln\left(\frac{[cyt]_{ox}}{[cyt]_{red}}\right)$ is linear with a slope of 1 and an intercept of $\frac{F}{RT}\left(E^{\ominus}_{cyt} - E^{\ominus}_{D}\right)$.

(b) Draw up the following table.

$\ln\left(\frac{[D_{ox}]_{eq}}{[D_{red}]_{eq}}\right)$	−5.882	−4.776	−3.661	−3.002	−2.593	−1.436	−0.6274
$\ln\left(\frac{[cyt_{ox}]_{eq}}{[cyt_{red}]_{eq}}\right)$	−4.547	−3.772	−2.415	−1.625	−1.094	−0.2120	−0.3293

The plot of $\ln\left(\frac{[D_{ox}]_{eq}}{[D_{red}]_{eq}}\right)$ against $\ln\left(\frac{[cyt_{ox}]_{eq}}{[cyt_{red}]_{eq}}\right)$ is shown in Figure 17.4. The intercept is −1.2124.

Hence,

$$E^{\ominus}_{cyt} = \frac{RT}{F} \times (-1.2124) + 0.237\,\text{V}$$

$$= 0.0257\,\text{V} \times (-1.2124) + 0.237\,\text{V}$$

$$= \boxed{+0.206\,\text{V}}$$

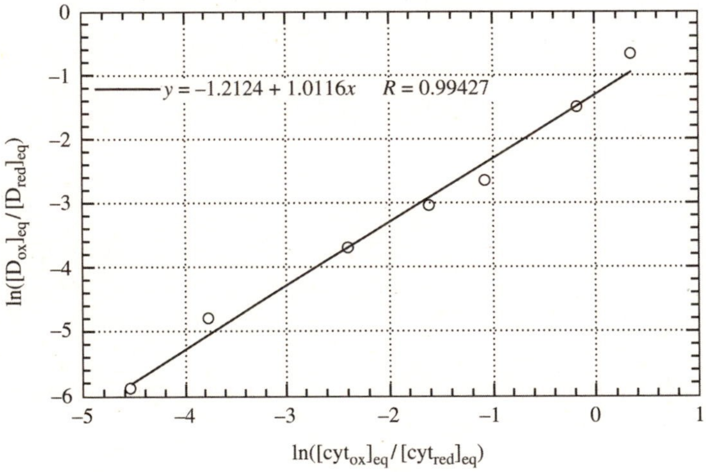

Figure 17.4

P17.39 (a) The cell reaction is $H_2(g) + \frac{1}{2}O_2(g) \rightarrow H_2O(l) \quad \nu = 2$

$$\Delta_r G^{\ominus} = \Delta_f G^{\ominus}(H_2O, l) = -237.13\ \text{kJ mol}^{-1}\ \text{[Table 14.6]}$$

$$E^{\ominus} = \frac{\Delta_r G^{\ominus}}{\nu F}[17.30] = \frac{+237.13\ \text{kJ mol}^{-1}}{(2) \times (96.485\ \text{kC mol}^{-1})} = \boxed{+1.23\ \text{V}}$$

(b) $C_6H_6(l) + \frac{15}{2}O_2(g) \rightarrow 6\,CO_2(g) + 3\,H_2O(l)$

$$\Delta_f G^{\ominus} = 6\,\Delta_f G^{\ominus}(CO_2, g) + 3\,\Delta_f G^{\ominus}(H_2O, l) - \Delta_f G^{\ominus}(C_6H_6, l)$$

$$= (6) \times (-394.36) + (3) \times (-237.13) - (+124.3.03)]\ \text{kJ mol}^{-1}\ \text{[Tables 14.5 and 14.6]}$$

$$= -3201.58\ \text{kJ mol}^{-1}$$

To find the value of v for this reaction, recognize that a carbon atom in benzene has a –1 oxidation state while a carbon atom in carbon dioxide has a +4 oxidation state. This is a change of +5 per carbon or +30 per benzene molecule. Thus, $v = 30$ for the reaction equation. Alternatively, to find v we break the cell reaction down into half-reactions as follows.

$$\text{R: } {}^{15}\!/\!_2\,O_2(g) + 30\ e^- + 30\ H^+(aq) \rightarrow 15\ H_2O(l)$$

$$\text{L: } 6\ CO_2(g) + 30\ e^- + 30\ H^+(aq) \rightarrow C_6H_6(l) + 12\ H_2O(l)$$

$$\text{R} - \text{L: } C_6H_6(l) + {}^{15}\!/\!_2\ O_2(g) \rightarrow 6\ CO_2(g) + 3\ H_2O(l)$$

Once again, we find that $v = 30$.

$$\text{Therefore, } E^{\ominus} = \frac{-\Delta G^{\ominus}}{vF}\ [17.30] = \frac{+3201.58\ \text{kJ mol}^{-1}}{(30)\times(96.485\ \text{kC mol}^{-1})} = \boxed{+1.11\ \text{V}}.$$

P17.41 A reaction proceeds spontaneously if its reaction Gibbs function is negative.

$$\Delta_r G = \Delta_r G^{\ominus} + RT \ln Q\ \ [17.9]$$

Note that under the given conditions, $RT = 1.58\ \text{kJ mol}^{-1}$.

(i) $\Delta_r G/(\text{kJ mol}^{-1}) = \{\Delta_r G^{\ominus}(\text{i}) - RT \ln p_{H_2O}\}/\text{kJ mol}^{-1} = -23.6 - 1.58\ \ln(1.3\times10^{-7})$
$= +1.5$

(ii) $\Delta_r G/(\text{kJ mol}^{-1}) = \{\Delta_r G^{\ominus}(\text{ii}) - RT \ln p_{H_2O}p_{HNO_3}\}/(\text{kJ mol}^{-1})$
$= -57.2 - 1.58 \ln[(1.3\times10^{-7})\times(4.1\times10^{10})]$
$= +2.0$

(iii) $\Delta_r G/(\text{kJ mol}^{-1}) = \{\Delta_r G^{\ominus}(\text{iii}) - RT \ln p_{H_2O}{}^2 p_{HNO_3}\}/(\text{kJ mol}^{-1})$
$= -85.6 - 1.58 \ln[(1.3\times10^{-7})^2\times(4.1\times10^{-10})]$
$= -1.3$

(iv) $\Delta_r G/(\text{kJ mol}^{-1}) = \{\Delta_r G^{\ominus}(\text{iv}) - RT \ln p_{H_2O}{}^3 p_{HNO_3}\}/(\text{kJ mol}^{-1})$
$= -112.8 - 1.58 \ln[(1.3\times10^{-7})^3\times(4.1\times10^{-10})]$
$= -3.5$

So both the dihydrate and trihydrate form spontaneously from the vapor. Does one convert spontaneously into the other? Consider the reaction

$$HNO_3\cdot 2H_2O(s) + H_2O(g) \rightleftharpoons HNO_3\cdot 3H_2O(s)$$

which may be considered as reaction(iv) – reaction(iii). $\Delta_r G$ for this reaction is

$$\Delta_r G = \Delta_r G(\text{iv}) - \Delta_r G(\text{iii}) = -2.2\ \text{kJ mol}^{-1}$$

We conclude that the dihydrate converts spontaneously to the $\boxed{\text{trihydrate}}$, the most stable solid (at least of the four we considered).

PART 5 Chemical dynamics

18 Molecular motion

Answers to discussion questions

D18.1 The assumptions of the kinetic model of gases are presented in text Section 18.1. The following critique emphasizes conditions for which the fundamental assumptions may not be applicable:

The gas consists of molecules of mass *m* in ceaseless random motion obeying the laws of classical mechanics. Critique: The laws of classical mechanics include the conservation of particulate total energy, linear momentum, and angular momentum. These laws are predicated on the assumption that gaseous matter exhibits the particulate behavior alone to the exclusion of electron wave behavior that is required to explain chemical reactions that may occur during particle collisions. Furthermore, the assumption of ceaseless random motion means that the simple kinetic model does not include a description of convective flow or diffusion. (This restriction is relieved by dividing the gas into very small adjacent volumes in which each separate volume is characterized by its own temperature, pressure, and composition.)

The size of the molecules is negligible, in the sense that the diameters of the molecules are much smaller than the average distance traveled between collisions. Critique: A molecular gas at low molar volumes occupies a significant percentage of available space. The negligible molecular volume restriction limits the applicability of the simple kinetic model to the limit of zero pressure and prevents formation of a description of the van der Waals constants. However, the restriction can be relieved by treating each molecule as a ball bearing or billiard ball of fixed radius a and diameter d (= $2a$). The resultant hard-sphere model, which is a kind of central-force model, allows for the estimation of the collision frequency and mean free path (see Section 8.1b and c); it does not account for collision effects from dipole forces that are not centered on the center-of-mass.

The molecules interact only through brief elastic collisions (collisions in which the total translational kinetic energy of the molecules is conserved). Critique: This assumption is inherent in the analysis of collision frequency, mean free path, and the collision frequency with walls and surfaces. The model does not account for unevenness of surfaces, which creates surface pockets that may trap a gas molecule during collision, or cause a gas molecule to reflect at an angle other than the collision angle. The successful description of the simple kinetic model to wall collisions and effusion does suggest that it describes these phenomena as an average of many possibilities. This basic assumption of the model does not account for **inelastic collisions** in which molecule collisions at high temperature can exchange energy between translational, vibrational, and rotational degrees of freedom. This complication becomes important when the mean translational energy is comparable to the energy difference between the ground and the first excited state.

D18.3 Diffusion is the migration of particles (molecules) down a concentration gradient. Diffusion can be interpreted at the molecular level as being the result of the random jostling of the molecules in a fluid. The motion of the molecules is the result of a series of short jumps in random directions, a so-called random walk.

In the random walk model of diffusion, although a molecule may take many steps in a given time, it has only a small probability of being found away from its starting point because some of the steps lead it away from the starting point but others lead it back. As a result, the net distance traveled increases only as the square root of the time. There is no net flow of molecules unless there is a concentration gradient in the fluid. Also there are just as many molecules moving in one direction as there are molecules moving in another. The rate at which the molecules spread out is proportional to the concentration gradient. The constant of proportionality is called the diffusion coefficient.

On the molecular level in a gas, thermal conduction occurs because of random molecular motions in the presence of a temperature gradient. Across any plane in the gas, there is a net flux of energy from the high-temperature side because molecules coming from that side carry a higher average energy per molecule across the plane than do those coming from the low-temperature side. In solids, the situation is more complex as energy transport occurs through quantized elastic waves (phonons) and also, in metals, by electrons. Conduction in liquids can occur by all the mechanisms mentioned.

At the molecular (ionic) level, electrical conduction in an electrolytic solution is the net migration of ions in any given direction. When a gradient in electrical potential exists in a conductivity cell there will be a greater flow of positive ions in the direction of the negative electrode than in the direction of the positive electrode; hence there is a net flow of positive charge toward the region of low electrical potential. Likewise, a net flow of negative ions in the direction of the positive electrode will occur. In metals, only negatively charged electrons contribute to the current.

To see the connection between the flux of momentum and the viscosity, consider a fluid in a state of **Newtonian flow**, which can be imagined as occurring by a series of layers moving past one another (Fig. 18.11 of the text). The layer next to the wall of the vessel is stationary, and the velocity of successive layers varies linearly with distance, z, from the wall. Molecules ceaselessly move between the layers and bring with them the x-component of linear momentum they possessed in their original layer. A layer is retarded by molecules arriving from a more slowly moving layer because the molecules have a low momentum in the x-direction. A layer is accelerated by molecules arriving from a more rapidly moving layer. We interpret the net retarding effect as the fluid's viscosity.

D18.5 In the **one-dimensional random walk model** (Section 18.11) a particle jumps a fixed distance λ to either the right or the left in each τ unit of time with the probability of a jump to the left being equal to the probability of a jump to the right and the sum of these two probabilities being equal to 1 because no other event is possible. This model concludes that the probability of a particle being at a distance x from the origin after a time t is

$$P(x)=\left(\frac{2\tau}{\pi t}\right)^{1/2}\mathrm{e}^{-x^2\tau/2t\lambda^2}\ [18.58] \quad \text{or} \quad P(n)=\left(\frac{2}{\pi N}\right)^{1/2}\mathrm{e}^{-n^2/2N}$$

where $t/\tau = N$ is the number of steps taken in time t and $x/\lambda = n$ is the net number of steps traveled from the origin.

In the **freely jointed chain** model of a random coil (see *Further Information* 18.2), the individual polymer residues occupy zero volume and connecting bonds are free to make any angle with respect to the previous one. In the **one-dimensional freely jointed chain**, the angle between neighbors is either 0° or 180° with all residues aligned in a non-realistic straight line. The probability that the ends of a one-dimensional freely jointed chain composed of N units of length l are at the distance nl is

$$P(n)=\left(\frac{2}{\pi N}\right)^{1/2}\mathrm{e}^{-n^2/2N}\ [18.73]$$

Since the above two equations are identical, the one-dimensional random walk model and the one-dimensional freely jointed chain are completely analogous: the number of steps in one model are analogous with the number of residue units in the other, and the number of steps traveled from the origin are analogous to the number of unit lengths separating the chain ends in the other. The number of left/right steps is even analogous to the number of 0°/180° residue orientations. However, whereas there would not seem to be a limitation on a step being either to the left or to the right, the 180° residue orientation unrealistically places one residue on another. Nonetheless, the concepts and mathematical forms of the one-dimensional freely jointed chain can be usefully modified and extended to more realistic random coil models.

D18.7 The **thermodynamic force** $\mathcal{F}$ is defined in eqn. 18.39.

$$\mathcal{F} = -\left(\frac{\partial \mu}{\partial x}\right)_{p,T} \quad [18.39]$$

This expression is a summary of the second thermodynamic law that molecules move in the direction that minimizes the chemical potential of the molecules when p and T are local constants. $\mathcal{F}$ is not one of the "real" forces, such as gravity or electromagnetism; it is the negative gradient of the chemical potential, which has a balance of terms involving enthalpy and entropy:

$$\mu_J = \left(\frac{\partial G}{\partial n_J}\right)_{p,T,n'} [16.2] = \left(\frac{\partial}{\partial n_J}\right)_{p,T,n'} (H - TS)$$

Thus, the thermodynamic force moves molecules so as to minimize the enthalpy, to which molecular interactions provide a great contribution, while simultaneously attempting to maximize entropy. Often one or the other of these tendencies predominates. For an ideal solution, the gradient of the molar enthalpy is zero, so the force represents the spontaneous tendency for molecules to disperse so that entropy is maximized.

D18.9 The processes by which ions are transported across hydrophobic biological membranes are described in a section on biochemistry, *Impact on biochemistry* 18.2. **Passive transport** involves the spontaneous tendency for a chemical species to move down concentration and membrane potential gradients. In **active transport**, the flow of a species is driven by an exergonic process, such as the hydrolysis of ATP. The passive transport of an ion across a membrane must be mediated by a carrier molecule that has favorable interactions with both the hydrophobic membrane and the hydrophilic ion. The hydrophobic character is expected to lie on the periphery of the carrier, the surface in contact with the membrane, with the hydrophilic character toward the interior, the surface of the ion channel, so that the ion and membrane are distinctly separate. The ion may be transported through a **channel former**, a protein that spans the membrane and creates a hydrophilic pore through which the ion can pass. **Ion channels** are proteins that are highly selective to the transported ion; their activity may be triggered by potential differences across the membrane or by the binding of an **effector** molecule. The activity of **ion pumps** is endergonic and therefore requires the coupling of an exergonic metabolic process. **Selectivity filtering** strips an ion of its hydration sphere and allows passage of ions of only a specific size.

Solutions to exercises

E18.1(a)

$$\bar{c} = \left(\frac{8RT}{\pi M}\right)^{1/2} [18.8] \quad \text{and} \quad c = \langle v^2 \rangle^{1/2} = \left(\frac{3RT}{M}\right)^{1/2} [18.3]$$

mean translational kinetic energy, $\langle E_K \rangle_m = \langle \tfrac{1}{2}mv^2 \rangle N_A = \tfrac{1}{2}mN_A \langle v^2 \rangle = \tfrac{1}{2}Mc^2 = \tfrac{3}{2}RT$.

The ratios of species 1 to species 2 at the same temperature are

$$\frac{\bar{c}_1}{\bar{c}_2} = \left(\frac{M_2}{M_1}\right)^{1/2} \quad \text{and} \quad \frac{\langle E_K \rangle_{m1}}{\langle E_K \rangle_{m2}} = 1$$

(a) $$\frac{\bar{c}_{H_2}}{\bar{c}_{Hg}} = \left(\frac{200.6}{2.016}\right)^{1/2} = \boxed{9.975}$$

(b) The mean translation kinetic energy is independent of molecular mass and depends on temperature alone! Consequently, because the mean translational kinetic energy for a gas is proportional to T, the ratio of mean translational kinetic energies for gases at the same temperature always equals $\boxed{1}$.

E18.2(a) $$c = \langle v^2 \rangle^{1/2} = \left(\frac{3RT}{M}\right)^{1/2} \quad [18.3]$$

$$c_{H_2} = \left\{\frac{3\times(8.3145\ \text{J mol}^{-1}\ \text{K}^{-1})\times(293\ \text{K})}{2.016\times10^{-3}\ \text{kg mol}^{-1}}\right\}^{1/2} \qquad (1\ \text{J} = 1\ \text{kg m}^2\ \text{s}^{-2})$$

$$= \boxed{1904\ \text{m s}^{-1}}$$

$$c_{O_2} = \left\{\frac{3\times(8.3145\ \text{J mol}^{-1}\ \text{K}^{-1})\times(293\ \text{K})}{32.00\times10^{-3}\ \text{kg mol}^{-1}}\right\}^{1/2}$$

$$= \boxed{478\ \text{m s}^{-1}}$$

E18.3(a) The Maxwell distribution of speeds is $f(v) = 4\pi\left(\frac{M}{2\pi RT}\right)^{3/2} v^2 e^{-Mv^2/2RT}$ [18.4].

The factor $M/2RT$ can be evaluated as

$$\frac{M}{2RT} = \frac{28.02\times10^{-3}\ \text{kg mol}^{-1}}{2\times(8.3145\ \text{J K}^{-1}\ \text{mol}^{-1})\times(400\ \text{K})} = 4.21\times10^{-6}\ \text{m}^{-2}\,\text{s}^2$$

Though $f(v)$ varies over the range 200 to 210 m s^{-1}, the variation is small over this small range and we estimate its value to be a constant equal to its value at the center of the range.

$$f(295\ \text{m s}^{-1}) = (4\pi)\times\left(\frac{4.21\times10^{-6}\ \text{m}^{-2}\ \text{s}^2}{\pi}\right)^{3/2}\times(205\ \text{m s}^{-1})^2\times e^{(-4.21\times10^{-6})\times(205)^2} = 6.86\times10^{-4}\ \text{m}^{-1}\ \text{s}$$

Therefore, the fraction of molecules in the specified range is

$$\int_{v_{low}}^{v_{high}} f\ dv\ [18.6] = f\int_{v_{low}}^{v_{high}} dv = f\times\Delta v = (6.86\times10^{-4}\ \text{m}^{-1}\ \text{s})\times(10\ \text{m s}^{-1}) = \boxed{6.86\times10^{-3}}$$

corresponding to 0.69 percent.

COMMENT. This is a rather small percentage and suggests that the approximation of constancy of $f(v)$ over the range is adequate. To test the approximation, f(200 m s^{-1}) and f(210 m s^{-1}) could be evaluated.

E18.4(a) $$c^* = \left(\frac{2RT}{M}\right)^{1/2} \quad [18.9]$$

$$= \left\{\frac{2\times(8.3145\ \text{J mol}^{-1}\ \text{K}^{-1})\times(293\ \text{K})}{44.01\times10^{-3}\ \text{kg mol}^{-1}}\right\}^{1/2} \qquad (1\ \text{J} = 1\ \text{kg m}^2\ \text{s}^{-2})$$

$$= \boxed{333\ \text{m s}^{-1}}$$

$$\bar{c} = \left(\frac{8RT}{\pi M}\right)^{1/2} \quad [18.8]$$

$$= \left\{\frac{8\times(8.3145\ \text{J mol}^{-1}\ \text{K}^{-1})\times(293\ \text{K})}{\pi\times(44.01\times10^{-3}\ \text{kg mol}^{-1})}\right\}^{1/2}$$

$$= \boxed{375\ \text{m s}^{-1}}$$

$$\bar{c}_{\text{rel}} = 2^{1/2}\bar{c} \quad [18.10]$$

$$= 2^{1/2}\times(375\ \text{m s}^{-1}) = \boxed{530\ \text{m s}^{-1}}$$

E18.5(a) (a)

$$\bar{c} = \left(\frac{8RT}{\pi M}\right)^{1/2} \quad [18.8]$$

$$= \left\{\frac{8\times(8.3145\ \text{J mol}^{-1}\ \text{K}^{-1})\times(298\ \text{K})}{\pi\times(28.02\times10^{-3}\ \text{kg mol}^{-1})}\right\}^{1/2}$$

$$= \boxed{475\ \text{m s}^{-1}}$$

(b)

$$\lambda = \frac{kT}{2^{1/2}\sigma p} \quad [18.14] = \frac{kT}{2^{1/2}\pi d^2 p} \quad [\sigma = \pi d^2;\ \text{see Fig. 18.9 of text}]$$

$$= \frac{(1.381\times10^{-23}\ \text{J K}^{-1})\times(298\ \text{K})}{2^{1/2}\pi\times(395\times10^{-12}\ \text{m})^2\times(1.01325\times10^5\ \text{Pa})} \quad [1\ \text{atm} = 1.01325\ \text{bar} = 1.01325\times10^5\ \text{Pa}]$$

$$= 1.84\times10^{-7}\ \text{m} \quad \left[1\ \text{Pa} = 1\ \text{J m}^{-3}\right] = \boxed{184\ \text{nm}}$$

(c)

$$z = \frac{\bar{c}}{\lambda} \quad [18.13]$$

$$= \frac{475\ \text{m s}^{-1}}{1.84\times10^{-7}\ \text{m}} = \boxed{2.58\times10^9\ \text{s}^{-1}}$$

E18.6(a) The volume and radius of a spherical container are related by

$$V = \tfrac{4}{3}\pi r^3 \quad \text{or} \quad r = (3V/4\pi)^{1/3}$$

Thus, $r = \left\{3\times(1.00^{-4}\ \text{m}^3)/4\pi\right\}^{1/3} = 0.0288\ \text{m}$.

The pressure at which the mean free path is comparable to the container diameter $2r$ is given by eqn. 18.14.

$$\lambda = \frac{kT}{2^{1/2}\sigma p} \quad [18.14] \quad \text{or} \quad p = \frac{kT}{2^{1/2}\sigma\lambda}$$

$$p = \frac{(1.381\times10^{-23}\ \text{J K}^{-1})\times(293\ \text{K})}{(2^{1/2})\times(0.36\times10^{-18}\ \text{m}^2)\times2\times(0.0288\ \text{m})} = \boxed{0.138\ \text{Pa}} \quad [1\ \text{J} = 1\ \text{Pa m}^3]$$

E18.7(a)

$$\lambda = \frac{kT}{2^{1/2}\sigma p} \quad [18.14]$$

$$= \frac{(1.381\times10^{-23}\ \mathrm{J\,K^{-1}})\times(217\ \mathrm{K})}{(2^{1/2})\times(0.43\times10^{-18}\ \mathrm{m^2})\times(0.050)\times(1.013\times10^{5}\ \mathrm{Pa})}$$

$$= \boxed{9.7\times10^{-7}\ \mathrm{m}}$$

E18.8(a) The collision frequency, z, is given by $z = 2^{1/2}\sigma\bar{c}p/kT$ [18.12b and 18.10]. Since the mean speed depends on temperature, not pressure, it can be calculated once and used in all parts of this exercise.

$$\bar{c} = \left(\frac{8RT}{\pi M}\right)^{1/2} \quad [18.8]$$

$$= \left\{\frac{8\times(8.3145\ \mathrm{J\,mol^{-1}\,K^{-1}})\times(298\ \mathrm{K})}{\pi\times(39.95\times10^{-3}\ \mathrm{kg\,mol^{-1}})}\right\}^{1/2}$$

$$= \boxed{397\ \mathrm{m\,s^{-1}}}$$

$$z = \frac{2^{1/2}\sigma\bar{c}p}{kT} \quad [18.12\mathrm{b\ and\ }18.10]$$

$$= \frac{2^{1/2}\times(0.36\times10^{-18}\ \mathrm{m^2})\times(397\ \mathrm{m\,s^{-1}})}{(1.381\times10^{-23}\ \mathrm{J\,K^{-1}})\times(298\ \mathrm{K})}\times\left(\frac{1.01325\times10^{5}\ \mathrm{Pa}}{1\ \mathrm{atm}}\right)\times p \quad [\text{Table 18.1}]$$

$$= (4.98\times10^{9}\ \mathrm{s^{-1}})\times(p/\mathrm{atm})$$

We need only to plug in the pressure in atm to calculate z at that pressure.

(a) $z = \boxed{5.0\times10^{10}\ \mathrm{s^{-1}}}$ when $p = 10$ atm

(b) $z = \boxed{5.0\times10^{9}\ \mathrm{s^{-1}}}$ when $p = 1$ atm

(c) $z = \boxed{5.0\times10^{3}\ \mathrm{s^{-1}}}$ when $p = 10^{-6}$ atm

E18.9(a)

$$A = (5.0\ \mathrm{mm})\times(4.0\ \mathrm{mm}) = 2.0\times10^{-5}\ \mathrm{m^2}$$

The collision frequency of the Ar gas molecules with surface area A equals $Z_W A$

$$Z_W A = \frac{p}{(2\pi MkT/N_A)^{1/2}}A \quad [18.15;\ m = M/N_A]$$

$$= \frac{(25\ \mathrm{Pa})\times(2.0\times10^{-5}\ \mathrm{m^2})}{\{2\pi(39.95\times10^{-3}\ \mathrm{kg\,mol^{-1}})\times(1.381\times10^{-23}\ \mathrm{J\,K^{-1}})\times(300\ \mathrm{K})/(6.022\times10^{23}\ \mathrm{mol^{-1}})\}^{1/2}}$$

$$= 1.2\times10^{19}\ \mathrm{s^{-1}}$$

The number of argon molecule collisions within A in time interval t equals $Z_W At$ if p does not change significantly during the period t.

$$Z_W At = (1.2\times10^{19}\ \mathrm{s^{-1}})\times(100\ \mathrm{s}) = \boxed{1.2\times10^{21}}$$

E18.10(a) If p does not change significantly during the period t, $t_{O_2} = \dfrac{(\text{rate of effusion})_{H_2}}{(\text{rate of effusion})_{O_2}} t_{H_2}$.

The rate of effusion at constant temperature, pressure, and molecule count is proportional to $M^{-1/2}$ [18.17]. Consequently, the above expression conveniently simplifies to

$$t_{O_2} = \left(\frac{M_{O_2}}{M_{H_2}}\right)^{1/2} t_{H_2}$$

$$= \left(\frac{32.00}{2.02}\right)^{1/2} \times (135\text{ s}) = \boxed{537\text{ s}}$$

E18.11(a) If p does not change significantly during the period t, the mass loss equals the effusion mass loss multiplied by the time period t: m_{loss} = (rate of effusion) × t × m = (rate of effusion) × t × M / N_A.

$$m_{\text{loss}} = \left(\frac{pA_0N_A}{(2\pi MRT)^{1/2}}\right)\times\left(\frac{Mt}{N_A}\right) \quad [18.17]$$

$$= pA_0t\times\left(\frac{M}{2\pi RT}\right)^{1/2}$$

$$= (0.735\text{ Pa})\times\left\{\pi(0.75\times10^{-3}\text{ m})^2\right\}\times(3600\text{ s})\times\left(\frac{0.300\text{ kg mol}^{-1}}{2\pi(8.3145\text{ J mol}^{-1}\text{ K}^{-1})\times(500\text{ K})}\right)^{1/2}$$

$$= \boxed{16\text{ mg}}$$

E18.12(a) The pressure of this exercise changes significantly during time period t, so it is useful to spend a moment finding an expression for $p(t)$. Mathematically, the rate of effusion is the derivative $-\mathrm{d}N/\mathrm{d}t$. Substitution of the perfect gas law for N, $N = pVN_A/RT$, where V and T are constants, reveals that the rate of effusion can be written as $-(N_AV/RT)\mathrm{d}p/\mathrm{d}t$. This formulation of the rate of effusion, along with eqn. 18.17, is used to find $p(t)$.

$$-\left(\frac{N_AV}{RT}\right)\frac{\mathrm{d}p}{\mathrm{d}t} = \frac{pA_0N_A}{(2\pi MRT)^{1/2}} \quad [18.17]$$

$$\frac{\mathrm{d}p}{\mathrm{d}t} = -\frac{pA_0}{V}\left(\frac{RT}{2\pi M}\right)^{1/2}$$

$$\frac{\mathrm{d}p}{p} = -\frac{\mathrm{d}t}{\tau} \quad \text{where} \quad \tau = \frac{V}{A_0}\left(\frac{2\pi M}{RT}\right)^{1/2}$$

$$\int_{p_0}^{p}\frac{\mathrm{d}p}{p} = -\frac{1}{\tau}\int_0^t \mathrm{d}t \quad \text{where } p_0 \text{ is the initial pressure}$$

$$\ln\frac{p}{p_0} = -\frac{t}{\tau} \quad \text{or} \quad p(t) = p_0\mathrm{e}^{-t/\tau}$$

The nitrogen gas and unknown gas data can be used to determine the relaxation time, τ, for each.

$$\tau_{N_2} = \frac{t_{N_2}}{\ln(p_0/p)_{N_2}} = \frac{45\text{ s}}{\ln(74/20)} = 34\text{ s}$$

$$\tau_{\text{unk}} = \frac{t_{\text{unk}}}{\ln(p_0/p)_{\text{unk}}} = \frac{152\text{ s}}{\ln(74/20)} = 116\text{ s}$$

This definition of τ shows that it is proportional to $M^{1/2}$. Since the ratio of the relaxation times cancels the constant of proportionality,

$$\left(\frac{M_{\text{unk}}}{M_{N_2}}\right)^{1/2} = \frac{\tau_{\text{unk}}}{\tau_{N_2}}$$

$$M_{\text{unk}} = \left(\frac{\tau_{\text{unk}}}{\tau_{N_2}}\right)^2 M_{N_2}$$

$$= \left(\frac{116}{34.\bar{4}}\right)^2 \times (28.02\ \text{g mol}^{-1}) = \boxed{31\bar{9}\ \text{g mol}^{-1}}$$

E18.13(a) In E18.12(a) it is shown that

$$\ln\frac{p}{p_0} = -\frac{t}{\tau} \quad \text{or} \quad p(t) = p_0 e^{-t/\tau} \quad \text{where} \quad \tau = \frac{V}{A_0}\left(\frac{2\pi M}{RT}\right)^{1/2}$$

The relaxation time, τ, of oxygen is calculated with the data.

$$\tau = \left(\frac{3.0\ \text{m}^3}{\pi(0.10\times10^{-3}\ \text{m})^2}\right) \times \left\{\frac{2\pi(32.00\times10^{-3}\ \text{kg mol}^{-1})}{(8.3145\ \text{J mol}^{-1}\ \text{K}^{-1})\times(298\ \text{K})}\right\}^{1/2} = 8.6\times10^5\ \text{s} = 10.\bar{0}\ \text{days}$$

The time required for the specified pressure decline is calculated with the above equation.

$$t = \tau\ln(p_0/p) = (10.\bar{0}\ \text{days})\times\ln(80/70) = \boxed{1.3\ \text{days}}$$

E18.14(a)

$$\kappa = \tfrac{1}{3}\lambda\bar{c}C_{V,\text{m}}[\text{A}]\ [18.24], \quad \bar{c} = \left(\frac{8RT}{\pi M}\right)^{1/2}\ [18.8]$$

and $$\lambda = \frac{kT}{2^{1/2}\sigma p}[18.14] = \frac{1}{2^{1/2}\sigma N_{\text{A}}[\text{A}]} \quad \left[\frac{p}{kT} = N_{\text{A}}[\text{A}]\right]$$

Therefore, $\kappa = \dfrac{\bar{c}\,C_{V,\text{m}}}{3\times2^{1/2}\sigma N_{\text{A}}}$.

For argon $M = 39.95\ \text{g mol}^{-1}$.

$$\bar{c} = \left(\frac{8\times(8.3145\ \text{J K}^{-1}\ \text{mol}^{-1})\times(298\ \text{K})}{\pi\times(39.95\times10^{-3}\ \text{kg mol}^{-1})}\right)^{1/2} = 397\ \text{m s}^{-1}$$

$$\kappa = \frac{(397\ \text{m s}^{-1})\times(12.5\ \text{J K}^{-1}\ \text{mol}^{-1})}{3\times2^{1/2}\times(0.36\times10^{-18}\ \text{m}^2)\times(6.022\times10^{23}\ \text{mol}^{-1})} = \boxed{5.4\times10^{-3}\ \text{J K}^{-1}\ \text{m}^{-1}\ \text{s}^{-1}}$$

COMMENT. This calculated value does not agree well with the value of κ listed in Table 18.2.

Question. Can the differences between the calculated and experimental values of κ be accounted for by the difference in temperature (298 K here, 273 K in Table 18.2)? If not, what might be responsible for the difference?

E18.15(a)

$$D = \tfrac{1}{3}\lambda\bar{c}\ [18.23], \quad \bar{c} = \left(\frac{8RT}{\pi M}\right)^{1/2}\ [18.8], \quad \text{and} \quad \lambda = \frac{kT}{2^{1/2}\sigma p}\ [18.14]$$

For argon $M = 39.95\ \text{g mol}^{-1}$ and $\sigma = 0.36\ \text{nm}^2$ [Table 18.1]. We draw a table (below) of mean speed, mean free path, and diffusion coefficient calculations at 293 K and various pressures.

The flux due to diffusion is $J = -D\left(\dfrac{d\mathcal{N}}{dz}\right)$ [18.20] where $\mathcal{N}$ is the number density and z is the direction of the pressure gradient. Dividing both sides by the Avogadro constant converts $\mathcal{N}$ to molar concentration c while converting the flux to number of moles per unit area per second. Thus,

$$J = -D\frac{dc}{dz} = -D\frac{d(n/V)}{dz} = -\frac{D}{RT}\frac{dp}{dz} \quad \text{[perfect gas law]}$$

The negative sign indicates flow from high pressure to low.

For a pressure gradient of 1.0 bar m^{-1} (= 1.0×10^5 Pa m^{-1}),

$$J = \frac{(1.0\times10^5\ \text{Pa m}^{-1})\times D}{(8.3145\ \text{J mol}^{-1}\ \text{K}^{-1})\times(293\ \text{K})} = 41\ \text{mol m}^{-2}\ \text{s}^{-1}\times D/(\text{m}^2\ \text{s}^{-1})$$

The mole flux due to diffusion toward low pressure is calculated with the above equation and added to the exercise summary table.

p / Pa	$\bar{c}$ / m s^{-1}	λ / m	D / m^2 s^{-1}	J / mol m^{-2} s^{-1}
(a) 1.00	394	7.95×10^{-3}	1.04	43
(b) 1.00×10^5	"	7.95×10^{-8}	1.04×10^{-5}	4.3×10^{-4}
(c) 1.00×10^7	"	7.95×10^{-10}	1.04×10^{-7}	4.3×10^{-6}

E18.16(a) The flux is

$$J = -\kappa\frac{dT}{dz}\ [18.21] = -\tfrac{1}{3}\lambda\bar{c}C_{V,m}\ [\text{A}]\ \frac{dT}{dz}\ [18.24] = -\frac{\tfrac{1}{3}\lambda\bar{c}C_{V,m}p}{RT}\frac{dT}{dz}\ \ [\text{perfect gas, } [\text{A}] = p/RT]$$

where the negative sign indicates flow toward lower temperature and

$$\bar{c} = \left(\frac{8RT}{\pi M}\right)^{1/2}\ [18.8] \quad \text{and} \quad \lambda = \frac{kT}{2^{1/2}\sigma p}\ [18.14] = \frac{RT}{2^{1/2}N_A\sigma p}$$

$$J = -\tfrac{1}{3}\left(\frac{RT}{2^{1/2}N_A\sigma p}\right)\times\left(\frac{8RT}{\pi M}\right)^{1/2}\times\left(\frac{C_{V,m}p}{RT}\right)\times\frac{dT}{dz}$$

$$= -\tfrac{2}{3}\times\left(\frac{RT}{\pi M}\right)^{1/2}\times\left(\frac{C_{V,m}}{N_A\sigma}\right)\times\frac{dT}{dz}$$

For argon, $M = 39.95$ g mol^{-1}, $\sigma = 0.36$ nm^2 [Table 18.1], and $C_{V,m} = 12.5$ J K^{-1} mol^{-1} [given in E18.14a].

$$J = -\tfrac{2}{3}\times\left(\frac{(8.3145\ \text{J K}^{-1}\ \text{mol}^{-1})\times(280\ \text{K})}{\pi\times(39.95\times10^{-3}\ \text{kg mol}^{-1})}\right)^{1/2}\times\left\{\frac{12.5\ \text{J K}^{-1}\ \text{mol}^{-1}}{(6.022\times10^{23}\ \text{mol}^{-1})\times(0.36\times10^{-18}\ \text{m}^2)}\right\}\times(10.5\ \text{K m}^{-1})$$

$$= \boxed{-0.055\ \text{J m}^{-2}\ \text{s}^{-1}}$$

Note: We have used the $C_{V,m}$ value provided in E18.14(a). It can, however, be calculated with the perfect gas relationship $C_{V,m} = C_{p,m} - R$ and the $C_{p,m}$ value provided in Table 14.6 (20.786 J K^{-1} mol^{-1}). This gives $C_{V,m} = 12.471$ J K^{-1} mol^{-1}.

E18.17(a) The thermal conductivity, κ, is a function of the mean free path, λ, which in turn is a function of the collision cross-section, σ. Hence, reversing the order, σ can be obtained from κ.

$$\kappa = \tfrac{1}{3}\lambda\bar{c}C_{V,m}[\text{A}]\ [18.24], \quad \bar{c} = \left(\frac{8RT}{\pi M}\right)^{1/2}\ [18.8]$$

and $\lambda = \dfrac{kT}{2^{1/2}\sigma p}$ [18.14] $= \dfrac{1}{2^{1/2}\sigma N_A[\mathrm{A}]}$ $\left[\dfrac{p}{kT} = N_A[\mathrm{A}]\right]$

Therefore, $\kappa = \dfrac{\bar{c}C_{V,\mathrm{m}}}{3\times 2^{1/2}\sigma N_A}$ or $\sigma = \dfrac{\bar{c}C_{V,\mathrm{m}}}{3\times 2^{1/2}N_A\kappa}$.

For neon at 273 K, $\bar{c} = \left\{\dfrac{8\times(8.3145\ \mathrm{J\ K^{-1}\ mol^{-1}})\times(273\ \mathrm{K})}{\pi\times(20.18\times10^{-3}\ \mathrm{kg\ mol^{-1}})}\right\}^{1/2} = 535\ \mathrm{m\ s^{-1}}$.

From Table 18.2, $\kappa = 0.0465\ \mathrm{J\ K^{-1}\ m^{-1}\ s^{-1}}$.

The $C_{V,\mathrm{m}}$ value is calculated with the perfect gas relationship $C_{V,\mathrm{m}} = C_{p,\mathrm{m}} - R$ and the $C_{p,\mathrm{m}}$ value provided in Table 14.6 (20.786 J K^{-1} mol^{-1}). This gives $C_{V,\mathrm{m}} = 12.472$ J K^{-1} mol^{-1}. Alternatively, the $C_{V,\mathrm{m}}$ value is calculated with the equipartition theorem: $C_{V,\mathrm{m}} = {}^{3}/_{2}R = 12.472$ J K^{-1} mol^{-1}.

$$\sigma = \frac{(535\ \mathrm{m\ s^{-1}})\times(12.472\ \mathrm{J\ K^{-1}\ mol^{-1}})}{3\times2^{1/2}\times(6.022\times10^{23}\ \mathrm{mol^{-1}})\times(0.0465\ \mathrm{J\ K^{-1}\ m^{-1}\ s^{-1}})}$$

$$= 5.62\times10^{-20}\ \mathrm{m^2} = \boxed{0.0562\ \mathrm{nm^2}}$$

The experimental value is $0.24\ \mathrm{nm^2}$.

Question. What approximations inherent in the equation used in the solution to this exercise are the likely cause of the factor of 4 difference between the experimental and calculated values of the collision cross-section for neon?

E18.18(a) The thermal energy flux ('heat' flux) is described by $J(\text{energy}) = -\kappa\dfrac{\mathrm{d}T}{\mathrm{d}z}$ [18.21] where the negative sign indicates flow toward lower temperature. This is the rate of energy transfer per unit area. The total rate of energy transfer across area A is

$$\frac{\mathrm{d}E}{\mathrm{d}t} = AJ(\text{energy}) = -\kappa A\frac{\mathrm{d}T}{\mathrm{d}z}$$

To calculate the temperature gradient with the given data, we assume that the gradient is in a steady state. Then, recognizing that temperature differences have identical magnitude in Celsius or Kelvin units,

$$\frac{\mathrm{d}T}{\mathrm{d}z} = \frac{\Delta T}{\Delta z} = \frac{\{(-15)-(28)\}\ \mathrm{K}}{1.0\times10^{-2}\ \mathrm{m}} = -4.3\times10^{3}\ \mathrm{K\ m^{-1}}$$

We now assume that the coefficient of thermal conductivity of the gas between the window panes is comparable to that of nitrogen given in Table 18.2: $\kappa \approx 0.0240\ \mathrm{J\ K^{-1}\ m^{-1}\ s^{-1}}$.

Therefore, the rate of outward energy transfer is

$$\frac{\mathrm{d}E}{\mathrm{d}t} \approx -(0.0240\ \mathrm{J\ K^{-1}\ m^{-1}\ s^{-1}})\times(1.0\ \mathrm{m^2})\times(-4.3\times10^{3}\ \mathrm{K\ m^{-1}})$$

$$\approx 10\bar{3}\ \mathrm{J\ s^{-1}} \quad \text{or} \quad \boxed{10\bar{3}\ \mathrm{W}}$$

A $10\bar{3}$ W heater is needed to balance this rate of heat loss.

E18.19(a) $\eta = \tfrac{1}{3}M\lambda\bar{c}p/RT$ [18.25, [A] $= p/RT$ for a perfect gas]

We begin by substituting kinetic theory relationships for λ [18.14] and $\bar{c}$ [18.8].

$$\eta = \tfrac{1}{3}M\times\left(\frac{RT}{2^{1/2}\sigma N_A p}\right)\times\left(\frac{8RT}{\pi M}\right)^{1/2}\times\left(\frac{p}{RT}\right) = \left(\frac{M}{3\times 2^{1/2}\sigma N_A}\right)\times\left(\frac{8RT}{\pi M}\right)^{1/2}$$

We now solve and compute the collision cross-section for neon at 273 K and 1 atm (1.01325×10^5 Pa).

From Table 18.2, $\eta = 298\times10^{-6}\ \text{P} = 298\times10^{-7}\ \text{kg m}^{-1}\ \text{s}^{-1}$.

$$\sigma = \left\{\frac{20.18\times10^{-3}\ \text{kg mol}^{-1}}{3\times2^{1/2}\times(6.022\times10^{23}\ \text{mol}^{-1})\times(298\times10^{-7}\ \text{kg m}^{-1}\ \text{s}^{-1})}\right\}\times\left\{\frac{8(8.3145\ \text{J K}^{-1}\ \text{mol}^{-1})\times(273\ \text{K})}{\pi(20.18\times10^{-3}\ \text{kg mol}^{-1})}\right\}^{1/2}$$

$$= 1.42\times10^{-19}\ \text{m}^2 = \boxed{0.142\ \text{nm}^2}$$

E18.20(a) Taking air to be composed of 20% oxygen and 80% nitrogen, we estimate the average molar mass of air to be $M = (0.20\times32 + 0.80\times28)\ \text{g mol}^{-1} = 29\ \text{g mol}^{-1}$.

$$\eta = \frac{1}{3}m\lambda\bar{c}N_A[\text{A}]\ [18.25,\ \text{with}\ M = mN_A] = \left(\frac{m}{3\sigma}\right)\times\left(\frac{4RT}{\pi M}\right)^{1/2}$$

$$= \left(\frac{(29)\times(1.6605\times10^{-27}\ \text{kg})}{(3)\times(0.40\times10^{-18}\ \text{m}^2)}\right)\times\left(\frac{(4)\times(8.3145\ \text{J K}^{-1}\ \text{mol}^{-1})\times T}{\pi\times(29\times10^{-3}\ \text{kg mol}^{-1})}\right)^{1/2}$$

$$= (7.7\times10^{-7}\ \text{kg m}^{-1}\ \text{s}^{-1})\times(T/\text{K})^{1/2}$$

(a) At $T = 273$ K, $\eta = 1.3\times10^{-5}\ \text{kg m}^{-1}\ \text{s}^{-1}$ or $\boxed{130\ \mu\text{P}}$.

(b) At $T = 298$ K, $\eta = \boxed{130\ \mu\text{P}}$.

(c) At $T = 1000$ K, $\eta = \boxed{240\ \mu\text{P}}$.

E18.21(a) We take the natural logarithm of eqn. 18.26, evaluate the resultant constant with a specific $(\eta, T)_0$ data pair, and solve for the activation energy, E_a.

$$\eta \propto e^{E_a/RT}\quad [18.26]$$
$$\ln\eta = \text{constant} + E_a/RT$$
$$\text{constant} = \ln\eta_0 - E_a/RT_0$$

Therefore,

$$E_a = \frac{R\ln(\eta/\eta_0)}{\left(\frac{1}{T}-\frac{1}{T_0}\right)}$$

$$= \frac{(8.3145\ \text{J K}^{-1}\ \text{mol}^{-1})\times\ln(1.002/0.7975)}{\left(\frac{1}{293\ \text{K}}-\frac{1}{303\ \text{K}}\right)} = \boxed{16.8\ \text{J mol}^{-1}}$$

E18 22(a) Molar ionic conductivity is related to mobility by

$$\lambda = zuF\quad [18.37]$$
$$= 1\times(7.91\times10^{-8}\ \text{m}^2\ \text{s}^{-1}\ \text{V}^{-1})\times(96485\ \text{C mol}^{-1})$$
$$= \boxed{7.63\times10^{-3}\ \text{S m}^2\ \text{mol}^{-1}}$$

E18.23(a) $s = u\mathcal{E}$ [18.35] and $\mathcal{E} = \dfrac{\Delta\phi}{l}$ [18.31].

Therefore,

$$s = u\left(\frac{\Delta\phi}{l}\right)$$

$$= (7.92\times10^{-8}\ \mathrm{m^2\ s^{-1}\ V^{-1}})\times\left(\frac{25.0\ \mathrm{V}}{7.00\times10^{-3}\ \mathrm{m}}\right)$$

$$= 2.83\times10^{-4}\ \mathrm{m\ s^{-1}} \quad \text{or} \quad \boxed{283\ \mu\mathrm{m\ s^{-1}}}$$

E18.24(a) The basis for the solution is Kohlrausch's law of independent migration of ions (eqn. 18.30). Switching counterions does not affect the mobility of the remaining other ion at infinite dilution.

$$\Lambda_\mathrm{m}^\circ = \nu_+\lambda_+ + \nu_-\lambda_- \quad [18.30]$$

$$\Lambda_\mathrm{m}^\circ(\mathrm{NaI}) = \lambda(\mathrm{Na^+}) + \lambda(\mathrm{I^-}) = 12.69\ \mathrm{mS\ m^2\ mol^{-1}}$$

$$\Lambda_\mathrm{m}^\circ(\mathrm{NaNO_3}) = \lambda(\mathrm{Na^+}) + \lambda(\mathrm{NO_3^-}) = 12.16\ \mathrm{mS\ m^2\ mol^{-1}}$$

$$\Lambda_\mathrm{m}^\circ(\mathrm{AgNO_3}) = \lambda(\mathrm{Ag^+}) + \lambda(\mathrm{NO_3^-}) = 13.34\ \mathrm{mS\ m^2\ mol^{-1}}$$

Hence,

$$\Lambda_\mathrm{m}^\circ(\mathrm{AgI}) = \Lambda_\mathrm{m}^\circ(\mathrm{AgNO_3}) + \Lambda_\mathrm{m}^\circ(\mathrm{NaI}) - \Lambda_\mathrm{m}^\circ(\mathrm{NaNO_3})$$

$$= (13.34 + 12.69 - 12.16)\ \mathrm{mS\ m^2\ mol^{-1}} = \boxed{13.87\ \mathrm{mS\ m^2\ mol^{-1}}}$$

Question. How well does this result agree with the value calculated directly from the data of Table 18.5?

E18.25(a) $u = \dfrac{\lambda}{zF}$ [18.37]; $z = 1$; $1\ \mathrm{S} = 1\ \Omega^{-1} = 1\ \mathrm{C\ V^{-1}\ s^{-1}}$

$$u(\mathrm{Li^+}) = \frac{3.87\ \mathrm{mS\ m^2\ mol^{-1}}}{9.6485\times10^4\ \mathrm{C\,mol^{-1}}} = 4.01\times10^{-5}\ \mathrm{mS\ C^{-1}\ m^2} = \boxed{4.01\times10^{-8}\ \mathrm{m^2\ V^{-1}\ s^{-1}}}$$

$$u(\mathrm{Na^+}) = \frac{5.01\ \mathrm{mS\ m^2\ mol^{-1}}}{9.6485\times10^4\ \mathrm{C\,mol^{-1}}} = \boxed{5.19\times10^{-8}\ \mathrm{m^2\ V^{-1}\ s^{-1}}}$$

$$u(\mathrm{K^+}) = \frac{7.35\ \mathrm{mS\ m^2\ mol^{-1}}}{9.6485\times10^4\ \mathrm{C\,mol^{-1}}} = \boxed{7.62\times10^{-8}\ \mathrm{m^2\ V^{-1}\ s^{-1}}}$$

E18.26(a) $D = \dfrac{uRT}{zF}$ [18.45]; $z = 2$; $1\ \mathrm{C\ V} = 1\ \mathrm{J}$

$$D = \frac{(8.19\times10^{-8}\ \mathrm{m^2\ V^{-1}\,s^{-1}})\times(8.3145\ \mathrm{J\ K^{-1}\ mol^{-1}})\times(298\ \mathrm{K})}{2\times9.6485\times10^4\ \mathrm{C\ mol^{-1}}} = \boxed{1.05\times10^{-9}\ \mathrm{m^2\ s^{-1}}}$$

E18.27(a) Eqn. 18.57, $\langle x^2\rangle = 2Dt$, gives the mean square distance traveled in any one dimension in time t. We need the distance traveled from a point in any direction. The distinction here is the distinction between the one-dimensional and three-dimensional diffusion. The mean square three-dimensional distance can be obtained from the one-dimensional mean square distance since motions in the three directions are independent. Since $r^2 = x^2 + y^2 + z^2$ [Pythagorean theorem],

$$\langle r^2\rangle = \langle x^2\rangle + \langle y^2\rangle + \langle z^2\rangle = 3\langle x^2\rangle \text{ [independent motion]}$$

$$= 3\times 2Dt \text{ [18.57 for } \langle x^2\rangle] = 6Dt$$

Therefore, $t = \dfrac{\langle r^2\rangle}{6D} = \dfrac{(5.0\times 10^{-3}\text{ m})^2}{(6)\times(6.73\times 10^{-10}\text{ m}^2\text{ s}^{-1})} = \boxed{6.2\times 10^3\text{ s}}$.

E18.28(a) $a = \dfrac{kT}{6\pi\eta D}$ [18.49]; $1\text{ P} = 10^{-1}\text{ kg m}^{-1}\text{ s}^{-1}$

$$a = \frac{(1.381\times 10^{-23}\text{ J K}^{-1})\times(298\text{ K})}{6\pi\times(1.00\times 10^{-3}\text{ kg m}^{-1}\text{ s}^{-1})\times(5.2\times 10^{-10}\text{ m}^2\text{ s}^{-1})} = 4.2\times 10^{-10}\text{ m} \quad\text{or}\quad \boxed{420\text{ pm}}$$

E18.29(a) The Einstein–Smoluchowski equation [18.59] relates the diffusion constant to the jump distance λ and jump time τ.

$$D = \frac{\lambda^2}{2\tau}\text{ [18.59]} \quad\text{so}\quad \tau = \frac{\lambda^2}{2D}$$

If the jump distance is about 1 molecular diameter, or 2 effective molecular radii, then the jump distance can be obtained by use of the Stokes–Einstein equation [18.49].

$$\lambda = 2a = 2\left(\frac{kT}{6\pi\eta D}\right)\text{ [18.49]} = \frac{kT}{3\pi\eta D}$$

$$\tau = \frac{1}{2D}\times\left(\frac{kT}{3\pi\eta D}\right)^2$$

$$= \frac{1}{2\times(2.03\times 10^{-9}\text{ m}^2\text{ s}^{-1})}\times\left\{\frac{(1.381\times 10^{-23}\text{ J K}^{-1})\times(298\text{ K})}{3\pi\times(0.891\times 10^{-3}\text{ kg m}^{-1}\text{ s}^{-1})\times(2.03\times 10^{-9}\text{ m}^2\text{ s}^{-1})}\right\}^2 \text{ [Table 18.4]}$$

$$= 1.44\times 10^{-11}\text{ s} = \boxed{14.4\text{ ps}}$$

COMMENT. In the strictest sense we are again dealing with three-dimensional diffusion here (cf. E18.27a and b). However, since we are assuming that only one jump occurs, it is probably an adequate approximation to use an equation derived for one-dimensional diffusion. For three-dimensional diffusion, the equation analogous to eqn. 18.59 is $\tau = \lambda^2/6D$.

Question. Can you derive the equation? Use an analysis similar to that described in the solution to E18.27.

E18.30(a) The diffusion equation solution for these boundary conditions is provided in eqn. 18.54 with

$$n_0 = (20.0\text{ g})\times\left(\frac{1\text{ mol sucrose}}{342.30\text{ g}}\right) = 5.84\times 10^{-2}\text{ mol sucrose}$$

$A = 5.0\text{ cm}^2$, $D = 5.216\times 10^{-9}\text{ m}^2\text{ s}^{-1}$, and $x = 10\text{ cm}$

$$c(x,t) = \frac{n_0}{A(\pi Dt)^{1/2}}e^{-x^2/4Dt}\text{ [18.54]}$$

$$c(10\text{ cm},t)=\frac{5.84\times10^{-2}\text{ mol}}{(5.0\times10^{-4}\text{ m}^2)\times\{\pi(5.216\times10^{-9}\text{ m}^2\text{ s}^{-1})\}^{1/2}t^{1/2}}e^{-(0.10\text{ m})^2/4\times(5.216\times10^{-9}\text{ m}^2\text{ s}^{-1})\times t}$$

$$=(9.12\times10^5\text{ mol m}^{-3})\times(t/\text{s})^{-1/2}e^{-(4.79\times10^5)/(t/\text{s})}$$

(a) $t = 10$ s:

$$c(10\text{ cm},10\text{ s})=(9.12\times10^5\text{ mol m}^{-3})\times(10)^{-1/2}e^{-(4.79\times10^5)/(10)}=\boxed{0.00\text{ mol dm}^{-3}}$$

(b) $t = 24 \times 3600\text{ s} = 8.64 \times 10^4$ s:

$$c(10\text{ cm},8.64\times10^4\text{ s})=(9.12\times10^5\text{ mol m}^{-3})\times(8.64\times10^4)^{-1/2}e^{-(4.79\times10^5)/(8.64\times10^4)}$$
$$=12.1\text{ mol m}^{-3}=\boxed{0.0121\text{ mol dm}^{-3}}$$

E18.31(a) For a random coil, the root mean square separation is

$$R_{\text{rms}}=N^{1/2}l\text{ [18.76] where }N\text{ is the number of jointed chain units of length }l$$
$$=(900)^{1/2}\times(1.05\text{ nm})=\boxed{381.5\text{ nm}}$$

E18.32(a) The repeating unit (monomer) of polyethylene is (—CH_2—CH_2—), which has a molar mass of 28 g mol^{-1}. The number of repeating units, N, is therefore

$$N=\frac{300\,000\text{ g mol}^{-1}}{28\text{ g mol}^{-1}}=1.07\times10^4$$

and $l = 2R$(C—C) [add half a bond-length on either side of the monomer].

Since R(C—C) = 154 pm, l = 308 pm.

Therefore,

$$R_{\text{c}}=Nl\text{ [18.75]}=(1.07\times10^4)\times(308\text{ pm})=3.30\times10^6\text{ pm}=\boxed{3.30\ \mu\text{m}}$$

$$R_{\text{rms}}=N^{1/2}l=(1.07\times10^4)^{1/2}\times(308\text{ pm})=3.19\times10^4\text{ pm}=\boxed{0.0319\ \mu\text{m}}$$

Solutions to problems

Solutions to numerical problems

P18.1 The time in seconds for a disk to rotate 360° is the inverse of the frequency. The time for it to advance 2° is $\left(\frac{2°}{360°}\right)\times\frac{1}{\nu}$. This is the time required for slots in neighboring disks to coincide along the atomic beam. For an atom to pass through all neighboring slots, it must have the speed v_x = disk spacing/alignment time $=(1.0\text{ cm})/\left\{\left(\frac{2°}{360°}\right)\times\frac{1}{\nu}\right\}=180\nu\text{ cm}=180\times(\nu/\text{Hz})\text{ cm s}^{-1}$.

Hence, the distributions of the x-component of velocity are as follows.

ν / Hz	20	40	80	100	120
v_x / cm s^{-1}	3600	7200	14400	18000	21600
I_{exp} (40 K)	0.846	0.513	0.069	0.015	0.002
I_{exp} (100 K)	0.592	0.485	0.217	0.119	0.057

Theoretically, the velocity distribution in the x-direction is

$$f(v_x)=\left(\frac{m}{2\pi kT}\right)^{1/2} e^{-mv_x^2/2kT} \quad [18.5 \text{ with } M/R=m/k]$$

Therefore, as $I \propto f$, $I \propto T^{-1/2}e^{-mv_x^2/2kT}$

$$\text{Since } mv_x^2/2kT=\frac{83.80\times(1.6605\times10^{-27}\text{ kg})\times\{1.80\times(\nu/\text{Hz})\text{ m s}^{-1}\}^2}{2\times(1.381\times10^{-23}\text{ J K}^{-1})\times T}=\frac{(1.63\times10^{-2})\times(\nu/\text{Hz})^2}{T/\text{K}},$$

we can write $I \propto (T/\text{K})^{-1/2}\,e^{-1.63\times10^{-2}\times(\nu/\text{Hz})^2/(T/\text{K})}$ and draw up the following table, obtaining the constant of proportionality by fitting I to the value at $T=40$ K, $\nu=80$ Hz.

ν / Hz	20	40	80	100	120
I_{calc} (40 K)	0.80	0.49	(0.069)	0.016	0.003
I_{calc} (100 K)	0.56	0.46	0.209	0.116	0.057

The calculated values are in fair agreement with the experimental data.

P18.3 $\kappa=\tfrac{1}{3}\lambda\bar{c}C_{V,\text{m}}[\text{A}]$ [18.24] and $\bar{c}=\left(\frac{8RT}{\pi M}\right)^{1/2}$ [18.8] $\propto T^{1/2}$

Hence, $\kappa \propto T^{1/2}C_{V,\text{m}}$ and $\frac{\kappa'}{\kappa}=\left(\frac{T'}{T}\right)^{1/2}\times\left(\frac{C'_{V,\text{m}}}{C_{V,\text{m}}}\right)$

The molar heat capacities at the two temperatures are estimated with the equipartition theorem. At 300 K there are three translational degrees of freedom and two rotational degrees of freedom, which gives $C_{V,\text{m}}\approx(3+2)\tfrac{1}{2}R=\tfrac{5}{2}R$. At 10 K the rotational degrees of freedom are not significantly populated, so there are three translational degrees of freedom alone, which gives $C_{V,\text{m}}\approx\tfrac{3}{2}R$.

Therefore, $\frac{\kappa'}{\kappa}=\left(\frac{300}{10}\right)^{1/2}\times\left(\frac{5}{3}\right)=\boxed{9.1}$.

P18.5 The time constant for the exponential mass loss is

$$\tau=\left(\frac{2\pi M}{RT}\right)^{1/2}\frac{V}{A}$$
$$=\left\{\frac{2\pi\times(137.33\times10^{-3}\text{ kg mol}^{-1})}{(8.3145\text{ J K}^{-1}\text{ mol}^{-1})\times(1573\text{ K})}\right\}^{1/2}\times\left(\frac{100\times10^{-6}\text{ m}^3}{0.10\times10^{-6}\text{ m}^2}\right)=8.12\text{ s}$$

The expression for the exponential pressure decay is

$p(t)=p_0e^{-t/\tau}$ or $t=\tau\ln(p_0/p)$ where p_0 is the initial pressure

Thus, the time required for the pressure to drop to 10% of the initial value is

$$t=(8.12\text{ s})\times\ln(10)=\boxed{18.9\text{ s}}$$

P18.7 We take the natural logarithm of eqn. 18.26,

$$\eta \propto e^{E_a/RT} \quad [18.26]$$
$$\ln\eta=\text{constant}+E_a/RT$$

and recognize that a plot of $\ln \eta$ against $1/T$ has a slope equal to E_a/R. Thus, a linear regression fit of $\ln \eta$ against $1/T$, shown in Figure 18.1, yields the slope from which we calculate E_a with the expression E_a = slope × R.

$$E_a = (1222\ \text{K}) \times (8.3145\ \text{J K}^{-1}\ \text{mol}^{-1}) = \boxed{10.2\ \text{kJ mol}^{-1}}$$

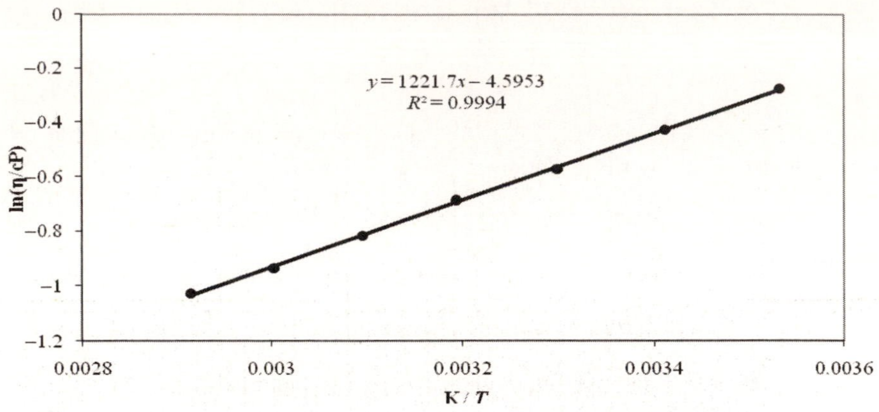

Figure 18.1

P18.9 The molar conductivity, Λ_m, is related to the conductivity, κ, by $\Lambda_m = \kappa/c$ [18.28], and the **Kohlrausch law** [18.29] indicates that molar conductivity is linear in $c^{1/2}$.

$$\Lambda_m = \Lambda_m{}^\circ - \mathcal{K}c^{1/2} \quad [18.29]$$

We draw a data table and calculate values for a plot of Λ_m against $c^{1/2}$.

c / (mol dm^{-3})	1.334	1.432	1.529	1.672	1.725
κ / (mS cm^{-1})	131	139	147	156	164
$c^{1/2}$ / (mol dm^{-3})$^{1/2}$	1.155	1.197	1.237	1.293	1.313
Λ_m / (mS m^2 mol^{-1})	9.82	9.71	9.61	9.33	9.51

The plot, shown in Figure 18.2, is linear and the linear regression fit yields the Kohlrausch parameters:

$$\Lambda_m{}^\circ = \boxed{12.78\ \text{mS m}^2\ \text{mol}^{-1}}$$

$$\mathcal{K} = \boxed{2.57\ \text{mS m}^2\ (\text{mol dm}^{-1})^{-3/2}}$$

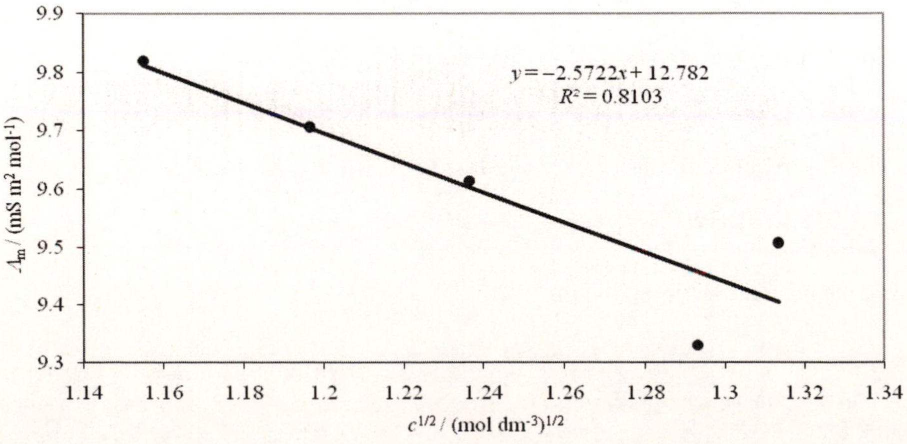

Figure 18.2

P18.11 The molar conductivity, Λ_m, is related to the conductivity, κ, by $\Lambda_m = \kappa/c$ [18.28] $= C/Rc$ where the cell constant is $C = 0.2063\ \text{cm}^{-1}$. The Kohlrausch law [18.29] indicates that molar conductivity is linear in $c^{1/2}$.

$$\Lambda_m = \Lambda_m{}^\circ - \mathcal{K}c^{1/2} \quad [18.29]$$

We draw a data table and calculate values for a plot of Λ_m against $c^{1/2}$.

c / (mol dm^{-3})	0.00050	0.0010	0.0050	0.010	0.020	0.050
R / Ω	3314	1669	342.1	174.1	89.08	37.14
$c^{1/2}$ / (mol dm^{-3})$^{1/2}$	0.0224	0.0316	0.0707	0.100	0.141	0.224
Λ_m / (mS m^2 mol^{-1})	12.45	12.36	12.06	11.85	11.58	11.11

The plot, shown in Figure 18.3, is linear, and the linear regression fit yields the intercept and slope. The intercept is the limiting molar conductivity and the slope is the negative of the Kohlrausch parameter $\mathcal{K}$:

$$\Lambda_m{}^\circ = \boxed{12.6\ \text{mS m}^2\ \text{mol}^{-1}}$$

$$\mathcal{K} = \boxed{6.66\ \text{mS m}^2\ (\text{mol dm}^{-1})^{-3/2}}$$

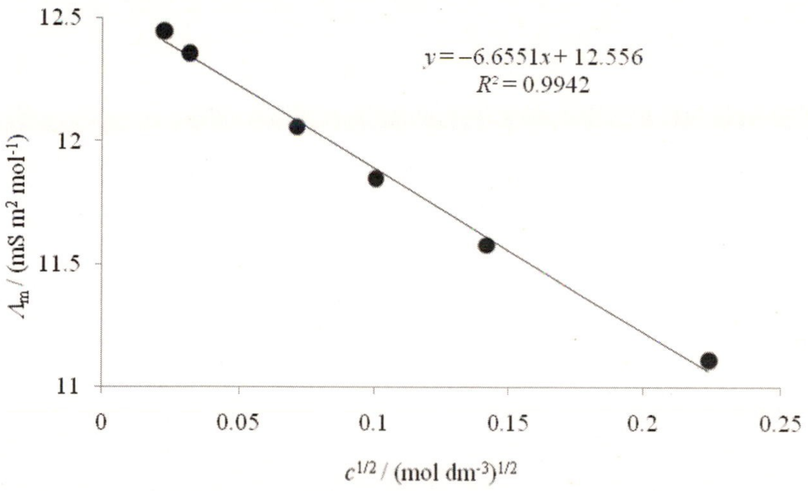

Figure 18.3

(a) $$\Lambda_m = \left((5.01+7.68) - 6.66\times(0.010)^{1/2}\right)\ \text{mS m}^2\ \text{mol}^{-1}$$

$$= \boxed{12.02\ \text{mS m}^2\ \text{mol}^{-1}}$$

(b) $$\kappa = c\Lambda_m = (10\ \text{mol m}^{-3})\times(12.02\ \text{mS m}^2\ \text{mol}^{-1}) = 120\ \text{mS m}^2\ \text{m}^{-3} = \boxed{120\ \text{mS m}^{-1}}$$

(c) $$R = \frac{C}{\kappa} = \frac{20.63\ \text{m}^{-1}}{120\ \text{mS m}^{-1}} = \boxed{172\ \Omega}$$

P18.13 The diffusion constant of a C_{60}^- ion in solution is related to the mobility of the ion and to its hydrodynamic radius a in eqn. 18.36, the Stokes formula:

$$a = \frac{ze}{6\pi\eta u} \quad [18.36]$$

$$= \frac{(1)\times(1.602\times10^{-19}\ \text{C})}{6\pi\left(0.93\times10^{-3}\ \text{kg m}^{-1}\text{s}^{-1}\right)\times\left(1.1\times10^{-8}\ \text{m}^2\ \text{V}^{-1}\ \text{s}^{-1}\right)} = 8.3\times10^{-10}\ \text{m} = \boxed{0.83\ \text{nm}}$$

This is substantially larger than the 0.5 nm van der Waals radius of a Buckminsterfullerene (C_{60}) molecule because the anion attracts a considerable hydration shell through the ion–dipole attraction to water molecules. The Stokes radius reflects the larger effective radius of the combined anion and its hydration shell.

P18.15 $\mathcal{F} = -\dfrac{RT}{c}\times\dfrac{\mathrm{d}c}{\mathrm{d}x}$ [18.40b] with the axis origin at the center of the tube.

$$RT = 2.48\times10^3\ \mathrm{J\ mol^{-1}} = 2.48\times10^3\ \mathrm{N\ m\ mol^{-1}}$$

$$c = c(x) = c_0\mathrm{e}^{-ax^2} \quad \text{where } c_0 = 0.100\ \mathrm{mol\ dm^{-3}} \quad \text{and} \quad a = 0.10\ \mathrm{cm^{-2}}$$

$$\frac{\mathrm{d}c}{\mathrm{d}x} = -2axc_0\mathrm{e}^{-ax^2} = -2axc$$

Thus, the thermodynamic force per mole is given by the expression

$$\mathcal{F} = 2aRTx = \boxed{(50.\ \mathrm{kN\ cm^{-1}\ mol^{-1}})x}$$

while the force per molecule is given by

$$\mathcal{F} = 2aRTx/N_\mathrm{A} = \boxed{(8.2\times10^{-23}\ \mathrm{kN\ cm^{-1}\ molecule^{-1}})x}$$

A plot of the absolute force per mole against x is shown in Figure 18.4. It demonstrates that mass is pushed by the thermodynamic force toward the ends of the tube where the concentration is lowest; a negative force pushes toward the left (i.e., $x < 0$), positive force pushes toward the right (i.e., $x > 0$).

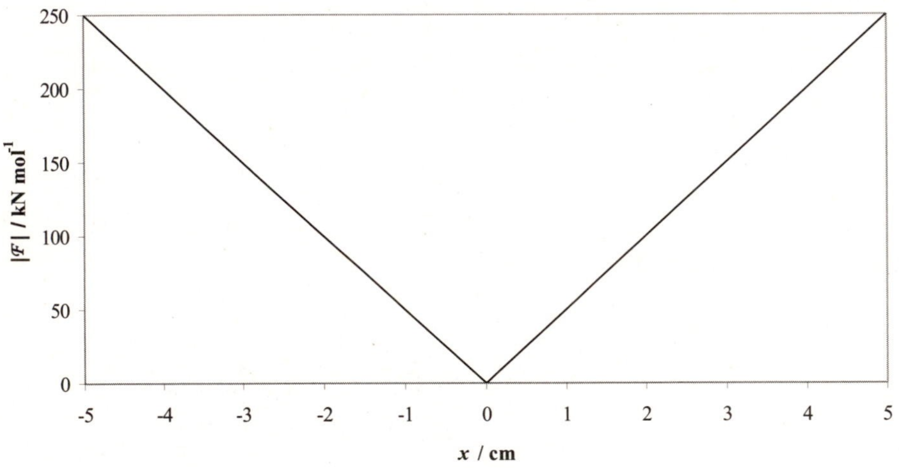

Figure 18.4

P18.17 $D = \dfrac{uRT}{zF}$ [18.45] and $a = \dfrac{ze}{6\pi\eta u}$ [18.36]

$$D = \frac{(8.3145\ \mathrm{J\ K^{-1}\ mol^{-1}})\times(298.15\ \mathrm{K})\times u}{96485\ \mathrm{C\ mol^{-1}}} = 2.569\times10^{-2}\ \mathrm{V}\times u$$

So $D/(\mathrm{cm^2\ s^{-1}}) = (2.569\times10^{-2})\times u/(\mathrm{cm^2\ s^{-1}\ V^{-1}})$

$$a = \frac{1.602\times10^{-19}\ \mathrm{C}}{(6\pi)\times(0.891\times10^{-3}\ \mathrm{kg\ m^{-1}\ s^{-1}})\times u}$$

$$= \frac{9.54\times10^{-18}\ \mathrm{C\ kg^{-1}\ m\ s}}{u} = \quad (1\ \mathrm{J} = 1\ \mathrm{C\ V},\ 1\ \mathrm{J} = 1\ \mathrm{kg\ m^2\ s^{-2}})$$

so $\quad a/\mathrm{m} = \dfrac{9.54\times10^{-14}}{u/\mathrm{cm^2\ s^{-1}\ V^{-1}}}$

and therefore $a/\mathrm{pm} = \dfrac{9.54\times10^{-2}}{u/\mathrm{cm^2\ s^{-1}\ V^{-1}}}$

We can now draw up the following table using data from Table 18.5.

	Li^+	Na^+	K^+	Rb^+
$u/(10^{-4}\ cm^2\ s^{-1}\ V^{-1})$	4.01	5.19	7.62	7.92
$D/10^{-5}\ cm^2\ s^{-1}$	1.03	1.33	1.96	2.04
a / pm	238	184	125	120

The ionic radii themselves (i.e. their crystallographic radii) are

	Li^+	Na^+	K^+	Rb^+
r_+ / pm	59	102	138	149

and it would seem that K^+ and Rb^+ have effective hydrodynamic radii that are smaller than their ionic radii. The effective hydrodynamic and ionic volumes of Li^+ and Na^+ are $\frac{4\pi}{3}\pi a^3$ and $\frac{4\pi}{3}\pi r_+^3$, respectively, so the volumes occupied by hydrating water molecules are

(a) Li^+: $\Delta V = \left(\frac{4\pi}{3}\right)\times(238^3 - 59^3)\times10^{-36}\ \text{m}^3 = 5.5\bar{6}\times10^{-29}\ \text{m}^3$

(b) Na^+: $\Delta V = \left(\frac{4\pi}{3}\right)\times(184^3 - 102^3)\times10^{-36}\ \text{m}^3 = 2.1\bar{6}\times10^{-29}\ \text{m}^3$

The volume occupied by a single H_2O molecule is approximately $\left(\frac{4\pi}{3}\right)\times(150\,\text{pm})^3 = 1.4\times10^{-29}\ \text{m}^3$. Therefore, Li^+ has about $\boxed{\text{four}}$ firmly attached H_2O molecules whereas Na^+ has only $\boxed{\text{one or two}}$ (according to this analysis).

P18.19 $c(r,t) = \dfrac{n_0}{8(\pi Dt)^{3/2}}\text{e}^{-r^2/4Dt}$ [18.55] where $n_0 = (10.0\ \text{g})\times(1\ \text{mol}/342.3\ \text{g}) = 0.0292\ \text{mol}$

Using $D = 5.22\times10^{-6}\ \text{cm}^2\ \text{s}^{-1} = 5.22\times10^{-8}\ \text{dm}^2\ \text{s}^{-1}$ and $r = 10$ cm, the working equation becomes

$$c(10\ \text{cm},t) = \frac{5.50\times10^7\ \text{mol dm}^{-3}}{(t/\text{s})^{3/2}}\text{e}^{-4.79\times10^6/(t/\text{s})}$$

(a) 1 hr = 3600 s

$$c(10\ \text{cm},1\ \text{hr}) = \frac{5.50\times10^7\ \text{mol dm}^{-3}}{(3600)^{3/2}}\text{e}^{-4.79\times10^6/(3600)} = \boxed{0.00\ \text{mol dm}^{-3}}$$

(b) 1 wk = 6.048×10^5 s

$$c(10\ \text{cm},1\ \text{wk}) = \frac{5.50\times10^7\ \text{mol dm}^{-3}}{(6.048\times10^5)^{3/2}}\text{e}^{-4.79\times10^6/(6.048\times10^5)} = \boxed{4.25\times10^{-5}\ \text{mol dm}^{-3}}$$

Solutions to theoretical problems

P18.21 The most probable speed of a gas molecule corresponds to the condition that the Maxwell distribution be a maximum (it has no minimum); hence we find it by setting the first derivative of the function equal to zero and solving for the value of v for which this condition holds.

$$f(v) = 4\pi\left(\frac{m}{2\pi kT}\right)^{3/2} v^2 e^{-mv^2/2kT}\ [18.4] = \text{const}\times v^2 e^{-mv^2/2kT} \quad [M/R = m/k]$$

$$\frac{\text{d}f(v)}{\text{d}v} = 0 \quad \text{when} \left(2 - \frac{mv^2}{kT}\right) = 0$$

So $\boxed{v(\text{most probable}) = c^* = \left(\frac{2kT}{m}\right)^{1/2} = \left(\frac{2RT}{M}\right)^{1/2}}$ [18.9].

The average kinetic energy corresponds to the average of $\tfrac{1}{2}mv^2$. The average is obtained by determining $\langle v^2\rangle = \int_0^\infty v^2 f(v)dv$ [18.7] $= 4\pi\left(\frac{m}{2\pi kT}\right)^{3/2}\int_0^\infty v^4 e^{-mv^2/2kT}\,dv$ [18.4].

The integral evaluates to $\frac{3\pi^{1/2}}{8}\left(\frac{m}{2kT}\right)^{-5/2}$. Then $\langle v^2\rangle = 4\pi\left(\frac{m}{2\pi kT}\right)^{3/2}\times\frac{3\pi^{1/2}}{8}\left(\frac{m}{2kT}\right)^{-5/2} = \frac{3kT}{m}$.

Thus, $\langle\varepsilon\rangle = \tfrac{1}{2}m\langle v^2\rangle = \tfrac{3}{2}kT$.

P18.23 We proceed as in *Justification* 18.2 except that to get the three-dimensional distribution, instead of taking a product of three one-dimensional distributions, we take a product of two one-dimensional distributions.

$$f(v_x, v_y)\mathrm{d}v_x\mathrm{d}v_y = f(v_x^2)f(v_y^2)\mathrm{d}v_x dv_y = \left(\frac{m}{2\pi kT}\right)e^{-mv^2/2kT}\mathrm{d}v_x\mathrm{d}v_y$$

where $v^2 = v_x^2 + v_y^2$. The probability $f(v)\mathrm{d}v$ that the molecules have a two-dimensional speed, v, in the range v to $v+\mathrm{d}v$ is the sum of the probabilities that it is in any of the area elements $\mathrm{d}v_x\mathrm{d}v_y$ in the circular shell of radius v. The sum of the area elements is the area of the circular shell of radius v and thickness $\mathrm{d}v$, which is $\pi(v+\mathrm{d}v)^2 - \pi v^2 = 2\pi v\,\mathrm{d}v$. Therefore,

$$\boxed{f(v) = \left(\frac{m}{kT}\right)ve^{-mv^2/2kT}} \quad \left[\frac{M}{R} = \frac{m}{k}\right]$$

The mean speed is determined as

$$\bar{c} = \int_0^\infty vf(v)\,\mathrm{d}v = \left(\frac{m}{kT}\right)\int_0^\infty v^2e^{-mv^2/2kT}\mathrm{d}v = \left(\frac{m}{kT}\right)\times\left(\frac{\pi^{1/2}}{4}\right)\times\left(\frac{2kT}{m}\right)^{3/2} \quad \text{[standard integral]}$$

$$= \boxed{\left(\frac{\pi kT}{2m}\right)^{1/2} \text{ or } \left(\frac{\pi RT}{2M}\right)^{1/2}}$$

P18.25 Rewriting eqn. 18.4 with $M/R = m/k$,

$$f(v) = 4\pi\left(\frac{m}{2\pi kT}\right)^{3/2}v^2e^{-mv^2/2kT}$$

The proportion of molecules with speeds less than c is

$$P = \int_0^c f(v)\,\mathrm{d}v = 4\pi\left(\frac{m}{2\pi kT}\right)^{3/2}\int_0^c v^2e^{-mv^2/2kT}\mathrm{d}v$$

Defining $a \equiv m/2kT$,

$$P = 4\pi\left(\frac{a}{\pi}\right)^{3/2}\int_0^c v^2e^{-av^2}\,\mathrm{d}v = -4\pi\left(\frac{a}{\pi}\right)^{3/2}\frac{\mathrm{d}}{\mathrm{d}a}\int_0^c e^{-av^2}\,\mathrm{d}v$$

Defining $\chi^2 \equiv av^2$. Then $\mathrm{d}v = a^{-1/2}\mathrm{d}\chi$ and

$$P = -4\pi\left(\frac{a}{\pi}\right)^{3/2}\frac{\mathrm{d}}{\mathrm{d}a}\left\{\frac{1}{a^{1/2}}\int_0^{ca^{1/2}}e^{-\chi^2}\,\mathrm{d}\chi\right\}$$

$$= -4\pi\left(\frac{a}{\pi}\right)^{3/2}\left\{-\frac{1}{2}\left(\frac{1}{a}\right)^{3/2}\int_0^{ca^{1/2}}e^{-\chi^2}\,\mathrm{d}\chi + \left(\frac{1}{a}\right)^{1/2}\frac{\mathrm{d}}{\mathrm{d}a}\int_0^{ca^{1/2}}e^{-\chi^2}\,\mathrm{d}\chi\right\}$$

Then we use $\int_0^{ca^{1/2}} e^{-\chi^2}\,d\chi = (\pi^{1/2}/2)\,\mathrm{erf}(ca^{1/2})$

$$\frac{d}{da}\int_0^{ca^{1/2}} e^{-\chi^2}\,d\chi = \left(\frac{dca^{1/2}}{da}\right)\times(e^{-c^2 a}) = \frac{1}{2}\left(\frac{c}{a^{1/2}}\right)e^{-c^2 a}$$

where we have used $\frac{d}{dz}\int_0^z f(y)dy = f(z)$.

Substituting and canceling, we obtain $P = \mathrm{erf}(ca^{1/2}) - (2ca^{1/2}/\pi^{1/2})e^{-c^2 a}$.

Now, $c = (3kT/m)^{1/2}$, so $ca^{1/2} = (3kT/m)^{1/2}\times(m/2kT)^{1/21/2} = (3/2)^{1/2}$, and

$$P = \mathrm{erf}\left(\sqrt{\frac{3}{2}}\right) - \left(\frac{6}{\pi}\right)^{1/2} e^{-3/2} = 0.92 - 0.31 = \boxed{0.61}$$

Therefore, (b) $\boxed{61\%}$ of the molecules have a speed less than the root mean square speed and (a) $\boxed{39\%}$ have a speed greater than the root mean square speed. (c) For the proportions in terms of the mean speed $\bar{c}$, replace c by $\bar{c} = (8kT/\pi m)^{1/2} = (8/3\pi)^{1/2}c$, so $\bar{c}a^{1/2} = 2/\pi^{1/2}$.

Then $P = \mathrm{erf}(\bar{c}a^{1/2}) - (2\bar{c}a^{1/2}/\pi^{1/2})\times(e^{-\bar{c}^2 a}) = \mathrm{erf}(2/\pi^{1/2}) - (4/\pi)e^{-4/\pi} = 0.889 - 0.356 = \boxed{0.533}$.

That is, $\boxed{53\%}$ of the molecules have a speed less than the mean, and $\boxed{47\%}$ have a speed greater than the mean.

P18.27

$$\langle v^n\rangle = \int_0^\infty v^n f(v)dv\ [18.7] = 4\pi\left(\frac{m}{2\pi kT}\right)^{3/2}\int_0^\infty v^{n+2}e^{-mv^2/2kT}\,dv\ [18.4]$$

The integral of the above expression must be handled in a manner that depends on whether n is odd or even. When n is an odd, positive integer (n = 1, 3, 5, ...), let $n + 2 = 2p + 1$ where p = 0, 1, 2, 3, ... and the integral is given by

$$\int_0^\infty v^{n+2}e^{-mv^2/2kT}\,dv = \int_0^\infty v^{2p+1}e^{-mv^2/2kT}\,dv = \tfrac{1}{2}p!\left(\frac{2kT}{m}\right)^{p+1}\ \text{[standard integral]}$$

$$= \tfrac{1}{2}\left(\frac{n+1}{2}\right)!\left(\frac{2kT}{m}\right)^{\frac{n+3}{2}}$$

Thus,

$$\langle v^n\rangle = 4\pi\left(\frac{m}{2\pi kT}\right)^{3/2}\times\tfrac{1}{2}\left(\frac{n+1}{2}\right)!\left(\frac{2kT}{m}\right)^{\frac{n+3}{2}}$$

$$= \boxed{\frac{2}{\pi^{1/2}}\left(\frac{n+1}{2}\right)!\left(\frac{2kT}{m}\right)^{n/2}}\ \text{when } n \text{ is an odd, positive integer}$$

When n is an even, positive integer (n = 0, 2, 4, ...), let $n + 2 = 2p$, where p = 0, 1, 2, 3, ... , and the integral is given by

$$\int_0^\infty v^{n+2}e^{-mv^2/2kT}\,dv = \int_0^\infty v^{2p}e^{-mv^2/2kT}\,dv = \frac{(2p-1)!!}{2^{p+1}}\left(\frac{2kT}{m}\right)^p\left(\frac{2\pi kT}{m}\right)^{1/2}\ \text{[standard integral]}$$

$$= \frac{\pi^{1/2}(n+1)!!}{2^{\frac{n+4}{2}}}\left(\frac{2kT}{m}\right)^{\frac{n+3}{2}}$$

Thus,

$$\langle v^n \rangle = 4\pi\left(\frac{m}{2\pi kT}\right)^{3/2} \times \frac{\pi^{1/2}(n+1)!!}{2^{\frac{n+4}{2}}}\left(\frac{2kT}{m}\right)^{\frac{n+3}{2}}$$

$$= \boxed{\frac{(n+1)!!}{2^{n/2}}\left(\frac{2kT}{m}\right)^{n/2}} \quad \text{when } n \text{ is an even, positive integer}$$

P18.29 The rate constant, k_r, for a transport process in which a molecule and its hydration sphere move a single step is governed by the activation energy for the step, E_a, where the general definition of activation energy is

$$E_a = RT^2\left(\frac{d\ln k_r}{dT}\right) \quad \text{[see Section 20.1]}$$

We expect that larger viscosities should retard the rate constant, so let us assume that k_r is inversely proportional to the viscosity. Then,

$$\ln k_r = \text{constant} - \ln(\eta)$$

$$\frac{d\ln k_r}{dT} = -\frac{1}{\eta}\frac{d\eta}{dT}$$

$$E_a = -\frac{RT^2}{\eta}\frac{d\eta}{dT}$$

$$\frac{d\eta}{\eta} = -\left(\frac{E_a}{R}\right)\frac{dT}{T^2}$$

In the case for which E_a is independent of temperature, the above working equation can be integrated. We choose the lower integration limit to be the viscosity at a reference temperature T_{ref}, η_{ref}. The upper integration limit is the viscosity at temperature T, η.

$$\int_{\eta_{ref}}^{\eta} \frac{d\eta}{\eta} = -\left(\frac{E_a}{R}\right)\int_{T_{ref}}^{T}\frac{dT}{T^2}$$

$$\ln\left(\frac{\eta}{\eta_{ref}}\right) = \left(\frac{E_a}{R}\right)\times\left(\frac{1}{T} - \frac{1}{T_{ref}}\right) \quad \text{or} \quad \boxed{\eta = Ae^{E_a/RT} \text{ where } A = \eta_{ref}e^{-E_a/RT_{ref}}}$$

Thus, we see that, when E_a is a constant, the pre-exponential factor A is a constant and $\eta \propto e^{E_a/RT}$ [18.26]. This demonstrates that the general definition $E_a = RT^2\left(\frac{d\ln k_r}{dT}\right)$ is compatible with eqn. [18.26] when the activation energy is a constant.

We explore the possibility that the Problem 18.8 empirical equation for the viscosity of water reflects an activation energy that has a dependence on temperature by applying the above working equation prior to the constancy assumption. The reference temperature is 20° C and from the *CRC Handbook* (71st ed., 1990–1991), η_{20} = 1002 μPa s = 1.002×10^{-3} kg m^{-1} s^{-1} (the value is not actually necessary in the following calculations).

$$\eta(\theta) = \eta_{20}10^{\{a(20-\theta/°\text{C})-b(20-\theta/°\text{C})^2\}/(\theta/°\text{C}+c)} \quad \text{[P18.8] where } a = 1.3272,\ b = 0.001053, \text{ and } c = 105$$

$$\eta(x) = \eta_{20}10^{f(x)} \text{ where } x = \theta/°\text{C and } f(x) = \{a(20-x)-b(20-x)^2\}/(x+c)$$

$$\frac{d\eta}{dx} = \eta_{20}\frac{d}{dx}(10^{f(x)}) = \eta_{20}10^{f(x)}\times\ln(10)\times\frac{d}{dx}(f(x)) = \eta(x)\times\ln(10)\times\frac{d}{dx}(f(x))$$

$$= \eta(x)\times\ln(10)\times\left\{\frac{-a+2b(20-x)}{x+c} - \frac{a(20-x)-b(20-x)^2}{(x+c)^2}\right\}$$

$$E_a = -\frac{RT^2}{\eta}\frac{d\eta}{dT} = -\frac{R\times(x+273.15)^2\,\text{K}}{\eta(x)}\frac{d\eta(x)}{d(x)}$$

$$= -\left\{R\times(x+273.15)^2\,\text{K}\right\}\times\ln(10)\times\left\{\frac{-a+2b(20-x)}{x+c} - \frac{a(20-x)-b(20-x)^2}{(x+c)^2}\right\}$$

This equation is used to make the plot of E_a against θ/° C (i.e., x) shown in Fig. 18.5. The activation energy drops from 17.5 kJ mol^{-1} at 20° C to 12.3 kJ mol^{-1} at 100° C. This decrease may be caused by the density decrease that occurs across this temperature range because the increased average intermolecular distance may cause a decrease in the hydrogen bond strength between water molecules. There may also be a decrease in the hydration sphere of a molecule, thereby making movement easier.

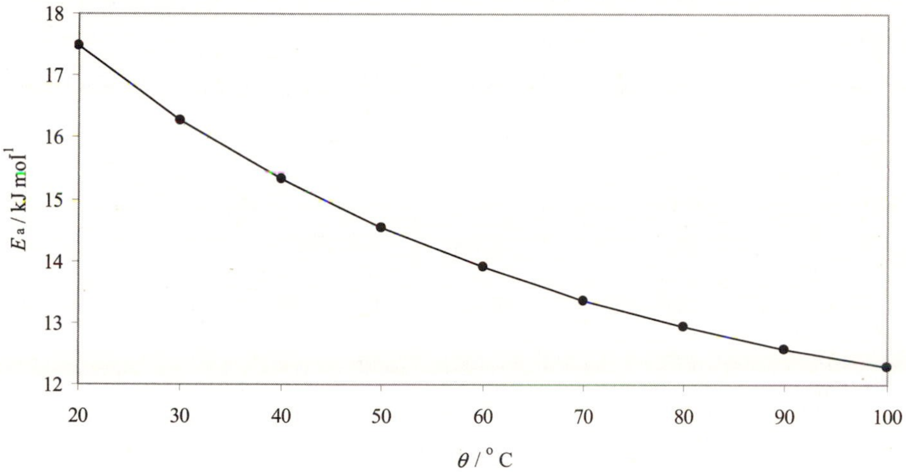

Figure 18.5

P18.31 The generalized diffusion equation is $\frac{\partial c}{\partial t} = D\frac{\partial^2 c}{\partial x^2} - v\frac{\partial c}{\partial x}$ [18.53].

We confirm that $c(x,t) = \frac{a}{t^{1/2}}e^{-b(x-x_0-vt)^2/t}$, where $a = \frac{c_0}{(4\pi D)^{1/2}}$ and $b = \frac{1}{4D}$, is a solution to the generalized diffusion equation by taking the partial derivative w/r/t time and both the first and second partials w/r/t position to find whether they are related by eqn. 18.53.

$$\frac{\partial c}{\partial t} = -\left(\frac{1}{2}\right)\times\left(\frac{a}{t^{3/2}}\right)e^{-b(x-x_0-vt)^2/t} + \left(\frac{a}{t^{1/2}}\right)\times\left\{\frac{b(x-x_0-vt)}{t}\right\}\times\left\{2v+\frac{(x-x_0-vt)}{t}\right\}e^{-b(x-x_0-vt)^2/t}$$

$$= -\frac{c}{2t} + \left\{\frac{b(x-x_0-vt)}{t}\right\}\times\left\{2v+\frac{(x-x_0-vt)}{t}\right\}\times c$$

$$\frac{\partial c}{\partial x} = \left(\frac{a}{t^{1/2}}\right)\times\left\{\frac{-2b(x-x_0-vt)}{t}\right\}e^{-b(x-x_0-vt)^2/t} = \left\{\frac{-2b(x-x_0-vt)}{t}\right\}\times c$$

$$\frac{\partial^2 c}{\partial x^2} = \frac{\partial}{\partial x}\frac{\partial c}{\partial x} = \frac{\partial}{\partial x}\left[\left\{\frac{-2b(x-x_0-vt)}{t}\right\}\times c\right]$$

$$= \frac{-2b}{t}\times c + \left\{\frac{-2b(x-x_0-vt)}{t}\right\}\times\frac{\partial c}{\partial x}$$

$$= \frac{-2b}{t}\times c + \left\{\frac{-2b(x-x_0-vt)}{t}\right\}\times\left\{\frac{-2b(x-x_0-vt)}{t}\right\}\times c$$

$$= \frac{-2b}{t}\times c + \left\{\frac{2b(x-x_0-vt)}{t}\right\}^2\times c$$

Thus,

$$D\frac{\partial^2 c}{\partial x^2}-v\frac{\partial c}{\partial x}=D\left\{\frac{-2b}{t}\times c+\left\{\frac{2b(x-x_0-vt)}{t}\right\}^2\times c\right\}-v\left\{\frac{-2b(x-x_0-vt)}{t}\right\}\times c$$

$$=\frac{-1}{2t}\times c+b\left\{\frac{(x-x_0-vt)}{t}\right\}^2\times c+v\left\{\frac{2b(x-x_0-vt)}{t}\right\}\times c\quad [D=1/4b]$$

$$=\frac{-c}{2t}+\left\{\frac{b(x-x_0-vt)}{t}\right\}\times\left\{2v+\frac{(x-x_0-vt)}{t}\right\}\times c$$

$$=\frac{\partial c}{\partial t}\text{ as required}$$

Initially the material is concentrated at $x=x_0$ because $c=0$ for $|x-x_0|>0$ when $t=0$ on account of the very strong exponential factor $\left(e^{-b(x-x_0)^2/t}\rightarrow 0\text{ more strongly than }\frac{1}{t^{1/2}}\rightarrow\infty\right)$.

The term $x_0 + vt$ in the concentration expression is the movement of the centroid due to fluid flow, x_{centroid}. To prepare a very general set of concentration profiles at a series of times without specifying x_0, v, or D, define z and C as follows.

$$C\equiv\frac{c}{c_0/(4\pi D\text{ h})^{1/2}}\quad\text{and}\quad z\equiv\frac{x-x_{\text{centroid}}}{(4D\text{ h})^{1/2}}$$

The hour (h) has been chosen for the unit because of the slow pace of diffusion activity. With these definitions, the concentration expression becomes

$$C(z,t)=\frac{1}{(t/\text{h})^{1/2}}e^{-z^2/(t/\text{h})}$$

Concentration profiles as C against t at various times (1, 5, and 20 h) are displayed in Figure 18.6.

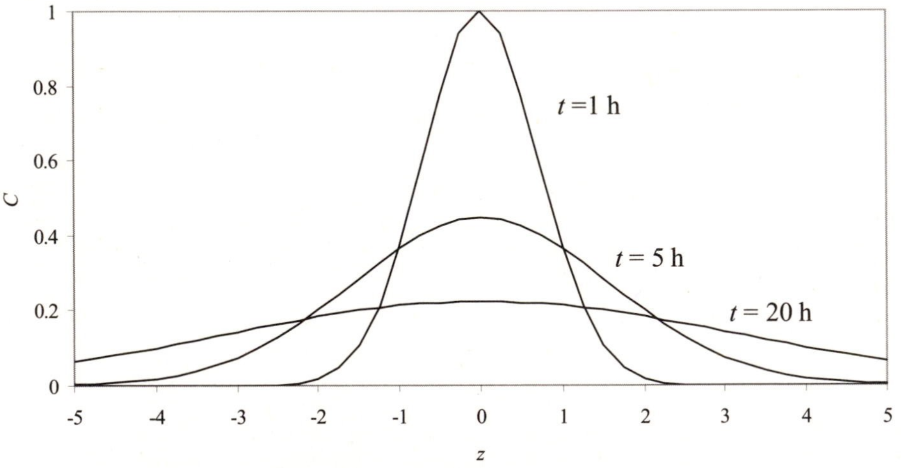

Figure 18.6

P18.33 $P(x)=\frac{N!}{\left\{\frac{1}{2}(N+s)\right\}!\left\{\frac{1}{2}(N-s)\right\}!\,2^N}$ [*Justification* 18.7], $s=\frac{x}{\lambda}$

$$P(6\lambda)=\frac{N!}{\left\{\frac{1}{2}(N+6)\right\}!\left\{\frac{1}{2}(N-6)\right\}!\,2^N}$$

(a) $N=4,\quad P(6\lambda)=\boxed{0}\quad [m!=\infty\text{ for }m<0]$

(b) $N=6,\quad P(6\lambda)=\frac{6!}{6!\,0!\,2^6}=\frac{1}{2^6}=\frac{1}{64}=\boxed{0.016}\quad [0!=1]$

(c) $N=12,\quad P(6\lambda)=\frac{12!}{9!\,3!\,2^{12}}=\frac{12\times11\times10}{3\times2\times2^{12}}=\boxed{0.054}$

P18.35

$$P=\frac{N!}{\{\tfrac{1}{2}(N+n)\}!\{\tfrac{1}{2}(N-n)\}!\,2^N}$$

The intermediate mathematical manipulations of *Justification* 18.7 begin with the above expression. Simplification of the expression proceeds by taking the natural logarithm of the expression, applying Stirling's approximation to each term that has the ln(x!) form, checking for term cancelations, and simplifying using basic logarithm properties.

Stirling's approximation: $\ln x! = \ln(2\pi)^{1/2}+(x+\tfrac{1}{2})\ln x - x$

Basic logarithm properties: $\ln(x\times y)=\ln x+\ln y$

$\ln(x/y)=\ln x-\ln y$

$\ln(x^y)=y\ln x$

Taking the natural logarithm and applying Stirling's formula gives

$$\begin{aligned}
\ln P &= \ln\left\{\frac{N!}{\{\tfrac{1}{2}(N+n)\}!\{\tfrac{1}{2}(N-n)\}!\,2^N}\right\}\\
&= \ln N! - \ln\left(\{\tfrac{1}{2}(N+n)\}!\right)-\ln\left(\{\tfrac{1}{2}(N-n)\}!\right)-\ln 2^N\\
&= \cancel{\ln(2\pi)^{1/2}}+(N+\tfrac{1}{2})\ln N-\cancel{N}\\
&\quad -\left[\cancel{\ln(2\pi)^{1/2}}+\{\tfrac{1}{2}(N+n)+\tfrac{1}{2}\}\ln\{\tfrac{1}{2}(N+n)\}-\tfrac{1}{2}(\cancel{N}+\cancel{n})\right]\\
&\quad -\left[\ln(2\pi)^{1/2}+\{\tfrac{1}{2}(N-n)+\tfrac{1}{2}\}\ln\{\tfrac{1}{2}(N-n)\}-\tfrac{1}{2}(\cancel{N}-\cancel{n})\right]-\ln 2^N
\end{aligned}$$

$$\begin{aligned}
\ln P &= (N+\tfrac{1}{2})\ln N-\ln(2\pi)^{1/2}-\ln 2^N\\
&\quad -\{\tfrac{1}{2}(N+n)+\tfrac{1}{2}\}\ln\{\tfrac{1}{2}(N+n)\}-\{\tfrac{1}{2}(N-n)+\tfrac{1}{2}\}\ln\{\tfrac{1}{2}(N-n)\}\\
&= \ln\left\{\frac{(N/2)^{N+\frac{1}{2}}}{\pi^{1/2}}\right\}-\tfrac{1}{2}\{N+n+1\}\ln\left\{\frac{N}{2}\left(1+\frac{n}{N}\right)\right\}-\tfrac{1}{2}\{N-n+1\}\ln\left\{\frac{N}{2}\left(1-\frac{n}{N}\right)\right\}\\
&= \ln\left\{\frac{(N/2)^{N+\frac{1}{2}}}{\pi^{1/2}}\right\}-\tfrac{1}{2}\{N+n+1\}\left\{\ln\left(\frac{N}{2}\right)+\ln\left(1+\frac{n}{N}\right)\right\}-\tfrac{1}{2}\{N-n+1\}\left\{\ln\left(\frac{N}{2}\right)+\ln\left(1-\frac{n}{N}\right)\right\}\\
&= \ln\left\{\frac{(N/2)^{N+\frac{1}{2}}}{\pi^{1/2}}\right\}-\tfrac{1}{2}\{N+\cancel{n}+1\}\ln\left(\frac{N}{2}\right)-\tfrac{1}{2}\{N+n+1\}\ln\left(1+\frac{n}{N}\right)\\
&\qquad -\tfrac{1}{2}\{N-\cancel{n}+1\}\ln\left(\frac{N}{2}\right)-\tfrac{1}{2}\{N-n+1\}\ln\left(1-\frac{n}{N}\right)\\
&= \ln\left\{\frac{(N/2)^{N+\frac{1}{2}}}{\pi^{1/2}}\right\}-\{N+1\}\ln\left(\frac{N}{2}\right)-\tfrac{1}{2}\{N+n+1\}\ln\left(1+\frac{n}{N}\right)-\tfrac{1}{2}\{N-n+1\}\ln\left(1-\frac{n}{N}\right)\\
&= \ln\left\{\frac{(N/2)^{N+\frac{1}{2}}}{(N/2)^{N+1}\pi^{1/2}}\right\}-\tfrac{1}{2}\{N+n+1\}\ln\left(1+\frac{n}{N}\right)-\tfrac{1}{2}\{N-n+1\}\ln\left(1-\frac{n}{N}\right)
\end{aligned}$$

$$\ln P=\ln\left(\frac{2}{\pi N}\right)^{\frac{1}{2}}-\tfrac{1}{2}\{N+n+1\}\ln\left(1+\frac{n}{N}\right)-\tfrac{1}{2}\{N-n+1\}\ln\left(1-\frac{n}{N}\right)$$

P18.37 For a three-dimensional, freely jointed chain of N repeating units each of length l, the probability that the ends lie in the range r to $r + dr$ is $f(r)dr$ where

$$f(r) = 4\pi\left(\frac{a}{\pi^{1/2}}\right)^3 r^2 e^{-a^2r^2} \qquad a = \left(\frac{3}{2Nl^2}\right)^{1/2} \qquad [18.74]$$

$$\langle R^2\rangle = \int_0^\infty r^2 f(r)\,dr$$

$$= 4\pi\left(\frac{a}{\pi^{1/2}}\right)^3 \int_0^\infty r^4 e^{-a^2r^2}\,dr = 4\pi\left(\frac{a}{\pi^{1/2}}\right)^3 \times \left(\frac{3\pi^{1/2}}{8a^5}\right) \text{ [standard integral]}$$

$$= \frac{3}{2a^2} = Nl^2$$

Thus,

$$R_{\text{rms}} = \langle R^2\rangle^{1/2} = \boxed{N^{1/2}l}$$

Solutions to applications

P18.39 (a) Each hydrogen atom contributes one electron and one proton to the "free" particle plasma. Consequently, the number of electrons in volume V equals the number of protons and the total number of free particles is given by $N = N_p + N_e = 2N_p$ or, alternatively, $n = n_p + n_e = 2n_p$. Since the perfect gas law is applicable and because the mass of an electron is negligible compared to the mass of a proton,

$$p = \frac{nRT}{V} = \frac{2n_pRT}{V} = \frac{2m_pRT}{VM_p} = \frac{2\rho_pRT}{M_p} \text{ where } \rho_p \text{ is the mass density of the protons}$$

$$= \frac{2\rho RT}{M_p} \text{ where total mass density, } \rho\text{, equals } \rho_p \text{ because electron mass is negligible}$$

$$= \frac{2(1.20\times10^3\text{ kg m}^{-3})\times(8.3145\text{ J mol}^{-1}\text{ K}^{-1})\times(3.6\times10^6\text{ K})}{1.0\times10^{-3}\text{ kg mol}^{-1}} = \boxed{7.2\times10^{13}\text{ Pa}}$$

(b) From *Justification* 18.1, $p = nM\langle v_x^2\rangle/V$ and $\langle v_x^2\rangle = \tfrac{1}{3}c^2$.

Thus,

$$p = \frac{m\langle v_x^2\rangle}{V} = \frac{mc^2}{3V} = \frac{2}{3}\left(\frac{\tfrac{1}{2}mc^2}{V}\right) = \boxed{\tfrac{2}{3}\rho_k} \text{ where } \rho_k = \frac{\tfrac{1}{2}mc^2}{V}$$

(c) $\rho_k = \tfrac{3}{2}p = \tfrac{3}{2}\times(7.2\times10^{13}\text{ Pa}) = 1.1\times10^{14}\text{ J m}^{-3} = \boxed{1.1\times10^2\text{ TJ m}^{-3}}$

This is about 1 billion times larger than the translational energy density of the Earth's atmosphere on a warm day.

(d) Each carbon atom contributes 6 electrons and 1 nucleus to the "free" particle plasma. Consequently, the number of electrons in volume V equals 6 times the number of carbon nuclei and the total number of free particles is given by $N = N_C + 6N_e = 7N_C$ or, alternatively, $n = n_C + 6n_e = 7n_C$. Since the perfect gas law is applicable and because the mass of the electrons is negligible compared to the mass of carbon nuclei,

$$p = \frac{nRT}{V} = \frac{7n_CRT}{V} = \frac{7m_CRT}{VM_C} = \frac{7\rho_cRT}{M_C} \text{ where } \rho_C \text{ is the mass density of carbon nuclei}$$

$$= \frac{7\rho RT}{M_C} \text{ where total mass density, } \rho\text{, equals } \rho_C \text{ because electron mass is negligible}$$

$$= \frac{7(1.20\times10^3\text{ kg m}^{-3})\times(8.3145\text{ J mol}^{-1}\text{ K}^{-1})\times(3500\text{ K})}{12\times10^{-3}\text{ kg mol}^{-1}} = \boxed{2.0\times10^{10}\text{ Pa}}$$

(e) $p=\dfrac{\rho RT}{M}$ where ρ is the mass density

$$=\frac{(1.20\times10^3\ \text{kg m}^{-3})\times(8.3145\ \text{J mol}^{-1}\ \text{K}^{-1})\times(3500\ \text{K})}{12.0\times10^{-3}\ \text{kg mol}^{-1}}=\boxed{2.9\times10^9\ \text{Pa}}$$

P18.41 Dry atmospheric air is 78.08% N_2, 20.95% O_2, 0.93% Ar, 0.03% CO_2, plus traces of other gases. Nitrogen, oxygen, and carbon dioxide contribute 99.06% of the molecules in a volume with each molecule contributing an average rotational energy equal to kT. The rotational energy density is given by

$$\rho_R=\frac{E_R}{V}=\frac{0.9906N\langle\varepsilon^R\rangle}{V}=\frac{0.9906\langle\varepsilon^R\rangle pN_A}{RT}$$

$$=\frac{0.9906\ kTpN_A}{RT}=0.9906p$$

$$=0.9906(1.013\times10^5\ \text{Pa})=0.1004\ \text{J cm}^{-1}$$

The total energy density (translational plus rotational) is

$$\rho_T=\rho_K+\rho_R=0.15\ \text{J cm}^{-3}+0.10\ \text{J cm}^{-3}=\boxed{0.25\ \text{J cm}^{-3}}$$

P18.43 $c(x,t)=c_0+(c_s-c_0)\{1-\text{erf}(\xi)\}$ where $\xi(x,t)=x/(4Dt)^{1/2}$ and $\text{erf}(\xi)=1-\frac{2}{\pi^{1/2}}\int_\xi^\infty e^{y^2}\,\text{d}y$

For $c(x,t)$ to be the correct solution of this diffusion problem, it must satisfy the boundary condition, the initial condition, and the diffusion equation (eqn. 18.50). At the boundary, $x=0$, $\xi=0$, and $\text{erf}(0)=1-\frac{2}{\pi^{1/2}}\int_0^\infty e^{-y^2}\,\text{d}y=1-\left(\frac{2}{\pi^{1/2}}\right)\times\left(\frac{\pi^{1/2}}{2}\right)=0.$

Thus, $c(0,t)=c_0+(c_s-c_0)\{1-0\}=c_s$. The boundary condition is satisfied. At the initial time ($t=0$), $\xi(x,0)=\infty$ and $\text{erf}(\infty)=1$. Thus, $c(x,0)=c_0+(c_s-c_0)\{1-1\}=c_0$. The initial condition is satisfied. We must find the analytical forms for $\partial c/\partial t$ and $\partial^2 c/\partial x^2$. If they are proportional with a constant of proportionality equal to D, then $c(x,t)$ satisfies the diffusion equation.

$$\frac{\partial c(x,t)}{\partial x}=D\left[\frac{1}{2}\frac{(c_s-c_0)x}{\sqrt{\pi}(Dt)^{3/2}}e^{-x^2/4Dt}\right]$$

$$\frac{\partial^2 c(x,t)}{\partial x^2}=\left[\frac{1}{2}\frac{(c_s-c_0)x}{\sqrt{\pi}(Dt)^{3/2}}e^{-x^2/4Dt}\right]$$

The constant of proportionality between the partials equals D and we conclude that the suggested solution satisfies the diffusion equation. Diffusion through alveoli sites (about 1 cell thick) of oxygen and carbon dioxide between lungs and blood capillaries (also about 1 cell thick) occurs through about 0.075 mm (the diameter of a red blood cell). So we will examine diffusion profiles for $0\le x\le 0.1$ mm. The largest distance suggests that the longest time that must be examined is estimated with eqn. 18.56.

$$t_{max}\simeq\frac{\pi x_{max}^2}{4D}=\frac{\pi(1\times10^{-4}\ \text{m})^2}{4(2.10\times10^{-9}\ \text{m}^2\ \text{s}^{-1})}=3.74\ \text{s}$$

Figure 18.7 shows oxygen concentration distributions for times between 0.01 s and 4.0 s. We set c_0 equal to zero and calculate c_s with Henry's law [16.26b].

$$b_{O_2} = \frac{p_{O_2}}{K_{O_2}} \text{ [16.26b]} = \frac{21 \text{ kPa}}{7.9\times10^4 \text{ kPa kg mol}^{-1}} \text{ [Table 16.1]} = 2.9\times10^{-4} \text{ mol kg}^{-1}$$

So $\quad c_s = 2.9 \times 10^{-4}$ mol dm^{-3}

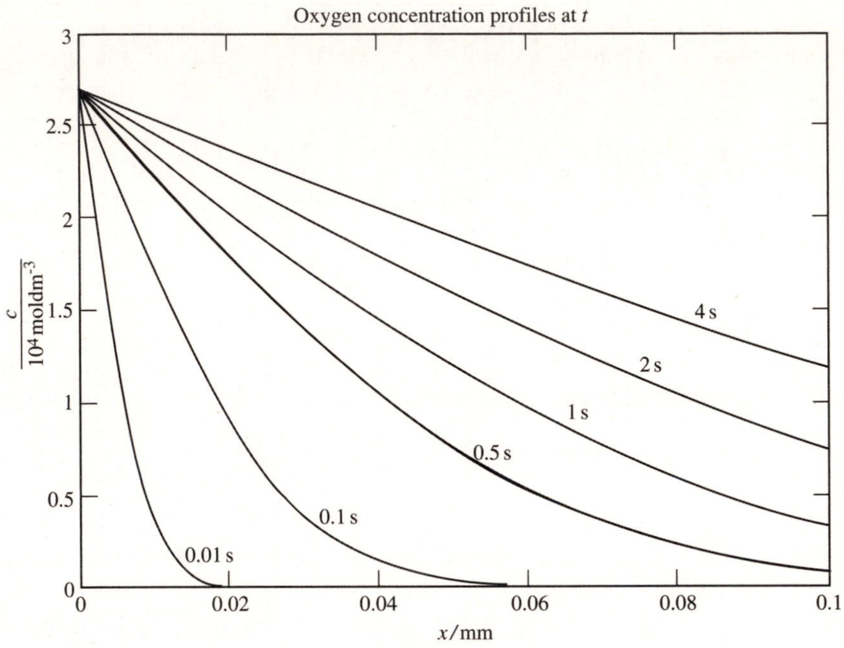

Figure 18.7

19 Chemical kinetics

Answers to discussion questions

D19.1 The time scales of atomic processes are rapid indeed: according to the following table, a nanosecond is an eternity. Note that the times given here are in some way typical values for times that may vary over two or three orders of magnitude. For example, vibrational wavenumbers can range from about 4400 cm^{-1} (for H_2) to 100 cm^{-1} and even lower (for I_2), with a corresponding range of associated times. Radiative decay rates of electronic states can vary even more widely: Times associated with phosphorescence can be in the millisecond and even second range. A large number of time scales for physical, chemical, and biological processes on the atomic and molecular scale are reported in Figure 2 of A. H. Zewail, "Femtochemistry: Atomic-Scale Dynamics of the Chemical Bond," *Journal of Physical Chemistry A* **104**, 5660 (2000).

Process	t / ns	Reference
Radiative decay of electronic excited state	1×10^{1}	Section 11.5
Molecular rotational motion	3×10^{-2}	$B \approx 1\ cm^{-1}$
Molecular vibrational motion	3×10^{-5}	$\tilde{\nu} \approx 1000\ cm^{-1}$
Proton transfer	0.3	Zewail 2000
Initial chemical reaction of vision*	2×10^{-4}	*Impact* 11.1
Energy transfer in photosynthesis†	1×10^{-3}	*Impact* 19.1
Electron transfer in photosynthesis	3×10^{-3}	*Impact* 19.1
Collision frequency in liquids	4×10^{-4}	Section 18.1(b)‡

*Photoisomerisation of retinal from 11-*cis* to all-*trans*

†Time from absorption until electron transfer to adjacent pigment

‡Use the formula for gas collision frequency at 300 K, the parameters for benzene from the data section, and the density of liquid benzene.

Radiative decay of excited electronic states can range from about 10^{-9} s to 10^{-4} s—even longer for phosphorescence involving "forbidden" decay paths. Molecular rotational motion takes place on a scale of 10^{-12} s to 10^{-9} s. Molecular vibrations are faster still, about 10^{-14} s to 10^{-12} s. Proton transfer reactions occur on a timescale of about 10^{-10} s to 10^{-9} s, although protons can hop from molecule to molecule in water even more rapidly (1.5×10^{-12} s, Section 18.7a). *Impact on biochemistry* 11.1 describes several events in vision, including the 200-fs photoisomerization that gets the process started. *Impact on biochemistry* 19.1 lists time scales of several energy-transfer and electron-transfer steps in photosynthesis. Initial energy transfer (to a nearby pigment) has a time scale of around 10^{-13} s to 5×10^{-12} s, with longer-range transfer (to the reaction center) taking about 10^{-10} s. Immediate electron transfer is also very fast (about 3 ps), with ultimate transfer (leading to oxidation of water and reduction of plastoquinone) taking from 10^{-10} s to 10^{-3} s. The mean time between collisions in liquids is similar to vibrational periods, around 10^{-13} s. One can estimate collision times in liquids very roughly by applying the expression for collisions in gases (Section 18.1b) to liquid conditions.

D19.3 The overall reaction order is the sum of the powers of the concentrations of all of the substances appearing in the experimental rate law for the reaction (eqn. 19.7); hence, it is the sum of the individual orders (exponents) associated with a given reactant (or, occasionally, product). Reaction order is an experimentally determined, not theoretical, quantity, although theory may attempt to predict or explain it. Molecularity is the number of reactant molecules participating in an elementary reaction. Molecularity has meaning only for an elementary reaction, but reaction order applies to any reaction. In general, reaction order bears no necessary relation to the stoichiometry of the reaction, with the exception of elementary reactions, where the order of the reaction corresponds to the number of molecules participating in the reaction, that is, to its molecularity. Thus, for an elementary reaction, overall order and molecularity are the same and are determined by the stoichiometry.

D19.5 The rate-determining step is not just the slowest step: it must be slow *and* be a crucial gateway for the formation of products. If a faster reaction can also lead to products, then the slowest step is irrelevant because the slow reaction can then be side-stepped. The rate-determining step is like a slow ferry crossing between two fast highways: the overall rate at which traffic can reach its destination is determined by the rate at which it can cross on the ferry.

If the first step in a mechanism is the slowest step with the highest activation energy, then it is rate-determining, and the overall reaction rate is equal to the rate of the first step because all subsequent steps are so fast that once the first intermediate is formed, it results immediately in the formation of products. Once over the initial barrier, the intermediates cascade into products. However, a rate-determining step may also stem from the low concentration of a crucial reactant or catalyst and need not correspond to the step with highest activation barrier. A rate-determining step arising from the low activity of a crucial enzyme can sometimes be identified by determining whether or not the reactants and products for the step are in equilibrium: if the reaction is not at equilibrium, the step may be slow enough to be rate-determining.

D19.7 Simple diagrams of Gibbs energy against reaction coordinate are useful for distinguishing between kinetic and thermodynamic control of a reaction. For the simple parallel reactions $R \rightarrow P_1$ and $R \rightarrow P_2$, shown in Figure 19.1 as Cases I and II, the product P_1 is thermodynamically favored because the Gibbs energy decreases to a greater extent for its formation. However, the rate at which each product appears does not depend on thermodynamic favorability. Rate constants depend on activation energy. In Case I the activation energy for the formation of P_1 is much larger than that for formation of P_2. At low and moderate temperature, the large activation energy may not be readily available and P_1 either cannot form or forms at a slow rate. The much smaller activation energy for P_2 formation is available, and consequently P_2 is produced even though it is not the thermodynamically favored product. This is kinetic control. In this case, $[P_2]/[P_1] = k_2/k_1 > 1$ [19.32].

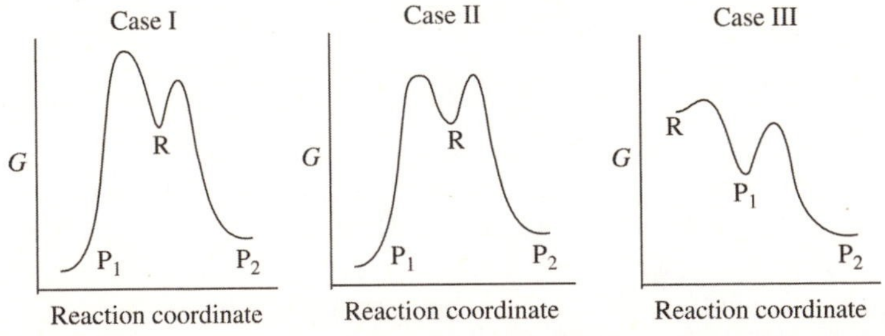

Figure 19.1

The activation energies for the parallel reactions are equal in Case II, and consequently the two products appear at identical rates. If the reactions are irreversible, $[P_2]/[P_1] = k_2/k_1 = 1$ at all times. The results are very different for reversible reactions. The activation energy for $P_1 \rightarrow R$ is much larger than that for $P_2 \rightarrow R$ and P_1 accumulates as the more rapid $P_2 \rightarrow R \rightarrow P_1$ occurs. Eventually the ratio $[P_2]/[P_1]$ approaches the equilibrium value.

$$\left(\frac{[P_2]}{[P_1]}\right)_{eq} = e^{-(\Delta G_2 - \Delta G_1)/RT} < 1$$

This is thermodynamic control.

Case III in Figure 19.1 represents an interesting consecutive reaction series $R \rightarrow P_1 \rightarrow P_2$. The first step has relatively low activation energy and P_1 rapidly appears. However, the relatively large activation energy for the second step is not available at low and moderate temperatures. By using low or moderate temperatures and short reaction times, it is possible to produce more of the thermodynamically less favorable P_1. This is kinetic control. High temperatures and long reaction times will yield the thermodynamically favored P_2.

The ratio of reaction products is determined by relative reaction rates in kinetic controlled reactions. Favorable conditions include short reaction times, lower temperatures, and irreversible reactions. Thermodynamic control is favored by long reaction times, higher temperatures, and reversible reactions. The ratio of products depends on the relative stability of products for thermodynamically controlled reactions.

D19.9 In the analysis of stepwise polymerization, the rate constant for the second-order condensation is assumed to be independent of the chain length and to remain constant throughout the reaction. It follows, then, that the degree of polymerization is given by

$$\langle N\rangle = 1 + k_r t[A]_0 \quad [19.49b]$$

Therefore, the average molar mass can be controlled by adjusting the initial concentration of monomer and the length of time that the polymerization is allowed to proceed.

Chain polymerization is a complicated radical chain mechanism involving initiation, propagation, and termination steps (see Section 19.8(b) for the details of this mechanism). The derivation of the overall rate equation utilizes the steady-state approximation and leads to the following expression for the average number of monomer units in the polymer chain:

$$\langle N\rangle = 2k_r[M][I]^{-1/2} \quad [19.58]$$

where $k_r = (1/2)k_p(fk_ik_t)^{-1/2}$, where k_p, k_i, and k_t are the rate constants for the propagation, initiation, and termination steps respectively, and where f is the fraction of radicals that successfully initiate a chain. We see that the average molar mass of the polymer is directly proportional to the monomer concentration and inversely proportional to the square root of the initiator concentration and to the rate constant for initiation. Therefore, the slower the initiation of the chain, the higher the average molar mass of the polymer.

D19.11 The shortening of the lifetime of an excited state is called quenching. Quenching effects may be studied by monitoring the emission from the excited state that is involved in the photochemical process. The addition of a quencher opens up an additional channel for the deactivation of the excited singlet state.

Three common mechanisms for bimolecular quenching of an excited singlet (or triplet) state are

Collisional deactivation	$S^* + Q \rightarrow S + Q$
Energy transfer	$S^* + Q \rightarrow S + Q^*$
Electron transfer	$S^* + Q \rightarrow S^+ + Q^-$ or $S^- + Q^+$

Collisional quenching is particularly efficient when Q is a heavy species, such as an iodide ion, which receives energy from S^* and then decays primarily by internal conversion to the ground state. Pure collisional quenching can be detected by the appearance of vibrational and rotational excitation in the spectrum of the acceptor.

In many cases, it is possible to prove that energy transfer is the predominant mechanism of quenching if the excited state of the acceptor fluoresces or phosphoresces at a characteristic wavelength. In a pulsed laser experiment, the rise in fluorescence intensity from Q^* with a characteristic time that is the same as the time for the decay of the fluorescence of S^* is often taken as an indication of energy transfer from S to Q.

Electron transfer can be studied by time-resolved spectroscopy (Section 11.7f). The oxidized and reduced products often have electronic absorption spectra distinct from those of their neutral parent compounds. Therefore, the rapid appearance of such known features in the absorption spectrum after excitation by a laser pulse may be taken as an indication of quenching by electron transfer.

Solutions to exercises

E19.1(a) Let the initial amount of ICl be n_{ICl} and the initial amount of H_2 be n_H; the initial amounts of I_2 and HCl are assumed to be zero. Thus, the initial total quantity of gas is $n_{ICl} + n_H$. Let the amount of I_2 formed at any given time be n. In that case, the amount of HCl is $2n$, that of H_2 is $n_H - n$, and the amount of ICl is $n_{ICl} - 2n$. At any given time, then, the total quantity of gas is

$$n_{total} = n_{ICl} - 2n + n_H - n + n + 2n = n_{ICl} + n_H = n_{initial}$$

Thus there is no change in the amount of gas during the course of the reaction. Since there is no change in volume or temperature either, there is $\boxed{\text{no change in pressure}}$.

COMMENT. Measuring the pressure would **not** be a practical way of monitoring the progress of this reaction.

E19.2(a) $$v = \frac{1}{\nu_J}\frac{d[J]}{dt} \ [19.3b] \qquad \text{so} \qquad \frac{d[J]}{dt} = \nu_J v$$

Rate of formation of C $= 3v = \boxed{8.1 \text{ mol dm}^{-3}\text{ s}^{-1}}$

Rate of formation of D $= v = \boxed{2.7 \text{ mol dm}^{-3}\text{ s}^{-1}}$

Rate of consumption of A $= v = \boxed{2.7 \text{ mol dm}^{-3}\text{ s}^{-1}}$

Rate of consumption of B $= 2v = \boxed{5.4 \text{ mol dm}^{-3}\text{ s}^{-1}}$

E19.3(a) $$v = \frac{1}{\nu_J}\frac{d[J]}{dt} \ [19.3b] = \frac{1}{2}\frac{d[C]}{dt} = \frac{1}{2}\times(2.7 \text{ mol dm}^{-3}\text{ s}^{-1}) = \boxed{1.3\bar{5} \text{ mol dm}^{-3}\text{ s}^{-1}}$$

Rate of formation of D $= 3v = \boxed{4.0\bar{5} \text{ mol dm}^{-3}\text{ s}^{-1}}$

Rate of consumption of A $= 2v = \boxed{2.7 \text{ mol dm}^{-3}\text{ s}^{-1}}$

Rate of consumption of B $= v = \boxed{1.3\bar{5} \text{ mol dm}^{-3}\text{ s}^{-1}}$

E19.4(a) The rate is expressed in mol dm^{-3} s^{-1}; therefore,

$$\text{mol dm}^{-3}\text{ s}^{-1} = [k_r] \times (\text{mol dm}^{-3}) \times (\text{mol dm}^{-3})$$

where $[k_r]$ denotes units of k_r, requires the units to be $\boxed{\text{dm}^3\text{ mol}^{-1}\text{ s}^{-1}}$.

(a) Rate of formation of A $= v = \boxed{k_r[A][B]}$
(b) Rate of consumption of C $= 3v = \boxed{3k_r[A][B]}$

E19.5(a) Given $\frac{d[C]}{dt} = k_r[A][B][C]$, the rate of reaction is [19.3b]

$$v = \frac{1}{\nu_J}\frac{d[J]}{dt} = \frac{1}{2}\frac{d[C]}{dt} = \boxed{\frac{1}{2}k_r[A][B][C]}$$

The units of k_r, $[k_r]$, must satisfy

$$\text{mol dm}^{-3}\text{ s}^{-1} = [k_r] \times (\text{mol dm}^{-3}) \times (\text{mol dm}^{-3}) \times (\text{mol dm}^{-3})$$

Therefore, $[k_r] = \boxed{\text{dm}^6\text{ mol}^{-2}\text{ s}^{-1}}$.

E19.6(a) (a) For a second-order reaction, denoting the units of k_r by $[k_r]$,

$$\text{mol dm}^{-3}\text{ s}^{-1} = [k_r] \times (\text{mol dm}^{-3})^2\text{; therefore } \boxed{[k_r] = \text{dm}^3\text{ mol}^{-1}\text{ s}^{-1}}$$

For a third-order reaction,

$$\text{mol dm}^{-3}\text{ s}^{-1} = [k_r] \times (\text{mol dm}^{-3})^3\text{; therefore } \boxed{[k_r] = \text{dm}^6\text{ mol}^{-2}\text{ s}^{-1}}$$

(b) For a second-order reaction,

$$\text{kPa s}^{-1} = [k_r] \times \text{kPa}^2;\ \text{therefore}\ \boxed{[k_r] = \text{kPa}^{-1}\ \text{s}^{-1}}$$

For a third-order reaction,

$$\text{kPa s}^{-1} = [k_r] \times \text{kPa}^3;\ \text{therefore}\ \boxed{[k_r] = \text{kPa}^{-2}\ \text{s}^{-1}}$$

E19.7(a) The rate law is

$$v = k_r[\text{A}]^a \propto p_\text{A}{}^a = \{p_{\text{A},0}(1-f)\}^a$$

where f is the fraction reacted. That is, concentration and partial pressure are proportional to each other. Thus we can write

$$\frac{v_1}{v_2} = \frac{p^a_{\text{A},1}}{p^a_{\text{A},2}} = \left(\frac{1-f_1}{1-f_2}\right)^a$$

Taking logarithms,

$$\ln\left(\frac{v_1}{v_2}\right) = a\ln\left(\frac{1-f_1}{1-f_2}\right)$$

so $$a = \frac{\ln\left(\frac{v_1}{v_2}\right)}{\ln\left(\frac{1-f_1}{1-f_2}\right)} = \frac{\ln\left(\frac{1.07}{0.76}\right)}{\ln\left(\frac{0.95}{0.80}\right)} = 1.9\overline{9}$$

Hence, the reaction is $\boxed{\text{second-order}}$.

COMMENT. Knowledge of the initial pressure is not required for the solution to this exercise. The ratio of pressures was computed using fractions of the initial pressure.

E19.8(a) Table 19.3 gives a general expression for the half-life of a reaction of the type $\text{A} \rightarrow \text{P}$ for orders other than 1:

$$t_{1/2} = \frac{2^{n-1}-1}{(n-1)k_r[\text{A}]_0^{n-1}} \propto [\text{A}]_0^{1-n} \propto p_0^{1-n}$$

where the proportionality constants may be functions of the reaction order, the rate constant, or even the temperature, but not a function of the concentration. Form a ratio of the half-lives at different initial pressures:

$$\frac{t_{1/2}(p_{0,1})}{t_{1/2}(p_{0,2})} = \left(\frac{p_{0,1}}{p_{0,2}}\right)^{1-n} = \left(\frac{p_{0,2}}{p_{0,1}}\right)^{n-1}$$

Hence, $$\ln\left(\frac{t_{1/2}(p_{0,1})}{t_{1/2}(p_{0,2})}\right) = (n-1)\ln\left(\frac{p_{0,2}}{p_{0,1}}\right)$$

or $$(n-1) = \frac{\ln\left(\frac{410\ \text{s}}{880\ \text{s}}\right)}{\ln\left(\frac{169\ \text{Torr}}{363\ \text{Torr}}\right)} = 0.999 \approx 1$$

Therefore, $\boxed{n = 2}$ in agreement with the result of Exercise 19.7(a).

E19.9(a) $$2\ \text{N}_2\text{O}_5 \rightarrow 4\ \text{NO}_2 + \text{O}_2 \qquad v = k_r[\text{N}_2\text{O}_5]$$

Therefore, rate of consumption of $\text{N}_2\text{O}_5 = 2v = 2k_r[\text{N}_2\text{O}_5]$.

$$\frac{d[\text{N}_2\text{O}_5]}{dt} = -2k_r[\text{N}_2\text{O}_5] \qquad \text{so} \qquad [\text{N}_2\text{O}_5] = [\text{N}_2\text{O}_5]_0 e^{-2k_r t}$$

Solve this for t:

$$t = \frac{1}{2k_r}\ln\frac{[N_2O_5]_0}{[N_2O_5]}$$

Therefore, the half-life is

$$t_{1/2} = \frac{1}{2k_r}\ln 2 = \frac{\ln 2}{(2)\times(3.38\times10^{-5}\ \mathrm{s^{-1}})} = \boxed{1.03\times10^4\ \mathrm{s}}$$

Since the partial pressure of N_2O_5 is proportional to its concentration,

$$p(N_2O_5) = p_0(N_2O_5)e^{-2k_rt}$$

(a) $$p(N_2O_5) = (500\ \mathrm{Torr})\times\left(e^{-(2\times3.38\times10^{-5}/\mathrm{s})\times(50\ \mathrm{s})}\right) = \boxed{498\ \mathrm{Torr}}$$

(b) $$p(N_2O_5) = (500\ \mathrm{Torr})\times\left(e^{-(2\times3.38\times10^{-5}/\mathrm{s})\times(20\times60\ \mathrm{s})}\right) = \boxed{461\ \mathrm{Torr}}$$

COMMENT. The half-life formula in Table 19.3 is based on a rate constant for the rate of change of the reactant; that is, the formula based on the assumption that

$$-\frac{d[A]}{dt} = k_r[A]$$

Our expression for the rate of consumption has $2k_r$ instead of k_r, and our expression for $t_{1/2}$ does likewise.

E19.10(a) The integrated rate law is

$$k_rt = \frac{1}{[B]_0-[A]_0}\ln\left(\frac{[B]/[B]_0}{[A]/[A]_0}\right)\quad [19.17]$$

(a) The stoichiometry of the reaction requires that when

$$\Delta[B] = (0.020 - 0.050)\ \mathrm{mol\ dm^{-3}} = -0.030\ \mathrm{mol\ dm^{-3}}$$

then $\Delta[A] = -0.030\ \mathrm{mol\ dm^{-3}}$, as well.

Thus $[A] = 0.075\ \mathrm{mol\ dm^{-3}} - 0.030\ \mathrm{mol\ dm^{-3}} = 0.045\ \mathrm{mol\ dm^{-3}}$

When $[B] = 0.020\ \mathrm{mol\ dm^{-3}}$. Therefore,

$$k_rt = \frac{1}{(0.050-0.075)\ \mathrm{mol\,dm^{-3}}}\ln\left(\frac{0.020/0.050}{0.045/0.075}\right)$$

$$k_r\times1.0\ \mathrm{h} = 16.\bar{2}\ \mathrm{dm^3\ mol^{-1}}$$

so $$k_r = \boxed{16.\bar{2}\ \mathrm{dm^3\ mol^{-1}\ h^{-1}}}\times\left(\frac{1\ \mathrm{h}}{3600\ \mathrm{s}}\right) = \boxed{4.5\times10^{-3}\ \mathrm{dm^3\ mol^{-1}\ s^{-1}}}$$

(b) The half-life with respect to A is the time required for [A] to fall to 0.0375 mol dm^{-3} (and [B] to 0.0125 mol dm^{-3}). We solve eqn. 19.17 for t.

$$t_{1/2}(A) = \left(\frac{1}{(16.\bar{2}\ \mathrm{dm^3\ mol^{-1}\ h^{-1}})\times(-0.030\ \mathrm{mol\,dm^{-3}})}\right)\times\ln\left(\frac{0.0125/0.050}{0.50}\right)$$

$$= 1.4\bar{2}\ \mathrm{h} = \boxed{5.1\times10^3\ \mathrm{s}}$$

Similarly, the half-life with respect to B is the time required for [B] to fall to 0.025 mol dm^{-3} (and [A] to 0.050 mol dm^{-3}).

$$t_{1/2}(B) = \left(\frac{1}{(16.\bar{2}\ \mathrm{dm^3\ mol^{-1}\ h^{-1}})\times(-0.030\ \mathrm{mol\,dm^{-3}})}\right)\times\ln\left(\frac{0.50}{0.050/0.075}\right)$$

$$= 0.59\ \mathrm{h} = \boxed{2.1\times10^3\ \mathrm{s}}$$

COMMENT. This exercise illustrates that there is no unique half-life for reactions other than those of the type A → P.

E19.11(a) The integrated second-order rate law for a reaction of the type A + B → products is

$$k_r t = \frac{1}{[B]_0 - [A]_0}\ln\left(\frac{[B]/[B]_0}{[A]/[A]_0}\right) \quad [19.17]$$

Introducing $[B] = [B]_0 - x$ and $[A] = [A]_0 - x$ and rearranging, we obtain

$$k_r t = \left(\frac{1}{[B]_0 - [A]_0}\right)\ln\left(\frac{[A]_0([B]_0 - x)}{([A]_0 - x)[B]_0}\right)$$

Solving for x yields, after some rearranging,

$$x = \frac{[A]_0[B]_0\left(e^{k_r([B]_0-[A]_0)t} - 1\right)}{[B]_0 e^{([B]_0-[A]_0)k_r t} - [A]_0} = \frac{(0.060)\times(0.110\,\text{mol dm}^{-3})\times\left(e^{(0.110-0.060)\times 0.11\times t/\text{s}} - 1\right)}{(0.110)\times e^{(0.110-0.060)\times 0.11\times t/\text{s}} - 0.060}$$

$$= \frac{(0.060\,\text{mol dm}^{-3})\times(e^{0.0055t/\text{s}} - 1)}{e^{0.0055t/\text{s}} - 0.55}$$

(a) After 20 s,

$$x = \frac{(0.060\,\text{mol dm}^{-3})\times(e^{0.0055\times 20} - 1)}{e^{0.0055\times 20} - 0.55} = 0.0122\,\text{mol dm}^{-3}$$

which implies that

$$[CH_3COOC_2H_5] = (0.110 - 0.0122)\,\text{mol dm}^{-3} = \boxed{0.098\,\text{mol dm}^{-3}}$$

(b) After 15 min = 900 s,

$$x = \frac{(0.060\,\text{mol dm}^{-3})\times(e^{0.0055\times 900} - 1)}{e^{0.0055\times 900} - 0.55} = 0.060\,\text{mol dm}^{-3}$$

so $$[CH_3COOC_2H_5] = (0.110 - 0.060)\,\text{mol dm}^{-3} = \boxed{0.050\,\text{mol dm}^{-3}}$$

E19.12(a) The rate of consumption of A is

$$-\frac{d[A]}{dt} = 2v = 2k_r[A]^2 \quad [\nu_A = -2]$$

which integrates to $\frac{1}{[A]} - \frac{1}{[A]_0} = 2k_r t$ [19.13b with k_r replaced by $2k_r$].

Therefore, $t = \frac{1}{2k_r}\left(\frac{1}{[A]} - \frac{1}{[A]_0}\right)$

$$t = \left(\frac{1}{2\times 4.30\times 10^{-4}\,\text{dm}^3\,\text{mol}^{-1}\,\text{s}^{-1}}\right)\times\left(\frac{1}{0.010\,\text{mol dm}^{-3}} - \frac{1}{0.210\,\text{mol dm}^{-3}}\right)$$

$$= \boxed{1.11\times 10^5\,\text{s}} = \boxed{1.28\text{ days}}$$

E19.13(a) The reactions whose rate constants are sought are the forward and reverse reactions in the following equilibrium.

$$NH_3(aq) + H_2O(l) \underset{k_r'}{\overset{k_r}{\rightleftharpoons}} NH_4^+(aq) + OH^-(aq)$$

The rate constants are related by

$$K_b = \frac{k_r}{k_r'} = \frac{[NH_4^+][OH^-]}{[NH_3]} = 1.78\times 10^{-5}\,\text{mol dm}^{-3}$$

where the concentrations are equilibrium concentrations. (We assign units to K_b, which technically is a pure number, to help us keep track of units in the rate constants. Keeping track of the units makes us realize that k_r is a pseudo first-order protonation of NH_3 in excess water, for water does not appear in the above expression.) We need one more relationship between the constants, which we can obtain by proceeding as in Example 19.4:

$$\frac{1}{\tau} = k_r + k_r'([NH_4^+]+[OH^-])$$

Substitute into this expression

$$k_r = K_b k_r' \quad \text{and} \quad [NH_4^+] = [OH^-] = (K_b[NH_3])^{1/2}$$

Hence, $\dfrac{1}{\tau} = K_b k_r' + 2k_r'(K_b[NH_3])^{1/2} = k_r'\{K_b + 2(K_b[NH_3])^{1/2}\}$

So the reverse rate constant is

$$k_r' = \frac{1}{\tau\{K_b + 2(K_b[NH_3])^{1/2}\}}$$

$$= \frac{1}{7.61\times10^{-9}\ \text{s}\{1.78\times10^{-5}\ \text{mol dm}^{-3} + 2(1.78\times10^{-5}\times0.15)^{1/2}\ \text{mol dm}^{-3}\}}$$

$$= \boxed{4.0\times10^{10}\ \text{dm}^{-3}\ \text{mol s}^{-1}}$$

and the forward constant is

$$k_r = K_b k_r' = 1.78\times10^{-5}\ \text{mol dm}^{-3}\times 4.0\times10^{10}\ \text{dm}^3\ \text{mol}^{-1}\ \text{s}^{-1} = \boxed{7.1\times10^5\ \text{s}^{-1}}$$

Recall that k_r is the rate constant in the pseudo first-order rate law.

$$-\frac{d[NH_3]}{dt} = k_r[NH_3]$$

Let us call k the rate constant in the bimolecular rate law.

$$-\frac{d[NH_3]}{dt} = k[NH_3][H_2O]$$

Setting these two expressions equal to each other yields

$$k = \frac{k_r}{[H_2O]} = \frac{7.1\times10^5\ \text{s}^{-1}}{(1000\ \text{g dm}^{-3})/(18.02\ \text{g mol}^{-1})} = \boxed{1.28\times10^4\ \text{dm}^3\ \text{mol}^{-1}\ \text{s}^{-1}}$$

E19.14(a) The rate of the overall reaction is

$$v = \frac{d[P]}{dt} = k_2[A][B]$$

However, we cannot have the concentration of an intermediate in the overall rate law.

(i) Assume a pre-equilibrium with

$$K = \frac{[A]^2}{[A_2]}, \quad \text{implying that} \quad [A] = K^{1/2}[A_2]^{1/2}$$

and $\quad v = \boxed{k_2K^{1/2}[A_2]^{1/2}[B]} = k_{\text{eff}}[A_2]^{1/2}[B]$

where $k_{\text{eff}} = k_2K^{1/2}$.

(ii) Apply the steady-state approximation:

$$\frac{d[A]}{dt} \approx 0 = 2k_1[A_2] - 2k_1'[A]^2 - k_2[A][B]$$

This is a quadratic equation in [A].

$$[\mathrm{A}]=\frac{-b\pm\sqrt{b^2-4ac}}{2a}=\frac{k_2[\mathrm{B}]\pm\sqrt{k_2{}^2[\mathrm{B}]^2+16k_1'k_1[\mathrm{A}_2]}}{-4k_1'}=\frac{k_2[\mathrm{B}]}{4k_1'}\left(\sqrt{1+\frac{16k_1'k_1[\mathrm{A}_2]}{k_2{}^2[\mathrm{B}]^2}}-1\right)$$

(In the last step, choose the sign that gives a positive quantity for [A].) Thus the rate law is

$$v=k_2[\mathrm{A}][\mathrm{B}]=\boxed{\frac{k_2{}^2[\mathrm{B}]^2}{4k_1'}\left(\sqrt{1+\frac{16k_1'k_1[\mathrm{A}_2]}{k_2{}^2[\mathrm{B}]^2}}-1\right)}$$

This is a perfectly good rate law, albeit a complicated one. It is not in typical power-law form, but it is a function of reactant concentrations only, with no intermediates. This law simplifies under certain conditions. If $16k_1k_1'[\mathrm{A}_2] >> k_2{}^2[\mathrm{B}]^2$, then

$$\sqrt{1+\frac{16k_1'k_1[\mathrm{A}_2]}{k_2{}^2[\mathrm{B}]^2}}-1\approx\frac{4\sqrt{k_1'k_1[\mathrm{A}_2]}}{k_2[\mathrm{B}]}$$

and $$v\approx\frac{k_2{}^2[\mathrm{B}]^2}{4k_1'}\times\frac{4\sqrt{k_1'k_1[\mathrm{A}_2]}}{k_2[\mathrm{B}]}=k_2[\mathrm{B}]\sqrt{\frac{k_1[\mathrm{A}_2]}{k_1'}}=\boxed{k_2K^{1/2}[\mathrm{A}_2]^{1/2}[\mathrm{B}]}$$

recovering the pre-equilibrium rate law. If, on the other hand, $16k_1k_1'[\mathrm{A}_2] << k_2{}^2[\mathrm{B}]^2$, we expand the square root

$$\sqrt{1+\frac{16k_1'k_1[\mathrm{A}_2]}{k_2{}^2[\mathrm{B}]^2}}\approx 1+\frac{8k_1'k_1[\mathrm{A}_2]}{k_2{}^2[\mathrm{B}]^2}$$

and $$v\approx\frac{k_2{}^2[\mathrm{B}]^2}{4k_1'}\left(1+\frac{8k_1'k_1[\mathrm{A}_2]}{k_2{}^2[\mathrm{B}]^2}-1\right)=\boxed{2k_1[\mathrm{A}_2]}$$

COMMENT. If the equilibrium is "fast," the latter condition will not be fulfilled. In fact, this special case amounts to having the first step rate-limiting. Note that the full (messy) steady-state approximation is less severe than either the pre-equilibrium or the rate-limiting step approximations, for it includes both as special cases.

E19.15(a) $$\frac{1}{k_\mathrm{r}}=\frac{k_\mathrm{a}'}{k_\mathrm{a}k_\mathrm{b}}+\frac{1}{k_\mathrm{a}p_\mathrm{A}}\ \text{[analogous to 19.46]}$$

Therefore, for two different pressures we have

$$\frac{1}{k_\mathrm{r}(p_1)}-\frac{1}{k_\mathrm{r}(p_2)}=\frac{1}{k_\mathrm{a}}\left(\frac{1}{p_1}-\frac{1}{p_2}\right)$$

so $$k_\mathrm{a}=\frac{\frac{1}{p_1}-\frac{1}{p_2}}{\frac{1}{k_\mathrm{r}(p_1)}-\frac{1}{k_\mathrm{r}(p_2)}}=\frac{\frac{1}{12\ \mathrm{Pa}}-\frac{1}{1.30\times10^3\ \mathrm{Pa}}}{\frac{1}{2.10\times10^{-5}\ \mathrm{s}^{-1}}-\frac{1}{2.50\times10^{-4}\ \mathrm{s}^{-1}}}=\boxed{1.9\times10^{-6}\ \mathrm{Pa}^{-1}\ \mathrm{s}^{-1}}$$

or $\boxed{1.9\ \mathrm{MPa}^{-1}\ \mathrm{s}^{-1}}$

E19.16(a) The degree of polymerization is [19.49b]

$$\begin{aligned}\langle N\rangle&=1+k_\mathrm{r}t[\mathrm{A}]_0\\&=1+(1.39\ \mathrm{dm}^{-3}\ \mathrm{mol}^{-1}\ \mathrm{s}^{-1})\times5.00\ \mathrm{h}\times3600\ \mathrm{s\ h}^{-1}\times1.00\times10^{-2}\ \mathrm{mol\ dm}^{-3}\\&=\boxed{251}\end{aligned}$$

The fraction condensed is related to the degree of polymerization by

$$\langle N\rangle = \frac{1}{1-p} \quad \text{so} \quad p = \frac{\langle N\rangle - 1}{\langle N\rangle} = \frac{251-1}{251} = \boxed{0.996}$$

E19.17(a) The kinetic chain length varies with concentration as

$$\nu = k_r[\mathrm{M}][\mathrm{I}]^{-1/2} \; [19.57]$$

so the ratio of kinetic chain lengths under different concentrations is

$$\frac{\nu_2}{\nu_1} = \frac{[\mathrm{M}]_2}{[\mathrm{M}]_1} \times \left(\frac{[\mathrm{I}]_1}{[\mathrm{I}]_2}\right)^{1/2} = \frac{1}{4.2} \times \left(\frac{1}{3.6}\right)^{1/2} = \boxed{0.125}$$

E19.18(a) Number of photons absorbed = ϕ^{-1} × number of molecules that react [19.59a]. Therefore,

$$\text{number absorbed} = \frac{(2.28\times10^{-3}\ \mathrm{mol}/2)\times(6.022\times10^{23}\ \mathrm{einstein}^{-1})}{2.1\times10^{2}\ \mathrm{mol\,einstein}^{-1}} = \boxed{3.3\times10^{18}}$$

E19.19(a) For a source of power output P at wavelength λ, the number of photons (n_λ) generated in a time t is the energy output divided by the energy per mole of photons

$$n_\lambda = \frac{Pt}{h\nu N_A} = \frac{P\lambda t}{hcN_A} = \frac{(100\ \mathrm{W})\times(45\ \mathrm{min})\times(60\ \mathrm{s\ min}^{-1})\times(490\times10^{-9}\ \mathrm{m})}{(6.626\times10^{-34}\ \mathrm{J\ s})\times(2.998\times10^{8}\ \mathrm{m\ s}^{-1})\times(6.022\times10^{23}\ \mathrm{mol}^{-1})}$$
$$= 1.11\ \mathrm{mol} = 1.11\ \mathrm{einstein}$$

If 40% of the incident photon flux is transmitted, then 60% is absorbed. Therefore,

$$\phi = \frac{\text{molecules decomposed}}{\text{photons absorbed}} = \frac{0.344\ \mathrm{mol}}{0.60\times1.11\ \mathrm{einstein}} = \boxed{0.52}$$

E19.20(a) The Stern–Volmer equation (eqn. 19.66) relates the ratio of fluorescence quantum yields in the absence and presence of quenching.

$$\frac{\phi_{\mathrm{f},0}}{\phi_{\mathrm{f}}} = 1 + \tau_0 k_{\mathrm{Q}}[\mathrm{Q}] = \frac{I_{\mathrm{f},0}}{I_{\mathrm{f}}}$$

The last equality reflects the fact that fluorescence intensities are proportional to quantum yields. Solve this equation for [Q]:

$$[\mathrm{Q}] = \frac{(I_{\mathrm{f},0}/I_{\mathrm{f}})-1}{\tau_0 k_{\mathrm{Q}}} = \frac{2-1}{(6.0\times10^{-9}\ \mathrm{s})\times(3.0\times10^{8}\ \mathrm{dm^3\ mol^{-1}\ s^{-1}})} = \boxed{0.56\ \mathrm{mol\ dm^{-3}}}$$

E19.21(a) The efficiency of resonance energy transfer is given by [19.67].

$$\eta_{\mathrm{T}} = 1 - \frac{\phi_{\mathrm{f}}}{\phi_{\mathrm{f},0}} = 0.10$$

Förster theory relates this quantity to the distance R between donor-acceptor pairs by

$$\eta_{\mathrm{T}} = \frac{R_0^{\,6}}{R_0^{\,6} + R^6} \; [19.68]$$

where R_0 is an empirical parameter listed in Table 19.6. Solving for the distance yields

$$R = R_0\left(\frac{1}{\eta_{\mathrm{T}}} - 1\right)^{1/6} = (4.9\ \mathrm{nm})\times\left(\frac{1}{0.10} - 1\right)^{1/6} = \boxed{7.1\ \mathrm{nm}}$$

Solutions to problems

Solutions to numerical problems

P19.1 A simple but practical approach is to make an initial guess at the order by observing whether the half-life of the reaction appears to depend on concentration. If it does not, the reaction is first-order; if it does, refer to Table 19.3 for an expression for the half-life of a reaction of the type $A \rightarrow P$ for orders other than 1:

$$t_{1/2} = \frac{2^{n-1}-1}{(n-1)k_r[A]_0^{n-1}} \propto [A]_0^{1-n}$$

Examination of the data shows that the first half-life is roughly 45 minutes and the second is about double the first. (Compare the 0–50 minute data to the 50–150 minute data.) That is, the half-life starting from *half* of the initial concentration is about *twice* the initial half-life, suggesting that the half-life is inversely proportional to initial concentration:

$$t_{1/2} \propto [A]_0^{-1} = [A]_0^{1-n} \text{ with } n = 2$$

Confirm this suggestion by plotting 1/[A] against time. A second-order reaction will obey

$$\frac{1}{[A]} = k_r t + \frac{1}{[A]_0} \quad [19.13b]$$

We draw up the following table ($A = NH_4CNO$).

t/min	0	20.0	50.0	65.0	150
m(urea)/g	0	7.0	12.1	13.8	17.7
m(A)/g	22.9	15.9	10.8	9.1	5.2
$[A]/(\text{mol dm}^{-3})$	0.381	0.265	0.180	0.152	0.0866
$[A]^{-1}/(\text{dm}^3\text{ mol}^{-1})$	2.62	3.78	5.56	6.60	11.5

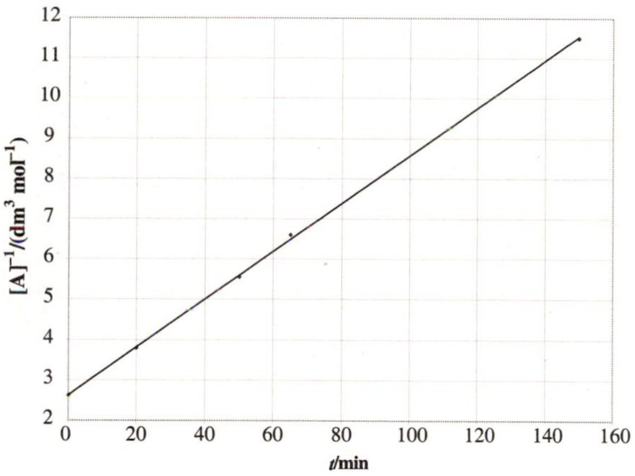

Figure 19.2

The data are plotted in Figure 19.2 and fit closely to a straight line. Hence, the reaction is indeed $\boxed{\text{second-order}}$. The rate constant is the slope: $\boxed{k_r = 0.059\bar{4}\text{ dm}^3\text{ mol}^{-1}\text{ min}^{-1}}$. To find [A] at 300 min, use eqn. 19.13(c):

$$[A] = \frac{[A]_0}{1+k_r t[A]_0} = \frac{0.381\text{ mol dm}^{-3}}{1+(0.059\bar{4})\times(300)\times(0.381)} = 0.048\bar{9}\text{ mol dm}^{-3}$$

The mass of NH_4CNO left after 300 minutes is

$$m = (0.048\bar{9}\text{ mol dm}^{-3})\times(1.00\text{ dm}^3)\times(60.06\text{ g mol}^{-1}) = \boxed{2.94\text{ g}}$$

P19.3 Use the procedure adopted in the solutions to problems 19.1 and 19.2: Is the half-life (or any other similarly defined "fractional life") constant, or does it vary over the course of the reaction? The data are not quite so clear-cut. The half-life appears to be approximately constant at about 10 minutes: in the interval 0–10 the initial concentration drops by just over half, while in the interval 2–12 the concentration drops by slightly less than half. Another measure would compare the fractional consumption in two equal time intervals. The fractional consumption in the 0–2 interval is about 1/6, while that in the 10–12 interval is about 1/10—suggesting that the fractional consumption is *not* constant over the time the reaction was monitored. We draw up the following table (A = nitrile) to examine both first-order and second-order plots. The former is a plot of $\ln\left(\dfrac{[A]}{[A]_0}\right)$ against time (eqn. 19.10b); the latter is a plot of 1/[A] against time.

$t/(10^3\ s)$	0	2.00	4.00	6.00	8.00	10.00	12.00
$[A]/(mol\ dm^{-3})$	1.50	1.26	1.07	0.92	0.81	0.72	0.65
$\dfrac{[A]}{[A]_0}$	1.00	0.840	0.713	0.613	0.540	0.480	0.433
$\ln\left(\dfrac{[A]}{[A]_0}\right)$	0	–0.174	–0.338	–0.489	–0.616	–0.734	–0.836
1/[A]	0.667	0.794	0.935	1.09	1.23	1.39	1.54

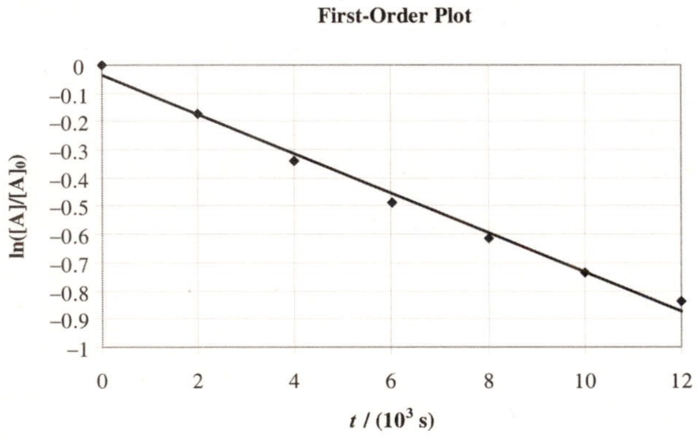

Figure 19.3(a)

The first-order plot (Figure 19.3a) is not bad: the correlation coefficient is 0.991. The corresponding first-order rate constant is $k_r = -\text{slope} = \boxed{7.0\times10^{-5}\ s^{-1}}$.

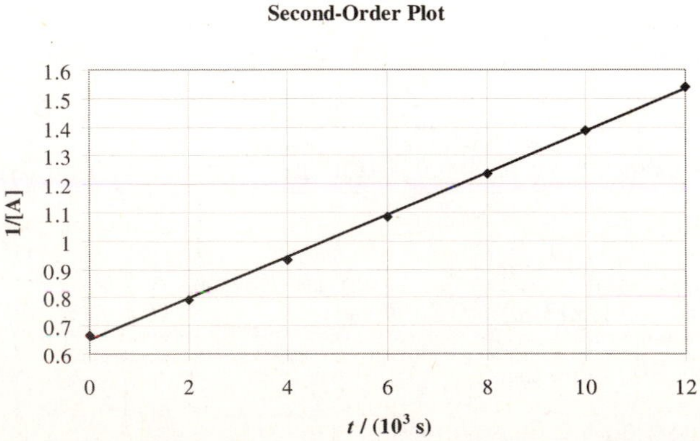

Figure 19.3(b)

The second-order plot (Figure 19.3b) looks even better: the correlation coefficient is 0.999. The corresponding second-order rate constant is $k_r = \text{slope} = \boxed{7.3\times10^{-5}\ dm^3\ mol^{-1}\ s^{-1}}$.

COMMENT. Based on the given data, the reaction appears to be closer to second-order than to first-order. (The reaction order need not be an integer.) This conclusion is a tenuous one, though, based on the assumption that the data contain little experimental error. The best course of action for an investigator seeking to establish the reaction order would be further experimentation, following the reaction over a wider range of concentrations.

P19.5 Since both reactions are first-order,

$$-\frac{d[A]}{dt} = k_1[A] + k_2[A] = (k_1 + k_2)[A]$$

so $$[A] = [A]_0 e^{-(k_1+k_2)t} \quad [19.10b \text{ with } k_r = k_1 + k_2]$$

We are interested in the yield of ketene, CH_2CO; call it K:

$$\frac{d[K]}{dt} = k_2[A] = k_2[A]_0 e^{-(k_1+k_2)t}$$

Integrating yields

$$\int_0^{[K]} d[K] = k_2[A]_0 \int_0^t e^{-(k_1+k_2)t} dt$$

$$[K] = \frac{k_2[A]_0}{k_1+k_2}(1 - e^{-(k_1+k_2)t}) = \frac{k_2}{k_1+k_2}([A]_0 - [A])$$

The percent yield is the amount of K produced compared to complete conversion; since the stoichiometry of reaction (2) is one-to-one, we can write

$$\%\ \text{yield} = \frac{[K]}{[A]_0} \times 100\% = \frac{k_2}{k_1+k_2}(1 - e^{-(k_1+k_2)t}) \times 100\%$$

which has its maximum value when the reaction reaches completion.

$$\text{max}\ \%\ \text{yield} = \frac{k_2}{k_1+k_2} \times 100\% = \frac{4.65\ s^{-1}}{(3.74 + 4.65)\ s^{-1}} \times 100\% = \boxed{55.4\%}$$

COMMENT. If we are interested in yield of the desired product (ketene) compared to the products of side reactions (products of reaction 1), it makes sense to define the conversion ratio, the ratio of desired product formed to starting material *reacted*, namely,

$$\frac{[K]}{[A]_0 - [A]}$$

which works out in this case to be independent of time.

$$\frac{[K]}{[A]_0 - [A]} = \frac{k_2}{k_1 + k_2}$$

If a substance reacts by parallel processes of the same order, then the ratio of the amounts of products will be constant and independent of the extent of the reaction, no matter what the order.

Question. Can you demonstrate the truth of the statement made in the comment?

P19.7 The stoichiometry of the reaction relates product and reaction concentrations as follows:

$$[A] = [A]_0 - 2[B]$$

When the reaction goes to completion, $[B] = [A]_0/2$; hence $[A]_0 = 0.624\ \text{mol dm}^{-3}$. We can therefore tabulate [A], and examine its half-life. We see that the half-life of A from its initial concentration is approximately 20 min, and that its half-life from the concentration at 20 min is also 20 min. This indicates a first-order reaction. We confirm this conclusion by plotting the data accordingly (in Figure 19.4), using

$$\ln\frac{[A]_0}{[A]} = k_A t \quad [19.10b]$$

which follows from

$$\frac{d[A]}{dt} = -k_A[A]$$

t / min	0	10	20	30	40	∞
[B] / (mol dm^{-3})	0	0.089	0.153	0.200	0.230	0.312
[A] / (mol dm^{-3})	0.624	0.446	0.318	0.224	0.164	0
$\ln\frac{[A]_0}{[A]}$	0	0.34	0.67	1.02	1.34	

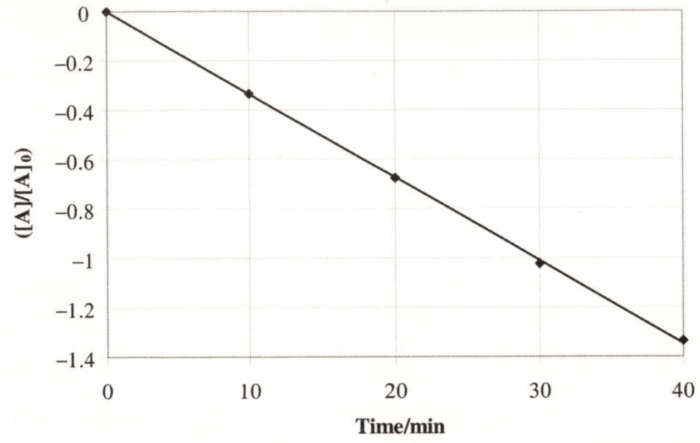

Figure 19.4

The points lie on a straight line, which confirms first-order kinetics. Since the slope of the line is -3.4×10^{-2} min^{-1}, we conclude that $k_A = 3.4\times10^{-2}$ min^{-1}. To express the rate law in the form $v = k_r[A]$, we note that

$$v = -\frac{1}{2}\frac{d[A]}{dt} = -\left(\frac{1}{2}\right)\times(-k_A[A]) = \frac{1}{2}k_A[A]$$

and hence $k_r = \frac{1}{2}k_A = \boxed{1.7\times10^{-2}\ \text{min}^{-1}}$.

P19.9 If the reaction is first-order, the concentrations obey

$$\ln\left(\frac{[A]}{[A]_0}\right) = -k_r t \quad [19.10b]$$

and, since pressures and concentrations of gases are proportional, the pressures should obey

$$\ln\frac{p_0}{p} = k_r t$$

and $\frac{1}{t}\ln\frac{p_0}{p}$ should be a constant. We test this by drawing up the following table.

p_0 / Torr	200	200	400	400	600	600
t / s	100	200	100	200	100	200
p / Torr	186	173	373	347	559	520
$10^4\left(\frac{1}{t/\text{s}}\right)\ln\frac{p_0}{p}$	7.3	7.3	7.0	7.1	7.1	7.2

The values in the last row of the table are virtually constant, so (in the pressure range spanned by the data) the reaction has first-order kinetics with $k_r = \boxed{7.2\times10^{-4}\ \text{s}^{-1}}$.

P19.11 Using spreadsheet software to evaluate eqn. 19.27(b), one can draw up a plot like that in Figure 19.5. The curves in this plot represent the concentration of the intermediate [I] as a function of time. They are labeled with the ratio k_a/k_b, where $k_b = 1\ s^{-1}$ for all curves and k_a varies. The thickest curve, labeled 10, corresponds to $k_a = 10\ s^{-1}$, as specified in part (a) of the problem. As the ratio k_a/k_b gets smaller (or, as the problem puts it, the ratio k_b/k_a gets larger), the concentration profile for I becomes lower, broader, and flatter; that is, [I] becomes more nearly constant over a longer period of time. This is the nature of the steady-state approximation, which becomes more and more valid as consumption of the intermediate becomes fast compared with its formation.

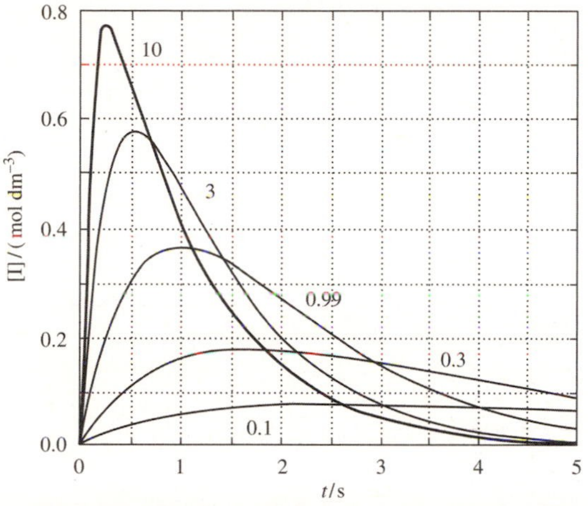

Figure 19.5

P19.13 (a) First, find an expression for the relaxation time, using Example 19.4 as a model.

$$\frac{d[A]}{dt} = -2k_a[A]^2 + 2k_a'[A_2]$$

Rewrite the expression in terms of a difference from equilibrium values, $[A] = [A]_{eq} + x$.

$$\frac{d[A]}{dt} = \frac{d([A]_{eq} + x)}{dt} = \frac{dx}{dt} = -2k_a([A]_{eq} + x)^2 + 2k_a'\left([A_2]_{eq} - \tfrac{1}{2}x\right)$$

$$\frac{dx}{dt} = -2k_a[A]_{eq}^{\ 2} - 4k_a[A]_{eq}x - 2k_a x^2 + 2k_a'[A_2]_{eq} - k_a' x$$

Neglect powers of x greater than x^1, and use the fact that at equilibrium the forward and reverse rates are equal,

$$k_a[A]_{eq}^{\ 2} = k_a'[A_2]_{eq}$$

to obtain

$$\frac{dx}{dt} \approx -(4k_a[A]_{eq} + k_a')x \quad \text{so} \quad \frac{1}{\tau} \approx 4k_a[A]_{eq} + k_a'$$

To get the desired expression, square the reciprocal relaxation time,

$$\text{(i)} \qquad \frac{1}{\tau^2} \approx 16k_a^{\ 2}[A]_{eq}^{\ 2} + 8k_a k_a'[A]_{eq} + (k_a')^2$$

introduce $[A]_{tot} = [A]_{eq} + 2[A_2]_{eq}$ into the middle term,

$$\frac{1}{\tau^2} \approx 16k_a^{\ 2}[A]_{eq}^{\ 2} + 8k_a k_a'([A]_{tot} - 2[A_2]_{eq}) + (k_a')^2$$

$$\approx 16k_a^{\ 2}[A]_{eq}^{\ 2} + 8k_a k_a'[A]_{tot} - 16k_a k_a'[A_2]_{eq} + (k_a')^2 = \boxed{8k_a k_a'[A]_{tot} + (k_a')^2}$$

and use the equilibrium condition again to see that the remaining equilibrium concentrations cancel each other.

COMMENT. Introducing $[A]_{tot}$ into just one term of eqn. (i) above is a permissible step but not a very systematic one. It is worth trying because of the resemblance between eqn. (i) and the desired expression: We would be finished if we could get $[A]_{tot}$ into the middle term and somehow get the first term to disappear! A more systematic but messier approach would be to express $[A]_{eq}$ in terms of the desired $[A]_{tot}$ by using the equilibrium condition and $[A]_{tot} = [A]_{eq} + 2[A_2]_{eq}$: Solve both of those equations for $[A_2]_{eq}$, set the two resulting expressions equal to each other, solve for $[A]_{eq}$ in terms of the desired $[A]_{tot}$, and substitute **that** expression for $[A]_{eq}$ everywhere in eqn. (i).

(b) Plot $\dfrac{1}{\tau^2}$ vs. $[A]_{tot}$

The resulting curve should be a straight line whose y-intercept is $(k_a')^2$ and whose slope is $8k_ak_a'$.

(c) Draw up the following table.

$[A]_{tot}$ / (mol dm^{-3})	0.500	0.352	0.251	0.151	0.101
τ / ns	2.3	2.7	3.3	4.0	5.3
$1/(\tau/\text{ns})^2$	0.189	0.137	0.092	0.062	0.036

The plot is shown in Figure 19.6.

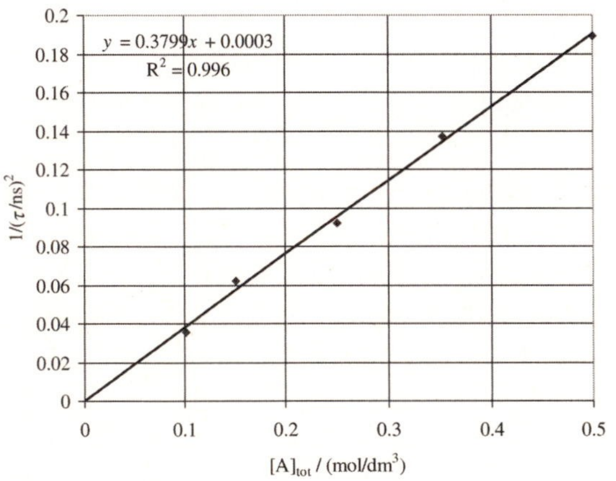

Figure 19.6

The y-intercept is 0.0003 ns^{-2} and the slope is 0.38 ns^{-2} dm^3 mol^{-1}, so

$$k_a' = \{3\times10^{-4}\times(10^{-9}\text{ s})^{-2}\}^{1/2} = (3\times10^{14}\text{ s}^{-2})^{1/2} = \boxed{1.\overline{7}\times10^{7}\text{ s}^{-1}}$$

$$k_a = \frac{0.38\times(10^{-9}\text{ s})^{-2}\text{ dm}^3\text{ mol}^{-1}}{8\times(1.\overline{7}\times10^{7}\text{ s}^{-1})} = \boxed{2.\overline{7}\times10^{9}\text{ dm}^3\text{ mol}^{-1}\text{ s}^{-1}}$$

and $$K = \frac{k_a/\text{dm}^3\text{ mol}^{-1}\text{ s}^{-1}}{k_a'/\text{s}^{-1}} = \frac{2.\overline{7}\times10^{9}}{1.\overline{7}\times10^{7}} = \boxed{1.\overline{6}\times10^{2}}$$

COMMENT. The data define a good straight line, as the correlation coefficient $R^2 = 0.996$ shows. The straight line appears to go through the origin, but the best-fit equation gives a small non-zero y-intercept. Inspection of the plot shows that several of the data points lie about as far from the fit line as the y-intercept does from zero. This suggests that the y-intercept has a fairly high relative uncertainty and so do the rate constants.

P19.15 (a) The fluorescence intensity is proportional to the concentration of fluorescing species, so

$$\frac{I_f}{I_0} = \frac{[S]}{[S]_0} = e^{-t/\tau_0}\ [19.62] \quad \text{so} \quad \ln\left(\frac{I_f}{I_0}\right) = -\frac{t}{\tau_0}$$

A plot of $\ln(I_f/I_0)$ against t should be linear with a slope equal to $-1/\tau_0$ (i.e., $\tau_0 = -1/\text{slope}$) and an intercept equal to zero. See Figure 19.7. The plot is linear, with slope $-0.150\ \text{ns}^{-1}$, so

$$\tau_0 = -1/(-0.14\overline{5}\ \text{ns}^{-1}) = \boxed{6.9\ \text{ns}}$$

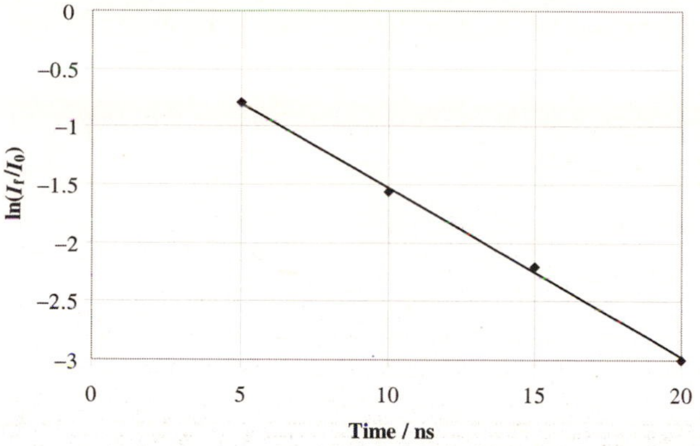

Figure 19.7

Alternatively, average the experimental values of $\frac{1}{t}\ln\left(\frac{I_f}{I_0}\right)$ and check that the standard deviation is a small fraction of the average (it is). The average equals $-1/\tau_0$ (i.e., $\tau_0 = -1/\text{average}$).

(b) The quantum yield for fluorescence is related to the rate constants for the various decay mechanisms of the excited state by

$$\phi_f = \frac{k_f}{k_f + k_{ISC} + k_{IC}}\ [19.64] = k_f\tau_0\ [19.63]$$

so $$k_f = \phi_f/\tau_0 = 0.70/(6.9\ \text{ns}) = \boxed{0.10\overline{1}\ \text{ns}^{-1}}$$

P19.17 Proceed as in Problem 19.15. In the absence of a quencher, a plot of $\ln I_f/I_0$ against t should be linear with a slope equal to $-1/\tau_0$. The plot is in fact linear with a best-fit slope of $-1.00\overline{4}\ \mu\text{s}^{-1}$. (See Figure 19.8)

$$\tau_0 = \frac{-1}{-1.00\overline{4}\ \mu\text{s}^{-1}} = 9.96\ \mu\text{s}$$

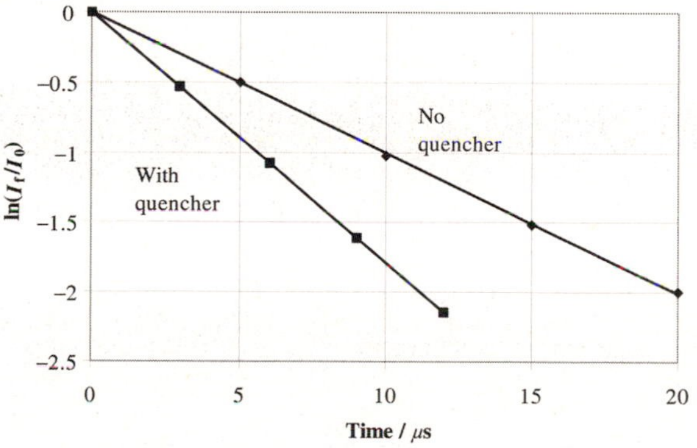

Figure 19.8

In the presence of a quencher, a graph of $\ln I_f/I_0$ against t is still linear but has a slope equal to $-1/\tau$. This plot is found to be linear with a regression slope equal to $-1.78\overline{8}\ \mu\text{s}^{-1}$.

$$\tau = \frac{-1}{-1.78\overline{8}\ \mu\text{s}^{-1}} = 5.59\ \mu\text{s}$$

The rate constant for quenching (i.e., for energy transfer to the quencher) can be obtained from

$$\frac{1}{\tau} = \frac{1}{\tau_0} + k_Q[\text{Q}] \quad \text{[Example 19.8]}$$

Thus $k_Q = \frac{\tau^{-1} - \tau_0^{-1}}{[N_2]} = \frac{RT(\tau^{-1} - \tau_0^{-1})}{p_{N_2}}$

$$= \frac{(0.08206\,\text{dm}^3\,\text{atm}\,\text{K}^{-1}\,\text{mol}^{-1})(300\,\text{K})(0.1788 - 0.1004)\times(10^{-6}\,\text{s})^{-1}}{9.74\times10^{-4}\,\text{atm}}$$

$$= \boxed{1.98\times10^{9}\,\text{dm}^3\,\text{mol}^{-1}\,\text{s}^{-1}}$$

Solutions to theoretical problems

P19.19 The rate of change of [A] is

$$\frac{d[A]}{dt} = -k_r[A]^n$$

Hence, $\int_{[A]_0}^{[A]} \frac{d[A]}{[A]^n} = -k_r \int_0^t dt = -k_r t$

Therefore, $k_r t = \left(\frac{1}{n-1}\right)\times\left(\frac{1}{[A]^{n-1}} - \frac{1}{[A]_0^{n-1}}\right)$

At $t = t_{1/2}$, $[A] = [A]_0/2$

$$k_r t_{1/2} = \left(\frac{1}{n-1}\right)\times\left(\frac{2^{n-1}}{[A]_0^{n-1}} - \frac{1}{[A]_0^{n-1}}\right) = \left(\frac{2^{n-1}-1}{n-1}\right)\times\left(\frac{1}{[A]_0^{n-1}}\right)$$

and $t_{1/2} = \boxed{\frac{2^{n-1}-1}{k_r(n-1)[A]_0^{n-1}}}$ [as in eqn. 19.15]

Now let $t_{1/3}$ be the time at which $[A] = [A]_0/3$. Substitute these expressions into the integrated rate law:

$$k_r t_{1/3} = \left(\frac{1}{n-1}\right)\times\left(\frac{3^{n-1}}{[A]_0^{n-1}} - \frac{1}{[A]_0^{n-1}}\right) = \left(\frac{3^{n-1}-1}{n-1}\right)\times\left(\frac{1}{[A]_0^{n-1}}\right)$$

and $t_{1/3} = \boxed{\frac{3^{n-1}-1}{k_r(n-1)[A]_0^{n-1}}}$

P19.21 $v = \frac{d[P]}{dt} = k_r[A][B]$

Let the initial concentrations be $[A]_0 = A_0$, $[B]_0 = B_0$, and $[P]_0 = 0$. Then when P is formed in concentration x, the concentration of A changes to $A_0 - 2x$ and that of B changes to $B_0 - 3x$. Therefore,

$$\frac{d[P]}{dt} = \frac{dx}{dt} = k_r(A_0 - 2x)(B_0 - 3x) \text{ with } x = 0 \text{ at } t = 0$$

$$\int_0^t k_r dt = \int_0^x \frac{dx}{(A_0 - 2x)\times(B_0 - 3x)}$$

Apply partial fractions decomposition to the integrand on the right.

$$\int_0^t k_r dt = \int_0^x \left(\frac{6}{2B_0 - 3A_0}\right)\times\left(\frac{1}{3(A_0 - 2x)} - \frac{1}{2(B_0 - 3x)}\right)dx$$

$$= \left(\frac{-1}{(2B_0 - 3A_0)}\right)\times\left(\int_0^x \frac{dx}{x - (1/2)A_0} - \int_0^x \frac{dx}{x - (1/3)B_0}\right)$$

$$k_r t = \left(\frac{-1}{(2B_0 - 3A_0)}\right) \times \left[\ln\left(\frac{x - \frac{1}{2}A_0}{-\frac{1}{2}A_0}\right) - \ln\left(\frac{x - \frac{1}{3}B_0}{-\frac{1}{3}B_0}\right)\right]$$

$$= \left(\frac{-1}{2B_0 - 3A_0}\right)\ln\left(\frac{(2x - A_0)B_0}{A_0(3x - B_0)}\right)$$

$$= \boxed{\left(\frac{1}{3A_0 - 2B_0}\right)\ln\left(\frac{(2x - A_0)B_0}{A_0(3x - B_0)}\right)}$$

P19.23 The rate equations are

$$\frac{d[A]}{dt} = -k_a[A] + k_a'[B]$$

$$\frac{d[B]}{dt} = k_a[A] - k_a'[B] - k_b[B] + k_b'[C]$$

$$\frac{d[C]}{dt} = k_b[B] - k_b'[C]$$

These equations are a set of coupled differential equations. Although it is not immediately apparent, they have a closed-form general solution; however, we are looking for the particular circumstances under which the mechanism reduces to the second form given. Since the reaction involves an intermediate, let us explore the result of applying the steady-state approximation to it. Then

$$\frac{d[B]}{dt} = k_a[A] - k_a'[B] - k_b[B] + k_b'[C] \approx 0$$

and $$[B] \approx \frac{k_a[A] + k_b'[C]}{k_a' + k_b}$$

Therefore, $$\frac{d[A]}{dt} = -\frac{k_a k_b}{k_a' + k_b}[A] + \frac{k_a' k_b'}{k_a' + k_b}[C].$$

This rate expression may be compared to that given in the text [Section 19.4] for a first-order reaction approaching equilibrium.

$$A \underset{k_r'}{\overset{k_r}{\rightleftharpoons}} C$$

Here $$k_r = \frac{k_a k_b}{k_a' + k_b} \quad \text{and} \quad k_r' = \frac{k_a' k_b'}{k_a' + k_b}$$

The solutions are $$[A] = \left(\frac{k_r' + k_r e^{-(k_r' + k_r)t}}{k_r' + k_r}\right) \times [A]_0 \quad [19.19]$$

and $[C] = [A]_0 - [A]$

Thus, the conditions under which the first mechanism given reduces to the second are the conditions under which the steady-state approximation holds, namely, when B can be treated as a steady-state intermediate.

P19.25 Let the forward rates be written as

$$r_1 = k_1[A] \qquad r_2 = k_2[B] \qquad r_3 = k_3[C]$$

and the reverse rates as

$$r_1' = k_1'[B] \qquad r_2' = k_2'[C] \qquad r_3' = k_3'[D]$$

The net rates are then

$$R_1 = k_1[A] - k_1'[B] \qquad R_2 = k_2[B] - k_2'[C] \qquad R_3 = k_3[C] - k_3'[D]$$

But $[A] = [A]_0$ and $[D]=0$, so the steady-state equations for the net rates of the individual steps are

$$k_1[A]_0 - k_1'[B] = k_2[B] - k_2'[C] = k_3[C]$$

From the second of these equations we find

$$[C] = \frac{k_2[B]}{k_2' + k_3}$$

After inserting this expression for [C] into the first of the steady-state equations, we obtain

$$[B] = \frac{k_1[A]_0 + k_2'[C]}{k_1' + k_2} = \frac{k_1[A]_0 + k_2'\left(\frac{k_2[B]}{k_2'+k_3}\right)}{k_1'+k_2}$$

which yields, on isolating [B],

$$[B] = [A]_0 \times \frac{k_1}{k_1' + k_2 - \left(\frac{k_2 k_2'}{k_2'+k_3}\right)}$$

Thus, at the steady state,

$$R_1 = R_2 = R_3 = [A]_0 k_1 \times \left(1 - \frac{k_1}{k_1' + k_2 - \left(\frac{k_2 k_2'}{k_2'+k_3}\right)}\right) = \boxed{\frac{k_1 k_2 k_3 [A]_0}{k_1' k_2' + k_1' k_3 + k_2 k_3}}$$

COMMENT. At steady state, not only are the net rates of reactions 1, 2, and 3 steady but so are the concentrations [B] and [C]. That is,

$$\frac{d[B]}{dt} = k_1[A]_0 - (k_1' + k_2)[B] + k_2'[C] \approx 0$$

and $$\frac{d[C]}{dt} = k_2[B] - (k_2' + k_3)[C] \approx 0$$

In fact, another approach to solving the problem is to solve *these* equations for [B] and [C].

P19.27 The number-average molar mass of the polymer is the average chain length times the molar mass of the monomer.

$$\langle M \rangle = \langle N \rangle M_1 = \frac{M_1}{1-p} \quad [19.49a]$$

The probability P_N that a polymer consists of N monomers is equal to the probability that it has $N-1$ reacted end groups and one unreacted end group. The former probability is p^{N-1}; the latter $1-p$. Therefore, the total probability of finding an N-mer is

$$P_N = p^{N-1}(1-p)$$

We need this probability to get at $\langle M^2 \rangle$, again using number averaging:

$$\langle M^2 \rangle = M_1^2 \langle N^2 \rangle = M_1^2 \sum_N N^2 P_N = M_1^2 (1-p) \sum_N N^2 p^{N-1}$$

$$= M_1^2(1-p)\frac{d}{dp} p \frac{d}{dp} \sum_N p^N = M_1^2(1-p)\frac{d}{dp} p \frac{d}{dp}(1-p)^{-1} = \frac{M_1^2(1+p)}{(1-p)^2}$$

Thus, $$\langle M^2 \rangle_N - \langle M \rangle_N^2 = M_1^2\left(\frac{1+p}{(1-p)^2} - \frac{1}{(1-p)^2}\right) = \frac{pM_1^2}{(1-p)^2}$$

and $$\boxed{\left(\langle M^2\rangle_N - \langle M\rangle_N^{\,2}\right)^{1/2} = \frac{p^{1/2}M_1}{1-p}}$$

The time dependence is obtained from

$$p = \frac{k_r t[\mathrm{A}]_0}{1+k_r t[\mathrm{A}]_0} \quad [19.48]$$

and $$\langle N\rangle = \frac{1}{1-p} = 1 + k_r t[\mathrm{A}]_0 \qquad [19.49\text{a and }19.49\text{b}]$$

Hence, $$\frac{p^{1/2}}{1-p} = p^{1/2}(1+k_r t[\mathrm{A}]_0) = \{k_r t[\mathrm{A}]_0(1+k_r t[\mathrm{A}]_0)\}^{1/2}$$

and $$\left(\langle M^2\rangle_N - \langle M\rangle_N^{\,2}\right)^{1/2} = \boxed{M_1\{kt[\mathrm{A}]_0(1+kt[\mathrm{A}]_0)\}^{1/2}}$$

P19.29 In termination by disproportionation, the chain carriers do not combine. The average number of monomers in a polymer molecule equals the average number in a chain carrier when it terminates, namely, the kinetic chain length, ν.

$$\langle N\rangle = \nu = \boxed{k_r[\cdot\mathrm{M}][\mathrm{I}]^{-1/2}}$$

COMMENT. Contrast this result to the reasoning before eqn. 19.58, in which the average number is the sum of the average numbers of a pair of combining radicals.

P19.31 The rates of the individual steps are

$$\mathrm{A} \rightarrow \mathrm{B} \qquad \frac{\mathrm{d}[\mathrm{B}]}{\mathrm{d}t} = I_a$$

$$\mathrm{B} \rightarrow \mathrm{A} \qquad \frac{\mathrm{d}[\mathrm{B}]}{\mathrm{d}t} = -k_r[\mathrm{B}]^2$$

In the photostationary state, $I_a - k_r[\mathrm{B}]^2 = 0$. Hence,

$$[\mathrm{B}] = \boxed{\left(\frac{I_a}{k_r}\right)^{1/2}}$$

This concentration can differ significantly from an equilibrium distribution because changing the illumination may change the rate of the forward reaction without affecting the reverse reaction. Contrast this situation to the corresponding equilibrium expression, in which $[\mathrm{B}]_{eq}$ depends on a ratio of rate constants for the forward and reverse reactions. In the equilibrium case, the rates of forward and reverse reactions cannot be changed *independently*.

Solutions to applications

P19.33 A simple but practical approach is to make an initial guess at the order by observing whether the half-life of the reaction appears to depend on concentration. If it does not, the reaction is first-order; if it does, it may be second-order. Examination of the data shows that the half-life is roughly 90 minutes, but it is not exactly constant. (Compare the 60–150-minute data to the 150–240-minute data; in both intervals the concentration drops by roughly half. Then examine the 30–120-minute interval, where the concentration drops by less than half.) If the reaction is first-order, it will obey

$$\ln\left(\frac{c}{c_0}\right) = -k_r t \quad [19.10\text{b}]$$

If it is second-order, it will obey

$$\frac{1}{c} = k_r t + \frac{1}{c_0} \quad [19.13b]$$

See whether a first-order plot of ln c versus time or a second-order plot of $1/c$ versus time has a substantially better fit. We draw up the following table.

t / min	30	60	120	150	240	360	480
c / (ng cm^{-3})	699	622	413	292	152	60	24
(ng cm^{-3}) / c	0.00143	0.00161	0.00242	0.00342	0.00658	0.0167	0.0412
ln {c/(ng cm^{-3})}	6.550	6.433	6.023	5.677	5.024	4.094	3.178

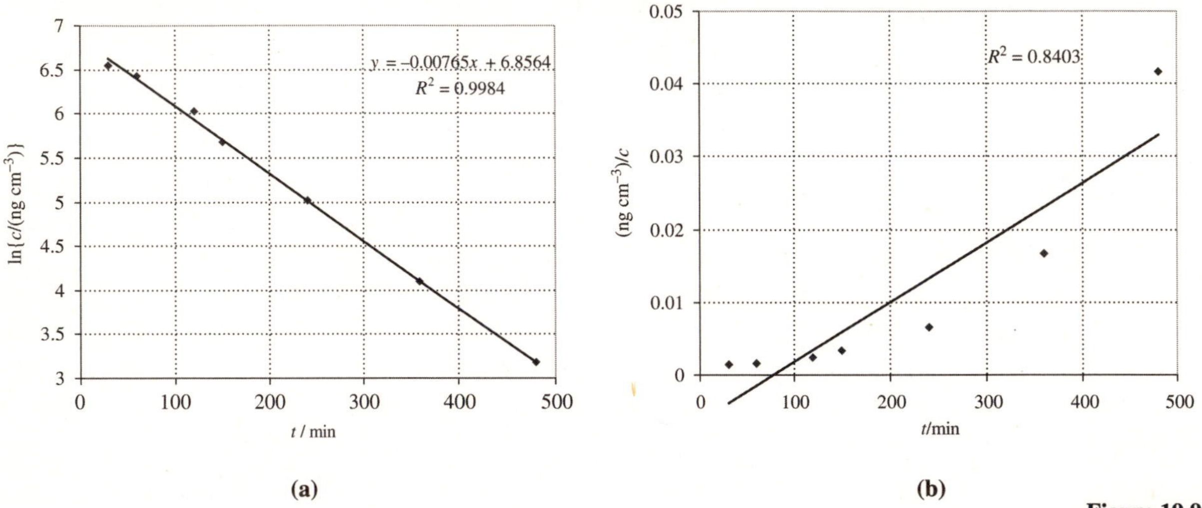

Figure 19.9

The data are plotted in parts (a) and (b), of Figure 19.9. The first-order plot fits closely to a straight line with just a hint of curvature near the outset. The second-order plot, conversely, is strongly curved throughout. Hence, the reaction is $\boxed{\text{first-order}}$. The rate constant is the slope of the first-order plot: $k_r = \boxed{0.00765\ \text{min}^{-1}} = \boxed{0.459\ \text{h}^{-1}}$. The half-life is (eqn. 19.11)

$$t_{1/2} = \frac{\ln 2}{k_r} = \frac{\ln 2}{0.459\ \text{h}^{-1}} = \boxed{1.51\ \text{h}} = \boxed{91\ \text{min}}$$

COMMENT. As noted in the problem, the drug concentration is a result of absorption and elimination of the drug, two processes with distinct rates. Elimination is characteristically slower, so the later data points reflect elimination only because absorption is effectively complete by then. The earlier data points, by contrast, reflect both absorption and elimination. It is therefore not surprising that the early points do not adhere so closely to the line so well defined by the later data.

P19.35 The initial rate is

$$v_0 = (3.6\times10^{6}\ \text{dm}^9\ \text{mol}^{-3}\ \text{s}^{-1})\times(5\times10^{-5}\,\text{mol dm}^{-3})^2\times(10^{-5.6}\ \text{mol dm}^{-3})^2$$
$$= \boxed{6\times10^{-14}\,\text{mol dm}^{-3}\ \text{s}^{-1}}$$

The half-life for a second-order reaction is

$$t_{1/2} = \frac{1}{k_{\text{eff}}[\text{HSO}_3^-]_0} \quad [19.13b]$$

where k_{eff} is the rate constant in the expression

$$-\frac{\text{d}[\text{HSO}_3^-]}{\text{d}t} = 2v = k_{\text{eff}}[\text{HSO}_3^-]^2$$

Comparison to the given rate law and rate constant shows

$$k_{\text{eff}} = 2k_r[\text{H}^+]^2 = 2(3.6\times10^6\ \text{dm}^9\ \text{mol}^{-3}\ \text{s}^{-1})\times(10^{-5.6}\ \text{mol dm}^{-3})^2$$
$$= 4.5\times10^{-5}\,\text{dm}^3\ \text{mol}^{-1}\ \text{s}^{-1}$$

and $$t_{1/2} = \frac{1}{(4.5\times10^{-5}\ \text{dm}^3\ \text{mol}^{-1}\ \text{s}^{-1})\times(5\times10^{-5}\,\text{mol dm}^{-3})} = \boxed{4.\bar{4}\times10^8\ \text{s} = 14\ \text{yr}}$$

P19.37 The rate of reaction is the rate at which ozone absorbs photons times the quantum yield. The rate at which ozone absorbs photons is the rate at which photons impinge on the ozone times the fraction of photons absorbed. That fraction is 1 – *T*, where *T* is the transmittance. *T* is related to the absorbance *A* by

$$A = -\log T = \varepsilon cl \quad \text{so} \quad 1 - T = 1 - 10^{-\varepsilon cl}$$

and $$1-T = 1-10^{-\{(260\,\text{dm}^3\ \text{mol}^{-1}\ \text{cm}^{-1})\times(8\times10^{-9}\ \text{mol dm}^{-3})\times(10^5\ \text{cm})} = 0.38$$

If we let *F* stand for the flux of photons (the rate at which photons impinge on our sample of ozone), then the rate of reaction is

$$v = \phi(1-T)F = (0.94)\times(0.38)\times\frac{(1\times10^{14}\ \text{cm}^{-2}\ \text{s}^{-1})\times(1000\,\text{cm}^3\ \text{dm}^{-3})}{(6.022\times10^{23}\ \text{mol}^{-1})\times(10^5\ \text{cm})}$$
$$= \boxed{5.9\times10^{-13}\ \text{mol dm}^{-3}\ \text{s}^{-1}}$$

20 Molecular reaction dynamics

Answers to discussion questions

D20.1 The parameter A is called the pre-exponential factor or the frequency factor. The parameter E_a is called the activation energy. Collectively, the two quantities are called the Arrhenius parameters.

Plots of ln k_r versus $1/T$ (called Arrhenius plots) yield a straight line for many reactions over temperature ranges of interest. In such plots the intercept of the line at $1/T = 0$ (at infinite temperature) is ln A and the slope is $-E_a/R$. For some reactions over some temperature ranges, though, the plot is not linear, and we describe the temperature dependence as non-Arrhenius. However, it is still possible to define an activation energy as

$$E_a = RT^2\left(\frac{d\ln k_r}{dT}\right) \quad [20.2]$$

This definition reduces to the earlier one (as the slope of a straight line) for a temperature-independent activation energy. The latter definition is more general, however, because it allows E_a to be obtained from the instantaneous slope of an Arrhenius plot even if it is not a straight line (i.e., the tangent of the curve at the temperature of interest). Non-Arrhenius behavior is sometimes a sign that quantum mechanical tunneling is playing a significant role in the reaction.

D20.3 Collision theory expresses a rate of reaction as a fraction of the rate of collision, on the assumption that reaction happens only between colliding molecules, and then only if the collision has enough energy and the proper orientation. So the rate of reaction is directly proportional to the rate of collision, and the expression of this rate comes directly from kinetic-molecular theory. The fraction of collisions energetic enough for reaction also comes from kinetic-molecular theory via the Boltzmann distribution of energy.

D20.5 In collision theory, the steric factor or P-factor is introduced as the probability that a collision energetic enough to lead to reaction actually does lead to reaction. This probability can be interpreted in various ways, such as the probability of proper orientation of colliding reactants. In the RRK model, the P-factor is the probability that sufficient energy is concentrated in a particular mode of motion that leads to reaction; thus, the complexity of a reacting molecule affects the steric P-factor. The RRK expression for the P factor is

$$P = \left(1-\frac{E^*}{E}\right)^{s-1} \quad [20.19a]$$

where E^* is the energy needed to break the bond-critical for reaction, E is the energy available within the molecule for distribution over various modes of motion, and s is the number of modes.

D20.7 The Eyring equation (eqn. 20.40) results from activated complex theory, which is an attempt to account for the rate constants of bimolecular reactions of the form $A+B \rightleftharpoons C^a \rightarrow P$ in terms of the formation of an activated complex. In the formulation of the theory, it is assumed that the activated complex and the reactants are in equilibrium, and the concentration of activated complex is calculated in terms of an equilibrium constant, which in turn is calculated from the partition functions of the reactants and a postulated form of the activated complex. It is further supposed that one normal mode of the activated complex, the one corresponding to displacement along the reaction coordinate, has a very low force constant. Displacement along this mode leads to products, provided that the complex enters a certain configuration of its atoms, known as the transition state. The derivation of the equilibrium constant from the partition functions leads to eqn. 20.39 and in turn to eqn. 20.40, the Eyring equation. See Section 20.6 for a more extensive discussion.

D20.9 An expression of the rate constant for electron transfer is given by eqn. 20.58:

$$k_{\mathrm{et}} = CH_{\mathrm{DA}}(r)^2 \mathrm{e}^{-\Delta^{\ddagger}G/RT}$$

This rate constant depends on the distance between donor and acceptor (r) through the function $H_{\mathrm{DA}}(r)^2$, which is given by eqn. 20.57; k_{et} decays exponentially with increasing r. The parameter β in eqns. 20.57 and 20.61 determines how strongly k_{et} depends on distance. This parameter changes with the transfer medium. The standard Gibbs energy of the electron-transfer process ($\Delta_r G^\circ$) affects the rate through the activation Gibbs energy and a quantity called the reorganization energy (λ), as shown in eqn. 20.59. In systems where the reorganization energy is constant, the dependence of ln k_{et} on $\Delta_r G^\circ$ is given by eqn. 20.62; it is an inverted parabola in which the maximum rate occurs when $\Delta_r G^\circ = \lambda$. A more thorough discussion can be found in Section 20.8.

D20.11 The saddle point on the potential energy surface corresponds to the transition state of a reaction. The saddle-point energy is the minimum energy required for reaction; it is the minimum energy for a path on the potential energy surface that leads from reactants to products. Because many paths on the surface between reactants and products do not pass through the saddle point, they necessarily pass through points of greater energy, so the activation energy can be greater than the saddle-point energy. Thus, the saddle-point energy is a lower limit to the activation energy.

D20.13 Attractive and repulsive potential-energy surfaces are discussed in Section 20.11(b). An attractive surface is one whose saddle point is closer to reactants than to products so that the transition state occurs early in the reaction. On such a surface, trajectories in which excess energy is translational tend to end in products, whereas trajectories in which the reactant is vibrationally excited tend not to cross the saddle point and end in products. Conversely, on a repulsive surface, the oscillatory motion of a trajectory that has excess vibrational energy in the reactant enhances the likelihood that the trajectory will end in products rather than simply reflect back to reactants.

Solutions to exercises

E20.1(a)

$$\ln k_r(T_1) = \ln A - \frac{E_a}{RT_1} \qquad \ln k_r(T_2) = \ln A - \frac{E_a}{RT_2} \quad [20.1\mathrm{a}]$$

Subtract these two expressions and solve for E_a:

$$E_a = \frac{R\ \ln\left(\dfrac{k_r(T_2)}{k_r(T_1)}\right)}{\left(\dfrac{1}{T_1}-\dfrac{1}{T_2}\right)} = \frac{(8.3145\,\mathrm{J\,K^{-1}\,mol^{-1}})\times\ln\left(\dfrac{2.67\times10^{-2}}{3.80\times10^{-3}}\right)}{\dfrac{1}{308\,\mathrm{K}}-\dfrac{1}{323\,\mathrm{K}}} = \boxed{106\,\mathrm{kJ\,mol^{-1}}}$$

For A, we rearrange eqn. 20.1(b):

$$A = k_r \times \mathrm{e}^{E_a/RT} = (3.80\times10^{-3}\,\mathrm{mol\,dm^{-3}\,s^{-1}})\times \mathrm{e}^{106\times10^3/(8.3145\times308)}$$

$$= \boxed{6.5\times10^{15}\,\mathrm{mol\,dm^{-3}\,s^{-1}}}$$

E20.2(a) Proceed as in Exercise 20.1(a):

$$E_a = \frac{R\ \ln\left(\dfrac{k_r(T_2)}{k_r(T_1)}\right)}{\left(\dfrac{1}{T_1}-\dfrac{1}{T_2}\right)} = \frac{(8.3145\ \mathrm{J\,K^{-1}\ mol^{-1}})\times\ln 3}{\dfrac{1}{297\,\mathrm{K}}-\dfrac{1}{322\,\mathrm{K}}} = \boxed{35\,\mathrm{kJ\,mol^{-1}}}$$

E20.3(a) Let the steps be

$$\mathrm{A+B} \rightleftharpoons \mathrm{I}\ (\text{fast: } k_a,\ k_a') \qquad \text{and} \qquad \mathrm{I}\rightarrow\mathrm{P} \qquad (k_b)$$

Then the rate of reaction is

$$v = \frac{\mathrm{d[P]}}{\mathrm{d}t} = k_b[\mathrm{I}]$$

Applying the pre-equilibrium approximation yields

$$\frac{[\mathrm{I}]}{[\mathrm{A}][\mathrm{B}]} = K = \frac{k_a}{k_a'} \qquad \text{so} \qquad [\mathrm{I}] = \frac{k_a[\mathrm{A}][\mathrm{B}]}{k_a'}$$

and $$v = \frac{k_a k_b[\mathrm{A}][\mathrm{B}]}{k_a'} = k_r[\mathrm{A}][\mathrm{B}] \qquad \text{with} \qquad k_r = \frac{k_a k_b}{k_a'}$$

Thus $E_a = E_a(a) + E_a(b) - E_a'(a)$ [20.4] $= (25 + 10 - 38)\ \mathrm{kJ\ mol^{-1}} = \boxed{-3\ \mathrm{kJ\ mol^{-1}}}$

COMMENT. Activation energies are rarely negative; however, some composite reactions are known to have small, negative activation energies.

E20.4(a) The collision frequency is

$$z = \sigma\bar{c}_{\mathrm{rel}}\mathcal{N}_A \ [20.9]$$

where $\bar{c}_{\mathrm{rel}} = \left(\dfrac{16kT}{\pi m}\right)^{1/2}$ [20.10], $\sigma = \pi d^2$ [20.7b] $= 4\pi R^2$, and $\mathcal{N}_A = \dfrac{p}{kT}$

Therefore, $$z = \sigma\mathcal{N}_A\left(\frac{16kT}{\pi m}\right)^{1/2} = 16pR^2\left(\frac{\pi}{mkT}\right)^{1/2}$$

$$= 16\times(120\times10^3\ \mathrm{Pa})\times(190\times10^{-12}\ \mathrm{m})^2$$

$$\times\left(\frac{\pi}{17.03\ \mathrm{u}\times1.661\times10^{-27}\ \mathrm{kg\ u^{-1}}\times1.381\times10^{-23}\ \mathrm{J\,K^{-1}}\times303\ \mathrm{K}}\right)^{1/2}$$

$$= \boxed{1.13\times10^{10}\ \mathrm{s^{-1}}}$$

The collision density is

$$Z = \frac{z\mathcal{N}_A}{2}\ [20.12\mathrm{a}] = \frac{z}{2}\left(\frac{p}{kT}\right) = \frac{1.13\times10^{10}\ \mathrm{s^{-1}}}{2}\left(\frac{120\times10^3\ \mathrm{Pa}}{1.381\times10^{-23}\ \mathrm{J\,K^{-1}}\times303\ \mathrm{K}}\right)$$

$$= \boxed{1.62\times10^{35}\ \mathrm{s^{-1}\ m^{-3}}}$$

For the percentage increase at constant volume, note that $\mathcal{N}_A$ is constant at constant volume, so the only constant-volume temperature dependence on z (and on Z) is in the speed factor.

$$z \propto T^{1/2} \quad \text{so} \quad \frac{1}{z}\left(\frac{\partial z}{\partial T}\right)_V = \frac{1}{2T} \quad \text{and} \quad \frac{1}{Z}\left(\frac{\partial Z}{\partial T}\right)_V = \frac{1}{2T}$$

Therefore $\dfrac{\delta z}{z} = \dfrac{\delta T}{Z} \approx \dfrac{\delta T}{2T} = \dfrac{1}{2}\left(\dfrac{10\text{ K}}{303\text{ K}}\right) = 0.017$

so both z and Z increase by about $\boxed{1.7\text{ percent}}$.

E20.5(a) The fraction of collisions having at least E_a along the line of flight can be inferred by dividing out of the collision-theory rate constant (eqn. 20.18) the factors that can be identified as belonging to the steric factor or collision rate $f = e^{-E_a/RT}$.

$$\text{(a) (i)} \frac{E_a}{RT} = \frac{20\times10^3\text{ J mol}^{-1}}{(8.3145\text{ J K}^{-1}\text{ mol}^{-1})\times(350\text{ K})} = 6.9 \quad \text{so} \quad f = e^{-6.9} = \boxed{1.04\times10^{-3}}$$

$$\text{(ii)} \frac{E_a}{RT} = \frac{20\times10^3\text{ J mol}^{-1}}{(8.3145\text{ J K}^{-1}\text{ mol}^{-1})\times(900\text{ K})} = 2.67 \quad \text{so} \quad f = e^{-2.67} = \boxed{0.069}$$

$$\text{(b) (i)} \frac{E_a}{RT} = \frac{100\times10^3\text{ J mol}^{-1}}{(8.3145\text{ J K}^{-1}\text{ mol}^{-1})\times(350\text{ K})} = 34.4 \quad \text{so} \quad f = e^{-34.4} = \boxed{1.19\times10^{-15}}$$

$$\text{(ii)} \frac{E_a}{RT} = \frac{100\times10^3\text{ J mol}^{-1}}{(8.3145\text{ J K}^{-1}\text{ mol}^{-1})\times(900\text{ K})} = 13.4 \quad \text{so} \quad f = e^{-13.4} = \boxed{1.57\times10^{-6}}$$

E20.6(a) A straightforward approach would be to compute $f = e^{-E_a/RT}$ at the new temperature and compare it to that at the old temperature. An approximate approach would be to note that f changes from $f_0 = e^{-E_a/RT}$ to $\exp\left(\dfrac{-E_a}{RT(1+x)}\right)$, where x is the fractional increase in the temperature. If x is small, the exponent changes from $-E_a/RT$ to approximately $-E_a(1-x)/RT$ and f changes from f_0 to

$$f \approx e^{-E_a(1-x)/RT} = e^{-E_a/RT}(e^{-E_a/RT})^{-x} = f_0 f_0^{-x}$$

Thus the new fraction is the old one times a factor of f_0^{-x}. The increase in f expressed as a percentage is

$$\frac{f-f_0}{f_0}\times100\% = \frac{f_0 f_0^{-x} - f_0}{f_0}\times100\% = (f_0^{-x}-1)\times100\%$$

(a) (i) $f_0^{-x} = (1.04\times10^{-3})^{-10/350} = 1.22$ and the percentage change is $\boxed{22\%}$

(ii) $f_0^{-x} = (0.069)^{-10/900} = 1.03$ and the percentage change is $\boxed{3\%}$

(b) (i) $f_0^{-x} = (1.19\times10^{-15})^{-10/350} = 2.7$ and the percentage change is $\boxed{170\%}$

(ii) $f_0^{-x} = (1.57\times10^{-6})^{-10/900} = 1.16$ and the percentage change is $\boxed{16\%}$

E20.7(a) $$k_r = P\sigma\left(\frac{8kT}{\pi\mu}\right)^{1/2} N_A e^{-E_a/RT} \quad [20.18]$$

We are not given a steric factor, so assume that $P = 1$.

$$k_r = 0.36\times(10^{-9}\,\text{m})^2\times\left(\frac{8\times(1.381\times10^{-23}\,\text{J K}^{-1})\times(650\,\text{K})}{\pi\times(3.32\times10^{-27}\,\text{kg})}\right)^{1/2}\times(6.022\times10^{23}\,\text{mol}^{-1})$$
$$\times\exp\left(\frac{-171\times10^{3}\,\text{J}}{(8.3145\,\text{J K}^{-1}\,\text{mol}^{-1})\times(650\,\text{K})}\right)$$
$$= \boxed{1.03\times10^{-5}\,\text{m}^3\,\text{mol}^{-1}\,\text{s}^{-1}} = \boxed{1.03\times10^{-2}\,\text{dm}^3\,\text{mol}^{-1}\,\text{s}^{-1}}$$

COMMENT. Assuming that $P = 1$ is done because it is convenient in the absence of additional information, not because it is likely to be accurate.

E20.8(a) The rate constant for a diffusion-controlled bimolecular reaction is

$$k_d = 4\pi R^* D N_A \quad [20.22]$$

where $D = D_A + D_B = 2\times(6\times10^{-9}\,\text{m}^2\,\text{s}^{-1}) = 1.2\times10^{-8}\,\text{m}^2\,\text{s}^{-1}$

$$k_d = 4\pi\times(0.5\times10^{-9}\,\text{m})\times(1.2\times10^{-8}\,\text{m}^2\,\text{s}^{-1})\times(6.022\times10^{23}\,\text{mol}^{-1})$$

$$k_d = \boxed{4.\bar{5}\times10^{7}\,\text{m}^3\,\text{mol}^{-1}\,\text{s}^{-1}} = \boxed{4.\bar{5}\times10^{10}\,\text{dm}^3\,\text{mol}^{-1}\,\text{s}^{-1}}$$

E20.9(a) The rate constant for a diffusion-controlled bimolecular reaction is

$$k_d = \frac{8RT}{3\eta}\ [20.25] = \frac{8\times(8.3145\,\text{J K}^{-1}\,\text{mol}^{-1})\times(298\,\text{K})}{3\eta} = \frac{6.61\times10^{3}\,\text{J mol}^{-1}}{\eta}$$

(a) For water, $\eta = 1.00\times10^{-3}\,\text{kg m}^{-1}\,\text{s}^{-1}$.

$$k_d = \frac{6.61\times10^{3}\,\text{J mol}^{-1}}{1.00\times10^{-3}\,\text{kg m}^{-1}\,\text{s}^{-1}} = \boxed{6.61\times10^{6}\,\text{m}^3\,\text{mol}^{-1}\,\text{s}^{-1}} = \boxed{6.61\times10^{9}\,\text{dm}^3\,\text{mol}^{-1}\,\text{s}^{-1}}$$

(b) For pentane, $\eta = 2.2\times10^{-4}\,\text{kg m}^{-1}\,\text{s}^{-1}$.

$$k_d = \frac{6.61\times10^{3}\,\text{J mol}^{-1}}{2.2\times10^{-4}\,\text{kg m}^{-1}\,\text{s}^{-1}} = \boxed{3.0\times10^{7}\,\text{m}^3\,\text{mol}^{-1}\,\text{s}^{-1}} = \boxed{3.0\times10^{10}\,\text{dm}^3\,\text{mol}^{-1}\,\text{s}^{-1}}$$

E20.10(a) The rate constant for a diffusion-controlled bimolecular reaction is

$$k_d = \frac{8RT}{3\eta} = \frac{8\times(8.3145\,\text{J K}^{-1}\,\text{mol}^{-1})\times(320\,\text{K})}{3\times(0.89\times10^{-3}\,\text{kg m}^{-1}\,\text{s}^{-1})}$$
$$= \boxed{8.0\times10^{6}\,\text{m}^3\,\text{mol}^{-1}\,\text{s}^{-1}} = \boxed{8.0\times10^{9}\,\text{dm}^3\,\text{mol}^{-1}\,\text{s}^{-1}}$$

Since this reaction is elementary bimolecular, it is second-order; hence,

$$t_{1/2} = \frac{1}{2k_d[\text{A}]_0} \quad [\text{Table 19.3, with } k_r = 2k_d \text{ because 2 atoms are consumed}]$$

so $$t_{1/2} = \frac{1}{2\times(8.0\times10^{9}\,\text{dm}^3\,\text{mol}^{-1}\,\text{s}^{-1})\times(1.5\times10^{-3}\,\text{mol dm}^{-3})} = \boxed{4.2\times10^{-8}\,\text{s}}$$

E20.11(a) The steric factor, P, is

$$P = \frac{\sigma^*}{\sigma} \quad [\text{Section 20.3(c)}]$$

The mean collision cross-section is $\sigma = \pi d^2$ with $d = (d_A + d_B)/2$.

Get the diameters from the collision cross-sections:

$$d_A = (\sigma_A/\pi)^{1/2} \qquad \text{and} \qquad d_B = (\sigma_B/\pi)^{1/2}$$

so $$\sigma = \frac{\pi}{4}\left\{\left(\frac{\sigma_A}{\pi}\right)^{1/2} + \left(\frac{\sigma_B}{\pi}\right)^{1/2}\right\}^2 = \frac{(\sigma_A^{1/2}+\sigma_B^{1/2})^2}{4} = \frac{\{(0.95\ \text{nm}^2)^{1/2}+(0.65\ \text{nm}^2)^{1/2}\}^2}{4}$$

$$= \boxed{0.79\ \text{nm}^2}$$

Therefore, $P = \dfrac{9.2\times10^{-22}\ \text{m}^2}{0.79\times(10^{-9}\ \text{m})^2} = \boxed{1.16\times10^{-3}}$.

E20.12(a) Since the reaction is diffusion-controlled, the rate-limiting step is bimolecular and therefore second-order; hence,

$$\frac{d[P]}{dt} = k_d[A][B]$$

where $k_d = 4\pi R^* D N_A [20.22] = 4\pi N_A R^*(D_A + D_B)$

$$= 4\pi N_A \times (R_A + R_B) \times \frac{kT}{6\pi\eta}\left(\frac{1}{R_A}+\frac{1}{R_B}\right)[20.24] = \frac{2RT}{3\eta}(R_A+R_B)\times\left(\frac{1}{R_A}+\frac{1}{R_B}\right)$$

$$k_d = \frac{2\times(8.3145\ \text{J K}^{-1}\ \text{mol}^{-1})\times(313\text{K})}{3\times(2.93\times10^{-3}\ \text{kg m}^{-1}\ \text{s}^{-1})}\times(655+1820)\times\left(\frac{1}{655}+\frac{1}{1820}\right)$$

$$= 3.04\times10^6\ \text{m}^3\ \text{mol}^{-1}\ \text{s}^{-1} = 3.04\times10^9\ \text{dm}^3\ \text{mol}^{-1}\ \text{s}^{-1}$$

Therefore, the initial rate is

$$\frac{d[P]}{dt} = (3.04\times10^9\ \text{dm}^3\ \text{mol}^{-1}\ \text{s}^{-1})\times(0.170\,\text{mol dm}^{-3})\times(0.350\,\text{mol dm}^{-3})$$

$$= \boxed{1.81\times10^8\ \text{mol dm}^{-3}\ \text{s}^{-1}}$$

COMMENT. If the approximation of eqn. 20.25 is used, $k_d = 2.37\times10^9\ \text{dm}^3\ \text{mol}^{-1}\ \text{s}^{-1}$. In this case the approximation results in a difference of about 25% compared to the expression used above.

E20.13(a) The enthalpy of activation for a bimolecular solution reaction is [Section 20.7(a)]

$$\Delta^{\ddagger}H = E_a - RT = (8681\ \text{K} - 303\ \text{K})\times 8.3145\ \text{J mol}^{-1}\ \text{K}^{-1} = \boxed{+69.7\ \text{kJ mol}^{-1}}$$

$$k_r = B\,e^{\Delta^{\ddagger}S/R}e^{-\Delta^{\ddagger}H/RT},\quad B = \left(\frac{kT}{h}\right)\times\left(\frac{RT}{p^{\ominus}}\right)[20.46] = \frac{kRT^2}{hp^{\ominus}}$$

$$= B\,e^{\Delta^{\ddagger}S/R}e^{-E_a/RT}e = A\,e^{-E_a/RT}$$

Therefore, $A = e\,B\,e^{\Delta^{\ddagger}S/R}$, implying that $\Delta^{\ddagger}S = R\left(\ln\dfrac{A}{B}-1\right)$

$$B = \frac{(1.381\times10^{-23}\,\text{J K}^{-1})\times(8.3145\,\text{J K}^{-1}\ \text{mol}^{-1})\times(303\text{K})^2}{6.626\times10^{-34}\,\text{J s}\times10^5\ \text{Pa}}$$

$$= 1.59\times10^{11}\,\text{m}^3\ \text{mol}^{-1}\ \text{s}^{-1} = 1.59\times10^{14}\,\text{dm}^3\ \text{mol}^{-1}\ \text{s}^{-1}$$

and hence $$\Delta^{\ddagger}S = R\left[\ln\left(\frac{2.05\times10^{13}\ \text{dm}^3\ \text{mol}^{-1}\ \text{s}^{-1}}{1.59\times10^{14}\ \text{dm}^3\ \text{mol}^{-1}\ \text{s}^{-1}}\right)-1\right]$$

$$= 8.3145\ \text{J K}^{-1}\ \text{mol}^{-1}\times(-3.05) = \boxed{-25\ \text{J K}^{-1}\ \text{mol}^{-1}}.$$

E20.14(a) The enthalpy of activation for a bimolecular solution reaction is [Section 20.7a]

$$\Delta^{\ddagger}H = E_a - RT = 8.3145\ \text{J K}^{-1}\ \text{mol}^{-1} \times (9134\ \text{K} - 303\ \text{K}) = \boxed{+73.4\ \text{kJ mol}^{-1}}$$

The entropy of activation is [Exercise 20.13a]

$$\Delta^{\ddagger}S = R\left(\ln\frac{A}{B} - 1\right)$$

with $B = \dfrac{kRT^2}{hp^{\ominus}} = 1.59 \times 10^{14}\ \text{dm}^3\ \text{mol}^{-1}\ \text{s}^{-1}$ [Exercise 20.13a]

Therefore, $\Delta^{\ddagger}S = 8.3145\ \text{J K}^{-1}\ \text{mol}^{-1} \times \left[\ln\left(\dfrac{7.78\times10^{14}}{1.59\times10^{14}}\right) - 1\right] = +4.9\ \text{J K}^{-1}\ \text{mol}^{-1}$.

Hence, $\Delta^{\ddagger}G = \Delta^{\ddagger}H - T\Delta^{\ddagger}S = \{73.4 - (303) \times (4.9 \times 10^{-3})\}\ \text{kJ mol}^{-1} = \boxed{+71.9\ \text{kJ mol}^{-1}}$.

E20.15(a) Use eqn. 20.47(a) to relate a bimolecular gas-phase rate constant to activation energy and entropy:

$$k_r = e^2 B e^{\Delta^{\ddagger}S/R} e^{-E_a/RT}$$

where $B = \left(\dfrac{kT}{h}\right) \times \left(\dfrac{RT}{p^{\ominus}}\right)$ [20.46] $= \dfrac{(1.381\times10^{-23}\ \text{J K}^{-1}) \times (338\ \text{K})^2 \times (8.3145\ \text{J mol}^{-1}\ \text{K}^{-1})}{(6.626\times10^{-34}\ \text{J s}) \times (10^5\ \text{Pa})}$

$$= 1.98\times10^{11}\ \text{m}^3\ \text{mol}^{-1}\ \text{s}^{-1}$$

Solve for the entropy of activation:

$$\Delta^{\ddagger}S = R\left(\ln\frac{k_r}{B} - 2\right) + \frac{E_a}{T}$$

The derivations in Section 20.7(a) are based on a k_r that contains concentration units, whereas the rate constant given here has pressure units. So k_r is not the given rate constant, but

$$k_r = 7.84 \times 10^{-3}\ \text{kPa}^{-1}\ \text{s}^{-1} \times RT$$

$$k_r = 7.84 \times 10^{-3} \times (10^3\ \text{Pa})^{-1}\ \text{s}^{-1} \times 8.3145\ \text{J K}^{-1}\ \text{mol}^{-1} \times 338\ \text{K} = 0.0220\ \text{m}^3\ \text{mol}^{-1}\ \text{s}^{-1}$$

Hence $\Delta^{\ddagger}S = 8.3145\ \text{J K}^{-1}\ \text{mol}^{-1} \times \left(\ln\dfrac{0.0220\ \text{m}^3\ \text{mol}^{-1}\ \text{s}^{-1}}{1.98\times10^{11}\ \text{m}^3\ \text{mol}^{-1}\ \text{s}^{-1}} - 2\right) + \dfrac{58.6\times10^3\ \text{J mol}^{-1}}{338\ \text{K}}$

$$= \boxed{-91\ \text{J K}^{-1}\ \text{mol}^{-1}}$$

E20.16(a) For a bimolecular gas-phase reaction,

$$\Delta^{\ddagger}S = R\left(\ln\frac{k_r}{B} - 2\right) + \frac{E_a}{T}\ \text{[Exercise 20.15(a)]} = R\left(\ln\frac{A}{B} - \frac{E_a}{RT} - 2\right) + \frac{E_a}{T} = R\left(\ln\frac{A}{B} - 2\right)$$

where $B = \dfrac{kRT^2}{hp^{\ominus}}$ [Exercise 20.13(a)]

For two structureless particles, the rate constant is

$$k_r = N_A \sigma^* \left(\frac{8kT}{\pi\mu}\right)^{1/2} e^{-\Delta E_0/RT}\ [20.42]$$

The activation energy is [20.2]

$$E_a = RT^2 \frac{d\ln k_r}{dT} = RT^2 \frac{d}{dT}\left(\ln N_A\sigma^* + \frac{1}{2}\ln\frac{8k}{\pi\mu} + \frac{1}{2}\ln T - \frac{\Delta E_0}{RT}\right)$$

$$= RT^2\left(\frac{1}{2T} + \frac{\Delta E_0}{RT^2}\right) = \Delta E_0 + \frac{RT}{2},$$

so the pre-factor is

$$A = k_r e^{E_a/RT} = N_A\sigma^*\left(\frac{8kT}{\pi\mu}\right)^{1/2} e^{-\Delta E_0/RT}(e^{\Delta E_0/RT}e^{1/2}) = N_A\sigma^*\left(\frac{8kT}{\pi\mu}\right)^{1/2} e^{1/2}$$

Hence,

$$\Delta^{\ddagger}S = R\left\{\ln N_A\sigma^*\left(\frac{8kT}{\pi\mu}\right)^{1/2} + \frac{1}{2} - \ln\frac{kRT^2}{p^{\ominus}h} - 2\right\} = R\left\{\ln\frac{\sigma^* p^{\ominus}h}{(kT)^{3/2}}\left(\frac{8}{\pi\mu}\right)^{1/2} - \frac{3}{2}\right\}$$

For identical particles,

$$\mu = m/2 = (65\ \text{u})(1.661\times10^{-27}\ \text{kg u}^{-1})/2 = 5.4\times10^{-26}\ \text{kg}$$

and hence,

$$\Delta^{\ddagger}S = 8.3145\ \text{J K}^{-1}\ \text{mol}^{-1}$$

$$\times\left\{\ln\frac{0.35\times(10^{-9}\ \text{m})^2\times10^5\ \text{Pa}\times6.626\times10^{-34}\ \text{J s}}{(1.381\times10^{-23}\ \text{J K}^{-1}\times300\ \text{K})^{3/2}}\left(\frac{8}{\pi\times5.4\times10^{-26}\ \text{kg}}\right)^{1/2} - \frac{3}{2}\right\}$$

$$= \boxed{-74\ \text{J K}^{-1}\ \text{mol}^{-1}}$$

E20.17(a) At low pressure, the reaction can be assumed to be bimolecular (see Chapter 19.7).

(a) $\Delta^{\ddagger}S = R\left(\ln\frac{A}{B} - 2\right)$ [Exercise 20.16(a)]

where $B = \frac{kRT^2}{hp^{\ominus}}$ [Exercise 20.13(a)] $= \frac{1.381\times10^{-23}\ \text{J K}^{-1}\times8.3145\ \text{J K}^{-1}\ \text{mol}^{-1}\times(298\ \text{K})^2}{6.626\times10^{-34}\ \text{J s}\times10^5\ \text{Pa}}$

$$= 1.54\times10^{11}\ \text{m}^3\ \text{mol}^{-1}\ \text{s}^{-1} = 1.54\times10^{14}\ \text{dm}^3\ \text{mol}^{-1}\ \text{s}^{-1}$$

Hence, $\Delta^{\ddagger}S = 8.3145\ \text{J K}^{-1}\ \text{mol}^{-1}\times\left(\ln\frac{4.6\times10^{12}\ \text{dm}^3\ \text{mol}^{-1}\ \text{s}^{-1}}{1.54\times10^{14}\ \text{dm}^3\ \text{mol}^{-1}\ \text{s}^{-1}} - 2\right)$

$$= \boxed{-46\ \text{J K}^{-1}\ \text{mol}^{-1}}$$

(b) The enthalpy of activation for a bimolecular gas-phase reaction is [Section 20.7(a)] $\Delta^{\ddagger}H = E_a - 2RT = 10.0\ \text{kJ mol}^{-1} - 2\times8.3145\ \text{J mol}^{-1}\ \text{K}^{-1}\times298\ \text{K} = \boxed{+5.0\ \text{kJ mol}^{-1}}$.

(c) The Gibbs energy of activation at 298 K is

$$\Delta^{\ddagger}G = \Delta^{\ddagger}H - T\Delta^{\ddagger}S = 5.0\ \text{kJ mol}^{-1} - (298\ \text{K})\times(-46\times10^{-3}\ \text{kJ K}^{-1}\ \text{mol}^{-1})$$

$$\Delta^{\ddagger}G = \boxed{+18.7\ \text{kJ mol}^{-1}}$$

E20.18(a) Use eqn. 20.51 to examine the effect of ionic strength on a rate constant:

$$\log k_r = \log k_r^{\circ} + 2A|z_Az_B|I^{1/2}$$

Hence, $\log k_r^{\circ} = \log k_r - 2A|z_Az_B|I^{1/2} = \log 12.2 - 2\times0.509\times|1\times(-1)|\times(0.0525)^{1/2} = 0.85$

and $k_r^{\circ} = \boxed{7.1\ \text{dm}^6\ \text{mol}^{-2}\ \text{min}^{-1}}$

E20.19(a) The rate constant for electron transfer is

$$k_{et} = C\{H_{DA}(r)\}^2 e^{-\Delta^{\ddagger}G/RT} \quad [20.58]$$

The reorganization energy, λ, appears in two of these factors:

$$\Delta^{\ddagger}G = \frac{(\Delta_r G^{\ominus} + \lambda)^2}{4\lambda} \quad [20.59] \text{ and } C = \frac{1}{h}\left(\frac{\pi^3}{\lambda RT}\right)^{1/2} \quad [20.60]$$

So $$k_{et} = \frac{H_{DA}{}^2}{h}\left(\frac{\pi^3}{\lambda RT}\right)^{1/2} \exp\left(\frac{-(\Delta_r G^{\ominus} + \lambda)^2}{4\lambda RT}\right) = \frac{H_{DA}{}^2}{h}\left(\frac{\pi^3}{\lambda kT}\right)^{1/2} \exp\left(\frac{-(\Delta_r G^{\ominus} + \lambda)^2}{4\lambda kT}\right)$$

depending on whether the energies are expressed in molar units or molecular units. The only unknown in this equation is λ. Isolating λ analytically is not possible; however, one can solve for it numerically using the root-finding command of a symbolic mathematics package, or graphically by plotting the right-hand side versus the (constant) left-hand side and finding the value of λ at which the two lines cross. Before we put in numbers, we must make sure to use compatible units. We recognize that $H_{DA}(r)$ and $\Delta_r G^{\ominus}$ are both given in molecular units, but that the former is really a wavenumber rather than an energy. So we choose to express all energies in molecular units, namely, joules:

$$H_{DA}(r) = hc \times 0.04 \text{ cm}^{-1} = (6.626 \times 10^{-34} \text{ J s}) \times (2.998 \times 10^{10} \text{ cm s}^{-1}) \times (0.04 \text{ cm}^{-1})$$

$$H_{DA}(r) = 8\times10^{-25} \text{ J}$$

$$\frac{H_{DA}{}^2}{h}\left(\frac{\pi^3}{kT}\right)^{1/2} = \frac{(8\times10^{-25} \text{ J})^2}{6.626\times10^{-34} \text{ J s}}\left(\frac{\pi^3}{1.381\times10^{-23} \text{ J K}^{-1}\times298 \text{ K}}\right)^{1/2} = 8\times10^{-5} \text{ J}^{0.5} \text{ s}^{-1}$$

$$\Delta_r G^{\ominus} = -0.185 \text{ eV} \times 1.602\times10^{-19} \text{ J eV}^{-1} = -2.96\times10^{-20} \text{ J}$$

and $$4kT = 4 \times (1.381\times10^{-23} \text{ J K}^{-1}) \times (298 \text{ K}) = 1.65\times10^{-20} \text{ J}$$

Thus $$37.5 = 8\times10^{-5}\left(\frac{\text{J}}{\lambda}\right)^{1/2} \exp\left(\frac{-(-2.96\times10^{-20} \text{ J} + \lambda)^2}{\lambda\times1.65\times10^{-20} \text{ J}}\right)$$

where $\lambda = \boxed{4\times10^{-21} \text{ J}}$ or $\boxed{2 \text{ kJ mol}^{-1}}$

E20.20(a) For the same donor and acceptor at different distances, eqn. 20.61 applies:

$$\ln k_{et} = -\beta r + \text{constant}$$

The slope of a plot of k_{et} versus r is $-\beta$. The slope of a line defined by two points is

$$\text{slope} = \frac{\Delta y}{\Delta x} = \frac{\ln k_{et,2} - \ln k_{et,1}}{r_2 - r_1} = -\beta = \frac{\ln 4.51\times10^4 - \ln 2.02\times10^5}{(1.23 - 1.11) \text{ nm}}$$

so $\beta = \boxed{12.\bar{5} \text{ nm}^{-1}}$

Solutions to problems

Solutions to numerical problems

P20.1 If the rate constant obeys the Arrhenius equation (eqn. 20.1a), a plot of $\ln k_r$ against $1/T$ should yield a straight line with slope $-E_a/R$ (eqn. 20.1b). Construct a table as follows.

K_r / s	θ/°C	10^3 K / T	ln k_r
2.46×10^{-3}	0	3.66	−6.01
0.0451	20	3.41	−3.10
0.576	40	3.19	−0.552

The points are plotted in Figure 20.1.

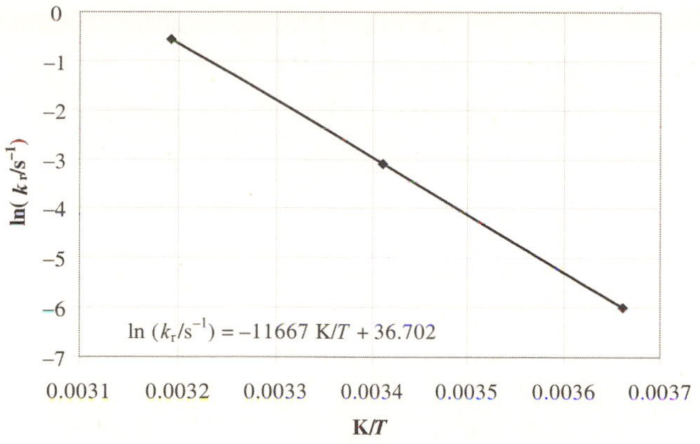

Figure 20.1

The best-fit straight line fits the three data points very well:

$$\ln(k_r/\text{s}^{-1}) = -1.17 \times 10^4\ \text{K}/T + 36.7$$

so $E_a = -(-1.17 \times 10^4\ \text{K}) \times (8.3145\ \text{J mol}^{-1}\ \text{K}^{-1}) = \boxed{9.70 \times 10^4\ \text{J mol}^{-1}} = \boxed{97.0\ \text{kJ mol}^{-1}}$

P20.3 The relation between the equilibrium constant and the rate constants is obtained from

$$\Delta_r G^\ominus = -RT \ln K = \Delta_r H^\ominus - T\Delta_r S^\ominus$$

so $$K = \frac{k_r}{k_r'} = \exp\left(\frac{-\Delta_r H^\ominus}{RT}\right)\exp\left(\frac{\Delta_r S^\ominus}{R}\right) = \left(\frac{A}{A'}\right)\exp\left(\frac{E_a' - E_a}{RT}\right)$$

Setting the temperature-dependent parts equal yields

$$\Delta_r H^\ominus = E_a - E_a' = [-4.2 - (53.3)]\ \text{kJ mol}^{-1} = -57.5\ \text{kJ mol}^{-1}$$

Setting the temperature-independent parts equal yields

$$\exp\left(\frac{\Delta_r S^\ominus}{R}\right) = \left(\frac{A}{A'}\right)$$

so $$\Delta_r S^\ominus = R\ln\left(\frac{A}{A'}\right) = (8.3145\ \text{J K}^{-1}\ \text{mol}^{-1})\ln\left(\frac{1.0\times10^9}{1.4\times10^{11}}\right) = -41.1\ \text{J K}^{-1}\ \text{mol}^{-1}$$

The thermodynamic quantities of the reaction are related to standard molar quantities, most of which can be looked up in thermodynamic data tables.

$$\Delta_r H^\ominus = \Delta_f H^\ominus(C_2H_6) + \Delta_f H^\ominus(\text{Br}) - \Delta_f H^\ominus(C_2H_5) - \Delta_f H^\ominus(\text{HBr})$$

so $$\Delta_f H^\ominus(C_2H_5) = \Delta_f H^\ominus(C_2H_6) + \Delta_f H^\ominus(\text{Br}) - \Delta_f H^\ominus(\text{HBr}) - \Delta_r H^\ominus$$

and $$\Delta_f H^\ominus(C_2H_5) = [(-84.68) + 111.88 - (-36.40) - (-57.5)]\ \text{kJ mol}^{-1} = \boxed{+121.1\ \text{kJ mol}^{-1}}$$

Similarly,

$$S_m^\ominus(C_2H_5) = [229.60 + 175.02 - 198.70 - (-41.1)]\ \text{J mol}^{-1}\ \text{K}^{-1} = \boxed{247.0\ \text{J K}^{-1}\ \text{mol}^{-1}}$$

Finally, $\Delta_f G^{\ominus}(C_2H_5) = \{-32.82 + 82.396 - (-53.45)\}\,\text{kJ mol}^{-1} - \Delta_r G^{\ominus}$

$$= 103.03\,\text{kJ mol}^{-1} - \Delta_r G^{\ominus}$$

but $\Delta_r G^{\ominus} = \Delta_r H^{\ominus} - T\Delta_r S^{\ominus} = -57.5\,\text{kJ mol}^{-1} - (298\,\text{K}) \times (-41.1 \times 10^{-3}\,\text{kJ K}^{-1}\,\text{mol}^{-1})$

$$= -45.3\,\text{kJ mol}^{-1}$$

so $\Delta_f G^{\ominus}(C_2H_5) = [103.03 - (-45.3)]\,\text{kJ mol}^{-1} = \boxed{+148.3\text{ kJ mol}^{-1}}$

P20.5 Draw up the following table as the basis of an Arrhenius plot.

T / K	600	700	800	1000
10^3 K / T	1.67	1.43	1.25	1.00
k_r / (cm^3 mol^{-1} s^{-1})	4.6×10^2	9.7×10^3	1.3×10^5	3.1×10^6
ln (k_r / cm^3 mol^{-1} s^{-1})	6.13	9.18	11.8	14.9

The points are plotted in Figure 20.2.

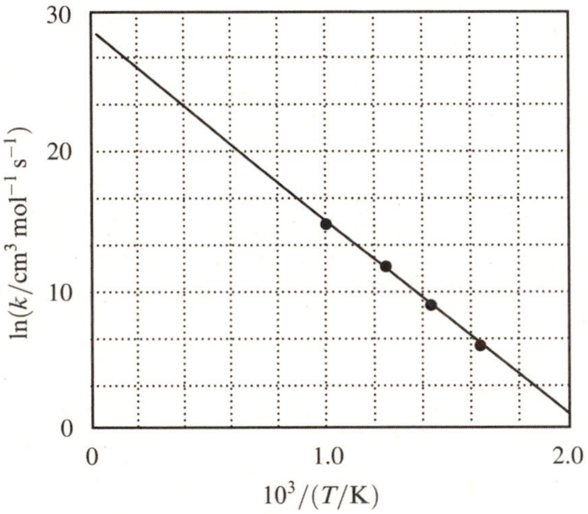

Figure 20.2

The least-squares intercept is at 28.3, which implies that

$$A / (\text{cm}^3\,\text{mol}^{-1}\,\text{s}^{-1}) = e^{28.3} = 2.0 \times 10^{12}$$

But comparison of eqn. 20.18 to the Arrhenius equation tells us that $A = N_A P\sigma \left(\frac{8kT}{\pi\mu}\right)^{1/2}$

so $$P = \frac{A}{N_A \sigma}\left(\frac{\pi\mu}{8kT}\right)^{1/2}$$

The reduced mass is

$$\mu = m(NO_2)/2 = 46\,\text{u} \times (1.661 \times 10^{-27}\,\text{kg u}^{-1}) / 2 = 3.8 \times 10^{-26}\,\text{kg}$$

so evaluating P in the center of the range of temperatures spanned by the data,

$$P = \frac{2.0 \times 10^{12} \times (10^{-2}\,\text{m})^3\,\text{mol}^{-1}\,\text{s}^{-1}}{(6.022 \times 10^{23}\,\text{mol}^{-1}) \times 0.60 \times (10^{-9}\,\text{m})^2} \times \left(\frac{\pi \times 3.8 \times 10^{-26}\,\text{kg}}{8 \times 1.381 \times 10^{-23}\,\text{J K}^{-1} \times 800\,\text{K}}\right)^{1/2}$$

$$= \boxed{6.5 \times 10^{-3}}.$$

$$\sigma^* = P\sigma = (6.5 \times 10^{-3}) \times (0.60\,\text{nm}^2) = \boxed{3.9 \times 10^{-3}\,\text{nm}^2} = \boxed{3.9 \times 10^{-21}\,\text{m}^2}$$

P20.7 Draw up the following table for an Arrhenius plot.

θ / °C	−24.82	−20.73	−17.02	−13.00	−8.95
T / K	248.33	252.42	256.13	260.15	264.20
10^3 K / T	4.027	3.962	3.904	3.844	3.785
$10^4\, k_r$ / s^{-1}	1.22	2.31	4.39	8.50	14.3
ln (k_r / s^{-1})	−9.01	−8.37	−7.73	−7.07	−6.55

The points are plotted in Figure 20.3.

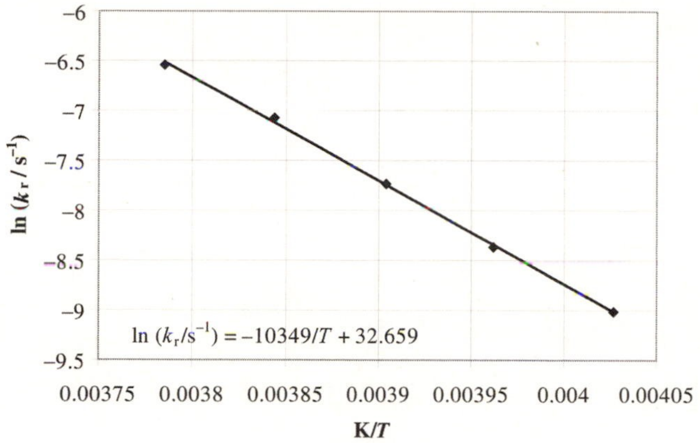

Figure 20.3

A least-squares fit of the data yields the intercept +32.7 at $1/T = 0$, which implies that

$\ln\left(\frac{A}{\text{s}^{-1}}\right) = 32.7$ and hence that $A = 1.53 \times 10^{14}\ \text{s}^{-1}$. The slope is

$$-1.035\times10^4\ \text{K} = E_a/R \qquad \text{and hence } \boxed{E_a = 86.0\ \text{kJ mol}^{-1}}$$

In solution $\Delta^{‡}H = E_a - RT$ [Section 20.7(a)], so at −20°C,

$$\Delta^{‡}H = 86.0\ \text{kJ mol}^{-1} - (8.3145\ \text{J mol}^{-1}\ \text{K}^{-1}) \times (253\ \text{K}) = \boxed{+83.9\ \text{kJ mol}^{-1}}$$

We assume that the reaction is first-order, for which, by analogy to Section 20.6(c),

$$K^{‡} = \frac{kT}{h\nu}\bar{K}^{‡} \qquad \text{and} \qquad k_r = k^{‡}K^{‡} = \nu \times \frac{kT}{h\nu} \times \bar{K}^{‡}$$

with $\Delta^{‡}G = -RT \ln \bar{K}^{‡}$

Therefore, $k_r = A\,\text{e}^{-E_a/RT} = \frac{kT}{h}\text{e}^{-\Delta^{‡}G/RT} = \frac{kT}{h}\text{e}^{\Delta^{‡}S/R}\,\text{e}^{-\Delta^{‡}H/RT}$.

We can identify $\Delta^{‡}S$ by writing

$$k_r = \frac{kT}{h}\text{e}^{\Delta^{‡}S/R}\,\text{e}^{-E_a/RT}\,\text{e} = A\,\text{e}^{-E_a/RT}$$

and hence obtain

$$\Delta^{‡}S = R\left[\ln\left(\frac{hA}{kT}\right) - 1\right]$$

$$= 8.3145\ \text{J K}^{-1}\ \text{mol}^{-1} \times \left[\ln\left(\frac{(6.626\times10^{-34}\ \text{J s})\times(1.53\times10^{14}\ \text{s}^{-1})}{(1.381\times10^{-23}\ \text{J K}^{-1})\times(253\ \text{K})}\right) - 1\right]$$

$$= \boxed{+19.6\ \text{J K}^{-1}\ \text{mol}^{-1}}.$$

Therefore, $\Delta^{‡}G = \Delta^{‡}H - T\Delta^{‡}S = 83.9\ \text{kJ mol}^{-1} - 253\ \text{K} \times 19.6\ \text{J K}^{-1}\ \text{mol}^{-1} = \boxed{+79.0\ \text{kJ mol}^{-1}}$.

P20.9 Figure 20.4 shows that $\log k_r$ is proportional to the ionic strength even when one of the reactants is a neutral molecule.

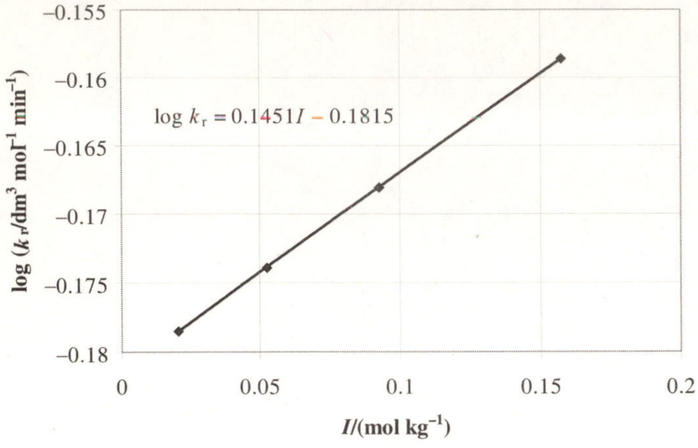

Figure 20.4

From the graph, the intercept at $I = 0$ is –0.182, so the limiting value of k_r is

$$k_r^\circ = 10^{-0.182} = \boxed{0.658\ \text{dm}^2\ \text{mol}^{-1}\ \text{min}^{-1}}$$

Compare the equation of the best-fit line to the logarithm of eqn. 20.49(b):

$$\log k_r = \log k_r^\circ - \log K_\gamma = \log k_r^\circ - \log \frac{\gamma_{C^\ddagger}}{\gamma_{I^-}\gamma_{H_2O_2}} = \log k_r^\circ + \log \frac{\gamma_{I^-}\gamma_{H_2O_2}}{\gamma_{C^\ddagger}}$$

which implies that $\log \frac{\gamma_{I^-}\gamma_{H_2O_2}}{\gamma_{C^\ddagger}} = 0.145I.$

If the Debye-Hückel limiting law holds (an approximation at best), the activity coefficients of I^- and the activated complex are equal, which would imply that $\log \gamma_{H_2O_2} = 0.145I.$

P20.11 According to the Debye-Hückel limiting law, the logarithms of ionic activity coefficients, γ, are proportional to $I^{1/2}$ (where I is ionic strength). Thus, a plot of $\log k_r$ versus $I^{1/2}$ should give a straight line whose y-intercept is $\log k_r^\circ$ and whose slope is $2Az_Az_B$, where z_A and z_B are charge numbers of the component ions of the activated complex (eqn. 20.51). The extended Debye–Hückel law has $\log \gamma$ proportional to $\left(\frac{I^{1/2}}{1+BI^{1/2}}\right)$, so it requires plotting $\log k_r$ versus $\left(\frac{I^{1/2}}{1+BI^{1/2}}\right)$, and it also has a slope of $2Az_Az_B$ and a y-intercept of $\log k_r^\circ$. The ionic strength in a 2:1 electrolyte solution is three times the molal concentration. The transformed data are in the following table and plots of $\log k_r$ are in Figure 20.5.

$[Na_2SO_4]$ / (mol kg^{-1})	0.2	0.15	0.1	0.05	0.025	0.0125	0.005
k_r / (dm$^{3/2}$ mol$^{-1/2}$ s^{-1})	0.462	0.430	0.390	0.321	0.283	0.252	0.224
$I^{1/2}$	0.775	0.671	0.548	0.387	0.274	0.194	0.122
$I^{1/2}(1+BI^{1/2})$	0.436	0.401	0.354	0.279	0.215	0.162	0.109
$\log k_r$	–0.335	–0.367	–0.409	–0.493	–0.548	–0.599	–0.650

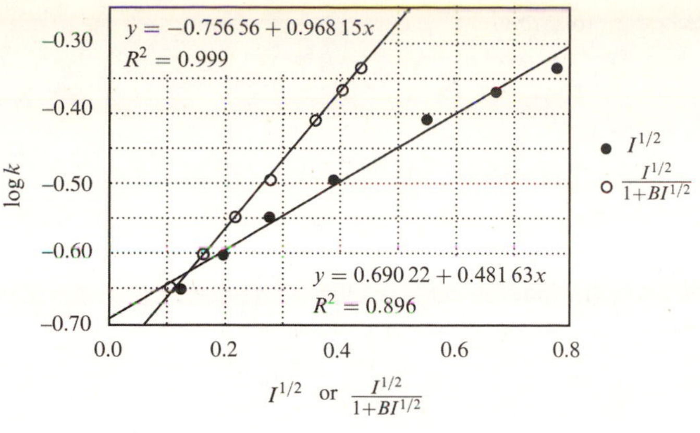

Figure 20.5

The line based on the limiting law appears curved. The zero-ionic-strength rate constant based on it is

$$k_r^\circ = 10^{-0.690}\ \mathrm{dm^{3/2}\ mol^{-1/2}\ s^{-1}} = 0.204\ \mathrm{dm^{3/2}\ mol^{-1/2}\ s^{-1}}$$

The slope is positive, so the complex must overcome repulsive interactions. The product of charges, however, works out to be 0.5, not easily interpretable in terms of charge numbers. The line based on the extended law appears straighter and has a better correlation coefficient. The zero-ionic-strength rate constant based on it is

$$k_r^\circ = 10^{-0.757}\ \mathrm{dm^{3/2}\ mol^{-1/2}\ s^{-1}} = 0.175\ \mathrm{dm^{3/2}\ mol^{-1/2}\ s^{-1}}$$

The product of charges works out to be 0.9, nearly 1, interpretable in terms of a complex of two univalent ions of the same sign.

P20.13 For a bimolecular gas-phase reaction,

$$\Delta^\ddagger H = E_a - 2RT \text{ [Section 20.7(a)]}$$

so $$\Delta^\ddagger H = 65.43\ \mathrm{kJ\ mol^{-1}} - 2\times 8.3145\ \mathrm{J\ mol^{-1}\ K^{-1}}\times 300\ \mathrm{K} = \boxed{+60.44\ \mathrm{kJ\ mol^{-1}}}$$

$$\Delta^\ddagger H = \Delta^\ddagger U + \Delta^\ddagger(pV)$$

so $$\Delta^\ddagger U = \Delta^\ddagger H - \Delta^\ddagger(pV) = \Delta^\ddagger H - RT\Delta\nu_{\mathrm{gas}}$$

$$\Delta^\ddagger U = 60.44\ \mathrm{kJ\ mol^{-1}} - (8.3145\ \mathrm{J\ K^{-1}\ mol^{-1}})\times(300\ \mathrm{K})\times(-1) = \boxed{+62.9\ \mathrm{kJ\ mol^{-1}}}$$

$$\Delta^\ddagger S = R\left(\ln\frac{A}{B} - 2\right) \text{ [Exercise 20.16(a)]}$$

where $$B = \frac{kRT^2}{hp^\ominus} = \frac{(1.381\times10^{-23}\ \mathrm{J\ K^{-1}})\times(300\ \mathrm{K})^2\times(8.3145\ \mathrm{J\ K^{-1}\ mol^{-1}})}{(6.626\times10^{-34}\ \mathrm{J\ s})\times(10^5\ \mathrm{Pa})}$$

$$= 1.56\times10^{11}\ \mathrm{m^3\ mol^{-1}\ s^{-1}} = 1.56\times10^{14}\ \mathrm{dm^3\ mol^{-1}\ s^{-1}}$$

so $$\Delta^\ddagger S = (8.3145\ \mathrm{J\ K^{-1}\ mol^{-1}})\times\left(\ln\frac{4.07\times10^5\ \mathrm{dm^3\ mol^{-1}\ s^{-1}}}{1.56\times10^{14}\ \mathrm{dm^3\ mol^{-1}\ s^{-1}}} - 2\right)$$

$$= \boxed{-181\ \mathrm{J\ K^{-1}\ mol^{-1}}}$$

$$\Delta^\ddagger G = \Delta^\ddagger H - T\Delta^\ddagger S = 60.44\ \mathrm{kJ\ mol^{-1}} - (300\ \mathrm{K})\times(-181\ \mathrm{J\ K^{-1}\ mol^{-1}})$$

$$\Delta^\ddagger G = \boxed{+114.7\ \mathrm{kJ\ mol^{-1}}}$$

P20.15 Estimate the bimolecular rate constant k_r for the reaction

$$\mathrm{Ru(bpy)_3^{3+} + Fe(H_2O)_6^{2+} \rightarrow Ru(bpy)_3^{2+} + Fe(H_2O)_6^{3+}}$$

by using the approximate Marcus cross-relation:

$$k_r \approx (k_{DD}k_{AA}K)^{1/2}$$

The standard cell potential for the reaction is

$$E^{\ominus} = E^{\ominus}_{red}\left(Ru(bpy)_3{}^{3+}\right) - E^{\ominus}_{red}\left(Fe(H_2O)_6{}^{3+}\right) = (1.26 - 0.77)\,V = 0.49\ V$$

so the equilibrium constant is

$$K = \exp\left(\frac{\nu F E^{\ominus}}{RT}\right) = \exp\left(\frac{(1)(96485\ C\ mol^{-1}\ s^{-1})(0.49\,V)}{(8.3145\ J\ K^{-1}\ mol^{-1})(298\,K)}\right) = 1.9\times10^{8}$$

The rate constant is approximately

$$k_r \approx \{(4.0\times10^{8}\ dm^3\ mol^{-1}\ s^{-1})(4.2\ dm^3\ mol^{-1}\ s^{-1})(1.9\times10^{8})\}^{1/2}$$

$$k_r \approx \boxed{5.7\times10^{8}\ dm^3\ mol^{-1}\ s^{-1}}$$

Solutions to theoretical problems

P20.17 We are to show that eqn. 20.29

$$[J]^* = [J]e^{-k_r t}$$

is a solution of eqn. 20.28,

$$\frac{\partial [J]^*}{\partial t} = D\frac{\partial^2 [J]^*}{\partial x^2} - k_r[J]^*$$

provided that [J] is a solution of

$$\frac{\partial [J]}{\partial t} = D\frac{\partial^2 [J]}{\partial x^2}$$

Evaluate the derivatives of $[J]^*$:

$$\frac{\partial [J]^*}{\partial t} = \frac{\partial [J]}{\partial t}e^{-k_r t} - k_r[J]e^{-k_r t} \qquad \text{and} \qquad \frac{\partial^2 [J]^*}{\partial x^2} = \frac{\partial^2 [J]}{\partial x^2}e^{-k_r t}$$

Use the fact that [J] is a solution in the absence of reaction

$$D\frac{\partial^2 [J]^*}{\partial x^2} = D\frac{\partial^2 [J]}{\partial x^2}e^{-k_r t} = \frac{\partial [J]}{\partial t}e^{-k_r t} = \frac{\partial [J]^*}{\partial t} + k_r[J]e^{-k_r t}$$

which gives us back eqn. 20.28, as required.

P20.19 The standard molar partition function is defined in Section 17.3(a) as the molecular partition function at standard pressure:

$$\frac{q_m^{\ominus T}}{N_A} = \frac{V_m^{\ominus}}{N_A\Lambda^3} = \frac{RT}{N_A p^{\ominus}\Lambda^3} = \frac{kT}{p^{\ominus}}\left(\frac{(T/K)^{1/2}(M/g\ mol^{-1})^{1/2}}{1749\ pm}\right)^3 \quad \text{[Table 13.1]}$$

For $T \approx 300$ K, $M \approx 50$ g mol^{-1}, $\frac{q_m^{\ominus T}}{N_A} \approx \boxed{1.4\times10^{7}}$

$$q^R(\text{nonlinear}) = \frac{1.0270}{\sigma}\times\frac{(T/K)^{3/2}}{(\tilde{A}\tilde{B}\tilde{C}/cm^{-3})^{1/2}} \quad \text{[Table 13.1]}$$

For $T \approx 300$ K, $\tilde{A} \approx \tilde{B} \approx \tilde{C} \approx 2\,\text{cm}^{-1}$, $\sigma \approx 2$, $q^{\text{R}}(\text{nonlinear}) \approx \boxed{900}$

$$q^{\text{R}}(\text{linear}) = \frac{0.6950}{\sigma} \times \frac{(T/\text{K})}{(\tilde{B}/\text{cm}^{-1})} \quad \text{[Table 13.1]}$$

For $T \approx 300$ K, $\tilde{B} \approx 1\,\text{cm}^{-1}$, $\sigma \approx 1$, $q^{\text{R}}(\text{linear}) \approx \boxed{200}$

Energies of most excited vibrational states in small molecules and of most excited electronic states are high enough to make $q^{\text{V}} \approx q^{\text{E}} \approx \boxed{1}$.

$$k_{\text{r}} = \frac{\kappa kT}{h}\bar{K}_{\text{c}}^{\ddagger}\,[20.40] = \left(\frac{\kappa kT}{h}\right) \times \left(\frac{RT}{p^{\ominus}}\right) \times \left(\frac{N_{\text{A}}\bar{q}_{\text{C}^{\ddagger}}^{\ominus}}{q_{\text{A}}^{\ominus}q_{\text{B}}^{\ominus}}\right)\text{e}^{-\Delta E_0/RT}\,[20.39] = A\text{e}^{-E_{\text{a}}/RT}$$

If A and B are structureless molecules, then we use estimates from above to evaluate.

$$\frac{q_{\text{A}}^{\ominus}}{N_{\text{A}}} = \frac{q_{\text{A}}^{\ominus\text{T}}}{N_{\text{A}}} \approx 1.4 \times 10^7 \approx \frac{q_{\text{B}}^{\ominus}}{N_{\text{A}}} = \frac{q_{\text{B}}^{\ominus\text{T}}}{N_{\text{A}}}$$

$$\frac{\bar{q}_{\text{C}^{\ddagger}}^{\ominus}}{N_{\text{A}}} = \frac{q_{\text{C}^{\ddagger}}^{\ominus\text{T}}q^{\text{R}}(\text{linear})}{N_{\text{A}}} \approx (2^{3/2}) \times (1.4 \times 10^7) \times (200) = 8 \times 10^9$$

(The factor of $2^{3/2}$ comes from $m_{\text{C}} = m_{\text{A}} + m_{\text{B}} \approx 2m_{\text{A}}$ and $q^{\text{T}} \propto m^{3/2}$.)

$$\frac{RT}{p^{\ominus}} = \frac{(8.3145\ \text{J K}^{-1}\ \text{mol}^{-1}) \times (300\ \text{K})}{10^5\ \text{Pa}} = 2.5 \times 10^{-2}\ \text{m}^3\ \text{mol}^{-1}$$

and $$\frac{\kappa kT}{h} \approx \frac{kT}{h} = \frac{(1.381 \times 10^{-23}\ \text{J K}^{-1}) \times (300\ \text{K})}{6.626 \times 10^{-34}\ \text{J s}} = 6.25 \times 10^{12}\ \text{s}^{-1}$$

Strictly speaking, E_{a} is not exactly the same as ΔE_0, but they are approximately equal—close enough for the purposes of estimating the order of magnitude of the pre-exponential factor. So once we identify $E_{\text{a}} \approx \Delta E_0$, we identify the pre-exponential factor with everything *other than* the exponential term. Therefore, the pre-exponential factor

$$A \approx \frac{(6.25 \times 10^{12}\ \text{s}^{-1}) \times (2.5 \times 10^{-2}\ \text{m}^3\ \text{mol}^{-1}) \times (8 \times 10^9)}{(1.4 \times 10^7)^2}$$

$$\approx 6.3 \times 10^6\ \text{m}^3\ \text{mol}^{-1}\ \text{s}^{-1} = \boxed{6.3 \times 10^9\ \text{dm}^3\ \text{mol}^{-1}\ \text{s}^{-1}}$$

According to collision theory [20.18],

$$A = P\sigma\left(\frac{8kT}{\pi\mu}\right)^{1/2} N_{\text{A}}$$

Take $\sigma \approx 0.5\ \text{nm}^2 = 5 \times 10^{-19}\ \text{m}^2$ as a typical value for small molecules:

$$A = P \times 5 \times 10^{-19}\ \text{m}^2 \times \left(\frac{8 \times (1.381 \times 10^{-23}\ \text{J K}^{-1}) \times (300\ \text{K})}{\pi \times (25\ \text{u}) \times (1.661 \times 10^{-27}\ \text{kg u}^{-1})}\right)^{1/2} \times 6.022 \times 10^{23}\ \text{mol}^{-1}$$

$$= 1.5 \times 10^8\ \text{m}^3\ \text{mol}^{-1}\ \text{s}^{-1} \times P.$$

The values are quite consistent, for they imply $P \approx 0.04$, which is certainly a plausible value. If A and B are non-linear triatomics, then

$$\frac{q_{\text{A}}^{\ominus}}{N_{\text{A}}} \approx (1.4 \times 10^7) \times (900) = 1.3 \times 10^{10} \approx \frac{q_{\text{B}}^{\ominus}}{N_{\text{A}}}$$

$$\frac{q_{\text{C}^{\ddagger}}^{\ominus}}{N_{\text{A}}} \approx (2^{3/2}) \times (1.4 \times 10^7) \times (900) = 3.6 \times 10^{10}$$

and $$A \approx \frac{(6.25\times10^{12}\,\mathrm{s}^{-1})\times(2.5\times10^{-2}\,\mathrm{m}^3\,\mathrm{mol}^{-1})\times(3.6\times10^{10})}{(1.3\times10^{10})^2}$$

$$\approx 33\,\mathrm{m}^3\,\mathrm{mol}^{-1}\,\mathrm{s}^{-1} = \boxed{3.3\times10^4\,\mathrm{dm}^3\,\mathrm{mol}^{-1}\,\mathrm{s}^{-1}}$$

Comparison to the expression from collision theory implies $\boxed{P = 2\times10^{-7}}$.

P20.21 The diffusion process described is unimolecular, hence first-order, and therefore analogous but not identical to the second-order case of Section 20.6. We can write

$$[\mathrm{A}^\ddagger] = K^\ddagger[\mathrm{A}] \quad \text{[analogous to 20.32]}$$

and $$-\frac{\mathrm{d}[\mathrm{A}]}{\mathrm{d}t} = k^\ddagger[\mathrm{A}^\ddagger] = \kappa\nu^\ddagger[\mathrm{A}^\ddagger] \approx \nu^\ddagger[\mathrm{A}^\ddagger] = \kappa\nu^\ddagger K^\ddagger[\mathrm{A}] = k_\mathrm{r}[\mathrm{A}] \qquad [20.33\text{–}20.35]$$

Thus, $$k_\mathrm{r} \approx \nu^\ddagger K^\ddagger = \nu^\ddagger\left(\frac{kT}{h\nu^\ddagger}\right)\times\left(\frac{\bar{q}^\ddagger}{q}\right)\mathrm{e}^{-\Delta E_0/RT}$$

where $\bar{q}^\ddagger$ and q are the (vibrational) partition functions at the top (missing one mode) and foot of the well, respectively. Let the y-direction be the direction of diffusion. Hence, for the activated atom the vibrational mode in this direction is lost, and

$\bar{q}^\ddagger = q_x^{\ddagger\mathrm{V}} q_z^{\ddagger\mathrm{V}}$ for the activated atom, and

$q = q_x^{\mathrm{V}} q_y^{\mathrm{V}} q_z^{\mathrm{V}}$ for an atom at the bottom of a well

For classical vibration, $q^\mathrm{V} \approx \dfrac{kT}{hc\tilde{\nu}}$[Table 13.1]$= \dfrac{kT}{h\nu}$.

Hence, $$k_\mathrm{r} = \frac{kT}{h}\left(\frac{(kT/h\nu^\ddagger)^2}{(kT/h\nu)^3}\right)\mathrm{e}^{-\Delta E_0/RT} = \boxed{\frac{\nu^3}{(\nu^\ddagger)^2}\mathrm{e}^{-\Delta E_0/RT}} \approx \frac{\nu^3}{(\nu^\ddagger)^2}\mathrm{e}^{-E_\mathrm{a}/RT}.$$

(a) If $\nu^\ddagger = \nu$, then $k_\mathrm{r} \approx \nu\mathrm{e}^{-E_\mathrm{a}/RT} = 10^{11}\,\mathrm{s}^{-1}\times\mathrm{e}^{-60000/(8.3145\times500)} = 5.4\times10^4\,\mathrm{s}^{-1}$

$$D = \frac{\lambda^2}{2\tau}[18.59] \approx \frac{1}{2}\lambda^2 k_\mathrm{r}\left[\tau = \frac{1}{k_\mathrm{r}}\text{(period for vibration with enough energy)}\right]$$

$$= \frac{1}{2}\times(316\times10^{-12}\,\mathrm{m})^2\times5.4\times10^4\,\mathrm{s}^{-1} = \boxed{2.7\times10^{-15}\,\mathrm{m}^2\,\mathrm{s}^{-1}}$$

(b) If $\nu^\ddagger = \nu/2$, then $k_\mathrm{r} \approx 4\nu\mathrm{e}^{-E_\mathrm{a}/RT} = 2.2\times10^5\,\mathrm{s}^{-1}$

$$D = 4\times(2.7\times10^{-15}\,\mathrm{m}^2\,\mathrm{s}^{-1}) = \boxed{1.1\times10^{-14}\,\mathrm{m}^2\,\mathrm{s}^{-1}}$$

P20.23 The change in intensity of the beam, dI, is proportional to the number of scatterers per unit volume, $\mathcal{N}$, the intensity of the beam, I, and the path length dL. The constant of proportionality must be the collision cross-section σ, the "target area" of each scatterer. Thus, σdL is the volume of scatterers to be encountered within the beam, and

$$\mathrm{d}I = -\sigma\mathcal{N}I\mathrm{d}L \quad \text{or} \quad \mathrm{d}\ln I = -\sigma\mathcal{N}\mathrm{d}L$$

If the incident intensity (at $L = 0$) is I_0 and the emergent intensity is I, we can write

$$\ln\frac{I}{I_0} = -\sigma\mathcal{N}L \quad \text{or} \quad \boxed{I = I_0\mathrm{e}^{-\sigma\mathcal{N}L}}$$

P20.25 $$\mathrm{A} + \mathrm{B} \rightarrow \mathrm{C}^\ddagger \rightarrow \mathrm{P}$$

$$k_\mathrm{r} = \left(\kappa\frac{kT}{h}\right)\times\bar{K}_\mathrm{c}^\ddagger[20.40] = \left(\kappa\frac{kT}{h}\right)\times\left(\frac{N_\mathrm{A}RT}{p^\ominus}\right)\frac{q_{\mathrm{C}^\ddagger}^\ominus}{q_\mathrm{A}^\ominus q_\mathrm{B}^\ominus}e^{-\Delta E_0/RT}[20.39]$$

We assume that the only factor that changes between the atomic and molecular case is the ratio of the partition functions. For collisions between atoms,

$$q_A^{\ominus} = q_A^{T} \approx 10^{26} \approx q_B^{\ominus} = q_B^{T}$$

$$q_C^{\ominus} = (q_C^{R})^2 q_C^{V} q_C^{T} \approx (10^{1.5})^2 \times (1) \times (10^{26}) = 10^{29}$$

$$k_r(\text{atoms}) \propto \frac{10^{29}}{10^{26} \times 10^{26}} = 10^{-23}$$

For collisions between nonlinear molecules,

$$q_A^{\ominus} = (q_A^{R})^3 (q_A^{V})^{3N-6} (q_A^{T}) \approx (10^{1.5})^3 \times (1) \times (10^{26}) = 3 \times 10^{30} \approx q_B^{\ominus}$$

$$q_C^{\ominus} = (q_C^{R})^3 (q_C^{V})^{3(N+N')-6} (q_C^{T}) \approx (10^{1.5})^3 \times (1) \times (10^{26}) = 3 \times 10^{30}$$

$$k_r(\text{molecules}) \propto \frac{3 \times 10^{30}}{1 \times 10^{61}} = 3 \times 10^{-31}$$

Therefore, $k_r(\text{atoms}) / k_r(\text{molecules}) \approx \dfrac{10^{-23}}{3 \times 10^{-31}} = \boxed{3 \times 10^{7}}$.

P20.27 For $D + A \rightarrow D^+ + A^-$

the rate constant is

$$k_r = Kk_{et}[20.55] = K\kappa\nu^{\ddagger} e^{-\Delta^{\ddagger}G/RT}[20.56]$$

(a) $$\Delta^{\ddagger}G = \frac{(\Delta_r G^{\ominus} + \lambda)^2}{4\lambda} \ [20.59]$$

so $$\Delta^{\ddagger}G_{DD} = \frac{(0 + \lambda_{DD})^2}{4\lambda_{DD}} = \frac{\lambda_{DD}}{4} \qquad \Delta^{\ddagger}G_{AA} = \frac{\lambda_{AA}}{4}$$

and $$\Delta^{\ddagger}G_{DA} = \frac{(\Delta_r G^{\ominus} + \lambda_{DA})^2}{4\lambda_{DA}} = \frac{(\Delta_r G^{\ominus})^2 + 2\Delta_r G^{\ominus}\lambda_{DA} + \lambda_{DA}{}^2}{4\lambda_{DA}}$$

(b) If $|\Delta_r G^{\ominus}| = \lambda_{DA}$, then $\Delta^{\ddagger}G_{DA} \approx \dfrac{\Delta_r G^{\ominus}}{2} + \dfrac{\lambda_{DA}}{4}$

Assume $\lambda_{DA} = \dfrac{\lambda_{AA} + \lambda_{DD}}{2} = 2(\Delta^{\ddagger}G_{AA} + \Delta^{\ddagger}G_{DD})$

Hence, $\Delta^{\ddagger}G_{DA} \approx \dfrac{\Delta_r G^{\ominus} + \Delta^{\ddagger}G_{AA} + \Delta^{\ddagger}G_{DD}}{2}$

(c) From the above expression based on eqns. 20.55 and 20.56, the rate constants for the self-exchange reactions are

$$k_{AA} = K_{AA}\kappa\nu^{\ddagger} e^{-\Delta^{\ddagger}G_{AA}/RT} \qquad \text{and} \qquad k_{DD} = K_{AA}\kappa\nu^{\ddagger} e^{-\Delta^{\ddagger}G_{DD}/RT}$$

(d) Compare these results to the rate constant for the reaction of interest:

$$k_r = K_{DA}\kappa\nu^{\ddagger} e^{-\Delta^{\ddagger}G_{DA}/RT} \approx K_{DA}\kappa\nu^{\ddagger} e^{-\Delta_r G/2RT} e^{-\Delta^{\ddagger}G_{AA}/2RT} e^{-\Delta^{\ddagger}G_{DD}/2RT}$$

(e) Thus $k_r = (k_{AA}k_{DD})^{1/2} \dfrac{K_{DA}}{(K_{AA}K_{DD})^{1/2}} e^{-\Delta_r G/2RT}$

The constants K_{DA} and so on are equilibrium constants for diffusive pairing, i.e., for steps like eqn. 20.53(a). It is reasonable to expect that $K_{DA} \approx (K_{AA}K_{DD})^{1/2}$, eliminating them from the expression. Finally, the equilibrium constant for the overall reaction is

$$e^{-\Delta_r G/RT} = K$$

so the exponential term in our expression is its square root. Therefore

$$\boxed{k_r \approx (k_{AA}k_{DD}K)^{1/2}}$$

Solutions to applications

P20.29 (a) The rate constant of a diffusion-limited reaction is

$$k_d = \frac{8RT}{3\eta}[20.25] = \frac{8\times(8.3145\text{ J K}^{-1}\text{ mol}^{-1})\times(298\text{ K})}{3\times(1.06\times10^{-3}\text{ kg m}^{-1}\text{ s}^{-1})}$$

$$= \boxed{6.23\times10^{6}\text{ m}^3\text{ mol}^{-1}\text{ s}^{-1}} = \boxed{6.23\times10^{9}\text{ dm}^3\text{ mol}^{-1}\text{s}^{-1}}$$

(b) The rate constant is related to the diffusion constants and reaction distance by

$$k_d = 4\pi R^* D N_A$$

so $$R^* = \frac{k_d}{4\pi D N_A} = \frac{(2.77\times10^{9}\text{ dm}^3\text{ mol}^{-1}\text{ s}^{-1})\times(10^{-3}\text{ m}^3\text{ dm}^{-3})}{4\pi\times(1\times10^{-9}\text{ m}^2\text{ s}^{-1})\times(6.022\times10^{23}\text{ mol}^{-1})}$$

$$= \boxed{4\times10^{-10}\text{ m}} = \boxed{0.4\text{ nm}}$$

P20.31 For a series of reactions with a fixed edge-to-edge distance and reorganization energy, the logarithm of the rate constant depends quadratically on the reaction free-energy:

$$\ln k_{et} = -\frac{(\Delta_r G^\ominus)^2}{4\lambda kT} - \frac{\Delta_r G^\ominus}{2kT} + \text{constant} \quad [20.62 \text{ in molecular units}]$$

Draw up the following table.

$\Delta_r G^\ominus$/eV	k_{et} / (10^6 s^{-1})	ln (k_{et}/s^{-1})
−0.665	0.657	13.4
−0.705	1.52	14.2
−0.745	1.12	13.9
−0.975	8.99	16.0
−1.015	5.76	15.6
−1.055	10.1	16.1

Plot ln k_{et} versus $\Delta_r G^\ominus$ (Figure 20.6).

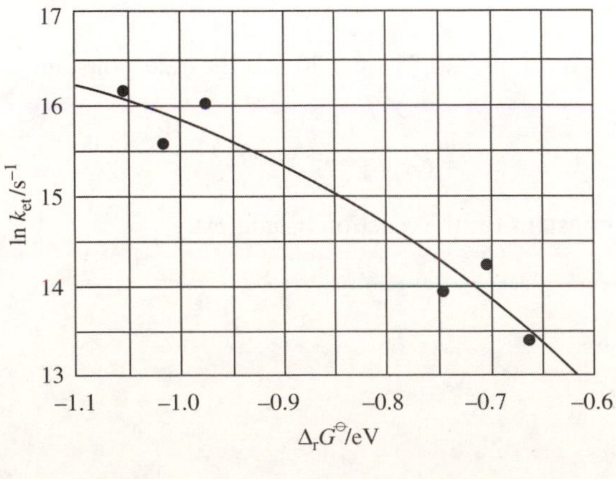

Figure 20.6

The least-squares quadratic fit equation is

$$\ln k_{\mathrm{ct}}/\mathrm{s}^{-1} = 3.23 - 21.1(\Delta_{\mathrm{r}}G^{\ominus}/\mathrm{eV}) - 8.48(\Delta_{\mathrm{r}}G^{\ominus}/\mathrm{eV})^2 \quad r^2 = 0.938$$

The coefficient of the quadratic term is

$$-\frac{1}{4\lambda kT} = -\frac{8.48}{\mathrm{eV}^2}$$

so $$\lambda = \frac{(\mathrm{eV})^2}{4(8.48)kT} = \frac{(1.602\times10^{-19}\ \mathrm{J\,eV}^{-1})(\mathrm{eV})^2}{2(8.48)(1.381\times10^{-23}\ \mathrm{J\,K}^{-1})(298\ \mathrm{K})} = \boxed{1.15\ \mathrm{eV}}$$

P20.33 The theoretical treatment of Section 20.8 applies only at relatively high temperatures. At temperatures above 130 K, the reaction in question is observed to follow a temperature dependence consistent with eqn. 20.58, namely, increasing rate with increasing temperature. Below 130 K, the temperature-dependent terms in eqn. 20.58 are replaced by Frank–Condon factors; that is, temperature-dependent terms are replaced by temperature-independent wavefunction overlap integrals.

P20.35 (a) The rate of reaction is

$$\begin{aligned} v &= k_{\mathrm{r}}[\mathrm{CH_4}][\mathrm{OH}] \\ &= (1.13\times10^{9}\ \mathrm{dm^3\ mol^{-1}\ s^{-1}})\times\exp\left(\frac{-14.1\times10^{3}\,\mathrm{J\ mol^{-1}}}{(8.3145\ \mathrm{J\,K^{-1}\ mol^{-1}})\times(263\ \mathrm{K})}\right) \\ &\quad\times(4.0\times10^{-8}\ \mathrm{mol\ dm^{-3}})\times(3.0\times10^{-15}\,\mathrm{mol\ dm^{-3}}) = \boxed{2.1\times10^{-16}\,\mathrm{mol\ dm^{-3}\ s^{-1}}}. \end{aligned}$$

(b) The mass is the amount consumed (in moles) times the molar mass; the amount consumed is the rate of consumption times the volume of the "reaction vessel" times the time.

$$\begin{aligned} m = MvVt &= (0.01604\ \mathrm{kg\ mol^{-1}})\times(2.1\times10^{-16}\ \mathrm{mol\ dm^{-3}\ s^{-1}}) \\ &\quad\times(4\times10^{21}\ \mathrm{dm^3})\times(365\times24\times3600\ \mathrm{s}) \\ &= \boxed{4.3\times10^{11}\ \mathrm{kg\ or\ 430\ Tg}}. \end{aligned}$$

21 Catalysis

Answers to discussion questions

D21.1 The **Michaelis–Menten mechanism** of enzyme activity models the enzyme with one active site, weakly and reversibly, binding a substrate in homogeneous solution. It is a three-step mechanism. The first and second steps are the reversible formation of the enzyme–substrate complex (ES). The third step is the decay of the complex into the product. The steady-state approximation is applied to the concentration of the intermediate (ES), and its use simplifies the derivation of the final rate expression. However, the justification for the use of the approximation with this mechanism is suspect. Both rate constants for the reversible step may not be as large (in comparison to the rate constant for the decay to products) as they need to be for the approximation to be valid. The mechanism clearly indicates that the simplest form of the rate law, $v = v_{max} = k_b[E]_0$, occurs when $[S]_0 \gg K_M$. In addition, the general form of the rate law does seem to match the principal experimental features of enzyme-catalyzed reactions. It provides a mechanistic understanding of both the turnover number and catalytic efficiency. The model may be expanded to include multisubstrate reactions and inhibition.

D21.3 As temperature increases we expect the rate of an enzyme-catalyzed reaction to increase. However, at a sufficiently high temperature, the enzyme **denatures** and a decrease in the reaction rate is observed. Temperature-related denaturation is caused by the action of vigorous vibrational motion, which destroys secondary and tertiary protein structure. Electrostatic, internal hydrogen bonding and van der Waals interactions that hold the protein in its active, folded shape are broken with the protein unfolding into a **random coil**. The active site and enzymatic activity are lost.

The rate of a particular enzyme-catalyzed reaction may also appear to decrease at high temperature in the special case in which an alternative substrate reaction, which has a relatively slow rate at low temperature, has the faster rate increase with increasing temperature. A temperature may be reached at which the alternative reaction predominates.

D21.5 Figure 21.1 is a sketch of the enzyme-catalyzed reaction rate against substrate concentration both with and without product inhibition. Inhibition reduces the reaction rate and lowers the maximum achievable reaction rate.

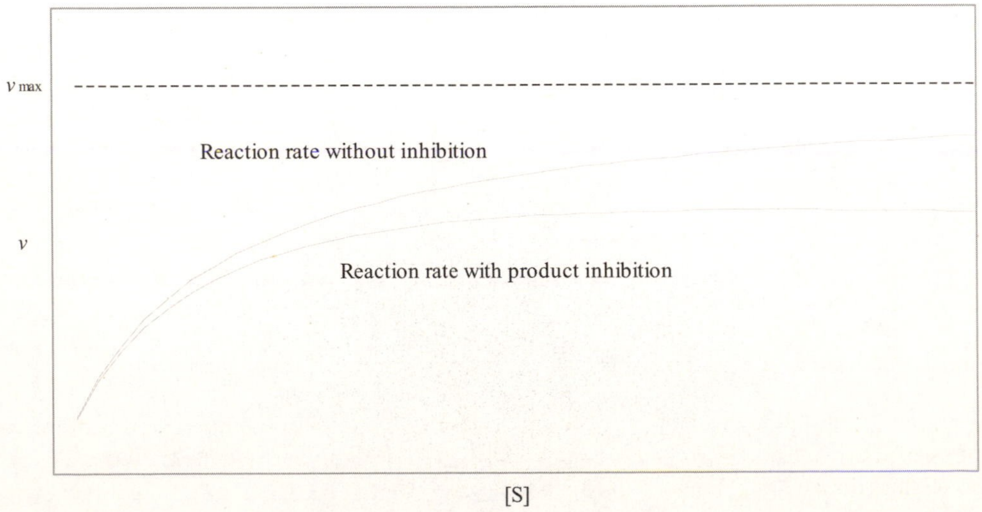

Figure 21.1

D21.7 **AFM**, atomic force microscopy, drags a sharp stylus attached to a cantilever across a surface and monitors the deflection of a laser beam from the back of the cantilever. Tiny changes in deflection indicate attraction to or repulsion from atoms on a sample surface. Since no current is involved, both conductive and nonconductive surfaces may be viewed. Surface damage is avoided by using a cantilever that has a very small spring constant. The method does not require a vacuum, and it has been applied in a liquid environment. Biological polymers may be viewed and nanometer resolutions have been achieved. However, an incorrect probe choice may cause image artifacts and distortions. Thermal drift of adsorbates may result in image distortions during relatively slow surface scans.

FIM, field-ionization microscopy, points a tip, with a point radius of about 50 nm, toward a fluorescent screen in a chamber containing about 1 mTorr to 1 nTorr of either hydrogen or helium. A positive 2–20 kV potential applied to the tip causes the hydrogen or helium gas adsorbate molecules to ionize and accelerate to the fluorescent screen. The image portrays the electrical characteristics of the tip surface and surface diffusion characteristics of the adsorbate are deduced. See the very interesting historical review of the technique in *C&EN* 83, no. 48 (November 28, 2005): 13–16.

LEED, low-energy electron diffraction, uses electrons with energies in the range 10–200 eV, which ensures diffraction from atoms only on or near the sample surface. Diffraction intensities depend on the vertical location of the atoms. The diffraction pattern is sharp if the surface is well-ordered for long distances compared with the wavelength of the incident electrons. Diffuse patterns indicate a poorly ordered surface or the presence of impurities. If the LEED pattern does not correspond to the pattern expected by extrapolation of the bulk to the surface, then either a reconstruction of the surface has occurred or there is order in the arrangement of an adsorbed layer. The interpretation of LEED data can be very complicated.

SAM, scanning Auger electron microscopy uses a focused 1–5 keV electron beam to probe and map surface composition to a resolution of about 50 nm. The high-energy impact causes the ejection of an electron from a low-lying orbital, and an upper electron falls into it. The energy this releases may result either in the generation of **X-ray fluorescence** or in the ejection of a second electron, the **Auger effect**. The emissions are used to identify chemical constituents at interfaces and surfaces of conducting and semiconducting materials to a depth of 1–5 nm.

SEM, scanning electron microscopy, uses magnetic fields to focus and scan a beam of electrons across a sample surface. Scattered electrons from a small, irradiated area are detected and the electrical signal is sent to a video screen. Resolution is typically between 1.5 and 3.0 nm. Nonconductive materials require a thin conductive coating to prevent electrical charging of the sample.

STM, scanning tunneling microscopy, reveals atomic details of surface and adsorbate structure. Surface chemical reactions can be viewed as they happen. The tip of the STM, which may end in a single atom, can also be used to manipulate adsorbed atoms on a surface, making possible the fabrication of complex and yet very tiny structures, such as nanometer-sized electronic devices. The method is based on the quantum mechanical tunneling effect in the presence of a bias voltage between the STM tip and sample surface. A piezoelectric scanner is used to position and move the tip in very close proximity to the surface, and the electrical current of tunneling generates an image of the surface topography with a resolution in the nanometer range. Images of surface electronic states may be generated. A host of very interesting STM images can be viewed at http://www.almaden.ibm.com/vis/stm/gallery.html.

D21.9 In the Langmuir–Hinshelwood mechanism of surface-catalyzed reactions, the reaction takes place by encounters between molecular fragments and atoms already adsorbed on the surface. We therefore expect the rate law to be second-order in the extent of surface coverage:

$$A + B \rightarrow P \qquad v = k\theta_A\theta_B$$

Insertion of the appropriate isotherms for A and B then gives the reaction rate in terms of the partial pressures of the reactants. For example, if A and B follow Langmuir isotherms (eqn. 21.11), and adsorb without dissociation, then it follows that the rate law is

$$v = \frac{kK_A K_B p_A p_B}{(1+K_A p_A + K_B p_B)^2}$$

The parameters K in the isotherms and the rate constant k are all temperature dependent, so the overall temperature dependence of the rate may be strongly non-Arrhenius (in the sense that the reaction rate is unlikely to be proportional to $\exp(-E_a/RT)$.

In the Eley–Rideal mechanism (ER mechanism) of a surface-catalyzed reaction, a gas phase molecule collides with another molecule already adsorbed on the surface. The rate of formation of product is expected to be proportional to the partial pressure, p_B of the non-adsorbed gas B and the extent of surface coverage, θ_A, of the adsorbed gas A. It follows that the rate law should be

$$A + B \rightarrow P \qquad v = kp_A\theta_B$$

The rate constant, k, might be much larger than for the uncatalyzed gas-phase reaction because the reaction on the surface has a low activation energy and the adsorption itself is often not activated.

If we know the adsorption isotherm for A, we can express the rate law in terms of its partial pressure, p_A. For example, if the adsorption of A follows a Langmuir isotherm in the pressure range of interest, then the rate law would be

$$v = \frac{kKp_A p_B}{1+Kp_A}$$

If A were a diatomic molecule that adsorbed as atoms, we would substitute the isotherm given in eqn. 21.13 instead.

According to eqn. 21.33, when the partial pressure of A is high (in the sense $Kp_A \gg 1$), there is almost complete surface coverage, and the rate is equal to kp_B. Now the rate-determining step is the collision of B with the adsorbed fragments. When the pressure of A is low ($Kp_A \ll 1$), perhaps because of its reaction, the rate is equal to $kKp_A p_B$. Now the extent of surface coverage is important in the determination of the rate.

In the Mars van Krevelen mechanism of catalytic oxidation, for example in the partial oxidation of propene to propenal, the first stage is the adsorption of the propene molecule with loss of a hydrogen to form the allyl radical, $CH_2{=}CHCH_2$. An O atom in the surface can now transfer to this radical, leading to the formation of acrolein (propenal, $CH_2{=}CHCHO$) and its desorption from the surface. The H atom also escapes with a surface O atom and goes on to form H_2O, which leaves the surface. The surface is left with vacancies and metal ions in lower oxidation states. These vacancies are attacked by O_2 molecules in the overlying gas, which then chemisorb as O_2^- ions reforming the catalyst. This sequence of events involves great upheavals of the surface, and some materials break up under the stress.

Solutions to exercises

E21.1(a) The fast, reversible step suggests the pre-equilibrium approximation:

$$K = \frac{[BH^+][A^-]}{[AH][B]} \quad \text{and} \quad [A^-] = \frac{K[AH][B]}{[BH^+]}$$

Thus, the rate of product formation is

$$\frac{d[P]}{dt} = k_b[AH][A^-] = \boxed{\frac{k_b K[AH]^2[B]}{[BH^+]}}$$

The application of the steady-state approximation to $[A^-]$ gives a similar but significantly different result:

$$\frac{d[A^-]}{dt} = k_a[AH][B] - k_a'[A^-][BH^+] - k_b[A^-][AH] = 0$$

Therefore, $[\mathrm{A^-}] = \dfrac{k_a[\mathrm{AH}][\mathrm{B}]}{k_a'[\mathrm{BH^+}] + k_b[\mathrm{AH}]}$

and the rate of formation of product is

$$\frac{\mathrm{d[P]}}{\mathrm{d}t} = k_b[\mathrm{AH}][\mathrm{A^-}] = \frac{k_a k_b[\mathrm{AH}]^2[\mathrm{B}]}{k_a'[\mathrm{BH^+}] + k_b[\mathrm{AH}]}$$

E21.2(a) Since $v = \dfrac{v_{\max}}{1 + K_M/[\mathrm{S}]_0}$ [21.4a], then

$$\begin{aligned} v_{\max} &= (1 + K_M/[\mathrm{S}]_0)v \\ &= (1 + 0.046/0.105) \times (1.04\ \mathrm{mmol\ dm^{-3}\ s^{-1}}) \\ &= \boxed{1.50\ \mathrm{mmol\ dm^{-3}\ s^{-1}}} \end{aligned}$$

E21.3(a)

$$\begin{aligned} k_{\mathrm{cat}} &= v_{\max}/[\mathrm{E}]_0 \quad [21.5] \\ &= (0.425\ \mathrm{mmol\ dm^{-3}\ s^{-1}})/(3.60\times10^{-6}\ \mathrm{mmol\ dm^{-3}}) = \boxed{1.18\times10^5\ \mathrm{s^{-1}}} \end{aligned}$$

$$\begin{aligned} \eta &= k_{\mathrm{cat}}/K_M \quad [21.6] \\ &= (1.18\times10^5\ \mathrm{s^{-1}})/(0.015\ \mathrm{mol\ dm^{-3}}) = \boxed{7.9\times10^6\ \mathrm{dm^3\ mol^{-1}\ s^{-1}}} \end{aligned}$$

Diffusion limits the catalytic efficiency, η, to a maximum of about 10^8–$10^9\ \mathrm{dm^3\ mol^{-1}\ s^{-1}}$. Since the catalytic efficiency of this enzyme is significantly smaller than the maximum, the enzyme is not "catalytically perfect."

E21.4(a) Eqn. 21.8 describes competitive inhibition as the case for which $\alpha = 1 + [\mathrm{I}]/K_I$ and $\alpha' = 1$. Thus,

$$v = \frac{v_{\max}}{1 + \alpha K_M/[\mathrm{S}]_0}$$

By setting the ratio $v([\mathrm{I}] = 0)/v([\mathrm{I}])$ equal to 2 and solving for α, we can subsequently solve for the inhibitor concentration that reduces the catalytic rate by 50%.

$$\frac{v([\mathrm{I}]=0)}{v([\mathrm{I}])} = \frac{1 + \alpha K_M/[\mathrm{S}]_0}{1 + K_M/[\mathrm{S}]_0} = 2$$

$$\begin{aligned} \alpha &= \frac{2(1 + K_M/[\mathrm{S}]_0) - 1}{K_M/[\mathrm{S}]_0} \\ &= \frac{2(1 + 3.0/0.10) - 1}{3.0/0.10} = 2.0\overline{3} \\ [\mathrm{I}] &= (\alpha - 1)K_I \\ &= 1.0\overline{3} \times (2.0\times10^{-5}\ \mathrm{mol\ dm^{-3}}) = \boxed{2.0\times10^{-5}\ \mathrm{mol\ dm^{-3}}} \end{aligned}$$

E21.5(a) The collision frequency, Z_W, of gas molecules with an ideally smooth surface area is given by eqn. 18.15.

$$Z_W = \frac{p}{(2\pi MkT/N_A)^{1/2}} \quad [18.15;\ m = M/N_A]$$

$$= \frac{p \times \{(\mathrm{kg\ m^{-1}\ s^{-2}})/\mathrm{Pa}\} \times (10^{-4}\ \mathrm{m^2/cm^2})}{\{2\pi \times (1.381\times10^{-23}\ \mathrm{J\ K^{-1}}) \times (298.15\ \mathrm{K}) \times (\mathrm{kg\ mol^{-1}})/(6.022\times10^{23}\ \mathrm{mol^{-1}})\}^{1/2}\{M/(\mathrm{kg\ mol^{-1}})\}^{1/2}}$$

$$= 4.825\times10^{17}\left(\frac{p/\mathrm{Pa}}{\{M/(\mathrm{kg\ mol^{-1}})\}^{1/2}}\right)\mathrm{cm^{-2}\ s^{-1}} \quad \text{at } 25°\mathrm{C}$$

(a) Hydrogen ($M = 0.002016$ kg mol^{-1})

(i) $p = 100$ Pa, $Z_W = \boxed{1.07\times10^{21}\ \text{cm}^{-2}\ \text{s}^{-1}}$

(ii) $p = 0.10$ μTorr $= 1.33\times10^{-5}$ Pa, $Z_W = \boxed{1.4\times10^{14}\ \text{cm}^{-2}\ \text{s}^{-1}}$

(b) Propane ($M = 0.04410$ kg mol^{-1})

(i) $p = 100$ Pa, $Z_W = \boxed{2.30\times10^{20}\ \text{cm}^{-2}\ \text{s}^{-1}}$

(ii) $p = 0.10$ μTorr $= 1.33\times10^{-5}$ Pa, $Z_W = \boxed{3.1\times10^{13}\ \text{cm}^{-2}\ \text{s}^{-1}}$

E21.6(a) $$A = \pi d^2/4 = \pi(1.5\ \text{mm})^2/4 = 1.77\times10^{-6}\ \text{m}^2$$

The collision frequency of the Ar gas molecules with surface area A equals Z_WA.

$$Z_WA = \frac{p}{(2\pi MkT/N_A)^{1/2}}A\ [18.15;\ m = M/N_A]$$
$$\begin{aligned}p &= (Z_WA)\times(2\pi MkT/N_A)^{1/2}/A\\ &= (4.5\times10^{20}\ \text{s}^{-1})\times\{2\pi(39.95\times10^{-3}\ \text{kg mol}^{-1})\times(1.381\times10^{-23}\ \text{J K}^{-1})\\ &\quad\times(425\ \text{K})/(6.022\times10^{23}\ \text{mol}^{-1})\times\}^{1/2}/(1.77\times10^{-6}\ \text{m}^2)\\ &= 1.3\times10^4\ \text{Pa} = \boxed{0.13\ \text{bar}}\end{aligned}$$

E21.7(a) The farther apart the atoms responsible for the pattern, the closer the spots appear in the pattern (see Example 21.3). Doubling the vertical separation between atoms of the unreconstructed face, which has LEED pattern (a), yields a reconstructed surface that gives LEED pattern (b).

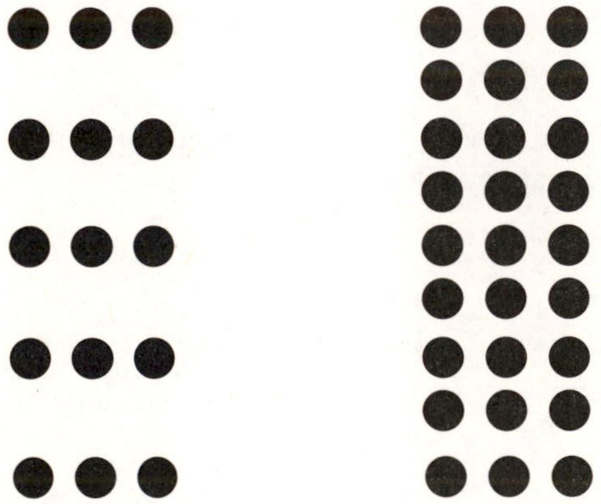

(a) (b)

E21.8(a) Let us assume that the nitrogen molecules are close-packed, as shown in Figure 21.2 as spheres, in the monolayer. Then one molecule occupies the parallelogram area of $2\sqrt{3}\,r^2$ where r is the radius of the adsorbed molecule. Furthermore, let us assume that the collision cross-section of Table 18.1 ($\sigma = 0.43\ \text{nm}^2 = 4\pi r^2$) gives a reasonable estimate of r: $r = (\sigma/4\pi)^{1/2}$. With these assumptions the surface area occupied by one molecule is

$$\begin{aligned}A_{\text{molecule}} &= 2\sqrt{3}\,(\sigma/4\pi) = \sqrt{3}\,\sigma/2\pi\\ &= \sqrt{3}\,(0.43\ \text{nm}^2)/2\pi = 0.12\ \text{nm}^2\end{aligned}$$

In this model, the surface area per gram of the catalyst equals $A_{\text{molecule}}N$ where N is the number of adsorbed molecules. N can be calculated with the 0°C data, a temperature that is so high compared to the boiling point of nitrogen that all molecules are likely to be desorbed from the surface as perfect gas.

$$N=\frac{pV}{kT}=\frac{(760\text{ Torr})\times(133.3\text{ Pa/Torr})\times(3.86\times10^{-6}\text{ m}^3)}{(1.381\times10^{-23}\text{ J K}^{-1})\times(273.15\text{ K})}=1.04\times10^{20}$$

$$A_{\text{molecule}}N=(0.12\times10^{-18}\text{ m}^2)\times(1.04\times10^{20})=\boxed{12\text{ m}^2}$$

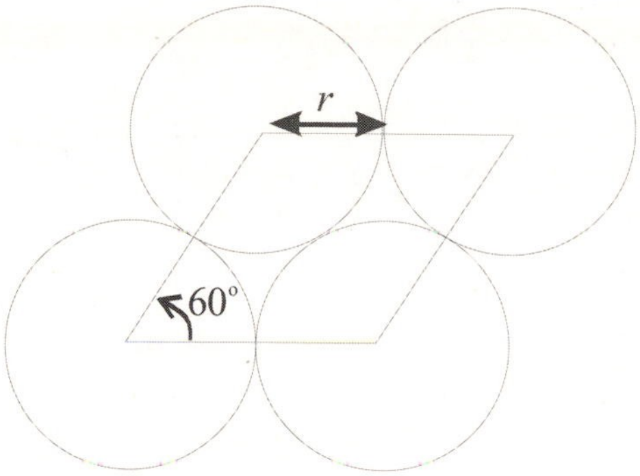

Figure 21.2

E21.9(a) $$\theta=\frac{V}{V_\infty}\,[21.9]=\frac{V}{V_{\text{mon}}}=\frac{Kp}{1+Kp}\,[21.11]$$

This rearranges to [Example 21.4]

$$\frac{p}{V}=\frac{p}{V_{\text{mon}}}+\frac{1}{KV_{\text{mon}}}$$

Hence, $\frac{p_2}{V_2}-\frac{p_1}{V_1}=\frac{p_2}{V_{\text{mon}}}-\frac{p_1}{V_{\text{mon}}}$

Solving for V_{mon}

$$V_{\text{mon}}=\frac{p_2-p_1}{(p_2/V_2-p_1/V_1)}=\frac{(760-145.4)\text{ Torr}}{(760/1.443-145.4/0.286)\text{ Torr cm}^{-3}}=\boxed{33.6\text{ cm}^3}$$

E21.10(a) The enthalpy of adsorption is typical of $\boxed{\text{chemisorption}}$ (Table 21.2) for which $\tau_0\approx10^{-14}$ s [21.23] because the adsorbate-substrate bond is stiff. The half-life for remaining on the surface is

$$t_{1/2}=\tau_0 e^{E_d/RT}\,[21.23]\approx(10^{-14}\text{ s})\times(e^{120\times10^3/(8.3145\times400)})\;[E_d\approx-\Delta_{\text{ad}}H]\approx\boxed{50\text{ s}}$$

E21.11(a) $$\frac{m_1}{m_2}=\frac{\theta_1}{\theta_2}=\frac{p_1}{p_2}\times\frac{1+Kp_2}{1+Kp_1}\;[21.9\text{ and }21.11]$$

which solves to

$$K=\frac{(m_1p_2/m_2p_1)-1}{p_2-(m_1p_2/m_2)}=\frac{(m_1/m_2)\times(p_2/p_1)-1}{1-(m_1/m_2)}\times\frac{1}{p_2}$$

$$=\frac{(0.44/0.19)\times(3.0/26.0)-1}{1-(0.44/0.19)}\times\frac{1}{3.0\text{ kPa}}=0.19\text{ kPa}^{-1}$$

Therefore,

$$\theta_1=\frac{(0.19\text{ kPa}^{-1})\times(26.0\text{ kPa})}{(1)+(0.19\text{ kPa}^{-1})\times(26.0\text{ kPa})}=\boxed{0.83}\;[21.11]\quad\text{and}\quad\theta_2=\frac{(0.19)\times(3.0)}{(1)+(0.19)\times(3.0)}=\boxed{0.36}$$

E21.12(a) $\theta = \dfrac{Kp}{1+Kp}$ [21.11], which implies that $p = \left(\dfrac{\theta}{1-\theta}\right)\dfrac{1}{K}$.

(a) $p = (0.15/0.85)/0.75\ \text{kPa}^{-1} = \boxed{0.24\ \text{kPa}}$

(b) $p = (0.95/0.05)/0.75\ \text{kPa}^{-1} = \boxed{25\ \text{kPa}}$

E21.13(a) $\theta = \dfrac{Kp}{1+Kp}$ [21.11], which implies that $K = \left(\dfrac{\theta}{1-\theta}\right)\times\left(\dfrac{1}{p}\right)$

Additionally, $\ln\left(\dfrac{K_2}{K_1}\right) = -\dfrac{\Delta_{\text{ad}}H}{R}\left(\dfrac{1}{T_2}-\dfrac{1}{T_1}\right)$ [17.28] $= \dfrac{\Delta_{\text{des}}H}{R}\left(\dfrac{1}{T_2}-\dfrac{1}{T_1}\right)$ $[\Delta_{\text{ad}}H = -\Delta_{\text{des}}H]$.

Since $\theta_2 = \theta_1$, $K_2/K_1 = p_1/p_2$ and

$$\ln\frac{p_1}{p_2} = \frac{\Delta_{\text{des}}H}{R}\left(\frac{1}{T_2}-\frac{1}{T_1}\right) = \left(\frac{10.2\ \text{kJ mol}^{-1}}{8.3145\ \text{J K}^{-1}\ \text{mol}^{-1}}\right)\times\left(\frac{1}{313\ \text{K}}-\frac{1}{298\ \text{K}}\right) = -0.197$$

which implies that $p_2 = (12\ \text{kPa})\times(\text{e}^{0.197}) = \boxed{15\ \text{kPa}}$.

E21.14(a) $\theta = \dfrac{Kp}{1+Kp}$ [21.11], which implies that $K = \left(\dfrac{\theta}{1-\theta}\right)\times\left(\dfrac{1}{p}\right)$

Additionally, $\ln\left(\dfrac{K_2}{K_1}\right) = -\dfrac{\Delta_{\text{ad}}H}{R}\left(\dfrac{1}{T_2}-\dfrac{1}{T_1}\right)$ [17.28] or $\Delta_{\text{ad}}H = -R\ln\left(\dfrac{K_2}{K_1}\right)\times\left(\dfrac{1}{T_2}-\dfrac{1}{T_1}\right)^{-1}$

Since $\theta_2 = \theta_1$, $K_2/K_1 = p_1/p_2$ and

$$\Delta_{\text{ad}}H = -R\ln\left(\frac{p_1}{p_2}\right)\times\left(\frac{1}{T_2}-\frac{1}{T_1}\right)^{-1}$$

$$= -(8.3145\ \text{J K}^{-1}\ \text{mol}^{-1})\times\ln\left(\frac{490\ \text{kPa}}{3.2\times10^3\ \text{kPa}}\right)\times\left(\frac{1}{250\ \text{K}}-\frac{1}{190\ \text{K}}\right)^{-1}$$

$$= \boxed{-12.\overline{4}\ \text{kJ mol}^{-1}}$$

E21.15(a) The desorption time for a given volume is proportional to the half-life of the absorbed species, and consequently the ratio of desorption times at two different temperatures is given by

$$t(2)/t(1) = t_{1/2}(2)/t_{1/2}(1) = \text{e}^{E_\text{d}/RT_2}/\text{e}^{E_\text{d}/RT_1}\ [21.23] = \text{e}^{E_\text{d}(1/T_2-1/T_1)/R}$$

Solving for the activation energy for desorption, E_d, gives

$$E_\text{d} = R\ln\{t(2)/t(1)\}(1/T_2-1/T_1)^{-1}$$

$$= (8.3145\ \text{J K}^{-1}\ \text{mol}^{-1})\times\ln\left(\frac{2.0\ \text{min}}{27\ \text{min}}\right)\times\left(\frac{1}{1978\ \text{K}}-\frac{1}{1856\ \text{K}}\right)^{-1}$$

$$= \boxed{65\overline{1}\ \text{kJ mol}^{-1}}$$

The desorption time, t, for the same volume at temperature t is given by

$$t = t(1)\text{e}^{E_\text{d}(1/T-1/T_1)/R} = (27\ \text{min})\exp\left\{(65\overline{1}\times10^3\ \text{J mol}^{-1})\times\left(\frac{1}{T}-\frac{1}{1856\ \text{K}}\right)/(8.3145\ \text{J K}^{-1}\ \text{mol}^{-1})\right\}$$

$$= (27\ \text{min})\exp\left\{(78.3)\times\left(\frac{1}{T/1000\ \text{K}}-\frac{1}{1.856}\right)\right\}$$

(a) At 298 K, $t = \boxed{1.6 \times 10^{97}\ \text{min}}$, which is about forever.

(b) At 3000 K, $t = \boxed{2.8 \times 10^{-6}\ \text{min}}$.

E21.16(a) The average time of molecular residence is proportional to the half-life of the absorbed species, and consequently the ratio of average residence times at two different temperatures is given by

$$t(2)/t(1) = t_{1/2}(2)/t_{1/2}(1) = e^{E_d/RT_2}/e^{E_d/RT_1}\ [21.23] = e^{E_d(1/T_2 - 1/T_1)/R}$$

Solving for the activation energy for desorption, E_d, gives

$$E_d = R\ln\{t(2)/t(1)\}(1/T_2 - 1/T_1)^{-1}$$
$$= (8.3145\ \text{J K}^{-1}\ \text{mol}^{-1}) \times \ln\left(\frac{3.49\ \text{s}}{0.36\ \text{s}}\right) \times \left(\frac{1}{2362\ \text{K}} - \frac{1}{2548\ \text{K}}\right)^{-1}$$
$$= \boxed{61\bar{1}\ \text{kJ mol}^{-1}}$$

E21.17(a) At 400 K, $t_{1/2} = \tau_0 e^{E_d/RT}\ [21.23] = (0.10\ \text{ps}) \times e^{0.301 E_d/\text{kJ mol}^{-1}}$.

At 1000 K, $t_{1/2} = \tau_0 e^{E_d/RT}\ [21.23] = (0.10\ \text{ps}) \times e^{0.120 E_d/\text{kJ mol}^{-1}}$.

(a) $E_d = 15\ \text{kJ mol}^{-1}$

$$t_{1/2}(400\ \text{K}) = (0.10\ \text{ps}) \times e^{0.301\times 15} = \boxed{9.1\ \text{ps}},\quad t_{1/2}(1000\ \text{K}) = (0.10\ \text{ps}) \times e^{0.120\times 15} = \boxed{0.60\ \text{ps}}$$

(b) $E_d = 150\ \text{kJ mol}^{-1}$

$[\text{ATP}]/(\mu\text{mol dm}^{-3})$	0.60	0.80	1.4	2.0	3.0
$v/(\mu\text{mol dm}^{-3}\ \text{s}^{-1})$	0.81	0.97	1.30	1.47	1.69
$1/\{[\text{ATP}]/(\mu\text{mol dm}^{-3})\}$	1.67	1.25	0.714	0.500	0.333
$1/\{v/(\mu\text{mol dm}^{-3}\ \text{s}^{-1})\}$	1.23	1.03	0.769	0.680	0.592

$$t_{1/2}(400\ \text{K}) = (0.10\ \text{ps}) \times e^{0.301\times 150} = \boxed{4.1\times 10^{6}\ \text{s}},\quad t_{1/2}(1000\ \text{K}) = (0.10\ \text{ps}) \times e^{0.120\times 150} = \boxed{6.6\ \mu\text{s}}$$

E21.18(a)

$$v = k_r\theta = \frac{k_r Kp}{1 + Kp}\ [21.28a]$$

(a) On gold, $\theta \approx 1$, and $v = k_r\theta \approx$ constant, a $\boxed{\text{zeroth-order}}$ reaction.

(b) On platinum, $\theta \approx Kp$ (as $Kp \ll 1$), so $v = k_r Kp$ and the reaction is $\boxed{\text{first-order}}$.

Solutions to problems

Solutions to numerical problems

P21.1 We draw up the table below, which includes data rows required for a Lineweaver–Burk plot ($1/v$ against $1/[\text{S}]_0$). The linear regression fit is summarized in Figure 21.3.

$$1/v_{max} = \text{intercept}\ [21.4b]$$

$$v_{max} = 1/\text{intercept} = 1/(0.433\ \mu\text{mol dm}^{-3}\ \text{s}^{-1}) = \boxed{2.31\ \mu\text{mol dm}^{-3}\ \text{s}^{-1}}$$

$$k_b = v_{max}/[\text{E}]_0\ [21.3b] = (2.31\ \mu\text{mol dm}^{-3}\ \text{s}^{-1})/(0.020\ \mu\text{mol dm}^{-3}) = \boxed{115\ \text{s}^{-1}}$$

$$k_{cat} = k_b\ [21.5] = \boxed{115\ \text{s}^{-1}}$$

$$K_M = v_{max} \times \text{slope}\ [21.4b] = (2.31\ \mu\text{mol dm}^{-3}\ \text{s}^{-1}) \times (0.480\ \text{s}) = \boxed{1.11\ \mu\text{mol dm}^{-3}}$$

$$\eta = k_{cat}/K_M\ [21.6] = (115\ \text{s}^{-1})/(1.11\ \mu\text{mol dm}^{-3}) = \boxed{104\ \text{dm}^{3}\ \mu\text{mol}^{-1}\ \text{s}^{-1}}$$

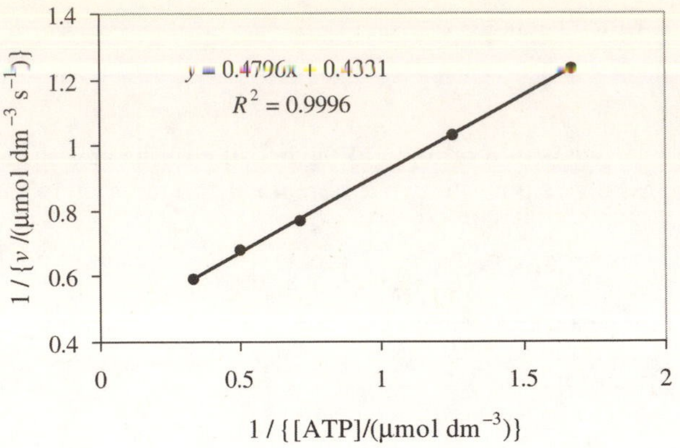

Figure 21.3

P21.3 (a) The dissociation equilibrium may be rearranged to give the following four relationships.

$$[E^-] = K_{E,a}[EH]/[H^+] \qquad [EH_2^+] = [EH][H^+]/K_{E,b}$$

$$[ES^-] = K_{ES,a}[ESH]/[H^+] \qquad [ESH_2] = [ESH][H^+]/K_{ES,b}$$

Mass balance provides an equation for [EH].

$$[E]_0 = [E^-]+[EH]+[EH_2^+]+[ES^-]+[ESH]+[ESH_2]$$
$$= \frac{K_{E,a}[EH]}{[H^+]}+[EH]+\frac{[EH][H^+]}{K_{E,b}}+\frac{K_{ES,a}[ESH]}{[H^+]}+[ESH]+\frac{[ESH][H^+]}{K_{ES,b}}$$

$$[EH] = \frac{[E]_0 - \left\{1+\dfrac{[H^+]}{K_{ES,b}}+\dfrac{K_{ES,a}}{[H^+]}\right\}[ESH]}{1+\dfrac{[H^+]}{K_{E,b}}+\dfrac{K_{E,a}}{[H^+]}}$$

$$= \frac{[E]_0 - c_1[ESH]}{c_2} \quad \text{where} \quad c_1 = 1+\frac{[H^+]}{K_{ES,b}}+\frac{K_{ES,a}}{[H^+]} \text{ and } c_2 = 1+\frac{[H^+]}{K_{E,b}}+\frac{K_{E,a}}{[H^+]}$$

The steady-state approximation provides an equation for [ESH].

$$\frac{d[ESH]}{dt} = k_a[EH][S]-k_a'[ESH]-k_b[ESH] = 0$$

$$[ESH] = \frac{k_a}{k_a'+k_b}[EH][S] = K_M^{-1}[EH][S]$$
$$= K_M^{-1}[S]\left\{\frac{[E]_0 - c_1[ESH]}{c_2}\right\}$$

$$[ESH] = \frac{K_M^{-1}[S][E]_0/c_2}{1+K_M^{-1}[S]c_1/c_2} = \frac{[E]_0/c_1}{1+K_M(c_2/c_1)/[S]}$$

The rate law becomes

$$v = d[P]/dt = k_b[ESH] = \frac{k_b[E]_0/c_1}{1+K_M(c_2/c_1)/[S]} = \frac{v'_{max}}{1+K_M'/[S]}$$

$$\text{where} \quad v'_{max} = k_b[E]_0\bigg/\left\{1+\frac{[H^+]}{K_{ES,b}}+\frac{K_{ES,a}}{[H^+]}\right\} = v_{max}\bigg/\left\{1+\frac{[H^+]}{K_{ES,b}}+\frac{K_{ES,a}}{[H^+]}\right\}$$

and $$K'_M = K_M \left\{1 + \frac{[H^+]}{K_{E,b}} + \frac{K_{E,a}}{[H^+]}\right\} \Big/ \left\{1 + \frac{[H^+]}{K_{ES,b}} + \frac{K_{ES,a}}{[H^+]}\right\}$$

(b) $v_{max} = 1.0\times10^{-6}$ mol dm^{-3} s^{-1} $\quad K_{ES,b} = 1.0\times10^{-6}$ $\quad K_{ES,a} = 1.0\times10^{-8}$

Figure 21.4 shows a plot of v'_{max} against pH. The plot indicates a maximum value of v'_{max} at pH = 7.0 for this set of equilibrium and kinetic constants. A formula for the pH of the maximum can be derived by finding the point at which $\frac{dv'_{max}}{d[H^+]} = 0$. This gives

$$[H^+]_{max} = (K_{ES,a}\,K_{ES,b})^{1/2} = \sqrt{(1.0\times10^{-8}\text{ mol dm}^{-3})(1.0\times10^{-6}\text{ mol dm}^{-3})} = 1.0\times10^{-7}\text{ mol dm}^{-3}$$

which corresponds to $\boxed{pH = 7.0}$.

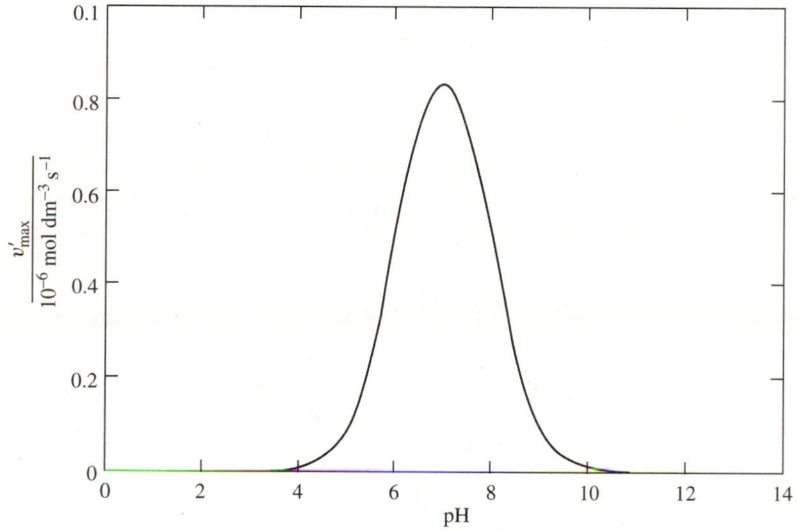

Figure 21.4

(c) $v_{max} = 1.0\times10^{-6}$ mol dm^{-3} s^{-1} $\quad K_{ES,b} = 1.0\times10^{-4}$ $\quad K_{ES,a} = 1.0\times10^{-10}$

Figure 21.5 shows a plot of v'_{max} against pH. The plot once again indicates a maximum value of v'_{max} at pH = 7.0 for this set of equilibrium and kinetic constants. However, the rate is high over a much larger pH range than appeared in part (b). This reflects the behavior of the term $1 + [H^+]/K_{ES,b} + K_{ES,a}/[H^+]$ in the denominator of the v'_{max} expression. When $K_{ES,b}$ is relatively large, large $[H^+]$ values (low pH) cause growth in the values of v'_{max}. However, when $K_{ES,a}$ is relatively small, very small $[H^+]$ values (high pH) cause a decline in the v'_{max} values.

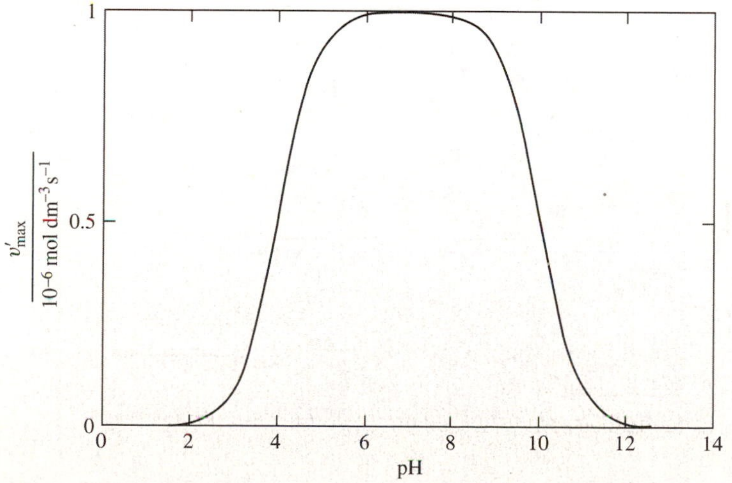

Figure 21.5

P21.5 Refer to Figure 21.6.

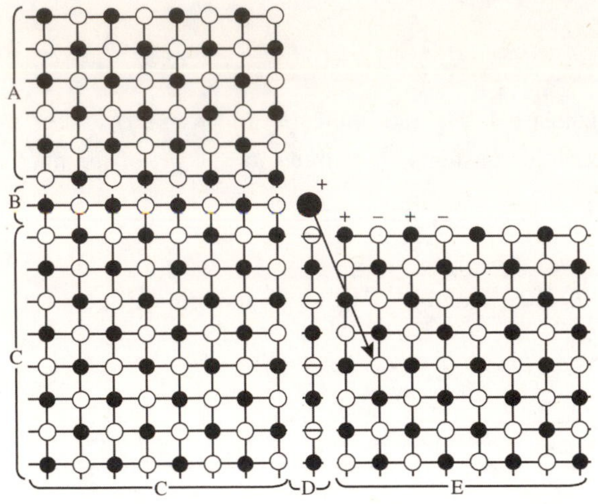

Figure 21.6

Evaluate the sum of $\pm 1/r_i$, where r_i is the distance from the ion i to the ion of interest, taking $+1/r$ for ions of like charge and $-1/r$ for ions of opposite charge. The array has been divided into five zones. Zones B and D can be summed analytically to give $-\ln 2 = -0.69$. The summation over the other zones, each of which gives the same result, is tedious because of the very slow convergence of the sum. Unless you make a very clever choice of the sequence of ions (grouping them so that their contributions almost cancel), you will find the following values for arrays of different sizes

10 × 10	20 × 20	50 × 50	100 × 100	200 × 200
0.259	0.273	0.283	0.286	0.289

The final figure is in good agreement with the analytical value, 0.289 259 7... .

(a) For a cation above a flat surface, the energy (relative to the energy at infinity and in multiples of $e^2/4\pi\varepsilon r_0$ where r_0 is the lattice spacing (200 pm)) is

Zone C + D + E = $0.29 - 0.69 + 0.29 = \boxed{-0.11}$, which implies an attractive state

(b) For a cation at the foot of a high cliff, the energy is

Zone A + B + C + D + E = $3 \times 0.29 + 2 \times (-0.69) = \boxed{-0.51}$, which is significantly more attractive

Hence, the latter is more likely the settling point (if potential energy considerations such as these are dominant).

P21.7 Refer to Figure 21.7.

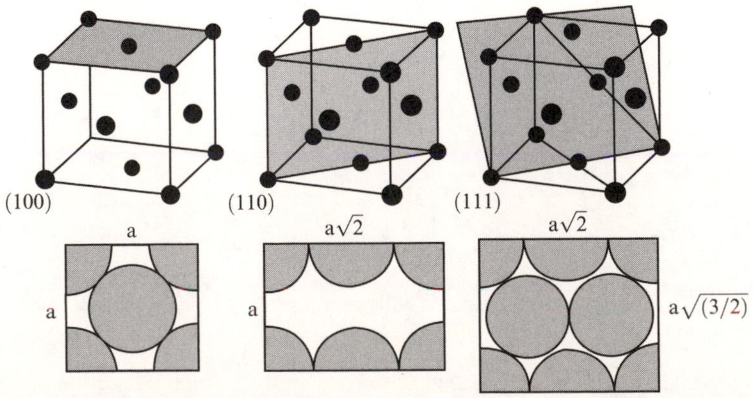

Figure 21.7

The (100) and (110) faces each expose two atoms, and the (111) face exposes four. The areas of the faces of each cell are (a) $(352\ \text{pm})^2 = 1.24\times10^{-15}\ \text{cm}^2$, (b) $\sqrt{2}\times(352\ \text{pm})^2 = 1.75\times10^{-15}\ \text{cm}^2$, and

(c) $\sqrt{3}\times(352\text{ pm})^2 = 2.15\times10^{-15}\text{cm}^2$. The numbers of atoms exposed per square centimetre, the surface number density, are therefore

(a) $$\frac{2}{1.24\times10^{-15}\text{cm}^2} = \boxed{1.61\times10^{15}\text{ cm}^{-2}}$$

(b) $$\frac{2}{1.75\times10^{-15}\text{cm}^2} = \boxed{1.14\times10^{15}\text{ cm}^{-2}}$$

(c) $$\frac{4}{2.15\times10^{-15}\text{cm}^2} = \boxed{1.86\times10^{15}\text{ cm}^{-2}}$$

The collision frequency, Z_W, of gas molecules with a surface is given by eqn. 18.15.

$$Z_W = \frac{p}{(2\pi MkT/N_A)^{1/2}} \quad [18.15;\ m = M/N_A]$$

$$= \frac{p\times\{(\text{kg m}^{-1}\text{ s}^{-2})/\text{Pa}\}\times(10^{-4}\text{ m}^2/\text{cm}^2)}{\{2\pi\times(1.381\times10^{-23}\text{ J K}^{-1})\times(298.15\text{ K})\times(\text{kg mol}^{-1})/(6.022\times10^{23}\text{ mol}^{-1})\}^{1/2}\{M/(\text{kg mol}^{-1})\}^{1/2}}$$

$$= 4.825\times10^{17}\left(\frac{p/\text{Pa}}{\{M/(\text{kg mol}^{-1})\}^{1/2}}\right)\text{cm}^{-2}\text{ s}^{-1} \quad \text{at } 25°\text{ C}$$

(a) Hydrogen ($M = 0.002016$ kg mol^{-1})
(i) $p = 100$ Pa, $Z_W = 1.07\times10^{21}$ cm^{-2} s^{-1}
(ii) $p = 0.10$ μTorr $= 1.33\times10^{-5}$ Pa, $Z_W = 1.4\times10^{14}$ cm^{-2} s^{-1}

(b) Propane ($M = 0.04410$ kg mol^{-1})
(i) $p = 100$ Pa, $Z_W = 2.30\times10^{20}$ cm^{-2} s^{-1}
(ii) $p = 0.10$ μTorr $= 1.33\times10^{-5}$ Pa, $Z_W = 3.1\times10^{13}$ cm^{-2} s^{-1}

The frequency of collisions per surface atom, Z, is calculated by dividing Z_W by the surface number densities for the different planes. We can therefore draw up the following table.

	Hydrogen		Propane	
Z/(atom^{-1} s^{-1})	100 Pa	10^{-7} Torr	100 Pa	10^{-7} Torr
(100)	6.6×10^5	8.7×10^{-2}	1.4×10^5	1.9×10^{-2}
(110)	9.4×10^5	1.2×10^{-1}	2.0×10^5	2.7×10^{-2}
(111)	5.8×10^5	7.5×10^{-2}	1.2×10^5	1.7×10^{-2}

P21.9 $$\frac{V}{V_{\text{mon}}} = \frac{cz}{(1-z)\{1-(1-c)z\}} \quad \left[21.15,\text{ BET isotherm, } z = \frac{p}{p^*}\right]$$

This rearranges to

$$\frac{z}{(1-z)V} = \frac{1}{cV_{\text{mon}}} + \frac{(c-1)z}{cV_{\text{mon}}}$$

A plot of the left-hand side, $z/(1-z)V$, against z should result in a straight line if the data obey the BET isotherm. Should it be linear, a linear regression fit of the plot yields values for the intercept and slope, which are related to c and V_{mon} by

$1/cV_{\text{mon}}$ = intercept and $(c-1)/cV_{\text{mon}}$ = slope

Solving for c and V_{mon} yields

$c = 1 +$ slope/intercept and $V_{\text{mon}} = 1/(c\times\text{intercept})$

We draw up the following tables.

(a) 0°C, $p^* = 429.6$ kPa

p / kPa	14.0	37.6	65.6	79.2	82.7	100.7	106.4
$10^3 z$	32.6	87.5	152.7	184.4	192.4	234.3	247.7
$\dfrac{10^3 z}{(1-z)(V/\text{cm}^3)}$	3.03	7.11	12.1	14.1	15.4	17.7	20.0

(b) 18°C, $p^* = 819.7$ kPa

p / kPa	5.3	8.4	14.4	29.2	62.1	74.0	80.1	102.0
$10^3 z$	6.5	10.2	17.6	35.6	75.8	90.3	97.8	124.4
$\dfrac{10^3 z}{(1-z)(V/\text{cm}^3)}$	0.70	1.06	1.74	3.27	6.35	7.58	8.08	10.1

The $z/(1-z)V$ against z points are plotted in Figure 21.8. It is apparent that the plots are linear, so we conclude that the data fit the BET isotherm. The linear regression fits are summarized in the figure.

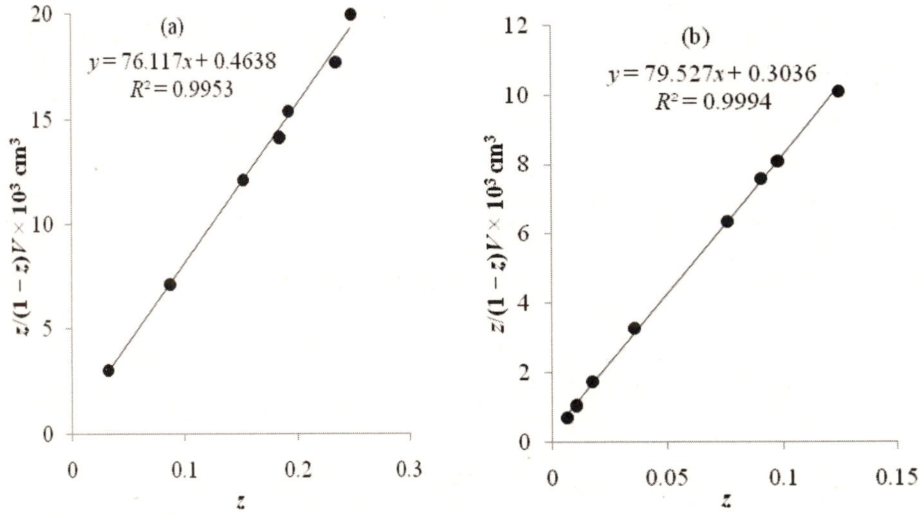

Figure 21.8

(a) Intercept = 0.4638×10^{-3} cm^{-3} and slope = 76.12×10^{-3} cm^{-3}

$c = 1 + \text{slope/intercept} = 1 + 76.12/0.4638 = \boxed{165}$

$V_{\text{mon}} = 1/(c \times \text{intercept}) = 1/(165 \times 0.4638 \times 10^{-3}\ \text{cm}^{-3}) = \boxed{13.1\ \text{cm}^3}$

(b) Intercept = 0.3036×10^{-3} cm^{-3} and slope = 79.53×10^{-3} cm^{-3}

$c = 1 + \text{slope/intercept} = 1 + 79.53/0.3036 = \boxed{263}$

$V_{\text{mon}} = 1/(c \times \text{intercept}) = 1/(263 \times 0.3036 \times 10^{-3}\ \text{cm}^{-3}) = \boxed{12.5\ \text{cm}^3}$

P21.11 $\theta = c_1 p^{1/c_2}$ [Freundlich isotherm, 21.19]

We adapt this isotherm to a liquid by noting that $w_a \propto \theta$ and replacing p by [A], the concentration of the acid. Then $w_a = c_1[\text{A}]^{1/c_2}$ (with c_1, c_2 modified constants), and hence $\log w_a = \log c_1 + \dfrac{1}{c_2} \times \log[\text{A}]$. We draw up the following table.

[A]/(mol dm^{-3})	0.05	0.10	0.50	1.0	1.5
log([A]/mol dm^{-3})	−1.30	−1.00	−0.30	−0.00	0.18
log(ω_a / g)	−1.40	−1.22	−0.92	−0.80	−0.72

These points are plotted in Figure 21.9. They fall on a reasonably straight line with slope 0.42 and intercept –0.80. Therefore, $c_2 = 1/0.42 = \boxed{2.4}$ and $c_1 = \boxed{0.16}$. (The units of c_1 are bizarre: $c_1 = 0.16\ \text{g mol}^{-0.42}\ \text{dm}^{1.26}$).

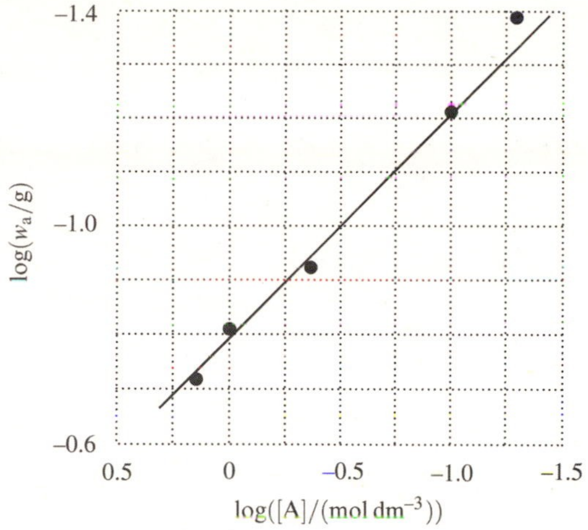

Figure 21.9

P21.13 Taking the natural logarithm of the isotherm $c_{\text{ads}} = Kc_{\text{sol}}^{1/n}$ gives

$$\ln c_{\text{ads}} = \ln K + (\ln c_{\text{sol}})/n$$

so a plot of $\ln c_{\text{ads}}$ versus $\ln c_{\text{sol}}$ would have a slope of $1/n$ and a y-intercept of $\ln K$. The transformed data and plot are shown in Figure 21.10.

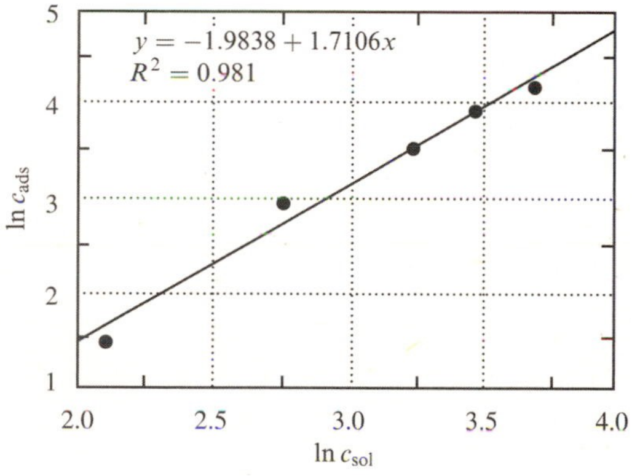

Figure 21.10

$c_{\text{sol}}/(\text{mg g}^{-1})$	8.26	15.65	25.43	31.74	40.00
$c_{\text{ads}}/(\text{mg g}^{-1})$	4.41	19.2	35.2	52.0	67.2
$\ln c_{\text{sol}}$	2.11	2.75	3.24	3.46	3.69
$\ln c_{\text{ads}}$	1.48	2.95	3.56	3.95	4.21

$$K = e^{-1.9838}\ \text{mg g}^{-1} = \boxed{0.138\ \text{mg g}^{-1}} \quad \text{and} \quad n = 1/1.71 = \boxed{0.58}$$

To express this information in terms of fractional coverage, the amount of adsorbate corresponding to monolayer coverage must be known. This saturation point, however, has no special significance in the Freundlich isotherm (i.e., it does not correspond to any limiting case).

P21.15 The Langmuir isotherm is

$$\theta = \frac{Kp}{1+Kp} = \frac{n}{n_\infty} \quad \text{so} \quad n(1+Kp) = n_\infty Kp \text{ and } \frac{p}{n} = \frac{p}{n_\infty} + \frac{1}{Kn_\infty}$$

So a plot of p/n against p should be a straight line with slope $1/n_\infty$ and y-intercept $1/Kn_\infty$. The transformed data and plot (Figure 21.11) follow.

p / kPa	31.00	38.22	53.03	76.38	101.97	130.47	165.06	182.41	205.75	219.91
n / (mol kg^{-1})	1.00	1.17	1.54	2.04	2.49	2.90	3.22	3.30	3.35	3.36
$\dfrac{p/n}{\text{kPa mol}^{-1}\text{ kg}}$	31.00	32.67	34.44	37.44	40.95	44.99	51.26	55.28	61.42	65.45

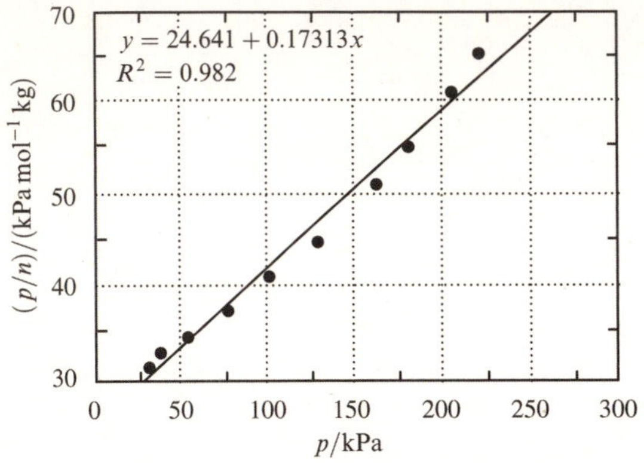

Figure 21.11

$$n_\infty = \frac{1}{0.17313\ \text{mol}^{-1}\ \text{kg}} = \boxed{5.78\ \text{mol kg}^{-1}}$$

The y-intercept is

$$b = \frac{1}{Kn_\infty} \quad \text{so} \quad K = \frac{1}{bn_\infty} = \frac{1}{(24.641\ \text{kPa mol}^{-1}\ \text{kg}) \times (5.78\ \text{mol kg}^{-1})}$$

$$K = 7.02 \times 10^{-3}\ \text{kPa}^{-1} = \boxed{7.02\ \text{Pa}^{-1}}$$

Solutions to theoretical problems

P21.17 (a) $A + P \rightarrow P + P$ autocatalytic step, $v = k[A][P]$

Let $[A] = [A]_0 - x$ and $[P] = [P]_0 + x$.

We substitute these definitions into the rate expression, simplify, and integrate.

$$v = -\frac{d[A]}{dt} = k[A][P]$$

$$-\frac{d([A]_0 - x)}{dt} = k([A]_0 - x)([P]_0 + x)$$

$$\frac{dx}{([A]_0 - x)([P]_0 + x)} = k\,dt$$

$$\frac{1}{[A]_0 + [P]_0}\left(\frac{1}{[A]_0 - x} + \frac{1}{[P]_0 + x}\right)dx = k\,dt$$

$$\frac{1}{[A]_0 + [P]_0}\int_0^x \left(\frac{1}{[A]_0 - x} + \frac{1}{[P]_0 + x}\right)dx = k\int_0^t dt$$

$$\frac{1}{[A]_0 + [P]_0}\left\{\ln\left(\frac{[A]_0}{[A]_0 - x}\right) + \ln\left(\frac{[P]_0 + x}{[P]_0}\right)\right\} = kt$$

$$\ln\left\{\left(\frac{[A]_0}{[P]_0}\right)\left(\frac{[P]_0+x}{[A]_0-x}\right)\right\}=k\left([A]_0+[P]_0\right)t$$

$$\ln\left\{\left(\frac{[A]_0}{[P]_0}\right)\left(\frac{[P]}{[A]_0+[P]_0-[P]}\right)\right\}=k\left([A]_0+[P]_0\right)t$$

$$\ln\left\{\left(\frac{1}{b}\right)\left(\frac{[P]}{[A]_0+[P]_0-[P]}\right)\right\}=at \quad \text{where} \quad a=k\left([A]_0+[P]_0\right) \quad \text{and} \quad b=\frac{[P]_0}{[A]_0}$$

$$\frac{[P]}{[A]_0+[P]_0-[P]}=be^{at}$$

$$[P]=\left([A]_0+[P]_0\right)be^{at}-be^{at}[P]$$

$$\left(1+be^{at}\right)[P]=[P]_0\left(1+\frac{[A]_0}{[P]_0}\right)be^{at}=[P]_0\left(1+\frac{1}{b}\right)be^{at}=[P]_0(b+1)e^{at}$$

$$\boxed{\frac{[P]}{[P]_0}=(b+1)\frac{e^{at}}{1+be^{at}}}$$

(b) See Figure 21.12(a).

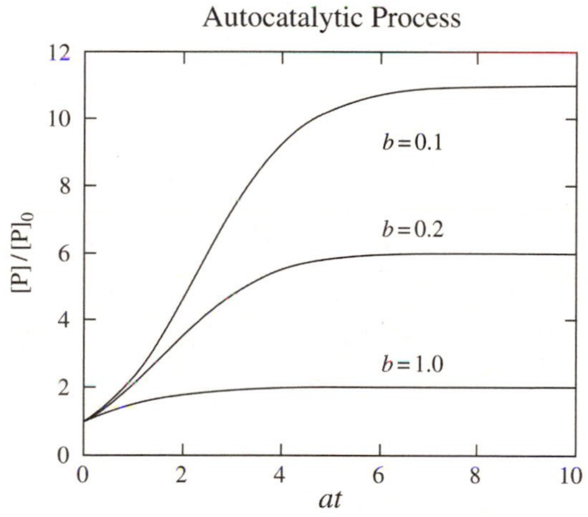

Figure 21.12(a)

The growth to [P] reaches a maximum at very long times. As $t\rightarrow\infty$, the exponential term in the denominator of $[P]/[P]_0 =(b+1)e^{at}/\left(1+be^{at}\right)$ becomes so large that the denominator becomes be^{at}. Thus, $\left([P]/[P]_0\right)_{max} =(b+1)e^{at}/\left(be^{at}\right) =(b+1)/b$ where $b=[P]_0/[A]_0$, and this maximum occurs as $t\rightarrow\infty$.

The autocatalytic curve $[P]/[P]_0=(b+1)e^{at}/\left(1+be^{at}\right)$ has a shape that is very similar to that of the first-order process $[P]/[A]_0=1-e^{-kt}$. However, $[P]_{max}$ = $[A]_0$ at $t\rightarrow\infty$ for the first-order process, whereas $[P]_{max} = (1 + 1/b)[P]_0$ for the autocatalytic mechanism. In a series of experiments at fixed $[A]_0$ and assorted $[P]_0$, only the autocatalytic mechanism will show variation in $[P]_{max}$. Another difference is that the autocatalytic curve is initially concave up, which gives an overall sigmoidal curve, whereas the first-order curve is concave down. See Figure 21.12(b).

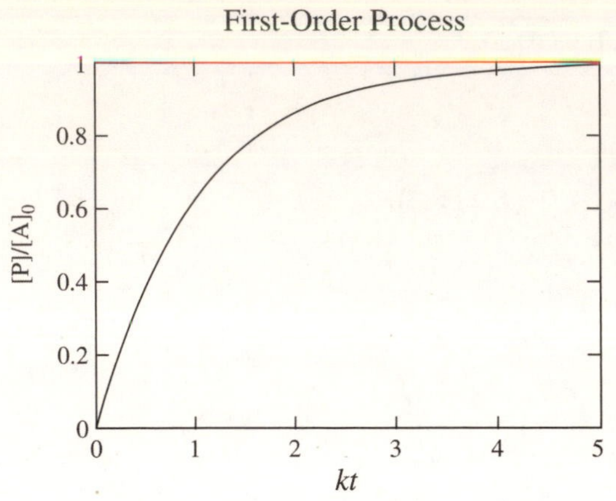

Figure 21.12(b)

(c) Let $[\mathrm{P}]_{v_{\max}}$ be the concentration of P at which the reaction rate is a maximum and let $t_{\max}$ be the corresponding time.

$$v = k[\mathrm{A}][\mathrm{P}] = k([\mathrm{A}]_0 - x)([\mathrm{P}]_0 + x)$$
$$= k\{[\mathrm{A}]_0[\mathrm{P}]_0 + ([\mathrm{A}]_0 - [\mathrm{P}]_0)x - x^2\}$$
$$\frac{\mathrm{d}v}{\mathrm{d}t} = k([\mathrm{A}]_0 - [\mathrm{P}]_0 - 2x)$$

The reaction rate is a maximum when $\mathrm{d}v/\mathrm{d}t = 0$. This occurs when

$$x = [\mathrm{P}]_{v_{\max}} - [\mathrm{P}]_0 = \frac{[\mathrm{A}]_0 - [\mathrm{P}]_0}{2} \quad \text{or} \quad \frac{[\mathrm{P}]_{v_{\max}}}{[\mathrm{P}]_0} = \frac{b+1}{2b}$$

Substitution into the final equation of part (a) gives

$$\frac{[\mathrm{P}]_{v_{\max}}}{[\mathrm{P}]_0} = \frac{b+1}{2b} = (b+1)\frac{\mathrm{e}^{at_{\max}}}{1+b\mathrm{e}^{at_{\max}}}$$

Solving for $t_{\max}$,

$$1 + b\mathrm{e}^{at_{\max}} = 2b\mathrm{e}^{at_{\max}}$$
$$\mathrm{e}^{at_{\max}} = b^{-1}$$
$$at_{\max} = \ln(b^{-1}) = -\ln(b)$$
$$t_{\max} = \boxed{-\frac{1}{a}\ln(b)}$$

(d) $$\frac{\mathrm{d}[\mathrm{P}]}{\mathrm{d}t} = k[\mathrm{A}]^2[\mathrm{P}]$$

$$[\mathrm{A}] = A_0 - x, \quad [\mathrm{P}] = P_0 + x, \quad \frac{\mathrm{d}[\mathrm{P}]}{\mathrm{d}t} = \frac{\mathrm{d}x}{\mathrm{d}t} = k(A_0 - x)^2(P_0 + x)$$

$$\int_0^x \frac{\mathrm{d}x}{(A_0 - x)^2(P_0 + x)} = kt$$

Solve the integral by partial fractions.

$$\frac{1}{(A_0 - x)^2(P_0 + x)} = \frac{\alpha}{(A_0 - x)^2} + \frac{\beta}{A_0 - x} + \frac{\gamma}{P_0 + x}$$
$$= \frac{\alpha(P_0 + x) + \beta(A_0 - x)(P_0 + x) + \gamma(A_0 - x)^2}{(A_0 - x)^2(P_0 + x)}$$

$$\left.\begin{aligned} P_0\alpha + A_0P_0\beta + A_0^2\gamma &= 1 \\ \alpha + (A_0 - P_0)\beta - 2A_0\gamma &= 0 \\ -\beta + \gamma &= 0 \end{aligned}\right\}$$

This set of simultaneous equations solves to

$$\alpha = \frac{1}{A_0 + P_0} \qquad \beta = \gamma = \frac{\alpha}{A_0 + P_0}$$

Therefore,

$$kt = \left(\frac{1}{A_0+P_0}\right)\int_0^x \left[\left(\frac{1}{A_0-x}\right)^2 + \left(\frac{1}{A_0+P_0}\right)\left(\frac{1}{A_0-x}+\frac{1}{P_0-x}\right)\right]\mathrm{d}x$$

$$= \left(\frac{1}{A_0+P_0}\right)\left\{\left(\frac{1}{A_0-x}\right) - \left(\frac{1}{A_0}\right) + \left(\frac{1}{A_0+P_0}\right)\left[\ln\left(\frac{A_0}{A_0-x}\right) + \ln\left(\frac{P_0+x}{P_0}\right)\right]\right\}$$

$$= \left(\frac{1}{A_0+P_0}\right)\left[\left(\frac{x}{A_0(A_0-x)}\right) + \left(\frac{1}{A_0+P_0}\right)\ln\left(\frac{A_0(P_0+x)}{(A_0-x)P_0}\right)\right]$$

Therefore with $y = \frac{x}{A_0}$ and $p = \frac{P_0}{A_0}$,

$$A_0(A_0+P_0)kt = \boxed{\left(\frac{y}{1-y}\right) + \left(\frac{1}{1-p}\right)\ln\left(\frac{p+y}{p(1-y)}\right)}$$

The maximum rate occurs at

$$\frac{\mathrm{d}v_{\mathrm{P}}}{\mathrm{d}t} = 0, \qquad v_{\mathrm{P}} = k[\mathrm{A}]^2[\mathrm{P}]$$

and hence at the solution of

$$2k\left(\frac{\mathrm{d}[\mathrm{A}]}{\mathrm{d}t}\right)[\mathrm{A}][\mathrm{P}] + k[\mathrm{A}]^2\frac{\mathrm{d}[\mathrm{P}]}{\mathrm{d}t} = 0$$

$$-2k[\mathrm{A}][\mathrm{P}]v_{\mathrm{P}} + k[\mathrm{A}]^2 v_{\mathrm{P}} = 0 \quad [\text{as } v_{\mathrm{A}} = -v_{\mathrm{P}}]$$

$$k[\mathrm{A}]([\mathrm{A}] - 2[\mathrm{p}])v_{\mathrm{P}} = 0$$

The rate is a maximum when $[\mathrm{A}] = 2[\mathrm{P}]$, which occurs at

$$A_0 - x = 2P_0 + 2x, \quad \text{or} \quad x = \tfrac{1}{3}(A_0 - 2P_0); \quad y = \tfrac{1}{3}(1-2p)$$

Substituting this condition into the integrated rate law gives

$$A_0(A_0+P_0)kt_{\max} = \left(\frac{1}{1+p}\right)\left(\frac{1}{2}(1-2p) + \ln\frac{1}{2p}\right)$$

or $$\boxed{(A_0+P_0)^2 kt_{\max} = \frac{1}{2} - p - \ln 2p}$$

(e) $$\frac{\mathrm{d}[\mathrm{P}]}{\mathrm{d}t} = k[\mathrm{A}][\mathrm{P}]^2$$

$$\frac{\mathrm{d}x}{\mathrm{d}t} = k(A_0 - x)(P_0 + x)^2 \quad [x = P - P_0]$$

$$kt = \int_0^x \frac{dx}{(A_0-x)(P_0+x)^2}$$

Integrate by partial fractions (as in part d).

$$kt = \left(\frac{1}{A_0+P_0}\right)\int_0^x \left\{\left(\frac{1}{P_0+x}\right)^2 + \left(\frac{1}{A_0+P_0}\right)\left[\frac{1}{P_0+x}+\frac{1}{A_0-x}\right]\right\} dx$$

$$= \left(\frac{1}{A_0+P_0}\right)\left\{\left(\frac{1}{P_0}-\frac{1}{P_0+x}\right)+\left(\frac{1}{A_0+P_0}\right)\left[\ln\left(\frac{P_0+x}{P_0}\right)+\ln\left(\frac{A_0}{A_0-x}\right)\right]\right\}$$

$$= \left(\frac{1}{A_0+P_0}\right)\left[\left(\frac{x}{P_0(P_0+x)}\right)+\left(\frac{1}{A_0+P_0}\right)\ln\left(\frac{(P_0+x)A_0}{P_0(A_0-x)}\right)\right]$$

Therefore, with $y = \frac{x}{[\mathrm{A}]_0}$ and $p = \frac{P_0}{A_0}$,

$$\boxed{A_0(A_0+P_0)kt = \left(\frac{y}{p(p+y)}\right)+\left(\frac{1}{1+p}\right)\ln\left(\frac{p+y}{p(1-y)}\right)}$$

The rate is maximum when

$$\frac{d\upsilon_\mathrm{P}}{dt} = 2k[\mathrm{A}][\mathrm{P}]\left(\frac{d[\mathrm{P}]}{dt}\right)+k\left(\frac{d[\mathrm{A}]}{dt}\right)[\mathrm{P}]^2$$

$$= 2k[\mathrm{A}][\mathrm{P}]\upsilon_\mathrm{P} - k[\mathrm{P}]^2\upsilon_\mathrm{P} = k[\mathrm{P}](2[\mathrm{A}]-[\mathrm{P}])\upsilon_\mathrm{P} = 0$$

—that is, at $[\mathrm{A}] = \tfrac{1}{2}[\mathrm{P}]$.

On substitution of this condition into the integrated rate law, we find

$$A_0(A_0+P_0)kt_{\max} = \left(\frac{2-p}{2p(1+p)}\right)+\left(\frac{1}{1+p}\right)\ln\frac{2}{p}$$

or $$(A_0+P_0)^2 kt_{\max} = \boxed{\frac{2-p}{2p}+\ln\frac{2}{p}}$$

P21.19 Assuming a rapid pre-equilibrium of E, S, and ES for eqn. 21.1 implies that

$$K = \frac{k_\mathrm{a}}{k'_\mathrm{a}} = \frac{[\mathrm{ES}]}{[\mathrm{E}][\mathrm{S}]} \quad \text{and} \quad [\mathrm{ES}] = K[\mathrm{E}][\mathrm{S}]$$

But the law of mass balance demands that $[\mathrm{E}] = [\mathrm{E}]_0 - [\mathrm{ES}]$, so $[\mathrm{ES}] = K([\mathrm{E}]_0 - [\mathrm{ES}])[\mathrm{S}]$, and solving for [ES], we find that

$$[\mathrm{ES}] = \frac{[\mathrm{E}]_0}{1+\dfrac{1}{K[\mathrm{S}]_0}}$$

where the free substrate concentration has been replaced by $[\mathrm{S}]_0$ because the substrate is typically in large excess relative to the enzyme. Now substitute the latter expression into the Michaelis–Menten rate law [21.1].

$$v = k_\mathrm{b}[\mathrm{ES}] = \frac{k_\mathrm{b}[\mathrm{E}]_0}{1+\dfrac{1}{K[\mathrm{S}]_0}} \quad \text{where } v_{\max} = k_\mathrm{b}[\mathrm{E}]_0$$

$$\boxed{v = \frac{v_{\max}}{1+\dfrac{1}{K[\mathrm{S}]_0}} \quad \text{(rate law based on rapid pre-equilibrium approximation)}}$$

With $K_M = (k_a' + k_b)/k_a$ eqn. 21.2 is

$$v = \frac{v_{max}}{1 + \dfrac{K_M}{[S]_0}} \quad \text{(rate law based on steady-state approximation)}$$

Inspection reveals that the two approximations are identical when $K_M = 1/K$, which implies that

$$(k_a' + k_b)/k_a = k_a'/k_a \text{ or } \boxed{k_a' \gg k_b}$$

P21.21 (a) We add to the Michaelis–Menten mechanism [21.1] the inhibition by the substrate

$$\text{SES} \rightleftharpoons \text{ES} + \text{S} \qquad K_I = [\text{ES}][\text{S}]/[\text{SES}]$$

where the inhibited enzyme, SES, forms when S binds to ES and thereby prevents the formation of product. This inhibition might possibly occur when S is at a very high concentration. Enzyme mass balance is written in terms of [ES], K_I, K_M (= [E][S]/[ES]), and [S]. (For practical purposes the free substrate concentration is replaced by $[S]_0$ because the substrate is typically in large excess relative to the enzyme.)

$$\begin{aligned}[\text{E}]_0 &= [\text{E}] + [\text{ES}] + [\text{SES}] \\ &= \frac{K_M[\text{ES}]}{[\text{S}]} + [\text{ES}] + \frac{[\text{ES}][\text{S}]}{K_I} \\ &= \left(1 + \frac{K_M}{[\text{S}]} + \frac{[\text{S}]}{K_I}\right)[\text{ES}]\end{aligned}$$

Thus,

$$[\text{ES}] = \frac{[\text{E}]_0}{\left(1 + \dfrac{K_M}{[\text{S}]} + \dfrac{[\text{S}]}{K_I}\right)}$$

and the expression for the rate of product formation becomes

$$v = k_b[\text{ES}] = \frac{v_{max}}{1 + \dfrac{K_M}{[\text{S}]_0} + \dfrac{[\text{S}]_0}{K_I}} \quad \text{where} \quad v_{max} = k_b[\text{E}]_0$$

The denominator term $[S]_0/K_I$ reflects a reduced reaction rate caused by inhibition as the concentration of S becomes very large.

(b) To examine the effect that substrate inhibition has on the double reciprocal, use the Lineweaver–Burk plot of $1/v$ against $1/[S]_0$ to take the inverse of the above rate expression and compare it to the uninhibited expression [21.4b]:

$$\frac{1}{v} = \frac{1}{v_{max}} + \left(\frac{K_M}{v_{max}}\right)\frac{1}{[\text{S}]_0} \quad [21.4b]$$

The inverse of the inhibited rate law is

$$\begin{aligned}\frac{1}{v} &= \frac{1}{v_{max}} + \left(\frac{K_M}{v_{max}}\right)\frac{1}{[\text{S}]_0} + \left(\frac{[\text{S}]_0^2}{v_{max}K_I}\right)\frac{1}{[\text{S}]_0} \\ &= \frac{1}{v_{max}} + \left(\frac{K_M}{v_{max}} + \frac{[\text{S}]_0^2}{v_{max}K_I}\right)\frac{1}{[\text{S}]_0}.\end{aligned}$$

The uninhibited and inhibited line shapes are sketched in Figure 21.13.

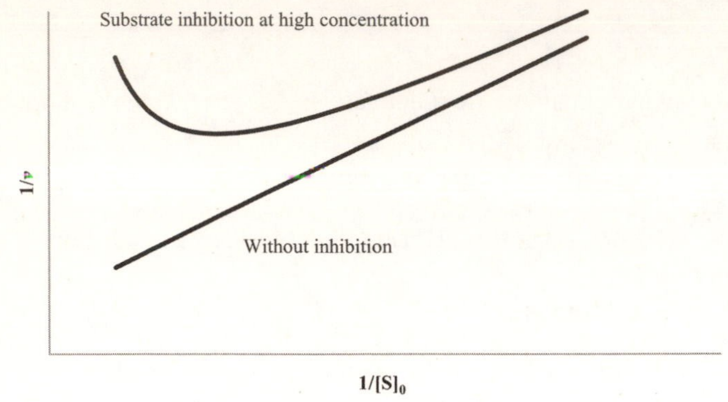

Figure 21.13

Comparing the two expressions, we see that the two curves match at high values of $1/[S]_0$. However, as the concentration of $[S]_0$ increases ($1/[S]_0$ decreases), the $1/v$ curve with inhibition curves upward because the reaction rate is decreasing.

P21.23 For association:

$$\frac{dR}{dt} = k_{on}a_0(R_{eq} - R) \quad [21.26]$$

$$\frac{dR}{R_{eq} - R} = k_{on}a_0 dt$$

$$\int_0^R \frac{dR}{R_{eq} - R} = \int_0^t k_{on}a_0 dt = k_{on}a_0 t$$

$$-\ln(R_{eq} - R)\,|_0^R = k_{on}a_0 t$$

$$-\ln\left(\frac{R_{eq} - R}{R_{eq}}\right) = k_{on}a_0 t$$

$$\frac{R_{eq} - R}{R_{eq}} = e^{-k_{on}a_0 t}$$

$$R = R_{eq}\left\{1 - e^{-k_{on}a_0 t}\right\}$$

$$\boxed{R = R_{eq}\left\{1 - e^{-k_r t}\right\} \text{ where } k_r = k_{on}a_0}$$

For dissociation:

$$\frac{dR}{dt} = -k_{off}R$$

$$\frac{dR}{R} = -k_{off}dt$$

$$\int_{R_{eq}}^R \frac{dR}{R} = -\int_0^t k_{off}dt$$

$$\ln\left(\frac{R}{R_{\text{eq}}}\right) = -k_{\text{off}}t$$

$$\boxed{R = R_{\text{eq}}e^{-k_r t} \text{ where } k_r = k_{\text{off}}}$$

Solutions to applications

P21.25 Equilibrium constants vary with temperature according to the van't Hoff equation [17.28], which can be written in the form

$$\frac{K_2}{K_1} = \mathrm{e}^{-\left[\frac{\Delta_{\text{ads}}H^{\ominus}}{R}\left(\frac{1}{T_2}-\frac{1}{T_1}\right)\right]}$$

or

$$\frac{K_2}{K_1} = \exp\left[\frac{160\times10^3\ \mathrm{J\,mol^{-1}}}{8.3145\ \mathrm{J\,K^{-1}\,mol^{-1}}}\left(\frac{1}{773\ \mathrm{K}} - \frac{1}{673\ \mathrm{K}}\right)\right] = \boxed{0.0247}$$

As measured by the equilibrium constant of absorption, NO is less strongly absorbed by a factor of 0.0247 at 500°C than at 400°C.

P21.27 (a) $q_{\text{water}} = k(\text{RH})^{1/n}$

With a power law regression analysis we find

$\boxed{k = 0.2289}$, standard deviation = 0.0068

$1/n = 1.6182$, standard deviation = 0.0093; $\boxed{n=0.6180}$

$R = 0.999508$

A linear regression analysis may be performed by transforming the equation to the following form by taking the logarithm of the Freundlich type equation.

$$\ln q_{\text{water}} = \ln k + \frac{1}{n}\ln(\text{RH})$$

$\ln k = -1.4746$, standard deviation = 0.0068; $\boxed{k = 0.2289}$

$\frac{1}{n} = 1.6183$, standard deviation = 0.0093; $\boxed{n = 0.6180}$

$R = 0.999508$

The two methods give exactly the same result because the software package for performing the power law regression performs the transformation to linear form for you. Both methods are actually performing a linear regression. The correlation coefficient indicates that 99.95% of the data variation is explained with the Freundlich type isotherm. The Freundlich fit hypothesis looks very good.

(b) The Langmuir isotherm model describes adsorption sites that are independent and equivalent. This assumption seems to be valid for the VOC case in which molecules interact very weakly. However, water molecules interact much more strongly through forces such as hydrogen bonding, and multilayers can readily form at the lower temperatures. The intermolecular forces of water apparently cause adsorption sites to become non-equivalent and dependent. In this particular case, the Freundlich-type isotherm becomes the better description.

(c) $r_{\text{VOC}} = 1 - q_{\text{water}}$ where $r_{\text{VOC}} \equiv q_{\text{VOC}} / q_{\text{VOC,RH=0}}$

$$r_{\text{VOC}} = 1 - k(\text{RH})^{1/n}$$
$$1 - r_{\text{VOC}} = k(\text{RH})^{1/n}$$

To determine the goodness-of-fit values for k and n, we perform a power law regression fit of $1 - r_{\text{VOC}}$ against RH. Results are

$\boxed{k = 0.5227}$, standard deviation = 0.0719

$\dfrac{1}{n} = 1.3749$, standard deviation = 0.0601; $\boxed{n = 0.7273}$

$R = 0.99620$

Since 99.62% of the variation is explained by the regression, we conclude that the hypothesis that $r_{\text{VOC}} = 1 - q_{\text{water}}$ may be very useful. The values of R and n differ significantly from those of part (a). It may be that water is adsorbing to some portions of the surface and VOC to others.